Local Drug Delivery System

Local Drug Delivery System

Editors

Silvia Tampucci
Daniela Monti

MDPI

Basel • Beijing • Wuhan • Barcelona • Belgrade • Novi Sad • Cluj • Manchester

Editors
Silvia Tampucci
Department of Pharmacy
University of Pisa
Pisa
Italy

Daniela Monti
Department of Pharmacy
University of Pisa
Pisa
Italy

Editorial Office
MDPI
St. Alban-Anlage 66
4052 Basel, Switzerland

This is a reprint of articles from the Special Issue published online in the open access journal *Pharmaceutics* (ISSN 1999-4923) (available at: www.mdpi.com/journal/pharmaceutics/special_issues/Local_Drug_Delivery).

For citation purposes, cite each article independently as indicated on the article page online and as indicated below:

Lastname, A.A.; Lastname, B.B. Article Title. *Journal Name* **Year**, *Volume Number*, Page Range.

ISBN 978-3-7258-0330-9 (Hbk)
ISBN 9978-3-7258-0329-3 (PDF)
doi.org/10.3390/books9978-3-7258-0329-3

Contents

Preface . . . . **vii**

Laura Müller, Christoph Rosenbaum, Adrian Rump, Michael Grimm, Friederike Klammt, Annabel Kleinwort, et al.
Determination of Mucoadhesion of Polyvinyl Alcohol Films to Human Intestinal Tissue
Reprinted from: *Pharmaceutics* **2023**, *15*, 1740, doi:10.3390/pharmaceutics15061740 . . . . **1**

Robert Mau, Thomas Eickner, Gábor Jüttner, Ziwen Gao, Chunjiang Wei, Nicklas Fiedler, et al.
Micro Injection Molding of Drug-Loaded Round Window Niche Implants for an Animal Model Using 3D-Printed Molds
Reprinted from: *Pharmaceutics* **2023**, *15*, 1584, doi:10.3390/pharmaceutics15061584 . . . . **15**

Himangsu Mondal, Ho-Joong Kim, Nijaya Mohanto and Jun-Pil Jee
A Review on Dry Eye Disease Treatment: Recent Progress, Diagnostics, and Future Perspectives
Reprinted from: *Pharmaceutics* **2023**, *15*, 990, doi:10.3390/ pharmaceutics15030990 . . . . **33**

Marco Uboldi, Cristiana Perrotta, Claudia Moscheni, Silvia Zecchini, Alessandra Napoli, Chiara Castiglioni, et al.
Insights into the Safety and Versatility of 4D Printed Intravesical Drug Delivery Systems
Reprinted from: *Pharmaceutics* **2023**, *15*, 757, doi:10.3390/pharmaceutics15030757 . . . . **54**

Xin Gao, Xingyan Fan, Kuan Jiang, Yang Hu, Yu Liu, Weiyue Lu, et al.
Intraocular siRNA Delivery Mediated by Penetratin Derivative to Silence Orthotopic Retinoblastoma Gene
Reprinted from: *Pharmaceutics* **2023**, *15*, 745, doi:10.3390/pharmaceutics15030745 . . . . **75**

Ruchi Tiwari and Kamla Pathak
Local Drug Delivery Strategies towards Wound Healing
Reprinted from: *Pharmaceutics* **2023**, *15*, 634, doi:10.3390/pharmaceutics15020634 . . . . **93**

Caterina Valentino, Barbara Vigani, Giuseppina Sandri, Franca Ferrari and Silvia Rossi
Current Status of Polysaccharides-Based Drug Delivery Systems for Nervous Tissue Injuries Repair
Reprinted from: *Pharmaceutics* **2023**, *15*, 400, doi:10.3390/pharmaceutics15020400 . . . . **132**

Rajesh Pradhan, Anuradha Dey, Rajeev Taliyan, Anu Puri, Sanskruti Kharavtekar and Sunil Kumar Dubey
Recent Advances in Targeted Nanocarriers for the Management of Triple Negative Breast Cancer
Reprinted from: *Pharmaceutics* **2023**, *15*, 246, doi:10.3390/pharmaceutics15010246 . . . . **158**

Brijesh Patel and Hetal Thakkar
Formulation Development of Fast Dissolving Microneedles Loaded with Cubosomes of Febuxostat: In Vitro and In Vivo Evaluation
Reprinted from: *Pharmaceutics* **2023**, *15*, 224, doi:10.3390/pharmaceutics15010224 . . . . **199**

Mahipal Reddy Donthi, Siva Ram Munnangi, Kowthavarapu Venkata Krishna, Ranendra Narayan Saha, Gautam Singhvi and Sunil Kumar Dubey
Nanoemulgel: A Novel Nano Carrier as a Tool for Topical Drug Delivery
Reprinted from: *Pharmaceutics* **2023**, *15*, 164, doi:10.3390/pharmaceutics15010164 . . . . **227**

Linda Maurizi, Jacopo Forte, Maria Grazia Ammendolia, Patrizia Nadia Hanieh, Antonietta Lucia Conte, Michela Relucenti, et al.
Effect of Ciprofloxacin-Loaded Niosomes on *Escherichia coli* and *Staphylococcus aureus* Biofilm Formation
Reprinted from: *Pharmaceutics* **2022**, *14*, 2662, doi:10.3390/pharmaceutics14122662 . . . . **255**

Noriaki Nagai, Fumihiko Ogata, Saori Deguchi, Aoi Fushiki, Saki Daimyo, Hiroko Otake, et al.
Design of a Transdermal Sustained Release Formulation Based on Water-Soluble Ointment Incorporating Tulobuterol Nanoparticles
Reprinted from: *Pharmaceutics* **2022**, *14*, 2431, doi:10.3390/pharmaceutics14112431 . . . . **273**

Mailine Gehrcke, Carolina Cristóvão Martins, Taíne de Bastos Brum, Lucas Saldanha da Rosa, Cristiane Luchese, Ethel Antunes Wilhelm, et al.
Novel Pullulan/Gellan Gum Bilayer Film as a Vehicle for Silibinin-Loaded Nanocapsules in the Topical Treatment of Atopic Dermatitis
Reprinted from: *Pharmaceutics* **2022**, *14*, 2352, doi:10.3390/pharmaceutics14112352 . . . . **288**

Zhiyu Jin, Yu Han, Danshen Zhang, Zhongqiu Li, Yongshuai Jing, Beibei Hu, et al.
Application of Intranasal Administration in the Delivery of Antidepressant Active Ingredients
Reprinted from: *Pharmaceutics* **2022**, *14*, 2070, doi:10.3390/pharmaceutics14102070 . . . . **311**

Silvia Tampucci, Giorgio Tofani, Patrizia Chetoni, Mariacristina Di Gangi, Andrea Mezzetta, Valentina Paganini, et al
Sporopollenin Microcapsule: Sunscreen Delivery System with Photoprotective Properties
Reprinted from: *Pharmaceutics* **2022**, *14*, 2041, doi:10.3390/pharmaceutics14102041 . . . . **345**

Preface

Site-specific drug delivery is among the main objectives for the optimization of pharmaceutical therapies. In particular, local drug delivery systems represent a way to avoid systemic administration, reducing the associated side effects and increasing patient compliance. In the design of local drug delivery systems, different strategies to overcome physiological barriers to obtain effective drug concentration at the target site without affecting adjacent tissues can be pursued. Physical and chemical enhancers, microneedles, and nanostructured drug delivery systems have been proposed as effective tools to influence drug release and partition in the target tissues. Innovative drug delivery systems could be based on natural or synthetic polymers that are biodegradable, endowed with stimuli-responsive behavior, and, if applicable, mucoadhesive properties.

Silvia Tampucci and Daniela Monti
Editors

pharmaceutics

Article

Determination of Mucoadhesion of Polyvinyl Alcohol Films to Human Intestinal Tissue

Laura Müller [1], Christoph Rosenbaum [1], Adrian Rump [1], Michael Grimm [1], Friederike Klammt [1], Annabel Kleinwort [2], Alexandra Busemann [2] and Werner Weitschies [1,*]

[1] Department of Biopharmaceutics and Pharmaceutical Technology, Institute of Pharmacy, University of Greifswald, Felix-Hausdorff-Str. 3, 17487 Greifswald, Germany
[2] Department of General, Visceral, Thoracic and Vascular Surgery, Greifswald University Medicine, Ferdinand-Sauerbruch-Str., 17457 Greifswald, Germany
* Correspondence: werner.weitschies@uni-greifswald.de; Tel.: +49-3834-420-4813

Abstract: The absorption of drugs with narrow absorption windows in the upper small intestine can be improved with a mucoadhesive drug delivery system such as enteric films. To predict the mucoadhesive behaviour in vivo, suitable in vitro or ex vivo methods can be performed. In this study, the influence of tissue storage and sampling site on the mucoadhesion of polyvinyl alcohol film to human small intestinal mucosa was investigated. Tissue from twelve human subjects was used to determine adhesion using a tensile strength method. Thawing of tissue frozen at −20 °C resulted in a significantly higher work of adhesion ($p = 0.0005$) when a low contact force was applied for one minute, whereas the maximum detachment force was not affected. When the contact force and time were increased, no differences were found for thawed tissue compared to fresh tissue. No change in adhesion was observed depending on the sampling location. Initial results from a comparison of adhesion to porcine and human mucosa suggest that the tissues are equivalent.

Keywords: mucoadhesion; site-specific application; intestinal application; ex vivo measurements; human intestinal mucosa

Citation: Müller, L.; Rosenbaum, C.; Rump, A.; Grimm, M.; Klammt, F.; Kleinwort, A.; Busemann, A.; Weitschies, W. Determination of Mucoadhesion of Polyvinyl Alcohol Films to Human Intestinal Tissue. *Pharmaceutics* **2023**, *15*, 1740. https://doi.org/10.3390/pharmaceutics15061740

Academic Editors: Daniela Monti and Silvia Tampucci

Received: 24 May 2023
Revised: 12 June 2023
Accepted: 13 June 2023
Published: 15 June 2023

1. Introduction

An ideal drug substance should be absorbed uniformly throughout the small intestine. Some drugs are poorly absorbed due to narrow absorption areas, also known as absorption windows. These are usually located in the upper part of the small intestine. Poor absorption may be caused by specific transport mechanisms such as active transport or active excretion. Several drug delivery approaches have been developed to overcome this challenge, such as mucoadhesive films, which can be a highly beneficial drug delivery system (DDS) for site-specific applications, such as in the upper intestine. Examples of drugs that are only absorbed in the upper small intestine are furosemide [1], acyclovir [2,3] and gabapentin [4]. Other possible drugs that could benefit from prolonged residence time through mucoadhesion are therapeutic peptides and proteins [5,6]. These macromolecules mostly have very low oral bioavailability due to their high molecular weight and vulnerable structure. The specific amino acid sequence essential for drug activity can be destroyed by the chemical, physical and proteolytic nature of the gastrointestinal tract [7]. An ideal DDS for peptides and proteins should protect and preserve the drug structure and release it at the highly vasculated specific absorption site [8].

To predict the adhesion of the dosage form in vivo during formulation development, appropriate in vitro methods can be useful. In vitro methods have the advantage of good reproducibility and avoidance of biological tissues. Many biomimetic materials have been described in literature to mimic and replace tissue. They include, for example, simple hydrogels such as gelatin [9] or agar gels [10] and more complex hydrogels such as HEMA-AGA hydrogels [11] or mucin compacts [12]. The disadvantage of these biomimetic materials is that they may not adequately represent the inter-individual variability of ex vivo and in vivo studies. This can lead to biased prediction of in vivo behavior by in vitro methods. Therefore, ex vivo methods using tissue can be very helpful to get an idea of the variability in vivo. Ideally, the tissue used should represent as closely as possible the application site of the DDS under development.

Tissues from animal sources are mainly used as ex vivo substrates, such as chicken pouch [13], porcine tissue [14] or bovine tissue [15]. Although animal tissues are often used in ex vivo studies, there is the ethical drawback that animals have to be slaughtered to obtain the tissue. Along with these ethical concerns, the choice of suitable animal tissue is another issue. When it comes to mucoadhesion studies in the small intestine, rodents are known to be poor model animals. Not only the anatomy and physiology have been found to be different from humans [16,17] but also the pH and water content [18]. Therefore, large animal models such as pigs are often used to study the small intestine. However, although the pig anatomy is quite similar to that of humans, there are still some differences. Mucus thickness and composition are known to influence mucoadhesion [19]. In pigs, the average thickness of the small intestinal mucus is about 26 to 31 μm [19], whereas in humans the gastroduodenal mucus layer is of variable thickness [20]. These differences may affect mucoadhesion and the in vitro-in vivo correlation of mucoadhesion studies. Therefore, the ideal tissue for mucoadhesion studies is potentially human tissue. Patients taking medicines are usually elderly people suffering from more than one disease [21]. Their gastrointestinal tract may further differ from that of animals used in animal models. Theoretically, tissue from the target patient population should ideally be used to obtain the most predictive results.

Despite the choice of tissue, tissue preparation and storage may also affect the outcome of studies. In previous studies, mucoadhesion was found to be higher on thawed porcine small intestine tissue than on fresh tissue [22]. As these results may not be applicable to human tissue, further mucoadhesion studies on human small intestinal mucosa are needed. To the best of our knowledge, there is no ex vivo mucoadhesion study on human intestinal tissue. In this study, several questions will be addressed, the first of which is whether tissue storage has an effect on mucoadhesion in two different test setups. Secondly, the effect on the sampling site was investigated. Finally, a comparison was made with results on porcine tissue obtained in previous studies [22] using the same methodology. The results should indicate that the choice and storage of the tissue and the experimental design of each mucoadhesion study are very important variables that need to be investigated in order to understand the underlying mechanisms of mucoadhesion and to achieve predictive results for respective DDS.

2. Materials and Methods

2.1. Study Materials

The water-soluble polyvinyl alcohol quality EMPROVE® ESSENTIAL PVA 18–88 (PVA 18–88, $M_W \approx$ 96,000 g/mol, Merck KGaA, Darmstadt, Germany) with a degree of hydrolysis of 88% was used as the mucoadhesive polymer for the preparation of the mucoadhesive films. Anhydrous glycerol (AppliChem GmbH, Darmstadt, Germany) was used as a plasticizer. The chemicals were dissolved in demineralized water.

2.2. Preparation of Mucoadhesive Films

Mucoadhesive films were prepared using the solvent casting technique on the day before the planned surgery. A total of 80.00 g demineralised water and 2.00 g anhydrous glycerol were mixed on a magnetic stirring plate (IKA® RCT basic, IKA®-Werke GmbH & CO. KG, Staufen, Germany) at 250 rpm. Then, 18.00 g ground PVA 18–88 was added at 500 rpm. The dispersion was heated to 85 °C under continuous magnetic stirring at 150 rpm for 1 h until a clear solution was obtained. The solution was centrifuged at 4400 rpm for 15 min to remove air bubbles (Centrifuge 5702 R, Eppendorf SE, Hamburg, Germany). The solution was cast on a liner at 12.0 mm/s with a coating knife set to 1000 μm (mtv messtechnik oHG, Erftstadt, Germany) using an automatic coating bench (Automatic Precision Film Applicator CX4, mtv messtechnik oHG, Erftstadt, Germany). The cast film was dried at room temperature.

2.3. Study Participants

A positive ethical vote was obtained from the Ethics Committee of the University Medicine of Greifswald for the mucoadhesion study on human tissue (Ethical Protocol No. BB 027/21, date of approval: 2 March2021). A total of 13 patients (10 male, 3 females; BMI = 24.5 ± 5.5 kg/m^2) aged 36 to 84 years (68 ± 13 years) was included. The patients suffered from various diseases of the gastrointestinal tract, such as cancer, sigmoid diverticulitis or Crohn's disease. The operations during which the samples for the study were taken were directly related to these diseases. Written informed consent was obtained from all subjects and included information about the tissue sampling, the experimental plan, the handling of personal data and the data protection laws of Germany. During medically necessary surgery for Whipple procedure (n = 4), right hemicolectomy (n = 4), ileostomy (n = 4) or pancreatectomy (n = 1), a portion of healthy small bowel was also removed for technical reasons. This tissue was the proximal jejunum (n = 5) or the distal ileum (n = 8). In addition to demographic data, premedication data were also collected from the study participants.

2.4. Mucoadhesion Study

Mucoadhesion was determined using the same texture analysis method described in a previous study [22]. Briefly, a texture analyser (TA plus, AMETEK Lloyd Instruments Ltd., Hampshire, UK) equipped with a 10 N load cell was used to measure the maximum detachment force (F_{max}) and the work of adhesion (WoA). Circular pieces (d = 14 mm, A ≈ 1.54 cm^2) of the PVA films were punched out using a punching iron. The films were attached to the upper probe using double-sided adhesive tape (tesa® Doppelseitiges Klebeband universal, tesa SE, Hamburg, Germany). The tissues were placed on the lower base of the apparatus. They were collected at the time of removal during surgery and transported to the laboratory in a polystyrene cooler filled with ice. To avoid direct contact, a bag filled with water was placed between the tissue placed in another bag and the ice. The time between the collection of the samples and the start of the experiments was a maximum of 30 min. The intestinal tissue was cut into four pieces, two of which were used immediately. The other two were placed in sealed PE bags and frozen at −20 °C in a freezer. After one week of storage, the tissues in the PE bags were thawed in a water bath at 37 °C.

F_{max} and WoA were measured in two settings based on a previous study [22]. In brief, a standard setting (setting A) was used as a starting point to investigate the influence of contact force, contact time and withdrawal speed on the mucoadhesion of PVA films to agar/mucin gels. Setting A was chosen on the basis of literature values. A low contact time and low contact force were used. In the following investigations, setting B was found to be the best compromise between the highest F_{max} and WoA and gel integrity. The experimental setups for both are described in Table 1. Each setting was performed on fresh and thawed tissue. The upper probe with the polymer film was lowered at a constant speed to the tissue from a distance of 5 cm until a specified contact force was detected. The probe remained in this position during the contact time and was then removed at a defined withdrawal speed. During removal from the tissue, load and machine extension were measured using NEXYGEN Plus software (AMETEK Lloyd Instruments Ltd., Hampshire, UK).

Table 1. Instrument settings for Setting A and B.

Variable	Setting A	Setting B
contact force [N]	0.1	0.35
contact time [s]	60	180
withdrawal speed [mm/s]	0.5	1.0

2.5. Statistical Analysis

F_{max} and WoA were calculated using Microsoft® Excel® 2019 (Microsoft Corporation, Redmond, WA, USA) and reported as individual data and medians. F_{max} was the maximum force measured during film detachment. WoA describes the area under the curve (AUC) and was calculated using the linear trapezoidal rule. Statistical analysis was performed using GraphPad Prism 5 (v. 5.01; GraphPad Software, Boston, MA, USA). F_{max} and WoA were tested for normal distribution using the D'Agostino and Pearson omnibus normality test. If the data were normally distributed and paired (e.g., derived from same subject), a paired t-test was used. Data that were not normally distributed were compared non-parametrically. A Wilcoxon signed rank test was used for paired data and a Mann-Whitney U test was used for unpaired data.

3. Results

A total of 13 subjects was initially part of the study. One subject (female, age = 84 years, BMI: 18 kg/m^2) had to be excluded during the study because the amount of tissue removed for clinical reasons of the main indication was too small to perform mucoadhesion measurements with a sufficient number of samples. Data from this subject are excluded below. The demographics of the subjects are shown in Table 2.

Table 2. Demographic data of the remaining 12 study participants.

Parameter	Median (Range)	Mean ± SD
sex	m = 10; f = 2	m = 83%; f = 17%
age/y	70 (36–80)	67 ± 12
height/m	1.76 (1.59–1.87)	1.74 ± 0.09
weight/kg	75.6 (43.0–105.0)	75.6 ± 16.8
BMI/kg/m^2	23.9 (16.4–37.6)	25.0 ± 5.4

In the remaining 12 subjects, both settings were to be performed on fresh and thawed tissue. In four subjects the tissue was too small to try both settings. Therefore, setting A with lower contact force, contact time and withdrawal speed was preferred on fresh and thawed tissue. The data have been checked for normal distribution. As not all data were normally distributed, a Gaussian distribution was not assumed for statistical comparison. The individual medians of the subjects can be found in Table S1.

3.1. Processing of the Tissue

The intestinal segments were divided into four parts, two of which were frozen and thawed for the experiments after a one-week storage period, and the other two were used fresh. As setting A was the preferred setting, a comparison of fresh and thawed tissue could be made for all 12 subjects.

Significant differences were found for WoA ($p = 0.0005$) in setting A (Figure 1), whereas no significant difference was found for F_{max} ($p = 0.6221$). For individual data, WoA was higher on thawed tissue than on fresh tissue. No clear trend can be seen for F_{max}.

For setting B, where a higher contact time and force are applied, no significant differences can be found for either WoA ($p = 0.7422$) or F_{max} ($p = 0.3125$) (Figure 2). In contrast to setting A, no trend can be seen in the individual data for either of the calculated mucoadhesion values. Overall, the measured and calculated results were higher in setting B than in setting A.

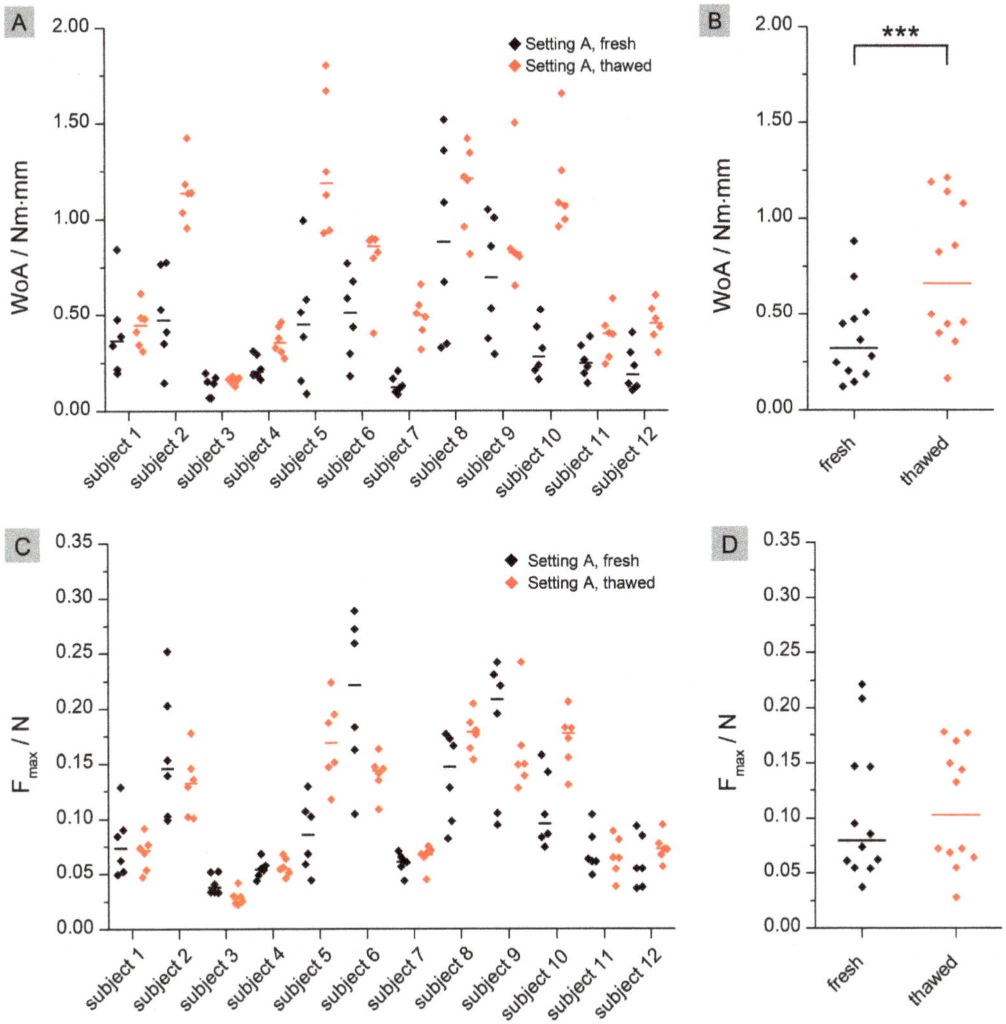

Figure 1. Results for Setting A. (**A**): Individual data of the calculated WoA (Nm × mm) with median (n = 12); black: fresh tissue; red: thawed tissue; (**B**): pooled medians of all subjects with median line; (**C**): Individual data of the calculated F_{max} (N) with median (n = 12); black: fresh tissue; red: thawed tissue; (**D**): pooled medians of all subjects with median line. Significant difference of WoA and F_{max} was checked by using a Wilcoxon signed rank test: *** ($p < 0.001$).

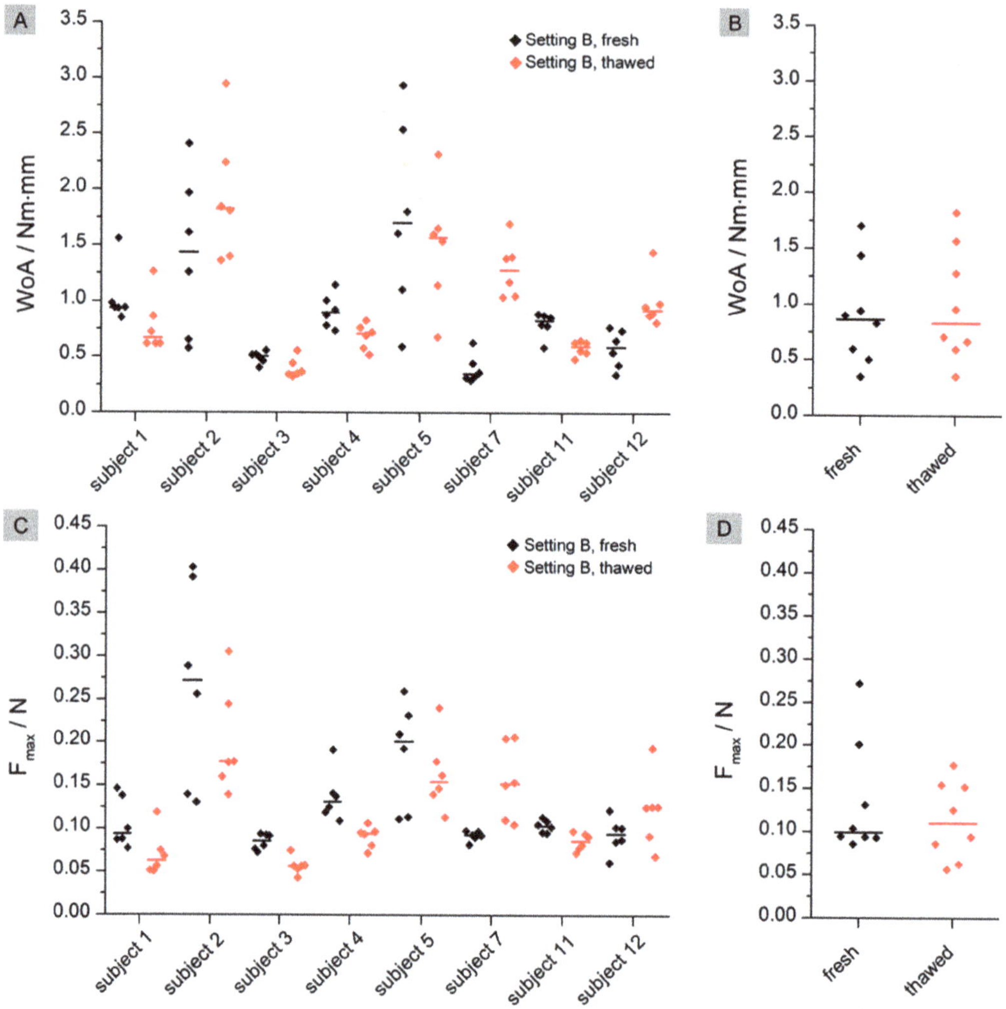

Figure 2. Results for Setting B. (**A**): Individual data of the calculated WoA (Nm × mm) with median (n = 8); black: fresh tissue; red: thawed tissue; (**B**): pooled medians of all subjects with median line; (**C**): Individual data of the calculated F_{max} (N) with median (n = 8); black: fresh tissue; red: thawed tissue; (**D**): pooled medians of all subjects with median line. Significant difference of WoA and F_{max} was checked by using a Wilcoxon signed rank test.

3.2. Comparison of Different Test Settings

Fresh and thawed tissues were also compared for both settings to investigate the influence of the test parameters. Setting A used a lower contact time, lower contact force and lower withdrawal speed. Only data from participants who were able to use both settings were included in the comparison, resulting in eight measurements.

As shown in Figure 3, the fresh tissue showed significant differences between setting A and setting B for WoA (p = 0.0078) and F_{max} (p = 0.0078). For both WoA and F_{max}, the median of each individual data set was significantly higher in setting B than in setting A.

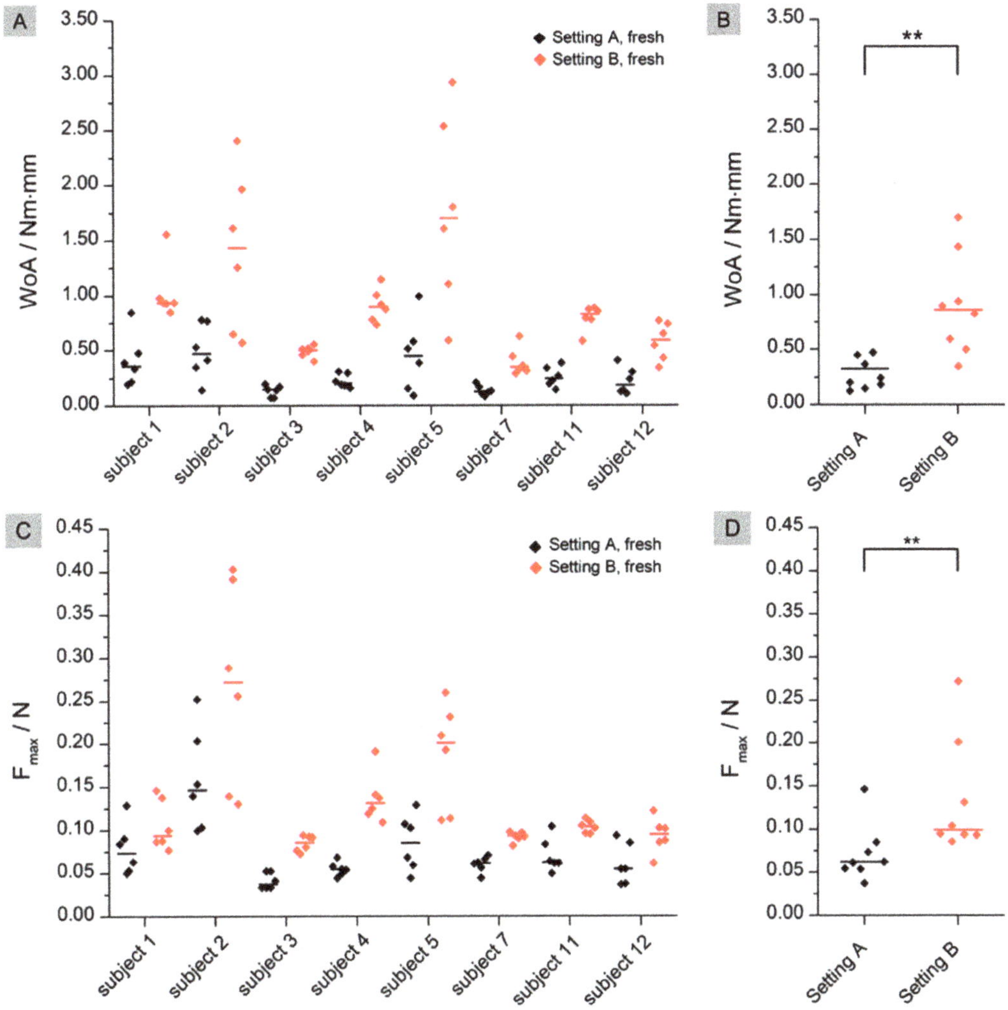

Figure 3. Comparison of the results for Setting A and B on fresh tissue. (**A**): Individual data of the calculated WoA (Nm × mm) with median (n = 8); black: Setting A; red: Setting B; (**B**): pooled medians of all subjects with median line; (**C**): Individual data of the calculated F_{max} (N) with median (n = 8); black: Setting A; red: Setting B; (**D**): pooled medians of all subjects with median line. Significant difference of WoA and F_{max} was checked by using a Wilcoxon signed rank test: ** ($p < 0.01$).

The same comparison was made for thawed tissue. As shown in Figure 4, statistically significant differences were also found for WoA ($p = 0.0078$), but not for F_{max} ($p = 0.3828$).

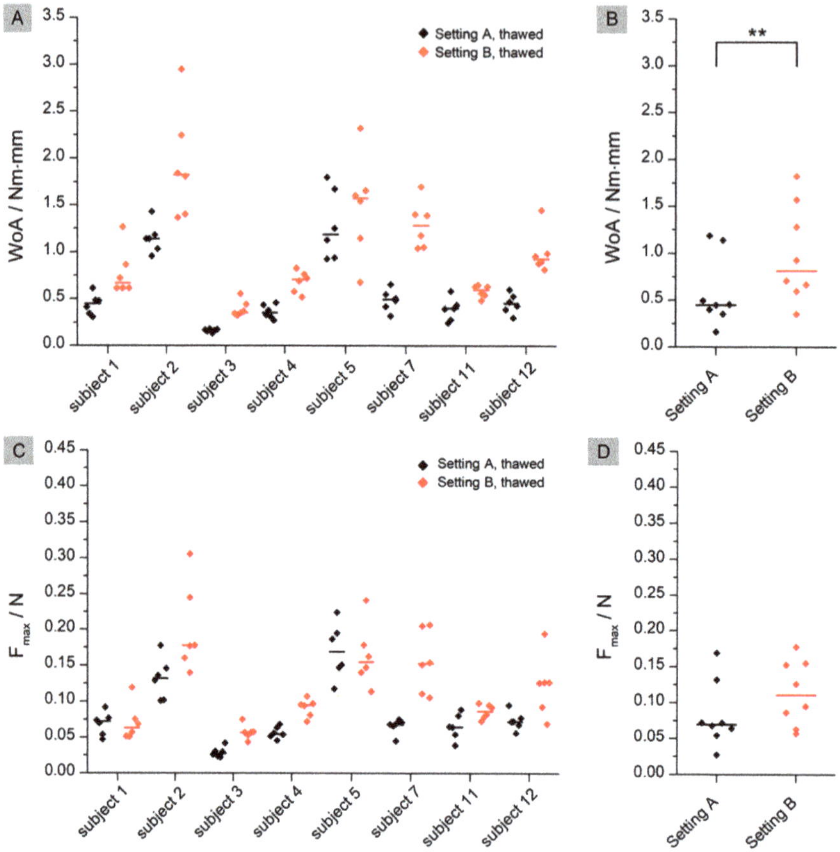

Figure 4. Comparison of the results for Setting A and B on thawed tissue. (**A**): Individual data of the calculated WoA (Nm × mm) with median (n = 8); black: Setting A; red: Setting B; (**B**): pooled medians of all subjects with median line; (**C**): Individual data of the calculated F_{max} (N) with median (n = 8); black: Setting A; red: Setting B; (**D**): pooled medians of all subjects with median line. Significant difference of WoA and F_{max} was checked by using a Wilcoxon signed rank test: ** ($p < 0.01$).

3.3. Sampling Location

Another issue was the importance of the sampling site, as there may be differences in adhesion in the proximal jejunum compared with the distal ileum. A Mann-Whitney U test was performed as the data were not paired and the number of samples was too small to assume a normal distribution. The test was performed on fresh and thawed tissue for setting A only, as the number of samples in this case was 12. No statistical differences were found for either WoA ($p_{fresh} = 0.5303$; $p_{thawed} = 0.2020$) or F_{max} ($p_{fresh} = 0.2677$; $p_{thawed} = 0.1490$).

3.4. Comparison of Mucoadhesion on Porcine Versus Human Intestinal Tissue

The data obtained in this study were compared with those of a previous study carried out on porcine small intestine tissue [22]. In the previous study, the identical test setup A was used to measure mucoadhesion. The only difference, apart from the origin of the tissue, was that the number of samples was much smaller (n = 3) compared to the new study (n = 12). Cleaned porcine tissue was used as a reference because, unlike the participants' tissue, it was not free of food residues due to the surgical specifications. As a result, the statistical analysis presented below may only give an indication of the difference. A Mann-

Whitney U test was performed to evaluate possible differences in WoA and F_{max} between fresh and thawed tissue. The results are shown in Figure 5.

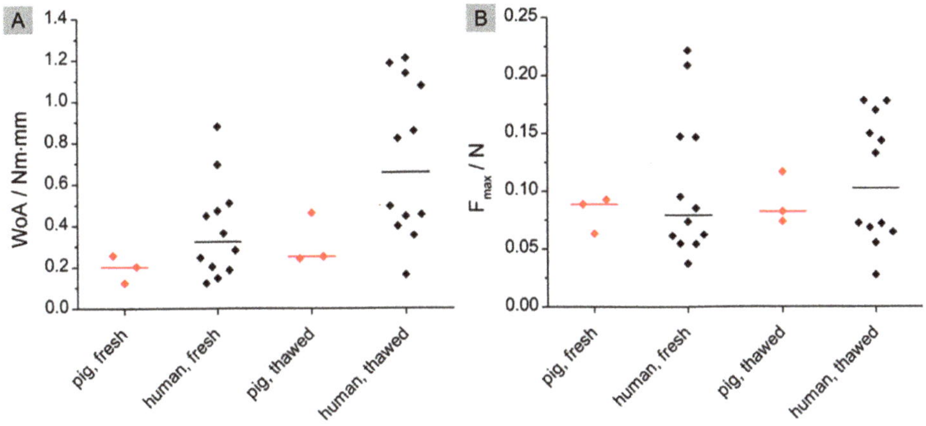

Figure 5. Comparison of the results for Setting A on fresh tissue of pigs (n = 3) and humans (n = 12). (**A**): Pooled medians of the calculated WoA (Nm × mm) with median line; red: pig; black: human; (**B**): Pooled medians of the calculated F_{max} (N) with median line; red: pig; black: human. Significant difference of WoA and F_{max} was checked by using a Mann-Whitney U test.

No significant differences could be found for the WoA ($p_{fresh} = 0.2790$; $p_{thawed} = 0.1296$) nor the F_{max} ($p_{fresh} = 0.9425$; $p_{thawed} = 0.9425$). The data presented in Figure 5A indicate a trend towards a slightly higher WoA on fasted human tissue compared to washed porcine tissue. However, the sample numbers of porcine tissues are too small to state this with certainty.

4. Discussion

Ex vivo mucoadhesion measurements of PVA-films on human small intestine tissue show that F_{max} and WoA are highly variable inter-individually and intra-individually. Possible influences on the measurement results were investigated. Statistical analysis of the mucoadhesion values in two settings and on tissues prepared in different ways showed that the WoA appears to be sensitive to storage and test parameters. WoA was significantly higher on thawed tissue in setting A, where a lower contact force is applied for a shorter time, but surprisingly not when a higher force is applied for a longer contact time, as in setting B. If tissue is frozen without a cryoprotectant, ice crystals may form. This depends on the rate at which the tissue is frozen. A slow freezing rate often results in the formation of sharp crystals that can damage tissue cells by perforating them [23]. In addition, cells can be further damaged by the osmotic pressure that can result from ice formation [23]. Signs of possible tissue damage were observed after storage of the respective tissue samples. The appearance of the tissue changed during storage. The macrostructure of the tissue appeared flatter than in the fresh condition. There was also some leakage of fluid from the tissue as can be seen in Figure 6.

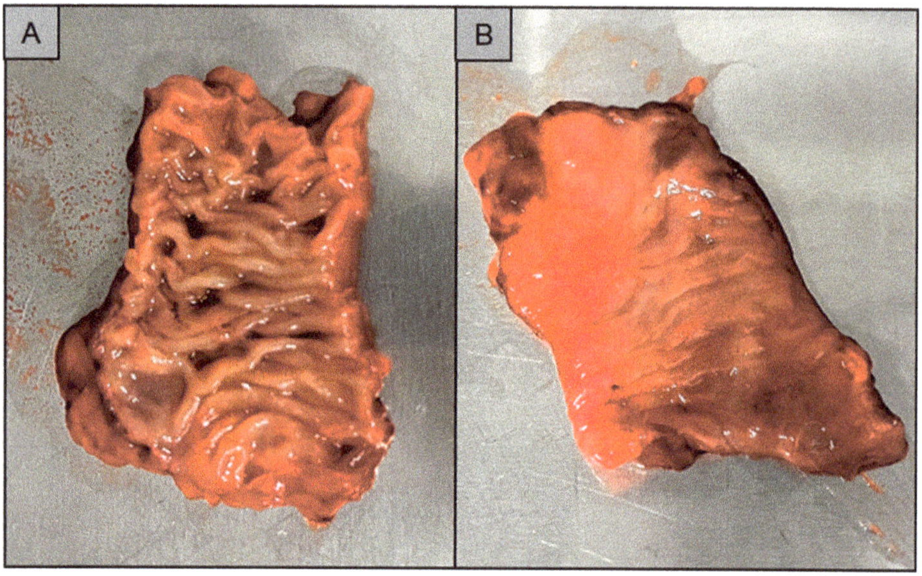

Figure 6. (**A**): Fresh tissue; (**B**): thawed tissue after a storage time of 7 days at T = −20 °C in a freezer.

The flattened structure of thawed tissue may explain the higher observed WoA. When a low force is applied in the fresh state, the mucoadhesive film may not be in contact with the entire tissue due to the macroscopically visible folded structure. The contact forces of 0.1 N and 0.35 N correspond to biorelevant pressures of 6.5 mbar and 22.7 mbar, respectively. These are within the physiological range of the small intestine as determined in telemetric studies with the SmartPill [24]. Higher contact forces may result in a flatter structure due to tissue compression and therefore more even contact between the film and the mucosa. As a result, WoA may increase when higher forces are applied (setting A versus setting B) or when the tissue loses structure due to thawing. No statistical differences can be found for WoA on fresh versus thawed tissue in setting B. A possible reason for this finding could be that not only the macrostructure but also the microstructure of the mucus changes during storage, as observed by Hägerström et al. [25]. Negatively charged glycoproteins called mucins, which make up approximately 0.5–5% of mucus [26], play an essential role in mucoadhesion. Mucoadhesive polymers can bind to mucins either through chemical bonds, such as ionic, covalent or secondary bonds, or through physical bonds. These include interpenetration and entanglement of polymer structures and mucin chains [27]. Polyvinyl alcohol, which was used in our study, is a non-ionic polymer. This group of polymers is known to bind to mucus through secondary chemical bonds such as hydrogen bonds and chain entanglements. Typically, the mucoadhesion of non-ionic polymers is lower than that of cationic polymers such as chitosan, which bind by electrostatic attraction to negatively charged mucins [28]. If the structure of the mucus changes during freezing and thawing [29], it may loosen, resulting in a looser structure that potentially facilitates interpenetration and chain entanglement. This may have a positive effect on the mucoadhesion of non-ionic polymers, as observed for thawed tissue in setting A (Figure 1B). The influence of a higher contact force (setting B) appears to have a greater effect on mucoadhesion than the thawing process, as there are no statistical differences between fresh and thawed tissue in setting B. However, the loss of the microstructure of mucus may influence the adhesion of charged polymers. When mucus hydrogels are frozen and thawed, there is a phase separation between the aqueous phase and the hydrogel former, resulting in a concentration of mucins. This in turn can lead to an increased number of possible electrostatic interactions, resulting in a higher mucoadhesive work. In contrast to WoA, F_{max} is not influenced by storage (Figures 1D and 2D). The question therefore arises as to whether WoA or F_{max} is the better surrogate for the measurement of mucoadhesion. In their paper, das Neves et al. [30] discussed whether WoA or F_{max} is more suitable for evaluating the mucoadhesion of semi-solids to bovine vaginal mucosa. WoA represents the sum of all adhesive forces, whereas F_{max} represents only the peak force during detachment. Therefore, the authors consider WoA to be the more accurate parameter for mucoadhesion. Da Silva et al. [12] also confirmed in their work that the WoA is more sensitive to changes in the test parameters and therefore the better surrogate for mucoadhesion studies in texture analyser studies. These results are confirmed by our study. For future mucoadhesion measurements with the texture analyser, it should be noted that both the test parameters and, in particular, the storage of the tissues have an influence on the measurement results, making it difficult to compare different studies.

In addition to the storage and test setup, the influence of the sampling site was investigated. No statistical differences were found between the results obtained from the proximal jejunum and the distal ileum. Another patient-specific parameter that may influence the results of the study is the amount of aqueous medium (e.g., mucus and/or bile acid) present on the tissue. A higher amount of water can cause faster hydration of the solid polymer in the mucoadhesive film. This effect is advantageous in the contact stage, as chain disentanglement of the former solid polymer occurs upon hydration [31]. The detrimental effect begins as soon as the polymer hydrogel is diluted. If the amount of water in the polymeric gel becomes too high, the cohesiveness of the gel will decrease. This results in a failure of adhesion within the gel as the test preparation is detached from the mucosa which is represented by lower values for F_{max} and WoA. To minimise the influence of intestinal fluids, the tissues can be washed [32] or wettened with a specified amount of liquid [33] prior to the experiment. The disadvantage of these methods is that the tissues may no longer represent the actual in vivo state.

Another patient-related factor to be considered is age. Intestinal morphology does not appear to change in older people [34,35]. There is some evidence that there may be changes in the structure of mucus with age. Elderman et al. [36] reported that the age of mice can influence the thickness of their colonic mucus, with older mice having a thinner layer of mucus compared to young mice. As mentioned above, mucus thickness may influence mucoadhesion, so it would be interesting to investigate age-related changes in mucus in humans. Other important changes that occur with ageing are increased illness and polymedication [21]. Drugs and inflammatory bowel diseases could also affect mucoadhesion, as they can affect pH and mucus [37]. It is important to note that there are many factors that can influence mucoadhesion in vivo, especially in the elderly. These factors are less likely to have a visible effect on ex vivo mucoadhesion studies as their influence may be small. However, when it comes to in vivo performance, they should be considered.

The final point evaluated in our study was the comparison of mucoadhesion to porcine versus human small intestinal tissue. The data for porcine tissue were taken from a previous study carried out in our laboratory [22] under the same conditions. It should be noted that the results of this comparison can only indicate a possible trend, as the number of samples for the porcine tissue was too small. This is related to the fact that the porcine experiments were aimed at a broader screening with more different setups and variables, thus limiting the sample size of measurements comparable to human ex vivo measurements from this study. No significant differences were found for either WoA or F_{max} on fresh and thawed porcine or human tissue. The individual data may suggest that the WoA is slightly higher on human mucosa compared to porcine mucosa, especially when thawed tissue is used. This again might lead to the conclusion that the WoA is the better surrogate for mucoadhesion as it seems to be more sensitive to changes in the tested mucosal sample.

Pigs are often used as model animals for studies involving the gastrointestinal tract because their gastrointestinal physiology is very similar to that of humans [16]. In addition, the availability of tissues is usually good, as pigs are common farm animals, and intestinal tissues are most often slaughterhouse waste. Jackson and Perkins reported that the mucoadhesion of cholestyramine on porcine gastric mucosa was found to be higher than with human mucosa [38]. They explained this result with a possibly thicker mucus layer in pigs compared to humans. This is contrary to the results obtained in our study, but the limited number of porcine tissue samples and a different mucosa may influence the outcome of these studies.

5. Conclusions

The purpose of this ex vivo study was to highlight the inter-individual variability of mucoadhesion to human small intestine tissue. The study data showed the range of individual results, highlighting the high variability of biological materials. The results show that an ideal mucoadhesive DDS should be able to demonstrate good adhesion despite the high inter-individual variability. PVA mucoadhesive films were used to investigate the test-related factors influencing this variability.

Storage is an important factor influencing the study results and should be considered when performing a mucoadhesion test. The WoA seems to be more sensitive to the storage of tissue when a force of 0.1 N is applied for 60 s. The effect on F_{max} is less pronounced. No statistical differences for both can be found if a higher force of 0.35 N is applied for 180 s. Comparing setting A (lower force and contact time) to setting B (higher force and contact time) shows that there is a significant difference in the measurement results of WoA and F_{max} on fresh tissue. Again, no difference was found on thawed tissue for the F_{max}. Therefore, WoA is assumed to be the better surrogate for mucoadhesion. The results show that the adhesion is dependent on both the test setup and the sample preparation. An ideal test setup and storage of the sample to which the dosage form is to adhere must be individually tested prior to each test.

Although the data available were limited, a comparison of mucoadhesion on porcine and human mucosa was made. The initial impression is that the two tissues are comparable. If further studies confirm the hypothesis that porcine intestinal tissue could replace human intestinal tissue, this would facilitate ex vivo mucoadhesion studies. Tissue of animal origin can be obtained in larger quantities and without the regulatory requirements of ex vivo human tissue studies.

Despite the test-related factors, the sampling site was examined as a patient-related factor. No difference was found between proximal jejunum and distal ileum. Other patient-related factors need to be investigated in the future, as mucoadhesion is a complex phenomenon and the understanding of all factors affecting mucoadhesion in vivo is still limited. Gender, mucus thickness, gastrointestinal fluids, diseases and medications may be other parameters to consider. A larger number of subjects would be needed to gain a deeper understanding of the physiological effects on mucoadhesion and to design an ideal DDS that is minimally affected by these variables. A simple way to address the variability and allow comparability between different test devices, tissue preparations and possible new innovative delivery forms could be to measure adhesion against a simple and reproducible manufacturable standard, such as a polyvinyl alcohol film.

Supplementary Materials: The following supporting information can be downloaded at: https://www.mdpi.com/article/10.3390/pharmaceutics15061740/s1, Table S1. Individual medians of F_{max} (N) and WoA (Nm × mm) of all subjects in both test setups (setting A and B) and different tissue preparation. All tests were performed in six replicates.

Author Contributions: Conceptualization, L.M., C.R., M.G., A.B. and W.W.; methodology, L.M. and C.R.; formal analysis, L.M., A.R. and M.G.; investigation, L.M. and F.K.; resources, A.K. and A.B.; data curation, L.M., A.K. and A.B.; writing—original draft preparation, L.M.; writing—review and editing, C.R., A.R., M.G., A.B. and W.W.; visualization, L.M.; supervision, A.B. and W.W.; project administration, M.G., A.B. and W.W. All authors have read and agreed to the published version of the manuscript.

Funding: This research received no external funding.

Institutional Review Board Statement: The study was approved by the Ethics Committee of the Greifswald University Medicine (Ethical Protocol No. BB 027/21).

Informed Consent Statement: Informed consent was obtained from all subjects involved in the study. Subjects are not identifiable in these data; nonetheless, written informed consent has been obtained from the subjects to publish this paper.

Data Availability Statement: Data are available within the paper and its supplementary material.

Acknowledgments: The authors would like to thank the team at the Department of General, Visceral, Thoracic and Vascular Surgery of the University Medicine of Greifswald for their excellent support. The graphical abstract of the manuscript has been created using BioRender.com.

Conflicts of Interest: The authors declare no conflict of interest.

References

1. Sultan, A.A.; El-Gizawy, S.A.; Osman, M.A.; el Maghraby, G.M. Colloidal Carriers for Extended Absorption Window of Furosemide. *J. Pharm. Pharmacol.* **2016**, *68*, 324–332. [CrossRef] [PubMed]
2. Shin, S.; Kim, T.H.; Lee, D.Y.; Chung, S.E.; Lee, J.B.; Kim, D.-H.; Shin, B.S. Development of a Population Pharmacokinetics-Based in Vitro-in Vivo Correlation Model for Drugs with Site-Dependent Absorption: The Acyclovir Case Study. *AAPS J.* **2020**, *22*, 27. [CrossRef] [PubMed]
3. Kharia, A.A.; Singhai, A.K. Development and Optimisation of Mucoadhesive Nanoparticles of Acyclovir Using Design of Experiments Approach. *J. Microencapsul.* **2015**, *32*, 521–532. [CrossRef] [PubMed]
4. Abouelatta, S.M.; Aboelwafa, A.A.; El-Gazayerly, O.N. Gastroretentive Raft Liquid Delivery System as a New Approach to Release Extension for Carrier-Mediated Drug. *Drug Deliv.* **2018**, *25*, 1161–1174. [CrossRef]
5. Sheng, J.; He, H.; Han, L.; Qin, J.; Chen, S.; Ru, G.; Li, R.; Yang, P.; Wang, J.; Yang, V.C. Enhancing Insulin Oral Absorption by Using Mucoadhesive Nanoparticles Loaded with LMWP-Linked Insulin Conjugates. *J. Control. Release* **2016**, *233*, 181–190. [CrossRef]
6. Banerjee, A.; Lee, J.; Mitragotri, S. Intestinal Mucoadhesive Devices for Oral Delivery of Insulin. *Bioeng. Transl. Med.* **2016**, *1*, 338–346. [CrossRef]
7. Amaral, M.; Martins, A.S.; Catarino, J.; Faísca, P.; Kumar, P.; Pinto, J.F.; Pinto, R.; Correia, I.; Ascensão, L.; Afonso, R.A.; et al. How Can Biomolecules Improve Mucoadhesion of Oral Insulin? A Comprehensive Insight Using Ex-Vivo, In Silico, and In Vivo Models. *Biomolecules* **2020**, *10*, 675. [CrossRef]
8. Dragan, E.S.; Dinu, M.V. Polysaccharides Constructed Hydrogels as Vehicles for Proteins and Peptides. A Review. *Carbohydr. Polym.* **2019**, *225*, 115210. [CrossRef]
9. Göbel, A.; da Silva, J.B.; Cook, M.; Breitkreutz, J. Development of Buccal Film Formulations and Their Mucoadhesive Performance in Biomimetic Models. *Int. J. Pharm.* **2021**, *610*, 121233. [CrossRef]
10. Alaei, S.; Omidi, Y.; Omidian, H. In Vitro Evaluation of Adhesion and Mechanical Properties of Oral Thin Films. *Eur. J. Pharm. Sci.* **2021**, *166*, 105965. [CrossRef]
11. Hall, D.J.; Khutoryanskaya, O.V.; Khutoryanskiy, V.V. Developing Synthetic Mucosa-Mimetic Hydrogels to Replace Animal Experimentation in Characterisation of Mucoadhesive Drug Delivery Systems. *Soft Matter* **2011**, *7*, 9620–9623. [CrossRef]
12. Bassi da Silva, J.; Ferreira, S.; Reis, A.; Cook, M.; Bruschi, M. Assessing Mucoadhesion in Polymer Gels: The Effect of Method Type and Instrument Variables. *Polymers* **2018**, *10*, 254. [CrossRef]
13. Ammar, H.O.; Ghorab, M.M.; Mahmoud, A.A.; Shahin, H.I. Design and In Vitro/In Vivo Evaluation of Ultra-Thin Mucoadhesive Buccal Film Containing Fluticasone Propionate. *AAPS PharmSciTech* **2017**, *18*, 93–103. [CrossRef]
14. Dalskov Mosgaard, M.; Strindberg, S.; Abid, Z.; Singh Petersen, R.; Højlund Eklund Thamdrup, L.; Joukainen Andersen, A.; Sylvest Keller, S.; Müllertz, A.; Hagner Nielsen, L.; Boisen, A. Ex Vivo Intestinal Perfusion Model for Investigating Mucoadhesion of Microcontainers. *Int. J. Pharm.* **2019**, *570*, 118658. [CrossRef]
15. Rençber, S.; Karavana, S.Y.; Yilmaz, F.F.; Eraç, B.; Nenni, M.; Gurer-Orhan, H.; Limoncu, M.H.; Güneri, P.; Ertan, G. Formulation and Evaluation of Fluconazole Loaded Oral Strips for Local Treatment of Oral Candidiasis. *J. Drug Deliv. Sci. Technol.* **2019**, *49*, 615–621. [CrossRef]
16. Ziegler, A.; Gonzalez, L.; Blikslager, A. Large Animal Models: The Key to Translational Discovery in Digestive Disease Research. *Cell Mol. Gastroenterol. Hepatol.* **2016**, *2*, 716–724. [CrossRef]
17. Kararli, T.T. Comparison of the Gastrointestinal Anatomy, Physiology, and Biochemistry of Humans and Commonly Used Laboratory Animals. *Biopharm. Drug Dispos.* **1995**, *16*, 351–380. [CrossRef]
18. McConnell, E.L.; Basit, A.W.; Murdan, S. Measurements of Rat and Mouse Gastrointestinal PH, Fluid and Lymphoid Tissue, and Implications for in-Vivo Experiments. *J. Pharm. Pharmacol.* **2010**, *60*, 63–70. [CrossRef]
19. Varum, F.J.O.; Veiga, F.; Sousa, J.S.; Basit, A.W. An Investigation into the Role of Mucus Thickness on Mucoadhesion in the Gastrointestinal Tract of Pig. *Eur. J. Pharm. Sci.* **2010**, *40*, 335–341. [CrossRef] [PubMed]
20. Allen, A.; Flemstrom, G.; Garner, A.; Kivilaakso, E. Gastroduodenal Mucosal Protection. *Physiol. Rev.* **1993**, *73*, 823–857. [CrossRef]

21. Pazan, F.; Wehling, M. Polypharmacy in Older Adults: A Narrative Review of Definitions, Epidemiology and Consequences. *Eur. Geriatr. Med.* **2021**, *12*, 443–452. [CrossRef]
22. Müller, L.; Rosenbaum, C.; Krause, J.; Weitschies, W. Characterization of an In Vitro/Ex Vivo Mucoadhesiveness Measurement Method of PVA Films. *Polymers* **2022**, *14*, 5146. [CrossRef] [PubMed]
23. Baraibar, M.A.; Schoning, P. Effects of Freezing and Frozen Storage on Histological Characteristics of Canine Tissues. *J. Forensic. Sci.* **1985**, *30*, 439–447. [CrossRef] [PubMed]
24. Koziolek, M.; Schneider, F.; Grimm, M.; Modeβ, C.; Seekamp, A.; Roustom, T.; Siegmund, W.; Weitschies, W. Intragastric PH and Pressure Profiles after Intake of the High-Caloric, High-Fat Meal as Used for Food Effect Studies. *J. Control. Release* **2015**, *220*, 71–78. [CrossRef] [PubMed]
25. Hägerström, H.; Edsman, K. Interpretation of Mucoadhesive Properties of Polymer Gel Preparations Using a Tensile Strength Method. *J. Pharm. Pharmacol.* **2010**, *53*, 1589–1599. [CrossRef] [PubMed]
26. Kulkarni, R.; Fanse, S.; Burgess, D.J. Mucoadhesive Drug Delivery Systems: A Promising Non-Invasive Approach to Bioavailability Enhancement. Part I: Biophysical Considerations. *Expert Opin. Drug Deliv.* **2023**, *20*, 395–412. [CrossRef] [PubMed]
27. Peppas, N.A.; Thomas, J.B.; McGinty, J. Molecular Aspects of Mucoadhesive Carrier Development for Drug Delivery and Improved Absorption. *J. Biomater. Sci. Polym. Ed.* **2009**, *20*, 1–20. [CrossRef]
28. Khutoryanskiy, V.V. Advances in Mucoadhesion and Mucoadhesive Polymers. *Macromol. Biosci.* **2011**, *11*, 748–764. [CrossRef] [PubMed]
29. Bayer, I.S. Recent Advances in Mucoadhesive Interface Materials, Mucoadhesion Characterization, and Technologies. *Adv. Mater. Interfaces* **2022**, *9*, 2200211. [CrossRef]
30. das Neves, J.; Amaral, M.H.; Bahia, M.F. Performance of an in Vitro Mucoadhesion Testing Method for Vaginal Semisolids: Influence of Different Testing Conditions and Instrumental Parameters. *Eur. J. Pharm. Biopharm.* **2008**, *69*, 622–632. [CrossRef]
31. Smart, J.D. The Basics and Underlying Mechanisms of Mucoadhesion. *Adv. Drug Deliv. Rev.* **2005**, *57*, 1556–1568. [CrossRef] [PubMed]
32. Srivastava, A.; Verma, A.; Saraf, S.; Jain, A.; Tiwari, A.; Panda, P.K.; Jain, S.K. Mucoadhesive Gastroretentive Microparticulate System for Programmed Delivery of Famotidine and Clarithromycin. *J. Microencapsul.* **2021**, *38*, 151–163. [CrossRef] [PubMed]
33. Baus, R.A.; Haug, M.F.; Leichner, C.; Jelkmann, M.; Bernkop-Schnürch, A. In Vitro-in Vivo Correlation of Mucoadhesion Studies on Buccal Mucosa. *Mol. Pharm.* **2019**, *16*, 2719–2727. [CrossRef]
34. Dumic, I.; Nordin, T.; Jecmenica, M.; Stojkovic Lalosevic, M.; Milosavljevic, T.; Milovanovic, T. Gastrointestinal Tract Disorders in Older Age. *Can. J. Gastroenterol. Hepatol.* **2019**, *2019*, 6757524. [CrossRef]
35. Lipski, P.S.; Bennett, M.K.; Kelly, P.J.; James, O.F. Ageing and Duodenal Morphometry. *J. Clin. Pathol.* **1992**, *45*, 450–452. [CrossRef] [PubMed]
36. Elderman, M.; Sovran, B.; Hugenholtz, F.; Graversen, K.; Huijskes, M.; Houtsma, E.; Belzer, C.; Boekschoten, M.; de Vos, P.; Dekker, J.; et al. The Effect of Age on the Intestinal Mucus Thickness, Microbiota Composition and Immunity in Relation to Sex in Mice. *PLoS ONE* **2017**, *12*, e0184274. [CrossRef]
37. Johansson, M.E.V. Mucus Layers in Inflammatory Bowel Disease. *Inflamm. Bowel. Dis.* **2014**, *20*, 2124–2131. [CrossRef]
38. Jackson, S.J.; Perkins, A.C. In Vitro Assessment of the Mucoadhesion of Cholestyramine to Porcine and Human Gastric Mucosa. *Eur. J. Pharm. Biopharm.* **2001**, *52*, 121–127. [CrossRef]

pharmaceutics

MDPI

Article

Micro Injection Molding of Drug-Loaded Round Window Niche Implants for an Animal Model Using 3D-Printed Molds

Robert Mau [1], Thomas Eickner [2], Gábor Jüttner [3], Ziwen Gao [4,5], Chunjiang Wei [4,5], Nicklas Fiedler [2], Volkmar Senz [2], Thomas Lenarz [4,5], Niels Grabow [2,6], Verena Scheper [4,5,*] and Hermann Seitz [1,6,*]

1 Microfluidics, Faculty of Mechanical Engineering and Marine Technology, University of Rostock, Justus-von-Liebig Weg 6, 18059 Rostock, Germany; robert.mau@uni-rostock.de
2 Institute for Biomedical Engineering, University Medical Center Rostock, Friedrich-Barnewitz-Straße 4, 18119 Rostock, Germany; thomas.eickner@uni-rostock.de (T.E.); nicklas.fiedler@uni-rostock.de (N.F.); volkmar.senz@uni-rostock.de (V.S.); niels.grabow@uni-rostock.de (N.G.)
3 Kunststoff-Zentrum in Leipzig gGmbH (KUZ), Erich-Zeigner-Allee 44, 04229 Leipzig, Germany; juettner@kuz-leipzig.de
4 Lower Saxony Center for Biomedical Engineering, Implant Research and Development (NIFE), Department of Otorhinolaryngology, Head and Neck Surgery, Hannover Medical School, Stadtfelddamm 34, 30625 Hannover, Germany; gao.ziwen@mh-hannover.de (Z.G.); wei.chunjiang@mh-hannover.de (C.W.); lenarz.thomas@mh-hannover.de (T.L.)
5 Cluster of Excellence "Hearing4all", Department of Otorhinolaryngology, Head and Neck Surgery, Hannover Medical School, Carl-Neuberg-Straße 1, 30625 Hannover, Germany
6 Department Life, Light & Matter, Interdisciplinary Faculty, University of Rostock, Albert-Einstein-Str. 25, 18059 Rostock, Germany
* Correspondence: scheper.verena@mh-hannover.de (V.S.); hermann.seitz@uni-rostock.de (H.S.); Tel.: +49-511-532-4369 (V.S.); +49-381-498-9090 (H.S.)

Abstract: A novel approach for the long-term medical treatment of the inner ear is the diffusion of drugs through the round window membrane from a patient-individualized, drug-eluting implant, which is inserted in the middle ear. In this study, drug-loaded (10 wt% Dexamethasone) guinea pig round window niche implants (GP-RNIs, ~1.30 mm × 0.95 mm × 0.60 mm) were manufactured with high precision via micro injection molding (μIM, T_{mold} = 160 °C, crosslinking time of 120 s). Each implant has a handle (~3.00 mm × 1.00 mm × 0.30 mm) that can be used to hold the implant. A medical-grade silicone elastomer was used as implant material. Molds for μIM were 3D printed from a commercially available resin (T_G = 84 °C) via a high-resolution DLP process (xy resolution of 32 μm, z resolution of 10 μm, 3D printing time of about 6 h). Drug release, biocompatibility, and bioefficacy of the GP-RNIs were investigated in vitro. GP-RNIs could be successfully produced. The wear of the molds due to thermal stress was observed. However, the molds are suitable for single use in the μIM process. About 10% of the drug load (8.2 ± 0.6 μg) was released after 6 weeks (medium: isotonic saline). The implants showed high biocompatibility over 28 days (lowest cell viability ~80%). Moreover, we found anti-inflammatory effects over 28 days in a TNF-α-reduction test. These results are promising for the development of long-term drug-releasing implants for human inner ear therapy.

Keywords: micro injection molding; 3D printing; rapid tooling; digital light processing; implant; drug delivery system; dexamethasone; anti-inflammatory; TNF-α; biocompatibility; inner ear therapy

Citation: Mau, R.; Eickner, T.; Jüttner, G.; Gao, Z.; Wei, C.; Fiedler, N.; Senz, V.; Lenarz, T.; Grabow, N.; Scheper, V.; et al. Micro Injection Molding of Drug-Loaded Round Window Niche Implants for an Animal Model Using 3D-Printed Molds. *Pharmaceutics* **2023**, *15*, 1584. https://doi.org/10.3390/pharmaceutics15061584

Academic Editors: Silvia Tampucci and Daniela Monti

Received: 14 April 2023
Revised: 16 May 2023
Accepted: 22 May 2023
Published: 24 May 2023

1. Introduction

There is increasing interest in novel concepts of medical treatment of the inner ear in order to treat disorders such as Menière's disease (MD) and idiopathic sudden sensorineural hearing loss (ISSHL). Pharmaceutical substances can pass from the middle to the inner ear via diffusion through the semipermeable round window membrane (RWM). The RWM is located deep in a recess, the round window niche (RWN), between the middle and inner ear (Figure 1). There are various pharmacological treatment methods and drug delivery strategies. For example, it is a common treatment to inject drugs or drug-laden gels directly into the middle ear cavity (intratympanically) by needle through the tympanic membrane [1]. However, such methods have a significant disadvantage. Large portions of the applied drug cannot diffuse through the RWM in the inner ear because it does not come into sufficient contact with it. Instead, much of the applied drug is absorbed by the mucosa of the middle ear or evacuated from the middle ear space by the Eustachian tube [2,3].

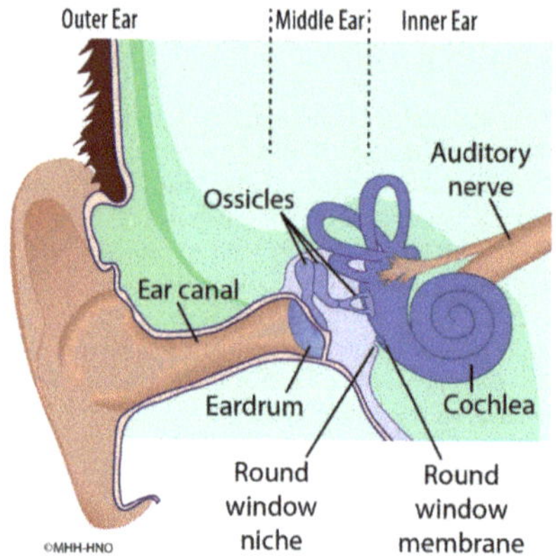

Figure 1. Scheme of the anatomic structures of outer ear, middle ear and inner ear. The round window membrane is located between the middle ear and the inner ear.

To overcome this drawback and to provide a more efficient and safe administration route for controlled drug release in the inner ear, we introduced a new concept of a patient-individualized, drug-loaded round window niche implant (RNI) [4,5]. Following our concept, an improved drug transport into the inner ear for several weeks might be obtained via drug diffusion from an RNI through the RWM. For that purpose, an RNI must meet the individual anatomical needs of a patient to fit precisely onto the RWM, and it should have dimensions of just a few millimeters in xyz directions. The RNI should be characterized by a soft and stretchable mechanical behavior, as we have already found in a prototype implantation study [5]. Moreover, inflammation-suppressing substances, such as glucocorticoids and especially dexamethasone (DEX), are promising for the drug load of an RNI. These substances have been proven to be promising in treating inner ear pathologies, including sudden sensorineural hearing loss (SSNHL) [6–9], Menière's disease [10–13] and acute tinnitus [14]. Moreover, DEX positively affects the preservation of residual hearing and the reduction of fibrosis after cochlear implant (CI) surgery [15]. Recently, we found preferable concentration ranges for different DEX formulations to ensure biocompatibility and bioefficacy [16].

For highly individualized and complex products, such as the RNI described, 3D printing technology (also referred to as additive manufacturing) offers promising opportunities for both time- and cost-efficient production. In general, additive manufacturing methods that use photopolymerization enable the highest resolutions in the 3D printing sector [17]. There are various 3D printing processes that use photopolymerization, such as the vat photopolymerization methods (e.g., Digital Light Processing (DLP), stereolithography (SLA), two-photon polymerization (2PP)) or the material jetting methods (e.g., PolyJet, Multi-Jet Modeling (MJM)). The technology of DLP 3D printing is one of the most widely used processes because it offers high printing speeds and low running costs [18,19]. However, the availability and validity of ready-to-use, medical-grade materials that can be processed with photopolymerizing additive manufacturing methods are still limiting factors. On the one hand, there is a range of commercially available biocompatible photosensitive resins, but the majority of these are for dental applications (e.g., surgical guides, retainers, aligners, temporary dentures) or hearing aids and are suitable only for temporary skin contact (30+ days) or short-term mucosal contact (not more than 24 h) [20,21]. On the other hand, there is a lack of bioresins for applications such as tissue-engineered human constructs [21]. Very few resins certified as hemocompatible are commercially available. In a recent review article on biocompatible 3D printing resins, Guttridge et al. report that they found only one material (PrintoDent GR-20, pro3dure medical GmbH, Iserlohn, Germany) that was tested and certified as hemocompatible according to ISO 10993-4 [20]. When dealing with photopolymers, in general, there are risks such as cytotoxic, mutagenic and allergic reactions resulting from incomplete photopolymerization [18]. Information about the intended use, certification and postprocessing is highly variable for commercially available photopolymers [20]. It is often necessary to develop a specific postprocessing treatment to ensure sufficient biocompatibility for the desired use [20]. Another limitation is the lack of biodegradable resins. As described by Bao et al., there has been considerable progress in the development of biodegradable medical devices or implants using vat photopolymerization techniques such as DLP. However, further advancements in both novel materials and photopolymerization 3D printing techniques are needed for the challenging translation process toward clinical applications [22]. In contrast, conventional and well-established manufacturing via injection molding (IM) technologies is suitable for various materials, e.g., composite materials, foamed materials, thermoplastic and thermosetting plastics, rubber and even metals [23–25]. The use of medical-grade polymers [26,27] and the applicability of IM for the manufacturing of drug delivery systems have been investigated and established for many years [28]. Moreover, the micro injection molding method (μIM) enables the highest resolution and precision [29–31].

The processes of IM and μIM are based on the utilization of molds. Conventional manufacturing of metal molds via milling is relatively cost- and time-consuming, especially when there is a need for complex geometries. Therefore, IM and μIM are not usually used to manufacture a small series or even individualized single parts. To address this limitation, 3D printing technology enables rapid tooling as a cost- and time-saving method for mold manufacturing [32,33]. High-resolution photopolymerizing 3D printing methods, such as material jetting, SLA, or DLP, enable high process resolution for rapid tooling applications in micromanufacturing applications [34]. For instance, DLP potentially enables a cost reduction of 80% to 90% compared to conventional mold manufacturing [32].

In this study, we demonstrate a promising way of saving costs and time with a high-precision μIM of individualized DEX-loaded RNI using a medical-grade soft material. For that purpose, a mean guinea pig round window niche implant (GP-RNI) is manufactured via μIM. With a view to future investigation and translation, the GP-RNI features an exemplary implant geometry suitable for a favored animal model. We use high-resolution DLP 3D printing for rapid tooling of molds for the μIM process. After manufacturing the drug release, the biocompatibility and bioefficacy of the GP-RNI are investigated. Our manufacturing process aims to combine the best of both worlds of 3D printing and μIM: a cost- and time-saving, high-precision rapid tooling of molds via high-resolution 3D printing and the accessibility of a wide range of (polymeric) medical-grade materials for implant manufacturing, as such materials are well-established for μIM.

2. Materials and Methods

2.1. Three-Dimensional Models of Guinea Pig Round Window Niche and Mold

Figure 2A shows a photograph of the anatomical structure of an exemplary guinea pig round window niche and Figure 2B illustrates a graphic of the digital 3D model of a mean GP-RNI. The model has approximate nominal dimensions of a length of 1.30 mm, a width of 0.95 mm and a height of 0.60 mm. The digital 3D model was established by reconstructing 3D volumes from microCT images (XtremeCTII, ScancoMedical AG, Brüttisellen, Switzerland) via 3D Slicer^{TM} software version 4.11 (Surgical Planning Laboratory, Brigham and Women's Hospital, Harvard Medical School, Boston, MA, USA) [35,36].

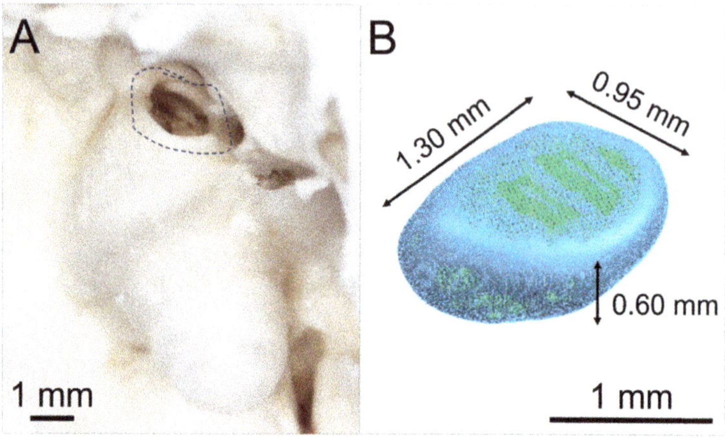

Figure 2. (**A**) Photograph of the anatomical structure of exemplary guinea pig round window niche (GP RWN, marked); (**B**) 3D model of mean guinea pig round window niche implant (GP-RNI). The approximate dimensions of the model are marked in the figure.

The 3D model of the mean GP-RNI was joined with a 3D model of a handle structure (Figure 3A (scheme), Figure 3B (final design)). The handle is an element to ensure good gripping and handling by forceps during the implantation process. This completed 3D model of the mean GP-RNI with a handle is designed to be manufactured via μIM. A sprue structure was added to the 3D model to obtain a flow path for the μIM material (Figure 3C). The sprue structure is removed from the μIM implant after the μIM process. Figure 3D shows the final 3D model of the mold (2 parts). It is based on the 3D model of the GP-RNI with a handle and the sprue. The 3D model of the mold will be used for rapid tooling of the mold via DLP 3D printing.

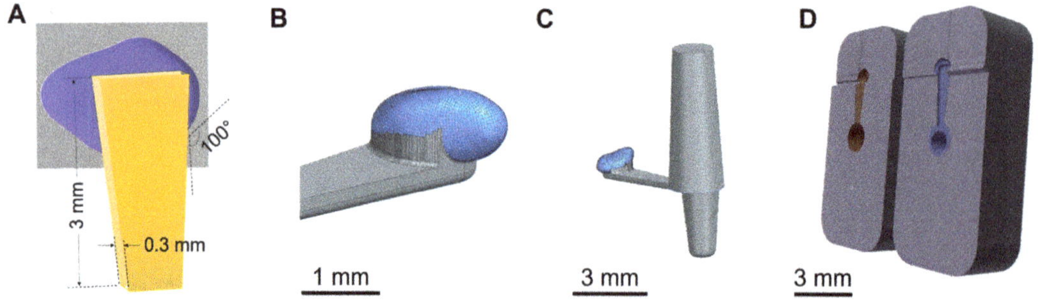

Figure 3. (**A**) Scheme of mean guinea pig round window niche implant (GP-RNI) (blue) with a handle (yellow, for a good gripping by forceps during the implantation process); (**B**) 3D model of mean GP-RNI (blue) with a handle (grey); (**C**) 3D model of GP-RNI (blue) with a handle for implantation (grey) and sprue for μIM (grey, conic); (**D**) 3D model of mold (2 parts, left: upper half, right: lower half) for μIM of GP-RNI with a handle (the negative halves of the implant with a handle and the sprue are colored orange and blue).

2.2. *Rapid Tooling of Molds via Digital Light Processing*

The 3D printing of the micro injection molds was performed with an Asiga Pro 4K45 (Asiga, Alexandria, Australia) using DLP technology. The xy resolution of the LED projector (UV light, $\lambda = 385$ nm, 4k-resolution-mode) was 32 μm and a build platform of 122 mm × 68 mm was installed. For the printing process, the photopolymeric resin Asiga PlasGRAY V2 (Asiga, Alexandria, Australia) [37] was used. The resin enabled a minimum layer thickness of 10 μm and enabled cured parts with a shore hardness of Shore D 82 and a glass transition temperature of T_G = 84 °C. It is designed for the manufacturing of highly detailed parts for dental, jewelry and design industries with high surface smoothness and quality. There were no reinforcing fillers, e.g., to enhance heat resistance for increased IM suitability, as the PlasGRAY V2 resin is a general purpose photopolymeric resin in DLP 3D printing.

Since a single mold is made of two parts (a lower half and an upper half), the parts of the same kind were 3D printed simultaneously. A total of 52 molds were manufactured. For that purpose, 52 lower halves of the molds (including supporting structure) were 3D printed simultaneously in 201 min in a first printing sequence. In addition, 52 upper halves of the molds (including the supporting structure) were 3D printed simultaneously in 166 min in a second printing sequence. All parts were built in a horizontal position (Figure 4). A base plate with a height of 0.4 mm was used as a supporting structure to ensure sufficient adhesion of the 3D-printed part with the built platform. The base plate was built using a layer resolution in z direction (layer height) of 100 μm per layer, exposed for 23.871 s per layer. Further layers of the mold parts were built using a layer height of 10 μm per layer, exposed for 0.498 s per layer. There was a material consumption of 0.417 mL per single mold.

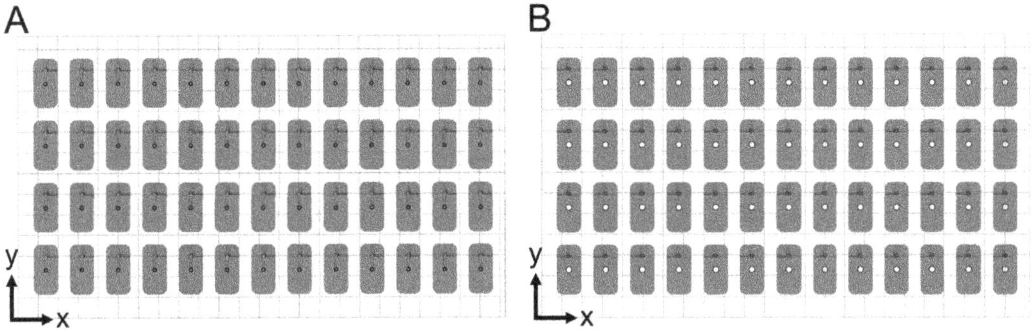

Figure 4. Placement of molds on build platform. Fifty-two of each lower (**A**) and upper (**B**) halves of the final molds were printed in the same orientation to eliminate possible differences in xy resolution.

Postprocessing steps were the washing of the printed parts in 98% isopropyl alcohol in an ultrasonic bath twice for 5 min, drying at room temperature for 30 min, and postcuring the mold halves for 2000 flashes on each side (total of 2 × 2000 flashes) in a UV curing unit Otoflash G171 (NK Optik GmbH, Baierbrunn, Germany).

2.3. Micro Injection Molding of Drug-Loaded Implants

For μIM investigations, a homogenous mixture of medical-grade silicone elastomer MED-4244 (NuSil Technology LLC, Radnor, PA, USA) containing 10 wt% DEX (powder, Sanofi SA, Paris, France) was prepared by manually stirring using a stainless steel laboratory scoop. MED-4244 is a two-part (10:1 *w*:*w*, part A: part B), pourable, translucent silicone elastomer and cures by heat via addition-cure chemistry. It can be used for implants that remain in the human body for a period of more than 29 days [38]. The stirring process was performed for several minutes until a uniform whitish coloration of the mixture appeared as an indicator of the homogenous distribution of the DEX powder in the liquid silicone. The homogeneity of the mixture and the absence of air bubbles in the mixture were evaluated by the naked eye. The preparation of the mixture was performed under an ambient atmosphere.

Micro injection molding (μIM) investigations were performed using a machine of type formicaPlast (Klöckner DESMA Elastomertechnik GmbH, Achim, Germany), modified as shown previously [29] and featuring an injection piston with a diameter of 3 mm. For μIM, a 3D-printed mold was inserted in the μIM machine using customized metallic housing as shown in Figure 5. Each mold was used only once for μIM. The μIM process was performed at a mold temperature of 160 °C and an injection flow rate of 4.2 mm^3/s. The mold temperature of 160 °C was necessary for the heat-driven curing process of the used silicone elastomer. The shot weight (inclusive sprue) was 8.2 mm^3. No extra holding pressure was applied. After injection of the prepared mixture of the silicone elastomer DEX mixture, there was a crosslinking time of 120 s.

Figure 5. Micro injection molding (μIM) machine equipped with an exemplary 3D-printed mold (see white circle). A metallic holder keeps the mold in position.

2.4. Drug Release

Drug release was investigated via high-performance liquid chromatography (HPLC). For that purpose, the masses of the GP-RNI (n = 3) were determined on a Kern 770 microbalance (KERN & Sohn, Balingen, Germany), as shown in Table 1.

Table 1. Masses of the GP-RNIs prior to the drug release testing, resulting in a mean ± standard deviation of 0.86 ± 0.04 mg.

Sample of GP-RNIs	Mass in Mg
1	0.92
2	0.83
3	0.83

The GP-RNIs were placed in 4 mL glass vials and stored at 37 °C in 2 mL isotonic saline (B.Braun, Melsungen, Germany) on a lab shaker (Heidolph, Schwabach, Germany) at 100 rpm. For sampling, the medium was exchanged completely after defined time periods of 0.25; 0.75; 1.5; 3; 6; 13; 24; 29; 101; 197 and 317 h and then every 7 days for an additional 6 weeks. The medium was subsequently mixed 1:1 (*v:v*) with methanol (Carl Roth, Karlsruhe, Germany) and distilled water (Ultrapure water system (Sartorius, Göttingen, Germany)) prior to the HPLC measurements. Quantification of DEX was performed on a

HPLC system (Knauer Wissenschaftlicher Gerätebau Dr. Ing. Herbert Knauer GmbH, Berlin, Germany) equipped with a Chromolith FastGrad RP-18e 50-2 column (Merck KGaA, Darmstadt, Germany). Methanol/Water 1:1 was used as the mobile phase in an isocratic chromatographic method at a flow rate of 0.8 mL/min. Detection occurred with a UV-Detector at the wavelength λ = 254 nm [39]. For calibration, DEX standards with concentrations of 0.1, 0.5, 1.0, 2.0, 5.0, 10 and 50 µg/mL were used.

2.5. Biocompatibility

For in vitro biocompatibility and bioefficacy (see Section 2.6.) studies, eluates were generated by incubating RNIs in 24-well plates (Nunc, Thermo Fisher Scientific, Waltham, MA, USA) with 500 µL saline (600 µL NaCl 0.9%, B. Braun, Melsungen, Germany) per well in an incubator (CB150; Binder, Tübingen, Germany; 37 °C, 5% CO_2, 95% humidity) for 1, 3, 7, 10, 14, 21 and 28 days. On the day of collecting the supernatant, all of the supernatants in each well was taken out and separated into two Eppendorf tubes, one for a biocompatibility test and the second for a bioefficacy test. The wells were refilled with fresh 500 µL saline and the sample was further incubated until the next sampling time point was reached. The supernatants were stored at −20 °C before processing. All experiments were performed in triplicate and repeated three times.

A 3-(4,5-dimethylthiazol-2-yl)-2,5-diphenyltetrazolium bromide (MTT) assay (PanReac AppliChem, Darmstadt, Germany) was performed to investigate the biocompatibility of the supernatant as previously described [16]. NIH/3T3 fibroblasts (mouse, ATCC-Number: CRL-1658, German Collection of Microorganisms and Cell Cultures GmbH, Braunschweig, Germany; passage 3 to 10) were seeded in Dulbecco's modified Eagle's medium (DMEM, Bio and Sell GmbH, Feucht, Germany) with 10% fetal calf serum (FCS, Bio and Sell GmbH, Feucht, Germany), penicillin and streptomycin (100 units/mL each) in a humidified atmosphere (5% CO_2/95% air, 37 °C) as shown before [16]. To perform the MTT assay, the fibroblasts were seeded in 96-well plates at a concentration of 1.5×10^4 cells/mL with 100 µL fresh culture medium. After an incubation time of 24 h, the culture medium was replaced by a fresh culture medium and culture supernatant of GP-RNI samples at a 1:1 ratio. Cells treated with 0.1% DMSO were used as a positive control (PC) for a toxic effect on the cells. The negative control cells (blank) were cultured in a pure complemented medium for regular cell proliferation (50 µL 0.9% NaCl + 50 µL cell culture medium). For validation that the experiments were performed successfully, the PC and blank conditions ran in parallel with every single experiment. All experiments were performed in duplicate and repeated three times.

After 24 h, the medium was removed, replaced by 50 µL 0.5 mg/mL MTT reagent, and incubated for two hours in a humidified atmosphere (5% CO_2/95% air, 37 °C). Subsequently, the MTT reagent medium was removed and replaced by 100 µL MTT solution (isopropanol) per well. To dissolve the formazan produced by MTT reduction, the MTT reagent medium was incubated for five minutes on a rotary shaker at room temperature of 21 °C. The optical density (OD) was determined at a wavelength of 570 nm utilizing a microplate reader (Gen5 2.06.Ink, BioTek Synergy™ H1HyBrid Reader, Santa Clara, CA, USA). The measurement of empty wells without cells was performed for a correction of the OD. The relative cell viability was calculated in percentage terms by dividing the empty-subtracted OD of the test groups by the empty-subtracted OD of the blank and multiplying the result by 100. Cell viability below 70% was judged as being cytotoxic, which is in accordance with ISO guideline 10993-5:2009 for the biological evaluation of medical devices. The normal distribution of data was checked (Kolmogorov–Smirnov Test) and an analysis of variance (ANOVA) and a Dunnett's Multiple Comparison Test was conducted using GraphPad Prism® version 8.4.3 (GraphPad Prism Software Inc., La Jolla, CA, USA). Means ± standard deviations of the data were reported and the statistical significance was considered at p-values less than 0.05.

2.6. *Bioefficacy*

A TNF-α-reduction test was performed for the investigation of the anti-inflammatory effect of the μIM-manufactured GP-RNI samples containing 10 wt% DEX as shown before [40]. It is assumed that the GP-RNI should release DEX into the supernatant (see Section 2.5. for supernatant sampling), and to be bioeffective, the released DEX should reduce the TNF-α-production of cells being stressed with lipopolysaccharide (LPS, Sigma-Aldrich, St. Louis, MO, USA). Experiments were performed in triplicate (n = 3) per plate (N = 3) for every condition: DC2.4 mouse cells (DCs) (Sigma-Aldrich, St. Louis, MO, USA, LOT:3093896) were cultured in 48-well plates in RPMI 1640 medium (Sigma-Aldrich, St. Louis, MO, USA), which was supplemented with non-essential amino acids (1 mmol/L, Sigma-Aldrich, St Louis, MO, USA) and 10% FCS (Bio & Sell GmbH, Feucht, Germany). The cells were cultivated for 24 h in an incubator and subsequently divided into negative control (NC), positive control (PC) and supernatant groups. All cells, except the NC, were stressed by adding 100 μL LPS to the medium (0.5 μg/mL). The PC and NC conditions were conducted in parallel with each single experiment for the validation that each experiment was performed successfully. The supernatants of the cultured GP-RNIs were added to the wells and the cells were incubated for an additional 24 h. After 24 h, the supernatant was collected and ELISA analysis was performed. ELISA kits (Boster Biological Technology, Pleasanton, CA, USA) were used in accordance with the manufacturer's instructions. Each supernatant was applied in dilution and as a replicate to the ELISA plate. The absorbance of OD was recorded at a wavelength of 450 nm utilizing a MicroPlate Reader (Gen5 2.06.Ink, BioTekSynergy™ H1HyBrid Reader, Santa Clara, CA, USA).

Because of the non-normal distribution of data (Kolmogorov–Smirnov Test), subsequently, each sampling time point was separately tested for relevant differences in TNF reduction compared to PC using the Mann–Whitney U Test. The Friedman test ($p < 0.0001$), followed by Dunn's Multiple Comparison Test, was run to detect differences between the dependent variable of the different time points. The data are reported as mean ± standard deviation. Statistical significance was considered at p-values less than 0.05.

3. Results and Discussion

3.1. *Rapid Tooling of Molds via Digital Light Processing*

Figure 6 shows both parts of a DLP 3D-printed mold after 3D printing but before finishing the parts by milling off the remnants of the supporting structure (Figure 6A,B, marked via white arrows). The upper half of the mold (Figure 6A) features the geometry of the upper half of the GP-RNI and the sprue. The sprue is a relatively large channel through which the liquid polymer material enters the mold. Furthermore, the lower half of the mold (Figure 6B) features the geometry of the GP-RNI and the form of the handle structure of the GP-RNI. The handle structure is used as a runner structure for the transport of liquid polymer inside the mold. Moreover, the lower half of the mold features a "cold slug" structure for the μIM process. The forms of the halves of the GP-RNI are of high precision because of the high resolution of the utilized DLP 3D printing process featuring a relatively low z-layer height of 10 μm per layer (Figure 6C,D).

3.2. *Micro Injection Molding of Implant Prototypes*

Figure 7 shows the finished and assembled 3D-printed halves of the mold. The parts fit closely. There is minimal clearance between the upper and lower halves before they are used for the μIM injection molding process (see Figure 7A). The edges of the parts are flush with each other. After the μIM process (single use, Figure 7B), the clearance between the upper and the lower halves is decreased. The upper half of the mold shows a crack as a sign of wear.

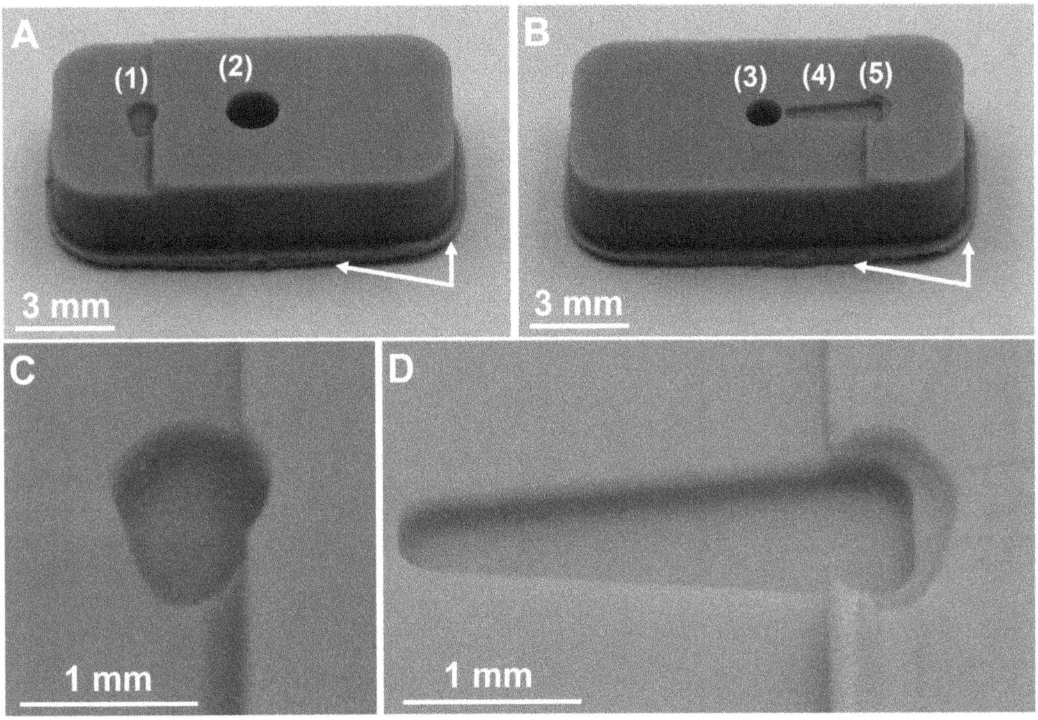

Figure 6. Photographs of DLP 3D-printed mold after 3D printing (but before finishing). The mold consists of two parts. (**A**) shows the upper half of mold featuring the form of the upper half of the GP-RNI (1) and the sprue (2). (**B**) shows the lower half of the mold featuring the form of the lower half of the GP-RNI (3) and the form of the handle structure of the GP-RNI (4). The mold is designed to use the handle structure as a runner structure for liquid μIM material during μIM process. Moreover, there is the "cold slug" structure (5) for μIM process. Remnants of the supporting structure are marked via white arrows. (**C**,**D**) show the forms of the GP-RNI in more detail. The small structures are 3D printed with a high grade of precision because a high-resolution DLP 3D printing process featuring a relatively low z-layer height of 10 μm per layer was used.

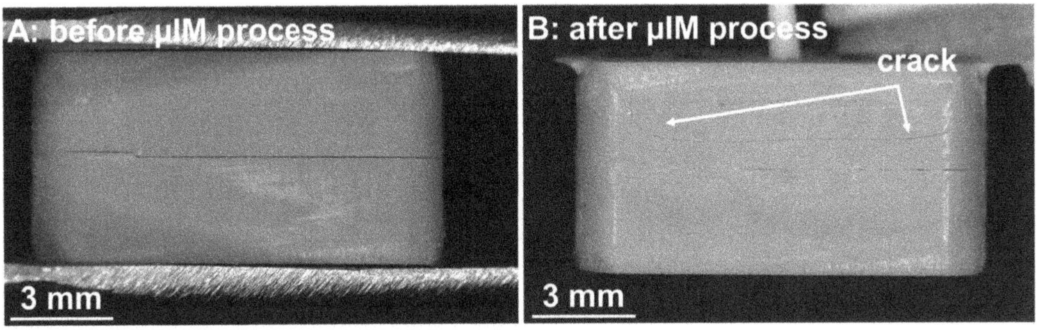

Figure 7. (**A**) Photograph of an assembled mold (made of upper and lower halves) before μIM process. Both parts of the mold fit well together. The edges of both halves are flush with each other. (**B**) A mold after μIM process. The upper half is cracked.

Figure 8 shows a separated mold after a finished μIM process. The mixture of the silicone elastomer (MED-4244) and the glucocorticoid DEX (10 wt%) was successfully injected and cured. Despite the decreased clearance between the parts (compare Figure 7A,B), the two halves were not significantly merged and the separation of the halves was easy. However, there are significant signs of wear all over the surface of both parts of the mold. Nevertheless, the separation of the GP-RNI from the mold worked well and did not cause any damage to the μIM implant.

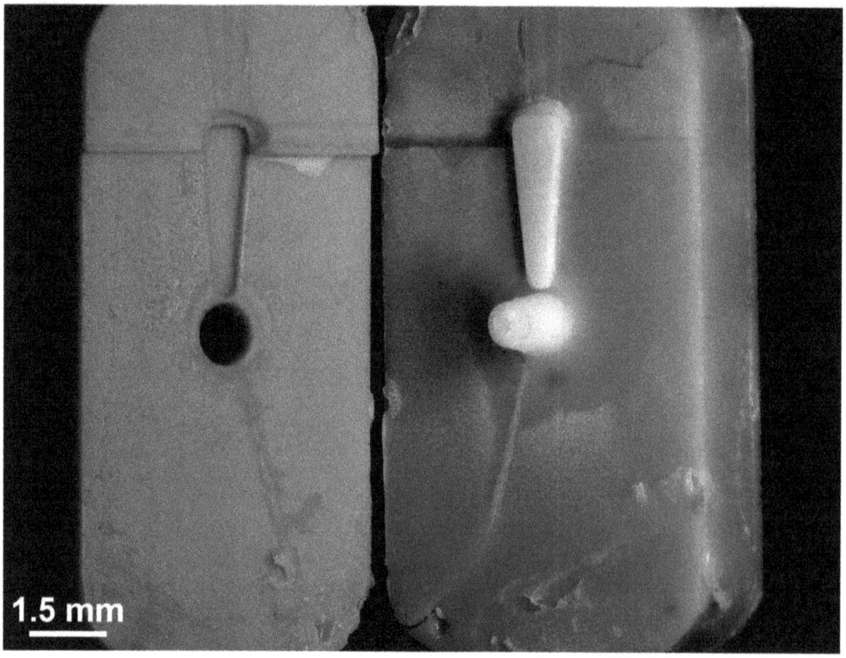

Figure 8. Photograph of a separated mold after the finished μIM process. The μIM material, a mixture of a medical-grade silicone elastomer (MED-4244, NuSil Technology) and DEX (10 wt%), was successfully injected and cured. It was separated from one part of the mold without any damage to the GP-RNI. There are significant signs of wear all over the surface of both parts of the mold.

Low thermal conductivity, low heat resistance and, consequently, low durability and high cycle times are known limitations of molds manufactured from available photopolymeric resins [32,34]. In this study, the combination of thermal and mechanical stress during the μIM process is most likely the reason for the wear of the DLP 3D-printed molds, as these are significant factors [32]. The glass transition temperature (T_G) of the photopolymer PlasGRAY V2, which was used for DLP 3D printing of the molds, is T_G = 84 °C [37]. The mold temperature during the performed μIM process was T_{mold} = 160 °C. When the mold temperature is similar to T_G, the polymer material may become brittle and prone to failure [32]. Zink et al. recommend keeping the mold temperature below the glass transition temperature of the mold material [41]. Following Zink et al., the parameter of the mold temperature significantly affects the applicability of 3D-printed polymeric molds. The mold material's mechanical properties are impaired as a function of temperature, especially when using temperatures above T_G. In another study by Martinho et al., the authors recommend keeping the mold temperature below 15 °C above T_G of the 3D-printed photopolymer resin [42]. In our study, we used a relatively high T_{mold} = 160 °C for a mold material with a T_G = 84 °C. The manufactured GP-RNIs were of proper quality. However, the molds showed significant wear because of thermal stress. Because of that, we used a single mold only once for our μIM process.

Due to economical manufacturing criteria, 3D-printed polymer molds can be well suited for low- and medium-volume injection molding production [32]. Commonly, for a medical implant, it is desirable to meet a patient's individual needs. Implants such as the described RNI benefit from high customization. Low durability of the molds and relatively high cycle times of an injection molding process might be acceptable. As a result, the described rapid tooling process chain based on 3D-printed molds and μIM has proven to be a promising manufacturing technique for the production of highly individualized single parts. Furthermore, there are promising strategies to overcome the limitations of heat transport and heat resistance, such as composite materials [32] and innovative cooling channels in the 3D-printed molds [41,43]. Even relatively simple methods such as tight-fitting metal mold holders can help to deal with low heat resistance of 3D-printed molds, as we showed in [44].

Figure 9A shows the top view and Figure 9B shows the bottom view of two GP-RNI. The implants are homogeneously colored and show no failures such as burns, black spots, short shots or deformations. There are a few flash failures around the contour of the implant, where the mold halves met (Figure 9C). Moreover, the implant's surface shows stair casing (Figure 9C) because the built resolution of the implant is limited to the DLP 3D-printing resolution of z = 10 μm per layer, which was used for mold manufacturing. Nevertheless, a layer height of z = 10 μm is a relatively low value compared to other methods, especially non-photopolymerzing 3D-printing methods [34]. Consequently, the utilized DLP technique allows μIM of implants with a relatively high resolution. An alternative photopolymerizing 3D-printing technique that enables a higher resolution and a lower staircase effect could be two-photon polymerization (2PP) [21]. However, in comparison to DLP, the use of 2PP would most likely lead to significantly higher 3D printing times. With our current state of knowledge, it is not clear which resolution is needed for therapeutically effective RNIs. In [45], we reported good fitting accuracy of prototypes of human RNIs with a z = 100 μm per layer, which were implanted in human cadaver RWN. We further found that a higher resolution, respectively a lower z-value per layer, leads to increased contour accuracy. The highest possible contour accuracy might be desirable. Increased contour accuracy most likely may lead to a better interface between the implant and the RWM. Consequently, a more effective drug transport into the inner ear by drug diffusion through the RWN might be achieved. Further investigations are needed at this point. At this point, the xy resolution of 32 μm and the z resolution of z = 10 μm (layer height) used here enable 3D printing of a relatively high resolution and a high contour accuracy.

We used the silicone elastomer MED-4244 since it is a soft and stretchable material [38]. Materials with such mechanical behavior are favorable for RNIs because of the beneficial tactile feedback and handling during implantation while also minimizing the likelihood of traumatizing sensitive structures such as the RWM during insertion [45]. Moreover, there is high potential to adapt our rapid tooling-based μIM manufacturing process to other medical-grade materials and biodegradable materials, as such materials are established for IM processes [26,27].

3.3. *Drug Release*

The release of DEX from the GP-RNI shows a two-phase progression with a burst release at the beginning, followed by a slower release (Figure 10A,B). The burst release occurred within the first 13 h (Figure 10C,D), during which a relatively large amount of DEX (about 0.7 μg in total) is released. The release of DEX from the GP-RNIs showed a diffusion-controlled mechanism, behaving like a matrix system [46,47]. As diffusion is dependent on the concentration gradient between the drug-releasing implant and the release medium, a faster diffusion results with a higher concentration gradient. The gradient remains maximal at the beginning of the release. A slower release phase follows with a linear slope. A release rate of about 1 µg/week is observed from the second week onwards. At the time of evaluation, this phase had not yet ended. After 6 weeks, about 10% of the DEX (8.2 ± 0.6 µg) had been released.

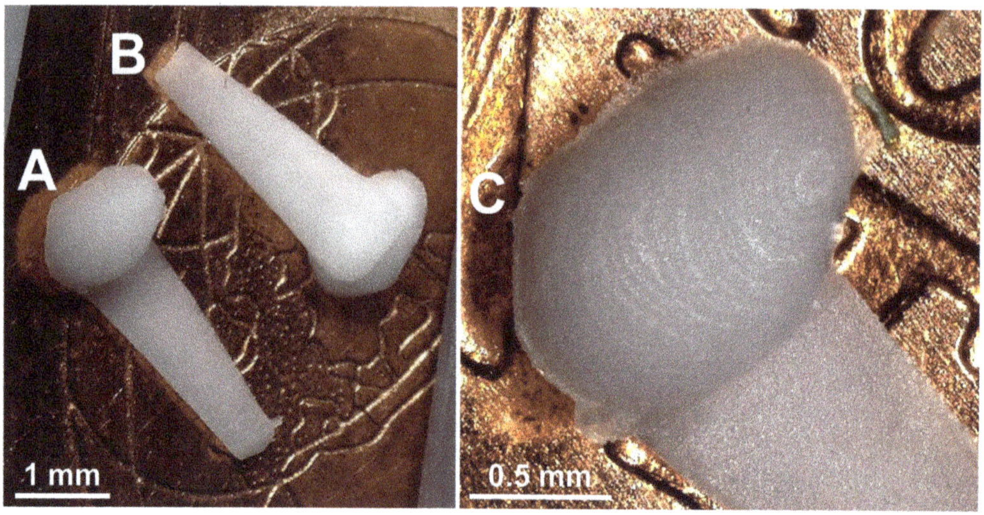

Figure 9. Photographs of two GP-RNI implants with handle structure manufactured via µIM under the usage of rapid tooling with high-resolution DLP 3D printing. (**A**) shows the top view and (**B**) shows the bottom view of the implant. The implants are manufactured from a medical-grade silicone elastomer (MED-4244) and are drug-loaded with the glucocorticoid DEX (10 wt%). The implants show no signs of burns, black spots, short shots or deformations. (**C**) marks a few flash failures around the implant contour where the mold halves met. Moreover, the implants show stair casing, as its resolution is limited to the DLP 3D printing z resolution of 10 µm/layer, which was used for mold manufacturing.

The drug release behavior found is promising, as the GP-RNI allows prolonged drug release over several weeks to months. Such long-release behavior can be beneficial compared to inner ear therapy methods [4]. Nevertheless, the drug release from GP-RNIs has to be tested in a more realistic scenario in vivo. By intratympanic application, the drug has to pass the RWM. This affects the drug concentration reached in the inner ear [48]. Further research on diffusion-based drug transport from the implant through the RWM is needed. Moreover, further research is needed to investigate whether the amount of DEX released is sufficient to achieve a positive therapeutic effect. It has been found in the literature that even relatively low DEX concentrations of 0.00118 mg/mL are effective [49]. However, the findings in the literature on therapeutic effective drug concentrations in cochlear pharmacotherapy are not consistent, and individual experimental parameters make it difficult to compare results [16]. In addition, the DEX concentrations must be chosen depending on the DEX formulation, as there may be significant differences in cytotoxicity [16].

The thermal stress from our µIM process might increase the risk of the degradation of the drug load in the processed material [48,50]. In our process, we used a mold temperature of T_{mold} = 160 °C and a crosslinking time of t = 2 min. These parameters should not lead to significant degradation of the drug load in the processed µIM material as DEX has a melting temperature of T = 262.4 °C. There is a rapid decomposition at higher temperatures [51], but process temperatures below are considered to be suitable. Farto-Vaamonde et al. successfully processed a DEX-loaded filament via extrusion-based 3D printing at extrusion temperatures of T = 220 °C [51]. In the work of Li et al., DEX was exposed to a temperature of T = 185 °C for a period of 5 min during hot melt extrusion, but the authors reported no significant drug degradation [52]. However, as Farto et al. recommend in [51], unnecessarily long heating periods at relatively high temperatures should be avoided as much as possible to prevent stability problems with the drug.

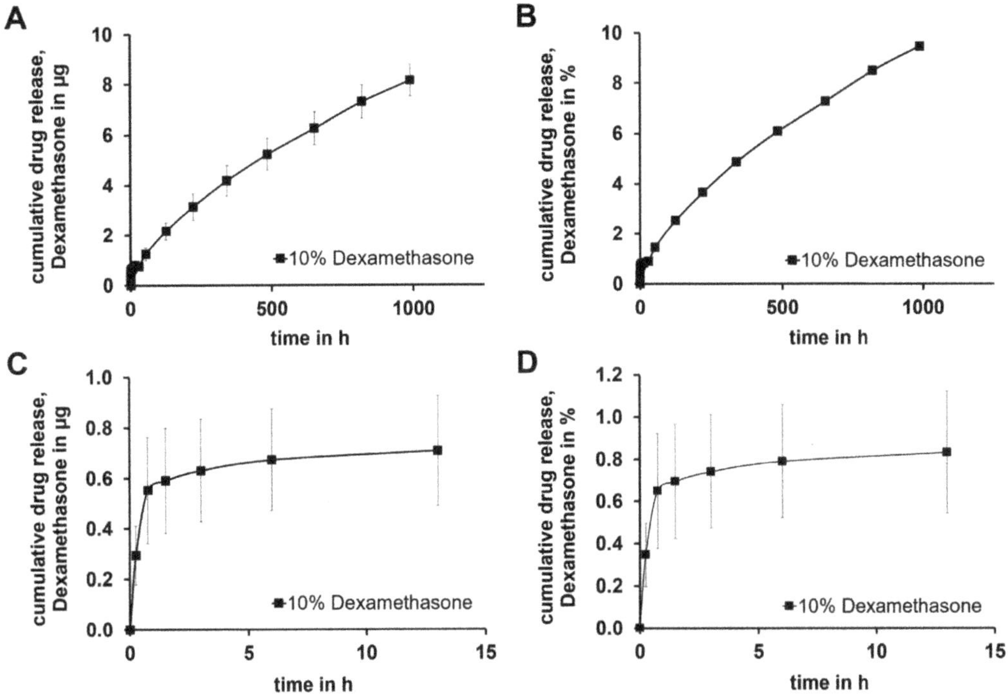

Figure 10. Diagrams show (**A**) cumulative drug release of DEX absolute amount of DEX for about 6 weeks, (**B**) relative DEX release for about 6 weeks, (**C**) enlarged representation of the complete release over the first 15 h, (**D**) enlarged representation of the relative release over the first 15 h; Diagrams (**B**,**D**) were normalized according to the calculated amount of DEX 10 % of the RNI-mass. Drug release testing continues.

Further glucocorticoids, such as prednisone or hydrocortisone (cortisol), offer therapeutic potential for inner ear therapy [53] and are promising for inclusion in further investigations. Prednisone melts and degrades at temperatures of 230–235 °C [54]. For hydrocortisone, the start of thermal degradation was found at a temperature of 225 °C [55]. Therefore, both of these drugs could be suitable for our process. The risk of drug degradation due to thermal stress should be taken into account, especially when using further, thermal sensitive components. For example, proteins such as the growth factors brain-derived neurotrophic factor (BDNF) and insulinlike growth factor 1 (IGF1) have been identified as potentially protective for hearing [56]. They might be promising for long-term inner ear therapy. However, our process is not suitable for the processing of proteins, as they can denature at temperatures far below T = 160 °C.

3.4. Biocompatibility and Bioefficacy

The results of in vitro biocompatibility (cell viability) and bioefficacy (TNF-α-reduction test) investigations are shown in Figure 11. Compared to the blank (100%), the cell viability of the PC, including the cytotoxic agent DMSO, was significantly reduced (12 ± 9%; $p < 0.001$), proving the successful experimental setup. The cell viability of all GP-RNI-supernatant samples (mean ± SD; day 1: 114 ± 19%; day 3: 80 ± 13%; day 7: 94 ± 16%; day 10: 85 ± 22%; day 14: 99 ± 9%; day 21: 102 ± 9%; day 28: 104 ± 13%) did not differ significantly from that of the blank (Figure 11A). The lowest cell viability was 80.44 ± 13.30%. It was found when the supernatant of sampling day 3 was applied. It is still clearly above the 70% of the blank, which is the mark for indicating cytotoxic potential.

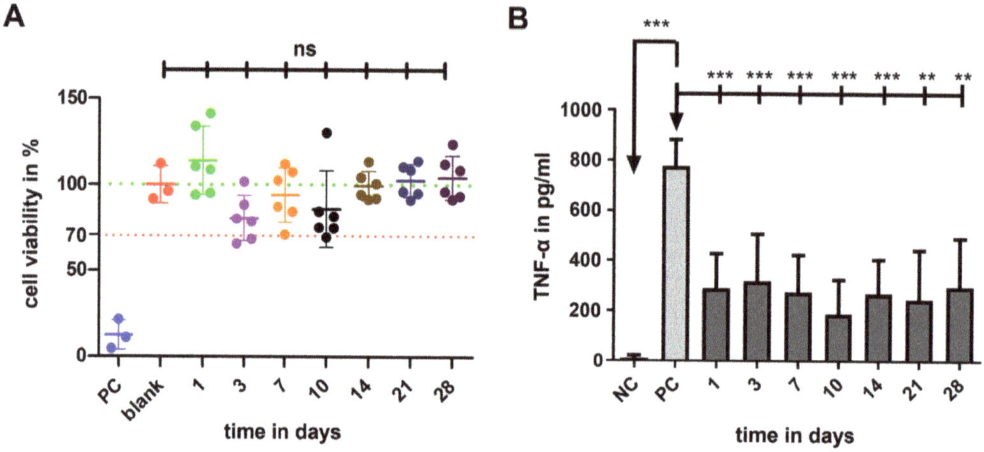

Figure 11. (**A**) Comparison of cell viability (CV in %) of fibroblasts treated with the supernatant of cultured GP-RNI sampled after various incubation times. The mean survival in blank is set as 100% (green line). PC illustrates the cytotoxicity of DMSO and the successful performance of the assay. The dotted line at 70% CV marks the toxicity level, based on the ISO guideline for biocompatibility testing of medical devices (ISO 10993-5:2009—8.5.1). The means of all data per time point were all in the safe range above 70% and statistical analysis did not report significant differences compared to the blank. Data are given as mean ± standard deviation with single experimental results included as dots (N = 3; n = 3); ns = not significant. (**B**) TNF-α amounts measured by ELISA in the supernatants of dendritic cells (DCs). TNF-α production is induced by the addition of 0.5 μg/mL LPS to the culture medium. This results in a high release of TNF-α in the PC when compared with the basic TNF-α level of unstressed cells in the NC. All tested RNI-eluates reduced the TNF-α amount in culture. Data are given as mean ± standard deviation and detected significances are marked with ** ($p < 0.01$) and *** ($p < 0.001$).

Cells without stress (NC) showed a very low basic level of TNF-α-production (11.58 ± 10.44 pg/mL), while this level significantly increased when LPS was added (PC, 776.2 ± 106 pg/mL). Compared to the PC, all tested supernatant reduced the TNF-α amount in the DC-cell-LPS-stress test significantly (Figure 11B). This anti-inflammatory effect was highest on the 10th day (188.2 ± 136.5 pg/mL) and lowest on the third day (318.9 ± 186.3 pg/mL). During the sampling period, the anti-inflammatory effect of the eluate varied. Data and results of statistical analyses are shown in the Appendix A (Table A1).

Our results show neither the usage of photopolymeric molds for μIM nor the drug load of DEX affect the biocompatibility of the used RNI material silicone elastomer MED-4244 critically. The exact amount of DEX released in the supernatant, which was used for biocompatibility investigations, is unknown at this point. Toxic effects of DEX are reported in the literature even for relatively low concentrations of 3 μM (0.00118 mg/mL, DEX: 392.46 g/mol) [49]. The authors report the start of toxic effects on outer hair cells by that drug concentration in vitro. However, as we reported previously [16], the findings in the literature concerning critical drug concentrations in cochlear pharmacotherapy are not consistent, or widespread, and are hard to compare as different individual experimental parameters must be considered. With an MTT assay, as we used here in this work, we found no significant toxic effects for DEX concentrations up to 2000 μM (0.784 mg/mL) [16]. With regard to the slow drug release behavior of the tested GP-RNI samples in isotonic saline, the DEX concentration in the supernatant should be far below 2000 μM. This supports our findings that the GP-RNIs containing DEX and that are made from 3D-printed molds are biocompatible.

We found significant anti-inflammatory effects during the whole 28-day course of the investigation. The anti-inflammatory effects of DEX are well known, and there are further potentials in terms of the protection of CI patients from hearing loss, fibrotic CI encapsulation and spiral ganglion degeneration [16]. In the literature, effective concentrations for DEX are found from 0.00118 mg/mL [49] to 24 mg/mL [57]. However, because of a wide variety of experimental parameters and treatment protocols, findings from the literature are hard to compare and there is a large variability between concentrations being toxic in vivo and those having a beneficial effect [16]. Further in vivo investigations need to show what DEX concentrations are needed in RNIs to receive specific therapeutic effects for inner ear therapy. Many studies highlight the therapeutic potentials of DEX for inner ear diseases, such as [15,58,59]. In addition to pure DEX, as we used here, there are other drug formulations, such as dexamethasone dihydrogen phosphate disodium (DPS). In a previous study, we found a slight tendency for DPS to be more effective in reducing TNF-α-production than other DEX formulations [16]. Moreover, other glucocorticoids, such as prednisone and hydrocortisone (cortisol), are promising for further investigations because they hold therapeutic potential for inner ear therapy [53].

4. Conclusions

We presented the high-precision manufacturing and analysis of drug-loaded implants for controlled drug delivery in the inner ear. In this study, mean guinea pig round window niche implants (GP-RNIs) were manufactured via micro injection molding (μIM) using molds manufactured via rapid tooling using a DLP 3D-printing process. A commercially available photopolymer resin was successfully used as the mold material. This photopolymer resin was primarily designed for highly detailed parts for dental, jewelry, and design. There was no need for reinforcing filler materials to improve material properties such as heat resistance. The 3D-printed molds were suitable for single use in our μIM process, which enables the individual manufacturing of highly patient-personalized implants.

A medical-grade silicone elastomer was drug-loaded with the glucocorticoid DEX (10 wt%) and successfully used for μIM of the implants. The μIM-manufactured implants showed high biocompatibility over a 28-day period. Moreover, we found anti-inflammatory effects over a 28-day period in a TNF-α-reduction test, which indicates high bioefficacy of the drug load. In vitro drug release investigations showed a burst release of about 0.7 μg DEX within the first 13 h in isotonic saline. A slower drug release phase follows with a linear slope. After 6 weeks, about 10% of the drug load of DEX (8.2 $\pm$ 0.6 μg) had been released. These results are promising for prolonged drug delivery of an RNI for inner ear therapy.

Further investigations will focus on in vivo testing of GP-RNIs. Drug transport through RWM in the inner ear and the therapeutic effect of GP-RNIs especially need to be examined. Moreover, further research is needed to adapt the presented rapid tooling-based μIM manufacturing process to other medical-grade materials, particularly biodegradable materials.

Author Contributions: Conceptualization, all authors; methodology, R.M., T.E., G.J. and V.S. (Verena Scheper); formal analysis, Z.G., C.W. and T.E.; investigation, R.M., T.E., G.J., Z.G., C.W., N.F. and V.S. (Verena Scheper); validation, R.M., T.E. and V.S. (Verena Scheper); writing—original draft preparation, R.M., T.E., N.F. and V.S. (Verena Scheper); writing—review and editing, H.S., T.E. and V.S. (Verena Scheper); visualization, R.M., T.E. and V.S. (Verena Scheper); supervision, H.S., V.S. (Verena Scheper), T.L. and N.G.; project administration, H.S., V.S. (Volkmar Senz), V.S. (Verena Scheper) and G.J.; funding acquisition, R.M., H.S., V.S. (Volkmar Senz), N.G., G.J., V.S. (Verena Scheper) and T.L. All authors have read and agreed to the published version of the manuscript.

Funding: The authors would like to thank the Federal Ministry of Education and Research of Germany (BMBF) for research funding of "RESPONSE—Partnership for Innovation in Implant Technology" (funding numbers 03ZZ0928A, 03ZZ0928D, 03ZZ0928J and 03ZZ0928L) in the program

"Twenty20—Partnership for Innovation". In addition, we want to thank the BMBF for funding of the DLP 3D printing device Asiga Pro 4K45 (funding number 03ZZ09X06).

Institutional Review Board Statement: Not applicable.

Informed Consent Statement: Not applicable.

Data Availability Statement: The raw data required to reproduce the findings of this article are available from the corresponding authors upon reasonable request.

Conflicts of Interest: The authors declare no conflict of interest.

Appendix A

Table A1. TNF-α concentration per time point of sampling and statistical analysis results. * ($p < 0.05$), ** ($p < 0.01$), *** ($p < 0.001$), ns = not significant.

	TNF-α in pg/mL (Mean ± Standard Deviation)						
	290 ± 138	318 ± 186	275 ± 148	188 ± 136	270 ± 134	246 ± 196	298 ± 192
Time point of sampling in days	**1**	**3**	**7**	**10**	**14**	**21**	**28**
1	-	ns	ns	***	ns	ns	ns
3	-	-	ns	***	ns	*	ns
7	-	-	-	**	ns	ns	ns
10	-	-	-	-	*	ns	***
14	-	-	-	-	-	ns	ns
21	-	-	-	-	-	-	ns
28	-	-	-	-	-	-	-

References

1. Nyberg, S.; Abbott, N.J.; Shi, X.; Steyger, P.S.; Alain, D. Delivery of therapeutics to the inner ear: The challenge of the blood-labyrinth barrier. *Sci. Transl. Med.* **2019**, *11*, eaao0935. [CrossRef] [PubMed]
2. Borden, R.C.; Saunders, J.E.; Berryhill, W.E.; Krempl, G.A.; Thompson, D.M.; Queimado, L. Hyaluronic acid hydrogel sustains the delivery of dexamethasone across the round window membrane. *Audiol. Neurootol.* **2011**, *16*, 1–11. [CrossRef] [PubMed]
3. Paulson, D.P.; Abuzeid, W.; Jiang, H.; Oe, T.; O'Malley, B.W.; Li, D. A novel controlled local drug delivery system for inner ear disease. *Laryngoscope* **2008**, *118*, 706–711. [CrossRef]
4. Matin, F.; Gao, Z.; Repp, F.; John, S.; Lenarz, T.; Scheper, V. Determination of the Round Window Niche Anatomy Using Cone Beam Computed Tomography Imaging as Preparatory Work for Individualized Drug-Releasing Implants. *J. Imaging* **2021**, *7*, 79. [CrossRef] [PubMed]
5. Mau, R.; Schick, P.; Matin-Mann, F.; Gao, Z.; Alcacer Labrador, D.; John, S.; Repp, F.; Lenarz, T.; Weitschies, W.; Scheper, V.; et al. Digital light processing and drug stability of Dexamethasone-loaded implant prototypes for medical treatment of the inner ear. *Trans. Addit. Manuf. Meets Med. Trans. AMMM* **2022**, *4*, 666. [CrossRef]
6. Plontke, S.K.; Löwenheim, H.; Mertens, J.; Engel, C.; Meisner, C.; Weidner, A.; Zimmermann, R.; Preyer, S.; Koitschev, A.; Zenner, H.-P. Randomized, double blind, placebo controlled trial on the safety and efficacy of continuous intratympanic dexamethasone delivered via a round window catheter for severe to profound sudden idiopathic sensorineural hearing loss after failure of systemic therapy. *Laryngoscope* **2009**, *119*, 359–369. [CrossRef]
7. Erdur, O.; Kayhan, F.T.; Cirik, A.A. Effectiveness of intratympanic dexamethasone for refractory sudden sensorineural hearing loss. *Eur. Arch. Otorhinolaryngol.* **2014**, *271*, 1431–1436. [CrossRef]
8. Li, X.; Chen, W.-J.; Xu, J.; Yi, H.-J.; Ye, J.-Y. Clinical Analysis of Intratympanic Injection of Dexamethasone for Treating Sudden Deafness. *Int. J. Gen. Med.* **2021**, *14*, 2575–2579. [CrossRef]
9. Berjis, N.; Soheilipour, S.; Musavi, A.; Hashemi, S.M. Intratympanic dexamethasone injection vs methylprednisolone for the treatment of refractory sudden sensorineural hearing loss. *Adv. Biomed. Res.* **2016**, *5*, 111. [CrossRef]
10. Albu, S.; Nagy, A.; Doros, C.; Marceanu, L.; Cozma, S.; Musat, G.; Trabalzini, F. Treatment of Meniere's disease with intratympanic dexamethazone plus high dosage of betahistine. *Am. J. Otolaryngol.* **2016**, *37*, 225–230. [CrossRef]
11. Atrache Al Attrache, N.; Krstulovic, C.; Pérez Guillen, V.; Morera Pérez, C.; Pérez Garrigues, H. Response Over Time of Vertigo Spells to Intratympanic Dexamethasone Treatment in Meniere's Disease Patients. *J. Int. Adv. Otol.* **2016**, *12*, 92–97. [CrossRef] [PubMed]
12. Silverstein, H.; Isaacson, J.E.; Olds, M.J.; Rowan, P.T.; Rosenberg, S. Dexamethasone inner ear perfusion for the treatment of Meniere's disease: A prospective, randomized, double-blind, crossover trial. *Am. J. Otol.* **1998**, *19*, 196–201. [PubMed]
13. Takeda, T.; Takeda, S.; Kakigi, A. Effects of Glucocorticoids on the Inner Ear. *Front. Surg.* **2020**, *7*, 596383. [CrossRef]

14. Elzayat, S.; El-Sherif, H.; Hegazy, H.; Gabr, T.; El-Tahan, A.-R. Tinnitus: Evaluation of Intratympanic Injection of Combined Lidocaine and Corticosteroids. *ORL J. Otorhinolaryngol. Relat. Spec.* **2016**, *78*, 159–166. [CrossRef]
15. Bas, E.; Bohorquez, J.; Goncalves, S.; Perez, E.; Dinh, C.T.; Garnham, C.; Hessler, R.; Eshraghi, A.A.; van de Water, T.R. Electrode array-eluted dexamethasone protects against electrode insertion trauma induced hearing and hair cell losses, damage to neural elements, increases in impedance and fibrosis: A dose response study. *Hear. Res.* **2016**, *337*, 12–24. [CrossRef] [PubMed]
16. Gao, Z.; Schwieger, J.; Matin-Mann, F.; Behrens, P.; Lenarz, T.; Scheper, V. Dexamethasone for Inner Ear Therapy: Biocompatibility and Bio-Efficacy of Different Dexamethasone Formulations In Vitro. *Biomolecules* **2021**, *11*, 1896. [CrossRef]
17. Chen, W.; Fernandez, C.S.; Xu, L.; Velliou, E.; Homer-Vanniasinkam, S.; Tiwari, M.K. High-resolution 3D printing for healthcare. In *3D Printing in Medicine*; Elsevier: Amsterdam, The Netherlands, 2023; pp. 225–271. ISBN 9780323898317.
18. Bahati, D.; Bricha, M.; El Mabrouk, K. Vat Photopolymerization Additive Manufacturing Technology for Bone Tissue Engineering Applications. *Adv. Eng. Mater.* **2023**, *25*, 2200859. [CrossRef]
19. Zhang, Q.; Weng, S.; Hamel, C.M.; Montgomery, S.M.; Wu, J.; Kuang, X.; Zhou, K.; Qi, H.J. Design for the reduction of volume shrinkage-induced distortion in digital light processing 3D printing. *Extrem. Mech. Lett.* **2021**, *48*, 101403. [CrossRef]
20. Guttridge, C.; Shannon, A.; O'Sullivan, A.; O'Sullivan, K.J.; O'Sullivan, L.W. Biocompatible 3D printing resins for medical applications: A review of marketed intended use, biocompatibility certification, and post-processing guidance. *Ann. 3d Print. Med.* **2022**, *5*, 100044. [CrossRef]
21. Ng, W.L.; Lee, J.M.; Zhou, M.; Chen, Y.-W.; Lee, K.-X.A.; Yeong, W.Y.; Shen, Y.-F. Vat polymerization-based bioprinting-process, materials, applications and regulatory challenges. *Biofabrication* **2020**, *12*, 22001. [CrossRef]
22. Bao, Y.; Paunović, N.; Leroux, J.-C. Challenges and Opportunities in 3D Printing of Biodegradable Medical Devices by Emerging Photopolymerization Techniques. *Adv. Funct. Mater.* **2022**, *32*, 2109864. [CrossRef]
23. Fu, H.; Xu, H.; Liu, Y.; Yang, Z.; Kormakov, S.; Wu, D.; Sun, J. Overview of Injection Molding Technology for Processing Polymers and Their Composites. *ES Mater. Manuf.* **2020**, *8*, 3–23. [CrossRef]
24. Dehghan-Manshadi, A.; Yu, P.; Dargusch, M.; StJohn, D.; Qian, M. Metal injection moulding of surgical tools, biomaterials and medical devices: A review. *Powder Technol.* **2020**, *364*, 189–204. [CrossRef]
25. Giboz, J.; Copponnex, T.; Mélé, P. Microinjection molding of thermoplastic polymers: A review. *J. Micromech. Microeng.* **2007**, *17*, R96–R109. [CrossRef]
26. Amellal, K.; Tzoganakis, C.; Penlidis, A.; Rempel, G.L. Injection molding of medical plastics: A review. *Adv. Polym. Technol.* **1994**, *13*, 315–322. [CrossRef]
27. de Melo, L.P.; Salmoria, G.V.; Fancello, E.A.; Roesler, C.R.d.M. Effect of Injection Molding Melt Temperatures on PLGA Craniofacial Plate Properties during In Vitro Degradation. *Int. J. Biomater.* **2017**, *2017*, 1256537. [CrossRef]
28. Zema, L.; Loreti, G.; Melocchi, A.; Maroni, A.; Gazzaniga, A. Injection Molding and its application to drug delivery. *J. Control. Release* **2012**, *159*, 324–331. [CrossRef]
29. Dormann, B.; Decker, C.; Juettner, G. High prescision micro molding injection of 2 component liquid silicone. In Proceedings of the Annual Technical Conference-ANTEC, Conference Proceedings, Orlando, FL, USA, 2–4 April 2012; pp. 1843–1846, ISBN 9781622760831.
30. Calaon, M.; Baruffi, F.; Fantoni, G.; Cirri, I.; Santochi, M.; Hansen, H.N.; Tosello, G. Functional Analysis Validation of Micro and Conventional Injection Molding Machines Performances Based on Process Precision and Accuracy for Micro Manufacturing. *Micromachines* **2020**, *11*, 1115. [CrossRef]
31. Wang, Y.; Weng, C.; Deng, Z.; Sun, H.; Jiang, B. Fabrication and performance of nickel-based composite mold inserts for micro-injection molding. *Appl. Surf. Sci.* **2023**, *615*, 156417. [CrossRef]
32. Lozano, A.B.; Álvarez, S.H.; Isaza, C.V.; Montealegre-Rubio, W. Analysis and Advances in Additive Manufacturing as a New Technology to Make Polymer Injection Molds for World-Class Production Systems. *Polymers* **2022**, *14*, 1646. [CrossRef]
33. Walsh, E.; ter Horst, J.H.; Markl, D. Development of 3D printed rapid tooling for micro-injection moulding. *Chem. Eng. Sci.* **2021**, *235*, 116498. [CrossRef]
34. Dempsey, D.; McDonald, S.; Masato, D.; Barry, C. Characterization of Stereolithography Printed Soft Tooling for Micro Injection Molding. *Micromachines* **2020**, *11*, 819. [CrossRef] [PubMed]
35. 3D Slicer Image Computing Platform. Available online: https://www.slicer.org/ (accessed on 20 May 2023).
36. Fedorov, A.; Beichel, R.; Kalpathy-Cramer, J.; Finet, J.; Fillion-Robin, J.-C.; Pujol, S.; Bauer, C.; Jennings, D.; Fennessy, F.; Sonka, M.; et al. 3D Slicer as an image computing platform for the Quantitative Imaging Network. *Magn. Reson. Imaging* **2012**, *30*, 1323–1341. [CrossRef] [PubMed]
37. LithoLabs GmbH. Product Information "Asiga PlasGRAY V2". Available online: https://litholabs.one/en/dental/resin/model/61/asiga-plasgray-v2 (accessed on 25 January 2023).
38. Avantor Inc., NuSil Technology LLM. MED-4244: Low Consistency Silicone Elastomer. Available online: https://www.avantorsciences.com/assetsvc/asset/en_US/id/29019092/contents/en_us_tds_nusimed-4244.pdf (accessed on 8 February 2023).
39. Bohl, A.; Rohm, H.W.; Ceschi, P.; Paasche, G.; Hahn, A.; Barcikowski, S.; Lenarz, T.; Stöver, T.; Pau, H.-W.; Schmitz, K.-P.; et al. Development of a specially tailored local drug delivery system for the prevention of fibrosis after insertion of cochlear implants into the inner ear. *J. Mater. Sci. Mater. Med.* **2012**, *23*, 2151–2162. [CrossRef] [PubMed]
40. Matin-Mann, F.; Gao, Z.; Schwieger, J.; Ulbricht, M.; Domsta, V.; Senekowitsch, S.; Weitschies, W.; Seidlitz, A.; Doll, K.; Stiesch, M.; et al. Individualized, Additively Manufactured Drug-Releasing External Ear Canal Implant for Prevention of Postoperative

Restenosis: Development, In Vitro Testing, and Proof of Concept in an Individual Curative Trial. *Pharmaceutics* **2022**, *14*, 1242. [CrossRef]
41. Zink, B.; Kovács, N.K.; Kovács, J.G. Thermal analysis based method development for novel rapid tooling applications. *Int. Commun. Heat Mass Transf.* **2019**, *108*, 104297. [CrossRef]
42. Martinho, P.G.; Pouzada, A.S. Alternative materials in moulding elements of hybrid moulds: Structural integrity and tribological aspects. *Int. J. Adv. Manuf. Technol.* **2021**, *113*, 351–363. [CrossRef]
43. Shinde, M.S.; Ashtankar, K.M.; Kuthe, A.M.; Dahake, S.W.; Mawale, M.B. Direct rapid manufacturing of molds with conformal cooling channels. *RPJ* **2018**, *24*, 1347–1364. [CrossRef]
44. Mau, R.; Jüttner, G.; Gao, Z.; Matin, F.; Alcacer Labrador, D.; Repp, F.; John, S.; Scheper, V.; Lenarz, T.; Seitz, H. Micro injection molding of individualised implants using 3D printed molds manufactured via digital light processing. *Curr. Dir. Biomed. Eng.* **2021**, *7*, 399–402. [CrossRef]
45. Mau, R.; Nazir, J.; Gao, Z.; Alcacer Labrador, D.; Repp, F.; John, S.; Lenarz, T.; Scheper, V.; Seitz, H.; Matin-Mann, F. Digital Light Processing of Round Window Niche Implant Prototypes for Implantation Studies. *Curr. Dir. Biomed. Eng.* **2022**, *8*, 157–160. [CrossRef]
46. Peppas, N.A.; Narasimhan, B. Mathematical models in drug delivery: How modeling has shaped the way we design new drug delivery systems. *J. Control. Release* **2014**, *190*, 75–81. [CrossRef] [PubMed]
47. Langer, R.S.; Peppas, N.A. Present and future applications of biomaterials in controlled drug delivery systems. *Biomaterials* **1981**, *2*, 201–214. [CrossRef] [PubMed]
48. Salt, A.N.; Plontke, S.K. Pharmacokinetic principles in the inner ear: Influence of drug properties on intratympanic applications. *Hear. Res.* **2018**, *368*, 28–40. [CrossRef]
49. Jia, H.; François, F.; Bourien, J.; Eybalin, M.; Lloyd, R.V.; van de Water, T.R.; Puel, J.-L.; Venail, F. Prevention of trauma-induced cochlear fibrosis using intracochlear application of anti-inflammatory and antiproliferative drugs. *Neuroscience* **2016**, *316*, 261–278. [CrossRef] [PubMed]
50. Domsta, V.; Seidlitz, A. 3D-Printing of Drug-Eluting Implants: An Overview of the Current Developments Described in the Literature. *Molecules* **2021**, *26*, 4066. [CrossRef]
51. Farto-Vaamonde, X.; Auriemma, G.; Aquino, R.P.; Concheiro, A.; Alvarez-Lorenzo, C. Post-manufacture loading of filaments and 3D printed PLA scaffolds with prednisolone and dexamethasone for tissue regeneration applications. *Eur. J. Pharm. Biopharm.* **2019**, *141*, 100–110. [CrossRef]
52. Li, D.; Guo, G.; Fan, R.; Liang, J.; Deng, X.; Luo, F.; Qian, Z. PLA/F68/dexamethasone implants prepared by hot-melt extrusion for controlled release of anti-inflammatory drug to implantable medical devices: I. Preparation, characterization and hydrolytic degradation study. *Int. J. Pharm.* **2013**, *441*, 365–372. [CrossRef]
53. Cortés Fuentes, I.A.; Videhult Pierre, P.; Engmér Berglin, C. Improving Clinical Outcomes in Cochlear Implantation Using Glucocorticoid Therapy: A Review. *Ear. Hear.* **2020**, *41*, 17–24. [CrossRef]
54. Ledeți, I.; Bengescu, C.; Cîrcioban, D.; Vlase, G.; Vlase, T.; Tomoroga, C.; Buda, V.; Ledeți, A.; Dragomirescu, A.; Murariu, M. Solid-state stability and kinetic study of three glucocorticoid hormones: Prednisolone, prednisone and cortisone. *J. Therm. Anal. Calorim.* **2020**, *141*, 1053–1065. [CrossRef]
55. Ayyoubi, S.; van Kampen, E.E.M.; Kocabas, L.I.; Parulski, C.; Lechanteur, A.; Evrard, B.; de Jager, K.; Muller, E.; Wilms, E.W.; Meulenhoff, P.W.C.; et al. 3D printed, personalized sustained release cortisol for patients with adrenal insufficiency. *Int. J. Pharm.* **2023**, *630*, 122466. [CrossRef]
56. Paulsen, A.J.; Cruickshanks, K.J.; Pinto, A.; Schubert, C.R.; Dalton, D.S.; Fischer, M.E.; Klein, B.E.K.; Klein, R.; Tsai, M.Y.; Tweed, T.S. Neuroprotective factors and incident hearing impairment in the epidemiology of hearing loss study. *Laryngoscope* **2019**, *129*, 2178–2183. [CrossRef] [PubMed]
57. Alexander, T.H.; Harris, J.P.; Nguyen, Q.T.; Vorasubin, N. Dose Effect of Intratympanic Dexamethasone for Idiopathic Sudden Sensorineural Hearing Loss: 24 mg/mL Is Superior to 10 mg/mL. *Otol. Neurotol.* **2015**, *36*, 1321–1327. [CrossRef] [PubMed]
58. Hütten, M.; Dhanasingh, A.; Hessler, R.; Stöver, T.; Esser, K.-H.; Möller, M.; Lenarz, T.; Jolly, C.; Groll, J.; Scheper, V. In vitro and in vivo evaluation of a hydrogel reservoir as a continuous drug delivery system for inner ear treatment. *PLoS ONE* **2014**, *9*, e104564. [CrossRef] [PubMed]
59. Connolly, T.M.; Eastwood, H.; Kel, G.; Lisnichuk, H.; Richardson, R.; O'Leary, S. Pre-operative intravenous dexamethasone prevents auditory threshold shift in a guinea pig model of cochlear implantation. *Audiol. Neurootol.* **2011**, *16*, 137–144. [CrossRef] [PubMed]

pharmaceutics

MDPI

Review

A Review on Dry Eye Disease Treatment: Recent Progress, Diagnostics, and Future Perspectives

Himangsu Mondal [1,†], Ho-Joong Kim [2,†], Nijaya Mohanto [1] and Jun-Pil Jee [1,*]

1 Drug Delivery Research Lab, College of Pharmacy, Chosun University, Gwangju 61452, Republic of Korea
2 Department of Chemistry, Chosun University, Gwangju 61452, Republic of Korea
* Correspondence: jee@chosun.ac.kr; Tel.: +82-62-230-6364; Fax: +82-62-222-5414
† These authors contributed equally to this work.

Abstract: Dry eye disease is a multifactorial disorder of the eye and tear film with potential damage to the ocular surface. Various treatment approaches for this disorder aim to alleviate disease symptoms and restore the normal ophthalmic environment. The most widely used dosage form is eye drops of different drugs with 5% bioavailability. The use of contact lenses to deliver drugs increases bioavailability by up to 50%. Cyclosporin A is a hydrophobic drug loaded onto contact lenses to treat dry eye disease with significant improvement. The tear is a source of vital biomarkers for various systemic and ocular disorders. Several biomarkers related to dry eye disease have been identified. Contact lens sensing technology has become sufficiently advanced to detect specific biomarkers and predict disease conditions accurately. This review focuses on dry eye disease treatment with cyclosporin A-loaded contact lenses, contact lens biosensors for ocular biomarkers of dry eye disease, and the possibility of integrating sensors in therapeutic contact lenses.

Keywords: dry eye; ocular; drug delivery; contact lens; cyclosporin A; biomarkers; biosensors

Citation: Mondal, H.; Kim, H.-J.; Mohanto, N.; Jee, J.-P. A Review on Dry Eye Disease Treatment: Recent Progress, Diagnostics, and Future Perspectives. *Pharmaceutics* **2023**, *15*, 990. https://doi.org/10.3390/pharmaceutics15030990

Academic Editors: Daniela Monti and Silvia Tampucci

Received: 15 February 2023
Revised: 11 March 2023
Accepted: 15 March 2023
Published: 19 March 2023

1. Introduction

Dry eye disease (DED) is an illness of the preocular tear film that occurs due to injury to the eye surface and is associated with signs of ocular discomfort. It is also known as keratoconjunctivitis sicca, sicca syndrome, keratitis sicca, dry eye syndrome (DES), xerophthalmia, dysfunctional tear syndrome, ocular surface disease, or simply dry eyes. "Sjögren's syndrome" is a form of DES where the eyes do not produce enough tears [1].

The International Dry Eye Workshop (2007) defined dry eye as a multifactorial disease of the tear film and ocular surface that results in symptoms of discomfort, visual disturbance, and tear film instability, with potential damage to the ocular surface. It is associated with ocular surface inflammation and increased tear film osmolarity [2]. A normal human tear film comprises water, lipids, mucins, electrolytes, proteins, and vitamins. Along with goblet cells, the meibomian and lacrimal glands produce tears that maintain a normal ocular environment by removing debris, lubricating the eye, and protecting the eye from infection [3]. Recently, it was found that dry eye is an inflammatory disease that shares many features with autoimmune diseases. The pathogenesis of dry eye may be due to stress on the ocular surface (infection, environmental factors, endogenous stress, genetic factors, and antigens) [4,5]. DES is a chronic disease, particularly among older people, but proper treatment decreases symptoms and, eventually, ocular damage [3].

The prevalence rate is 5–50%, which may be up to 75% in adults over 40 years old, with women being the most affected [6]. For younger adults aged 18–45 years, only 2.7% may develop DED [7]. It has an economic impact, ranging from $687 to $1267 annually, depending on the severity of the disease. DED costs the US economy approximately $3.8 billion [7,8]. These costs include prescription drugs, over-the-counter products, and punctual plug placement [9].

The eye is a complex organ with various anatomical and physiological barriers (Figure 1). These complex structures make ocular drug delivery challenging for scientists [10]. Available ophthalmic dosage forms include eye drops, ointments, and suspensions. Among these preparations, eye drops hold the major part (>90%) [11]. A healthy human has a 7–30 μL tear with a 0.5~2.2 μL/min turnover rate. The restoration time of the tear film is 2–3 min [12]. Thus, approximately 1–5% of the applied drug reaches the intended tissue. The ineffectiveness of this treatment necessitates repeated application at higher concentrations, which affects the patient's normal routine [13]. Numerous innovative drug delivery techniques have been developed to enhance drug residence duration and cross the cornea, but all have limitations [14,15]. Therefore, a unique drug delivery method can enhance patient compliance while increasing drug bioavailability and reducing systemic exposure to improve clinical outcomes [13]. Drug delivery using contact lenses is an intriguing field of research because of its exclusive features, such as prolonged wear, simple end of therapy (by removing the contact lens), and greater bioavailability (>50%) than eye drops, as established by numerous studies [16,17]. Patients' high compliance may be achieved with drug-loaded contact lenses by excluding multiple drug dosing, particularly for contact lenses worn mainly for vision improvement [18,19].

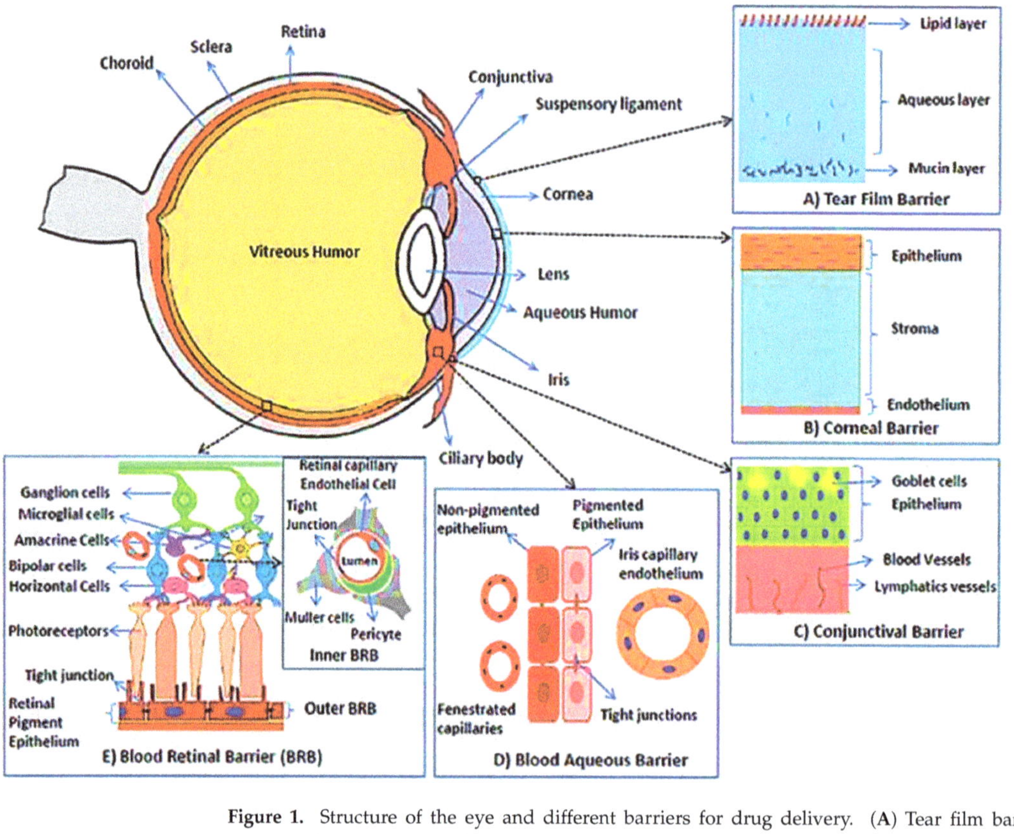

Figure 1. Structure of the eye and different barriers for drug delivery. (**A**) Tear film barrier. (**B**) Corneal barrier. (**C**) Conjunctival barrier. (**D**) Blood-aqueous barrier. (**E**) Blood-retinal barrier (BRB). Reprinted with permission from reference [12]. Copyright 2017 Elsevier.

As the solitary sensory organ of the human optical system, the human eye contains a wealth of important chemical, physical, and biological biomarkers related to human health. Consequently, it has become an important research topic, propelling the rapid development of soft electronic systems for eye research [20]. Biosensor-integrated contact lenses can be a good choice for pliable and wearable therapeutic devices [21]. They have sensing gears to track eye conditions, such as intraocular pressure (IOP) and tear fluid constitution [22]. The biosensing functionalities of contact lenses have become possible with advancements in device downsizing for microcircuits, microsensors, and other microscale devices [20]. There are two major groups of sensors for sensing tear fluid: chemical (biomolecules, metabolites, and electrolytes) and physiological (wrinkling behavior, tear production, IOP, and temperature) sensors [23,24]. Electrochemical sensing has a higher sensitivity and temporal resolution than fluorescence-based sensing using colorimetric assays [25,26]. In the case of DED, the tear production rate can be determined colorimetrically using microfluidic cells with a coloring dye if implanted in contact lenses [27].

This review aimed to (1) evaluate the current treatments for DED, (2) assess the potential of contact lens-based drug delivery as a new treatment for DED, covering the features of the current technologies and their pros and cons, (3) describe various biosensor technologies that can identify pathological eye characteristics, and (4) provide a future perspective of biosensor-fused contact lens-based drug delivery.

2. Treatments of Dry Eye Disease

As DED is a chronic disease, the treatment may take a long time to get a positive result, which can be achieved in different ways (Table 1). The treatment involves a hierarchy approach according to disease severity and information related to (subclinical) inflammation of the ocular surface, meibomian gland dysfunction, and/or associated systemic disease [28]. Cigarette smoking, air conditioning, and dry heating air increase the risk of DED, which must be avoided [5]. Currently, many drugs for DED treatments are in clinical trials (Table 2).

Table 1. Dry eye disease treatments.

Medication	Description	Mechanism of Action	References
Artificial tears	Polyvinyl alcohol, povidone, hydroxypropyl guar, cellulose derivatives, and hyaluronic acid	Increase tear film stability. Reduce ocular surface stress. Improve contrast sensitivity and the optical quality of the surface.	[28–33]
Topical corticosteroids (loteprednol 0.5%)	Unpreserved corticosteroid eyedrops, instilled over a period of 2 to 4 weeks, improve the symptoms and clinical signs of moderate to severe dry eye disease.	Corticosteroids act by the induction of phospholipase A2 inhibitory proteins and inhibiting the release of arachidonic acid.	[34–36]
Cyclosporin A (CsA)	Topical application of CsA leads to increased production of tear fluid, possibly via local release of parasympathetic neuro transmitters. CsA eyedrops 0.05% (Restasis) were approved for the topical treatment of dry eye by the FDA in 2002.	CsA is an immunosuppressant that inhibits the calcineurin–phosphatase pathway by complex formation with cyclophilin, and thus reduces the transcription of T-cell-activating cytokines such as interleukin-2 (IL-2).	[37–41]
Tacrolimus/ pimecrolimus	Appear to be as effective as CsA and are used in patients who cannot tolerate CsA	Inhibition of interleukin-2 gene transcription, nitric oxide synthase activation, cell degranulation, and apoptosis.	[5]
Tetracyclines	Bacteriostatic antibiotics with anti-inflammatory effect.	They reduce the synthesis and activity of matrix metalloproteinases, the production of interleukin-1 (IL-1) and tumor necrosis factor, collagenase activity, and B-cell activation.	[5,42,43]
Macrolides	Azithromycin 1% has been successfully used to improve meibomian gland function and symptoms, a reduction in bacterial colonization of the eyelid margins, and normalization of the meibomian gland secretion lipid profile.	Inhibition of bacterial protein biosynthesis by preventing peptidyltransferase from adding the growing peptide attached to tRNA to the next amino acid and also inhibiting bacterial ribosomal translation.	[44–46]

Table 1. *Cont.*

Medication	Description	Mechanism of Action	References
Omega fatty acids	Omega-3 and omega-6 are essential fatty acids for ocular surface homeostasis.	Omega-3 fatty acids work by blocking pro-inflammatory eicosanoids and reducing cytokines through anti-inflammatory activity.	[47]
Eyelid hygiene	Hot compresses, eyelid warming masks or goggles, infrared heaters, and eyelid massage improve eyelid margin morphology with a reduction in blocked meibomian gland excretory ducts, and an increase in tear film stability and lipid layer thickness of the tear film.		[48–51]
Punctal plugs	Temporary occlusion of the tear ducts by small collagen or silicone plugs (punctal plugs) is effective in patients with severe aqueous-deficient dry eye disease.		[36,52,53]
Lifitegrast (Xidra)	The U.S. Food and Drug Administration approved Xiidra (lifitegrast ophthalmic solution) for the treatment of signs and symptoms of dry eye disease, on Monday, 11 July 2016. Xiidra is the first medication in a new class of drugs, called lymphocyte function-associated antigen 1 (LFA-1) antagonist, approved by the FDA for dry eye disease. Xiidra is manufactured by Shire US Inc., of Lexington, Massachusetts.	Lifitegrast blocks the interaction of cell surface proteins LFA-1 and intercellular adhesion molecule-1 (ICAM-1), and is believed to inhibit T-cell-mediated inflammation in DED.	[54,55]
Vitamin A	Vitamin A is an essential nutrient present naturally in tear film of healthy eyes. Vitamin A plays an important role in production of the mucin layer, the most innermost lubricating layer of tear film that is crucial for a healthy tear film. Vitamin A deficiency leads to loss of mucin layer and goblet cell atrophy.	Vitamin A drops protect the eyes from free radicals, toxins, allergens, and inflammation.	[1,28,56]
Vitamin E	Vitamin E is a fat-soluble antioxidant that prevents the oxidation of fatty acids by reactive oxygen species. The retina is a lipid-rich environment and is bombarded by ultraviolet radiation. In cell culture, vitamin E has been found to enhance the antioxidant ability of lutein to protect retinal pigment epithelial cells from acrolein-induced oxidation.		[57,58]

Table 2. Clinical trials of DED drugs [59]. Copyright 2022 Elsevier.

Functions	Drug	Stage
A mucin-like glycoprotein	Lacritin	Phase II
	Lubricin	Phase II
Anti-inflammatory and/or immunosuppressive	Loteprednol etabonate 0.25% suspension	FDA-approved
	OCS-O_2	Phase II
	A higher concentration of Cyclosporine	Phase III
	Tacrolimus (0.03%) eye drops	Phase IV
	Rapamycin (sirolimus)	Phase I
	EBI-005	Phase III
	Resolvin E1 analogues	Phase II
Biological components	Albumin 5%	Phase II
	Estradiol	Phase II
	N-acetylcysteine	Phase II
	Thymosin b4	Phase II
	Amniotic membrane extract eye drops	Phase I/II
	Mesenchymal stem cells	Phase I/II
Mucin secretagogues	Tavilermide (MIM-D3, 1% or 5%)	Phase II
	Ecabet sodium	Phase III
	Mycophenolate mofetil	Phase II
Other's products	15(s)-HETE or Icomucret	Phase III/II
	Visomitin (SkQ1)	Phase II/III
	Tivanisiran (SYL1001)	Phase III

2.1. Weakness of Existing Treatments

The available dosage forms for DED are primarily eye drops or emulsions. After application, they tend to rapidly enter the nasolacrimal duct with a low turnover rate and short restoration time. The drug is then eliminated through lymphatic flow and conjunctival blood. As a result, only 1–5% of the drug administered is available for absorption by the target tissue. The bioavailability of lipophilic and hydrophilic drugs in eye drops is less than 5% and 0.5%, respectively. Frequent dosing with highly concentrated drugs [60] is required to overcome this limitation. This may be responsible for poor patient compliance, especially for chronic ocular disease, such as DES [10,61]. The fraction of the delivered drug that enters the systemic circulation escapes first-pass metabolism and enters all major organs, with potential side effects [60].

For 18.2–80% of patients, there is a chance of microbial contamination when administering eye drops due to contact between their face, eyes, or hands and the dispensing tip of the eye drop container. The delivery of the same number of drops is not always possible, and approximately 11.3–60.6% of patients fail to do this. Moreover, the amount of drug expelled from the eye drop bottle is not constant and depends on the force applied on the bottle surface, which may lead to dose variations [10,62]. Therefore, it is not possible to deliver the correct amount of drug irrespective of careful handling of conventional eye drops, and that might fail to gain patients' satisfaction and may lead to poor clinical outcomes [10,63].

2.2. Contact Lenses as an Alternative DED Treatment

An innovative dosage form is required to overcome the existing limitations of DED treatment. In this regard, contact lenses are a suitable candidate. Contact lenses have undergone numerous advancements and modifications (Figure 2).

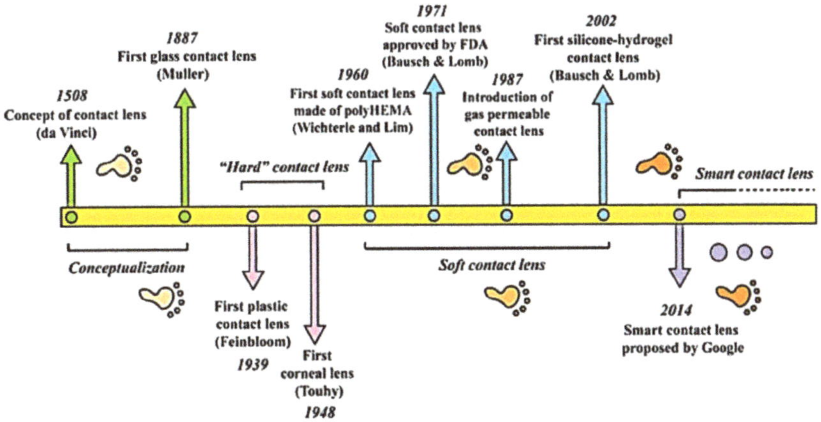

Figure 2. The milestones of contact lens development. Reprinted with permission from reference [24]. Copyright 2021 John Wiley and Sons.

Contact lenses, which are thin, curved, plastic lenses, are worn to protect the eye or correct vision. In 1965, Sedlacek first introduced contact lenses as vehicles for ocular drug delivery. Hydrogels and silicone hydrogels are suitable materials for drug-laden contact lenses. 2-hydroxyethyl methacrylate (HEMA) is polymerized to obtain hydrogel contact lenses with low oxygen permeability [64]. The introduction of various hydrophilic monomers amplifies the water content and subsequent oxygen permeation. HEMA-based contact lenses can be worn for less than 6 days [65]. However, silicone hydrogel lenses can be worn for 29 days and have high oxygen permeability [66,67]. Compared with eye drops, contact lenses extend the residence time of drugs in the eye from 2 min to 30 min, resulting in improved drug bioavailability in the cornea [68]. Potential side effects are also reduced as drug exposure in the systemic circulation is minimized [69].

Therapeutic or drug-eluting contact lenses can be excellent substitutes for treating eye diseases. With increased residence time of the drug in front of the cornea, the bioavailability of the drug also increases up to approximately 50%, which in turn can increase drug efficacy and abate systemic side effects [70,71]. This platform may increase the patient's compliance with single drug administrations, especially for vision correction patients who wear contact lenses [10,18]. Millions of people suffer from different types of ocular diseases. There are three purposes for developing a therapeutic contact lens to provide the loaded drug for slow and longer release periods. First, a "comfortable lens" ensures longer wearing of contact lenses for dry eyes. Second, "patients' compliance" by providing an easier way to maintain treatment effectiveness. Third, "bandage lenses" for managing postoperative complications, corneal wound healing, and viral corneal erosion. Anti-inflammatory or antimicrobial drugs may be incorporated into the lens, which releases the drug throughout the treatment period (1–30 days) [11]. Contact lens-based delivery of drugs for the treatment of DED are listed in Table 3.

The first technique for drug loading to the contact lens was soaking the contact lens in the drug solution, but within 1–3 h, almost all the loaded drug was released. Several procedures have been introduced to design drug-loaded contact lenses (Figure 3), including the diffusion of vitamin E barriers, molecular imprinting, prolonged drug release by ionic interactions, drug-loaded implants, colloidal micro-and nanoparticles with drugs, and supercritical fluid technology, to avoid this limitation [17].

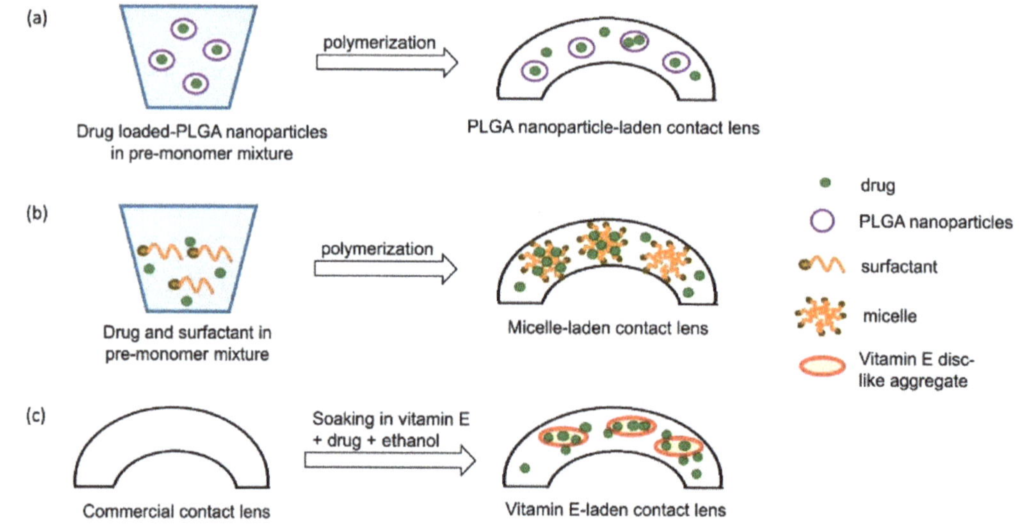

Figure 3. Hydrophobic drug controlled release techniques from soft contact lenses. (**a**) Nanoparticles loaded contact lenses. (**b**) Contact lenses with micelles. (**c**) Vitamin E fused contact lenses. Reprinted with permission from reference [72]. Copyright 2020 Elsevier.

2.3. Contact Lens-Based Drug Delivery for DED

Dry eye treatment aims to relieve symptoms, restore the ocular surface and tear film, enhance visual perception and quality of life, and correct causal defects. As dry eye disease is multifactorial, several therapies have been suggested for its management, including drug delivery through contact lenses. Cyclosporin A (CsA) can be loaded into the contact lens for DED treatment.

Table 3. DED treatment with contact lens drug delivery systems.

Drug Molecules	Contact Lens Type	Drug Loading Method	Duration	References
Cyclosporine A	Hydrogel and silicone hydrogel	Soaking	1 day (hydrogel) 15 days (silicone hydrogel). Pre-soaking with vit. E increases time release to 30 days	[73]
Hyaluronic acid	Hydrogel and silicone hydrogel	Soaking	1 h	[74]
Phospholipids	Silicone hydrogel	Soaking	10 h	[75]
Dexamethasone	Silicone hydrogel	Soaking	2 weeks–3 months	[76]
Dexamethasone	Silicone hydrogel	Soaking	7 days	[77,78]
Ap_4A (Secretagoge)	Hydrogel and silicone hydrogel	Soaking	5–6 h	[41,79]
Betaine (Osmoprotectant)	Silicone hydrogel	Soaking	10 h	[41,80]
Polyvinilpyrrolidone	Hydrogel	Polymerization	30 days	[81]
Hyaluronic acid	Hydrogel and silicone hydrogel	Polymerization	21 days (hydrogel), 49 days (silicone hydrogel)	[82]
Polyvinilpyrrolidone	Hydrogel	Polymerization	30 days	[81]
Diclofenac	Hydrogel	Copolymerization	7 days	[83]
Dexamethasone	Hydrogel	Copolymerization	50 h	[84]
Diclofenac	Hydrogel	Copolymerization	14 days	[85]
Cyclosporine A	Hydrogel	Nanoparticles (Brij surfactants)	20–30 days	[86–88]
Dexamethasone	Hydrogel	Nanoparticles/soaking	50 h	[89]
Hydroxypropyl methylcellulose	Silicone hydrogel	Molecular imprinting	60 days	[90]
Hyaluronic acid	Hydrogel and silicone hydrogel	Molecular imprinting	24 h	[91]
Diclofenac	Hydrogel	Molecular imprinting	6 days	[79]

2.3.1. Advantages of CsA

The underlying pathogenesis of DED lies in the infiltration of T-cells in the conjunctiva tissue, as well as the presence of proteases and cytokines in the tear fluid, which was the primary reason for presenting the use of immunomodulatory agents, such as corticosteroids, doxycycline, and CsA, and to treat DED. The FDA has authorized CsA emulsion for the treatment of dry eye, and clinical trials have confirmed CsA's effectiveness and safety. CsA appears to be a viable therapy for DED [92].

Tolypocladium inflatum is a CsA-producing fungus [93]. The natural product CsA is an immunosuppressant drug, and its immunosuppressive activity was first observed in 1976 [94,95]. It works by inhibiting the calcineurin–phosphatase pathway via complex formation with cyclophilin and thus reduces the transcription of T cell-activating cytokines, such as interleukin-2 (IL-2) and IL-4. It was the first drug approved by the US FDA for DED by topical application [37]. Additionally, CsA reduces the expression of proinflammatory chemokines and cytokines, such as IL-1β, IL-6, tumor necrosis factor-alpha (TNF-α), vascular cell adhesion molecule 1, and intercellular adhesion molecule 1, in benzalkonium chloride-mediated DED [96,97]. It also shelters the conjunctival epithelial cells of humans through anti-apoptotic activity, increases the density of conjunctival goblet cells, and improves the integrity of the corneal surface through immunomodulatory actions [98,99]. Goblet and epithelial cell apoptosis are related to a decrease in interferon-γ expression, resulting from CsA reduction in T-cell involvement and activation [100,101].

CsA for DED treatment is advantageous over other corticosteroids in several ways. First, the effect is reversible after therapy. Second, it has a very low systemic absorption rate. Third, no critical side effects were observed. These pharmacokinetic parameters are crucial because long-term treatment is essential for chronic illnesses, such as dry eye. Moreover, the advantages of CsA begin after 30 days of treatment, and a 90 days course of treatment appears to be required [92].

2.3.2. In Vitro CsA Release from Contact Lenses

CsA is a cyclic peptide of non-ribosomal origin consisting of 11 amino acids with a single d-amino acid. The cyclic structure consists of hydrogen bonds. The low aqueous solubility of CsA is due to this property and, thus, different cellular absorptions [102]. Several studies have demonstrated different techniques and construction materials for incorporating CsA into contact lenses (Table 4). CsA was loaded into silicone hydrogel contact lenses and hydrophilic poly-HEMA (p-HEMA) lenses by simple soaking. As CsA is a highly lipophilic drug, it has a higher affinity for lipophilic silicone-rich phases than the hydrophilic p-HEMA phase. Therefore, the partition coefficient is higher in silicone hydrogel contact lenses than in hydrophilic p-HEMA lenses. The in vitro release durations of CsA from silicone hydrogel contact lenses and hydrophilic p-HEMA contact lenses were approximately 15 days and 1 day, respectively [41]. In addition, the vitamin E barrier is used to slow and extend drug release (Figure 3c) with minimal impact on vital lens properties, such as light refraction and oxygen permeability [17,77]. The in vitro release of CsA was prolonged for approximately 1 month with a 20% loading of vitamin E into the silicone hydrogel contact lenses.

Table 4. CsA delivery from contact lenses for the management of DED.

Dosage Form	Contact Lens Material	Loading Method	Drug Release Duration	References
Contact lens	hydroxyethyl methacrylate (HEMA), cholesterol-hyaluronate (C-HA) micelle	mixing	12 days	[103]
Contact lens	poly-hydroxyethyl methacrylate (p-HEMA), Brij 97, Brij 78 and Brij 700	mixing	Brij 97—20 days, Brij 78—50 days, Brij 700—20 days	[86]
Silicone contact lenses	ethylene glycol dimethacrylate (EGDMA)	soaking	2 weeks, with vitamin E—1 month	[73]
Contact lens	poly-hydroxy ethyl methacrylate (p-HEMA), Brij 98	mixing	25 days	[87]
Contact lens	poly (2-hydroxyethyl methacrylate) (p-HEMA), Brij 97	mixing	20 days	[73]
Contact lens	graphene oxide	soaking	-	[104]

The use of surfactant in the p-HEMA contact lens polymerizing mixture forms micellar aggregates by creating hydrophobic sites inside the gel, where the hydrophobic CsA may preferentially partition. Kapoor et al. (2009) developed Brij surfactant-laden p-HEMA hydrogel contact lenses using CsA. Brij 97 and Brij 98 surfactants showed slow and longer CsA release of approximately 20 days. Brij 78 surfactant seemed most capable of delaying the release of CsA from p-HEMA contact lenses for more than 30 days (Figure 4) [41,86]. Peng and Chauhan (2011) showed that the incorporation of vitamin E extended the release of CsA for more than 30 days (Figure 5) [73]. Desai et al. (2021) developed a contact lens with graphene oxide to incorporate the hydrophobic drug CsA by the soaking method and evaluated its in vitro drug release. They demonstrated that the increased drug uptake did not change the optical or swelling characteristics [104].

2.3.3. In Vivo Biological Activity of CsA Contact Lenses

Mun et al. (2019) showed that CsA release from the contact lens in the rabbit eye produced a significant DED improvement (Figure 6). The DED was induced in rabbits with 3-concanavalin A injections (Con A, sigma L7647). The contact lens was able to release CsA for up to 7 days. Corneal immunofluorescence staining for MMP9 (a DED marker) was performed to confirm the treatment outcomes (Figure 7) [103]. A decrease in the MMP9 intensity was observed for the right eyes (OD) treated with eye drops and CsA/C-HA micelle CL compared to control group (OS). Desai et al. (2022) observed that rabbits recovered quickly from DED with a CsA graphene contact lens and a higher amount of CsA in the corneal fluid for a long time [104].

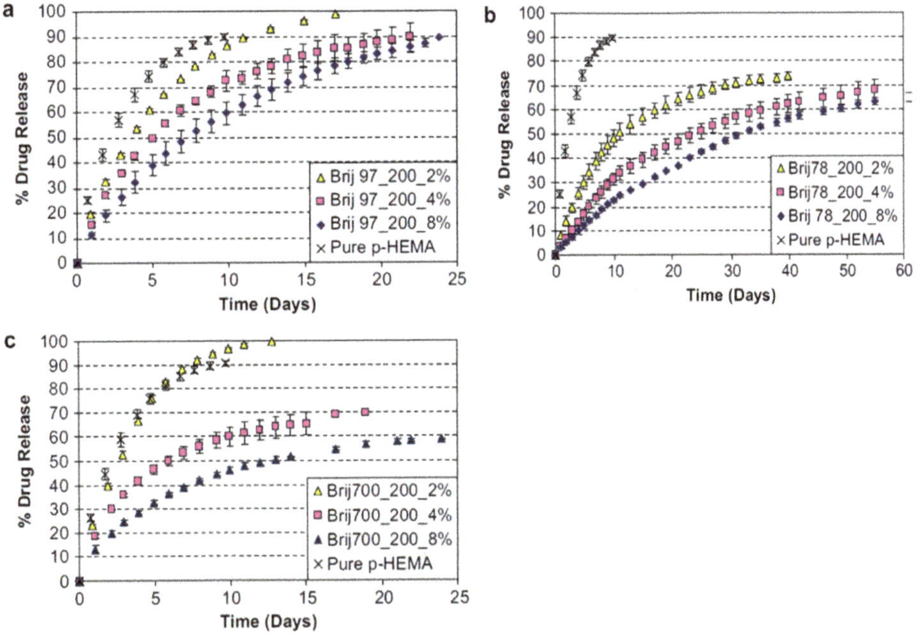

Figure 4. Influence of surfactant on cumulative release of CsA (50 mg/contact lens). The contact lenses consisted with different surfactants: (**a**) Brij 97, (**b**) Brij 78, and (**c**) Brij 700. Adapted from reference [86]. Copyright 2008 Elsevier.

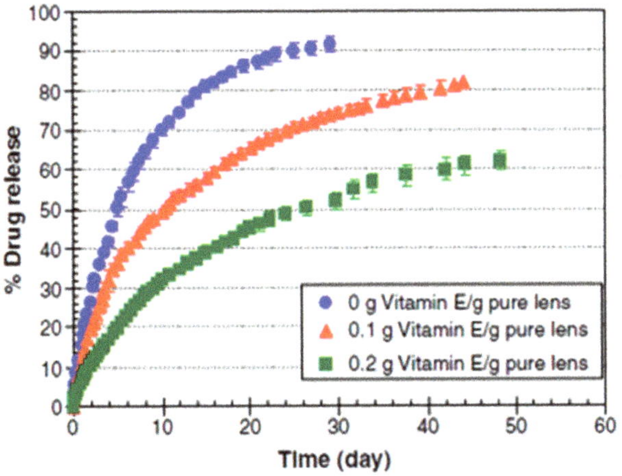

Figure 5. Effect of vitamin E on cumulative release of CsA from ACUVUE® OASYS™ contact lenses. Adapted from reference [73]. Copyright 2011 Elsevier.

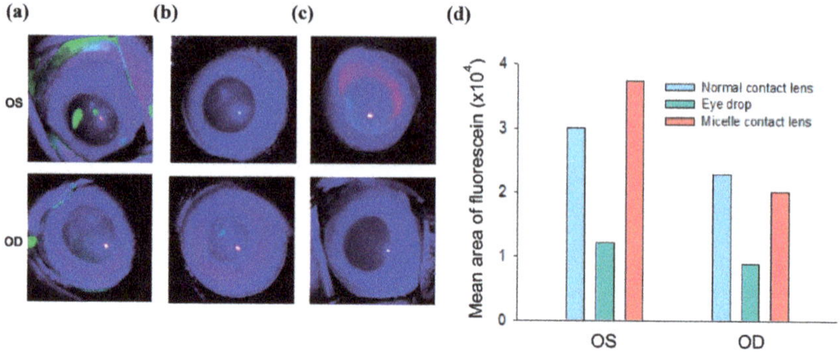

Figure 6. Corneal inflammation inspection by corneal fluorescein staining followed by (**a**) normal contact lens wear in OD, (**b**) CsA eye drop administration, and (**c**) CsA/C-HA micelle contact lens wear in OD. (**d**) The ROI (region of interest—ROI) values for fluorescein staining (oculus dexter—OD, oculus sinister—OS). In case of micelle CL, the OS did not contain any CsA, so the fluorescein staining showed increased intensity than the OD. Oculus dexter is abbreviated as OD, meaning right eye, and oculus sinister is abbreviated as OS, meaning left eye. Reprinted with permission from reference [103]. Copyright the Royal Society of Chemistry.

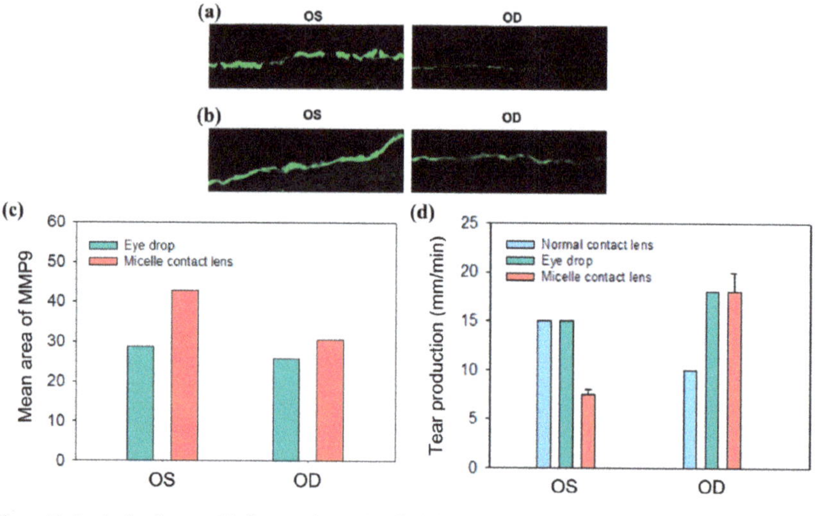

Figure 7. Analysis of corneal inflammation using the MMP9 DED marker (corneal immunofluorescein staining, green color) after (**a**) CsA eye drop application, and (**b**) CsA/C-HA micelle contact lens wear. (**c**) ROI values for MMP9. (**d**) Tear production by the contact lens-wearing (OD) and control (OS) eyes at day 7. Tear production was increased in the treated OD groups. Oculus dexter is abbreviated as OD, meaning right eye, and oculus sinister is abbreviated as OS, meaning left eye. Reprinted with permission from reference [103]. Copyright the Royal Society of Chemistry.

3. Challenges of Contact Lenses in Drug Delivery

Scientists have succeeded in prolonging drug release using contact lenses; however, critical lens properties, such as oxygen permeability, swelling, ion permeability, optical transparency, tensile strength, issues during monomer extraction, high burst release, sterilization, and storage, are yet to be addressed [105]. Corneal damage and infections related to the extended wearing of contact lenses for chronic diseases, such as DED, must be considered [13,91,106]. Contact lenses integrated with drug delivery systems are classified as combination medical products, which may delay the approval of new materials. The polymerization process and the chemical and physical properties of the generated hydrogel may be influenced by drug properties (the physical and chemical). Bacterial resistance and ocular toxicity may arise from long-term drug release from contact lenses [107].

A major limitation of contact-lens drug delivery is the initial burst release of the drug, which can lead to potential systemic toxicities. Researchers have employed various technologies, such as p-HEMA hydrogel, implants, molecular imprinting, use of vitamin E, and incorporation of drug-loaded nanoparticles, to minimize it [108]. Recently, bioelectronics and biosensors have been intensely utilized to monitor health conditions in real-time for chronic diseases and coronavirus disease 2019 (COVID-19). These multifunctional sensors process physiological signals to digital data without disturbing normal biological activities and can dramatically enhance therapeutic outcomes. These small devices can help deliver medicines more precisely, and personalization is possible [109]. Biosensors can be integrated with contact lenses to monitor disease conditions continuously via specific biomarkers for ocular diseases, improving lens properties and minimizing drug related systemic toxicities.

4. Biosensors Integrated Contact Lens

A biosensor is a diagnostic device used to sense a chemical substance that combines a biological element with a physicochemical indicator (Figure 8) [110–113]. The delicate biological components, such as enzymes, tissue, antibodies, organelles, cell receptors, and nucleic acids, are biologically obtained elements that bind with, interact with, or distinguish the analyte under experiment. The detector transforms one signal into another. It works through different mechanisms, for example, piezoelectric, electrochemiluminescence optical, and electrochemical, through the reaction of the analyte and biological sample to measure and quantify. A reader is connected to the display to show the results simply [99].

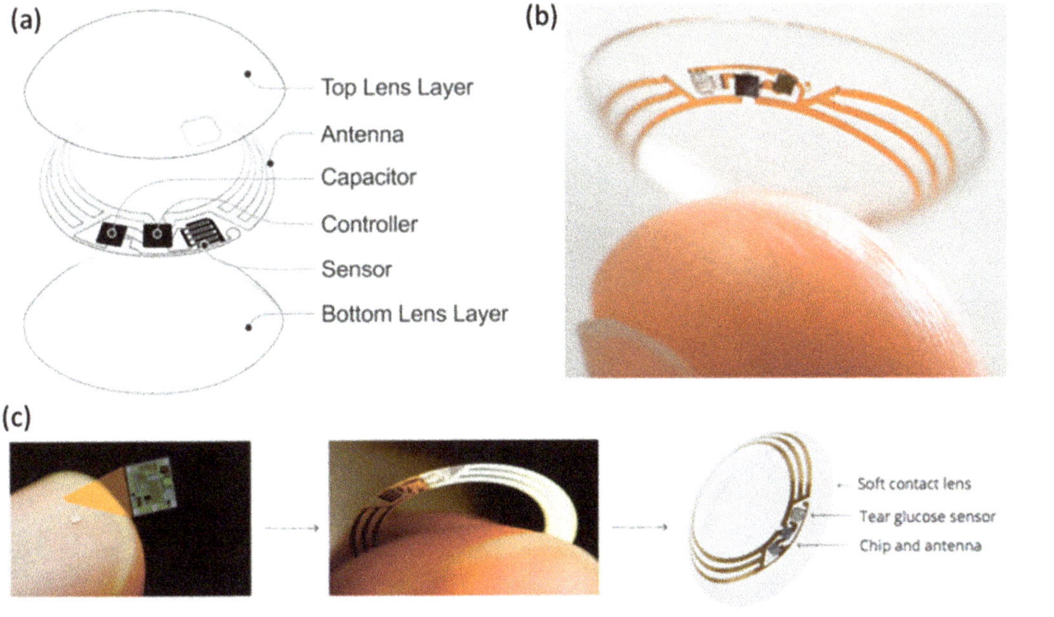

Figure 8. The contact lens sensor for tear glucose measurement, developed by Google and Novartis. (**a**) A diagram of the contact lens sensors. (**b**) The model contact lens sensor. (**c**) The wireless chip for the sensor. Reprinted with permission from reference [114]. Copyright 2021 John Wiley and Sons.

The human eye carries important chemical, physical, and biological data related to human health. Hence, it appears to be a vital research target that drives the rapid growth of soft electronic systems for diagnosing various diseases of the eye and other organs. As wearable and flexible medical devices, contact lenses have a significant capacity to support the analysis and treatment of ocular diseases [20].

4.1. Tear Film Biomarkers for DED

The concentration of the tear fluid constituents was related to its concentration in the blood (Table 5). DED is a multifactorial inflammatory disease characterized by tear film instability, ocular discomfort, visual disturbances, inflammation, and increased tear osmolarity. Research on tear film biomarkers has been increasing to identify diagnostic tools for DED or monitor treatment outcomes in clinical trials. In the last 5 years, numerous studies have been conducted on tear fluid biomarkers (Table 6) in DED [115]. Physicians face difficulty selecting suitable treatment options because of the lack of sufficient tools for observing and monitoring patient responses. Recent studies have investigated chemokines, cytokines, growth factors, neuromodulators, mucins, and lipids to find protein profiles that can be suitable biomarkers for DED [116]. Aluru and colleagues, found that lysozyme proline-rich protein 4 is downregulated in several types of DED, suggesting that this protein is a potential biomarker for DES [117]. Zhou and colleagues reported the upregulation of ∝-1 acid glycoprotein 1, ∝-enolase, calgranulin B, calgranulin A, and calgizzarin. Furthermore, four downregulated proteins, including lipocalin-1, prolactin-inducible protein, lysozyme, and lactoferrin, were also found in DED patients. Recently, different research groups have anticipated other proteins related to DED, regardless of the probable biomarkers under investigation [118]. For instance, malate dehydrogenase 2 activity increased, but mucin (MUC)5AC activity decreased. On the other hand, neuromediators, such as neuropeptide Y (NPY), calcitonin gene-related peptide (CGRP), nerve growth factor (NGF), and vasoactive intestinal peptide, have been identified as possible biomarkers for DED due to their association in clinical studies [119]. In DED, NGF levels were increased, while NPY and CGRP were found to decrease in tear fluid. Chhadva et al. (2015), reported higher tear serotonin concentrations in patients with DED [120].

Table 5. The concentrations of major analytes in tears and their relative concentrations in the blood. Reprinted with permission from reference [114]. Copyright John Wiley and Sons.

Analyte	Tear Fluid Concentration [mM]	Blood Concentration [mM]	Diagnostic Application
Glucose	0.013–0.051	3.3–6.5	Diabetes management
Lactate	2.0–5.0	0.36–0.75	Ischemia, sepsis, liver disease, and cancer
Na^+	120–165	130–145	Hyper/hyponatremia
K^+	20–40	3.5–5.0	Hyper/hypokalemia and an indicator of ocular disease
Ca^{2+}	0.4–1.1	2.0–2.6	Hyper/hypocalcemia
Mg^{2+}	0.5–0.9	0.7–1.1	Hyper/hypomagnesemia
Cl^-	118–135	95–125	Hyper/hypochloremia
HCO_3^-	20–26	24–30	Respiratory quotient indicator
Urea	3.0–6.0	3.3–6.5	Renal function
Pyruvate	0.05–0.35	0.1–0.2	Genetic disorders of mitochondrial energy metabolism
Ascorbate	0.22–1.31	0.04–0.06	Diabetes
Total Protein	≈7 g/L	≈70 g/L	Dry eye conditions, ocular insult, and inflammation
Dopamine	0.37	475×10^{-9}	Glaucoma

Table 6. DED biomarkers identified in tear fluid.

Types of Biomarker Molecule	Biomarkers	References
Proteins	Lysozyme, lactoferrin, lysozyme proline-rich protein 4 (LPRR4), calgranulin A/S100 A8, lysozyme proline-rich protein 3 (LPRR3), nasopharyngeal carcinoma-associated PRP 4, α-1 antitrypsin α-enolase, α-1 acid glycoprotein 1, S100 A4, S100 A11 (calgizzarin), S100 A9/calgranulin B, lipocalin-1 (LCN-1), mammaglobin B, lipophilin A, beta-2 microglobulin (B2M), S100A6, annexin A1 annexin A11, cystatin S (CST4), phospholipase A2-activating protein (PLAA), transferrin, defensin-1, clusterin, lactotransferrin, cathepsin S, anti-SS-A, anti-SS-B, anti-α-fodrin antibodies, malate dehydrogenase (MDH) 2, palate lung nasal clone-PLUNC	[115,121]
Mucins	(MUC)5AC	[122]
Neuromediators	Nerve growth factor (NGF), calcitonin gene related peptide (CGRP), neuropeptide Y (NPY), vasointestinal peptide (VIP), serotonin, substance P	[119]
Cytokines/chemokines	Interleukin-1(IL-1), interleukin-2 (IL-2), interleukin-5 (IL-5), interleukin-6 (IL-6), interleukin 8 (IL-8) or chemokine (C-X-C motif) ligand 8 (CXCL8), interleukin-10 (IL-10), interleukin-12 (IL-12), interleukin-16 (IL-16), interleukin-33 (IL-33), GCSF, monocyte chemoattractant protein 1 (MCP1)/chemokine (C-C motif) ligand 2 (CCL2), MIP5/chemokine (C-C motif) ligand 15 (CCL15), C-X-C motif chemokine 5 (CXCL5 or ENA78), soluble interleukin-1 receptor Type I (sIL-1RI), soluble interleukin-6 receptor (sIL-6R), soluble gp130 (sgp130), soluble vascular endothelial growth factor receptor 1 (sVEGFR1), soluble epidermal growth factor receptor (sEGFR), soluble tumor necrosis factor receptor I (sTNFR I), interleukin-17A (IL-17A), interleukin-21 (IL-21), interleukin-22 (IL-22), interleukin-1 receptor antagonist (IL-1RA), chemokine (C-X-C motif) ligand 9 (CXCL9)/monokine induced by gamma interferon (MIG), interferon-inducible T-cell alpha chemoattractant (I-TAC)/C-X-C motif chemokine 11 (CXCL11), C–X–C motif chemokine 10 (CXCL10)/interferon γ-induced protein 10 kDa (IP-10), ligand 4 (CCL4)/macrophage inflammatory protein-1β (MIP-1β), chemokine (C-C motif) ligand 5 (also CCL5)/regulated on activation, normal T cell expressed and secreted (RANTES), epidermal growth factor (EGF), tumor necrosis factor alpha (TNF-α), interferon gamma (IFNγ), matrix metallopeptidase 9 (MMP-9), macrophage inflammatory protein-1 alpha (MIP-1α/CCL3), vascular endothelial growth factor (VEGF), fractalkine	[115,123,124]
Lipids	(O-acyl) ω-hydroxy fatty acids (OAHFAs), lysophospholipids, PUFA-containing diacylglyceride species, hexanoyl-lysine (HEL), 4-hydroxy-2-nonenal (HNE), malondialdehyde (MDA)	[125]
Metabolites	Cholesterol, N-acetylglucosamine, glutamate, creatine, amino-n-butyrate, choline, acetylcholine, arginine, phosphoethanolamine, glucose, phenylalanine	[126]
Tear solutes	Osmolarity	[127]

The levels of chemokines and cytokines in tears play an important role in DED. Many studies have been conducted to identify a complete tear profile. Some inflammatory chemokine/cytokine levels (such as TNF-α, IL-1, IL-1RA, IL-6, metalloproteinase (MMP)-9, IL-8/C-X-C motif ligand (CXCL) 8, IL-17A, IL-22, interferon-γ, IP-10/CXCL10, MIG/CXCL9, I-TAC/CXCL11, macrophage inflammatory protein (MIP)-1β/CCL4, MIP-1α/CCL3, and RANTES/CCL5) are remarkably elevated in tears of patients with DED [115], while endothelial growth factor is decreased [128–130] with an increase in disease severity. Measurement of MMP-9 in tears has been proposed as a delicate technique for DED severity determination [131,132]. Some researchers have found that MMP-9 increases in the tear fluid of patients with DED [124,133–135]. Several studies have been conducted to identify the properties of tear lipids secreted by the meibomian gland in patients with DED. Compositional differences in the DED patient reflex tear metabolomic profile were revealed for N-acetylglucosamine, cholesterol, creatine, glutamate, amino-n-butyrate, acetylcholine, choline, arginine, glucose, phosphoethanolamine, and phenylalanine levels [126]. Willshire C et al. found that the basal tear osmolarity increases in DED compared to that in the control group [127]. Therefore, it may be a useful marker for DED. Moreover, correlated biomarkers in tears, such as cytokine profiles, have been anticipated for the initial diagnosis of the COVID-19 [136,137].

4.2. Contact Lens Sensors for Sensing of Tear Fluid Biomarkers in DED

Tear osmolarity is the only clinically established parameter directly associated with dry eye severity [103]. Human tears' chemical components (biomarkers) include proteins, electrolytes, lipids, urea, L-lactic acid, cholesterol, ascorbic acid, and many metabolites. If their concentrations in tears are known, then their concentrations in the blood could be correlated. Therefore, concurrently analyzing their concentrations in tears provides important physiological data that can improve treatment outcomes and anticipation of some illnesses [20]. Identifying appropriate biomarkers for specific diseases is a major challenge and an ongoing process. Once a biomarker is identified, it is tested for biosensor applications that vary from functionally integrated (contact lens) to on-chip sensors (Figure 9) [116]. Currently, it is possible to measure multiple parameters, such as the glucose level and IOP, using a single contact lens that integrates many sensors [138]. The number of illnesses that can be tracked and identified with contact lens biosensors will increase as sensing technology advances. Contact lenses can naturally gather tear components during wear and may be examined thereafter. It would be feasible to detect the existence and progression of certain diseases by combining the detection of certain biomarkers, such as cancer or dry eye [85]. Some electrochemical sensors have already been developed to identify several biomarkers (Table 7) in tear fluid to monitor the condition of patients with DED. Thus, integrating these sensors into a therapeutic contact lens can continuously track DED progress.

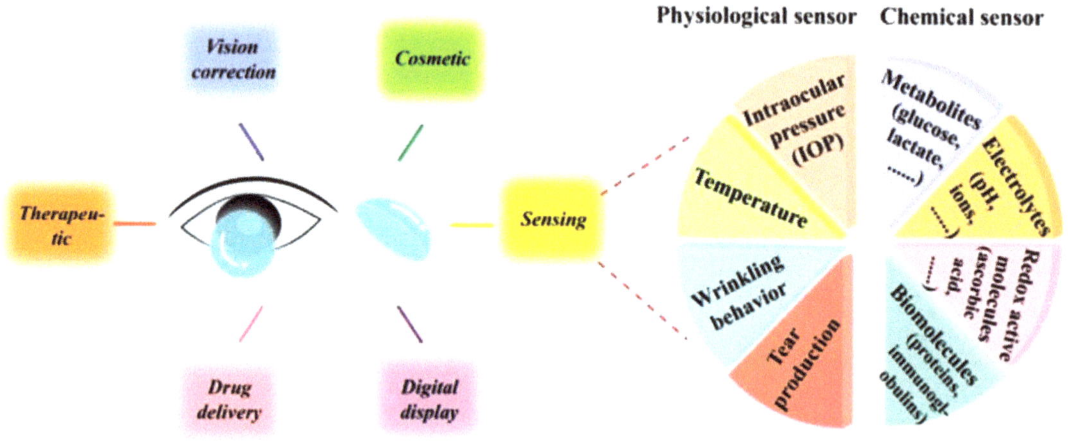

Figure 9. Sensing capability of contact lens sensors. Reprinted with permission from reference [24]. Copyright 2021 John Wiley and Sons.

Table 7. Electrochemical sensors that can sense vital biomarkers of DED in tears. Adapted from reference [24]. Copyright John Wiley and Sons.

Biomarkers	Sensor Type
Glucose	Enzymatic biosensor; amperometric
Osmolarity	Impedimetric
MMP-9	Electrochemical immunosensors
Urea	Voltammetric
Serum	Electrochemical immunosensors
TNF-α	Electrochemical aptamer sensor
Mucins	Electrochemical immunosensors

5. The Future Perspective of Biosensor Fused Contact Lens

The concept of biosensors in contact lenses is recent and unique. The implementation of biosensors in a contact lens can measure specific parameters even during sleep, enabling the analysis of the pattern of disease conditions at night or during sleeping hours [138]. Most contact lenses can only detect one biomarker in the eye, such as glucose, lactic acid, K^+, or Ca^{2+}. The detection of multiple chemical components in real-time increases the biomedical utility of contact lenses [139]. Most existing sensory systems lack the ability to power themselves. As natural sunlight is readily available for energy conversion, flexible photovoltaic self-powered technology can replace standard power supply modes in contact lenses. Photovoltaics will be a future trend in inflexible and stretchable electronics because of these and other characteristics [140]. More chips and interconnects must be added to the device to increase the performance and multifunctionality of contact lenses. The use of transparent materials, such as graphene, carbon nanowires, and indium tin oxide, will make this work easier. The shrinking of chips incorporated in the system for data storage, data transfer, and circuit powering has become increasingly significant, driving researchers and industrial suppliers to produce next-generation chips with multiplexed capabilities. Furthermore, when the entire circuit was scaled down, the sensitivity of the device was considerably diminished, particularly with respect to the size of the sensing electrodes. One possible solution to these issues is to use active sensors, such as field-effect transistors and complementary metal-oxide-semiconductor sensors, which have remarkable sensitivity and are sub-micrometers in size [20,141]. The concentration of tear biomarkers is low, which necessitates the use of highly sensitive biosensors. In addition, the development of sensitive biosensors is very expensive. Tear makeup varies significantly across and between individuals. The lag time caused by biomarker diffusion from the tear fluid to the implanted biosensor can affect the treatment outcomes. When the biosensor is attached to a contact lens, it becomes thicker and can cause patient discomfort [142].

Contact lenses have shown enormous potential in biomedicine owing to their features, such as real-time and non-invasive diagnostics and drug delivery. Multifunctional and integrated contact lenses can record physiological data regarding eye problems more efficiently than earlier approaches, reducing the need for human illness treatment. It offers great promise as an everyday medical device for the reliable measurement of ocular response to ophthalmic drugs and surgical procedure evaluation. Contact lenses represent technical and material advances that will pave the way for the next generation of precision medicine-based products [100]. Soon, it will be possible to combine biosensors in a medicated contact lens that will release the drug and monitor the overall disease condition simultaneously. No study has reported the electrical control of drug release from contact lenses using simultaneous biometric analysis [143]. Future biosensors may control drug release from the contact lens in response to a patient's need.

6. Conclusions

Drug-loaded contact lenses are a promising option for treating chronic ocular diseases. In the last few years, a number of innovative contact lens drug-delivery systems have been established that increase the drug-loading capacity and control the drug-release rate. Treatment of DED with a CsA-loaded contact lens has been successful in animal models. Further studies are required to confirm its feasibility in clinical trials. Contact lenses are not limited to drug-delivery devices, as they can also be used as a diagnostic tool. Contact lens sensor technology has gained popularity over the last decade, primarily owing to developments in the downsizing of electrical circuits and the discovery of several significant biomarkers in tear fluid. This sensor platform offers various advantages, including its non-invasive and constant biomarker-measuring properties. However, significant advancements in specificity, sensitivity, biocompatibility, integration with readout circuitry, and repeatability are still being made for such platforms to achieve feasibility. For example, a self-powered biosensor significantly simplifies the sensor layout. Furthermore, a better understanding of the relationship between illness and ocular biomarker concentration is necessary to develop practical multifunctional contact lens biosensors. This might pave the way for personalized therapeutic contact lenses with biosensors.

Author Contributions: H.M., H.-J.K. and J.-P.J.: conceptualization, methodology, data curation, writing original draft preparation; N.M., H.-J.K. and J.-P.J.: validation, writing review and editing. All authors have read and agreed to the published version of the manuscript.

Funding: This work was performed with financial support from the research funds provided by Chosun University in 2021.

Conflicts of Interest: The authors declare no conflict of interest.

References

1. Phadatare, S.P.; Momin, M.; Nighojkar, P.; Askarkar, S.; Singh, K.K. A Comprehensive Review on Dry Eye Disease: Diagnosis, Medical Management, Recent Developments, and Future Challenges. *Adv. Pharm.* **2015**, *2015*, 1–12. [CrossRef]
2. Lemp, M.A.; Baudouin, C.; Baum, J.; Dogru, M.; Foulks, G.N.; Kinoshita, S.; Laibson, P.; McCulley, J.; Murube, J.; Pflugfelder, S.C.; et al. The Definition and Classification of Dry Eye Disease: Report of the Definition and Classification Subcommittee of the International Dry Eye WorkShop (2007). *Ocul. Surf.* **2007**, *5*, 75–92. [CrossRef]
3. Marshall, L.L.; Roach, J.M. Treatment of Dry Eye Disease. *Consult. Pharm.* **2016**, *31*, 96–106. [CrossRef] [PubMed]
4. Saha, S.; Shilpi, J.A.; Mondal, H.; Hossain, F.; Anisuzzman, M.; Hasan, M.M.; Cordell, G.A. Ethnomedicinal, Phytochemical, and Pharmacological Profile of the Genus Dalbergia L.(Fabaceae). *Phytopharmacology* **2013**, *4*, 291–346.
5. Messmer, E.M. Pathophysiology, Diagnosis and Treatment of Dry Eye. *Dtsch. Arzteblatt Int.* **2015**, *112*, 71–82. [CrossRef] [PubMed]
6. Stapleton, F.; Alves, M.; Bunya, V.Y.; Jalbert, I.; Lekhanont, K.; Malet, F.; Na, K.S.; Schaumberg, D.; Uchino, M.; Vehof, J.; et al. TFOS DEWS II Epidemiology Report. *Ocul. Surf.* **2017**, *15*, 334–365. [CrossRef] [PubMed]
7. Farrand, K.F.; Fridman, M.; Stillman, I.Ö.; Schaumberg, D.A. Prevalence of Diagnosed Dry Eye Disease in the United States among Adults Aged 18 Years and Older. *Am. J. Ophthalmol.* **2017**, *182*, 90–98. [CrossRef] [PubMed]
8. Bielory, L.; Syed, B.A. Pharmacoeconomics of Anterior Ocular Inflammatory Disease. *Curr. Opin. Allergy Clin. Immunol.* **2013**, *13*, 537–542. [CrossRef]
9. Clayton, J.A. Dry Eye. *N. Engl. J. Med.* **2018**, *378*, 2212–2223. [CrossRef]
10. Choi, S.W.; Kim, J. Therapeutic Contact Lenses with Polymeric Vehicles for Ocular Drug Delivery: A Review. *Materials* **2018**, *11*, 1125. [CrossRef] [PubMed]
11. Kumar, A.; Jha, G. Drug Delivery through Soft Contact Lenses: An Introduction. *Chron. Young Sci.* **2011**, *2*, 3. [CrossRef]
12. Janagam, D.R.; Wu, L.; Lowe, T.L. Nanoparticles for Drug Delivery to the Anterior Segment of the Eye. *Adv. Drug Deliv. Rev.* **2017**, *122*, 31–64. [CrossRef] [PubMed]
13. Maulvi, F.A.; Shaikh, A.A.; Lakdawala, D.H.; Desai, A.R.; Pandya, M.M.; Singhania, S.S.; Vaidya, R.J.; Ranch, K.M.; Vyas, B.A.; Shah, D.O. Design and Optimization of a Novel Implantation Technology in Contact Lenses for the Treatment of Dry Eye Syndrome: In Vitro and in Vivo Evaluation. *Acta Biomater.* **2017**, *53*, 211–221. [CrossRef]
14. Carvalho, I.M.; Marques, C.S.; Oliveira, R.S.; Coelho, P.B.; Costa, P.C.; Ferreira, D.C. Sustained Drug Release by Contact Lenses for Glaucoma Treatment-a Review. *J. Control. Release Off. J. Control. Release Soc.* **2015**, *202*, 76–82. [CrossRef]
15. Sahoo, S.K.; Dilnawaz, F.; Krishnakumar, S. Nanotechnology in Ocular Drug Delivery. *Drug Discov. Today* **2008**, *13*, 144–151. [CrossRef]
16. Jung, H.J.; Abou-Jaoude, M.; Carbia, B.E.; Plummer, C.; Chauhan, A. Glaucoma Therapy by Extended Release of Timolol from Nanoparticle Loaded Silicone-Hydrogel Contact Lenses. *J. Control. Release Off. J. Control. Release Soc.* **2013**, *165*, 82–89. [CrossRef]
17. Peng, C.C.; Burke, M.T.; Carbia, B.E.; Plummer, C.; Chauhan, A. Extended Drug Delivery by Contact Lenses for Glaucoma Therapy. *J. Control. Release* **2012**, *162*, 152–158. [CrossRef] [PubMed]
18. Jung Jung, H.; Chauhan, A. Ophthalmic Drug Delivery by Contact Lenses. *Expert Rev. Ophthalmol.* **2012**, *7*, 199–201. [CrossRef]
19. Hsu, K.H.; Gause, S.; Chauhan, A. Review of Ophthalmic Drug Delivery by Contact Lenses. *J. Drug Deliv. Sci. Technol.* **2014**, *24*, 123–135. [CrossRef]
20. Ma, X.; Ahadian, S.; Liu, S.; Zhang, J.; Liu, S.; Cao, T.; Lin, W.; Wu, D.; de Barros, N.R.; Zare, M.R.; et al. Smart Contact Lenses for Biosensing Applications. *Adv. Intell. Syst.* **2021**, *3*, 2000263. [CrossRef]
21. Chen, G.Z.; Chan, I.S.; Leung, L.K.K.; Lam, D.C.C. Soft Wearable Contact Lens Sensor for Continuous Intraocular Pressure Monitoring. *Med. Eng. Phys.* **2014**, *36*, 1134–1139. [CrossRef]
22. Huang, J.F.; Zhong, J.; Chen, G.P.; Lin, Z.T.; Deng, Y.; Liu, Y.L.; Cao, P.Y.; Wang, B.; Wei, Y.; Wu, T.; et al. A Hydrogel-Based Hybrid Theranostic Contact Lens for Fungal Keratitis. *ACS Nano* **2016**, *10*, 6464–6473. [CrossRef]
23. Mohanto, N.; Khatun, A.; Begum, J.A.; Parvin, M.M.; Siddiqui, M.S.I.; Begum, S.; Parvin, R.; Islam, M.R.; Chowdhury, E.H. Trehalose Improves PPR Vaccine Virus Stability in Diluent. *Bangladesh J. Vet. Med. BJVM* **2019**, *17*, 117–123. [CrossRef]
24. Liu, H.; Yan, X.; Gu, Z.; Xiu, G.; Xiao, X. Electrochemical Sensing in Contact Lenses. *Electroanalysis* **2021**, *34*, 227–236. [CrossRef]
25. Moreddu, R.; Elsherif, M.; Adams, H.; Moschou, D.; Cordeiro, M.F.; Wolffsohn, J.S.; Vigolo, D.; Butt, H.; Cooper, J.M.; Yetisen, A.K. Integration of Paper Microfluidic Sensors into Contact Lenses for Tear Fluid Analysis. *Lab. Chip* **2020**, *20*, 3970–3979. [CrossRef]

26. Moreddu, R.; Wolffsohn, J.S.; Vigolo, D.; Yetisen, A.K. Laser-Inscribed Contact Lens Sensors for the Detection of Analytes in the Tear Fluid. *Sens. Actuators B Chem.* **2020**, *317*, 128183. [CrossRef]
27. Ray, T.R.; Ivanovic, M.; Curtis, P.M.; Franklin, D.; Guventurk, K.; Jeang, W.J.; Chafetz, J.; Gaertner, H.; Young, G.; Rebollo, S.; et al. Soft, Skin-Interfaced Sweat Stickers for Cystic Fibrosis Diagnosis and Management. *Sci. Transl. Med.* **2021**, *13*, eabd8109. [CrossRef]
28. Pflugfelder, S.C.; Geerling, G.; Kinoshita, S.; Lemp, M.A.; McCulley, J.; Nelson, D.; Novack, G.N.; Shimazaki, J.; Wilson, C. Management and Therapy of Dry Eye Disease: Report of the Management and Therapy Subcommittee of the International Dry Eye WorkShop (2007). *Ocul. Surf.* **2007**, *5*, 163–178. [CrossRef]
29. Lee, S.-Y.; Tong, L. Lipid-Containing Lubricants for Dry Eye. *Optom. Vis. Sci.* **2012**, *89*, 1654–1661. [CrossRef]
30. Messmer, E.M. Konservierungsmittel in Der Ophthalmologie. *Ophthalmologe* **2012**, *109*, 1064–1070. [CrossRef]
31. Raj, G.B.; Surendra, P.; Kim, D. Cellulose and its derivatives for application in 3D printing of pharmaceuticals. *J. Pharm. Investig.* **2021**, *51*, 1–22. [CrossRef]
32. Hasan, N.; Cao, J.; Lee, J.; Kim, H.; Yoo, J.-W. Development of Clindamycin-Loaded Alginate/Pectin/Hyaluronic Acid Composite Hydrogel Film for the Treatment of MRSA-Infected Wounds. *J. Pharm. Investig.* **2021**, *51*, 597–610. [CrossRef]
33. Kim, M.-H.; Nguyen, D.-T.; Kim, D.-D. Recent Studies on Modulating Hyaluronic Acid-Based Hydrogels for Controlled Drug Delivery. *J. Pharm. Investig.* **2022**, *52*, 397–413. [CrossRef]
34. Marsh, P.; Pflugfelder, S.C. Topical Nonpreserved Methylprednisolone Therapy for Keratoconjunctivitis Sicca in Sjogren Syndrome. *Ophthalmology* **1999**, *106*, 811–816. [CrossRef]
35. Pflugfelder, S.C.; Maskin, S.L.; Anderson, B.; Chodosh, J.; Holland, E.J.; De Paiva, C.S.; Bartels, S.P.; Micuda, T.; Proskin, H.M.; Vogel, R. A Randomized, Double-Masked, Placebo-Controlled, Multicenter Comparison of Loteprednol Etabonate Ophthalmic Suspension, 0.5%, and Placebo for Treatment of Keratoconjunctivitis Sicca in Patients with Delayed Tear Clearance. *Am. J. Ophthalmol.* **2004**, *138*, 444–457. [CrossRef]
36. Dursun, D.; Ertan, A.; Bilezikçi, B.; Akova, Y.A.; Pelit, A. Ocular Surface Changes in Keratoconjunctivitis Sicca with Silicone Punctum Plug Occlusion. *Curr. Eye Res.* **2003**, *26*, 263–269. [CrossRef]
37. Matsuda, S.; Koyasu, S. Mechanisms of Action of Cyclosporine. *Immunopharmacology* **2000**, *47*, 119–125. [CrossRef]
38. Moscovici, B.K.; Holzchuh, R.; Chiacchio, B.B.; Santo, R.M.; Shimazaki, J.; Hida, R.Y. Clinical Treatment of Dry Eye Using 0.03% Tacrolimus Eye Drops. *Cornea* **2012**, *31*, 945–949. [CrossRef] [PubMed]
39. Sanz-Marco, E.; Udaondo, P.; García-Delpech, S.; Vazquez, A.; Diaz-Llopis, M. Treatment of Refractory Dry Eye Associated with Graft versus Host Disease with 0.03% Tacrolimus Eyedrops. *J. Ocul. Pharmacol. Ther.* **2013**, *29*, 776–783. [CrossRef]
40. Auw-Hädrich, C.; Reinhard, T. Treatment of Chronic Blepharokeratoconjunctivitis with Local Calcineurin Inhibitors. *Ophthalmologe* **2009**, *106*, 635–638. [CrossRef]
41. Guzman-Aranguez, A.; Fonseca, B.; Carracedo, G.; Martin-Gil, A.; Martinez-Aguila, A.; Pintor, J. Dry Eye Treatment Based on Contact Lens Drug Delivery: A Review. *Eye Contact Lens* **2016**, *42*, 280–288. [CrossRef]
42. Yoo, S.E.; Lee, D.C.; Chang, M.H. The Effect of Low-Dose Doxycycline Therapy in Chronic Meibomian Gland Dysfunction. *Korean J. Ophthalmol. KJO* **2005**, *19*, 258–263. [CrossRef] [PubMed]
43. Solomon, A.; Rosenblatt, M.; Li, D.Q.; Liu, Z.; Monroy, D.; Ji, Z.; Lokeshwar, B.L.; Pflugfelder, S.C. Doxycycline Inhibition of Interleukin-1 in the Corneal Epithelium. *Investig. Ophthalmol. Vis. Sci.* **2000**, *41*, 2544–2557. [CrossRef] [PubMed]
44. Sadrai, Z.; Hajrasouliha, A.R.; Chauhan, S.; Saban, D.R.; Dastjerdi, M.H.; Dana, R. Effect of Topical Azithromycin on Corneal Innate Immune Responses. *Investig. Ophthalmol. Vis. Sci.* **2011**, *52*, 2525–2531. [CrossRef]
45. Foulks, G.N.; Borchman, D.; Yappert, M.; Kim, S.H.; McKay, J.W. Topical Azithromycin Therapy for Meibomian Gland Dysfunction: Clinical Response and Lipid Alterations. *Cornea* **2010**, *29*, 781–788. [CrossRef]
46. Haque, R.M.; Torkildsen, G.L.; Brubaker, K.; Zink, R.C.; Kowalski, R.P.; Mah, F.S.; Pflugfelder, S.C. Multicenter Open-Label Study Evaluating the Efficacy of Azithromycin Ophthalmic Solution 1% on the Signs and Symptoms of Subjects with Blepharitis. *Cornea* **2010**, *29*, 871–877. [CrossRef]
47. Barabino, S.; Rolando, M.; Camicione, P.; Ravera, G.; Zanardi, S.; Giuffrida, S.; Calabria, G. Systemic Linoleic and γ-Linolenic Acid Therapy in Dry Eye Syndrome with an Inflammatory Component. *Cornea* **2003**, *22*, 97–101. [CrossRef] [PubMed]
48. Guillon, M.; Maissa, C.; Wong, S. Eyelid Margin Modification Associated with Eyelid Hygiene in Anterior Blepharitis and Meibomian Gland Dysfunction. *Eye Contact Lens* **2012**, *38*, 319–325. [CrossRef] [PubMed]
49. Matsumoto, Y.; Dogru, M.; Goto, E.; Ishida, R.; Kojima, T.; Onguchi, T.; Yagi, Y.; Shimazaki, J.; Tsubota, K. Efficacy of a New Warm Moist Air Device on Tear Functions of Patients with Simple Meibomian Gland Dysfunction. *Cornea* **2006**, *25*, 644–650. [CrossRef]
50. Olson, M.C.; Korb, D.R.; Greiner, J.V. Increase in Tear Film Lipid Layer Thickness Following Treatment with Warm Compresses in Patients with Meibomian Gland Dysfunction. *Eye Contact Lens* **2003**, *29*, 96–99. [CrossRef] [PubMed]
51. Purslow, C. Evaluation of the Ocular Tolerance of a Novel Eyelid-Warming Device Used for Meibomian Gland Dysfunction. *Contact Lens Anterior Eye* **2013**, *36*, 226–231. [CrossRef] [PubMed]
52. Cohen, E.J. Punctal Occlusion. *Arch. Ophthalmol.* **1999**, *117*, 389–390. [CrossRef] [PubMed]
53. Tai, M.C.; Cosar, C.B.; Cohen, E.J.; Rapuano, C.J.; Laibson, P.R. The Clinical Efficacy of Silicone Punctal Plug Therapy. *Cornea* **2002**, *21*, 135–139. [CrossRef]
54. Commissioner, O. of the FDA Approves New Medication for Dry Eye Disease. Available online: https://www.fda.gov/news-events/press-announcements/fda-approves-new-medication-dry-eye-disease (accessed on 13 December 2022).

55. Donnenfeld, E.D.; Perry, H.D.; Nattis, A.S.; Rosenberg, E.D. Lifitegrast for the Treatment of Dry Eye Disease in Adults. *Expert Opin. Pharmacother.* **2017**, *18*, 1517–1524. [CrossRef]
56. Bron, A.J.; Mengher, L.S. The Ocular Surface in Keratoconjunctivitis Sicca. *Eye Basingstoke* **1989**, *3*, 428–437. [CrossRef]
57. McCusker, M.M.; Durrani, K.; Payette, M.J.; Suchecki, J. An Eye on Nutrition: The Role of Vitamins, Essential Fatty Acids, and Antioxidants in Age-Related Macular Degeneration, Dry Eye Syndrome, and Cataract. *Clin. Dermatol.* **2016**, *34*, 276–285. [CrossRef]
58. Feng, Z.; Liu, Z.; Li, X.; Jia, H.; Sun, L.; Tian, C.; Jia, L.; Liu, J. α-Tocopherol Is an Effective Phase II Enzyme Inducer: Protective Effects on Acrolein-Induced Oxidative Stress and Mitochondrial Dysfunction in Human Retinal Pigment Epithelial Cells. *J. Nutr. Biochem.* **2010**, *21*, 1222–1231. [CrossRef]
59. Nagai, N.; Otake, H. Novel Drug Delivery Systems for the Management of Dry Eye. *Adv. Drug Deliv. Rev.* **2022**, *191*, 114582. [CrossRef] [PubMed]
60. Singh, R.B.; Ichhpujani, P.; Thakur, S.; Jindal, S. Promising Therapeutic Drug Delivery Systems for Glaucoma: A Comprehensive Review. *Ther. Adv. Ophthalmol.* **2020**, *12*, 251584142090574. [CrossRef] [PubMed]
61. Ramadan, A.A.; Eladawy, S.A.; El-Enin, A.S.M.A.; Hussein, Z.M. Development and Investigation of Timolol Maleate Niosomal Formulations for the Treatment of Glaucoma. *J. Pharm. Investig.* **2020**, *50*, 59–70. [CrossRef]
62. Davis, S.A.; Sleath, B.; Carpenter, D.M.; Blalock, S.J.; Muir, K.W.; Budenz, D.L. Drop Instillation and Glaucoma. *Curr. Opin. Ophthalmol.* **2018**, *29*, 171–177. [CrossRef] [PubMed]
63. Gupta, R.; Patil, B.; Shah, B.M.; Bali, S.J.; Mishra, S.K.; Dada, T. Evaluating Eye Drop Instillation Technique in Glaucoma Patients. *J. Glaucoma* **2012**, *21*, 189–192. [CrossRef] [PubMed]
64. Fonn, D.; Bruce, A.S. A Review of the Holden-Mertz Criteria for Critical Oxygen Transmission. *Eye Contact Lens* **2005**, *31*, 247–251. [CrossRef] [PubMed]
65. McMahon, T.T.; Zadnik, K. Twenty-Five Years of Contact Lenses: The Impact on the Cornea and Ophthalmic Practice. *Cornea* **2000**, *19*, 730–740. [CrossRef]
66. Brennan, N.A.; Coles, M.L.C.; Comstock, T.L.; Levy, B. A 1-Year Prospective Clinical Trial of Balafilcon A (PureVision) Silicone-Hydrogel Contact Lenses Used on a 30-Day Continuous Wear Schedule. *Ophthalmology* **2002**, *109*, 1172–1177. [CrossRef]
67. Stapleton, F.; Stretton, S.; Papas, E.; Skotnitsky, C.; Sweeney, D.F. Silicone Hydrogel Contact Lenses and the Ocular Surface. *Ocul. Surf.* **2006**, *4*, 24–43. [CrossRef]
68. Dixon, P.; Ghosh, T.; Mondal, K.; Konar, A.; Chauhan, A.; Hazra, S. Controlled Delivery of Pirfenidone through Vitamin E-Loaded Contact Lens Ameliorates Corneal Inflammation. *Drug Deliv. Transl. Res.* **2018**, *8*, 1114–1126. [CrossRef]
69. Maulvi, F.A.; Soni, T.G.; Shah, D.O. A Review on Therapeutic Contact Lenses for Ocular Drug Delivery. *Drug Deliv.* **2016**, *23*, 3017–3026. [CrossRef]
70. Bengani, L.; Chauhan, A. Are Contact Lenses the Solution for Effective Ophthalmic Drug Delivery? *Future Med. Chem.* **2012**, *4*, 2141–2143. [CrossRef]
71. Li, C.C.; Chauhan, A. Modeling Ophthalmic Drug Delivery by Soaked Contact Lenses. *Ind. Eng. Chem. Res.* **2006**, *45*, 3718–3734. [CrossRef]
72. Torres-Luna, C.; Fan, X.; Domszy, R.; Hu, N.; Wang, N.S.; Yang, A. Hydrogel-Based Ocular Drug Delivery Systems for Hydrophobic Drugs. *Eur. J. Pharm. Sci.* **2020**, *154*, 105503. [CrossRef] [PubMed]
73. Peng, C.C.; Chauhan, A. Extended Cyclosporine Delivery by Silicone-Hydrogel Contact Lenses. *J. Control. Release* **2011**, *154*, 267–274. [CrossRef]
74. Scheuer, C.A.; Fridman, K.M.; Barniak, V.L.; Burke, S.E.; Venkatesh, S. Retention of Conditioning Agent Hyaluronan on Hydrogel Contact Lenses. *Contact Lens Anterior Eye J. Br. Contact Lens Assoc.* **2010**, *33* (Suppl. S1), S2–S6. [CrossRef]
75. Pitt, W.G.; Jack, D.R.; Zhao, Y.; Nelson, J.L.; Pruitt, J.D. Loading and Release of a Phospholipid from Contact Lenses. *Optom. Vis. Sci. Off. Publ. Am. Acad. Optom.* **2011**, *88*, 502–506. [CrossRef] [PubMed]
76. Kim, J.; Conway, A.; Chauhan, A. Extended Delivery of Ophthalmic Drugs by Silicone Hydrogel Contact Lenses. *Biomaterials* **2008**, *29*, 2259–2269. [CrossRef] [PubMed]
77. Peng, C.C.; Kim, J.; Chauhan, A. Extended Delivery of Hydrophilic Drugs from Silicone-Hydrogel Contact Lenses Containing Vitamin E Diffusion Barriers. *Biomaterials* **2010**, *31*, 4032–4047. [CrossRef] [PubMed]
78. Kim, J.; Peng, C.C.; Chauhan, A. Extended Release of Dexamethasone from Silicone-Hydrogel Contact Lenses Containing Vitamin E. *J. Control. Release* **2010**, *148*, 110–116. [CrossRef]
79. Dominguez-Godinez, C.O.; Martin-Gil, A.; Carracedo, G.; Guzman-Aranguez, A.; González-Méijome, J.M.; Pintor, J. In Vitro and in Vivo Delivery of the Secretagogue Diadenosine Tetraphosphate from Conventional and Silicone Hydrogel Soft Contact Lenses. *J. Optom.* **2013**, *6*, 205–211. [CrossRef]
80. Hsu, K.-H.; de la Jara, P.L.; Ariyavidana, A.; Watling, J.; Holden, B.; Garrett, Q.; Chauhan, A. Release of Betaine and Dexpanthenol from Vitamin E Modified Silicone-Hydrogel Contact Lenses. *Curr. Eye Res.* **2015**, *40*, 267–273. [CrossRef]
81. Yañez, F.; Concheiro, A.; Alvarez-Lorenzo, C. Macromolecule Release and Smoothness of Semi-Interpenetrating PVP-PHEMA Networks for Comfortable Soft Contact Lenses. *Eur. J. Pharm. Biopharm. Off. J. Arbeitsgemeinschaft Pharm. Verfahrenstechnik EV* **2008**, *69*, 1094–1103. [CrossRef]

82. Weeks, A.; Subbaraman, L.N.; Jones, L.; Sheardown, H. Physical Entrapment of Hyaluronic Acid during Synthesis Results in Extended Release from Model Hydrogel and Silicone Hydrogel Contact Lens Materials. *Eye Contact Lens* **2013**, *39*, 179–185. [CrossRef] [PubMed]
83. dos Santos, J.-F.R.; Alvarez-Lorenzo, C.; Silva, M.; Balsa, L.; Couceiro, J.; Torres-Labandeira, J.-J.; Concheiro, A. Soft Contact Lenses Functionalized with Pendant Cyclodextrins for Controlled Drug Delivery. *Biomaterials* **2009**, *30*, 1348–1355. [CrossRef]
84. Bengani, L.C.; Chauhan, A. Extended Delivery of an Anionic Drug by Contact Lens Loaded with a Cationic Surfactant. *Biomaterials* **2013**, *34*, 2814–2821. [CrossRef]
85. Tieppo, A.; Pate, K.M.; Byrne, M.E. In Vitro Controlled Release of an Anti-Inflammatory from Daily Disposable Therapeutic Contact Lenses under Physiological Ocular Tear Flow. *Eur. J. Pharm. Biopharm. Off. J. Arbeitsgemeinschaft Pharm. Verfahrenstechnik EV* **2012**, *81*, 170–177. [CrossRef] [PubMed]
86. Kapoor, Y.; Thomas, J.C.; Tan, G.; John, V.T.; Chauhan, A. Surfactant-Laden Soft Contact Lenses for Extended Delivery of Ophthalmic Drugs. *Biomaterials* **2009**, *30*, 867–878. [CrossRef] [PubMed]
87. Kapoor, Y.; Chauhan, A. Drug and Surfactant Transport in Cyclosporine A and Brij 98 Laden P-HEMA Hydrogels. *J. Colloid Interface Sci.* **2008**, *322*, 624–633. [CrossRef]
88. Kapoor, Y.; Chauhan, A. Ophthalmic Delivery of Cyclosporine A from Brij-97 Microemulsion and Surfactant-Laden p-HEMA Hydrogels. *Int. J. Pharm.* **2008**, *361*, 222–229. [CrossRef]
89. Andrade-Vivero, P.; Fernandez-Gabriel, E.; Alvarez-Lorenzo, C.; Concheiro, A. Improving the Loading and Release of NSAIDs from PHEMA Hydrogels by Copolymerization with Functionalized Monomers. *J. Pharm. Sci.* **2007**, *96*, 802–813. [CrossRef]
90. White, C.J.; McBride, M.K.; Pate, K.M.; Tieppo, A.; Byrne, M.E. Extended Release of High Molecular Weight Hydroxypropyl Methylcellulose from Molecularly Imprinted, Extended Wear Silicone Hydrogel Contact Lenses. *Biomaterials* **2011**, *32*, 5698–5705. [CrossRef] [PubMed]
91. Ali, M.; Byrne, M.E. Controlled Release of High Molecular Weight Hyaluronic Acid from Molecularly Imprinted Hydrogel Contact Lenses. *Pharm. Res.* **2009**, *26*, 714–726. [CrossRef]
92. Kymionis, G. Treatment of Chronic Dry Eye: Focus on Cyclosporine. *Clin. Ophthalmol.* **2008**, *2*, 829. [CrossRef] [PubMed]
93. Bushley, K.E.; Raja, R.; Jaiswal, P.; Cumbie, J.S.; Nonogaki, M.; Boyd, A.E.; Owensby, C.A.; Knaus, B.J.; Elser, J.; Miller, D.; et al. The Genome of Tolypocladium Inflatum: Evolution, Organization, and Expression of the Cyclosporin Biosynthetic Gene Cluster. *PLoS Genet.* **2013**, *9*, e1003496. [CrossRef]
94. Borel, J.F.; Feurer, C.; Magnée, C.; Stähelin, H. Effects of the New Anti-Lymphocytic Peptide Cyclosporin A in Animals. *Immunology* **1977**, *32*, 1017–1025. [PubMed]
95. Mondal, H.; Saha, S.; Awang, K.; Hossain, H.; Ablat, A.; Islam, K.; Jahan, I.A.; Sadhu, S.; Hossain, G.; Shilpi, J.; et al. Central-Stimulating and Analgesic Activity of the Ethanolic Extract of Alternanthera Sessilis in Mice. *BMC Complement. Altern. Med.* **2014**, *14*, 398. [CrossRef] [PubMed]
96. Bang, S.P.; Yeon, C.Y.; Adhikari, N.; Neupane, S.; Kim, H.; Lee, D.C.; Son, M.J.; Lee, H.G.; Kim, J.-Y.; Jun, J.H. Cyclosporine A Eyedrops with Self-Nanoemulsifying Drug Delivery Systems Have Improved Physicochemical Properties and Efficacy against Dry Eye Disease in a Murine Dry Eye Model. *PLoS ONE* **2019**, *14*, e0224805. [CrossRef]
97. Periman, L.M.; Mah, F.S.; Karpecki, P.M. A Review of the Mechanism of Action of Cyclosporine a: The Role of Cyclosporine a in Dry Eye Disease and Recent Formulation Developments. *Clin. Ophthalmol.* **2020**, *14*, 4187–4200. [CrossRef]
98. Gao, J.; Sana, R.; Calder, V.; Calonge, M.; Lee, W.; Wheeler, L.A.; Stern, M.E. Mitochondrial Permeability Transition Pore in Inflammatory Apoptosis of Human Conjunctival Epithelial Cells and T Cells: Effect of Cyclosporin A. *Investig. Ophthalmol. Vis. Sci.* **2013**, *54*, 4717–4733. [CrossRef]
99. Jones, L.; Downie, L.E.; Korb, D.; Benitez-Del-Castillo, J.M.; Dana, R.; Deng, S.X.; Dong, P.N.; Geerling, G.; Hida, R.Y.; Liu, Y.; et al. TFOS DEWS II Management and Therapy Report. *Ocul. Surf.* **2017**, *15*, 575–628. [CrossRef] [PubMed]
100. Pflugfelder, S.C.; De Paiva, C.S.; Moore, Q.L.; Volpe, E.A.; Li, D.-Q.; Gumus, K.; Zaheer, M.L.; Corrales, R.M. Aqueous Tear Deficiency Increases Conjunctival Interferon-γ (IFN-γ) Expression and Goblet Cell Loss. *Investig. Ophthalmol. Vis. Sci.* **2015**, *56*, 7545–7550. [CrossRef] [PubMed]
101. Ghasemi, H.; Djalilian, A. Topical Calcineurin Inhibitors: Expanding Indications for Corneal and Ocular Surface Inflammation. *J. Ophthalmic Vis. Res.* **2019**, *14*, 398. [CrossRef]
102. Mandal, A.; Gote, V.; Pal, D.; Ogundele, A.; Mitra, A.K. Ocular Pharmacokinetics of a Topical Ophthalmic Nanomicellar Solution of Cyclosporine (Cequa®) for Dry Eye Disease. *Pharm. Res.* **2019**, *36*, 36. [CrossRef]
103. Mun, J.; Mok, J.W.; Jeong, S.; Cho, S.; Joo, C.K.; Hahn, S.K. Drug-Eluting Contact Lens Containing Cyclosporine-Loaded Cholesterol-Hyaluronate Micelles for Dry Eye Syndrome. *RSC Adv.* **2019**, *9*, 16578–16585. [CrossRef]
104. Desai, D.T.; Maulvi, F.A.; Desai, A.R.; Shukla, M.R.; Desai, B.V.; Khadela, A.D.; Shetty, K.H.; Shah, D.O.; Willcox, M.D.P. In Vitro and in Vivo Evaluation of Cyclosporine-Graphene Oxide Laden Hydrogel Contact Lenses. *Int. J. Pharm.* **2022**, *613*, 121414. [CrossRef]
105. Bowman, F.W. The Sterility Testing of Pharmaceuticals. *J. Pharm. Sci.* **1969**, *58*, 1301–1308. [CrossRef]
106. Yu, F.; Liu, X.; Zhong, Y.; Guo, X.; Li, M.; Mao, Z.; Xiao, H.; Yang, S. Sodium Hyaluronate Decreases Ocular Surface Toxicity Induced by Benzalkonium Chloride-Preserved Latanoprost: An in Vivo Study. *Investig. Ophthalmol. Vis. Sci.* **2013**, *54*, 3385–3393. [CrossRef]

107. Elshaer, A.; Ghatora, B.; Mustafa, S.; Alany, R.G. Contact Lenses as Drug Reservoirs & Delivery Systems: The Successes & Challenges. *Ther. Deliv.* **2014**, *5*, 1085–1100. [CrossRef]
108. Desai, A.R.; Maulvi, F.A.; Desai, D.M.; Shukla, M.R.; Ranch, K.M.; Vyas, B.A.; Shah, S.A.; Sandeman, S.; Shah, D.O. Multiple Drug Delivery from the Drug-Implants-Laden Silicone Contact Lens: Addressing the Issue of Burst Drug Release. *Mater. Sci. Eng. C* **2020**, *112*, 110885. [CrossRef] [PubMed]
109. Guo, S.; Wu, K.; Li, C.; Wang, H.; Sun, Z.; Xi, D.; Zhang, S.; Ding, W.; Zaghloul, M.E.; Wang, C.; et al. Integrated Contact Lens Sensor System Based on Multifunctional Ultrathin MoS2 Transistors. *Matter* **2021**, *4*, 969–985. [CrossRef]
110. Banica, F.-G. *Chemical Sensors and Biosensors: Fundamentals and Applications*; John Wiley & Sons: Hoboken, NJ, USA, 2012. [CrossRef]
111. Khalilian, A.; Khan, M.R.R.; Kang, S.W. Highly Sensitive and Wide-Dynamic-Range Side-Polished Fiber-Optic Taste Sensor. *Sens. Actuators B Chem.* **2017**, *249*, 700–707. [CrossRef]
112. Dincer, C.; Bruch, R.; Costa-Rama, E.; Fernández-Abedul, M.T.; Merkoçi, A.; Manz, A.; Urban, G.A.; Güder, F. Disposable Sensors in Diagnostics, Food, and Environmental Monitoring. *Adv. Mater.* **2019**, *31*, 1806739. [CrossRef]
113. Turner, A.; Karube, I.; Wilson, G.S. *Biosensors: Fundamentals and Applications*, 1st ed.; Oxford University Press: Oxford, UK; New York, NY, USA, 1987; ISBN 0198547242.
114. Farandos, N.M.; Yetisen, A.K.; Monteiro, M.J.; Lowe, C.R.; Yun, S.H. Contact Lens Sensors in Ocular Diagnostics. *Adv. Healthc. Mater.* **2015**, *4*, 792–810. [CrossRef]
115. Hagan, S.; Martin, E.; Enríquez-de-Salamanca, A. Tear Fluid Biomarkers in Ocular and Systemic Disease: Potential Use for Predictive, Preventive and Personalised Medicine. *EPMA J.* **2016**, *7*, 15. [CrossRef]
116. Tseng, R.C.; Chen, C.-C.; Hsu, S.-M.; Chuang, H.-S. Contact-Lens Biosensors. *Sensors* **2018**, *18*, 2651. [CrossRef]
117. Aluru, S.V.; Agarwal, S.; Srinivasan, B.; Iyer, G.K.; Rajappa, S.M.; Tatu, U.; Padmanabhan, P.; Subramanian, N.; Narayanasamy, A. Lacrimal Proline Rich 4 (LPRR4) Protein in the Tear Fluid Is a Potential Biomarker of Dry Eye Syndrome. *PLoS ONE* **2012**, *7*, e51979. [CrossRef] [PubMed]
118. Zhou, L.; Beuerman, R.W.; Choi, M.C.; Shao, Z.Z.; Xiao, R.L.; Yang, H.; Tong, L.; Liu, S.; Stern, M.E.; Tan, D. Identification of Tear Fluid Biomarkers in Dry Eye Syndrome Using ITRAQ Quantitative Proteomics. *J. Proteome Res.* **2009**, *8*, 4889–4905. [CrossRef] [PubMed]
119. Lambiase, A.; Micera, A.; Sacchetti, M.; Cortes, M.; Mantelli, F.; Bonini, S. Alterations of Tear Neuromediators in Dry Eye Disease. *Arch. Ophthalmol. Chic. Ill 1960* **2011**, *129*, 981–986. [CrossRef]
120. Chhadva, P.; Lee, T.; Sarantopoulos, C.D.; Hackam, A.S.; McClellan, A.L.; Felix, E.R.; Levitt, R.C.; Galor, A. Human Tear Serotonin Levels Correlate with Symptoms and Signs of Dry Eye. *Ophthalmology* **2015**, *122*, 1675–1680. [CrossRef] [PubMed]
121. Boehm, N.; Funke, S.; Wiegand, M.; Wehrwein, N.; Pfeiffer, N.; Grus, F.H. Alterations in the Tear Proteome of Dry Eye Patients–a Matter of the Clinical Phenotype. *Investig. Ophthalmol. Vis. Sci.* **2013**, *54*, 2385–2392. [CrossRef]
122. Argüeso, P.; Balaram, M.; Spurr-Michaud, S.; Keutmann, H.T.; Dana, M.R.; Gipson, I.K. Decreased Levels of the Goblet Cell Mucin MUC5AC in Tears of Patients with Sjögren Syndrome. *Investig. Ophthalmol. Vis. Sci.* **2002**, *43*, 1004–1011.
123. Boehm, N.; Riechardt, A.I.; Wiegand, M.; Pfeiffer, N.; Grus, F.H. Proinflammatory Cytokine Profiling of Tears from Dry Eye Patients by Means of Antibody Microarrays. *Investig. Ophthalmol. Vis. Sci.* **2011**, *52*, 7725–7730. [CrossRef] [PubMed]
124. López-Miguel, A.; Tesón, M.; Martín-Montañez, V.; Enríquez-De-Salamanca, A.; Stern, M.E.; González-García, M.J.; Calonge, M. Clinical and Molecular Inflammatory Response in Sjögren Syndrome-Associated Dry Eye Patients Under Desiccating Stress. *Am. J. Ophthalmol.* **2016**, *161*, 133–141.e2. [CrossRef]
125. Lam, S.M.; Tong, L.; Reux, B.; Duan, X.; Petznick, A.; Yong, S.S.; Khee, C.B.S.; Lear, M.J.; Wenk, M.R.; Shui, G. Lipidomic Analysis of Human Tear Fluid Reveals Structure-Specific Lipid Alterations in Dry Eye Syndrome. *J. Lipid Res.* **2014**, *55*, 299–306. [CrossRef]
126. Galbis-Estrada, C.; Martinez-Castillo, S.; Morales, J.M.; Vivar-Llopis, B.; Monleón, D.; Díaz-Llopis, M.; Pinazo-Durán, M.D. Differential Effects of Dry Eye Disorders on Metabolomic Profile by 1H Nuclear Magnetic Resonance Spectroscopy. *BioMed Res. Int.* **2014**, *2014*, 542549. [CrossRef]
127. Willshire, C.; Buckley, R.J.; Bron, A.J. Estimating Basal Tear Osmolarity in Normal and Dry Eye Subjects. *Contact Lens Anterior Eye J. Br. Contact Lens Assoc.* **2018**, *41*, 34–46. [CrossRef] [PubMed]
128. Pflugfelder, S.C.; Jones, D.; Ji, Z.; Afonso, A.; Monroy, D. Altered Cytokine Balance in the Tear Fluid and Conjunctiva of Patients with Sjögren's Syndrome Keratoconjunctivitis Sicca. *Curr. Eye Res.* **1999**, *19*, 201–211. [CrossRef]
129. Ohashi, Y.; Ishida, R.; Kojima, T.; Goto, E.; Matsumoto, Y.; Watanabe, K.; Ishida, N.; Nakata, K.; Takeuchi, T.; Tsubota, K. Abnormal Protein Profiles in Tears with Dry Eye Syndrome. *Am. J. Ophthalmol.* **2003**, *136*, 291–299. [CrossRef] [PubMed]
130. Lam, H.; Bleiden, L.; de Paiva, C.S.; Farley, W.; Stern, M.E.; Pflugfelder, S.C. Tear Cytokine Profiles in Dysfunctional Tear Syndrome. *Am. J. Ophthalmol.* **2009**, *147*, 198–205.e1. [CrossRef]
131. Chotikavanich, S.; de Paiva, C.S.; Li, D.Q.; Chen, J.J.; Bian, F.; Farley, W.J.; Pflugfelder, S.C. Production and Activity of Matrix Metalloproteinase-9 on the Ocular Surface Increase in Dysfunctional Tear Syndrome. *Investig. Ophthalmol. Vis. Sci.* **2009**, *50*, 3203–3209. [CrossRef]
132. Aragona, P.; Aguennouz, M.; Rania, L.; Postorino, E.; Sommario, M.S.; Roszkowska, A.M.; De Pasquale, M.G.; Pisani, A.; Puzzolo, D. Matrix Metalloproteinase 9 and Transglutaminase 2 Expression at the Ocular Surface in Patients with Different Forms of Dry Eye Disease. *Ophthalmology* **2015**, *122*, 62–71. [CrossRef] [PubMed]

133. Tesón, M.; González-García, M.J.; López-Miguel, A.; Enríquez-de-Salamanca, A.; Martín-Montañez, V.; Benito, M.J.; Mateo, M.E.; Stern, M.E.; Calonge, M. Influence of a Controlled Environment Simulating an In-Flight Airplane Cabin on Dry Eye Disease. *Investig. Ophthalmol. Vis. Sci.* **2013**, *54*, 2093–2099. [CrossRef] [PubMed]
134. López-Miguel, A.; Tesón, M.; Martín-Montañez, V.; Enríquez-de-Salamanca, A.; Stern, M.E.; Calonge, M.; González-García, M.J. Dry Eye Exacerbation in Patients Exposed to Desiccating Stress under Controlled Environmental Conditions. *Am. J. Ophthalmol.* **2014**, *157*, 788–798.e2. [CrossRef]
135. Mondal, H.; Hossain, H.; Awang, K.; Saha, S.; Mamun-Ur-Rashid, S.; Islam, K.; Rahman, M.S.; Jahan, I.A.; Rahman, M.M.; Shilpi, J.A. Anthelmintic Activity of Ellagic Acid, a Major Constituent of Alternanthera Sessilis against Haemonchus Contortus. *Pak. Vet. J.* **2015**, *35*, 58–62.
136. Burgos-Blasco, B.; Güemes-Villahoz, N.; Santiago, J.L.; Fernandez-Vigo, J.I.; Espino-Paisán, L.; Sarriá, B.; García-Feijoo, J.; Martinez-de-la-Casa, J.M. Hypercytokinemia in COVID-19: Tear Cytokine Profile in Hospitalized COVID-19 Patients. *Exp. Eye Res.* **2020**, *200*, 108253. [CrossRef]
137. Shinn, J.; Kwon, N.; Lee, S.A.; Lee, Y. Smart PH-Responsive Nanomedicines for Disease Therapy. *J. Pharm. Investig.* **2022**, *52*, 427–441. [CrossRef]
138. Kim, J.; Kim, M.; Lee, M.-S.; Kim, K.; Ji, S.; Kim, Y.-T.; Park, J.; Na, K.; Bae, K.-H.; Kyun Kim, H.; et al. Wearable Smart Sensor Systems Integrated on Soft Contact Lenses for Wireless Ocular Diagnostics. *Nat. Commun.* **2017**, *8*, 14997. [CrossRef]
139. Yin, R.; Xu, Z.; Mei, M.; Chen, Z.; Wang, K.; Liu, Y.; Tang, T.; Priydarshi, M.K.; Meng, X.; Zhao, S.; et al. Soft Transparent Graphene Contact Lens Electrodes for Conformal Full-Cornea Recording of Electroretinogram. *Nat. Commun.* **2018**, *9*, 2334. [CrossRef]
140. Park, S.; Heo, S.W.; Lee, W.; Inoue, D.; Jiang, Z.; Yu, K.; Jinno, H.; Hashizume, D.; Sekino, M.; Yokota, T.; et al. Self-Powered Ultra-Flexible Electronics via Nano-Grating-Patterned Organic Photovoltaics. *Nature* **2018**, *561*, 516–521. [CrossRef]
141. Hong, G.; Lieber, C.M. Author Correction: Novel Electrode Technologies for Neural Recordings. *Nat. Rev. Neurosci.* **2019**, *20*, 376. [CrossRef]
142. Phan, C.M.; Subbaraman, L.; Jones, L.W. The Use of Contact Lenses as Biosensors. *Optom. Vis. Sci.* **2016**, *93*, 419–425. [CrossRef]
143. Keum, D.H.; Kim, S.K.; Koo, J.; Lee, G.H.; Jeon, C.; Mok, J.W.; Mun, B.H.; Lee, K.J.; Kamrani, E.; Joo, C.K.; et al. Wireless Smart Contact Lens for Diabetic Diagnosis and Therapy. *Sci. Adv.* **2020**, *6*, 1–13. [CrossRef]

pharmaceutics

Article

Insights into the Safety and Versatility of 4D Printed Intravesical Drug Delivery Systems

Marco Uboldi [1], Cristiana Perrotta [2], Claudia Moscheni [2], Silvia Zecchini [2], Alessandra Napoli [2], Chiara Castiglioni [3], Andrea Gazzaniga [1], Alice Melocchi [1,*] and Lucia Zema [1]

[1] Sezione di Tecnologia e Legislazione Farmaceutiche "Maria Edvige Sangalli", Dipartimento di Scienze Farmaceutiche, Università degli Studi di Milano, via Giuseppe Colombo 71, 20133 Milano, Italy
[2] Dipartimento di Scienze Biomediche e Cliniche, Università degli Studi di Milano, via Giovanni Battista Grassi 74, 20157 Milano, Italy
[3] Dipartimento di Chimica, Materiali e Ingegneria Chimica "Giulio Natta", Politecnico di Milano, piazza Leonardo da Vinci 32, 20133 Milan, Italy
* Correspondence: alice.melocchi@unimi.it; Tel.: +39-02-50324654

Abstract: This paper focuses on recent advancements in the development of 4D printed drug delivery systems (DDSs) for the intravesical administration of drugs. By coupling the effectiveness of local treatments with major compliance and long-lasting performance, they would represent a promising innovation for the current treatment of bladder pathologies. Being based on a shape-memory pharmaceutical-grade polyvinyl alcohol (PVA), these DDSs are manufactured in a bulky shape, can be programmed to take on a collapsed one suitable for insertion into a catheter and re-expand inside the target organ, following exposure to biological fluids at body temperature, while releasing their content. The biocompatibility of prototypes made of PVAs of different molecular weight, either uncoated or coated with Eudragit®-based formulations, was assessed by excluding relevant in vitro toxicity and inflammatory response using bladder cancer and human monocytic cell lines. Moreover, the feasibility of a novel configuration was preliminarily investigated, targeting the development of prototypes provided with inner reservoirs to be filled with different drug-containing formulations. Samples entailing two cavities, filled during the printing process, were successfully fabricated and showed, in simulated urine at body temperature, potential for controlled release, while maintaining the ability to recover about 70% of their original shape within 3 min.

Keywords: 3D printing; cytotoxicity; controlled release; fused deposition modeling; local delivery; retentive systems; shape memory polymers

Citation: Uboldi, M.; Perrotta, C.; Moscheni, C.; Zecchini, S.; Napoli, A.; Castiglioni, C.; Gazzaniga, A.; Melocchi, A.; Zema, L. Insights into the Safety and Versatility of 4D Printed Intravesical Drug Delivery Systems. *Pharmaceutics* **2023**, *15*, 757. https://doi.org/10.3390/pharmaceutics15030757

Academic Editor: Dimitrios A. Lamprou

Received: 31 January 2023
Revised: 20 February 2023
Accepted: 22 February 2023
Published: 24 February 2023

1. Introduction

Over the years, various strategies have been investigated to improve the local treatment of bladder diseases, having as the ultimate target the achievement and maintenance of effective levels of drugs at the target site [1–5]. In this respect, avoiding repeated catheterizations, which are responsible for a dramatic decrease in patient compliance towards intravesical administration, and improving adherence as well as penetration of the administered drug into the urothelium still represent the main challenges to be overcome. Liquid formulations able to undergo an increase in their viscosity once at the target site, through the formation of gels at body temperature, were recently proven able to ensure long-lasting residence in the bladder coupled with controlled release [6–8]. Drug delivery systems (DDSs) either capable of floating into the urine or of avoiding early elimination from the target site during physiological urination thanks to a swift expansion were also proposed [9–11] They were generally designed to be administered via catheter and, once exhausted, to be removed in the same way or to be spontaneously eliminated following solubilization, erosion and rupture phenomena. In more detail, the so-called expandable systems could be retained in the desired organ either following an increase in their size or a controlled variation in the relevant geometry. Notably, the resulting DDSs should not damage the bladder walls or interfere with their physiological contraction.

Expandable retentive systems can be classified based on the process driving the relevant increase in spatial encumbrance, which may rely on the removal of an external constraint of a mechanical nature (e.g., exit of the system from the catheter) or on the shape memory effect (SME) provided by so-called smart materials [12,13]. The latter mechanism consists in the recovery of an original shape obtained under manufacturing, triggered by an external stimulus of a non-mechanical nature, such as a change in temperature, moisture or light [14–17]. By way of example, the first category entails the LiRIS system, for the controlled release of lidocaine, and a S-shaped device, manufactured via stereolithography 3D printing and proposed by Xu and colleagues [9,18–21].

As far as applications relying on SME are concerned, various systems have been described over years, especially for biomedical applications (e.g., scaffolds, hemostatic plugs and devices for cellular surgery) [22–25]. In this respect, the advent of 3D printing has further prompted research into shape memory polymers (SMPs) [26–29]. Indeed, during this process, the final item is manufactured, layer-by-layer, reproducing a shape previously designed through computer-aided design software [30,31]. As a consequence, besides offering high flexibility and geometric freedom, it would allow the modification, in real-time, of the product in order to fulfill specific needs, all features that would be particularly interesting for R&D and customization purposes. Among the 3D printing techniques available, fused deposition modeling (FDM) has emerged as one of the most studied in pharmaceutics, probably in view of the limited cost of the equipment and its ease of use [32–34]. During FDM, polymer wires, generally known as filaments and manufactured by hot melt extrusion (HME), are fed into the printhead. Here, the filament is heated and extruded through a nozzle on a build plate. The reciprocating movement of the printhead and of the build plate ensures the deposition of the molten material layer-by-layer until the product is completed from the bottom up.

Focusing on the use of SMPs for the development of organ-retentive systems by FDM, the programmed shape-shifting of a pharmaceutical-grade poly(vinyl alcohol) (PVA) was recently leveraged to develop drug-embedded matrix-type DDSs for prolonged maintenance and release into hollow muscular organs, including the bladder and the stomach [35,36]. Although water-induced SME of PVA was already described in the material-related literature, especially upon chemical modification of the polymer or relevant blending with other compounds, in this case the shape changes were demonstrated to mainly depend on contact with body temperature [37–41]. In more detail, samples having different original shapes, endowed with such spatial encumbrance as to avoid rapid emptying through the sphincters of the selected organs, were produced by HME and FDM. In this respect, modifications occurring on a 3D material configuration over time, triggered by an external stimulus of a non-mechanical nature and resulting in macroscopic shape changes, has been associated with the concept of 4D printing, in which time represented the fourth dimension [15,42–44]. Indeed, the resulting PVA-based prototypes turned out able to take on, after production, a temporary collapsed shape and to quickly recover the original one in the desired environment. As the temporary shape would ease administration, it has been conceived according to the particular features of the route selected for reaching the target organ. The possibility of using film-coating to improve mechanical resistance and timescale of release provided by the matrix-like specimens, without impairing their working mechanism, was also demonstrated [45,46].

In the present work, a further step in the development of expandable bladder-retentive DDSs based on the smart behavior of PVA was undertaken. This was done to enable novel therapeutic approaches towards urothelial bladder cancer and many other disabling pathological conditions affecting this organ (e.g., interstitial cystitis, infections), thus reducing dropouts and providing patients with more personalized, effective and tolerated treatments. In more detail, a preliminary biocompatibility study involving the evaluation of cytotoxicity on bladder cancer and human monocytic cell lines was carried out on the materials employed so far for the manufacturing of uncoated and coated PVA-based prototypes. Moreover, taking advantage of the rapid prototyping capabilities of FDM, the feasibility of an improved design for an intravesical delivery system was considered, entailing 4D printed specimens provided with internal cavities. This evolution would make it possible to overcome the limitations related to the thermal stability of drugs embedded in the PVA-based material, which needs to be processed at temperatures ≥180 °C, while enhancing the versatility of the DDS proposed in terms of formulations to be conveyed and achievable release performance. Indeed, the reservoir units could not only be employed for the administration of separate doses of active molecules that are mutually incompatible, but also filled with new formulations, for instance, graphene-based nanoparticles already under development [47].

2. Materials and Methods

2.1. Materials

Prototype manufacturing and physio-technological characterization: PVA05 and PVA48 (Gohsenol™ EG 05P and 48P, Mitsubishi Chemical, Tokio, Japan); glycerol (Pharmagel, Milan, Italy; GLY); methacrylic acid copolymers, Eudragit® RS 100 and RL 100 (Evonik, Essen, Germany); ready-to-use dispersion of methacrylic acid copolymers, Eudragit® NE (Evonik, Essen, Germany); triethyl citrate (TEC; Sigma Aldrich, Darmstadt, Germany); ethanol (Sigma Aldrich, Darmstadt, Germany); acetaminophen for direct compression (Rhodia, Milan, Italy; AAP); PLA filament (TreeDfilaments, Milan, Italy); simulated urine fluids (NaCl 13.75 g/L; $MgSO_4$ 1.69 g/L; $MgCl_2$ 0.83 g/L; $CaCl_2$ 0.67 g/L, KCl 0.38 g/L and urea 17.40 g/L; pH 7.50).

In vitro studies: L-Glutamine, penicillin-streptomycin, Trypsin-EDTA, RPMI1940, Dulbecco's phosphate saline buffer (PBS) w/o calcium and magnesium and fetal bovine serum (FBS) (Euroclone, Milan, Italy). Minimum Essential Medium Eagle (EMEM), 3-[4,5-dimethylthiazol-2-yl]-2,5-diphenyl tetrazolium bromide (MTT), dimethylsulfoxide (DMSO), Phorbol 12-myristate 13-acetate (PMA), lipopolysaccharides (LPS) from Escherichia coli, paraformaldehyde (PFA), TritonX-100, Fluoroshield mounting medium, Epirubicin and Mitomycin C (Sigma-Aldrich, Darmstadt, Germany). Trypan blue stain, iScript gDNA clear cDNA synthesis kit, PureZOL RNA isolation reagent, 4′,6-diamidine-2′-phenylindole dihydrochloride (DAPI) nuclear staining dye and Universal SYBR Green supermix (Bio-Rad, Hercules, CA, USA). Ki67 antibody (Abcam, Cambridge, UK). Goat serum (GS), goat anti-rabbit IgG (H + L) cross-adsorbed secondary antibody Alexa Fluor 647, MitoTracker™ Orange and Fluorescein phalloidin (ThermoFisher, Milan, Italy).

2.2. Methods

2.2.1. Preparation of PVA-Based Formulations

PVA05 and PVA48 powders were kept in an oven at 40 °C for 24 h prior to use. Relevant formulations containing 15% by weight of GLY (calculated on the dry polymer) were prepared by kneading. Either PVA05 or PVA48 was placed in a mortar, and the liquid plasticizer was added dropwise under continuous mixing. The resulting mixtures were oven-dried at 40 °C for 8 h. Afterwards, aggregates were ground by means of a blade mill, and the <250 μm powder fraction was recovered.

2.2.2. HME

HME was carried out taking advantage of a twin-screw extruder (Haake™ MiniLab II, Thermo Scientific, Milwaukee, WI, USA) equipped with counter-rotating screws and a custom-made aluminum circular die (ø = 1.80 mm), as previously described [48]. The extrusion temperature and screw speed were set at 180 °C and at 100 rpm, respectively, while the maximum torque registered was approximately 150 N·cm. Extruded rods were cut into 50 mm-long samples that were employed, as such or after coating, for in vitro toxicity studies. PVA05-based rods were also employed to feed the FDM printer. In this case, they were manually pulled and forced to pass through a caliper set at 1.80 mm and connected to the extruder. This was done to counteract possible swelling phenomena and to calibrate the rod diameter, thus enhancing the yield of filaments suitable for 3D printing (i.e., 1.75 ± 0.05 mm). After cooling, the filament diameter was verified every 5 cm in length, and portions out of specifications were discarded. Indeed, filaments with a diameter greater than 1.80 mm were unsuitable for printing.

2.2.3. 3D Printing

I-shaped prototypes were fabricated using a Kloner3D 240® Twin printer (Kloner3D, Florence, Italy) using computer-aided design (CAD) files purposely developed, as described in the Results and Discussion section. These were designed using Autodesk® Autocad® 2016 software version 14.0 (Autodesk Inc., San Francisco, CA, USA), saved in .stl format and imported to the 3D printer software (Simplify 3D, Milan, Italy). The printing parameters set for printing the PVA-based formulation are reported in Table 1.

Table 1. Operating parameters set for the PVA-based formulation.

Parameter	Value
Nozzle diameter	0.5 mm
Printing temperature	200 °C
Build plate temperature	50 °C
Extrusion flow	100% of the maximum flow
Printing speed	23 mm/s
Retraction length	2.00 mm
Retraction speed	20 mm/s
Layer height	0.10 mm
Infill	100% or 50%,
Infill geometry	Rectilinear
Number of top/bottom	2
Number of perimeters	1

The printing process was interrupted at a specific height (i.e., 25th layer) to enable manual filling of the system cavities with a previously weighted (≈20 mg each cavity; analytical balance, Gibertini, Milan, Italy) amount of free-flowing AAP powder, selected as the drug tracer.

Using a commercial PLA filament as received, FDM was also employed to fabricate (i) a trapdoor tool to improve manual filling of the system cavities during the relevant fabrication and (ii) templates intended to make programming of samples in the desired temporary U shape easier and more reproducible (Figure 1). The printing parameters set for printing the PLA filament are reported in Table 2.

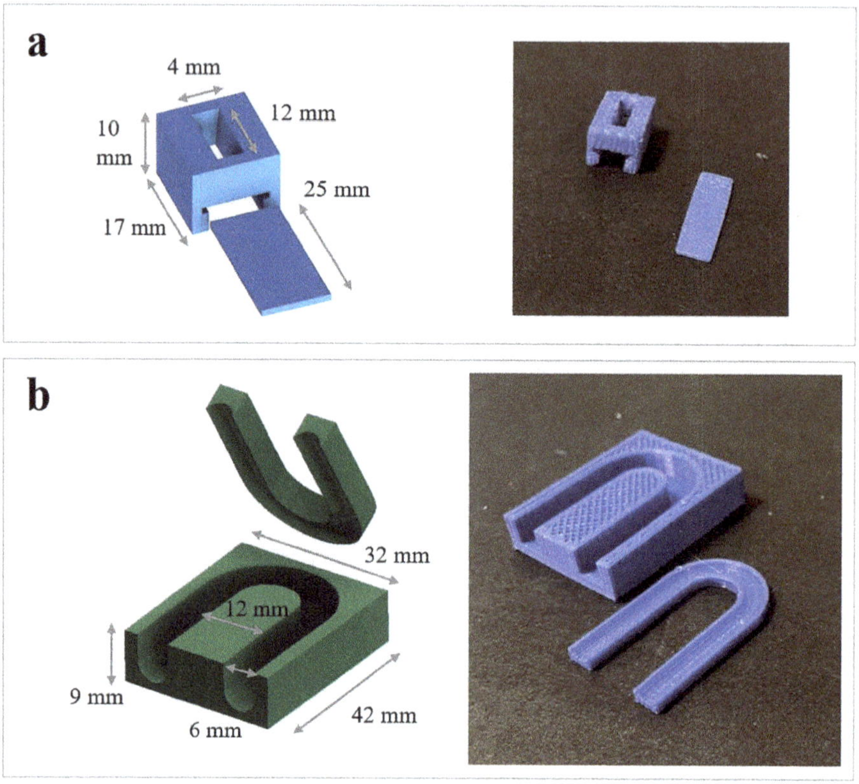

Figure 1. Digital models with dimensional details of (**a**) the trapdoor tool and (**b**) the template used for programming the temporary shape, together with photographs of the actual printed objects.

Table 2. Operating parameters set for the PLA filament.

Parameter	Value
Nozzle diameter	0.5 mm
Printing temperature	200 °C
Build plate temperature	40 °C
Extrusion flow	100% of the maximum flow
Printing speed	65 mm/s
Retraction length	2.40 mm
Retraction speed	45 mm/s
Layer height	0.20 mm
Infill	75%
Infill geometry	Honeycomb
Number of top/bottom	3
Number of perimeters	2

2.2.4. Film-Coating

Extruded rods and printed I-shaped prototypes were coated with (i) an ethanolic solution (final concentration 30% weight/volume) containing Eudragit® RS and RL (mixed in a 3:1 ratio by weight) and TEC as the plasticizer (15% by weight on the dry polymeric blend) and (ii) a 30% ready-to-use aqueous suspension of Eudragit® NE. While the former samples were referred as Eudragit® RS/RL-coated, the latter were identified as Eudragit® NE-coated. For simplicity reasons, within the Figures, they were identified as either RS/RL or NE.

Film-coating was performed by means of an in-house assembled machinery previously described, but adapting the procedure [45]. In this respect: (i) samples were inserted in the mandrels of the equipment for half of their length; (ii) the orientation of the spray gun was modified to assume an angle of 120° with respect to the horizontal plane, cutting the cylindrical samples on their major axis, thus enabling coating of the lateral surface and of the free end of the sample at the same time; (iii) the coating process was carried out for overall 4 min, being paused after 2 min to allow extraction of the specimen from the mandrel and its 180° rotation. This way, the prototypes were flipped, thus enabling coating of the half of the sample that was previously fixed in the mandrel. At the end of the process, coated specimens were maintained for 2 h in a ventilated oven set at 40 °C.

2.2.5. Physio-Technological Characterization

All the specimens were characterized for weight (n = 6; analytical balance, Gibertini, Milan, Italy). The thickness of the coating layer was also evaluated (n = 6). For this purpose, each sample was cut in six positions, i.e., one to four along its length and five and six on the ends (Figure 2). Notably, the positions in which the specimens were cut were selected to avoid the areas of the printed samples intended for drug filling. Photographs of each cross-section were acquired using a digital microscope (Digital Microscope AM-413T, Dino-Lite, Milan, Italy; resolution = 1.3 megapixel − 1280 × 1024) and processed by a dedicated software (ImageJ, Milan, Italy) to measure the thickness of the coating at six different points (L_1–L_6) along the circumference of each cut surface.

The SME was evaluated as previously described [35] using a purposely developed shape memory cycle, first involving the programming of the temporary shape and then recovery of the original one. The programming phase was carried out by heating the I-shaped samples up to 55 °C (i.e., at least 20 °C above their T_g) (oven, VWR, Milan, Italy). By means of the purposely printed template (see Figure 1b), which was also stored at 55 °C, the specimens were programmed to take on the temporary U shape. This step was manually performed. In more detail, the prototype was bent and positioned at the bottom part of the template (i.e., that resembling a U-shaped cavity), which was then closed by the relevant cover. Finally, the entire assembly maintaining the sample in the desired temporary configuration was cooled at −20° C for at least 8 h (Freezer, VWR, Milan, Italy). Recovery of the original shape was triggered upon immersion of the deformed specimens (n = 3) into 100 mL of unstirred simulated urine fluid, prepared as reported by Sherif and colleagues [49]. The latter was kept at 37 ± 0.5 °C, using a thermoregulated bath. The recovery process was monitored using a digital camera positioned at 10 cm above the samples (GoPro Hero Session, San Mateo, CA, USA). The photographs collected were processed by means of a specific software (ImageJ, Milan, Italy) to measure the variation of the angle between the two arms (α) of the samples so as to quantify the recovery of the original shape over time. Indeed, a recovery index (RI) versus time curves were then built, with RI calculated as follows:

$$RI = \frac{\alpha - \alpha_p}{\pi - \alpha_p} \quad (1)$$

where α_p is the angle obtained in the programming phase (angles in rad).

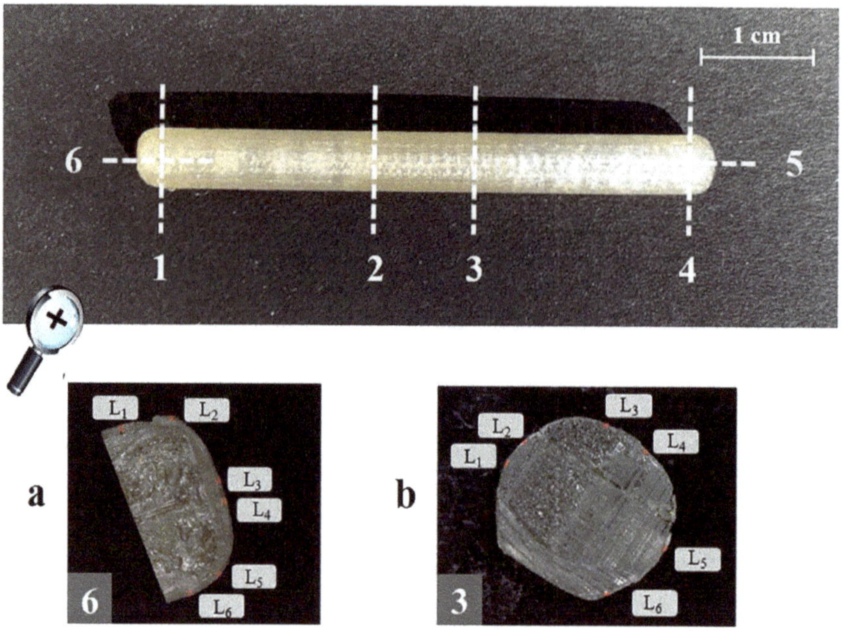

Figure 2. Outline of the positions in which each I-shaped sample was cut, together with photographs of the resulting cross-sections (types a and b). By way of example, details relevant to the thickness measurements (L_1–L_6) of the coating layer taken at position 3 and 6 are highlighted.

Uncoated and coated 3D printed prototypes, the inner cavities of which were filled with AAP during relevant fabrication, were tested for release by means of a USP38 dissolution apparatus 2 (10 rpm, 37 ± 0.5 °C; Distek, North Brunswick Township, NJ, USA; n = 3). A total of 400 mL of the abovementioned simulated urine fluids were used as the dissolution medium. The apparatus was connected to a pump (IPC Ismatec™, Thermo Fisher Scientific, Milan, Italy) for automatic collection of fluid samples and to a spectrophotometer for relevant assay (Lambda 35, Perkin Elmer, Milan, Italy; 1 mm cuvette path length, 248 nm λ_{max}). In this respect, AAP was selected as the drug tracer in light of its safety of use and based on the availability of a routine spectroscopic assay already developed. The amount of drug released at each time point was determined from a dedicated calibration curve (y = 6.43072x, R^2 = 0.9999; from 0.0125 to 0.40 mg/mL). Besides selecting the suitable range of drug concentrations to be tested during the initial set-up phase, the presence of excipients (i.e., PVA and GLY) in the dissolution medium was demonstrated not to affect the spectroscopic AAP determination. By linear interpolation of the release data immediately before and after the time point of interest, times to 10% and 90% release (i.e., $t_{10\%}$ and $t_{90\%}$, respectively) were calculated. While $t_{10\%}$ defined the lag phase, $t_{90\%}$ was used to calculate the pulse time (i.e., $t_{90\%}$–$t_{10\%}$).

2.2.6. In vitro Toxicity Studies

Cell Culture

In vitro studies were performed using the human bladder cancer HT1376 and the human monocytic THP-1 cell lines. HT1376 cells were obtained by American Type Culture Collection (ATCC), while THP-1 cells were kindly provided by Dr. Irma Saulle, Department of Pathophysiology and Transplantation, Università degli Studi di Milano. HT1376 cells were routinely cultured as a monolayer in Minimum Essential Medium Eagle supplemented with 10% heat inactivated FBS, 1% penicillin/streptomycin and 1% L-Glutamine. THP-1 cells were maintained in suspension in RPMI1640 supplemented with 1% L-glutamine, 1% streptomycin/penicillin and 20% FBS. For differentiation into macrophages, cells were seeded in 6-well plates at a confluence of 6×10^5 cells/well and treated for 24 h incubation with 50 ng/mL phorbol 12-myristate 13-acetate (PMA) [50,51]. For polarization toward a proinflammatory phenotype, macrophages were incubated with 250 ng/mL of LPS for 48 h.

Cell Viability, Proliferation and Death

Viability of HT1376 cells was assessed by MTT assay [52–54]. Cells were seeded into 6-well plates (2×10^5 cells per well) and incubated for 24 h. Then, the specimens (4 mm in length) were placed in direct contact with the cells or onto a transwell insert (0.4 μm pores) into the culture medium. Cells cultured in the medium without adding the specimens were taken as the negative control, while cells cultured in the presence of a solution (1 μM) of the chemotherapeutic drug epirubicin were used as the positive control. After 24–48 h of incubation, a MTT dye working solution was added to each well (final concentration 0.5 mg/mL). After 3 h of incubation, the supernatant was removed and replaced by 100 μL/well of DMSO. The absorbance (A) values of each well were recorded at 560 nm on an automatic plate reader (Glomax, Multi Detection System microplate reader, Promega, Milan, Italy). The relative viability versus the untreated control cells was calculated as follows:

$$\text{Relative viability } (\%) = \frac{A_{\text{exposed group}}}{A_{\text{control}}} \times 100 \tag{2}$$

Cell proliferation and death were assessed by immunofluorescence [55–57]. Cells treated as previously described were incubated with staining solution containing MitoTracker® fluorescent probe for 45 min in the dark, to analyze cell death. Then, cells were fixed in 4% PFA in 0.1 M PBS (pH 7.4) for 15 min, permeabilized with 0.1% TritonX-100 in PBS for 5 min and incubated in blocking buffer containing 5% normal goat serum and 0.1% TritonX-100 in PBS for 1 h. The primary antibody against Ki67, a proliferation marker, and Alexa Fluor conjugated secondary antibody were diluted in blocking buffer and incubated at 4 °C overnight and for 2 h at room temperature, respectively. Fluorescein phalloidin was used for cytoskeleton (actin) detection and incubated together with the secondary antibody. Nuclei were counterstained with DAPI Nuclear Staining Dye for 10 min at room temperature. Confocal imaging was performed with a Leica TCS SP8 AOBS microscope system with oil immersion objective 40X/1.30 (Leica, Heerbrugg, Switzerland). Image acquisitions were controlled by the Leica LAS AF software (Leica, Heerbrugg, Switzerland). Image analysis was performed with the ImageJ software (ImageJ, Milan, Italy).

Cytokine Analysis by Real-Time PCR

The analysis of the mRNA expression of cytokines was performed as previously described [54,57]. Total RNA from THP-1 derived macrophages was extracted with the PureZol RNA Isolation Reagent (Bio-Rad, Hercules, CA, USA), according to the manufacturer's protocol. First-strand cDNA was generated from 1 μg of total RNA using iScript Reverse Transcription Supermix (Bio-Rad, Hercules, CA, USA). A set of primer pairs (Eurofins Genomics, Milan, Italy) was designed to hybridize to unique regions of the appropriate gene sequence (Table 3). PCR was performed using SsoAdvanced Universal SYBR Green Supermix and the CFX96 Touch Real-Time PCR Detection System (Bio-Rad, Hercules, CA, USA). The fold change was determined relative to the control after normalizing to GAPDH and Rpl32 (internal standard) through the use of the formula $2^{-\Delta\Delta CT}$.

Table 3. List of primers designed for PCR.

Gene	Primer Sequences
IL-6	F: 5′-GGCACTGGCAGAAAACAACC-3′ R: 5′-GCAAGTCTCCTCATTGAATCC-3′
IL-1β	F: 5′-TTCGACACATGGGATAACGAGG-3′ R: 5′-TTTTTGCTGTGAGTCCCGGAG-3′
TNF	F: 5′-CCCAGGGACCTCTCTCTAATCA-3′ R: 5′-GCTACAGGCTTGTCACTCGG-3′
GAPDH	F: 5′-TGAGGTCAATGAAGGGGTC-3′ R: 5′-GTGAAGGTCGGAGTCAACG 3′
RPL32	R: 5′-TTAAGCGTAACTGGCGGAAAC-3′ F: 5′-AAACATTGTGAGCGATCTCGG-3′

Statistical Analysis

Statistical significance of raw data between the groups was evaluated using one-way ANOVA followed by Tukey post-tests (multiple comparisons). The analysis was carried out by using GraphPad Prism software package (GraphPad Software, San Diego, CA, USA). The results are expressed as means ± SEM of the indicated n values. p values ≤ 0.05 were considered statistically significant.

3. Results and Discussion

3.1. Cytotoxic Evaluation of PVA-Based Samples

In order to preliminarily evaluate the safety impact of the expandable bladder-retentive DDS under development, which involves pharmacopeial-grade materials of established safety by oral intake but processed through new hot melt technologies, a cytotoxicity study was carried out according to a predefined protocol. HME prototypes, based on two different PVA grades, uncoated and coated with Eudragit® RS/RL and NE formulations, were considered. The analyses were carried out in a cancer cell line of bladder origin (i.e., HT1376). The reason for the choice of this type of cells was twofold: (i) they are a good model of bladder carcinoma, widely used to evaluate efficacy of anticancer treatments [58]; (ii) they still maintain epithelial features (Figure 3a) and therefore may be considered a valuable model for any epithelium [59]. According to ISO 10993-Biological evaluation of medical devices Part 5: Tests for in vitro cytotoxicity, the prototypes were incubated either in direct physical contact with cultured cells or placed onto a transwell insert into the culture medium to allow an indirect contact with the cells. Cytotoxicity was determined by quantifying cell viability (i.e., the measure of the proportion of live, healthy cells within a population), cell proliferation (i.e., the assessment of dividing cells) and cell death (i.e., the evaluation of cells committed to death or already dead) [53]. First, cell viability upon contact with uncoated PVA prototypes was investigated by the MTT assay, using the chemotherapeutic drug epirubicin (1 μM) as positive control of toxicity [53,60]. After 24 h of incubation, none of the PVA-based specimens in either culturing condition (i.e., direct and indirect contact) caused a significant reduction of cell viability when compared with untreated control cells (Figure 3b). As expected, epirubicin showed a high toxicity by decreasing cell viability by nearly 50%. Then, cell proliferation was assessed by staining the cells for the proliferation marker Ki67, a nuclear nonhistone protein that is expressed in proliferating cells and absent in quiescent cells [61]. The percentage of Ki67+ cells in the specimen treated samples was similar to that of the control, therefore confirming no effect on cell proliferation (Figure 3c,d).

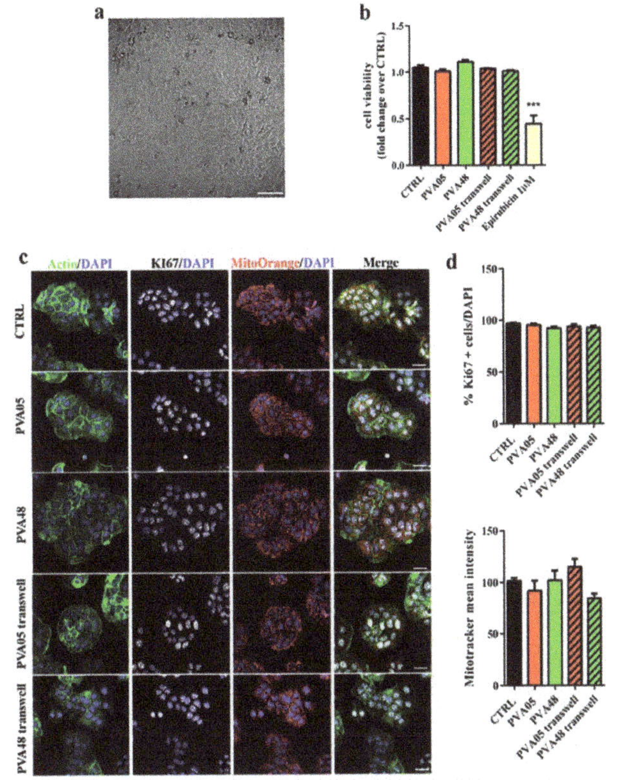

Figure 3. Cytotoxicity evaluation of uncoated PVA05- and PV48-based specimens. (**a**) Brightfield microscopy image of HT1376 cells (100 μm scale bar); (**b**) cell viability of HT1376 cells exposed to different samples for 24 h ($n = 6$). Untreated cells (CTRL) and epiribicin-treated cells were used as negative and positive controls, respectively (*** $p < 0.0001$ versus CTRL); (**c**) confocal microscope images of HT1376 cells stained with phalloidin to detect actin (green), Ki67 (white), Mitotracker Orange (red) and DAPI (blue) (20 μm scale bar); (**d**) upper panel: graph of the % of Ki67+ cells over total cells counted by DAPI staining; bottom panel: graph of the mean fluorescent intensity of Mitotracker Orange ($n = 3$).

A reliable and sensitive indicator of cell stress and apoptosis (e.g., programmed cell death) is the dissipation of the mitochondrial membrane potential. For this purpose, MitoTracker® Orange fluorescent and potentiometric dye, which accumulates in mitochondria within living cells but not in dying cells, was employed. No differences in the mean fluorescence intensity were observed among the samples analyzed. Moreover, no signs of apoptosis, e.g., cell and nucleus shrinkage, or condensed chromatin were highlighted [62] (Figure 3c). Taken together, these results indicated that the specimens were non-toxic to the cells, which is consistent with previous data on the good biocompatibility of PVA composites that enforced its use for different biomedical applications [63].

The same experimental protocol was then applied to prototypes coated with both Eudragit® RS/RL and Eudragit® NE formulations, which were maintained in both direct and indirect contact with the cells for 24 and 48 h (Figures 4 and 5). Because no differences between PVA05 and PVA48 coated specimens were highlighted, only data relevant to the lower molecular grade polymer are reported in the following Figures. As for the uncoated specimens, no signs of toxicity, potentially caused by the presence of the prototypes, was found either after 24 or 48 h of incubation. Cells were metabolically active and healthy (Figures 4a and 5a) and maintained their ability to proliferate, and no evidence of apoptosis was observed (Figures 4b,c and 5b,c). This was almost expected, as Eudragits® polymers are generally regarded as inactive, nontoxic and nonirritant materials [64]. Of note, this is the first report of the safety of the combination of PVA/Eudragit® in a model of bladder epithelial cells. Recently, PVA-based hydrogel beads coated with Eudragit® and orally administered have been tested in vivo, demonstrating the biocompatibility of the combination of such materials [65].

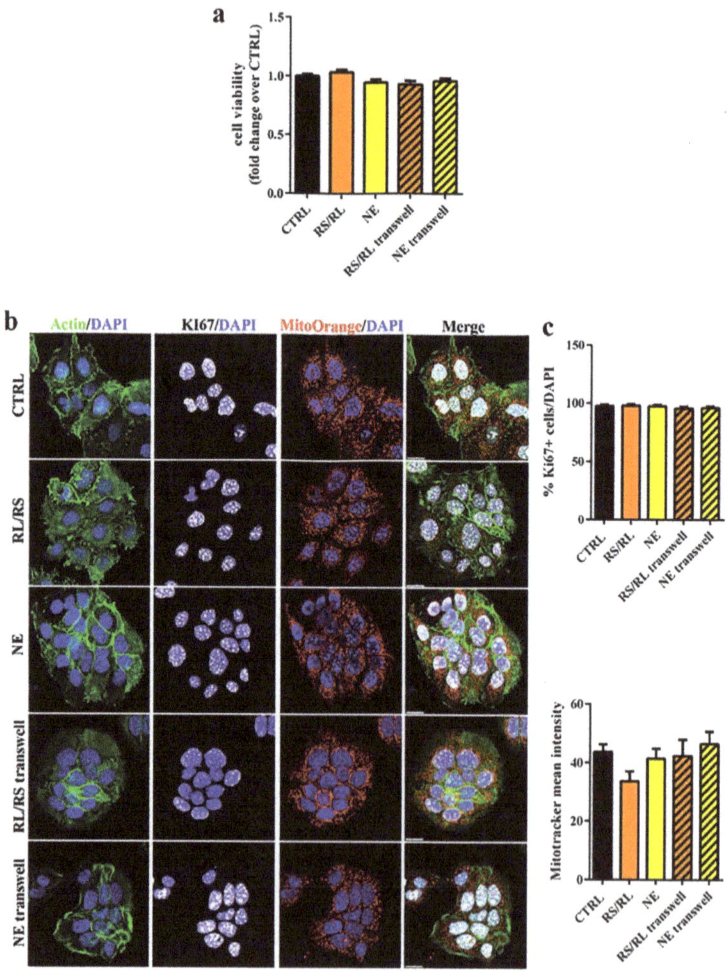

Figure 4. Cytotoxicity evaluation of Eudragit® RS/RL- and Eudragit® NE-coated specimens at 24 h of incubation. (**a**) Cell viability of HT1376 cells exposed to different samples for 24 h (n = 6). Untreated cells (CTRL) were used as negative control; (**b**) confocal microscope images of HT1376 cells stained with phalloidin to detect actin (green), Ki67 (white), Mitotracker Orange (red) and DAPI (blue) (20 µm scale bar); (**c**) upper panel: graph of the % of Ki67+ cells over total cells counted by DAPI staining; bottom panel: graph of the mean fluorescent intensity of Mitotracker Orange (n = 3).

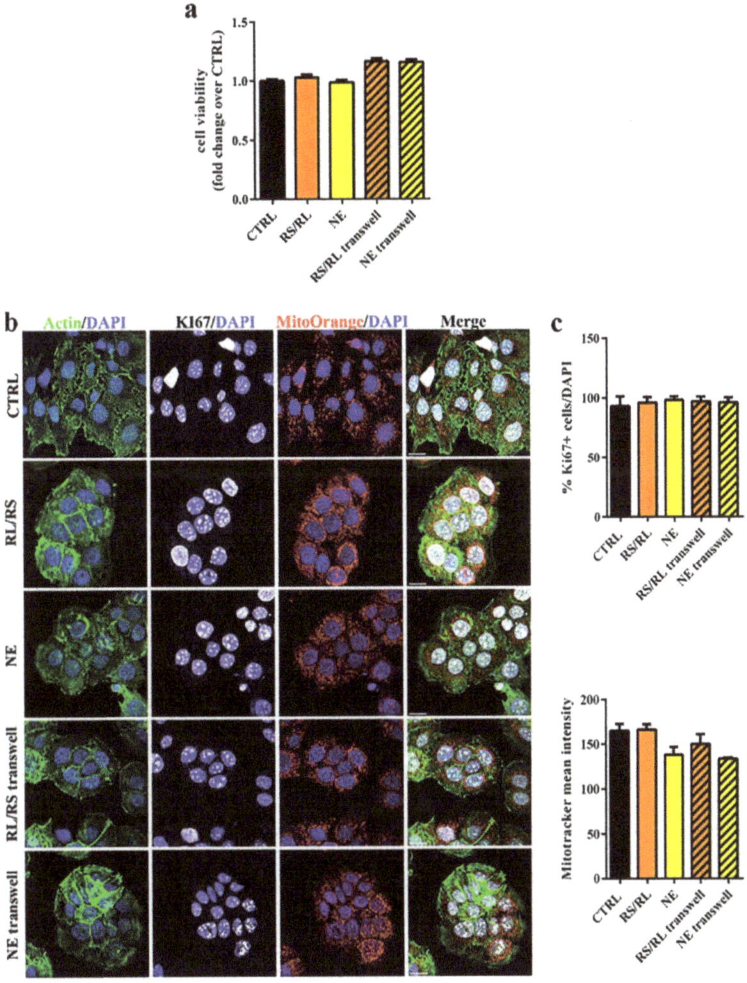

Figure 5. Cytotoxicity evaluation of Eudragit® RS/RL- and Eudragit® NE-coated specimens at 48 h of incubation. (**a**) Cell viability of HT1376 cells exposed to different samples for 48 h ($n = 6$). Untreated cells (CTRL) were used as negative; (**b**) confocal microscope images of HT1376 cells stained with phalloidin to detect actin (green), Ki67 (white), Mitotracker Orange (red) and DAPI (blue) (20 µm scale bar); (**c**) upper panel: graph of the % of Ki67+ cells over total cells counted by DAPI staining; bottom panel: graph of the mean fluorescent intensity of Mitotracker Orange ($n = 3$).

Finally, the inflammatory potential of uncoated and coated PVA-based specimens was investigated analyzing the expression of proinflammatory cytokines (IL1beta, IL6 and TNF alpha) by monocyte-derived THP-1 macrophages [66]. Macrophages treated with LPS, capable of inducing polarization toward an inflammatory phenotype and stimulating the expression of proinflammatory cytokines, were used as positive control (Figure 6). None of the devices tested was able to modify cytokine expression compared to the untreated control, suggesting that PVA-based specimens with or without coating did not affect the macrophage inflammatory response.

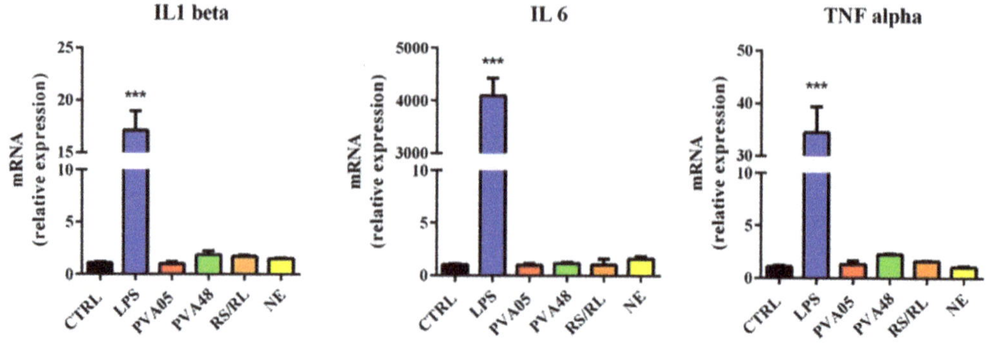

Figure 6. Proinflammatory cytokine expression in THP-1-derived macrophages exposed to uncoated PVA05- and PV48-based specimens and to Eudragit® RS/RL- and Eudragit® NE-coated ones. THP-1 cells differentiated into macrophage by the incubation with PMA for 24 h were exposed to the different samples for 48 h. Untreated macrophages (CTRL) and LPS treated macrophages were used as negative and positive controls, respectively. IL1 beta, IL6 and TNF alpha expression was analyzed by RT-PCR. Values are expressed as mean ± SEM ($n \geq 3$) normalized versus CTRL (*** $p < 0.001$ versus CTRL).

Similar findings were obtained by Omata and collaborators demonstrating that PVA-based coated particles were biocompatible and nontoxic and did not induce cytokine production by macrophages [67]. On the contrary, Strehl and co-investigators noticed an increase in the production of several cytokines, comparable to an acute inflammatory process, in human macrophages stimulated with PVA-coated nanoparticles [68]. This discrepancy might be explained by the differences in phagocytosis observed for the two types of particles, the former not being internalized as opposed to the latter. Phagocytosis is indeed a critical factor in macrophage activation [69,70].

3.2. New Configuration of the PVA-Based DDS

A second objective of the work was to demonstrate the feasibility of a different configuration for the PVA-based DDS under development, entailing internal cavities for drug filling and still exhibiting the SME ensuring bladder retention. To this end, samples characterized by a rather simple I shape were selected on account of the expected ease of fabrication and programming. Indeed, they were already demonstrated to be suitable screening tools for evaluating geometric and formulation changes (e.g., application of coatings) during development of PVA-based matrix-like prototypes. Moreover, the effectiveness of their shape recovery performance turned out independent of the original/temporary shapes considered [35,36,46].

As a first attempt, an I-shaped item characterized by the presence of a single internal cavity that would occupy most of its length was conceived for FDM manufacturing (Figure 7a). While the printing of samples starting from PVA05-based filaments was successful, during the programming of the temporary U shape and recovery of the original I shape, they showed a tendency to collapse and break at the curvature. Such a behavior was independent of the wall thickness (up to 1.5 mm) considered for the samples and was associated with the limited amount of polymeric material over which mechanical stresses could be released during U bending and subsequent opening of the arms of the specimens. For this reason, a new design was proposed, entailing a solid central polymeric portion separating two independent cavities, also named as compartments (Figure 7b). The presence of two separated cavities could also improve the versatility of the delivery system, offering more filling options. Inner compartments were designed with inward-facing ends conical in shape to increase the volume of the full central portion. The pseudo-circular section (5 mm in diameter) and the rounded external edges the system was provided with were intended to allow easy positioning of the latter in a catheter considered of medium size in clinical practice (i.e., external diameter greater than 16 Ch). The cross-section of the prototypes was flattened in correspondence with the portion resting on the printing plate, thus leading to a contact surface of approximately 3 mm in width. This detail improved the adherence of the first layers during the FDM process, limiting the chances of relevant detachment and reducing the number of printing failures. Moreover, to ensure loading of drug-containing formulations into the system cavities within a single process, printing was interrupted at the 25th layer. In this respect, the first trials were carried out having the operator fill each compartment by volume and then restarting the FDM process. However, with the aim of improving the consistency obtained with the volume-dependent filling, a trapdoor tool was designed and printed (see Figure 1a). It allowed for the accurate weighting of the powder and the easy transfer of the latter into the system cavities. Indeed, the trapdoor consisted in a chamber equipped with a removable base: in the closed configuration, the trapdoor could be placed on an analytical balance and filled at will from its top opening. When positioned on top of the prototype under fabrication, the base of the chamber could be manually removed, enabling the previously weighted powder to freely flow into the cavities, still maintaining an overall process time below 15 min.

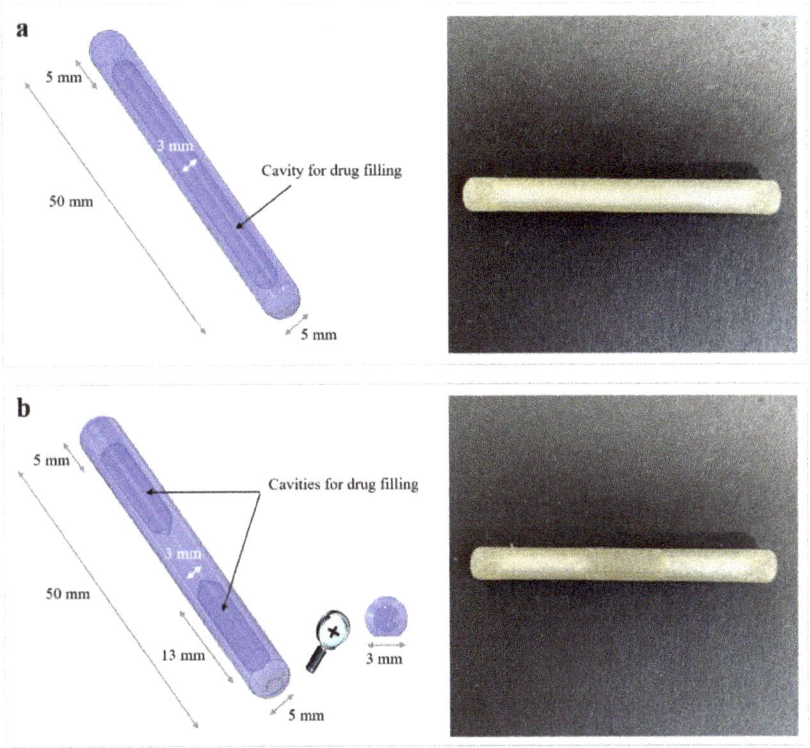

Figure 7. Digital models with dimensional and geometric details of (**a**) single and (**b**) two-compartment I-shaped prototypes, together with photographs of the actual printed and filled samples.

Based on the two-compartment design, prototypes with 100% and 50% infill, i.e., in principle characterized by different porosity, were printed. These proved to be quite reproducible in terms of the final weight of the device, and the variability of such parameters was reduced through the use of the semi-manual loading procedure (i.e., relying on the trapdoor tool). Based on these considerations, the weight variability observed was mainly attributed to the layer deposition mechanism typical of the 3D printing process (Table 4). Moreover, in view of the experience already gained with drug-embedded matrix-like DDSs, both 100 and 50% infill samples were coated with a low-permeable film based on Eudragit® NE, which could foster changes in the release performance. Considering the new configuration devised, the coating should have been layered over the entire external surface of the I-shaped specimens, in order to avoid undesired differences in the system surface exposed to aqueous fluids and thus of rate of dissolution/erosion of the polymeric walls (i.e., reducing the risk of uncontrolled penetration of water and early opening of the cavities). In this respect, a semi-automated coating procedure was preliminarily employed using the lab-scale equipment already developed for I-shaped specimens. As expected, based on the similar external dimensions and because the mass of the specimen would not affect its ability to rotate during the coating process, weight gain and coating thickness of all the samples turned out to be reproducible and independent of the infill percentage (Table 5).

Table 4. Weight data relevant to uncoated I-shaped samples printed by setting different infill.

	Weight, mg (CV)	
	Manually Filled	**Trapdoor-Filled**
100% Infill	841.85 (9.52)	838.63 (5.02)
50% Infill	723.44 (10.11)	736.11 (5.61)

Table 5. Thickness and weight gain data relevant to coated I-shaped samples printed by setting different infill.

	Thickness, μm (CV)						Overall Thickness, μm (CV)	Weight Gain, % (CV)
	1	**2**	**3**	**4**	**5**	**6**		
100% Infill	56.43 (6.36)	55.37 (7.78)	55.75 (6.61)	51.70 (8.72)	53.27 (9.25)	55.98 (6.37)	54.75 (7.64)	6.73 (6.84)
50% Infill	54.71 (7.35)	58.03 (9.06)	52.30 (6.23)	57.05 (7.59)	54.85 (9.63)	54.54 (9.93)	55.24 (8.53)	6.22 (7.01)

When tested for shape recovery, both empty samples and those filled with the powder tracer exhibited the desired behavior regardless of their design features (i.e., infill percentage and presence of the coating) (Figure 8). In particular, neither alteration nor collapse of the compartments occurred when programming the temporary shape or during the recovery process (by way of example, see photographs in Figure 9). Moreover, when dealing with coated samples, no visible damage to the external film was highlighted. After only 3 min of contact with simulated urine at 37 °C, all prototypes were able to recover ≥70% of their original shape (Figure 8). The presence of the Eudragit® NE-based coating seemed to slightly promote shape recovery by reducing the time required for its completion and increasing the relevant efficiency (i.e., higher RIs achieved sooner). The latter result was consistent with the data previously collected with matrix-like PVA-based prototypes and was associated with the flexibility of such a film, acting as a sort of rubbery envelope during shape recovery [46].

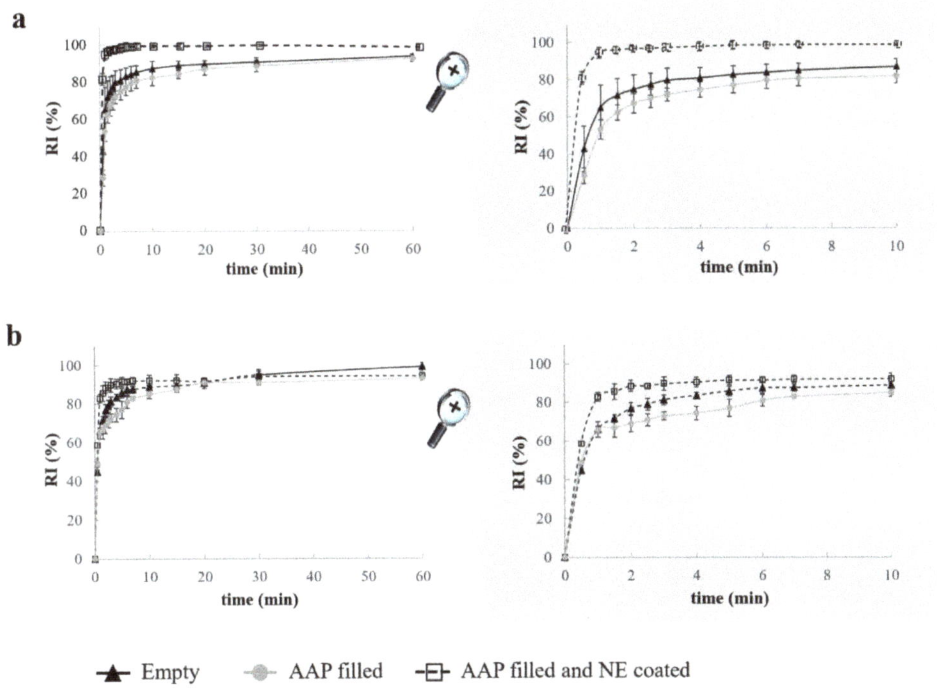

Figure 8. RI versus time curves relevant to different I-shaped samples (i.e., having empty reservoir units, being filled with the selected powder tracer and tested as such or after relevant coating) printed with (**a**) 100% and (**b**) 50% infill.

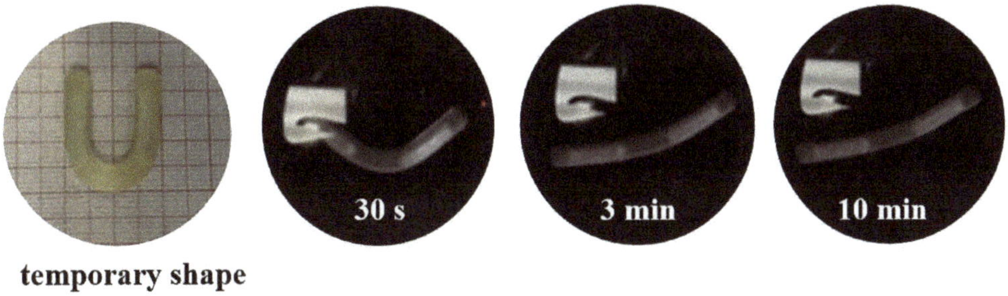

Figure 9. Photographs of an I sample, printed with 100% infill and filled with the drug tracer, after programming of the temporary U shape and during the recovery experiments.

As expected, based on the new configuration and composition of the system, uncoated PVA-based prototypes pointed out a pulsatile release performance, characterized by a lag phase prior to release (i.e., $t_{10\%}$) (Figure 10) [71–73]. The duration of the lag phase was determined by the hydration, erosion and dissolution of the swellable/soluble PVA-based walls surrounding the drug-filled compartments, at the end of which opening of the systems occurred. Accordingly, the erosion/dissolution of the PVA walls was completed faster when these were lighter (i.e., printed setting 50% infill), resulting in lower values for both lag time and pulse time of the relevant prototypes with respect to the 100% infill samples.

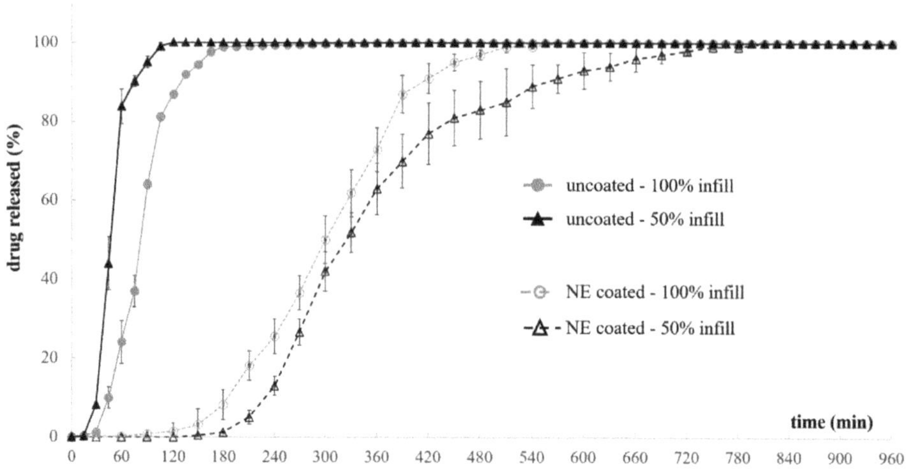

Figure 10. Release profiles relevant to uncoated and coated prototypes printed with different infill.

Dealing with coated samples for which interaction with aqueous fluids was mediated by the presence of a poorly-permeable film, the lag time increased four- to six-fold. Indeed, the hydration and swelling of PVA occurred more slowly with respect to the uncoated samples. However, when a threshold value was reached, the volume expansion of the hydrated polymer resulted in the formation of small openings in the coating layer along the entire length of all samples. In this respect, the systems printed with 100% infill, i.e., denser and with enhanced swelling capacity, were characterized by the shortest lag phase. On the other hand, 50% infill prototypes exhibited a reduced breaking ability associated with PVA expansion, which was responsible for a reduction in the number of openings in the coating layer. The expansion was also shown to occur later, as highlighted by the greater $t_{10\%}$ value. As a consequence of the reduction in the rate of swelling and of erosion/dissolution of the PVA walls, the aqueous fluid penetrating through the openings and the swollen polymeric matrix inside the cavities might also dissolve the conveyed drug, thus promoting its diffusion outward, even before the effective opening of the reservoir cavities. This phenomenon might explain why the overall duration of release from the coated samples turned out longer with respect to the uncoated ones.

4. Conclusions

The availability of organ-retentive DDSs conceived to remain and release their content for a prolonged period of time into the bladder would be highly advantageous from the patient perspective, as it might reduce the number of instillations and thus of catheterizations the patient would undergo over time. While improving compliance, life-quality and relevant expectancy, this approach might also limit healthcare and social expenses by acting on administration-related costs, entailing, for instance, hospitalizations, consumables, disposal operations, commitment of hospital personnel and management of inflammations/secondary infections. In addition, retentive systems could widen the number of available treatments for bladder pathologies by implementing new therapeutic approaches combining active ingredients and involving modified time and rate of release. In this respect, the expandable intravesical DDS already proposed as a matrix-like structure for prolonged release of active molecules was here further improved to be equipped with internal cavities for extemporaneous, independent and personalized filling. The new configuration would also enable programmed release of specific drug quantities at different times.

By taking advantage of the PVA SME, this work confirmed the application potential of 4D printing in the development of DDSs intended for retention in hollow muscular organs, especially towards more complex structures (e.g., multi-layer and hollow systems). Finally, preliminary biocompatibility studies highlighted the safety of the materials used, which was particularly promising in view of the next development steps.

Author Contributions: M.U.: software; investigation; data curation; writing—original draft; visualization. C.P.: conceptualization; methodology; data curation; writing—original draft; funding acquisition. C.M.: conceptualization; resources; writing—review and editing. S.Z.: methodology; validation; visualization. A.N.: formal analysis; investigation. C.C.: conceptualization; writing—review and editing. A.G.: resources; writing—review and editing. A.M.: conceptualization; methodology; writing—original draft; supervision. L.Z.: conceptualization; writing—review and editing; project administration; funding acquisition. All authors have read and agreed to the published version of the manuscript.

Funding: This work was partially supported by "Università degli Studi di Milano, Linea 3-Bando Straordinario per Progetti Interdipartimentali (Bando SEED 2019)".

Data Availability Statement: Data are available upon request.

Acknowledgments: The authors thank Marco Coazzoli for his technical support in preliminary trials.

Conflicts of Interest: The authors declare no conflict of interest.

References

1. Farokhzad, O.C.; Dimitrakov, J.D.; Karp, J.M.; Khademhosseini, A.; Freeman, M.R.; Langer, R. Drug delivery systems in urology-getting "smarter". *Urology* **2006**, *68*, 463–469. [CrossRef] [PubMed]
2. GuhaSarkar, S.; Banerjee, R. Intravesical drug delivery: Challenges, current status, opportunities and novel strategies. *J. Control. Release* **2010**, *148*, 147–159. [CrossRef] [PubMed]
3. Lee, S.H.; Choy, Y.B. Implantable devices for sustained, intravesical drug delivery. *Int. Neurourol. J.* **2016**, *20*, 101–106. [CrossRef] [PubMed]
4. Sarfraz, M.; Qamar, S.; Rehman, M.U.; Tahir, M.A.; Ijaz, M.; Ahsan, A.; Asim, M.H.; Nazir, I. Nano-formulation based intravesical drug delivery systems: An overview of versatile approaches to improve urinary bladder diseases. *Pharmaceutics* **2022**, *14*, 1909. [CrossRef]
5. Zacchè, M.M.; Srikrishna, S.; Cardozo, L. Novel targeted bladder drug-delivery systems: A review. *Res. Rep. Urol.* **2015**, *7*, 169–178. [CrossRef]
6. de Lima, C.S.A.; Varca, J.P.R.O.; Alves, V.M.; Nogueira, K.M.; Cruz, C.P.C.; Rial-Hermida, M.I.; Kadłubowski, S.S.; Varca, G.H.C.; Lugão, A.B. Mucoadhesive polymers and their applications in drug delivery systems for the treatment of bladder cancer. *Gels* **2022**, *8*, 587. [CrossRef]
7. Gugleva, V.; Michailova, V.; Mihaylova, R.; Momekov, G.; Zaharieva, M.M.; Najdenski, H.; Petrov, P.; Rangelov, S.; Forys, A.; Trzebicka, B.; et al. Formulation and evaluation of hybrid niosomal in situ gel for intravesical co-delivery of curcumin and gentamicin sulfate. *Pharmaceutics* **2022**, *14*, 747. [CrossRef]
8. Guo, P.; Wang, L.; Shang, W.; Chen, J.; Chen, Z.; Xiong, F.; Wang, Z.; Tong, Z.; Wang, K.; Yang, L.; et al. Intravesical in situ immunostimulatory gel for triple therapy of bladder cancer. *ACS ACS Appl. Mater. Interfaces* **2020**, *12*, 54367–54377. [CrossRef]
9. Cima, M.J.; Lee, H.; Daniel, K.; Tanenbauma, L.M.; Mantzavinou, A.; Spencer, K.C.; Ong, Q.; Sy, J.C.; Santini, J., Jr.; Schoellhammer, C.M.; et al. Single compartment drug delivery. *J. Control. Release* **2014**, *190*, 157–171. [CrossRef]
10. Wang, L.H.; Shang, L.; Shan, D.Y.; Che, X. Long-term floating control-released intravesical preparation of 5-fluorouracil for the local treatment of bladder cancer. *Drug Dev. Ind. Pharm.* **2017**, *43*, 1343–1350. [CrossRef]
11. Zhu, G.; Zhang, Y.; Wang, K.; Zhao, X.; Lian, H.; Wang, W.; Wang, H.; Wu, J.; Hu, Y.; Guo, H. Visualized intravesical floating hydrogel encapsulating vaporized perfluoropentane for controlled drug release. *Drug Deliv.* **2016**, *23*, 2820–2826. [CrossRef]
12. Maroni, A.; Melocchi, A.; Zema, L.; Foppoli, A.; Gazzaniga, A. Retentive drug delivery systems based on shape memory materials. *J. Appl. Polym. Sci.* **2020**, *137*, 48798. [CrossRef]
13. Palugan, L.; Cerea, M.; Cirilli, M.; Moutaharrik, A.; Maroni, A.; Zema, L.; Melocchi, A.; Uboldi, M.; Filippin, I.; Foppoli, A.; et al. Intravesical drug delivery approaches for improved therapy of urinary bladder diseases. *Int. J. Pharm. X* **2021**, *3*, 100100. [CrossRef] [PubMed]
14. Behl, M.; Lendlein, A. Shape memory polymers. *Mater. Today* **2007**, *10*, 20–28.
15. Melocchi, A.; Uboldi, M.; Cerea, M.; Foppoli, A.; Maroni, A.; Moutaharrik, S.; Palugan, L.; Zema, L.; Gazzaniga, A. Shape memory materials and 4D printing in pharmaceutics. *Adv. Drug Deliv. Rev.* **2021**, *173*, 216–237. [CrossRef] [PubMed]
16. Zhao, W.; Liu, L.; Zhang, F.; Leng, J.; Liu, Y. Shape memory polymers and their composites in biomedical applications. *Mater. Sci. Eng. C Mater. Biol. Appl.* **2019**, *97*, 864–883. [CrossRef]

17. Wischke, C.; Behl, M.; Lendlein, A. Drug-releasing shape-memory polymers-the role of morphology, processing effects, and matrix degradation. *Expert Opin. Drug Deliv.* **2013**, *10*, 1193–1205. [CrossRef]
18. Giesing, D.; Lee, H.; Daniel, K.D. Drug Delivery Systems and Methods for Treatment of Bladder Cancer with Gemcitabine. U.S. Patent US 2015/0250717 A1, 10 September 2015.
19. Lee, H.; Daniel, K.D. Intravesical drug delivery devices and methods including elastic polymer-drug matrix systems. U.S. Patent WO 2015/200752 Al, 30 December 2015.
20. Nickel, J.C.; Jain, P.; Shore, N.; Anderson, J.; Giesing, D.; Lee, H.; Kim, G.; Daniel, K.; White, S.; Larrivee-Elkins, C.; et al. Continuous intravesical lidocaine treatment for interstitial cystitis/bladder pain syndrome: Safety and efficacy of a new drug delivery device. *Sci. Transl. Med.* **2012**, *4*, 143a100. [CrossRef]
21. Xu, X.; Goyanes, A.; Trenfield, S.J.; Diaz-Gomez, L.; Alvarez-Lorenzo, C.; Gaisford, S.; Basit, A.W. Stereolithography (SLA) 3D printing of a bladder device for intravesical drug delivery. *Mater. Sci. Eng. C* **2021**, *120*, 111773. [CrossRef]
22. Peterson, G.I.; Dobrynin, A.V.; Becker, M.L. Biodegradable shape memory polymers in medicine. *Adv. Healthc. Mater.* **2017**, *6*, 1700694. [CrossRef]
23. Sun, L.; Huang, W.M. Thermo/moisture responsive shape-memory polymer for possible surgery/operation inside living cells in future. *Mat. Des.* **2010**, *31*, 2684–2689. [CrossRef]
24. Wong, Y.S.; Salvekar, A.V.; Zhuang, K.D.; Liu, H.; Birch, W.R.; Tay, K.H.; Huang, W.M.; Venkatraman, S.S. Bioabsorbable radiopaque water-responsive shape memory embolization plug for temporary vascular occlusion. *Biomaterials* **2016**, *102*, 98–106. [CrossRef] [PubMed]
25. Xiao, R.; Huang, W.M. Heating/solvent responsive shape-memory polymers for implant biomedical devices in minimally invasive surgery: Current status and challenge. *Macromol. Biosci.* **2020**, *20*, 2000108. [CrossRef] [PubMed]
26. Rahmatabadi, D.; Aberoumand, M.; Soltanmohammadi, K.; Soleyman, E.; Ghasemi, I.; Baniassadi, M.; Abrinia, K.; Zolfagharian, A.; Bodaghi, M.; Baghani, M. A New strategy for achieving shape memory effects in 4D printed two-layer composite structures. *Polymers* **2022**, *14*, 5446. [CrossRef] [PubMed]
27. Rahmatabadi, D.; Aberoumand, M.; Soltanmohammadi, K.; Soleyman, E.; Ghasemi, I.; Baniassadi, M.; Abrinia, K.; Bodaghi, M.; Baghani, M. 4D Printing-encapsulated polycaprolactone–thermoplastic polyurethane with high shape memory performances. *Adv. Eng. Mater.* **2023**, *2022*, 2201309. [CrossRef]
28. Soleyman, E.; Aberoumand, M.; Soltanmohammadi, K.; Rahmatabadi, D.; Ghasemi, I.; Baniassadi, M.; Abrinia, K.; Baghani, M. 4D printing of PET-G via FDM including tailormade excess third shape. *Manuf. Lett.* **2022**, *33*, 1–4. [CrossRef]
29. Soleyman, E.; Aberoumand, M.; Rahmatabadi, D.; Soltanmohammadi, K.; Ghasemi, I.; Baniassadi, M.; Abrinia, K.; Baghani, M. Assessment of controllable shape transformation, potential applications, and tensile shape memory properties of 3D printed PETG. *J. Mater. Res. Technol.* **2022**, *18*, 4201–4215. [CrossRef]
30. Elkasabgy, N.A.; Mahmoud, A.A.; Maged, A. 3D printing: An appealing route for customized drug delivery systems. *Int. J. Pharm.* **2020**, *588*, 119732. [CrossRef]
31. Patel, S.K.; Khoder, M.; Peak, M.; Alhnan, M.A. Controlling drug release with additive manufacturing-based solutions. *Adv. Drug Deliv. Rev.* **2021**, *174*, 369–386. [CrossRef]
32. Bandari, S.; Nyavanandi, D.; Dumpa, N.; Repka, M.A. Coupling hot melt extrusion and fused deposition modeling: Critical properties for successful performance. *Adv. Drug Deliv. Rev.* **2021**, *172*, 52–63. [CrossRef]
33. Krueger, L.; Miles, J.A.; Popat, A. 3D printing hybrid materials using fused deposition modelling for solid oral dosage forms. *J. Control. Rel.* **2022**, *351*, 444–455. [CrossRef]
34. Parulski, C.; Jennotte, O.; Lechanteur, A.; Evrard, B. Challenges of fused deposition modeling 3D printing in pharmaceutical applications: Where are we now? *Adv. Drug Deliv. Rev.* **2021**, *175*, 113810. [CrossRef]
35. Melocchi, A.; Inverardi, N.; Uboldi, M.; Baldi, F.; Maroni APandini, S.; Briatico-Vangosa, F.; Zema, L.; Gazzaniga, A. Retentive device for intravesical drug delivery based on water-induced shape memory response of poly(vinyl alcohol): Design concept and 4D printing feasibility. *Int. J. Pharm.* **2019**, *559*, 299–311. [CrossRef]
36. Melocchi, A.; Uboldi, M.; Inverardi, N.; Briatico-Vangosa, F.; Baldi, F.; Pandini, S.; Scalet, G.; Auricchio, F.; Cerea, M.; Foppoli, A.; et al. Expandable drug delivery system for gastric retention based on shape memory polymers: Development via 4D printing and extrusion. *Int. J. Pharm.* **2019**, *571*, 118700. [CrossRef]
37. Fang, Z.Q.; Kuang, Y.D.; Zhou, P.P.; Ming, S.Y.; Zhu, P.H.; Liu, Y.; Ning, H.L.; Chen, G. Programmable shape recovery process of water-responsive shape memory poly(vinyl alcohol) by wettability contrast strategy. *ACS Appl. Mater. Interfaces* **2017**, *9*, 5495–5502. [CrossRef] [PubMed]
38. Inverardi, N.; Scalet, G.; Melocchi, A.; Uboldi, M.; Maroni, A.; Zema, L.; Gazzaniga, A.; Auricchio, F.; Briatico-Vangosa, F.; Baldi, F.; et al. Experimental and computational analysis of a pharmaceutical-grade shape memory polymer applied to the development of gastroretentive drug delivery systems. *J. Mech. Behav. Biomed Mater.* **2021**, *124*, 104814. [CrossRef] [PubMed]
39. Lin, L.; Guo, Y. Enhanced shape memory property and mechanical property of polyvinyl alcohol by carbon black. *J. Biomater. Tissue Eng.* **2019**, *9*, 76–81. [CrossRef]
40. Uboldi, M.; Melocchi, A.; Moutaharrik, S.; Palugan, L.; Cerea, M.; Foppoli, A.; Maroni, A.; Gazzaniga, A.; Zema, L. Administration strategies and smart devices for drug release in specific sites of the upper GI tract. *J. Control. Release* **2022**, *348*, 537–552. [CrossRef] [PubMed]

41. Wang, W.; Lai, H.; Cheng, Z.; Kang, H.; Wang, Y.; Zhang, H.; Wang, J.; Liu, Y. Water-induced poly(vinyl alcohol)/carbon quantum dot nanocomposites with tunable shape recovery performance and fluorescence. *J. Mater. Chem. B* **2018**, *6*, 7444–7450. [CrossRef] [PubMed]
42. Afzali Naniz, M.; Askari, M.; Zolfagharian, A.; Afzali Naniz, M.; Bodaghi, M. 4D printing: A cutting-edge platform for biomedical applications. *Biomed. Mat.* **2022**, *17*, 062001. [CrossRef]
43. Pingale, P.; Dawre, S.; Dhapte-Pawar, V.; Dhas, N.; Rajput, A. Advances in 4D printing: From stimulation to simulation. *Drug Deliv. Transl. Res.* **2023**, *13*, 164–188. [CrossRef] [PubMed]
44. Pourmasoumi, P.; Moghaddam, A.; Nemati Mahand, S.; Heidari, F.; Salehi Moghaddam, Z.; Arjmand, M.; Kühnert, I.; Kruppke, B.; Wiesmann, H.-P.; Khonakdar, H.A. A review on the recent progress, opportunities, and challenges of 4D printing and bioprinting in regenerative medicine. *J. Biomater. Sci. Polym. Ed.* **2023**, *34*, 108–146. [CrossRef]
45. Uboldi, M.; Melocchi, A.; Moutaharrik, S.; Cerea, M.; Gazzaniga, A.; Zema, L. Dataset on a small-scale film-coating process developed for self-expanding 4D printed drug delivery devices. *Coatings* **2021**, *11*, 1252. [CrossRef]
46. Uboldi, M.; Pasini, C.; Pandini, S.; Baldi, F.; Briatico-Vangosa, F.; Inverardi, N.; Maroni, A.; Moutaharrik, S.; Melocchi, A.; Gazzaniga, A.; et al. Expandable drug delivery systems based on shape memory polymers: Impact of film coating on mechanical properties and release and recovery performance. *Pharmaceutics* **2022**, *14*, 2814. [CrossRef] [PubMed]
47. Hu, K.; Brambilla, L.; Sartori, P.; Moscheni, C.; Perrotta, C.; Zema, L.; Bertarelli, C.; Castiglioni, C. Development of tailored graphene nanoparticles: Preparation, sorting and structure assessment by complementary techniques. *Molecules* **2023**, *28*, 565. [CrossRef]
48. Melocchi, A.; Parietti, F.; Maroni, A.; Foppoli, A.; Gazzaniga, A.; Zema, L. Hot-melt extruded filaments based on pharmaceutical grade polymers for 3D printing by fused deposition modeling. *Int. J. Pharm.* **2016**, *509*, 255–263. [CrossRef]
49. Sherif, A.Y.; Mahrou, G.M.; Alanazi, F.K. Novel in-situ gel for intravesical administration of ketorolac. *Saudi Pharm. J.* **2018**, *26*, 845–851. [CrossRef]
50. Auwerx, J. The human leukemia cell line, THP-1: A multifacetted model for the study of monocyte-macrophage differentiation. *Experientia* **1991**, *47*, 22–31. [CrossRef]
51. Liu, X.; Yin, S.; Chen, Y.; Wu, Y.; Zheng, W.; Dong, H.; Bai, Y.; Qin, Y.; Li, J.; Feng, S.; et al. 1LPS-induced proinflammatory cytokine expression in human airway epithelial cells and macrophages via NF-κB, STAT3 or AP-1 activation. *Mol. Med. Rep.* **2018**, *17*, 5484–5491.
52. Bizzozero, L.; Cazzato, D.; Cervia, D.; Assi, E.; Simbari, F.; Pagni, F.; De Palma, C.; Monno, A.; Verdelli, C.; Querini, P.R.; et al. Acid sphingomyelinase determines melanoma progression and metastatic behaviour via the microphtalmia-associated transcription factor signalling pathway. *Cell Death Differ.* **2014**, *21*, 507–520. [CrossRef]
53. Perrotta, C.; Buonanno, F.; Zecchini, S.; Giavazzi, A.; Proietti Serafini, F.; Catalani, E.; Guerra, L.; Belardinelli, M.C.; Picchietti, S.; Fausto, A.M.; et al. Climacostol reduces tumour progression in a mouse model of melanoma via the p53-dependent intrinsic apoptotic programme. *Sci. Rep.* **2016**, *6*, 7281. [CrossRef] [PubMed]
54. Perrotta, C.; Cervia, D.; Di Renzo, I.; Moscheni, C.; Bassi, M.T.; Campana, L.; Martelli, C.; Catalani, E.; Giovarelli, M.; Zecchini, S.; et al. Nitric oxide generated by tumor-associated macrophages is responsible for cancer resistance to cisplatin and correlated with syntaxin 4 and acid sphingomyelinase inhibition. *Front. Immunol.* **2018**, *9*, 1186. [CrossRef] [PubMed]
55. Cervia, D.; Assi, E.; De Palma, C.; Giovarelli, M.; Bizzozero, L.; Pambianco, S.; Di Renzo, I.; Zecchini, S.; Moscheni, C.; Vantaggiato, C.; et al. Essential role for acid sphingomyelinase-inhibited autophagy in melanoma response to cisplatin. *Oncotarget* **2016**, *7*, 24995–25009. [CrossRef]
56. Coazzoli, M.; Napoli, A.; Roux-Biejat, P.; De Palma, C.; Moscheni, C.; Catalani, E.; Zecchini, S.; Conte, V.; Giovarelli, M.; Caccia, S.; et al. Acid sphingomyelinase downregulation enhances mitochondrial fusion and promotes oxidative metabolism in a mouse model of melanoma. *Cells* **2020**, *9*, 848. [CrossRef]
57. Roux-Biejat, P.; Coazzoli, M.; Marrazzo, P.; Zecchini, S.; Di Renzo, I.; Prata, C.; Napoli, A.; Moscheni, C.; Giovarelli, M.; Barbalace, M.C.; et al. Acid sphingomyelinase controls early phases of skeletal muscle regeneration by shaping the macrophage phenotype. *Cells* **2021**, *10*, 3028. [CrossRef] [PubMed]
58. Bernardo, C.; Costa, C.; Palmeira, C.; Pinto-Leite, R.; Oliveira, P.; Freitas, R.; Amado, F.; Santos, L.L. What we have learned from urinary bladder cancer models. *J. Cancer. Metastasis Treat.* **2016**, *2*, 51–58.
59. Baumgart, E.; Cohen, M.S.; Silva Neto, B.; Jacobs, M.A.; Wotkowicz, C.; Rieger-Christ, K.M.; Biolo, A.; Zeheb, R.; Loda, M.; Libertino, J.A.; et al. Identification and prognostic significance of an epithelial-mesenchymal transition expression profile in human bladder tumors. *Clin. Cancer. Res.* **2007**, *13*, 1685–1694. [CrossRef] [PubMed]
60. Onrust, S.V.; Wiseman, L.R.; Goa, K.L. Epirubicin: A review of its intravesical use in superficial bladder cancer. *Drugs Aging* **1999**, *15*, 307–333. [CrossRef]
61. Tian, Y.; Ma, Z.; Chen, Z.; Li, M.; Wu, Z. Clinicopathological and prognostic value of Ki-67 expression in bladder cancer: A systematic review and meta-analysis. *PLoS ONE* **2016**, *11*, e0158891. [CrossRef]
62. Saraste, A.; Pulkki, K. Morphologic and biochemical hallmarks of apoptosis. *Cardiovasc. Res.* **2000**, *45*, 528–537. [CrossRef]
63. Gaaz, T.S.; Sulong, A.B.; Akhtar, M.N.; Kadhum, A.A.; Mohamad, A.B.; Al-Amiery, A.A. Properties and Applications of polyvinyl alcohol, halloysite nanotubes and their nanocomposites. *Molecules* **2015**, *20*, 22833–22847. [CrossRef] [PubMed]
64. Patra Ch, N.; Priya, R.; Swain, S.; Jena, G.K.; Panigrahi, K.C.; Ghose, D. Pharmaceutical significance of Eudragit: A review. *Future J. Pharm. Sci.* **2017**, *3*, 33–45. [CrossRef]

65. Rehman, S.; Ranjha, N.M.; Shoukat, H.; Madni, A.; Ahmad, F.; Raza MRJameel, Q.A.; Majeed, A.; Ramzan, N. Fabrication, evaluation, in vivo pharmacokinetic and toxicological analysis of pH-sensitive Eudragit S-100-coated hydrogel beads: A promising strategy for colon targeting. *AAPS Pharm. Sci. Tech.* **2021**, *22*, 209. [CrossRef] [PubMed]
66. Sharif, O.; Bolshakov, V.N.; Raines, S.; Newham, P.; Perkins, N.D. Transcriptional profiling of the LPS induced NF-κB response in macrophages. *BMC Immunol.* **2007**, *8*, 1. [CrossRef] [PubMed]
67. Omata, S.; Sawae, Y.; Murakami, T. Effect of poly (vinyl alcohol) (PVA) wear particles generated in water lubricant on immune response of macrophage. *Biosurf. Biotribol.* **2015**, *1*, 71–79. [CrossRef]
68. Strehl, C.; Gaber, T.; Maurizi, L.; Hahne, M.; Rauch, R.; Hoff, P.; Häupl, T.; Hofmann-Amtenbrink, M.; Poole, A.R.; Hofmann, H.; et al. Effects of PVA coated nanoparticles on human immune cells. *Int. J. Nanomed.* **2015**, *10*, 3429–3445. [CrossRef]
69. Chikaura, H.; Nakashima, Y.; Fujiwara, Y.; Komohara, Y.; Takeya, M.; Nakanishi, Y. Effect of particle size on biological response by human monocyte-derived macrophages. *Biosurf. Biotribol.* **2016**, *2*, 18–25. [CrossRef]
70. Green, T.R.; Fisher, J.; Stone, M.; Wroblewski, B.M.; Ingham, E. Polyethylene particles of a 'critical size' are necessary for the induction of cytokines by macrophages in vitro. *Biomaterials* **1998**, *19*, 2297–2302. [CrossRef]
71. Lévi, F.; Okyar, A. Circadian clocks and drug delivery systems: Impact and opportunities in chronotherapeutics. *Expert Opin. Drug Deliv.* **2011**, *8*, 1535–1541. [CrossRef]
72. Maroni, A.; Zema, L.; Curto, M.D.D.; Loreti, G.; Gazzaniga, A. Oral pulsatile delivery: Rationale and chronopharmaceutical formulations. *Int. J. Pharm.* **2010**, *398*, 1–8. [CrossRef]
73. Melocchi, A.; Uboldi, M.; Briatico-Vangosa, F.; Moutaharrik, S.; Cerea, M.; Foppoli, A.; Maroni, A.; Palugan, L.; Zema, L.; Gazzaniga, A. The Chronotopic™ system for pulsatile and colonic delivery of active molecules in the era of precision medicine: Feasibility by 3D printing via fused deposition modeling (FDM). *Pharmaceutics* **2021**, *13*, 759. [CrossRef] [PubMed]

pharmaceutics

MDPI

Article

Intraocular siRNA Delivery Mediated by Penetratin Derivative to Silence Orthotopic Retinoblastoma Gene

Xin Gao [1,†], Xingyan Fan [1,†], Kuan Jiang [1,2], Yang Hu [1], Yu Liu [1], Weiyue Lu [1,3] and Gang Wei [1,3,4,*]

1 Key Laboratory of Smart Drug Delivery, Ministry of Education & Department of Pharmaceutics, School of Pharmacy, Fudan University, Shanghai 201203, China
2 Department of Pharmacology, School of Basic Medical Sciences & State Key Laboratory of Molecular Engineering of Polymers, Fudan University, Shanghai 200032, China
3 The Institutes of Integrative Medicine of Fudan University, Shanghai 200040, China
4 Shanghai Engineering Research Center of ImmunoTherapeutics, Shanghai 201203, China
* Correspondence: weigang@shmu.edu.cn; Tel.: +86-21-5198-0091; Fax: +86-21-5198-0090
† These authors contributed equally to this work.

Abstract: Gene therapy brings a ray of hope for inherited ocular diseases that may cause severe vision loss and even blindness. However, due to the dynamic and static absorption barriers, it is challenging to deliver genes to the posterior segment of the eye by topical instillation. To circumvent this limitation, we developed a penetratin derivative (89WP)-modified polyamidoamine polyplex to deliver small interference RNA (siRNA) via eye drops to achieve effective gene silencing in orthotopic retinoblastoma. The polyplex could be spontaneously assembled through electrostatic and hydrophobic interactions, as demonstrated by isothermal titration calorimetry, and enter cells intactly. In vitro cellular internalization revealed that the polyplex possessed higher permeability and safety than the lipoplex composed of commercial cationic liposomes. After the polyplex was instilled in the conjunctival sac of the mice, the distribution of siRNA in the fundus oculi was significantly increased, and the bioluminescence from orthotopic retinoblastoma was effectively inhibited. In this work, an evolved cell-penetrating peptide was employed to modify the siRNA vector in a simple and effective way, and the formed polyplex interfered with intraocular protein expression successfully via noninvasive administration, which showed a promising prospect for gene therapy for inherited ocular diseases.

Keywords: gene delivery; intraocular drug delivery; siRNA; penetratin derivative; nonviral vector; noninvasive administration

Citation: Gao, X.; Fan, X.; Jiang, K.; Hu, Y.; Liu, Y.; Lu, W.; Wei, G. Intraocular siRNA Delivery Mediated by Penetratin Derivative to Silence Orthotopic Retinoblastoma Gene. *Pharmaceutics* **2023**, *15*, 745. https://doi.org/10.3390/pharmaceutics15030745

Academic Editors: Daniela Monti, Silvia Tampucci and Gabriele Candiani

Received: 10 November 2022
Revised: 30 January 2023
Accepted: 21 February 2023
Published: 23 February 2023

1. Introduction

According to a recent epidemiological survey, nearly 295 million people worldwide suffer from moderate to severe visual impairments, which cause about 43.3 million blindness [1]. Even worse, some inherited blinding diseases, for example X-linked retinoschisis and Leber congenital amaurosis, still lack effective treatment regimens. Fortunately, gene therapy brings a ray of hope for these diseases [2]. The regulation of gene expression by nucleic acid agents has become one of the most promising strategies for the treatment of inherited blinding diseases, and more than 40 kinds of gene therapeutics are currently under clinical evaluation [3].

With its unique anatomy and immune-privilege characteristics, the eye is an ideal organ for topical administration and gene therapy [4,5]. Eye drops are convenient and can avoid the interference of a complicated physiological environment on gene delivery, as encountered by systemic administration. However, due to the dynamic and static absorption barriers of the eye, nucleic acid agents are obliged to be applied via subretinal or intravitreal injection, which requires rigorous medical conditions and is accompanied by poor patient compliance [6,7]. In late 2017, voretigene neparvovec-rzyl (Luxturna), a breakthrough in gene therapy, was approved by the U.S. Food and Drug Administration for the treatment of inherited retinal dystrophy. The human retinal pigment epithelial 65 kDa protein (*hRPE65*) genome is successfully encoded into the recombinant adenovirus-associated viruses (rAAV) vector, and then injected into the retina of patients to enable gene expression. This would allow the normally functioning photoreceptors to survive, and thus restore visual perception [2,8]. Although gene regulation provides a substantial therapeutic advance, intraocular injection may cause various side effects with an incidence higher than 5%, and even permanent vision loss [8]. Therefore, noninvasive delivery via topical instillation has great potential to meet the urgent clinical needs for the development of gene therapy [9].

Gene delivery is typically mediated by viral or nonviral vectors [10]. Viral vectors possess the advantages of high transfection efficiency, easy purification, and a strong ability to combine with genes [11]. However, they are usually afflicted with potential immunotoxicity [12], insertional mutagenesis, and limited loading capability, which hinders them from broadly clinical application [13]. On the contrary, the most prominent benefit of nonviral vectors is that they can alleviate the immune deficiency of viral vectors and regulate the loading capability of genes [14]. More importantly, nonviral vectors could be applied for noninvasive administration, making delivery more flexible and feasible, convenient, and safe [15]. However, the transfection efficiency of nonviral vectors is relatively low [16]. Thus, improvement of the ability of nonviral vectors to compress, deliver, and transfect genes via a simple and effective approach would be immensely beneficial to the clinical translation of gene therapy.

Retinoblastoma is an intraocular malignancy mostly affecting infants and children caused by hereditary gene mutations. Current clinical interventions include intravenous, intraarterial, or intravitreal chemotherapy, surgeries, and radiotherapy [17]. Some attempts have also been made to apply nano-drug delivery systems loaded with cytotoxic anti-tumor drugs or genes to treat retinoblastoma [18]. In the present work, we aimed to develop a nonviral vector that would deliver small interference RNA (siRNA) into the eye via topical instillation to manage retinoblastoma. siRNAs are characterized by high molecular weight, abundant negative charges, strong hydrophilicity, and poor stability in vitro and in vivo [19], so a variety of cationic polymers that can compress and neutralize genes are widely utilized to form polyplexes with siRNA through electrostatic interactions [20]. Polyamidoamine (PAMAM) dendrimer is one of the most commonly used gene carriers [21] due to its unique three-dimensional configuration, good biocompatibility, permeability, and stability. The 3rd generation PAMAM (PA) was chosen to pre-compress siRNA herein because of its low toxicity, but correspondingly, its delivery efficacy is relatively weak. Previously, we found that based on wild-type cell-penetrating peptide (CPP) penetratin, more powerful absorption enhancers, for example Q8W and N9W-penetratin (89WP), could be derived for ocular delivery [22]. Hence, 89WP was further employed to modify the primary polyplex composed of siRNA and 3rd generation PAMAM (siRNA/PA), forming a topical delivery system siRNA/PA/89WP to improve absorption of siRNA in the posterior segment of the eye. We recently used wild-type penetratin and 5th generation PAMAM to compress antisense oligonucleotides (ASOs) [23]. In order to modify PAMAM with the positively charged penetratin, negatively charged hyaluronic acid had to be introduced into the complex to implement layer-by-layer self-assembly via electrostatic interaction [24]. In this study, we tried to develop a simpler polyplex by direct modification with the evolved penetratin derivative to achieve more efficient intraocular delivery of siRNA. We hypothesized that 89WP was able to self-assemble with the siRNA/PA polyplex without the involvement of negatively charged hyaluronic acid. The assembling mechanism of this siRNA delivery system was investigated, and the efficacy of gene interference was compared with the commercially available transfection agent Lipofectamine 2000 (Lipo) using an orthotopic retinoblastoma model.

2. Materials and Methods

2.1. Materials

Small interference RNA targeting the exogenetic luciferase gene, whose sense sequence is 5′-GCACUCUGAUUGACAAAUATT-3′ and antisense sequence is 5′-UAUUUGUCAAU CAGAGUGCTT-3′, was synthesized by GenePharma (Jiangsu, China). The 5′ terminus of siRNA was labeled with 5-carboxyfluorescein (siRNA-FAM) and Cy5 (siRNA-Cy5) in the respective experiment. The penetratin-derived peptide 89WP and 5-carboxyfluorescein-labeled 89WP (89WP-FAM) were synthesized by China Peptides (Shanghai, China), whose amino acid sequence was RQIKIWFWWRRMKWKK. The 3rd generation PAMAM (MW 6.9 kDa) was purchased from Sigma-Aldrich (St. Louis, MO, USA). Lipofectamine 2000 (Lipo, Sydney, Australia) was obtained from Thermo Fisher Scientific (Waltham, MA, USA), and luciferin, the substrate of luciferase, was purchased from Sciencelight (Shanghai, China). All other chemicals used were of analytic grade.

2.2. Cell Lines and Animals

Human retinal glial cells (WERI-Rb-1) and retinal pigment epithelium cells (ARPE-19) were provided by the Cell Bank, Chinese Academy of Science (Shanghai, China). WERI-Rb-1-luci cells that express luciferase were built using lentiviral vectors encoding firefly luciferase (luci) to infect WERI-Rb-1 cells [25]. RPMI 1640 Dulbecco's modified Eagle medium containing nutrient mixture F-12 (DMEM), fetal bovine serum (FBS), 0.25% trypsin-EDTA, and penicillin-streptomycin was obtained from Gibco (Carlsbad, CA, USA). Cell counting kit-8 (CCK-8) was purchased from Meilun (Dalian, China). Cell culture plates and flasks were purchased from Cellvis (Mountain View, CA, USA).

Male nude mice (18–20 g) and male ICR mice (18–20 g) were purchased from the Experimental Animal Center of Fudan University and kept under specified conditions. All animal experiments were performed in accordance with the protocols and guidelines approved by the Ethics Committee of Fudan University (2021-06-YL-WG-81). Animals were acclimatized to laboratory conditions for 1 week before the experiments.

2.3. Preparation of siRNA Polyplexes

Preparation of the polyplex siRNA/PA/89WP involved two steps. First, the primary polyplex siRNA/PA was prepared by mixing an equal volume of 130 μg/mL PAMAM solution with 100 μg/mL siRNA solution, followed by vortexing for 30 s and then incubation for 30 min under static conditions. The ratio of nitrogen in PAMAM to phosphorus in the siRNA (N/P) of the primary polyplex in the mixed solution was 2:1, which was optimized by polyacrylamide gel electrophoresis. Second, an equal volume of 89WP solution was added to the siRNA/PA solution at different charge ratios of 89WP to siRNA (5:1, 8:1, 10:1, 15:1, 20:1, and 30:1), followed by vortex and incubation, as described above.

For preparation of the lipoplex siRNA/Lipo, an aliquot of 300 μg/mL siRNA solution was mixed with an equal volume of Lipofectamine 2000 solution according to the instructions, followed by vortexing for 30 s and then incubation for 30 min. All the above operations were performed at room temperature.

2.4. Characterization of siRNA Polyplexes

The particle size and zeta potential of the polyplexes were measured by dynamic light scattering (DLS, Malvern Instruments, Malvern, UK) at room temperature. The size distribution was presented by intensity. Each sample was replicated for 3 times. The refractive index was 1.59, and the detector angle was 173°. The morphology of the polyplexes was observed under a transmission electron microscope (FEI, Hillsboro, OR, USA). Briefly, 5 μL of each sample was added onto glow-discharged carbon-coated grids for 5 min, and then the remaining liquid was dried in an oven at 37 °C. The samples were visualized under a microscope operating at an accelerating voltage of 200 keV in bright-field image mode.

2.5. Cellular Uptake

In qualitative observation, the ARPE-19 cells in the logarithmic growth phase were inoculated in 24-well plates with 20,000 cells in each well ($n = 3$) and incubated for 24 h with 5% CO_2 at 37 °C in a Heraeus incubator (Kendro, San Francisco, CA, USA). The cells were then cultured in serum-free medium with various formulations at a final concentration of 1 µg/mL siRNA. After 4 h incubation, the cells were washed with 0.2% (m/v) heparin sodium solution three times, fixed with 4% (m/v) paraformaldehyde solution for 5 min, and then moisturized with 50% (v/v) glycerol. After that, the cells were then observed under an inverted fluorescence microscope (DMI4000B, LEICA, Germany). FAM fluorescence was excited at a 488 nm wavelength using an argon laser, and the emission was detected at 520 nm.

For quantitative analysis of cellular uptake of different formulations, ARPE-19 cells and WERI-Rb-1 cells in the logarithmic growth phase were inoculated in 12-well plates with 50,000 cells in each well ($n = 3$), as mentioned above. After being incubated with various formulations at a final concentration of 1 µg/mL siRNA, the cells were trypsinized with 0.25% trypsin-EDTA, collected in Eppendorf tubes, and centrifuged at 1000 rpm for 5 min. The precipitated cells were subsequently suspended, washed twice in 200 µL phosphate buffer saline (PBS), and analyzed using a flow cytometer (Beckman, Brea, CA, USA). The percentage of FAM-positive cells in total viable cells was defined as the uptake efficiency. The effects of various inhibitors on cellular uptake were also assessed. The applied concentrations and function mechanisms of the inhibitors are listed in Table S1.

2.6. In Vitro Gene Silencing

WERI-Rb-1-luci cells, which express luciferase in logarithmic growth phase and good condition, were inoculated in 24-well plates with 100,000 cells in each well ($n = 3$) and incubated for 24 h with 5% CO_2 at 37 °C. The cells were then administered various polyplexes at a final concentration of 1 µg/mL siRNA and incubated in serum-free medium for 4 h. After further cultivation in complete medium for 20 h, an equal volume of 0.15 mg/mL luciferin solution was added to each well for bioluminescence observation under an IVIS Spectrum system (Cailper PerkinElemer, Waltham, MA, USA).

2.7. Cell Viability Assay

ARPE-19 cells and WERI-Rb-1 cells in logarithmic growth phase and good condition were inoculated in 96-well plates at a density of 5000 cells per well ($n = 3$), and the edges of the plates were filled with sterile PBS. Following incubation at 37 °C and 5% CO_2 for 24 h, various polyplex formulations were added at a final concentration of 1 µg/mL siRNA and further incubated for 6 h. Afterwards, 10 µL CCK-8 solution was added to each well directly. The cells were incubated for an additional 2 h, and the absorbance value of each well was measured on a micro-plate reader (Bio-Tek, Shoreline, WA, USA) at a detection wavelength of 450 nm. The percentage of cell *viability* was calculated compared with that of untreated cells.

2.8. Isothermal Titration Calorimetry

An aliquot of 200 µL siRNA solution (70 µg/mL) or primary polyplex siRNA/PA solution (160 µg/mL) was filled into the sample cell of isothermal titration calorimeter (MicroCal iTC200, GE, USA), and about 40 µL of PA solution (240 µg/mL) or 89WP solution (2.80 mg/mL) was inhaled into the titration injector, respectively. The duration of each injection was 4 s, and the interval between each injection was 150 s. In order to ensure complete mixing in a few seconds, the injector stirred the solution in the sample cell at a rate of 1000 rpm. The system temperature was set at 25 °C. Titration was carried out first using a sample solution and then pure water as a control. Calorimetric data were analyzed using Origin software.

2.9. Intracellular Co-Localization

ARPE-19 cells in the logarithmic growth phase were seeded in confocal dishes with a density of 5000 cells/well and incubated for 24 h with 5% CO_2 at 37 °C. After administration with various polyplexes at a final concentration of 1 μg/mL siRNA, the cells were further cultured in serum-free medium for 4 h. Then, the cells were washed with 0.2% (m/v) heparin sodium solution three times, fixed with 4% (m/v) paraformaldehyde solution for 5 min, stained with 1 μg/mL 4′,6-diamidino-2-phenylindole (DAPI) solution for 10 min, and moisturized with 50% (v/v) glycerol. After that, the cells were observed with a laser scanning confocal microscope (Carl Zeiss, Jena, Germany).

2.10. In Vivo Retinal Distribution

Twenty mice were randomly divided into four groups, namely control group, naked siRNA group, polyplex siRNA/PA/89WP group, and lipoplex siRNA/Lipo group. In the control group (n = 2), 5 μL normal saline was instilled into the conjunctival sac of the right eyes of mice, while in the treatment groups (n = 6), the same volume of various formulations containing 2 μg siRNA labeled with FAM was applied. After topical administration, mice were sacrificed at 0.5, 1, 2, 4, 6, and 8 h by injection with a lethal dose of pentobarbital sodium (150 mg/kg). The eyeballs were harvested and fixed overnight with 4% (m/v) paraformaldehyde PBS solution, followed by preparation of frozen slices, which were observed with a laser scanning confocal microscope (Carl Zeiss, Germany) and analyzed using ZEN software.

2.11. In Vivo Gene Silencing

Nude mice were anesthetized by intraperitoneal injection with 40 mg/kg pentobarbital sodium solution. The right eye of each mouse was inoculated with 20,000 WERI-Rb-1-luci cells in the logarithmic growth phase by intravitreal injection, and the left eye was set as the control. The bioluminescence intensity of orthotopic retinoblastoma was observed regularly. The day of initial tumorigenesis (the 7th day after inoculation) was recorded as day 0, and on the same day, administration was also started.

Twelve nude mice were randomly divided into 4 groups, namely control group, naked siRNA group, siRNA/PA/89WP group, and siRNA/Lipo group. From day 0, the mice were given 5 μL various formulations containing 1.5 μg siRNA, except for the control group given normal saline, by topical instillation 3 times a day for 15 consecutive days.

On days 0, 3, 5, 7, 9, 11, 13, and 15, the bioluminescence of the inoculated eyes was determined. After intraperitoneal injection with 150 mg/kg luciferin, each mouse was anesthetized with isoflurane and subjected to in vivo bioluminescence imaging. The bioluminescence intensity of the retinoblastoma was semi-quantified to observe the variation. On day 15, after the last bioluminescence determination, all mice were sacrificed, and their right eyeballs were harvested for hematoxylin-eosin staining to observe the integrity of the cornea.

2.12. Statistical Analysis

The statistical significance of the quantitative data was analyzed via multiple comparisons of one-way or two-way ANOVA or t-test corrected by GraphPad Prism software. One asterisk (*) represents a significant difference ($p < 0.05$), two (**), three (***), and four (****) asterisks represent a highly significant difference ($p < 0.01$, $p < 0.001$, and $p < 0.0001$, respectively) and ns represents no significant difference ($p > 0.05$).

3. Results

3.1. Formation and Characterization of siRNA Polyplexes

As seen from the polyacrylamide gel electrophoresis (Figure S1), when the N/P ratio of 3rd generation PAMAM to siRNA was higher than 2:1, PAMAM could basically compress the siRNA and shield the negative charge. Therefore, in the primary polyplex siRNA/PA, the N/P ratio was chosen as 2:1. Then, 89WP was added to the primary polyplex at different ratios, forming siRNA/PA/89WPn, where n represents the charge ratio of 89WP to siRNA. As shown in Figure 1A,B, the primary polyplex siRNA/PA had a particle size of about 60 nm and exhibited electric neutrality. When 89WP was gradually added, both the particle size and zeta potential showed a tendency to increase, but they were lower than those of the lipoplex siRNA/Lipo (about 200 nm and 30 mV, respectively). The fluctuation in particle sizes of different polyplex formulations was mainly affected by two factors: the compression efficiency of siRNA and the composition of the polyplex. With the increase in the charge ratios, more peptide molecules were introduced into the polyplexes to improve the compression of siRNA. Simultaneously, these peptide molecules involved in the formation of the polyplexes also led to a slight increase in particle size. A smaller size means easier internalization by cells [26], and lower zeta potential implies better biocompatibility than siRNA/Lipo. The morphology of polyplexes siRNA/PA, siRNA/PA/P15, and lipoplex siRNA/Lipo observed by TEM was presented as a solid spherical shape, revealing that PAMAM and 89WP could compress siRNA tightly (Figure 1C). The particle sizes observed by TEM were consistent with those measured by DLS, as shown in Figure 1A. The particle size of polyplex siRNA/PA/P15 remained stable at 4 °C for at least 48 h (Figure S2).

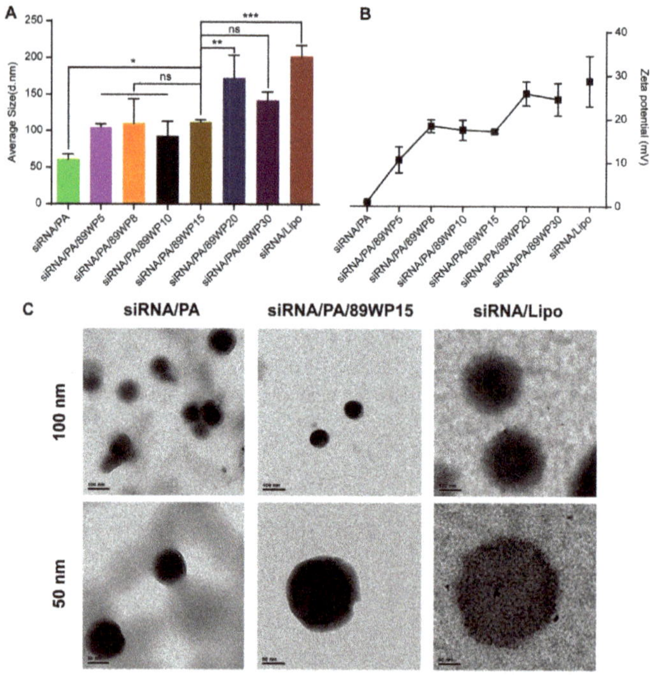

Figure 1. Characterization of siRNA polyplexes. (**A**) The particle sizes of various polyplex formulations measured using DLS. (**B**) Zeta potential of various polyplex formulations. (**C**) Morphology of the polyplexes siRNA/PA, siRNA/PA/P15, and lipoplex siRNA/Lipo observed by TEM. The scale bars are 100 nm for the first row and 50 nm for the second. Data are presented as mean ± SD (n = 3).

3.2. *Optimization of the Ratio of 89WP to siRNA*

Cellular uptake of various formulations was evaluated via qualitative observation using an inverted fluorescence microscope (Figure 2A). It was found that the fluorescence of siRNA-FAM in the co-incubated ARPE-19 cells increased gradually with the proportion of 89WP, and when the charge ratio of 89WP to siRNA ranged between 15:1 and 20:1, the highest cellular uptake occurred. For the cells treated by siRNA/Lipo prepared according to the instructions of Lipofectamine 2000, the fluorescence was almost as bright as the polyplexes containing 89WP, but the cells adhered to each other and looked in poor condition.

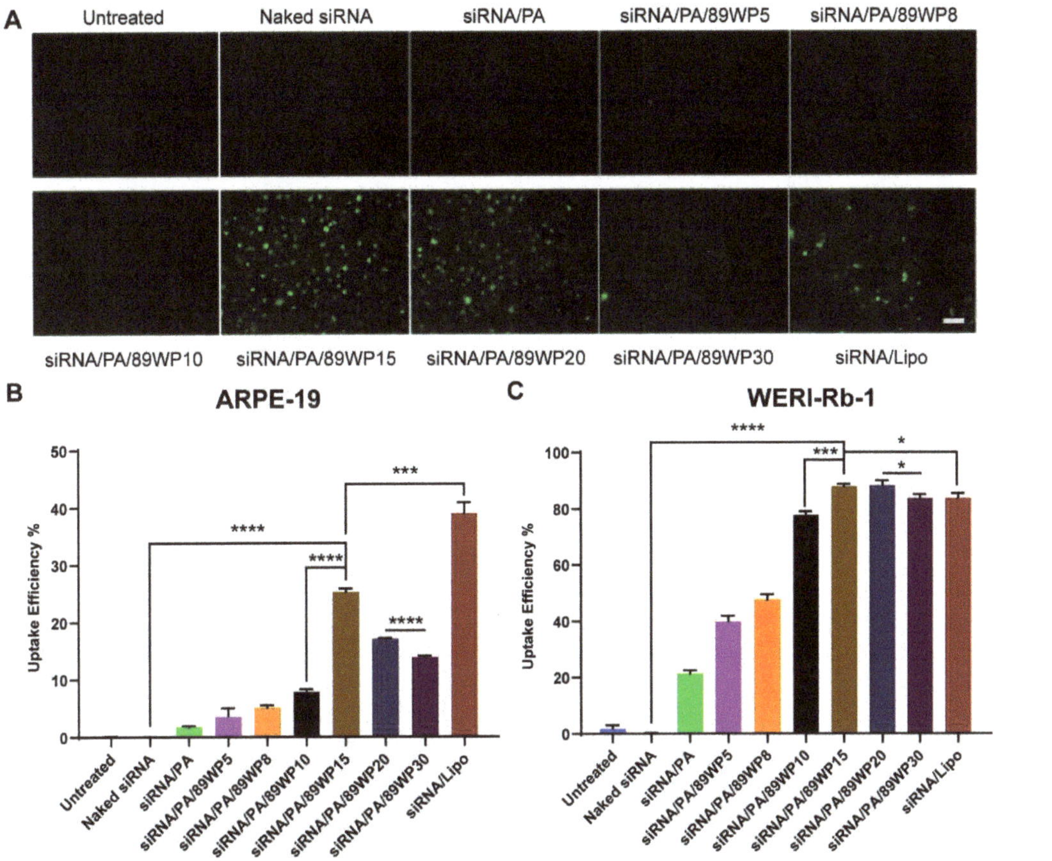

Figure 2. Optimization of the ratio of 89WP to siRNA via qualitative observation and quantitative flow cytometry evaluation. (**A**) The uptake of various siRNA polyplex formulations in ARPE-19 cells was qualitatively observed using an inverted fluorescence microscope (Scale bar, 100 μm). Flow cytometry was used to quantitatively investigate the uptake percentage of the siRNA polyplexes at different charge ratios of 89WP to siRNA in ARPE-19 (**B**) and WERI-Rb-1 (**C**) cells. The numbers behind 89WP in each polyplex formulation represent the charge ratios of 89WP to siRNA. Data are presented as mean ± SD (n = 3).

Flow cytometry was used to quantitatively investigate the uptake efficiency of siRNA polyplexes in normal ARPE-19 cells and tumor WERI-Rb-1 cells. The naked siRNA could barely be internalized alone by both cells (Figure 2B,C) due to its large molecular weight, strong hydrophilicity, and electronegativity. After being compressed by PAMAM, the uptake efficiencies of the primary polyplex siRNA/PA in ARPE-19 and WERI-Rb-1 cells were about 3% and 20%, respectively, which were substantially improved compared to the naked siRNA. With the addition of the peptide 89WP, the uptake efficiency in these two cell lines further increased with the proportion of 89WP, and peaked when the charge ratio of 89WP to siRNA reached 15:1. The uptake efficiency of siRNA/PA/89WP15 in ARPE-19 cells was 8.3 times higher than that of primary polyplex siRNA/PA (Figure 2B). Particularly in WERI-Rb-1 cells, the uptake efficiency of siRNA/PA/89WP15 reached 90%, which was 4.5 times higher than that of primary polyplex siRNA/PA, and was also significantly higher than that of siRNA/Lipo (Figure 2C), indicating that 89WP could greatly promote siRNA internalization. When the charge ratio of 89WP to siRNA further increased, the quantitative cellular uptake trended to decrease slightly, which might reflect the potential cytotoxicity caused by the polyplex formulations. It is worth noting that the uptake efficiencies of all the polyplexes and lipoplexes in the WERI-Rb-1 cells were much higher than in the ARPE-19 cells. We speculate that these differences were mainly due to the inherent endocytic capability of various cell lines to nanoparticles. This result is consistent with a previous report by Patiño et al. [27], who found that tumoral human breast epithelial cells (SKBR-3) mainly internalized positively charged microparticles, with a 3-fold higher efficiency than normal human breast epithelial cells (MCF-10A). In contrast, an opposed cellular uptake effect was observed for the negatively charged microparticles. The authors attributed the completely different endocytic capabilities of the two cell lines to their tumorigenic or non-tumorigenic nature. Since our polyplexes were all positively charged, they were more apt to be endocytosed by tumor cell WERI-Rb-1.

Bioluminescence imaging was conducted for qualitative (Figure 3A) and semi-quantitative (Figure 3B) evaluations of the interference effects on luciferase expression in WERI-Rb-1-luci cells using various siRNA polyplexes. No gene interference was observed in the cells treated with naked siRNA or the primary polyplex siRNA/PA. When the charge ratio of 89WP to siRNA reached 10:1, the expression of luciferase was basically inhibited (Figure 3B), which further proved the absorption-enhancing effect of 89WP.

CCK-8 was used to investigate the toxicity of various formulations to ARPE-19 and WERI-Rb-1 cells (Figure 3C,D). It was found that when the charge ratio of 89WP to siRNA was lower than 20:1, the survival rate of both cells was higher than 80%, and the biosafety of these formulations was acceptable. In the normal cell ARPE-19, there was no significant difference in the viabilities between the cells treated with polyplex siRNA/PA/89WP15 and the negative control naked siRNA, indicating that the carrier materials 3rd generation PAMAM and 89WP at this dose did not cause significant toxicity to the cells. In contrast, the positive control siRNA/Lipo was highly toxic to both cells, which reduced cell viability to below 70%, and the toxicity was significantly higher than that of siRNA/PA/89WP15. Considering the internalization, interference effect, and toxicity, siRNA/PA/89WP15 was chosen as the final polyplex formulation, also known as siRNA/PA/89WP hereafter.

3.3. Interactions between Components of siRNA Polyplexes

Isothermal titration calorimetry (ITC) was implemented to determine the interactions between PAMAM and siRNA and between 89WP and primary polyplex siRNA/PA. The ITC titration curves are shown in Figure 4, and the fitted thermodynamic parameters are presented in Table 1. When PAMAM was mixed with siRNA, the equilibrium binding constant (K_b) value was $3.63 \pm 1.25 \times 10^7$ M^{-1}, accompanied by an exothermic process ($\Delta H < 0$) and reduced entropy ($\Delta S < 0$). The corresponding Gibbs free energy change (ΔG), calculated via the equation $\Delta G = \Delta H - T\Delta S$, was also negative. Because under neutral pH, PA is positively charged, while siRNA is negatively charged, they could form stable primary polyplexes under room temperature due to electrostatic interaction. The thermodynamic parameters revealed that the mixed system became more orderly and that the primary polyplexes were spontaneously formed.

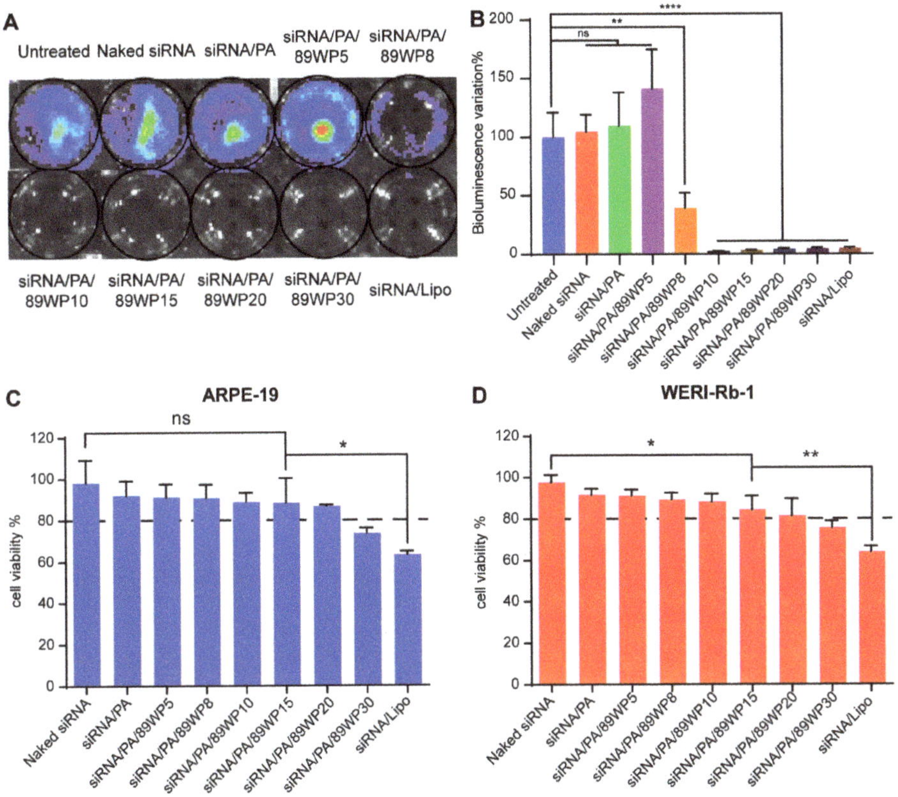

Figure 3. Optimization of the ratio of 89WP to siRNA by in vitro gene silencing effect and cell viability. Bioluminescence imaging was conducted to (**A**) qualitatively and (**B**) semi-quantitatively evaluate the interference effect of luciferase expression in WERI-Rb-1-luci cells using different siRNA polyplex formulations. Data were presented as mean ± SD (*n* = 4), and significance was compared with the untreated group. CCK-8 was used to evaluate the cell viability of different formulations to (**C**) ARPE-19 and (**D**) WERI-Rb-1 cells. Data were presented as mean ± SD (*n* = 3), and significance was compared with siRNA/PA/89WP15 group.

Table 1. Thermodynamic parameters of siRNA binding to PAMAM and 89WP binding to siRNA/PA.

Sample	K_b ($\times 10^5$ M^{-1})	ΔH (kcal/mol)	TΔS (kcal/mol)	ΔG (kcal/mol)
PA + siRNA	363 ± 125	−71.93 ± 2.02	−61.69	−10.24 ± 2.02
89WP + siRNA/PA	1.85 ± 0.90	1.04 ± 0.08	8.22	−7.18 ± 0.08

However, the ITC titration curve of 89WP and siRNA/PA in Figure 4B showed a completely different tendency from that of PA and siRNA as shown in Figure 4A. According to the corresponding thermodynamic parameters in Table 1, the K_b value was $1.85 \pm 0.90 \times 10^5$ M^{-1}, which was two orders of magnitude lower than that calculated from PA and siRNA, but there was still interaction between 89WP and siRNA/PA. The mixing process of 89WP and siRNA/PA was endothermic (ΔH > 0) with increased entropy change (ΔS > 0), suggesting that in addition to the formation of a polyplex, there might be some free 89WP in the system. The calculated Gibbs free energy change was also negative (ΔG < 0) and had a similar value to that obtained during formation of the primary polyplex siRNA/PA. These results indicated that 89WP and siRNA/PA could spontaneously form stable polyplexes at room temperature, perhaps via an entropy-driven process characterized by hydrophobic interaction [28].

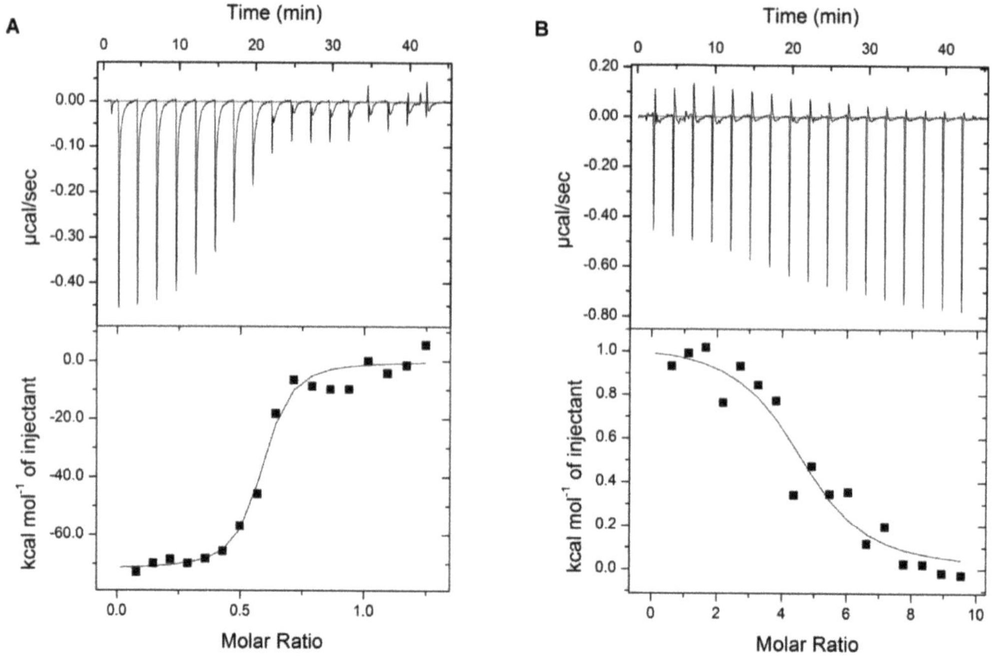

Figure 4. Interactions between the components of siRNA polyplexes. Isothermal titration curves for the interaction between siRNA and PAMAM (**A**), and for the interaction between 89WP and siRNA/PA (**B**).

3.4. Integrity of siRNA Polyplexes during Cellular Uptake

Laser confocal microscopy was used to investigate the co-localization of siRNA and 89WP after the polyplexes were internalized by ARPE-19 cells. As shown in Figure 5, the cell nuclei stained with DAPI are blue, the siRNA labeled with Cy5 is red, and the 89WP labeled with FAM is green. Naked siRNA-Cy5 could not enter the cells; however, when compressed by PAMAM and forming primary polyplex siRNA-Cy5/PA, a low level of internalization could be observed. These results are consistent with observations in cellular uptake studies. In contrast, the peptide 89WP-FAM could be well internalized by the cells and distributed around the nuclei, demonstrating that 89WP itself has strong cellular permeability. Moreover, both siRNA-Cy5/PA/89WP and siRNA/PA/89WP-FAM were labeled with single fluorescent dye, and could be well internalized by the cells, indicating that 89WP could not only penetrate the cell membrane barrier by itself, but also facilitate the polyplexes to enter the cells. In the cells treated with double-labeled polyplex siRNA-Cy5/PA/89WP-FAM, the yellow particles were the co-localized siRNA-Cy5 and 89WP-FAM, illustrating that 89WP and siRNA co-existed in the polyplex, which could enter the cells in its entirety. The remaining green particles were free 89WP, further proving that there are not only intact polyplexes but also free 89WP in the system, as revealed by the ITC determination. For the lipoplex siRNA-Cy5/Lipo, visible cellular internalization also occurred, and the uptake efficiency was comparable to that of siRNA-Cy5/PA/89WP.

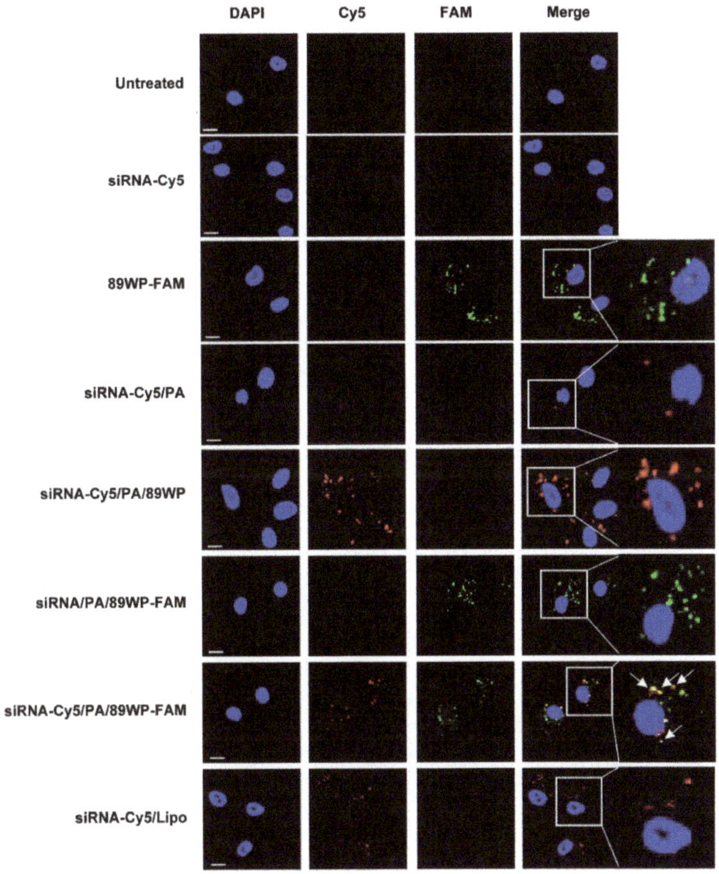

Figure 5. Co-localization of the polyplex components in cells. The siRNA labeled with Cy5 (siRNA-Cy5, red) and 89WP labeled with FAM (89WP-FAM, green) formed polyplexes siRNA-Cy5/PA/89WP, siRNA/PA/89WP-FAM, siRNA-Cy5/PA/89WP-FAM, and lipoplex siRNA-Cy5/Lipo. The distribution and co-localization (yellow) of siRNA-Cy5 and 89WP-FAM in ARPE-19 cells were observed under a confocal microscope. Scale bar, 20 μm.

3.5. Cellular Uptake Pathway of siRNA Polyplexes

Flow cytometry diagrams of WERI-Rb-1 and ARPE-19 cells internalizing siRNA polyplexes, in which siRNA was labeled with Cy5 and/or 89WP was labeled with FAM, are shown in Figure 6A and Figure S3A, respectively. In each diagram, the negative quadrant in the lower left corner represents those cells without uptake, the Cy5 quadrant in the upper left corner represents those cells internalizing siRNA-Cy5 only, the FAM quadrant in the lower right corner represents those cells internalizing 89WP-FAM only, and the Cy5 & FAM quadrant in the upper right corner represents those cells simultaneously internalizing both 89WP-FAM and siRNA-Cy5. It could be seen that the cells treated with single fluorescence-labeled polyplexes siRNA-Cy5/PA/89WP and siRNA/PA/89WP-FAM were distributed in their respective quadrants in the flow cytometry diagrams, indicating that the fluorescence signals of Cy5 and FAM dyes did not interfere with each other. In the cells treated with siRNA-Cy5/PA/89WP-FAM, the proportion of Cy5 and FAM double-positive cells reached more than 98% for ARPE-19 cells and 84% for WERI-Rb-1 cells, indicating that siRNA and 89WP almost simultaneously entered the same cells, implying that the polyplexes were internalized intactly.

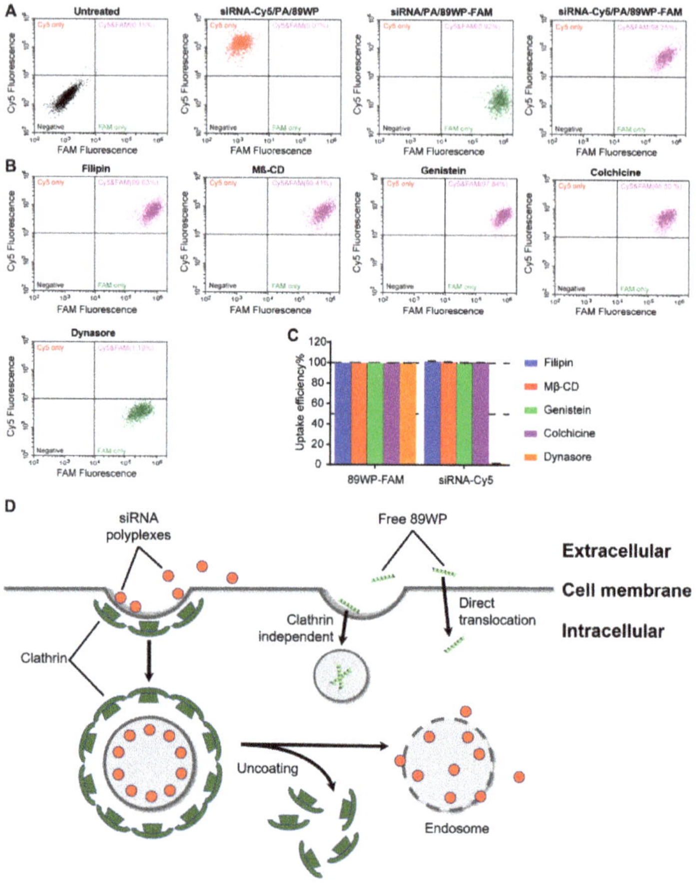

Figure 6. Cellular uptake pathway of siRNA polyplexes in ARPE-19 cells. (**A**) Flow cytometry diagrams of ARPE-19 cells treated with polyplexes siRNA-Cy5/PA/89WP, siRNA/PA/89WP-FAM, and siRNA-Cy5/PA/89WP-FAM, respectively. (**B**) Flow cytometry diagrams of ARPE-19 cells treated with polyplex siRNA-Cy5/PA/89WP-FAM in the presence of various cellular uptake inhibitors. (**C**) Effects of the inhibitors on the uptake efficiency of siRNA-Cy5/PA/89WP-FAM by ARPE-19 cells (n = 3). (**D**) Endocytosis pathway of the polyplex siRNA/PA/89WP mediated by clathrin.

The uptake efficiency was defined as the percentage of cellular uptake with an inhibitor compared to that without an inhibitor. When treated with the uptake inhibitors filipin, Mß-CD, genistein, and colchicine, no significant decrease in uptake efficiencies was observed in either ARPE-19 (Figure 6B,C) or WERI-Rb-1 (Figure S3B,C) cells, indicating that the main pathway for cellular uptake of the polyplex siRNA/PA/89WP might involve neither cholesterol- nor caveolin-mediated endocytosis nor pinocytosis. Although the uptake efficiencies of 89WP-FAM in both cells treated with siRNA-Cy5/PA/89WP-FAM remained unaffected in the presence of dynasore, those of siRNA-Cy5 were almost completely inhibited. Accordingly, the cellular uptake of polyplex siRNA/PA/89WP may be mainly via clathrin-mediated endocytosis (Figure 6D). In addition to the intact polyplex, free 89WP also existed in the delivery system, which was virtually not influenced by dynasore, implying that it was internalized via different pathways.

The uptake of fluorescence-labeled peptide 89WP-FAM and primary polyplex siRNA-Cy5/PA in ARPE-19 and WERI-Rb1 cells revealed that these inhibitors did not affect cellular uptake of the peptide alone, but that of polyplex siRNA-Cy5/PA was significantly restrained (Figure S4). This result further confirmed that the polyplexes with or without 89WP modification entered cells mainly through clathrin-mediated endocytosis, which was different from the free peptide 89WP.

3.6. In Vivo Retinal Distribution of siRNA Polyplexes

After topical instillation of the siRNA formulations labeled with FAM, the fluorescence intensity in the mouse retina was kept at a low level within 8 h in the naked siRNA group, indicating that the naked siRNA alone was difficult to absorb into the eye due to its hydrophilicity and relatively large molecular size. In contrast, the retinal fluorescence intensity of siRNA-FAM/PA/89WP and siRNA-FAM/Lipo groups increased remarkably and peaked at 2 h after administration. Then, the fluorescence intensity decreased gradually with time, and 6 h later, it was comparable with that of the siRNA-FAM group (Figure 7A). According to quantitative analysis (Figure 7B), the fluorescence intensity of siRNA-FAM/PA/89WP group was 2 to 3 times higher than that of the naked siRNA-FAM group at 0.5 h, 1 h, 2 h, and 4 h, and 1.5 times higher ($p < 0.0001$) than that of the siRNA-FAM/Lipo group at 2 h and 4 h. These results provided direct evidence that due to the ocular permeability of 89WP, siRNA polyplex could be absorbed into the eye and distributed in the retina, and the amount of siRNA accumulated in the ocular fundus was significantly higher than the lipoplex siRNA/Lipo.

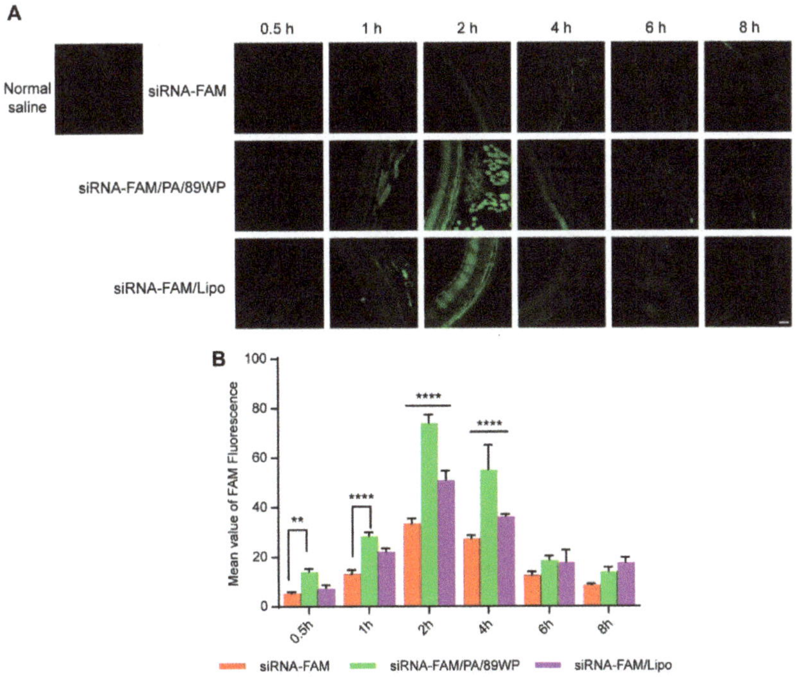

Figure 7. In vivo retinal distribution of siRNA formulations. All fluorescence-labeled naked siRNA-FAM, polyplex siRNA-FAM/PA/89WP, and lipoplex siRNA-FAM/Lipo contained 2 μg siRNA and were instilled into the conjunctival sac of mice. After the eyeballs were harvested at different time points, the green fluorescence intensity in the retina was observed under a confocal fluorescence microscope (**A**) and semi-quantitatively analyzed using ZEN software (**B**). Data are presented as mean ± SD ($n = 3$). Scale bar, 50 μm. Significance was compared with siRNA-FAM/PA/89WP group.

3.7. In Vivo Gene Silencing Efficacy

The siRNA formulations were administered as eye drops three times a day to evaluate gene silencing efficiency in retinoblastoma-bearing nude mice, and the eyes were subjected to in vivo bioluminescence imaging at set time points (Figure 8A). From Figure 8B, it could be found that the bioluminescence intensity increased significantly for the groups treated with normal saline and naked siRNA, indicating rapid proliferation of WERI-Rb-1-luci cells in the eyes. In contrast, the polyplex siRNA/PA/89WP could effectively interfere with luciferase gene expression of the intraocular WERI-Rb-1-luci cells, thereby inhibiting bioluminescence intensity at a low level similar to that at the beginning of tumorigenesis (day 0). Semi-quantitative analysis of the change in bioluminescence intensity was also conducted. The data showed that on day 15, the inhibition effect of siRNA/PA/89WP on tumor bioluminescence was 4 times better compared to that of the siRNA/Lipo group ($p < 0.05$), or 5 times and 6 times better compared to those of the groups treated with normal saline ($p < 0.001$) and naked siRNA ($p < 0.01$), respectively (Figure 8C,D). These results revealed that the polyplex siRNA/PA/89WP could efficiently penetrate across the ocular absorption barriers and delivered siRNA into the tumor cells in the posterior segment of the eye. After consecutive administration for 15 days, the corneas recovered from all the treatment groups still maintained normal morphology (Figure 8E), indicating that the siRNA formulations were safe for ocular tissues.

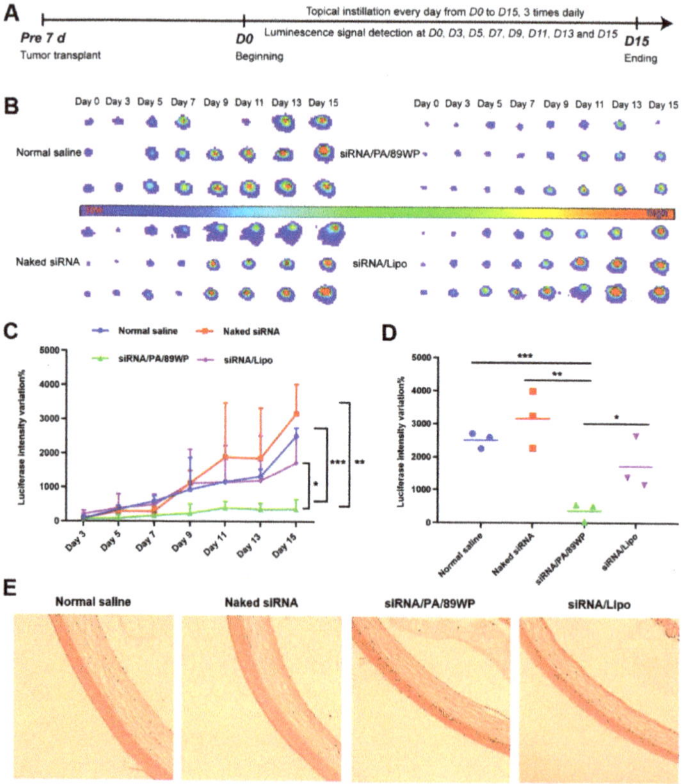

Figure 8. In vivo gene silencing efficacy. The eyes of nude mice were inoculated with WERI-Rb-1-luci cells, and the mice were divided into 4 groups, administered with normal saline, naked siRNA, siRNA/PA/89WP, and siRNA/Lipo, respectively. Each siRNA formulation contained 1.5 µg siRNA and was applied three times daily via topical instillation. (**A**) Time schedule of tumor transplant, treatment, and evaluation of tumor-bearing mice. (**B**) Bioluminescence imaging of the eyes. (**C**) Semi-

quantitative evaluation of changes in luciferase expression. (**D**) Comparison of changes in luciferase expression on day 15. (**E**) The HE-stained cornea after administration for 15 days. Significance was compared with siRNA/PA/89WP group, and $n = 3$.

4. Discussion

Ocular gene therapy, through altering gene expression to treat genetic diseases, has great potential for a variety of degenerative retinal syndromes [29]. Currently, there are still some technical barriers in this field that urgently need to be solved, such as seeking effective therapeutic genes and developing efficient delivery vectors [19,30,31]. Although ocular gene therapy has made remarkable progress in recent years, it is still in its infancy, and vast amounts of research are in the exploratory stages. Notably, in gene delivery, the efficiency and safety of most delivery vectors cannot meet clinical requirements, which has become a technical bottleneck of gene therapy [5].

In the present study, we developed an ocular gene delivery system consisting of a cationic polymer (PAMAM) and an optimized cell-penetrating peptide (89WP). PAMAM is a synthetic dendritic polymer rich in cationic functional groups that provides a reliable binding ability with nucleic acids [32]. The binding ability increases along with the generation of dendritic molecules [33], while cytotoxicity also increases [34,35]. In order to balance the capability of gene condensation and the potential cytotoxicity caused by the vector, the 3rd generation PAMAM was selected in this work to compress siRNA and form the primary polyplex siRNA/PA by simple physical complexation.

We previously screened out penetratin from several of the most commonly reported cell-penetrating peptides based on the ex vivo permeability in excised cornea and in vivo ocular distribution [36]. In a subsequent study, we optimized wild-type penetratin via amino acid mutation to further improve its permeability, and obtained a series of derivative peptides, which could be used as potential ocular absorption enhancers with low toxicity [22]. Among these derivatives, 89WP showed more distribution in the eyes, especially in the retina, after topical instillation. Therefore, we chose 89WP to facilitate the intraocular delivery of siRNA. It seemed that the N/P ratio of 2:1 was a critical value for the 3rd generation PAMAM to compress siRNA, because the formed primary polyplex siRNA/PA exhibited a nearly neutral zeta potential (Figure 1B), revealing the negatively charged siRNA was just neutralized by PAMAM. Actually, there were still some unbound or loosely bonded siRNAs in the mixed solution, according to the results of gel electrophoresis (Figure S1). Under this circumstance, the addition of positively charged 89WP would further compress siRNA together with PAMAM, forming ternary complexes with larger particle sizes and higher zeta potential compared to siRNA/PA. To modify the polyplex with penetratin, we previously introduced low molecular weight hyaluronic acid (HA) as an electronegative linker between the positively charged polyplex and penetratin to implement layer-by-layer self-assembly via electrostatic interaction [24]. In the present work, by virtue of thermodynamic analysis, we found that 89WP and siRNA/PA could spontaneously form a polyplex without the help of HA. This finding makes the siRNA polyplex simpler in structure and easier in preparation.

Both ARPE-19 and WERI-Rb-1 are retina-related cells, and here were chosen to predict the absorption of polyplex formulations in the posterior segment of eyes. In vitro evaluations showed that the polyplexes exhibited a much stronger uptake and silencing effect in both normal cells ARPE-19 and tumor cells WERI-Rb-1. With the aid of PAMAM and 89WP, the polyplexes were able to enter the cells through the clathrin-mediated pathway. Interestingly, the internalization process of the polyplex was different from that of peptide 89WP, which was not liable to be affected by uptake inhibitors, probably due to their direct binding and interaction with negatively charged cell membranes [37,38].

The interaction between the components of polyplex siRNA/PA/89WP and its integrity after cellular uptake was also investigated. The thermodynamic parameters revealed that a strong electrostatic interaction between PA and siRNA facilitated the formation of the primary polyplex. In contrast, when 89WP was mixed with siRNA/PA, entropy-driven hydrophobic interaction played the leading role, as indicated by negative ΔG and positive ΔS [28], consequently resulting in a system containing polyplex siRNA/PA/89WP and free peptide 89WP. This could explain why 89WP-FAM was still internalized in the cells when inhibited by dynasore. The confocal results showed that siRNA overlapped with 89WP in the cells, indicating that the polyplexes were internalized as an intact form and entered the cells via the clathrin-mediated endocytosis pathway. Besides the intact polyplexes, there was some free 89WP in the formulation. Even if the internalization of polyplex siRNA/PA/89WP was suppressed by dynasore, the free 89WP might enter cells through direct bonding with a negatively charged cytomembrane.

Fluorescence-labeled polyplexes were then instilled in the eyes of the mice. We found that after 2 h, the fluorescence intensity of polyplex siRNA-FAM/PA/89WP reached the maximum in the retina, then began to gradually attenuate, and almost totally eliminated after 6 h. The ocular pharmacokinetics revealed that the eye drops of siRNA polyplexes could be absorbed into the posterior segment within a short time after administration. Consequently, bioluminescence expressed in WERI-Rb-1-luci tumor cells could be effectively inhibited by topically applied siRNA/PA/89WP, and the interference effect was significantly better than that of the lipoplex siRNA/Lipo. The tumor-inhibiting efficiency was also noninferior to the polyplex constructed with the 5th generation PAMAM and wild-type penetratin [23], which may be attributed to the improved permeability of 89WP across the ocular absorption barriers and its hydrophobic interaction with siRNA/PA. More importantly, the present polyplex was composed of lower generation PAMAM, which was favorable for reducing the potential risk of safety for ocular application. These results provided solid proof that via noninvasive administration siRNA polyplex could distribute in the fundus oculi and therefore was competent to play the role of gene therapy in related diseases.

5. Conclusions

In this work, we successfully constructed a noninvasive intraocular gene delivery system, in which the 3rd generation PAMAM dendrimer compressed siRNA to form the primary polyplex siRNA/PA and then the penetratin derivative 89WP was introduced to self-assemble the ternary complex system targeting fundus oculi. The formed polyplex siRNA/PA/89WP was able to enter the ocular cells in an intact form through clathrin-mediated endocytosis, and was much safer to ocular cells than the siRNA lipoplex composed of commercialized cationic liposome. When topically applied in vivo, the polyplex siRNA/PA/89WP could reach the posterior segment of eyes in a short time, and perform an interference effect on the intraocular tumor without impairing anterior ocular tissues. Therefore, this simple and safe delivery system could be a useful tool for gene therapy for hereditary intraocular diseases.

Supplementary Materials: The following supporting information can be downloaded at: https://www.mdpi.com/article/10.3390/pharmaceutics15030745/s1. Table S1. Cellular uptake inhibitors and their function mechanisms. Figure S1. Polyacrylamide gel electrophoresis of primary polyplexes siRNA/PA. The ratio is N/P of 3rd generation PAMAM to siRNA. Figure S2. Time-dependent particle size variation of siRNA/PA/89WP incubated in 10 mM PBS at 4 °C (n = 3). Figure S3. Cellular uptake pathway of the siRNA polyplexes in WERI-Rb-1 cells. (A) Flow cytometry diagrams of WERI-Rb-1 cells treated with polyplexes siRNA-Cy5/PA/89WP, siRNA/PA/89WP-FAM, and siRNA-Cy5/PA/89WP-FAM, respectively. (B) Flow cytometry diagrams of WERI-Rb-1 cells treated with polyplex siRNA-Cy5/PA/89WP-FAM in the presence of various cellular uptake inhibitors. (C) Effects of the inhibitors on uptake efficiency of siRNA-Cy5/PA/89WP-FAM by WERI-Rb-1 cells (n = 3). Figure S4. Effects of various inhibitors on uptake efficiency of single fluorescence-labeled peptide (89WP-FAM) and primary polyplex (siRNA-Cy5/PA) in ARPE-19 (A) and WERI-Rb-1 (B) cells (n = 3).

Author Contributions: Conceptualization, G.W. and W.L.; methodology, G.W., X.F., X.G., K.J. and Y.H.; formal analysis, K.J. and Y.L.; investigation, X.G., X.F., K.J. and Y.H.; resources, G.W.; data curation, X.G. and X.F.; writing—original draft, X.G.; writing—review & editing, G.W., X.F., Y.L. and W.L.; visualization, X.F., supervision, G.W.; project administration, G.W. and W.L.; funding acquisition, G.W. All authors have read and agreed to the published version of the manuscript.

Funding: This research was funded by the Shanghai Science and Technology Program (21ZR1407100 and 21S11905300), the National Natural Science Fund of China (82273864 and 81573358), and the Development Project of Shanghai Peak Disciplines-Integrative Medicine (20180101).

Institutional Review Board Statement: The animal study protocol was approved by the Ethics Committee of Fudan University (protocol code 2021-06-YL-WG-81 and date of approval: June 2021).

Informed Consent Statement: Not applicable.

Data Availability Statement: All relevant data is contained in the article.

Acknowledgments: The isothermal titration calorimetry experiment platform was provided by National Facility for Protein Science in Shanghai Zhangjiang Lab.

Conflicts of Interest: The authors declare no conflict of interest.

References

1. Bourne, R.; Steinmetz, J.D.; Flaxman, S.; Briant, P.S.; Taylor, H.R.; Resnikoff, S.; Casson, R.J.; Abdoli, A.; Abu-Gharbieh, E.; Afshin, A.; et al. Trends in prevalence of blindness and distance and near vision impairment over 30 years: An analysis for the Global Burden of Disease Study. *Lancet Glob. Health* **2021**, *9*, e130–e143. [CrossRef]
2. Lloyd, A.; Piglowska, N.; Ciulla, T.; Pitluck, S.; Johnson, S.; Buessing, M.; O'Connell, T. Estimation of impact of RPE65-mediated inherited retinal disease on quality of life and the potential benefits of gene therapy. *Br. J. Ophthalmol.* **2019**, *103*, 1610–1614. [CrossRef] [PubMed]
3. Prado, D.A.; Acosta-Acero, M.; Maldonado, R.S. Gene therapy beyond luxturna: A new horizon of the treatment for inherited retinal disease. *Curr. Opin. Ophthalmol.* **2020**, *31*, 147–154. [CrossRef] [PubMed]
4. Garoon, R.B.; Stout, J.T. Update on ocular gene therapy and advances in treatment of inherited retinal diseases and exudative macular degeneration. *Curr. Opin. Ophthalmol.* **2016**, *27*, 268–273. [CrossRef] [PubMed]
5. Solinis, M.A.; del Pozo-Rodriguez, A.; Apaolaza, P.S.; Rodriguez-Gascon, A. Treatment of ocular disorders by gene therapy. *Eur. J. Pharm. Biopharm.* **2015**, *95*, 331–342. [CrossRef] [PubMed]
6. Falavarjani, K.G.; Nguyen, Q.D. Adverse events and complications associated with intravitreal injection of anti-VEGF agents: A review of literature. *Eye* **2013**, *27*, 787–794. [CrossRef]
7. Huang, M.; Zhang, M.; Zhu, H.; Du, X.; Wang, J. Mucosal vaccine delivery: A focus on the breakthrough of specific barriers. *Acta Pharm. Sin. B* **2022**, *12*, 3456–3474. [CrossRef]
8. Darrow, J.J. Luxturna: FDA documents reveal the value of a costly gene therapy. *Drug Discov. Today* **2019**, *24*, 949–954. [CrossRef]
9. Shan, X.; Gong, X.; Li, J.; Wen, J.; Li, Y.; Zhang, Z. Current approaches of nanomedicines in the market and various stage of clinical translation. *Acta Pharm. Sin. B* **2022**, *12*, 3028–3048. [CrossRef]
10. Zakeri, A.; Kouhbanani, M.A.J.; Beheshtkhoo, N.; Beigi, V.; Mousavi, S.M.; Hashemi, S.A.R.; Karimi Zade, A.; Amani, A.M.; Savardashtaki, A.; Mirzaei, E.; et al. Polyethylenimine-based nanocarriers in co-delivery of drug and gene: A developing horizon. *Nano Rev. Exp.* **2018**, *9*, 1488497. [CrossRef]
11. Zhang, T.; Guo, W.; Zhang, C.; Yu, J.; Xu, J.; Li, S.; Tian, J.H.; Wang, P.C.; Xing, J.F.; Liang, X.J. Transferrin-Dressed Virus-like Ternary Nanoparticles with Aggregation-Induced Emission for Targeted Delivery and Rapid Cytosolic Release of siRNA. *ACS Appl. Mater. Interfaces* **2017**, *9*, 16006–16014. [CrossRef]
12. Reichel, F.F.; Dauletbekov, D.L.; Klein, R.; Peters, T.; Ochakovski, G.A.; Seitz, I.P.; Wilhelm, B.; Ueffing, M.; Biel, M.; Wissinger, B.; et al. AAV8 Can Induce Innate and Adaptive Immune Response in the Primate Eye. *Mol. Ther.* **2017**, *25*, 2648–2660. [CrossRef] [PubMed]
13. Campbell, J.P.; McFarland, T.J.; Stout, J.T. Ocular Gene Therapy. *Dev. Ophthalmol.* **2016**, *55*, 317–321. [PubMed]
14. Yan, Y.; Liu, X.Y.; Lu, A.; Wang, X.Y.; Jiang, L.X.; Wang, J.C. Non-viral vectors for RNA delivery. *J. Control. Release* **2022**, *342*, 241–279. [CrossRef]
15. Liu, J.; Shui, S.L. Delivery methods for site-specific nucleases: Achieving the full potential of therapeutic gene editing. *J. Control. Release* **2016**, *244*, 83–97. [CrossRef] [PubMed]
16. Xu, L.; Anchordoquy, T. Drug delivery trends in clinical trials and translational medicine: Challenges and opportunities in the delivery of nucleic acid-based therapeutics. *J. Pharm. Sci.* **2011**, *100*, 38–52. [CrossRef]
17. Ancona-Lezama, D.; Dalvin, L.A.; Shields, C.L. Modern treatment of retinoblastoma: A 2020 review. *Indian J. Ophthalmol.* **2020**, *68*, 2356–2365.

18. Bhavsar, D.; Subramanian, K.; Sethuraman, S.; Krishnan, U.M. Management of retinoblastoma: Opportunities and challenges. *Drug Deliv.* **2016**, *23*, 2488–2496. [CrossRef]
19. Wang, Y.; Ji, X.; Ruan, M.; Liu, W.; Song, R.; Dai, J.; Xue, W. Worm-Like Biomimetic Nanoerythrocyte Carrying siRNA for Melanoma Gene Therapy. *Small* **2018**, *14*, e1803002. [CrossRef]
20. Dai, J.; Zou, S.; Pei, Y.; Cheng, D.; Ai, H.; Shuai, X. Polyethylenimine-grafted copolymer of poly(l-lysine) and poly(ethylene glycol) for gene delivery. *Biomaterials* **2011**, *32*, 1694–1705. [CrossRef]
21. Abedi-Gaballu, F.; Dehghan, G.; Ghaffari, M.; Yekta, R.; Abbaspour-Ravasjani, S.; Baradaran, B.; Dolatabadi, J.E.N.; Hamblin, M.R. PAMAM dendrimers as efficient drug and gene delivery nanosystems for cancer therapy. *Appl. Mater. Today* **2018**, *12*, 177–190. [CrossRef] [PubMed]
22. Jiang, K.; Gao, X.; Shen, Q.; Zhan, C.; Zhang, Y.; Xie, C.; Wei, G.; Lu, W. Discerning the composition of penetratin for safe penetration from cornea to retina. *Acta Biomater.* **2017**, *63*, 123–134. [CrossRef]
23. Tai, L.; Liu, C.; Jiang, K.; Chen, X.; Wei, G.; Lu, W.; Pan, W. Noninvasive delivery of oligonucleotide by penetratin-modified polyplexes to inhibit protein expression of intraocular tumor. *Nanomedicine* **2017**, *13*, 2091–2100. [CrossRef] [PubMed]
24. Tai, L.; Liu, C.; Jiang, K.; Chen, X.; Feng, L.; Pan, W.; Wei, G.; Lu, W. A novel penetratin-modified complex for noninvasive intraocular delivery of antisense oligonucleotides. *Int. J. Pharm.* **2017**, *529*, 347–356. [CrossRef] [PubMed]
25. Jiang, K.; Hu, Y.; Gao, X.; Zhan, C.; Zhang, Y.; Yao, S.; Xie, C.; Wei, G.; Lu, W. Octopus-like Flexible Vector for Noninvasive Intraocular Delivery of Short Interfering Nucleic Acids. *Nano Lett.* **2019**, *19*, 6410–6417. [CrossRef]
26. Choi, J.S.; Cao, J.; Naeem, M.; Noh, J.; Hasan, N.; Choi, H.K.; Yoo, J.W. Size-controlled biodegradable nanoparticles: Preparation and size-dependent cellular uptake and tumor cell growth inhibition. *Colloids Surf. B Biointerfaces* **2014**, *122*, 545–551. [CrossRef]
27. Patino, T.; Soriano, J.; Barrios, L.; Ibanez, E.; Nogues, C. Surface modification of microparticles causes differential uptake responses in normal and tumoral human breast epithelial cells. *Sci. Rep.* **2015**, *5*, 11371. [CrossRef]
28. Varese, M.; Guardiola, S.; Garcia, J.; Giralt, E. Enthalpy- versus Entropy-Driven Molecular Recognition in the Era of Biologics. *Chembiochem* **2019**, *20*, 2981–2986. [CrossRef] [PubMed]
29. Rodrigues, G.A.; Shalaev, E.; Karami, T.K.; Cunningham, J.; Slater, N.K.H.; Rivers, H.M. Pharmaceutical Development of AAV-Based Gene Therapy Products for the Eye. *Pharm. Res.* **2018**, *36*, 29. [CrossRef]
30. Wang, Y.; Mo, L.; Wei, W.; Shi, X. Efficacy and safety of dendrimer nanoparticles with coexpression of tumor necrosis factor-alpha and herpes simplex virus thymidine kinase in gene radiotherapy of the human uveal melanoma OCM-1 cell line. *Int. J. Nanomed.* **2013**, *8*, 3805–3816. [CrossRef]
31. Vasconcelos, A.; Vega, E.; Perez, Y.; Gomara, M.J.; Garcia, M.L.; Haro, I. Conjugation of cell-penetrating peptides with poly(lactic-co-glycolic acid)-polyethylene glycol nanoparticles improves ocular drug delivery. *Int. J. Nanomed.* **2015**, *10*, 609–631.
32. Shcharbin, D.; Shakhbazau, A.; Bryszewska, M. Poly(amidoamine) dendrimer complexes as a platform for gene delivery. *Expert Opin. Drug Deliv.* **2013**, *10*, 1687–1698. [CrossRef] [PubMed]
33. Pavan, G.M.; Posocco, P.; Tagliabue, A.; Maly, M.; Malek, A.; Danani, A.; Ragg, E.; Catapano, C.V.; Pricl, S. PAMAM dendrimers for siRNA delivery: Computational and experimental insights. *Chemistry* **2010**, *16*, 7781–7795. [CrossRef] [PubMed]
34. Shao, N.; Su, Y.; Hu, J.; Zhang, J.; Zhang, H.; Cheng, Y. Comparison of generation 3 polyamidoamine dendrimer and generation 4 polypropylenimine dendrimer on drug loading, complex structure, release behavior, and cytotoxicity. *Int. J. Nanomed.* **2011**, *6*, 3361–3372.
35. Jevprasesphant, R.; Penny, J.; Jalal, R.; Attwood, D.; McKeown, N.B.; D'Emanuele, A. The influence of surface modification on the cytotoxicity of PAMAM dendrimers. *Int. J. Pharm.* **2003**, *252*, 263–266. [CrossRef] [PubMed]
36. Liu, C.; Tai, L.; Zhang, W.; Wei, G.; Pan, W.; Lu, W. Penetratin, a potentially powerful absorption enhancer for noninvasive intraocular drug delivery. *Mol. Pharm.* **2014**, *11*, 1218–1227. [CrossRef]
37. Kauffman, W.B.; Fuselier, T.; He, J.; Wimley, W.C. Mechanism Matters: A Taxonomy of Cell Penetrating Peptides. *Trends Biochem. Sci.* **2015**, *40*, 749–764. [CrossRef]
38. Reid, L.M.; Verma, C.S.; Essex, J.W. The role of molecular simulations in understanding the mechanisms of cell-penetrating peptides. *Drug Discov. Today* **2019**, *24*, 1821–1835. [CrossRef] [PubMed]

pharmaceutics

Review

Local Drug Delivery Strategies towards Wound Healing

Ruchi Tiwari [1] and Kamla Pathak [2,*]

[1] Pranveer Singh Institute of Technology (Pharmacy), Kanpur 208020, Uttar Pradesh, India
[2] Faculty of Pharmacy, Uttar Pradesh University of Medical Sciences, Etawah 206130, Uttar Pradesh, India
* Correspondence: kamlapathak5@gmail.com

Abstract: A particular biological process known as wound healing is connected to the overall phenomena of growth and tissue regeneration. Several cellular and matrix elements work together to restore the integrity of injured tissue. The goal of the present review paper focused on the physiology of wound healing, medications used to treat wound healing, and local drug delivery systems for possible skin wound therapy. The capacity of the skin to heal a wound is the result of a highly intricate process that involves several different processes, such as vascular response, blood coagulation, fibrin network creation, re-epithelialisation, collagen maturation, and connective tissue remodelling. Wound healing may be controlled with topical antiseptics, topical antibiotics, herbal remedies, and cellular initiators. In order to effectively eradicate infections and shorten the healing process, contemporary antimicrobial treatments that include antibiotics or antiseptics must be investigated. A variety of delivery systems were described, including innovative delivery systems, hydrogels, microspheres, gold and silver nanoparticles, vesicles, emulsifying systems, nanofibres, artificial dressings, three-dimensional printed skin replacements, dendrimers and carbon nanotubes. It may be inferred that enhanced local delivery methods might be used to provide wound healing agents for faster healing of skin wounds.

Keywords: wound; physiology of wound healing; strategies towards wound healing; local delivery systems

Citation: Tiwari, R.; Pathak, K. Local Drug Delivery Strategies towards Wound Healing. *Pharmaceutics* **2023**, *15*, 634. https://doi.org/10.3390/pharmaceutics15020634

Academic Editors: Daniela Monti and Silvia Tampucci

Received: 9 December 2022
Revised: 30 January 2023
Accepted: 11 February 2023
Published: 13 February 2023

1. Introduction

One of the most crucial characteristics of simple bacteria to complex multicellular creatures is an undamaged outer coating. The largest organ in the human body, the skin serves a variety of purposes. Skin is susceptible to a range of external variables because of its exposure to the environment, which can lead to various skin injuries and damage. A series of physiological events can heal cuts or injuries because skin has remarkable regenerating abilities. The epidermis of the skin is central to several vital organs, including apocrine glands, eccrine sweat glands, and hair follicles with pilosebaceous units [1]. A wound is the result of the "disruption of normal anatomic structure and function," claims the Wound Healing Society. Depending on how long a wound takes to heal, attention might be given to the acute and chronic kinds. Acute wounds are typically treated effectively with a good possibility of success within a few weeks and are primarily caused by mechanical trauma or surgical procedures [2]. The location, size, depth, and type of an acute wound all affect its nature. The chain of events necessary for wound healing can be disrupted by a variety of disease processes, leading to chronic, non-healing wounds that cause the patient great suffering and need a tremendous number of resources from the medical system. The coagulation cascade, inflammatory pathways, and the cellular components of the immune system are all activated during wound healing, which causes a significant modification of all skin compartments [3]. For expediting in vivo wound healing and minimising scar formation, cellular scaffolds containing fibroblasts, keratinocytes, stem/progenitor cells, or reprogrammed cells have shown promising outcomes.

Depending on the type of wound and the patient, several wound treatments are used, but they all often start with cleaning, involve taking antibiotics, and involve choosing the right dressing [4]. The dressing should encourage autolytic debridement and be non-toxic, non-allergenic, non-adherent, and non-toxic. For deep wounds or after surgery for patients with implanted medical devices, effective local administration of antibacterial chemicals is crucial. Surface-adherent cells may release growth factors that may then affect the timing and result of the creation of a scar as skin cells move into a fibrin-rich temporary matrix to form a scar. A polymer applied therapeutically to the local site of interest may help such local wound-healing activities [5].

When serious cuts are treated with typical wound care techniques, permanent scars are created. As a result, considerable effort has gone into creating alternative therapies that bring back the natural skin's capacity for regeneration. Cell treatment, growth factor delivery, gene delivery, and other techniques have all been utilised to speed up the healing of non-healing wounds. Drugs and genes can have their half-lives extended, their bioavailability increased, their pharmacokinetics optimised, and their dose frequency decreased using drug delivery systems at the nano, micro, and macroscales [6]. If just extracellular delivery is necessary, microparticle-mediated distribution would have a more long-lasting therapeutic impact since the release kinetics would be slowed by the reduced surface-to-volume ratio. Protein and nucleic acid treatments, on the other hand, would require nanoparticle-mediated transport in order to reach the intracellular targets. Nanotechnology has a fantastic chance to improve currently used medical therapies, standard care, and wound management [7]. It has been discovered that nano drug delivery methods are non-toxic, completely compatible with skin, and favourably generate a helpful moist environment for activating and accelerating the wound healing process.

When comparing different drug delivery methods, tissue-engineered scaffolds are particularly pertinent to wound healing since they can act as a depot for medicines [8]. Additionally, they can serve as wound dressings to physically protect the wounds. Therefore, drug-incorporated scaffolds hold great promise for speeding the healing of chronic wounds through a combination of mechanisms. By acting through immune regulation, paracrine actions, and differentiation into epidermal and dermal cells to restore the injured skin, stem cell-based treatments show increasing promise [9]. The application of particular stem cells has emerged as a promising strategy to overcome the drawbacks of conventional treatments, offering the potential to improve the healing of chronic wounds as well as speed up the process of wound healing for acute wounds [10]. There are several different antibiotic formulations that can be used topically on wound sites. Medicine delivery devices are particularly crucial because when a drug is given systemically, the poor vasculature in the wound bed can prevent the drug from effectively reaching the healing tissue. Additionally, complicated drug delivery systems that can transport the active ingredients in the right quantity to the right area are needed due to the side effects of some medications, the short half-lives of biological components, and the dynamic nature of the wound environment [11].

The physiology of wound healing as well as current advancements in wound healing techniques, with a focus on local medication delivery elements, are critically examined in this review article.

2. Physiology of Wound Healing Process

Skin wound healing is an intriguing biological process that has served mammals well throughout evolution. Skin wound healing is a crucial phase for survival that culminates in wound closure because of its essential roles as a physical, chemical, and bacterial barrier. The process of cellular, humoral, and molecular mechanisms involved in skin wound healing is dynamic, tightly regulated, and can take years to complete [12]. The epithelialisation of skin wounds depends on the details of the lesion, including its location, depth, size, microbial contamination, patient-related medical problems, genetics, and epigenetics. Following a skin injury, the exposed sub-endothelium, collagen, and tissue factor will stimulate platelet aggregation, which in turn causes degranulation and releases chemotactic factors (chemokines) and growth factors to form the clot [12]. By following all of the aforementioned steps, successful haemostasis will be achieved.

A skin wound can heal completely through regeneration or repair. Skin repair shows an unspecific kind of healing in which the lesion heals by fibrosis and scar formation, in contrast to regeneration, which depicts the specific substitution of the tissue, such as the superficial epidermis, mucosa, or foetal skin. Unfortunately, the latter is the primary mode of adult skin wound healing [13]. The pathophysiology of chronic wounds is still poorly understood; however, it is known that they remain in the inflammation stage of the healing process rather than progressing further. Critical obstacles to the physiologic healing of chronic wounds include impaired vascularisation and the resulting hypoxia, the inability to move on to the healing phase, extended and exacerbated inflammation, and the incapacity of immune cells to manage bacterial infection. Normal wound healing can be hampered by prolonged wound-healing phases or overly aggressive reactions of the organism to the damage [14]. The majority of chronic wounds heal through fibrosis, which produces an excessive quantity of connective tissue rather than regeneration. Additionally, persistent inflammation is followed by fibrosis, and fibrosis-healing wounds have been discovered to have higher levels of pro-inflammatory mediators (such as TGF-β) (Figure 1). Unnecessary fibroblast proliferation, neovascularisation, and increased collagen and fibronectin synthesis are all results of poorly controlled growth factor activity. Additionally, a wound contracts too much and for too long, causing fibrotic scar tissue to grow. Keloids and hypertrophic scars are two different types of pathological scars that can result from an injury. In this regard, there is active current study on the change from the inflammatory to the proliferative stage of wound repair. Three to five phases that overlap in both time and location can be artificially created to represent the various stages of wound healing [12–16].

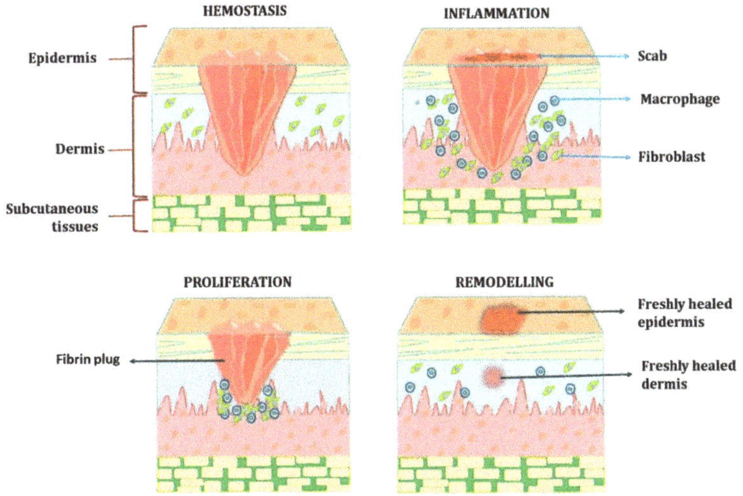

Figure 1. Physiology of wound healing. Haemostasis: The immediate response to a surgical injury is the vasoconstriction of blood vessels at the point of injury. Inflammatory process: brings nutrients to the area of surgery, removes debris and bacteria, and provides stimuli for wound repair. Proliferation: involves processes of angiogenesis, granulation tissue production, collagen deposition, and epithelialisation. Remodelling: Type 1 collagen replaces the type 3 collagen found in granulation tissue, and a scar forms.

2.1. Haemostasis and Coagulation: Vascular Mechanism

The thrombogenic sub-endothelium is initially exposed to platelets by haemorrhage into the wound. When the skin is wounded, bleeding typically occurs to help remove microorganisms and/or antigens from the wound. The main goal of the vascular mechanism is to stop exsanguination in order to maintain the integrity of the circulatory system and ensure important organs' abilities to function unharmed despite the injury. The long-term provision of a matrix for the invasive cells that are required in the later stages of healing is the second goal. Bleeding causes haemostasis to be activated, which occurs by exudate components such as clotting factors [17]. Vasoactive chemicals such as serotonin and catecholamines work through particular endothelium receptors to constrict nearby blood vessels. Leukocytes, red blood cells, and plasma proteins can enter the body by way of smaller arteries when they are triggered to vasodilate. A local perfusion failure with subsequent oxygen deprivation, increased glycolysis, and pH alterations is explained by the life-saving vasoconstriction with clot formation. Following vasoconstriction, there is a vasodilation during which thrombocytes infiltrate the temporary wound matrix [18]. Hyperaemia, a localised redness, and wound oedema are additional signs of vasodilation. Platelets are essential to this stage, as well as the overall healing process, because, in addition to establishing early haemostasis, they also release a number of cytokines, hormones, and chemokines that initiate the subsequent stages of healing. In order to prevent further blood loss, the coagulation cascade is activated alongside haemostatic processes through intrinsic and extrinsic routes, causing platelet aggregation and clot formation. The presence of fibrinogen in the exudate triggers the clotting mechanism, which causes the exudates (blood devoid of cells and platelets) to coagulate (Figure 2). This, along with the construction of a fibrin network, results in a clot in the wound, which stops bleeding [19]. The fibrin, fibronectin, vitronectin, and thrombospondin molecules found in the blood clot serve as the provisional matrix, a scaffold for the migration of leukocytes, keratinocytes, fibroblasts, and endothelial cells as well as a source of growth factors. Additionally, this stage is where the inflammatory process starts. This stage is occasionally referred to as the "lag phase" because the organism must coordinate the recruitment of numerous cells and components for the healing process while the wound lacks mechanical strength [18–20].

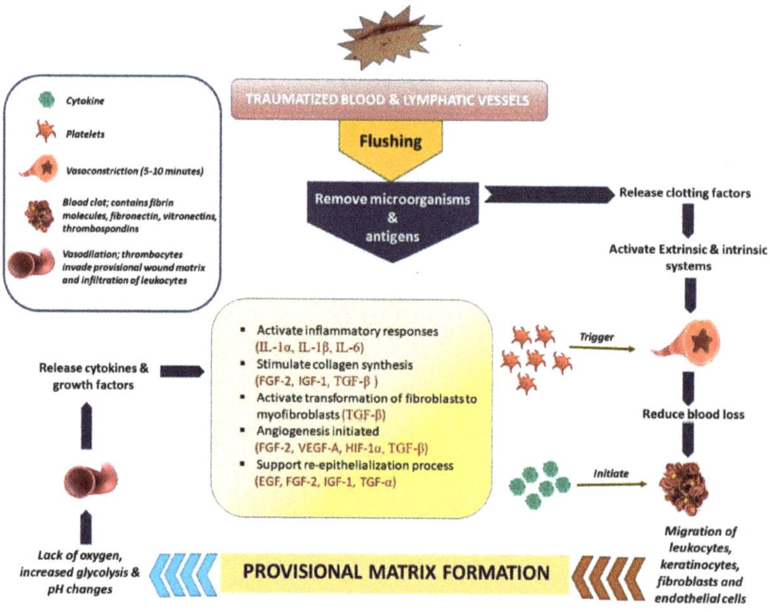

Figure 2. Vascular mechanism for Haemostasis and coagulation. This mechanism is regulated by a dynamic equilibrium between endothelial cells, thrombocytes, coagulation, and fibrinolysis, which also influences the amount of fibrin deposited at the wound site and affects how quickly the reparative processes proceed.

2.2. Inflammation: Cellular Mechanism

The inflammatory phase starts practically immediately after haemostasis and lasts for around 3 days, often starting as soon as a few minutes after damage and lasting up to 24 h. Through the production of histamine and serotonin, the release of protein-rich exudate into the wound produces vasodilation, allowing phagocytes to enter the wound and devour dead cells (necrotic tissue). It starts the complement cascade and sets off molecular processes that allow neutrophils, whose primary job it is to fight infection, to infiltrate the wound site. To remove bacteria, foreign objects, and injured tissue, the neutrophils must first perform the vital process of phagocytosis [21]. Within the first 24 h, many neutrophils arrive on the scene as the initial leukocytes. Macrophages quickly follow neutrophils because they are drawn to the consequences of neutrophil death. In the wound, phagocytic cells such as macrophages and other lymphocytes start to sweep away debris and microorganisms.

About 48 h after the injury, these macrophages infiltrate and remain there until the inflammatory phase is over. Long considered the main cell in the healing of wounds, macrophages appear to coordinate the most crucial stages of healing (Figure 3) [22]. Recent studies have looked at their role in both poor healing and scarring, despite the fact that they are essential to normal recovery. Re-epithelialisation, granulation tissue development, angiogenesis, wound cytokine production, and wound contracture are intricate processes involving macrophages. The succeeding procedures depend on phagocytotic activity because acute wounds with a bacterial imbalance are unable to heal. Inflammation is brought on as circulating monocytes enter the tissue and quickly mature into adult macrophages. By activating type-1 macrophages (M1), phagocytosis eliminates pathogens, foreign objects, necrotic neutrophils, and wound dermis from the diseased area. They generate a number of cytokines and proinflammatory mediators. Mastocytes are sensitive to tissue damage and play a crucial role in the healing process by secreting a number of cytokines that promote the recruitment of white blood cells [17–22]. T cells enter the wound site and control a variety of processes. Monocytes, cytokines, macrophages, corneocytes, big granular lymphocytes, T-lymphocytes, basophils, granulocytes, and vascular endothelial cells are all involved in wound regeneration. IL-1, TNF-α, IL-6, VPF, TGF- β, and IGF-1 are among the cytokines produced by monocytes, which develop into macrophages. Along with corneocytes, fibroblasts, granulocytes, and vascular endothelial cells, neutrophils, such as T lymphocytes and basophils, are important makers of TNF-α, IL-10, and other chemokines. VEGF, IGF-1, and TGF- β are all produced by macrophages [12–14]. As a result, each of the cells mentioned above participates in the intricate process of tissue repair. Subsequent repair mechanisms of an adult depend on cell and tissue movements that are caused by growth factor and cytokine signals, which are provided by the inflammatory response to injury. Epithelial cells and fibroblasts travel to the injured location during the migration phase to replace lost and damaged tissue. These cells quickly spread over the wound beneath the dried scab (clot), regenerating from the borders and thickening the epithelium [23].

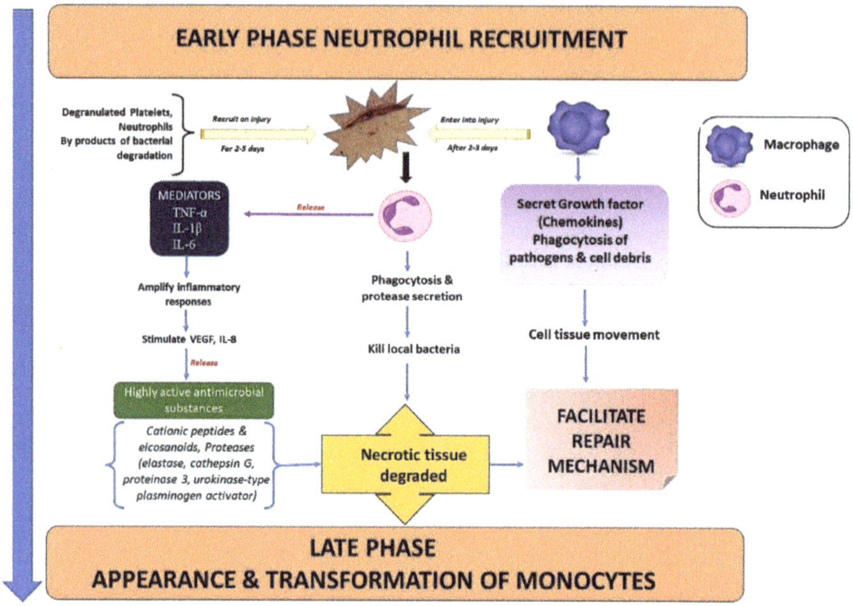

Figure 3. Inflammatory phase following a cutaneous incision. Inflammatory cell invasion: Early inflammatory phase activates the complement cascade and initiates molecular events while in the late phase, macrophages appear in the wound and continue the process of phagocytosis.

2.3. *Proliferation*

Because the proliferative stage of wound healing is highly metabolic with an increased demand for oxygen and nutrients, the restoration of blood flow is essential. For phagocytes undergoing a respiratory burst to effectively combat pathogens, the presence of oxygen in cutaneous wounds is also essential.

2.3.1. Epithelialisation

The multiplication and inflow of keratinocytes close to the wound's leading edge indicate epithelialisation. Re-epithelialisation begins during the proliferative phase of wound healing, roughly 16–24 h after injury, and continues through the second and third phases. It is characterised by fibroblast migration and the deposit of a recently created extracellular matrix, which takes the place of the temporary fibrin and fibronectin network [18]. Capillary budding and the synthesis of the extracellular matrix are also involved in the repair phase to fill in the gaps left by the debridement of the wound. Hair follicle and apocrine gland bulbs contain stem cells that start to develop into keratinocytes, repopulate the stratum basale, and move over the edge of the wound. They connect close to the inner edge of the wound once they come into contact with the mesenchyme of the extracellular matrix (ECM), and they start to lay down a new basement membrane. Keratinocytes are crucial for maintaining the barrier as well as for its repair after injury through a process called epithelialisation. Undifferentiated keratinocytes convert into differentiated non-dividing cells during differentiation as they move upward to eventually give rise to the cornified envelope [24]. The process of differentiation is mediated by three main mitogen-activated protein kinase (MAPK) pathways, which are activated by a variety of stimuli, including calcium influx, epidermal growth factor (EGF), and TNF-α. K6, K16, and K17 keratins are upregulated in migrating keratinocytes, which is thought to augment the viscoelastic characteristics of moving cells. In addition to producing paracrine and autocrine signals that are directed at surrounding keratinocytes, activated keratinocytes also alert fibroblasts, endothelial cells, melanocytes, and lymphocytes. In order to coordinate the activity of neighbouring cell types in the repair of damaged tissue, these responses are crucial [25].

2.3.2. Angiogenesis

In response to tissue injury, the complicated mechanism of angiogenesis is heavily controlled by signals from the ECM and serum. Activated macrophages, the epidermis, and soft tissue wounds can all produce angiogenesis. Gelatinase A is released by endothelial cells exposed to thrombin, and it aids in the local disintegration of the basement membrane, an essential first step in angiogenesis. Many soluble factors, most notably VEGF-A, positively influence the initiation of angiogenesis. Due to its powerful angiogenesis and vasopermeability activity, VEGF, a growth factor belonging to the PDGF family, was initially called the vasopermeability factor [22]. The VEGF family in mammals consists of five members (VEGF-A, -B, -C, and -D and placenta growth factor). In the wound, thrombin increases cellular receptors for VEGF. Due to a hypoxic gradient between injured and healthy tissue, expression of the HIF-1 gene causes the synthesis of VEGF. Nitric oxide generation by endothelial cells is also influenced by hypoxia [25]. To increase local blood flow, nitric oxide encourages angiogenesis and vasodilation. The most important proangiogenic factor in wound healing is VEGF-A, which is produced in response to hypoxia. A powerful proangiogenic mediator, VEGF-A also raises vascular permeability, which adds to wound oedema. Other factors, such as cardiac ankyrin repeat protein, as well as VEGF-A, FGF-2, PDGF, TGF-β family members, and other factors also encourage wound angiogenesis. Numerous highly technical depletion investigations on skin have shown that macrophages are a significant contributor to overall proangiogenic stimulation. Thus, it appears that in the healing wound, inflammation and the subsequent angiogenic response are related [26].

2.3.3. Granulation Tissue Formation

Granulation tissue normally grows from the wound's base and can cover any size wound. Chronic wound formation can be caused by any mistakes in the granulation tissue creation process. Fibroblasts, freshly sprouting blood vessels, and immature collagen make up granulation tissue (collagen type III) [25]. In this stage, some fibroblasts will also start differentiating into myofibroblasts, which can contract to close wound edges that are protruding from the body. TGF-β and PDGF, which are generated by inflammatory cells and platelets, entice fibroblasts and myofibroblasts from the surrounding tissue to move into the wound [27,28]. Through the creation of granulation tissue brought on by hypoxia, increased lactate, and different growth factors, healing through secondary intention is accomplished. Epithelisation over this granulation is a necessary step in healing by secondary intention, followed by substantial remodelling [29]. Any medicine that prevents the growth of new blood vessels may hinder the healing of wounds. In addition to the collagen matrix, fibrinogen, fibronectin, and hyaluronic acid, macrophages, proliferating fibroblasts, and vascularised stroma also make up the acute granulation tissue that takes the place of the fibrin-based provisional matrix. The blood vessels become less dense as collagen builds up, and the granulation tissue gradually reaches maturity to form a scar [30].

2.4. Remodelling Phase

Through cell apoptosis, the production of granulation tissue is stopped. Cellular connective tissue is generated during tissue maturation or remodelling, and the newly formed epithelium is strengthened. From a few months to around two years, cellular granular tissue transforms into an acellular mass. Fibroblasts and macrophages are important players in remodelling. Several growth factors, including TGF-β, PDGF, and FGF, which are stimulated during tissue injury and repair, control remodelling. Although the function of growth factors in the development of scars is not entirely known, TGF-β is assumed to be significant [26]. By boosting the development of tissue inhibitors of metalloproteinase and causing an increase in collagen deposition, this factor is known to decrease pro-collagenase production and enzyme activity. The collagen that was put down during proliferation is eventually replaced by a more stable interwoven type III collagen during remodelling as the water content of the wound decreases. In order to maintain a precise balance between synthesis and degradation and promote normal healing, regulatory mechanisms carefully govern the remodelling of an acute wound. Concurrent with the development of the granulation tissue, the extracellular matrix begins to be synthesised during the proliferative and remodelling phases. The proliferative phase's collagen III is now being replaced by the more robust collagen I. Later, the myofibroblasts generate wound contractions through various collagen attachments and aid in reducing the surface area of the scar [31,32].

3. Wound Healing Strategies

3.1. Cellular Activity Initiators

DNA synthesis is promoted in fibroblast cells by secretions from healthy wounds. Conversely, the same fibroblasts are inhibited by the secretions from long-lasting, nonhealing wounds, such as leg ulcers. Interestingly, heating the fluid contents of a chronic wound denatures them, removing the inhibitory impact and restoring fibroblast growth. As before, fibroblasts from chronic wounds have the worst response to growth factors than fibroblasts from acute wounds, suggesting that the fibroblasts in chronic wounds are also harmed. The most promising biomarkers are proteases and cytokines. Traditional medicines such as honey, curcumin, and tannin have been studied using modern pharmaceutical practices to learn how they affect cellular activity. One would anticipate that cytokine release, which represents neutrophil and macrophage activity, would increase in an effort to trigger a fibroblast response if fibroblasts stop responding. The pro-inflammatory cytokines IL-1 and TNF-α were found in higher concentrations in non-healing wounds than in healing wounds. When the healing starts, the levels significantly decrease. An electromechanical coupling bio-nanogenerator made of extremely discrete piezoelectric fibres was created by Tong et al. [33]. By using the inherent force of the cell, it can produce a surface piezopotential up to millivolts, providing in situ electrical stimulation for the living cells. Additionally, the three dimensional structure of bionanogenerator encourages growth of ECM. Bio-nanogenerators successfully support cell viability and development as a result, but more crucially, they maintain the cell's unique functional expression.

Several methods were used to modulate macrophages, including blocking IL-1 or TNF-α, inhibiting the inflammasome pharmacologically, neutralising MCP-1, and chelating iron with desferrioxamine. Sulphated hyaluronic acid is internalised by macrophages after being identified by CD44 and the scavenger receptors CD36 and LOX-1. Most notably, it prevents the phosphorylation of the transcription factors including pNFkB, pSTAT1, and IRF5 that are involved in M1-like activation states and the production of pro-inflammatory genes. Sulphated hyaluronic acid regulates macrophage activation in vivo. Inhibiting the secretion of growth factors by macrophages, blunting the immune system's response to presented antigens, blocking the conversion of membrane phospholipids to arachidonic acid, and reducing vascular permeability are a few of the ways that steroids modify the inflammatory process at various stages of the cascade of wound healing.

By altering fibroblast activity and proliferation, anti-fibrotic medications such as mitomycin C and 5-FU stop the formation of scars. By disrupting pyrimidine metabolism, the anti-proliferative activity of 5-FU is mediated. By preventing the production of thymidine nucleotides, it prevents DNA synthesis, leading to cell death. It has long-lasting effects on Tenon's fibroblasts and can effectively limit fibroblast development [34]. By reducing to an alkylating agent, mitomycin C is activated and subsequently works by cross-linking DNA. Mitomycin C can impede not just DNA replication but also mitosis and protein synthesis. Hypermongone C, a polycyclic polyprenylated acylphloroglucinol, was shown by Ehsan et al. [35] to have the ability to speed up wound closure by simultaneously boosting fibroblast proliferation and migration, encouraging angiogenesis, and inhibiting pro-inflammatory cytokines. This substance comes from the Hypericum plant family, which has long been used to cure wounds. There are now more opportunities for the combination of therapeutic strategies in multifunctional ECM-based wound dressings thanks to ECM-based materials that have already been investigated for the delivery of antimicrobials and sustained release of angiogenic and pro-fibrotic growth factors in wound healing. Proteomic and microbiome analyses, gene sequencing, high-resolution imaging, and single-cell laser capture are some of the new technologies that may offer comprehensive information that helps define macrophage subtypes with distinct activation profiles in physiologically healing wounds and recognise their dysregulation in chronic wounds. Improvements in collagen production and angiogenesis have been reported using microspheres carrying FGF-10 [36].

3.2. Collagen Synthesis Activators

In addition to providing resident cells with structural support, collagens, which are found in the dermis as fibrillar proteins, also control resident and inflammatory cell function. Since collagen plays a crucial role in the healing of wounds, chemicals that alter the molecular processes that cause collagen synthesis have been recognised as effective wound-healing medications. In in vitro and in physiological settings, the collagenase from the bacteria *Clostridium histolyticum* hydrolyses triple-helical collagen utilising synthetic peptides as substrates. The anti-collagenase activity of phytoconstituents and crude extracts from natural resources has received extensive study. Numerous phytoconstituents found in plants, including polyphenols with collagenase inhibitory activity, such as flavonoids, terpenoids, glycosides, vitamin E, vitamin C, phenolic acids, and tannins, are abundant. Reduced collagen production and stability, slower re-epithelialisation, and an elevated susceptibility to infection have all been linked to vitamin A insufficiency [37].

Ascorbic acid (vitamin C), is crucial for the manufacture of collagen. The collagen that is produced in scurvy is unhydroxylated, relatively unstable, and prone to collagenolysis. A lack of vitamin C causes fibroblasts to create unstable collagen, which offers a flimsy foundation for repair. Although it is generally known that animals with vitamin C deficiencies take longer to repair wounds, it is unclear whether oral vitamin C supplements speed up the healing process. A lack of vitamin K impairs the generation of the clotting factors (factors II, VII, IX, and X), which leads to bleeding disorders, the formation of haematomas, and consequent negative effects on wound healing [38]. In order to transport oxygen, iron is needed. The immune system and other mineral systems depend on substances such as copper and zinc. Lack of zinc causes the production of granulation tissue to be disrupted. *Aloe vera* is a natural substance that is still frequently used and acknowledged in domestic and professional settings as a tool for treating wounds. Aloe vera is primarily recognised for its ability to lessen pain in burn wounds, but it also helps wounds produce more collagen.

Collagenase and dexpanthenol-containing ointment formulations have been used to speed up re-epithelialisation, reduce fibroblast proliferation, and rebuild the ECM. Due to their advantageous effects, topical preparations of growth factors have recently been investigated in wound healing. Growth factors have a low bioavailability, which limits their use because they are quickly removed from the wound site. Emerging strategies that topically apply growth factors are nanoparticle-encapsulated and have improved stability and bioavailability to the wound area aim to address this limitation. Rats with parenchymal lung lesions experienced better wound healing and an increase in the presence of immature collagen after intraperitoneal glutamine treatment [39]. Dietary glutamine supplementation increased the collagen density in colonic anastomoses in rats, indicating that fibroblasts can synthesise enough glycine for collagen production, whereas they need a source of extracellular glutamine. Glutamine availability can also control collagen mRNA expression in fibroblasts [40].

3.3. Angiogenesis Activators

By acting as a chemoattractant for neutrophils, macrophages, and fibroblasts, TGF-β promotes the development of granulation tissue. As a result, TGF-β is a crucial regulator of angiogenesis during the healing of wounds because it controls cell division, migration, capillary tube formation, and ECM deposition. Through the production of particular proteins, gene augmentation brings about the return of normal cellular function. By delivering DNA or mRNA into the target cells, one can enhance genes. Exciting new choices for treating chronic wounds will be made possible in the following ten years by wound dressings that contain sustained nucleic acid delivery systems for promoting angiogenesis and therapy that targets the underlying morbidities. Reactive oxygen species (ROS)-scavenging hydrogel and oxygen-release microspheres were combined to create a sustained oxygenation system by Ya et al. [41]. In diabetic wounds, the continuous release of oxygen increased keratinocyte and dermal fibroblast survival and migration, encouraged the development of angiogenic growth factors and angiogenesis, and reduced the expression of pro-inflammatory cytokines. The pace of wound closure was greatly accelerated by these effects. With its interactions with a number of immune and non-immune cells, C1q is a well-known starter of the complement classical route and induces complement activation-independent activities. One of the probable targets of C1q, which binds to receptors found on cell surfaces and promotes inflammation, are endothelial cells. C1q has a special and hitherto unknown role in promoting angiogenesis through the globular heads. The ability of C1q to stimulate the growth of new blood vessels in both in vitro and in vivo models of wound healing provided evidence for its angiogenic action [42]. In order to promote the healing of wounds, Jin et al. created temperature-responsive nanobelt fibres that contain vitamin E. A correct matrix elasticity that encourages mesenchymal stem cell adhesion and angiogenesis was given by the cross-linked collagen sheets created by the nanoparticles [43]. Additionally, the scaffold encouraged endothelial cells to form tubes. The therapeutic potential of nanoparticle formulation results in the stimulation of angiogenesis. The flexibility of collagen sheets was improved by praseodymium-cobaltite nanoparticle cross-linking for the pro-angiogenic and stem cell differentiation ability [44].

3.4. Cytokine and Growth Factor Activators

Small, secreted proteins called cytokines influence not only the activity of immune cells but also that of other cells. Interleukins, lymphokines, and several related signalling molecules, such as TNF-α, interferons, and others, are among them. Through the activation of cell surface TGF-β serine/threonine type I and type II receptors and the activation of a Smad3-dependent signal, active TGF-β1 induces the fast chemotaxis of neutrophils and monocytes to the wound site. Leukocytes and fibroblasts that have expressed TGF- β1 are then stimulated to produce additional cytokines, such as TNF-α, IL-1β, and PDGF, as well as chemokines, which are all part of a cytokine cascade. Such factors serve to maintain the inflammatory cell response by influencing neutrophil and monocyte recruitment and activation. TGF-β and other cytokines that activate their corresponding cell surface receptors cause intracellular signalling pathways to be activated, which, in turn, causes target cell populations to respond phenotypically and functionally. NF-κB, early-growth response 1 (EGR1), Smads, and MAPK are some of the upstream signalling cascades involved in acute tissue injury. These cascades activate many cognate target genes, including adhesion molecules, coagulation factors, cytokines, and growth factors. The platform upon which circulating leukocyte-expressing counter-adhesion molecules (integrins, selectins, and Ig superfamily members) tether allows them to sense the microenvironment and react to chemotactic signals at the site of tissue injury. This is accomplished by cytokine-induced enhancement of adhesion molecules (VCAM-1, ELAM-1, and ICAM-1) on the endothelium. In response to various chemotactic cues, transmigration from within to outside the artery wall is made possible by interactions between adhesion molecules on blood leukocytes and endothelium. Numerous chemokines are generated in addition to the chemotactic action of TGF-β1 for neutrophils and monocytes to attract leukocytes to the site of tissue injury. Depending on where the cysteine residues are located, several families of related molecules serve as representations of chemokines [45].

Growth factors such as PDGF, TGF-β, and EGF are secreted when platelets degranulate and release alpha granules. PDGF plays a crucial role in luring neutrophils to the wound site to eliminate contaminated germs, coupled with proinflammatory cytokines such as IL-1. A number of pro-inflammatory cytokines (IL-1 and IL-6) and growth factors (FGF, EGF, TGF-β, and PDGF) are released by macrophages to support the development of granulation tissue. One of the first substances to be created in response to skin lesions is a substance known as a proinflammatory cytokine, which controls immune cell actions during epithelialisation. TNF-α, IL-1, IL-6, and IL17 are the main proinflammatory cytokines that play a role in the inflammation phase of wound healing. They also play a role in the epithelialisation phase by promoting cell proliferation and differentiation and mobilising local stem/progenitor cells. Cytokine modulators are a new class of medicinal drugs that prevent fibrogenesis. Scarring and fibrosis of the skin are frequently the results of excessive fibrogenesis [46]. As a different strategy to prevent fibrosis, the modulatory effects of natural compounds such as terpenes and honey should be taken into consideration. Mitomycin P modulators such as buckwheat and acacia honey should be taken into account as substances reducing scarring and encouraging re-epithelisation. Terpenoids are frequently present in essential oils and serve as cytokine suppressors, increasing the production of IL-10 and the anti-inflammatory cytokines TNF- and IL-1 [47].

3.5. Antimicrobials

There is a lot of debate about the application of topical antibiotics to wounds. Topical antimicrobial agents are described as substances that can eliminate, suppress, or lessen the number of bacteria. These substances include disinfectants, antiseptics, and antibiotics. Topical antimicrobial medicines are essential to topical burn care because they are used to prevent and control infection. The ideal topical preventive antimicrobial agent would be able to enter necrotic tissue without being absorbed by the body, have a broad spectrum of activity, a lengthy duration of action, have low toxicity, and have several other qualities [48].

A renewed interest in silver-based medications is a result of concerns about bacterial resistance. For thousands of years, silver has been utilised in medicine for its antibacterial properties. A good environment for wound healing can be created by using topical antimicrobials that do not impede epithelial outgrowth and deliver a high concentration of active components to devitalised, devascularised, and, perhaps, necrotic wounds. Topical antibacterial use may reduce the requirement for extensive debridement and subsequent grafting as well as wound deepening. Although microorganisms are present in every wound, the majority do not become infected and heal properly. In these circumstances, the immune system of the host and the bioburden of the wound are in equilibrium [49]. NF-κB is nuclear translocated as a result of toll-like receptor (TLR) stimulation through intracellular signalling from adapter proteins, which, in conjunction with mitogen-activated protein kinases, triggers the transcription of a variety of inflammatory cytokines, chemokines, antimicrobial peptides, and costimulatory factors. According to wound specialists, there is a threshold over which antimicrobial intervention is necessary when bacteria loads are more than or equal to 104 CFU/g. Silver compounds are among the antibacterial agents used in burn treatment. In cases when surgery is either not possible or would not be the first option right away, such as in cases of facial burns, silver sulphadiazine is frequently utilised and acts on burn eschar to restrict the area of non-viable tissue [50–55]. Since silver is a natural broad-spectrum antibiotic, there has not been any bacterial resistance to its treatments yet. There are numerous types of silver, including silver oxide, silver nitrate, silver sulphate, silver salt, silver zeolite, silver sulfadiazine, and silver nanoparticles. When silver cations come into touch with liquid, they are freed from their carrier dressings. Depending on the dressing employed, there are significant differences in the pace, duration, and peak level of silver released. Once discharged, silver kills germs in a variety of ways. The healing of both acute wounds and chronic wounds is currently aided by various forms of silver. Antiseptic and antibiotic dressings are the two primary categories into which antimicrobial dressings can be divided.

Due to their good efficacy and tolerability among the various antimicrobial agents available, iodophor-based formulations such as povidone iodine have remained well-liked after decades of usage for antisepsis and wound healing applications. Povidone iodine has been reported as having a wide range of activity, the capacity to penetrate biofilms, a lack of related resistance, anti-inflammatory qualities, low cytotoxicity, and good tolerability. In clinical practice, no adverse effects on wound healing have been noted. Another antimicrobial agent that penetrates burned tissue is cerium nitrate. It has a wide range of activity against Gram-positive and Gram-negative bacteria, as well as fungal species, and is highly effective when used in conjunction with silver sulphadiazine [56]. Povidone iodine is fully hazardous to keratinocytes and fibroblasts at concentrations greater than 0.004 and 0.05%, respectively [57–74]. According to in vitro testing, cadexomer iodine is not harmful to fibroblasts at doses up to 0.45%. At doses between 0.2 and 0.001%, chlorhexidine also exhibits dose-dependent toxicity to fibroblasts. Antiseptics are primarily used to avoid infection, reinfection, and probable disruption of wound healing. Antiseptic therapy also has the secondary purpose of promoting wound healing by stimulating cell growth and regeneration. These effects have been shown for polyhexamethylene biguanide, in addition to pure microbicidal action (polyhexanide). Antiseptics also have additional beneficial benefits, such as wound cleaning, which can aid in debridement. Effective antiseptics for local wounds include polyhexanide and octenidine dihydrochloride. Nitrofuran and natrium fusidate are other antibacterial substances (Table 1) [48–86].

Table 1. Commonly used antimicrobial agents with target species.

Class	Name	Wound Dressing Material	Tested Strains	Administration	References
Macrolides	Clarithromycin Erythromycin	PVA hydrogels	*Pseudomonas aeruginosa* *Staphylococcus aureus*	Oral/Systemic/ Topical/ ophthalmic	[58]
Tetracycline	Tetracycline Chlortetracycline	Cotton fabric coated with chitosan-Poly (vinyl pyrrolidone)-PEG	*E. coli* *S. aureus*	Oral/Topical	[59]
		Scaffolds with Collagen Microsphere with gelatin	*E. coli* *S. aureus*	Oral/Systemic/ Topical	[61]
Aminoglycosides	Streptomycin Neomycin	Wafers and film based on polymer Polyox/carrageenan	*E. coli* *S. aureus* *P. aeruginosa*	Oral/Topical/Systemic oral/topical	[66] [67]
		poly(styrene sulfonic acid-co-maleic acid) (PSSA-MA)/polyvinyl alcohol (PVA) ion exchange nanofibres	*E. coli* *S. aureus*		
Fluoroquinolones	Norfloxacin Ciprofloxacin	Films and nanofibre mats of povidone Electrospun fibers based on thermoresponsive polymer poly(N-isopropylacrylamide), poly(L-lactic acid-co-ϵ-caprolactone) Hydrogels from 2-hydroxyethyl methacrylate/citraconic anhydride-modified collagen Films and nanofibre mats of povidine	*E. coli* *Bacillus subtilis* *E. coli* *S. aureus* *S. aureus* *E.coli* *Bcillus subtilis*	Topical Topical Topical	[69] [70] [71]

The lipid–protein complex produced from damaged skin that is responsible for the severe immunosuppression linked to significant cutaneous burns is believed to be bound and denatured by cerium. Additionally, cerium nitrate hardens burn eschar, which is supposed to inhibit bacterial entry and maintain a moist wound. A broad-spectrum antibacterial agent, silver is efficient against yeast, bacteria, fungus, and viruses. When taken at the right concentration, it has also been demonstrated to be effective against vancomycin-resistant enterococci and Methicillin-resistant *Staphylococcus aureus* (MRSA) [81–86]. Additionally, silver is believed to lessen wound irritation and speed up recovery. The local wound environment affects the amount of silver required to have a bacteriostatic or bacteriocidal impact. It has shown to be effective against pathogens on the surface, but it might not have an impact on bacteria that have travelled a long way into the wound bed. So, when colonisation or critical colonisation is detected, silver may be utilised to assist in lowering the bacterial count in mild wound infections.

3.6. Stem Cell-Based Therapy

A promising new method in the area of regenerative medicine is stem cell-based therapy. The ability of stem cells to self-renew and specialise into distinct cell types is essential for physiologic tissue renewal and regeneration after injury, and is of great interest to biologists. Adult mesenchymal stem cells, embryonic stem cells, and, the more recently studied, induced pluripotent stem cells are the main sources of stem cells that are used for skin regeneration and wound healing among the several types of stem cells [57]. In essence, keratinocytes are produced from stem cells found in the skin's basal layer, which differentiate for three to six weeks before becoming corneocytes, which in turn create the stratum corneum layer. Keratin, a protein that plays a significant structural role in the stratum corneum, is one of the many proteins that keratinocytes generate. Along with proteins, the stratum corneum's complex lipid and cell membranes act as a significant barrier against bacteria and dehydration. The migration of fully developed cuboidal basal keratinocytes with large nuclei, phospholipid membranes, and organelles starts from the basal layer after roughly every 28 days [80–87].

A greater build-up of keratin and lipids develops throughout this turnover process and proceeds through terminal differentiation to form a stratum corneum. The dermo-epidermal junction, a basement membrane that separates the epidermal and dermal layers, is where the hemidesmosomes, which serve as cell adhesion molecules, are attached. Keratinocytes from the basal layer, ECM components, basal lamina, filaments, and anchoring fibrils make up the intricate dermo-epidermal junction structure. In order to reduce the possibility of the epidermis separating from the dermis layer, the dermo-epidermal junction must be restored during the wound healing process. The stem cell-based therapy for chronic wound healing makes use of a variety of processes, including growth factor interactions and activities, inflammation control, and immune process stimulation, for speeding vascularisation and re-epithelialisation. The ability of stem cell-based wound therapy to produce pro-regenerative cytokines and growth factors to encourage skin regeneration during the treatment of chronic wounds is largely responsible for its therapeutic potential [88,89].

3.7. Herbal Alternatives Acting as Activators for Wound Healing Factors

Wound healing may benefit greatly from a variety of plants or chemicals derived from plants that contain high concentrations of antioxidants and have anti-inflammatory, immunomodulatory, and antibacterial characteristics. Antioxidants help tissues recover from injury and can speed up wound healing. The antioxidant activity of flavonoids, anthraquinones, and naphthoquinones is strong. Shikonin, alkanin, lawsone, emodin, epigallocatechin-3-gallate, ellagic acid, and a few herbal extracts have strong antioxidant effects by scavenging ROS, preventing lipid peroxidation, and boosting intracellular antioxidant enzyme activities such as superoxide dismutase, catalase, and glutathione peroxidase [3,8]. Additionally, herbal medication encourages angiogenesis, fibroblast cell growth, and the production of temporary ECM.

Herbal extracts and other natural products' immunomodulatory and anti-inflammatory properties hasten the healing of wounds. It is true that plants, and the chemical substances derived from them, aid in healing and treatment. Some herbal remedies stimulate the expression of VEGF and TGF-β, both of which are crucial in stimulating angiogenesis, granulation tissue development, and the deposition of collagen fibres. Other herbal medicines used in wound dressings act as inhibitors of the expression of the proteins TNF-α, IL-1, and inducible nitric oxide synthase, resulting in the induction of antioxidant and anti-inflammatory properties during different stages of the wound healing process. In the healing of cutaneous wounds, curcumin promotes the proliferation of fibroblasts, the development of granulation tissue, and the deposition of collagen [14,15]. Cinnamon bark has some therapeutic characteristics, including anti-inflammatory, anti-diabetes, anti-ulcer, antimicrobial, and hypoglycaemic effects. It can also help with diabetic and infected wounds. In addition to the above-mentioned qualities, cinnamon is known to contain considerable amounts of polyphenols, which may improve an animal's ability to absorb glucose. Cinnamaldehyde, 2-hydroxycinnamaldehyde, and quercetin are anti-inflammatory properties of cinnamon components that can speed up wound healing. *Aloe vera* gel's antioxidant qualities, which are attributable to certain components including indoles and alkaloids, have positive benefits on the healing of wounds [90]. Burns, ulcers, and surgical wounds remain as first-line conditions for *Aloe vera* treatment. Numerous organic bioactive substances, such as pyrocatechol, saponins, acemannan, anthraquinones, glycosides, oleic acid, phytol, and simple and complex water-soluble polysaccharides, are found in *Aloe vera. Aloe vera* has a variety of wound-healing mechanisms, most of which are attributable to raising the level of lysyl oxidase and the turnover rate of the collagen in the tissue. IL-1, IL-1, IL-6, TNF-α, PGE2, and nitrous oxide are only a few of the proinflammatory mRNAs that are transcribed when acemannan, a primary mucopolysaccharide from *Aloe vera*, is consumed. Reactive oxygen species and endogenous mitogen inhibitors are bound and captured by mesoglycan molecules, which facilitates phagocytosis.

Glycans prolong the activity of released cytokines, growth factors, and other bioactives. Through the action of cyclin D1 and AKT/mTOR signal pathways, topically administered acemannan has been shown to drastically shorten the time to wound closure in a rat wound healing model. It is well known that *Anethum graveolens* has antibacterial, anti-diabetic, and anti-inflammatory qualities that can speed up wound healing. Major constituents of dill essential oil include cis-carvone, limonene, phellandrene, and anethofuran. Burns, blisters, herpes, cuts, wounds, skin infections, and insect bites are just a few of the skin conditions that eucalyptus oil is traditionally used to treat [91–93]. *Securigerasecuridaca* is well known for its antibacterial properties, and helps infected wounds heal faster [94]. Antioxidant, anti-inflammatory, antidiabetic, cancer-preventive, antibacterial, antiviral, antimalarial, hypotensive, immunostimulatory, and hepatoprotective properties are all present in *Andrographis paniculata* extracts. In one study, it was found that treatment with a 10% aqueous leaf extract of *Andrographis paniculate* considerably improved the rate of wound closure in rats [95]. The polysaccharides diosgenin, yamogenin, gitogenin, tigogenin, and neotigogens are found in the seeds of *Trigonella foenum-graecum*. Saponins have steroidal actions that can reduce bodily inflammation. Fenugreek also contains mucilage, volatile oils, flavonoids, and amino acid alkaloids, which are bioactive components. 4-Hydroxyisoleucine is the other substance in fenugreek that is active. Fenugreek is said to release an anti-inflammatory chemical into the area of a cut, which reduces inflammation. Fenugreek's antibacterial qualities may also enhance its anti-inflammatory effects. According to a study, the antibacterial characteristics of flavonoids and triterpenoids may help the wound healing process. Antioxidant properties of fenugreek are thought to hasten wound healing. The topical use of the fenugreek seed significantly increased the kinetics of wound contraction and epithelialisation [96].

In vivo studies have demonstrated that *Arctium lappa* root extract dramatically improves dermal ECM metabolism, affects glycosaminoglycan turnover, and minimises the appearance of wrinkles in human skin. *Arctium lappa* is also known to influence the Wnt/-catenin signalling pathway, which is recognised to be a significant regulator of wound healing, by regulating cell adhesion and gene expression in canine dermal fibroblasts. In diabetic rats, *Astragalus propinquus* and *Rehmanniaglutinosa* improve angiogenesis and reduce tissue oxidative stress to promote diabetic wound healing and post-ischaemic neovascularisation. In human skin fibroblasts, *Astragalus propinquus* and *Rehmanniaglutinosa* stimulated enhanced ECM deposition by activating the TGF-β signalling pathway. TNF-α and TGF-β1 levels were seen to be higher two days after the damage and to decrease as the wound healed. IL-10, on the other hand, was shown to be raised after 14 days, concurrent with wound healing [97]. Topical therapy with ethanolic *Ampelopsis japonica* increased re-epithelisation, granulation tissue development, vascularisation, and collagen deposition when compared to wounds treated with Vaseline or silver sulfadiazine.

Pharmacological benefits of *Angelica sinensis* include immunological modulator, anti-inflammatory, and anticancer properties. In human skin fibroblasts, extracts from Angelica sinensis have been demonstrated to promote cell proliferation, collagen secretion, and cell mobility while also activating an antiapoptotic mechanism. Additionally, extracts have been demonstrated to promote calcium fluxes and glycolysis, enhancing cell viability during tissue healing. Numerous studies have found conflicting results on *Angelica sinensis'* impact on the development of new blood vessels, raising questions about the plant's role in angiogenesis [98]. By changing the expression of connective tissue growth factor (CTGF) and smooth muscle actin in vivo, extracts from the *Calendula officinalis* flower promote the formation of granulation tissue in excisional wounds of BALB/c mice. *Blumea balsamifera*, also known as kakoranda in Ayurveda, is used to cure rheumatism, coughs, fevers, and pains. Eczema, dermatitis, skin damage, bruising, beriberi, lumbago, menorrhagia, rheumatism, and skin damage are all treated using leaf extracts that are administered topically [99]. It has been discovered that homoisoflavonoids extracted from *Caesalpinia sappan* exhibit anti-inflammatory and antiallergic properties as well as the ability to block viral neuraminidase activity. The main catechin, (-)-epigallocatechin-3-gallate (EGCG), promotes keratinocyte growth and differentiation. By altering TGF-signalling, lowering MMP-1 and MMP-2 expression, and reducing the synthesis of type I collagen in human dermal fibroblasts, EGCG inhibits TGF- receptors. These characteristics suggested that EGCG may have anti-scarring capabilities [100].

Entada phaseoloides are high in saponins and tannins and are used topically to treat skin lesions as well as analgesic, bacteriocide, haemostatic, and anticancer agents. In diabetic cutaneous wounds, vitamin E controlled inflammation and oxidative stress. Additionally, vitamin E enhanced the anti-oxidative enzymes superoxide dismutase (SOD), glutathione peroxidase (GPX), and catalase (CAT) that are in charge of purging ROS and oxidised macromolecules from damaged tissues. By encouraging the growth of new blood vessels as well as re-epithelialisation, matrix deposition, and collagen synthesis, vitamin E had positive impacts on the wound healing process' later stages (the proliferation and remodelling phases) [101]. Vasodilatation, blood lipid regulation, reduced inflammation; antioxidant, anti-cancer, anti-bacterial, anti-allergic, anti-ageing, and immunomodulatory potential are all benefits of *Panax ginseng* [102].

4. Localised Delivery Systems for Wound Healing

4.1. Microspheres/Microcarriers

A sort of injectable scaffold is a microsphere, often known as a microcarrier. Microspheres offer enough room for cell development and have good surface area-to-volume ratios. To further improve the transport of cells and bioactive compounds, microspheres with functional architectures (such as hollow or core-shell) can be easily customised and manufactured. As a result, functional microspheres have received a lot of attention recently as a new kind of injectable scaffold. Microspheres are typically referred to as spheres with a diameter between 1 and 1000 m. The diameter of most cells when they adhere and spread on a biomaterial is greater than 20 μm; hence, small-sized microspheres (<20 μm in diameter) are not suited as cellular carriers. Microspheres should therefore be between 20 and 200 μm in size when used as an injectable biomaterial for tissue engineering [103]. The development and uses of functional microspheres, including macroporous microspheres, nanofibrous microspheres, hollow microspheres, core–shell structured microspheres, and surface-modified functional microspheres, are covered after a brief introduction to the biomaterials and techniques for microsphere fabrication. Li and Wang explained the perspectives and directions for functional microspheres as injectable cell carriers, which are offered as a final step in the advancement of tissue regeneration [104].

Li et al. provide an explanation for the developments in functional microspheres, including the kinds of biomaterials used to make microspheres, the manufacturing processes for functional microspheres, and the uses of functional microspheres for the regeneration of bone, cartilage, the dentin–pulp complex, neural tissue, cardiac tissue, and skin. The use of microspheres has substantially increased as a result of the addition of additional structures and functions [105]. The core–shell configuration, for instance, can be easily used to combine cells in a microsphere with a regulated growth factor supply. Within a microsphere, there are numerous areas for cell migration and proliferation due to the hollow structure and macropores on the surface. Additionally, by incorporating nanofibrous architecture into microspheres, an ECM-like microenvironment is created that can direct tissue regeneration. This replicates the structure of natural ECM. Before using functional microspheres from a bench to a patient's bedside, however, there is still considerable work to be undertaken.

The following description highlights some of the major obstacles to the creation of functioning microspheres [91–97]. In order to include biomimetic elements in functional microspheres, newer methods must first be developed. Phase separation is currently the only efficient way to make nanofibrous microspheres. While electrospinning and electrospray are used to create nanofibrous microparticles, their sizes and geometries are not well regulated. The traditional layer-by-layer self-assembly technique has trouble producing spheres smaller than a micrometre. Although one of the most efficient ways to create functional microspheres is using microfluidics, it is still difficult to create nanofibrous microspheres with this technology. More bio-inspired functional microspheres will be developed as a result of new fabrication methods that use modern microsphere fabrication techniques [103]. The exact control of bioactive chemical release from microspheres is also required. Currently, the majority of microsphere systems release bioactive molecules over a period of several days to a week, which is insufficient for the regeneration of many tissues. Additionally, there is a significant initial burst release from microspheres, and limiting the burst release is essential to enhancing therapeutic efficacy. The first burst release from functional microspheres is predicted to be tuned and reduced using methods and procedures including novel chemical and physical interaction factors in addition to traditional production parameters, such as polymer concentration and cross-linking time. The third one concerns the mechanical capabilities of microspheres [104].

The majority of microspheres are made from polymeric materials and typically have a porous structure; hence, their compressive moduli are quite low. Even though the mechanical strength is increased by adding inorganic elements such as hyaluronic acid, the microspheres still cannot be employed in load-bearing situations. In order to broaden the use of functional microspheres, research is being conducted to improve the mechanical properties of microspheres. Fourth, more evidence is required to support microspheres' in vivo stability. To avoid diffusion of the microspheres to nearby faulty areas, microspheres are typically chemically cross-linked [105,106].

4.2. Inorganic Nanoparticles

Nanomaterials have received a lot of recent attention because of their improved efficacy and broad-spectrum antibacterial potential. Numerous metallic and metal oxide nanoparticles, such as Ag, Fe, Cu, and Au, as well as TiO_2, ZnO, and Fe_3O_4, are being thoroughly researched for the treatment of infectious disorders. Considering that both of their counterparts are typical of low molecular weight, the two most popular systems of metal nanoparticles (NPs), gold (AuNPs) and silver (AgNPs) provide excellent methods for delivering pharmaceuticals. These nanoparticles have advantages over commonly used antibiotics because of their distinctive features, including size, shape, surface charge, dispersion, and chemical composition. Additionally, these synthetic antibiotics frequently only work against a certain type of bacterium or bacterial family, rendering them useless against a diverse range of bacterial species [107]. Currently, antibiotic efficiency has decreased mostly as a result of uncontrolled, excessive dosage, and prolonged use, which has favoured the establishment of multi-drug-resistant bacterial strains or "superbugs" such as MRSA-3. Nanomaterials have received a lot of recent attention because of their improved efficacy and broad-spectrum antibacterial potential.

Additional benefits of these materials include targeted drug delivery, solubility, improved cellular internalisation, tissue/cell selectivity, compatibility with tissue/tumour imaging, minimal adverse effects, etc. Due to their effectiveness in combating bacteria, silver nanoparticles have received the most attention among all those studied. However, little is known about the methods by which these nanoparticles destroy pathogens. It has been suggested that Ag^+ ion release occurs from nanoparticles. However, the low stability and extreme cytotoxicity of the majority of silver nanoparticle (AgNP) forms limit their applicability in mammalian cells.

AgNPs must be engineered with various compositions to improve their properties and make them suitable for therapeutic application in order to overcome their current constraints. If these nanoparticles pass the required tests by regulatory agencies for better antibacterial efficacy and minimum cytotoxicity for patients, they may prove to be effective antibiotic alternatives or may be used in conjunction with antibiotics [106–108]. The antibacterial activity of metallic nanoparticles has been increased by a number of methods, including encasing silver nanoparticles in micelles, covering silver nanoparticles in gold, and capping gold nanoparticles with 5-amindole or sodium borohydride N-heterocyclic molecules.

A bimetallic nanoparticle with silver and gold was recently developed. Complex carbohydrates on the caps of these bimetallic nanoparticles improve their stability and other characteristics. These bimetallic NPs can be used for in vivo antimicrobial activities since they have considerably better antibacterial activities and do not harm mammalian cells. The use of silver to improve wound healing dates back to ancient times when it was utilised as an ion. Unlike silver ions (Ag^+) or other forms of silver, which are used as nanoparticles, silver metal (Ag) has no known medical applications. They can impede the healing process since many bacteria, viruses, fungi, and even yeast is cytotoxic to them. This ion was used as part of a wound dressing that contained silver ions specifically to treat severe wounds [109]. In general, silver has numerous beneficial qualities, including broad-spectrum antibacterial activities. Silver-based lotions and other ointments are also available, and silver nanoparticle products can be used in a variety of therapeutic ways to help accelerate wound regeneration. The use of silver as a common bandage has been linked to decreased inflammation foci and scarring, possibly preventing bacterial development, and boosting the healing process, and possibly enhancing remodelling in the wound area, according to multivariate retrospective analyses. Therefore, using nanocomposites that are submerged in silver molecules improves the healing process by directly expressing collagen and certain growth factors that result in re-epithelialisation, neovascularisation, and the deposition of collagen fibres. Additionally, silver nanoparticles can cause the fibroblast to differentiate into myofibroblast, which is in charge of contracting the wound and quickening the healing process, and in a similar way, they can cause keratinocytes to be stimulated, proliferate, and move to the necessary area. For better wound healing, silver nanoparticles encourage keratinocyte migration from the edge into the core of the lesion. According to certain researchers, antimicrobial peptide–AgNP composite has been examined for its ability to speed up the remodelling process without having any negative effects on the dermal tissues. Resonance scattering dark-field microscopy makes use of gold nanoparticles to detect microbial cells and their byproducts, bio-image tumour cells, identify receptors on their surface, and analyse endocytosis [110].

Due to their chemical characteristics, optical stability, and simplicity of surface modification, gold nanoparticles (AuNPs) have been researched for medicinal applications such as wound healing. Before using gold nanoparticles for wound healing, they must fuse or have their surfaces modified with other biomolecules. For instance, the effectiveness of AuNPs to promote healing is increased by the addition of polysaccharide peptides. The application of gold nanoparticles to skin wounds boosted angiopoietin, VEGF, and collagen expression while decreasing mitomycin P and TGF-β1 levels [111]. It demonstrated decreased bacterial load and aided in recovery. Due to reduced blood circulation, systemically administered antibiotics may have trouble reaching injured skin tissue, rendering them ineffective for lowering bacterial numbers in granulation wounds. Recently, there has been a lot of fascinating study on the antibacterial properties of AuNPs, which makes them appropriate for possible co-use with antibiotics. Particularly in the 18 wt% composite group, the substance accelerated wound closure by increasing angiogenesis and fibroblast proliferation without inducing cell damage. According to a recent ex vivo permeation study, AuNPs can be effective in the treatment of burns as well since they can speed up the healing process and prevent microbial colonisation while being transdermally active [112].

4.3. Hydrogel

Insoluble hydrophilic materials known as hydrogels are created using synthetic polymers such as poly (methacrylates) and polyvinyl pyrrolidone. Complex hydrophilic organic cross-linked polymers called hydrogels have a base that is 80–90% water. These hydrogels can provide water to the wound site and aid in keeping it moist, which promotes quicker wound healing. These are created into contact lenses, drug delivery systems, wound dressings, electrodes, and sensors. These gels can be found as fixed flexible sheets or free-flowing amorphous gels. They have a limited capacity for fluid absorption through swelling, but they can also contribute moisture to a dry wound, aiding in autolytic debridement and maintaining a moist, thermally insulated wound environment [113].

Additionally, they have been demonstrated to increase granulation and epithelialisation, lower wound bed temperature, and have a relaxing and cooling effect. They have been shown to be a less efficient bacterial barrier than occlusive dressings and are permeable to gas and water. These dressings are primarily used to moisten dry wound beds and to soften and remove slough and necrotic wound debris. Due to their high-water concentration, they are unable to absorb substantial drainage; they absorb very slowly and are consequently useless on bleeding wounds; and they typically require a secondary dressing. They can be applied to a range of wounds, including vascular ulcers, pressure ulcers, and partial and full-thickness wounds [114]. Maceration is a potential problem since the skin around open wounds must be shielded from excessive moisture.

Hydrogels can be used in conjunction with topical drugs or antibacterial agents, which is one of their advantages. Infected wounds should not be treated with hydrogels in their fixed state. Hydrogels must be coated with additional dressings and left on for up to three days. They transfer oxygen and moisture vapour, but the type of secondary dressing employed affects how permeable they are to bacteria and fluids. Until an equilibrium condition is attained, these systems may swell in water and keep their original shape. The process of hydration, which is related to the presence of chemical groups such as -OH, -COOH, -$CONH_2$, and -CONH-, as well as the existence of capillary regions and variations in osmotic pressure, is one of the interactions that contribute to the water sorption by hydrogels [115]. These are hydrophilic polymer networks that can absorb 10–20% of their dry weight in water as well as thousands of times that amount. These can dissolve and deteriorate or they can be chemically durable. When the polymer networks are bound together by molecular entanglements and/or secondary forces such as ionic and H-bonding, they are referred to as "reversible" or "physical" gels. When hydrogels have covalently bonded networks, they are referred to as "permanent" or "chemical" gels [116].

High-intensity radiation, freeze–thaw, or chemical processes can all be used to create hydrogels. Radiation, such as gamma rays, electron beams, X-rays, or ultraviolet light, is thought to be the most suited way for the creation of hydrogels since it allows for simple processing control and eliminates the need for potentially dangerous initiators or cross-linkers. Additionally, sterilisation and possible formation are both possible with irradiation. However, the mechanical strength of the hydrogels produced using this approach is subpar. Today, hydrogels are made using a freeze–thaw process to give them good strength and stability without the need for additional cross-linkers and initiators.

The hydrogels' limited swelling and thermal stability, as well as their opaque appearance, are the principal drawbacks of freeze–thawing. The application of hydrogel appears to considerably stimulate wound healing as compared to the standard gauze therapy. In order to create hydrogel wound dressings, a variety of natural and synthetic polymers with good biocompatibility are used [117].

There are several different types of hydrogels, including interpenetrating polymeric hydrogels, copolymer hydrogels, multipolymer hydrogels, and homopolymer hydrogels. In contrast to copolymer hydrogels, which are created by the cross-linking of two co-monomer units, one of which must be hydrophilic, homopolymer hydrogels are cross-linked networks of a single type of hydrophilic monomer unit. The cross-linking of more than three monomers results in the formation of multipolymer hydrogels. Finally, the swelling of a first network in a monomer and the reaction of the latter to generate a second intermeshing network structure results in interpenetrating polymeric hydrogels. It has been established that combining a natural polymer with a synthetic polymer appears to be a successful way to create materials with the necessary mechanical and thermal properties. It is also a quick process for producing the right forms, such as films, sponges, and hydrogels, to make a variety of biomedical devices.

For instance, chronic non-healing wounds are known to have an environment that is highly alkaline, whereas the healing process is more effective in an acidic environment. Therefore, manufactured dermal patches that can measure the pH of the wound continuously are essential for guiding point-of-care treatments and monitoring the healing process of chronic wounds. Since they are porous and permeable to oxygen and gas, hydrogels offer a lot of promise for use in biomedicine. They may be suggested as materials for both burn skin treatment and wound dressings. The hydrogel membrane was created based on a polyvinyl alcohol hydrogel, which may absorb wound exudates and release water, medications, or biomolecules (such as growth factors or antibiotics), creating the ideal environment for the healing of wounds. The epidermal sensor can also measure the ambient temperature, which allows it to deliver useful biological data regarding the state of the wound [118–120].

4.4. Vesicles Delivery System

Vesicular systems, which can be further divided into liposomes, ultra-deformable liposomes, and ethosomes, are composed of amphiphilic molecules because they have polar or hydrophilic regions and non-polar or lipophilic regions. It has been demonstrated that vesicular systems, including liposomes, niosomes, transferosomes, penetration enhancer-containing vesicles, and ethosomes, can improve the therapeutic activity of medications used to treat wounds. They can lengthen the shelf life of hydrophilic and hydrophobic medications and lessen major side effects including skin irritation, and act as a depot for controlled drug release. They can also improve the penetration of such medications into the skin. The two types of vesicular systems are hard vesicles, such as liposomes and niosomes, and flexible or ultra-flexible vesicles, such as transferosomes.

According to reports, rigid vesicles are ineffective for transdermal drug delivery because they stay on the stratum corneum's outer layer and do not thoroughly penetrate the skin. After topical administration, liposomes can cause a variety of reactions. The majority of efforts have been concentrated on the topical treatments' antibacterial action, which has fallen short due to the rising rate of antibiotic resistance. They can inhibit systemic absorption, maximise side effects, and provide a localising impact as well as tailored distribution to skin appendages. They can also improve drug deposition within the skin at the site of action. Additionally, these vesicles were crucial in the healing of wounds [121]. Antibiotics entrapped in liposomes exhibit reduced toxicity and more target specificity along with increased efficacy in treatment of bacterial infections and thus improve its pharmacokinetics and pharmacodynamics. Increased action against external pathogens that are resistant, as well as increased activity against intracellular pathogens, is also an attractive feature. Due to their occlusive action on the stratum corneum, lipid nanoparticles may be more appropriate in burn wounds and chronic wounds since they can prevent transepidermal water loss and maintain the lesion moisture. Additionally, compared to vesicular systems, nanostructured lipid carriers are offered as superior nano-delivery methods. They have great stability, low toxicity, high drug-loading capacity, and sustained drug release, which helps speed up wound healing and cuts down the number of drugs administered [122].

4.4.1. Conventional Liposomes in Wound Healing

Because each pathophysiology differs, distinct skin wounds may require different treatments. According to the extent of the burn, acute burn wounds not only cause harm to the skin's structures but also to all of the body's systems because of the leakage of plasma into interstitial spaces. Additionally, the compromised skin barrier makes people more vulnerable to bacterial infections. In order to maintain skin functionality, it is crucial to prevent infections and promote re-epithelialisation while treating burn wounds. The majority of efforts have been concentrated on the topical treatments' antibacterial action, which has fallen short due to the rising rate of antibiotic resistance. In order to combat infections and promote skin regeneration in burn wounds, a variety of nanosized lipid-based delivery systems, including liposomes, transferosomes, ethosomes, and lipid nanoparticles, have been investigated. The results are encouraging [122,123].

New treatments for chronic wounds have been made possible by advancements in vesicular drug delivery systems, which have decreased the cost, toxicity, and number of applications while enhancing the half-life and bioavailability of the medications. The lowered membrane permeability of madecassoside, a highly powerful substance used to heal wounds, was likewise outperformed by liposome encapsulation. A formulation with a high entrapment efficiency, excellent long-term stability, and small particle size was discovered. Furthermore, double-emulsion liposomes enhance transdermal penetration and wound healing, despite the fact that liposomes are non-toxic, biodegradable, and skin-compatible. Curcumin and quercetin have also been included in the liposomes. Polyphenols quercetin and curcumin have antioxidant and anti-inflammatory properties that are helpful for wound healing [124].

4.4.2. Ultra-Deformable Liposomes or Transferosomes in Wound Healing

As a result of the creation of new vesicular systems called ultra-deformable liposomes, elastic vesicles, or transferosomes, conventional liposomes are currently used less frequently as transdermal delivery systems. Vesicles' suppleness allows them to deform and enter skin pores that are much smaller than their diameters as intact vesicles. Ultra-deformable liposomes enter undamaged skin and penetrate deeply, allowing the systemic circulation to absorb them. Here, a transdermal hydration gradient allows the ultra-deformable liposomes to pass through the intact stratum corneum and enter the epidermis. This is a result of the vesicles' high degree of deformability, which is brought on by the presence of surfactants, also referred to as "edge activator" molecules. With the ability to solubilise and fluidise epidermal lipids, edge activators have a significant impact on the exceptional deformability of transferosomes and increase their permeability capacity. Recently, consideration has been given for their capacity to traverse the stratum corneum among vesicular transferosomes. Transferosomes can transmit low bioavailability medications via the skin, according to numerous studies. Due to their deformability and ability to resist dry environments, transferosomes can pass to deeper skin layers undamaged. Transferosomes are therefore typically used for transdermal medication delivery in addition to their potential for topical distribution due to their capacity to penetrate deeply through epidermal layers and reach systemic circulation without the risk of vesicle rupture [121,123].

4.4.3. Ethosomes and Phytosomes in Wound Healing

Drugs that are both extremely hydrophobic and highly hydrophilic now have easier access to the deep skin layers thanks to ethosomes. Depending on how much ethanol is present, the ethosomes can range in size from 103 to 200 nm. The second generation of liposomes can be introduced as ethosomes. As a result, the vesicular systems' storage stability is a major challenge. However, ethosomes have more stability than regular liposomes. Several studies have looked into the utility of ethosomes in the treatment of wounds and have found that ethosomes transfer active ingredients to the skin more effectively than liposomes or conventional formulations. Recent research has shown that natural ingredients and antibacterial agents are more effective when they are encapsulated while treating burn wounds. In a different study, the ability of ethosomes loaded with silver sulfadiazine, a topical antibiotic regarded as the gold standard in burn wounds, to speed up the healing process and decrease bacterial infections in second-degree burns, was examined in vivo and in vitro. High entrapment efficiency is offered by phytosomes, which can be employed to distribute phytoconstituents topically for wound healing. The preparation was discovered to be risk-free and to have considerable antioxidant and wound-healing properties [123–125].

4.5. Emulsifying Drug Delivery System

The oral distribution of such medications is frequently linked to inadequate bioavailability, considerable intra- and inter-subject variability, and a lack of dose proportionality. Approximately 40% of novel drug candidates have poor water solubility. The use of surfactants, lipids, permeation enhancers, micronisation, salt formation, cyclodextrins, nanoparticles, and solid dispersions are just a few of the formulation strategies that have been used to tackle these issues. Recently, lipid-based formulations have received a lot of attention, with a focus on self-emulsifying drug delivery systems to increase the oral bioavailability of lipophilic medications. The fact that self-emulsifying drug delivery systems offer a significant interfacial area for the partitioning of the medication between oil and water is another benefit they have over straightforward oily solutions. Therefore, these systems may provide an improvement in the pace and amount of absorption as well as more consistent plasma concentration profiles for lipophilic medicines with dissolution-limited oral absorption [126].

By combining water and the non-ionic surfactant Tween 20 (Polysorbate 20), Ghosh et al. created a cinnamon oil microemulsion. Oil and surfactant were consumed in a 1:4 ratio. The microemulsion was discovered to be kinetically stable and generated with droplets that were around 5.79 nm in diameter. Ponto et al. studied that a microemulsion with antibacterial properties promotes wound healing in Wistar rats. In order to increase the solubility and bioavailability of hydrophobic/lipophilic medicines such as curcumin, self-emulsifying drug delivery systems are crucial substitute vehicles. Self-emulsifying drug delivery systems are physically stable isotropic mixes of oil/lipids, surfactants, and co-surfactants/co-solvents that have a great deal of potential as therapeutic drug delivery systems. To achieve the required therapeutic objectives, the self-emulsifying drug delivery systems formulations represent the hydrophobic drug in a solubilised state (nanoglobules) in the target location [127]. Through a number of methods, including drug solubilisation, droplet size reduction, improved membrane permeability, and protection of pharmaceuticals from chemical and enzymatic degradation, the self-emulsifying drug delivery systems formulations can increase drug bioavailability.

Different self-emulsifying drug delivery systems formulations were created and tested with an eye toward skin applications, including self-emulsifying drug delivery systems for the topical delivery of mangosteen peel and self-emulsifying drug delivery systems for the transdermal drug delivery of curcumin, *Opuntia ficusindica* fixed oil, *Piper cubeba* essential oil, and *Opuntia ficusindica* fixed oil (*Garcinia Mangostana* L.). The absorption of self-emulsifying drug delivery systems by the fibroblasts was found to be linearly dependent and dose-dependent. When curcumin is synthesised into nanocarriers, which have the least amount of cytotoxicity when compared to free drug solution, the cellular uptake of curcumin is facilitated [125–128].

4.6. *Nanofiber/Film/Membrane*

One of the most significant areas of science and technology in the twenty-first century that can be likened to the industrial revolution is nanotechnology. Soon, nanotechnology will have a substantial impact on the global economy and industry, as well on as the technology used by humans and their way of life. In the past, different materials such as animal fat, plant fibres, honey, etc., have been used to cover wounds (Table 2). Polymers have been employed in several research studies to make films for use as wound dressings. With two external dimensions that are identical in size at the nanoscale (about 100 nm) and a third dimension that is noticeably larger, nanofibres are one of the most fascinating classes of nanomaterials. Depending on the medication's solubility in the polymer solution, a polymer solution (polymer + particular solvent) is first created, after which a predetermined proportion of the drug is added, resulting in either a homogenous solution or a suspension [126–131]. This combination is electrospun to create nanofibres that contain a solid polymer–drug complex.

Table 2. Localised delivery systems of several drugs.

Formulation	Drug	Administration	Outcome	References
Nanofibre	Gentamicin sulphate (GS)	Topical application	It promotes cell adhesion and proliferation to scaffolds, and ultimately tissue regeneration and promotes healing process.	[129]
Nanofibre	Ferulic acid	Topically applied every day	Increased migration of cells to the wound site to fill the gap and increased proliferation causing rapid wound healing.	[130]
Nanofibre	Berberine	Topical treatment	Exhibited antibacterial activity against Gram-positive and Gram-negative bacterium. Animal studies on the STZ-induced diabetic rats demonstrated that the CA/Gel/Beri dressing enhanced the wound healing process.	[131]
Nanofibre	Peppermint	Topical dressing	Accelerated response and less inflammation and nanofibres showed potent wound healing activity for diabetic ulcers.	[132]
Nanofibre	Beta-glucan (βG)	Topically applies once a day	Enhanced maturation of granulation tissue and better healing process.	[133]
Nanofibre	Huangbai liniment (compound phellodendron liquid, CPL)	Topical treatment	It was also found that composite nanofibre membrane could reduce wound inflammation, down-regulate the expression of IL-1β and TNF-α inflammatory genes, and facilitate wound healing.	[134]
Nanofibre	Gentamicin salt (GEN)	Topical dressing	It has excellent antibacterial properties against Gram-negative E. coli, which is due to the unique properties of silver nanoparticles for antibacterial activity, and this composite has a good release profile for wound healing.	[135]
Nanofibre	Poly (caprolactone) (PCL)	Topical dressing once every day	It significantly promoted adhesion, proliferation and induced angiogenesis, collagen deposition, and re-epithelialisation in the wound sites of diabetic mice model, as well as inhibited inflammation reaction.	[136]
Hydrogel	Curcumin	Daily topical treatment	Shorten inflammatory process, prevents infection and re-epithelisation and promotes wound closer.	[137]

Table 2. *Cont.*

Formulation	Drug	Administration	Outcome	References
Liposome	Citicoline/chitosan	Topical treatment	Chitosan-coated liposomes containing citicoline have emerged as a potential approach for promoting the healing process in diabetic rats. However, the therapeutic effectiveness of the suggested approach in diabetic patients needs to be investigated.	[138]
Liposome	Curcumin	Topically applied once a day for 18 days	Curcumin-loaded liposomes in lysine–collagen hydrogel was found to be the most effective of the three formulations in promoting wound healing. Hence, this formulation can serve as a prototype for further development and has great potential as a smart wound dressing for the treatment of surgical wounds.	[139]
Liposome	DangguiBuxue	Topically applied	Remarkably accelerates wound closure, enhances hydroxyproline content in wound granulation tissue, promotes cutaneous wound healing by reducing the inflammatory response and improving fresh granulation tissue formation, and significantly increases the density of blood vessels and cell proliferation.	[140]
Nanoparticle	Silver	Topical treatment	Rapid healing and improved cosmetic appearance via reduction in wound inflammation and modulation of fibrogenic.	[141]
Nanoparticle	Zinc oxide (ZnO_2)	Topical dressing	Had excellent anti-bacterial activity and rapid wound healing.	[142]
Nanomembrane	Triphala		Triphala PCL shows good broad spectrum of antimicrobial activity and biocompatibility and helps control wound infection and enhanced healing due to antioxidants of Triphala.	[143]
Nanomembrane	Chitosan	Topical treatment	This nanomembrane serves as an excellent microenvironment for cell adhesion, migration, proliferation, and differentiation. An in vivo experiment with this nanomembrane was also conducted, showing that it has a great capability for stem cell delivery for skin tissue reconstruction.	[144]
Hydrogel	Glycosaminoglycan	Topical application	Promotion of tissue proliferation and regeneration of vascular vessels.	[138]
Deformable liposome	Curcumin	Daily topical treatment	CDLs in hydrogel preserved hydrogel's bioadhesiveness to a higher degree than both NDLs and ADLs. In addition, CDLs-in-hydrogel enabled the most sustained skin penetration of curcumin and hence facilitates wound healing.	[145]
Liposomal ointment	Retinoic acid and growth factors	Topical application	Liposomal ointment on deep partial-thickness burn model stimulated wound closure ($p < 0.001$), promoted skin appendage formation and increased collagen production, thus improving healing quality.	[146]

Table 2. *Cont.*

Formulation	Drug	Administration	Outcome	References
Hydrogel nanoparticle	Copper (Cu)	Topical treatment	CuNP-comprised hydrogels exhibited a significant decrease in bacterial activity and promoted effective wound closure with negligible toxicity in our histological evolution.	[147]
Nanogel	Cerium oxide	Topical application	Showed significant antibacterial properties even at low absorptions and is effective at damage and scar production.	[148]
Nanoparticle	Thrombin	Topical treatment	The proportionate improvement in skin tensile strength after treatment with bound thrombin suggests that the novel thrombin conjugates may lessen surgical difficulties.	[149]
Microneedle	Trichostatin A, histone deacetylase 4	Topical	The microneedle-mediated Trichostatin A patch has been shown to improve the healing of diabetic wounds by reducing inflammation, promoting tissue regeneration, and inhibiting histone deacetylase 4.	[150]
Metal–organic framework microneedle patch	Nitric oxide	Topical application	Delivering nitric oxide molecules more precisely and deeply into the wound site may be made possible by the integrated microneedle's porous shape, increased specific surface area, and enough mechanical strength.	[151]
Microneedle	Curcumin nanodrugs/new Indocyanine Green/hyaluronic acid	Topical treatment	The two-layered microneedles platform has the potential to be used as a competitive technique for the treatment of melanoma since it can simultaneously remove the tumour and speed up wound healing.	[152]

A wide range of polymers can be used to create nanofibres. For dressing nanofibres, there are only three types of polymers now available: natural polymers, synthetic polymers, and mixed polymers. Natural polymers are appropriate for use in biomedical applications due to their wide range of benefits, including biocompatibility, non-toxicity, biodegradability, antibacterial properties, and desirable mechanical structure. In the procedure, the solvent evaporates. On the other hand, nanofibres are made of a polymeric base, which makes up most of the fibre's composition, and a bioactive molecule (such as a protein, hormone, or medication), or another type of polymer, but in a lesser amount than the base polymer. Currently, the three primary techniques for producing nanofibres are electrospinning, the phase-separation method, and the self-assembly method. The method that produces nanofibres most frequently is electrospinning. Depending on the electrospinning technique employed, various types of nanofibres can be produced. Today, the commercialisation of electrospinning equipment is advancing quickly. The most popular electrospinning methods include bubble electrospinning, melt electrospinning, coaxial electrospinning, self-bundling electrospinning, and nano-spider electrospinning [132–144]. Fibre electrospinning processes can be used to create polymer nanofibres (50–1000 nm) (wet or hot melt electro-spinning). Numerous excellent characteristics of nanofibres include their substantial surface area, the potential for surface functionalisation, tunable porosity, a broad range of material options, and outstanding mechanical performance. The advantageous characteristics of electrospun nanofibres, such as mechanical stability, high porosity, high surface area to volume ratio, and ability to exchange water, oxygen, and nutrients, encourage good cell attachment, differentiation, and proliferation. They enable the monitoring of infection and healing markers due to their capacity to mimic the structure and operation of ECM [145]. The nanofibres are excellent candidates for a variety of biomedical applications, such as tissue-engineered scaffolds (such as skin, cartilage, bone, and blood vessels), dressings for wound healing, biomedical devices, biosensors, and drug delivery systems, because of their exceptional capabilities. Electrospun nanofibre mats offer a native extracellular matrix-like structure with high interconnected porosity (60–90%), great absorbencies, and a water absorbance capacity of 18–21% more than films made of the same polymers [146]. These characteristics, along with balanced moisture and gas permeability, create an environment that is suitable for preventing exogenous infection, cell migration and proliferation, haemostasis, exudate absorption, and cell respiration in wounds. Electrospun nanofibres can control the proliferation, migration, differentiation, and synthesis of native extracellular matrix in skin cells.

Tissue engineering and wound healing are two of the most important and intriguing biomedical uses of nanofibres. Nanofibres have been utilised in the treatment of diabetic ulcers and wounds to aid in wound healing, haemostasis, skin regeneration, and wound dressing. Nanofibre keeps the wound surface moist while healing because it can hold more moisture in its structure. As a result, the nanofibres cannot adhere to the surface of the wound. Additionally, the porous nanofibre network makes it simpler for oxygen to diffuse into the wound area. To remove toxins from the blood of individuals with kidney failure, wearable blood purification systems may integrate the nanofibre membrane. Nanofibre scaffolds hold great promise for wound healing. These scaffolds are used in the treatment of diabetic ulcers, skin rejuvenation, wound dressings to encourage healing, and haemostasis [147,148]. Nanofibres keep the wound area moist while it heals because of their capacity to hold more moisture within their structures. As a result, the scaffold cannot adhere to the damaged surface. Furthermore, oxygen may easily diffuse to the location of the wound thanks to the porous networks. Nanofibres have the power to regulate a range of skin cell responses, including proliferative, migratory, differentiating, and extracellular matrix accumulation. Their microstructure ideally complements the ECM structure, which encourages cellular proliferation, adhesion, and growth. While creating a larger surface area than bulk materials, nanofibre arrangements have increased porosity, which is advantageous for cell activities.

The polymers used to create scaffolds for wound dressing include collagen, poly-vinyl pyrrolidone, polyacrylic acid, polyvinyl alcohol, gelatin, chitosan, silk fibroin, polyesters, and poly-urethane [149]. The goal of a wound dressing is to quickly achieve haemostasis, and it should also have strong antibacterial properties to guard against bacterial infections from the environment. The ability of electrospinning to construct nanofibrous membranes for wound dressings that can provide a moist environment surrounding the wound region to facilitate healing has drawn a great deal of interest. In order to create composite nanofibre membranes that carry a reservoir of biogenic AgNPs for use as a wound dressing, Bardania et al. used *T. polium* extract as a reducing agent. This method of "green synthesis", which does not use external stabilisers or reducing agents, produced AgNPs quickly, cheaply, and effectively [76,82,109].

4.7. Foam Dressings

Foam dressings are permeable to both gases and water vapour and have a polyurethane foundation. The outer layer's hydrophobic qualities shield it from liquids while allowing gaseous exchange and the passage of water vapour. Silicone-based rubber foam, or silicone, conforms to the shape of wounds. Depending on the thickness of the wound, foam can absorb varied amounts of wound drainage. There are foam dressings that are both adhesive and non-adhesive. Lower leg ulcers, mild to heavily exuding wounds, and granulating wounds can all benefit from foam dressings. They offer both thermal insulation and excellent absorption thanks to their hydrophilic characteristics [153]. These incredibly adaptable dressings should be used on moderate-to-heavy exudative wounds, partial- and full-thickness wounds that are granulating, or slough-coated donor sites, ostomy sites, mild burns, and diabetic ulcers. Due to their capacity to cause wounds to become even drier, they are not advised in dry or eschar-covered wounds and vascular ulcers. Due to their high absorbency and moisture vapour permeability, they are typically utilised as main dressings for absorption, and supplementary dressings are not necessary. The disadvantage of foam dressings is that they need to be changed frequently and are not appropriate for low exudating wounds, dry wounds, or dry scars since they need exudates to heal. They can stay in place for up to 4 to 7 days, but once they become saturated with exudates, they need to be changed. When removed, they are non-traumatic due to their makeup. They can also be applied to infected wounds if they are changed daily [39]. Numerous papers have examined various foam dressing forms, mostly because of their propensity for liquid and/or permeability to moisture vapour. These characteristics may make some foam more suitable for treating weakly exudative wounds as opposed to lesions that drain excessively [154]. The primary limitation of this kind of dressing is that it requires a second dressing, such as an elastic bandage or a film, to adhere to the wound. Foam dressings are used for deep wounds that have produced a lot of exudates as well as for long-term wounds such as venous ulcers. They may also be applied under compression bandages [149,153].

4.8. Biological Dressings

In order to restore the wound healing process and incorporate active biological agents to assist the wound healing process, advanced biological therapies are currently emerging in ischaemic wound therapies. The use of active biological agents, such as plant-derived active biomolecules with antioxidant, antibacterial, or anti-inflammatory properties, may be used in biological wound-healing therapies, which aim to aid the restoration of the body's natural repair mechanisms. Biological dressings stop contamination, heat loss, protein and electrolyte loss, and evaporative water loss [15,22]. Bioactive wound dressings can be created using naturally occurring biomaterials with endogenous activity or materials that release bioactive chemicals. Chitin, chitosan, hydrocolloids, alginate, and derivatives of organic biopolymers are a few examples of these biomaterials. Biological dressings made from animal collagen are known as collagen dressings. The fibroblast-produced protein collagen plays a part in each step of wound healing. The extracellular matrix is primarily made up of collagen, which gives it strength [29]. This category includes dressings made from collagen sourced from bovine, porcine, or avian sources. These products are all intended to hasten the healing process. Protease activity is decreased by both collagen and oxidised regenerated cellulose (ORC), which is usually coupled with collagen. Proteases have a variety of roles in the healing of wounds, albeit their levels typically decline over time. There is some proof that collagen dressings for venous leg ulcers are at the very least comparable to other modern wound dressings [36].

A glycoaminoglycan component of the ECM, hyaluronic acid has distinct biological and physical properties. Like collagen, hyaluronic acid is biocompatible, biodegradable, and naturally immune-suppressive. During the proliferative stage of wound healing, chitosan encourages the development of granulation tissue. Biological dressings are said to be more effective than other forms of dressings when compared to other dressings [37,38]. Alginates and chitosan have also been used successfully in a haemostatic dressing.

In a rat full-thickness wound model, it was demonstrated that sponges based on carboxymethyl chitosan, alginates, and a Chinese medication could quickly induce haemostasis and sustain wound closure. Lyophilisation and spray-drying procedures were used to create multi-resorbable haemostatic dressings made of chitosan, sodium/calcium alginate, and/or carboxymethyl cellulose. This kind of treatment was initially applied as a biological dressing to cover extensive burn wounds. Additionally, it has been used to heal chronic wounds, such as venous stasis ulcers [39]. Allografts from cadavers serve as a substrate for the formation of granulation tissue. The cadaveric skin is attached, an antibiotic/antimicrobial dressing is placed over the graft, and a compression dressing is used to lessen swelling once the wound bed has been prepped as previously mentioned. To enhance the physical and biological properties of tissue engineering products, such as wound dressings, chemical cross-linking is frequently utilised in the manufacturing process. Sodium trimetaphosphate, a secure and non-toxic cross-linking agent, can be used to improve the functional qualities of various polysaccharides, including starch, cellulose, and xanthan [48]. In patients with an ischaemic wound, biological wound dressing or topical agent therapy may hasten wound healing, increase limb salvage, be reasonably priced, and offer potential safety with nontoxic low-risk therapy. Therefore, patients with ischaemic wounds should also receive local wound care using biological dressings as an adjuvant therapy. To support the effectiveness and long-term results of these biological dressings in patients with ischaemic wounds, additional randomised studies are required [49].

4.9. Charcoal Dressings

By absorbing gases produced by bacteria, activated charcoal dressings serve the primary purpose of reducing wound odour. They can absorb odour molecules due to their huge surface area and function as a deodorising agent. Wound odour is very subjective in nature since it is challenging to define and quantify. Leg ulcerations and diverse fungating lesions are the wounds most frequently linked to odour generation. Numerous aerobic bacteria as well as anaerobes including *Bacteroides* and *Clostridium* species are among the organisms usually linked to malodorous wounds [51–53]. Eliminating the problematic organism is the best strategy for treating wound smells. Antibiotics taken systemically may be successful, but it may be challenging to obtain an adequate concentration of the drug at the infection site. In numerous studies to date, topical treatments such as metronidazole, clindamycin, honey, and sugar have demonstrated promise in this area. Activated charcoal dressings are frequently used in malodorous wounds; however, odour control has not received as much attention in the literature as wound healing has. According to current clinical experiences, it is evident that these dressings can reduce wound odour, but there are not any hard facts on just charcoal ingredients [56–58].

4.10. Three-Dimensional Skin Substitutes

Recently, several tissue engineering technologies have become available; these technologies take a fundamentally unique and new approach. Among these, three-dimensional free-form fabrication—often referred to as three-dimensional bioprinting—offers several benefits over traditional skin tissue engineering. An innovative method for designing and engineering human organs and tissues is three-dimensional bioprinting, a flexible automated on-demand platform for the free-form production of complex living constructions. Here, we use human skin as a representative example to show the potential of three-dimensional bioprinting for tissue engineering. In order to simulate the epidermis, dermis, and dermal matrix of the skin, keratinocytes, fibroblasts, and collagen were employed as constituent cells [154]. This method has enormous potential for the creation of three-dimensional skin tissue since it can dispense living cells, soluble components, and phase-changing hydrogels in a desired pattern while preserving very high cell viability.

The biomimetic mechanical cues that support vascularisation, alignment of fibrous proteins in the ECM, integration of dermal and epidermal components, and adhesion between these layers are absent from the existing skin grafts and their production techniques. When the biomechanics of the repaired tissue at the wound site are compared to those of the healthy tissue nearby wounds, this problem may become more difficult. Therefore, to prevent the separation of the layers during application, a regenerative skin scaffold should take biomimetic mechanical cues into account. By adjusting the physicomechanical characteristics of each layer, the scaffold can offer tailored microenvironments for various cell types [155]. The use of autologous epidermal sheets as a kind of skin replacement has progressed into the use of more sophisticated bilayered cutaneous tissue-designed skin substitutes. However, their regular use for restoring normal skin anatomy is constrained by insufficient vascularisation, rigid drug/growth factor loading, and the inability to regenerate skin appendages such as hair follicles [156].

Recent developments in cutting-edge science from stem cell biology, nanotechnology, and other vascularisation techniques have given researchers a huge head start in creating and modifying tissue-engineered skin substitutes for better skin regeneration and wound healing. The creation of scaffolds for tissue engineering, films and membranes for wound healing, artificial tissues, and even artificial organs, all include the use of three-dimensional printing. Superior flexibility, regulated porosity, and reproducibility are the key benefits of the scaffolds made using this method. One of the current synthetic alternatives imitates the layer of skin made up of keratinocytes and fibroblasts on the collagen matrix. Only the dermal components with fibroblasts on the collagen matrix are present in the cellular matrix. Focused application, which stimulates wound healing and directs the healing of severe burns without scars, yielded encouraging results.

Bioengineered organisms can adjust to their surroundings and release the growth factors and cytokines used in dressings. Both venous leg ulcers and diabetic foot ulcers can be treated using bioengineered dressings. The additive manufacturing method of three-dimensional bioprinting offers a viable method for creating biocompatible artificial skins by carefully layering growth factors, biomaterials, and living cells on top of one another. This automated technology is a versatile tool that, in terms of accuracy and usefulness, is ideal for therapeutic usage. Through the exact deposition of many cells and biomaterials, the bioprinted skin analogues closely replicate the native skin's architecture and heterogenicity [157]. The bioprinted skin constructions need to meet several requirements in terms of their compositional and functional qualities. The ability to transmit nutrients and wound exudates should be the first feature of bioprinted skin substitutes. The ability to precisely deposit various skin cells, such as keratinocytes, fibroblasts, adipocytes, melanocytes, Langerhans cells, etc., at certain layers and places is another requirement for bioprinted analogues. Finally, the bioprinted structure needs to be strong, biocompatible, biodegradable, and able to withstand the external forces and pressures that are present under in vivo circumstances. Despite revolutionising wound treatment, skin replacements have several drawbacks. They necessitate lengthy cell culture processes that extend production times. Most of the vascularisation is inadequate. Scarring occurs at the graft margins, which exacerbates existing functional and aesthetic issues. Infection is always a possibility. Since they only have fibroblasts and keratinocytes, they are unable to develop differentiated structures including sweat and sebum glands, hair follicles that cycle, pigmentation, sensory innervation, and motor innervation, as well as the epidermal barrier and dermal–epidermal junction [39,55–155].

Porosity, degradation, and mechanical properties should closely resemble the native skin structure. Onto a scaffold made of hyaluronic acid in three dimensions, autologous cultured fibroblasts are sown. Apligraf, a skin substitute for venous ulcers made from keratinocytes and fibroblast-seeded collagen, has received FDA approval [156,157]. Commercially available skin substitutes include IntegraTM artificial skin, which is made of a collagen/chondroitin-6-sulphate matrix coated with a thin silicone sheet, and AllodermTM, which is made of normal human fibroblasts with all cellular components removed. LaserskinTM, BiobraneTM, BioseedTM, and Hyalograft3-DTM are a few further alternatives.

4.11. Dendrimers

Dendrimers are a class of nanoscale (1–100 nm) three-dimensional globular macromolecules with numerous arms branching out from a central core. They have unique structural characteristics, including high levels of branching, multivalency, globular architecture, and well-defined molecular weight, making them promising drug delivery scaffolds. The distinctive structure of dendrimers makes them an ideal nanomaterial for the administration of medications to target certain tissues or molecules with solubility challenges. This is performed by trapping the drugs inside of their void spaces, branches, or outside functional groups. An expanding field of research involves cross-linking collagen with functionalised nanoparticles to produce scaffolds for use in wound healing. Due to their spherical structure, dendrimers can engage via hydrogen bonding, lipophilicity, and charge interactions with tiny medicines, metals, or imaging moieties that can fit within their branches. In addition to their structure, dendrimers may be the best drug delivery vehicles for many therapeutic treatments because of their size and lipophilicity, which allow them to easily permeate cell membranes. These distinguishing qualities have increased interest in using dendrimers nanoparticles for wound healing in research [158].

In order to build an efficient nanoparticle-mediated scaffold for tissue engineering and wound healing applications, Vedhanayagam et al. showed that nanoparticle form is a critical component that needs to be investigated. The re-epithelialisation and collagen deposition in damaged tissue are accelerated more quickly by the spherical shape of zinc oxide triethoxysilane poly(amidoamine) dendrimer generation 1 nanoparticles than by other shapes. The highly branched 3D structures known as antimicrobial peptide dendrimers (AMPDs) have a central core and a high density of flexible surface groups for possible molecule attachment. Their capacity to penetrate the cell membrane and the presence of functional groups of amino acid residues are what cause them to have bactericidal effects [159].

Mannose-decorated globular lysine dendrimers have the ability to reduce inflammation by targeting and reprogramming macrophages to the M2 phenotype, which is characterised by significant mannose receptor clustering on the cell surface and the elongated shape; increased production of TGF-1, IL-4, and IL-10; decreased secretion of IL-1, IL-6, and tumour necrosis factor (TNF); and increased ability to induce fibroblast proliferation. These results show that M2 macrophage polarisation can be directed by mannose-decorated globular lysine dendrimers, which may be helpful in the treatment of injuries and inflammation [160].

The development of an antisense delivery method based on dendrimers has assisted in the development of an antisense therapy strategy to treat bacterial infections. Dendrimers themselves might function as powerful antibacterial substances. In contrast, those with metal cores can produce active antimicrobial agents such as metal ions and ROS, which can kill bacteria. For example, those with positively charged surfaces typically have significant interactions with negatively charged bacterial cell membranes. AgNPs may benefit from the alternate template that dendrimers can offer. Dendrimers have a stronger antibacterial action when combined with silver than silver has by itself. In computerised tomography or immune-sensor coatings, poly-(amido-amine) dendrimers have been utilised as contrast agents. These highly branched dendritic molecules have a restricted size range, a well-defined globular structure, and a relatively large molecular mass [161].

Among the family of monodisperse, highly branched units with a clearly defined structure are dendrimers such as poly(amidoamine). Their surface contains a cationic primary amine group, which enables them to participate in chemical bonding. For the treatment of wounds, poly-(amido-amine) dendrimers are primarily used as a vector for the transport of various hydrophilic or hydrophobic medicines and genes. The hesperidin-loaded dendrimer was found to be biocompatible and suitable for use in wound healing after a haemolysis investigation [162]. Tumour suppression applications have demonstrated the anti-angiogenic properties of polycationic dendrimers. Angiogenesis and neovascularisation, however, are important components for skin regeneration following serious burn wounds because newly created blood vessels aid in the healing process by improving nutrition and oxygen supply to regenerating tissues. Therefore, using polycationic antimicrobial peptide dendrimers could prevent angiogenesis and endanger the process of skin regeneration [163].

Dendrimers are incorporated into gelatin nanofibres through covalent conjugation, which not only increases the capacity of nanofibre construction for drug loading but also offers a great deal of versatility for creating multipurpose electrospun dressing materials. Since dendrimer–gelatin nanofibre constructs are designed to treat a variety of wounds, including chronic wounds, burns, and skin malignancies, they can be customised to offer cutting-edge therapies [164,165]. Dendrimers have a variety of uses in biomedicine, but their toxicity has also been described in order to analyse the restrictions on their use. The structure of a chemical determines its toxicity as well as all of its other characteristics. A dendrimer's toxicity can either be increased or decreased depending on specific components (core, branch, and surface groups) [158,161].

4.12. Carbon Nanotubes

The carbon nanotubes are a class of stiff, stable, hollow nanomaterials with a variety of special physical, chemical, and mechanical properties that have been widely used as catalyst supports, nanowires, electronic components, and more recently in the fields of biomedical engineering and medical chemistry [166]. A needle-like structure with a significant surface area, carbon nanotubes are an allotropic form of carbon with nanoscale dimensions and μm lengths. The carbon atoms join to form sheets of graphite, which is made up of six-membered carbon atom rings. Graphite then spirals into tubes. Carbon nanotubes can be divided into single-wall and multiwall varieties based on the number of graphite layers present. CNTs have tensile strengths up to 63 gigapascals, which is around 50 times stronger than steel, and elastic moduli between 1.0 and 1.8 terapascals. With the goal of boosting drug delivery, regulating drug release, and improving therapeutic activity, carbon nanotubes are now being functionalised by various pharmacologically active compounds, with some degree of success [167]. The degradation products of carbon nanotubes can be eliminated by the functional tissues of the human body, making them non-toxic and safe for consumption [168]. The fibroblasts, which are crucial to the cell renewal system and the healing process of open wounds, may come into contact with the carbon nanotubes employed in the biomedical field and found in the environment as they pass through the skin or open wounds [166]. Proteins and receptors found in cell membranes can cling tightly to carbon nanotubes. Carbon nanotubes and cells must adhere to one another [166].

The ability of organic compounds with a carbon component to improve the antibacterial activity of polymers is well known. Because they can function as a conductive bridge over the insulating bilayer, carbon nanotubes' antimicrobial activity can be increased following functionalisation with -OH and -COOH functional groups or hybridisation with metallic compounds. After being incorporated into chitosan biopolymers, multiple-walled carbon nanotubes boost cell survival and proliferation with few negative side effects. Using the solution casting technique, Liu et al. created a bio-nanocomposite film-based wound healing dressing. The film's ability to absorb water is crucial to preventing tissue dehydration, limiting the growth of microorganisms, and safeguarding wound maceration. It is anticipated that the multiple-walled carbon nanotubes' hydrophilic nature and water-holding capacity will increase the bio-nanocomposite film's ability to absorb water. Through chemical interactions, the presence of multiple-walled carbon nanotubes increases ROS formation [169]. The physical properties of fibrous proteins found in the ECM, such as collagen and elastin, are approximated by the greater diameter of multiple-walled carbon nanotubes compared to single-walled carbon nanotubes.

According to molecular research conducted by Khalid et al., bacterial cellulose functionalised with multiwalled carbon nanotubes displayed a lower than control level of pro-inflammatory cytokines IL-1 and TNF- and a higher level of VEGF expression, which may have favoured a quicker healing process [170]. Carbon nanotubes may pass through a variety of physiological barriers and have substantial tissue penetration. Isoniazid/chitosan/carbc nanotube nanoparticles were created by Chen et al. and have been shown to dramatically speed up the healing of tuberculosis ulcers [171]. Experimental tests demonstrated by Zhao et al. revealed that the antibacterial gel made of nano-silver and multiple empty carbon nanotubes has a superior anti-infective impact on burn wounds and can significantly lessen the frequency of dressing changes [172]. By affecting cell spreading, adhesion, migration, and survival, exposure to carbon nanotubes causes inflammation, genotoxicity, and inhibits dermal fibroblasts' capacity to heal wounds. While concurrently harming the cytoskeleton and upsetting actin stress fibres in NIH 3T3 murine fibroblasts and human dermal fibroblasts, multiwalled carbon nanotubes reduced DNA synthesis and the levels of adhesion-related genes. Skin allergies are made worse, and keratinocyte cytotoxicity is also produced, by low doses of topical multiwalled carbon nanotubes [166–169].

4.13. Microneedle Drug Delivery Systems

The efficacy of traditional single-drug therapies is subpar, and penetration depth limits the effectiveness of drug delivery [173]. In the realm of wound healing, microneedle dressings with transdermal drug delivery capabilities have been crucial [174]. Additionally, the microstructure of microneedles allows for efficient medication administration to the target location while preventing overly strong skin and patch adherence. Additionally, temperature-sensitive hydrogel has been used to encapsulate vascular endothelial growth factor (VEGF) in the chitosan microneedle array micropores. As a result, the temperature increase brought on by the inflammatory response at the site of wounds can be used to controllably achieve the smart release of the medications [175]. The adaptable approach of the microneedle patch has been presented and has achieved several outstanding successes in the fields of disease therapy, biosensing, skin vaccination, and wound healing. Microneedles can efficiently deliver the desired active pharmaceuticals due to their better loading capacity and well-designed microstructures when compared to those used in conventional drug delivery systems. However, the microneedle that is so frequently used today is typically made from synthetic polymer materials that were created by difficult chemical synthesis using harsh experimental processes and environmentally hazardous organic reagents. This raises the danger of side effects. Additionally, the development of microneedle-based iatrotechnics is constrained by the acquisition of loaded active pharmaceuticals typically through a period of brutal elimination and extremely stringent clinical studies [176]. Wang et al. created a three-dimensional origami microneedle patch with extremely tiny needle structures, microfluidic channels, and numerous functionalities that was said to be able to detect biomarkers, distribute medications in a controlled manner, and monitor motions to speed up wound healing [177]. Exosomes can only partially reach the injury site through passive diffusion, according to Liu et al.'s explanation of the potential of microneedle delivery of exosomes for transdermal application [178]. As a result, the therapeutic effects and clinical applications are significantly diminished. An antibacterial and angiogenesis-promoting double-layer microneedle patch was reported by Gao et al. for the treatment of diabetic wounds. Tetracycline hydrochloride, an antibacterial medication filled with hyaluronic acid, serves as the tip of the double-layer microneedle, while deferoxamine, an angiogenic medication, serves as the substrate [179]. The transdermal route has been used to administer bioactives using microneedles. The bioactives can pass through the epidermis with the assistance of these microneedle devices. Microneedles are created via in situ polymerisation utilizing a mould-based approach using biomaterials such as chitosan, hyaluronic acid, and maltose [180]. In the realm of wound management, new individualised and programmable microneedle wound dressings are extremely valuable. With its straightforward, efficient, and safe qualities, the microneedle-mediated drug delivery system can also offer a novel method for the treatment of diabetic wounds. It has a wide range of applications in relevant biomedical sectors. The low stiffness of the microneedles for insertion into human skin poses a challenge for the translation of this method. The constraints and high cost of microfabrication technology have restricted the development of microneedle-based systems utilising standard subtractive techniques [181,182].

5. Future Prospects

The alignment of fibrous proteins, cell type, ECM composition, water content, and mechanical qualities vary between skin layers. However, the skin's natural ability to regenerate is constrained to treating relatively small wounds, necessitating the use of topical medications in some or most cases. Therefore, because the healing process entails complicated events to re-establish functional tissues following injury, big skin wounds present a problem. Due to the existence of several cells and chemicals orchestrating a process, wound healing has always been the most difficult problem. Any problem may hinder the healing process and cause an acute wound to become chronic. The difficulty in keeping immune cells in their naive immune state is currently a major barrier to their incorporation in the skin, and it is probably due to their ineffective incorporation within the skin strata as well as the absence of other functionally significant cells. Adult stem cells have emerged as a promising therapy for promoting scarless wound healing among the numerous techniques that have so far been used in the treatment of skin ulcers. Mesenchymal stem cells have drawn more attention than other adult stem cells due to their capacity for immunomodulation and tissue regeneration. Intense research is still being conducted on the creation of novel and efficient therapies for wound care. Future topical therapy for wound care can be evaluated using comparative effectiveness research as a technique. Over the past ten years, a number of novel techniques and items have been developed, helping to meet the ongoing demand for advancements in wound care. Numerous other techniques, including hyperbaric oxygen, growth hormones, biologic dressings, skin substitutes, and regenerative materials, have also demonstrated effectiveness in speeding up the healing of wounds.

Author Contributions: Conceptualization, R.T. and K.P.; validation, R.T.; resources, R.T. and K.P.; writing—original draft preparation, R.T.; writing—review and editing, K.P.; supervision, K.P. All authors have read and agreed to the published version of the manuscript.

Funding: This research received no external funding.

Institutional Review Board Statement: Not applicable.

Informed Consent Statement: Not applicable.

Data Availability Statement: Not applicable.

Conflicts of Interest: The authors declare no conflict of interest.

Abbreviations

AgNPs: Silver nanoparticles; AMPDs: Antimicrobial peptide dendrimers; AuNPs: Gold nanoparticles; CAT: Catalase; CNTs: Carbon nanotubes; CTGF: Connective tissue growth factor; ECM: Extracellular matrix; EGCG: Epigallocatechin-3-gallate; EGF: Epidermal growth factor; EGR1: Early-growth response 1; ELAM-1:Endothelial leukocyte adhesion molecule 1; FGF-2: Fibroblast growth factor -2; GPX: Glutathione peroxidase; HIF-1: Hypoxia-Inducible Factor; ICAM-1: Intercellular adhesion molecule 1; IGF-1: Insulin-like Growth Factor-1; IL-1: Interleukin-1; IRF5: Interferon regulatory factor 5; LOX-1: Lectin-like oxLDL (oxidised low-density lipoprotein); MAPK: Mitogen-activated protein kinase; MCP-1: Monocyte Chemoattractant Protein-1; MMP-1: Matrix metalloproteinase-1; MRSA: Methicillin-resistant *Staphylococcus aureus*; mTOR: mammalian target of rapamycin; ORC: Oxidised regenerated cellulose; PDGF: Platelet-derived growth factor; pNFκB: Phospho-Nuclear Factor Kappa B; PPAP: Polyprenylated acylphloroglucinol; pSTAT1:Phosphosignal transducer and activator of transcription 1; ROS: Reactive oxygen species; SMAD: Suppressor of Mothers against Decapentaplegic; SOD: Superoxide dismutase; TGF- β: Transforming Growth Factor-beta; TLR: Toll-like receptors; TNF-α: Tumour Necrosis Factor Alpha; VCAM-1:Vascular cell adhesion molecule 1; VEGF: Vascular endothelial growth factor; VPF: Vascular Permeability Factor.

References

1. Hassanshahi, A.; Hassanshahi, M.; Khabbazi, S.; Hosseini-Khah, Z.; Peymanfar, Y.; Ghalamkari, S.; Su, Y.W.; Xian, C.J. Adipose-derived stem cells for wound healing. *J. Cell. Physiol.* **2019**, *234*, 7903–7914. [CrossRef]
2. Kim, H.S.; Sun, X.; Lee, J.H.; Kim, H.W.; Fu, X.; Leong, K.W. Advanced drug delivery systems and artificial skin grafts for skin wound healing. *Adv. Drug Deliv. Rev.* **2019**, *146*, 209–239. [CrossRef] [PubMed]
3. Ryall, C.; Duarah, S.; Chen, S.; Yu, H.; Wen, J. Advancements in Skin Delivery of Natural Bioactive Products for Wound Management: A Brief Review of Two Decades. *Pharmaceutics* **2022**, *14*, 1072. [CrossRef] [PubMed]
4. Martin, P.; Nunan, R. Cellular and molecular mechanisms of repair in acute and chronic wound healing. *Br. J. Dermatol.* **2015**, *173*, 370–378. [CrossRef]
5. Han, G.; Ceilley, R. Chronic Wound Healing: A Review of Current Management and Treatments. *Adv. Ther.* **2017**, *34*, 599–610. [CrossRef] [PubMed]

6. Yuan, Z.; Zhang, K.; Jiao, X.; Cheng, Y.; Zhang, Y.; Zhang, P.; Zhang, X.; Wen, Y. A controllable local drug delivery system based on porous fibers for synergistic treatment of melanoma and promoting wound healing. *Biomater. Sci.* **2019**, *7*, 5084–5096. [CrossRef] [PubMed]
7. Elviri, L.; Bianchera, A.; Bergonzi, C.; Bettini, R. Controlled local drug delivery strategies from chitosan hydrogels for wound healing. *Expert Opin. Drug Deliv.* **2017**, *14*, 897–908. [CrossRef]
8. Gorain, B.; Pandey, M.; Leng, N.H.; Yan, C.W.; Nie, K.W.; Kaur, S.J.; Marshall, V.; Sisinthy, S.P.; Panneerselvam, J.; Molugulu, N.; et al. Advanced drug delivery systems containing herbal components for wound healing. *Int. J. Pharm.* **2022**, *617*, 121617. [CrossRef]
9. Veith, A.P.; Henderson, K.; Spencer, A.; Sligar, A.D.; Baker, A.B. Therapeutic strategies for enhancing angiogenesis in wound healing. *Adv. Drug Deliv. Rev.* **2019**, *146*, 97–125. [CrossRef]
10. Guo, S.; Dipietro, L.A. Factors affecting wound healing. *J. Dent. Res.* **2010**, *89*, 219–229. [CrossRef]
11. Rubalskii, E.; Ruemke, S.; Salmoukas, C.; Aleshkin, A.; Bochkareva, S.; Modin, E.; Mashaqi, B.; Boyle, E.C.; Boethig, D.; Rubalsky, M.; et al. Fibrin glue as a local drug-delivery system for bacteriophage PA5. *Sci. Rep.* **2019**, *9*, 2091. [CrossRef] [PubMed]
12. Jang, M.J.; Bae, S.K.; Jung, Y.S.; Kim, J.C.; Kim, J.S.; Park, S.K.; Suh, J.S.; Yi, S.J.; Ahn, S.H.; Lim, J.O. Enhanced wound healing using a 3D printed VEGF-mimicking peptide incorporated hydrogel patch in a pig model. *Biomed. Mater.* **2021**, *16*, 045013. [CrossRef] [PubMed]
13. Shedoeva, A.; Leavesley, D.; Upton, Z.; Fan, C. Wound Healing and the Use of Medicinal Plants. *Evid. Based Complement. Altern. Med.* **2019**, *2019*, 2684108. [CrossRef] [PubMed]
14. Ibrahim, N.; Wong, S.K.; Mohamed, I.N.; Mohamed, N.; Chin, K.Y.; Ima-Nirwana, S.; Shuid, A.N. Wound Healing Properties of Selected Natural Products. *Int. J. Environ. Res. Public Health* **2018**, *15*, 2360. [CrossRef] [PubMed]
15. Kumari, A.; Raina, N.; Wahi, A.; Goh, K.W.; Sharma, P.; Nagpal, R.; Jain, A.; Ming, L.C.; Gupta, M. Wound-Healing Effects of Curcumin and Its Nanoformulations: A Comprehensive Review. *Pharmaceutics* **2022**, *14*, 2288. [CrossRef]
16. Fatehi, P.; Abbasi, M. Medicinal plants used in wound dressings made of electrospun nanofibers. *J. Tissue Eng. Regen. Med.* **2020**, *14*, 1527–1548. [CrossRef]
17. Wang, Y.; Malcolm, D.W.; Benoit, D.S.W. Controlled and sustained delivery of siRNA/NPs from hydrogels expedites bone fracture healing. *Biomaterials* **2017**, *139*, 127–138. [CrossRef]
18. Barrientos, S.; Stojadinovic, O.; Golinko, M.S.; Brem, H.; Tomic-Canic, M. Growth factors and cytokines in wound healing. *Wound Repair Regen.* **2008**, *16*, 585–601. [CrossRef]
19. Kolimi, P.; Narala, S.; Nyavanandi, D.; Youssef, A.A.A.; Dudhipala, N. Innovative Treatment Strategies to Accelerate Wound Healing: Trajectory and Recent Advancements. *Cells* **2022**, *11*, 2439. [CrossRef]
20. Saghazadeh, S.; Rinoldi, C.; Schot, M.; Kashaf, S.S.; Sharifi, F.; Jalilian, E.; Nuutila, K.; Giatsidis, G.; Mostafalu, P.; Derakhshandeh, H.; et al. Drug delivery systems and materials for wound healing applications. *Adv. Drug Deliv. Rev.* **2018**, *127*, 138–166. [CrossRef]
21. Wang, W.; Lu, K.J.; Yu, C.H.; Huang, Q.L.; Du, Y.Z. Nano-drug delivery systems in wound treatment and skin regeneration. *J. Nanobiotechnol.* **2019**, *17*, 82. [CrossRef] [PubMed]
22. Farahani, M.; Shafiee, A. Wound Healing: From Passive to Smart Dressings. *Adv. Healthc. Mater.* **2021**, *10*, e2100477. [CrossRef] [PubMed]
23. Kuffler, D.P. Photobiomodulation in promoting wound healing: A review. *Regen. Med.* **2016**, *11*, 107–122. [CrossRef] [PubMed]
24. Chereddy, K.K.; Vandermeulen, G.; Préat, V. PLGA based drug delivery systems: Promising carriers for wound healing activity. *Wound Repair Regen.* **2016**, *24*, 223–236. [CrossRef]
25. Sorg, H.; Tilkorn, D.J.; Hager, S.; Hauser, J.; Mirastschijski, U. Skin Wound Healing: An Update on the Current Knowledge and Concepts. *Eur. Surg. Res.* **2017**, *58*, 81–94. [CrossRef] [PubMed]
26. Gantwerker, E.A.; Hom, D.B. Skin: Histology and physiology of wound healing. *Clin. Plast. Surg.* **2012**, *39*, 85–97. [CrossRef]
27. Velnar, T.; Bailey, T.; Smrkolj, V. The wound healing process: An overview of the cellular and molecular mechanisms. *J. Int. Med. Res.* **2009**, *37*, 1528–1542. [CrossRef]
28. Kaiser, P.; Wächter, J.; Windbergs, M. Therapy of infected wounds: Overcoming clinical challenges by advanced drug delivery systems. *Drug Deliv. Transl. Res.* **2021**, *11*, 1545–1567. [CrossRef]
29. Boateng, J.S.; Matthews, K.H.; Stevens, H.N.; Eccleston, G.M. Wound healing dressings and drug delivery systems: A review. *J. Pharm. Sci.* **2008**, *97*, 2892–2923. [CrossRef]
30. Broughton, G.; Janis, J.E.; Attinger, C.E. Wound healing: An overview. *PlastReconstr. Surg.* **2006**, *117* (Suppl. 7), 1e-S–32e-S. [CrossRef]
31. Eming, S.A.; Martin, P.; Tomic-Canic, M. Wound repair and regeneration: Mechanisms, signaling, and translation. *Sci. Transl. Med.* **2014**, *6*, 265sr6. [CrossRef] [PubMed]
32. Ud-Din, S.; Bayat, A. Non-animal models of wound healing in cutaneous repair: In silico, in vitro, ex vivo, and in vivo models of wounds and scars in human skin. *Wound Repair Regen.* **2017**, *25*, 164–176. [CrossRef] [PubMed]
33. Masoumpour, M.B.; Nowroozzadeh, M.H.; Razeghinejad, M.R. Current and Future Techniques in Wound Healing Modulation after Glaucoma Filtering Surgeries. *Open Ophthalmol. J.* **2016**, *10*, 68–85. [CrossRef] [PubMed]
34. Cabourne, E.; Clarke, J.C.; Schlottmann, P.G.; Evans, J.R. Mitomycin C versus 5-Fluorouracil for wound healing in glaucoma surgery. *Cochrane Database Syst. Rev.* **2015**, *2015*, CD006259. [CrossRef] [PubMed]

35. Moghadam, S.E.; MoridiFarimani, M.; Soroury, S.; Ebrahimi, S.N.; Jabbarzadeh, E. Hypermongone C Accelerates Wound Healing through the Modulation of Inflammatory Factors and Promotion of Fibroblast Migration. *Molecules* **2019**, *24*, 2022. [CrossRef] [PubMed]
36. Torregrossa, M.; Kakpenova, A.; Simon, J.C.; Franz, S. Modulation of macrophage functions by ECM-inspired wound dressings—A promising therapeutic approach for chronic wounds. *Biol. Chem.* **2021**, *402*, 1289–1307. [CrossRef]
37. Chattopadhyay, S.; Raines, R.T. Review collagen-based biomaterials for wound healing. *Biopolymers* **2014**, *101*, 821–833. [CrossRef]
38. Vivcharenko, V.; Wojcik, M.; Palka, K.; Przekora, A. Highly Porous and Superabsorbent Biomaterial Made of Marine-Derived Polysaccharides and Ascorbic Acid as an Optimal Dressing for Exuding Wound Management. *Materials* **2021**, *14*, 1211. [CrossRef]
39. Westby, M.J.; Dumville, J.C.; Soares, M.O.; Stubbs, N.; Norman, G. Dressings and topical agents for treating pressure ulcers. *Cochrane Database Syst. Rev.* **2017**, *6*, CD011947. [CrossRef]
40. Berger, M.M.; Binz, P.A.; Roux, C.; Charrière, M.; Scaletta, C.; Raffoul, W.; Applegate, L.A.; Pantet, O. Exudative glutamine losses contribute to high needs after burn injury. *J. Parenter. Enter. Nutr.* **2022**, *46*, 782–788. [CrossRef]
41. Guan, Y.; Niu, H.; Liu, Z.; Dang, Y.; Shen, J.; Zayed, M.; Ma, L.; Guan, J. Sustained oxygenation accelerates diabetic wound healing by promoting epithelialization and angiogenesis and decreasing inflammation. *Sci. Adv.* **2021**, *7*, eabj0153. [CrossRef] [PubMed]
42. Bossi, F.; Tripodo, C.; Rizzi, L.; Bulla, R.; Agostinis, C.; Guarnotta, C.; Munaut, C.; Baldassarre, G.; Papa, G.; Zorzet, S.; et al. C1q as a unique player in angiogenesis with therapeutic implication in wound healing. *Proc. Natl. Acad. Sci. USA* **2014**, *111*, 4209–4214. [CrossRef] [PubMed]
43. Jin, L.; Guo, X.; Gao, D. R-responsive MXene nanobelts for wound healing. *NPG Asia Mater.* **2021**, *13*, 24. [CrossRef]
44. Vijayan, V.; Sreekumar, S.; Singh, F.; Govindarajan, D.; Lakra, R.; Korrapati, P.-S.; Kiran, M.S. Praseodymium-Cobaltite-Reinforced Collagen as Biomimetic Scaffolds for Angiogenesis and Stem Cell Differentiation for Cutaneous Wound Healing. *ACS Appl. Bio Mater.* **2019**, *2*, 3458–3472. [CrossRef] [PubMed]
45. Ridiandries, A.; Tan, J.T.M.; Bursill, C.A. The Role of Chemokines in Wound Healing. *Int. J. Mol. Sci.* **2018**, *19*, 3217. [CrossRef] [PubMed]
46. Xiao, T.; Yan, Z.; Xiao, S.; Xia, Y. Proinflammatory cytokines regulate epidermal stem cells in wound epithelialization. *Stem Cell Res. Ther.* **2020**, *11*, 232. [CrossRef]
47. Heinrich, P.-C.; Behrmann, I.; Haan, S.; Hermanns, H.-M.; Müller-Newen, G.; Schaper, F. Principles of interleukin (IL)-6-type cytokine signalling and its regulation. *Biochem. J.* **2003**, *15*, 374. [CrossRef]
48. Mi, F.L.; Wu, Y.B.; Shyu, S.S.; Schoung, J.Y.; Huang, Y.B.; Tsai, Y.H.; Hao, J.Y. Control of wound infections using a bilayer chitosan wound dressing with sustainable antibiotic delivery. *J. Biomed. Mater. Res.* **2002**, *59*, 438–449. [CrossRef]
49. Tamahkar, E.; Özkahraman, B.; Süloğlu, A.K.; İdil, N.; Perçin, I. A novel multilayer hydrogel wound dressing for antibiotic release. *J. Drug Deliv. Sci. Technol.* **2020**, *58*, 101536. [CrossRef]
50. Sabitha, M.; Rajiv, S. Preparation and characterization of ampicillin-incorporated electrospun polyurethane scaffolds for wound healing and infection control. *Polym. Eng. Sci.* **2015**, *55*, 541–548. [CrossRef]
51. Ye, S.; Jiang, L.; Wu, J.; Su, C.; Huang, C.; Liu, X.; Shao, W. Flexible amoxicillin-grafted bacterial cellulose sponges for wound dressing: In vitro and in vivo evaluation. *ACS Appl. Mater. Interfaces* **2018**, *10*, 5862–5870. [CrossRef]
52. Basha, M.; AbouSamra, M.M.; Awad, G.A.; Mansy, S.S. A potential antibacterial wound dressing of cefadroxil chitosan nanoparticles in situ gel: Fabrication, in vitro optimization and in vivo evaluation. *Int. J. Pharm.* **2018**, *544*, 129–140. [CrossRef] [PubMed]
53. Nikdel, M.; Rajabinejad, H.; Yaghoubi, H.; Mikaeiliagah, E.; Cella, M.A.; Sadeghianmaryan, A.; Ahmadi, A. Fabrication of cellulosic nonwoven material coated with polyvinyl alcohol and zinc oxide/mesoporous silica nanoparticles for wound dressing purposes with cephalexin delivery. *ECS J. Solid State Sci. Technol.* **2021**, *10*, 057003. [CrossRef]
54. Rădulescu, M.; Holban, A.-M.; Mogoantă, L.; Bălşeanu, T.-A.; Mogos-anu, G.-D.; Savu, D.; Popescu, R.C.; Fufă, O.; Grumezescu, A.M.; Bezirtzoglou, E.; et al. Fabrication, Characterization, and Evaluation of Bionanocomposites Based on Natural Polymers and Antibiotics for Wound Healing Applications. *Molecules* **2016**, *21*, 761. [CrossRef]
55. Bakadia, B.M.; Boni, B.O.O.; Ahmed, A.A.Q.; Zheng, R.; Shi, Z.; Ullah, M.W.; Lamboni, L.; Yang, G. In Situ Synthesized Porous Bacterial Cellulose/Poly (vinyl alcohol)-Based Silk Sericin and Azithromycin Release System for Treating Chronic Wound Biofilm. *Macromol. Biosci.* **2022**, *1*, 2200201. [CrossRef] [PubMed]
56. Ciftci, F.; Ayan, S.; Duygulu, N.; Yilmazer, Y.; Karavelioglu, Z.; Vehapi, M.; ÇakırKoç, R.; Sengor, M.; Yılmazer, H.; Ozcimen, D.; et al. Selenium and clarithromycin loaded PLA-GO composite wound dressings by electrospinning method. *Int. J. Polym. Mater. Polym. Biomater.* **2022**, *13*, 71. [CrossRef]
57. de Souza, R.F.B.; de Souza, F.C.B.; Moraes, Â.M. Polysaccharide-based membranes loaded with erythromycin for application as wound dressings. *Appl. Polym. Sci.* **2016**, *10*, 133. [CrossRef]
58. Alavarse, A.C.; de Oliveira Silva, F.W.; Colque, J.T.; da Silva, V.M.; Prieto, T.; Venancio, E.C.; Bonvent, J.J. Tetracycline hydrochloride-loaded electrospun nanofibers mats based on PVA and chitosan for wound dressing. *Mater. Sci. Eng. C* **2017**, *1*, 77. [CrossRef]
59. Khampieng, T.; Wnek, G.-E.; Supaphol, P. Electrospun DOXY-h loaded-poly(acrylic acid) nanofiber mats:In vitro drug release and antibacterial properties investigation. *J. Biomater. Sci. Polym. Ed.* **2014**, *25*, 1292–1305. [CrossRef]
60. Akota, I.; Alvsaker, B.; Bjørnland, T. The effect of locally applied gauze drain impregnated with chlortetracycline ointment in mandibular third-molar surgery. *Acta Odontol. Scand.* **1998**, *56*, 25–29. [CrossRef]
61. Abbott, P.V.; Hume, W.R.; Pearman, J.W. Antibiotics and endodontics. *Aust. Dental. J.* **1990**, *35*, 50–60. [CrossRef]

62. Michalska-Sionkowska, M.; Kaczmarek, B.; Walczak, M.; Sionkowska, A. Antimicrobial activity of new materials based on the blends of collagen/chitosan/hyaluronic acid with gentamicin sulfate addition. *Mater. Sci. Eng. C* **2018**, *1*, 86. [CrossRef]
63. Anjum, A.; Sim, C.H.; Ng, S.F. Hydrogels containing antibiofilm and antimicrobial agents beneficial for biofilm-associated wound infection: Formulation characterizations and In vitro study. *AAPS PharmSciTech* **2018**, *19*, 1219–1230. [CrossRef]
64. Ahire, J.J.; Robertson, D.D.; van Reenen, A.J.; Dicks, L.M.T. Polyethylene oxide (PEO)-hyaluronic acid (HA) nanofibers with kanamycin inhibits the growth of Listeria monocytogenes. *Biomed. Pharmacother.* **2017**, *86*, 143–148. [CrossRef]
65. Nitanan, T.; Akkaramongkolporn, P.; Rojanarata, T.; Ngawhirunpat, T.; Opanasopit, P. Neomycin-loaded poly (styrene sulfonic acid-co-maleic acid) (PSSA-MA)/polyvinyl alcohol (PVA) ion exchange nanofibers for wound dressing materials. *Int. J. Pharm.* **2013**, *1*, 448. [CrossRef]
66. Denkbaş, E.U.R.B.; Öztürk, E.; Özdem&unknownr, N.; Agalar, C. Norfloxacin-loaded chitosan sponges as wound dressing material. *J. Biomater. Appl.* **2004**, *18*, 291–303. [CrossRef]
67. Contardi, M.; Heredia-Guerrero, J.A.; Perotto, G.; Valentini, P.; Pompa, P.P.; Spanò, R.; Goldonic, L.; Bertorelli, R.; Athanassiou, A.; Bayera, I.S. Transparent ciprofloxacin-povidone antibiotic films and nanofiber mats as potential skin and wound care dressings. *Eur. J. Pharm. Sci.* **2017**, *104*, 133–144. [CrossRef]
68. Li, H.; Williams, G.R.; Wu, J.; Wang, H.; Sun, X.; Zhu, L.M. Poly (N-isopropylacrylamide)/poly (l-lactic acid-co-ϵ-caprolactone) fibers loaded with ciprofloxacin as wound dressing materials. *Mater. Sci. Eng. C* **2017**, *1*, 79. [CrossRef]
69. Pamfil, D.; Vasile, C.; Tarţău, L.; Vereştiuc, L.; Poiată, A. pH-Responsive 2-hydroxyethyl methacrylate/citraconic anhydride–modified collagen hydrogels as ciprofloxacin carriers for wound dressings. *J. Bioact. Compat. Polym.* **2017**, *32*, 355–381. [CrossRef]
70. Pásztor, N.; Rédai, E.; Szabó, Z.I.; Sipos, E. Preparation and Characterization of Levofloxacin-Loaded Nanofibers as Potential Wound Dressings. *Acta Med. Marisiensis* **2017**, *1*, 63. [CrossRef]
71. Singh, B.; Dhiman, A. Designing bio-mimetic moxifloxacin loaded hydrogel wound dressing to improve antioxidant and pharmacology properties. *RSC Adv.* **2015**, *5*, 44666–44678. [CrossRef]
72. Kurczewska, J.; Pecyna, P.; Ratajczak, M.; Gajęcka, M.; Schroeder, G. Halloysite nanotubes as carriers of vancomycin in alginate-based wound dressing. *Saudi Pharm. J.* **2017**, *1*, 25. [CrossRef]
73. Amiri, N.; Ajami, S.; Shahroodi, A.; Jannatabadi, N.; Darban, S.A.; Bazzaz, B.S.F.; Pishavar, E.; Kalalinia, F.; Movaffagh, J. Teicoplanin-loaded chitosan-PEO nanofibers for local antibiotic delivery and wound healing. *Int. J. Biol. Macromol.* **2020**, *162*, 645–656. [CrossRef]
74. Rolston, K.V.I.; Dholakia, N.; Ho, D.H.; LeBlanc, B.; Dvorak, T.; Streeter, H. In-vitro activity of ramoplanin (a novel lipoglycopeptide), vancomycin, and teicoplanin against gram-positive clinical isolates from cancer patients. *J. Antimicrob. Chemother.* **1996**, *38*, 265–269. [CrossRef]
75. Habash, M.B.; Park, A.J.; Vis, E.C.; Harris, R.J.; Khursigara, C.M. Synergy of silver nanoparticles and aztreonam against Pseudomonas aeruginosa PAO1 biofilms. *Antimicrob. Agents Chemother.* **2014**, *58*, 5818–5830. [CrossRef]
76. Jones, R.N. Critical assessment of the newer non-quinolone oral antimicrobial agents. *Antimicrob. Newsl.* **1989**, *6*, 53–60. [CrossRef]
77. Bauernfeind, A.; Schweighart, S.; Chong, Y. Extended broad spectrum β-lactamase in Klebsiella pneumoniae including resistance to cephamycins. *Infection* **1989**, *17*, 316–321. [CrossRef]
78. Teaima, M.H.; Elasaly, M.K.; Omar, S.A.; El-Nabarawi, M.A.; Shoueir, K.R. Wound healing activities of polyurethane modified chitosan nanofibers loaded with different concentrations of linezolid in an experimental model of diabetes. *J. Drug Deliv. Sci. Technol.* **2022**, *67*, 102982. [CrossRef]
79. Mohammed, A.A.; Ali, M.A.; Ahmed, O.S. To evaluate safety and efficacy of tedizolid phosphate in the management of several skin infections. *Int. J. Res. Pharm.* **2018**, *1*, 41–49. [CrossRef]
80. Dou, J.L.; Jiang, Y.W.; Xie, J.Q.; Zhang, X.G. New is old, and old is new: Recent advances in antibiotic-based, antibiotic-free and ethnomedical treatments against methicillin-resistant Staphylococcus aureus wound infections. *Int. J. Mol. Sci.* **2016**, *17*, 617. [CrossRef]
81. Fajardo, A.R.; Lopes, L.C.; Caleare, A.O.; Britta, E.A.; Nakamura, C.V.; Rubira, A.F.; Muniz, E.C. Silver sulfadiazine loaded chitosan/chondroitin sulfate films for a potential wound dressing application. *Mater. Sci. Eng. C Mater. Biol. Appl.* **2013**, *33*, 588–595. [CrossRef] [PubMed]
82. Hasselmann, J.; Kühme, T.; Acosta, S. Antibiotic prophylaxis with trimethoprim/sulfamethoxazole instead of cloxacillin fails to improve inguinal surgical site infection rate after vascular surgery. *Eur. J. Vasc. Endovasc. Surg.* **2015**, *49*, 129–134. [CrossRef]
83. Gjorevski, N.; Nikolaev, M.; Brown, T.E.; Mitrofanova, O.; Brandenberg, N.; DelRio, F.W.; Yavitt, F.M.; Liberali, P.; Anseth, K.S.; Lutolf, M.P. Tissue geometry drives deterministic organoid patterning. *Science* **2022**, *375*, eaaw9021. [CrossRef] [PubMed]
84. ValadanTahbaz, S.; Azimi, L.; Asadian, M.; Lari, A.R. Evaluation of synergistic effect of tazobactam with meropenem and ciprofloxacin against multi-drug resistant Acinetobacter baumannii isolated from burn patients in Tehran. *GMS Hyg. Infect. Control* **2019**, *14*, Doc08. [CrossRef]
85. Yang, M.; Hu, Z.; Hu, F. Nosocomial meningitis caused by Acinetobacter baumannii: Risk factors and their impact on patient outcomes and treatments. *Future Microbiol.* **2012**, *7*, 787–793. [CrossRef]
86. NourianDehkordi, A.; MirahmadiBabaheydari, F.; Chehelgerdi, M.; RaeisiDehkordi, S. Skin tissue engineering: Wound healing based on stem-cell-based therapeutic strategies. *Stem Cell Res. Ther.* **2019**, *10*, 111. [CrossRef]
87. Gonzales, K.A.U.; Fuchs, E. Skin and its regenerative powers: An alliance between stem cells and their niche. *Dev. Cell* **2017**, *43*, 387–401. [CrossRef]

88. Azari, Z.; Nazarnezhad, S.; Webster, T.J.; Hoseini, S.J.; Brouki Milan, P.; Baino, F.; Kargozar, S. Stem cell-mediated angiogenesis in skin tissue engineering and wound healing. *Wound Repair Regen.* **2022**, *30*, 421–435. [CrossRef]
89. Kamath, J.V.C.; Rana, A.C.; Chowdhury, A.R. Pro-healing effect of Cinnamomum zeylanicum bark. *Phytother. Res.* **2003**, *17*, 970–972. [CrossRef]
90. Liang, J.; Cui, L.; Li, J.; Guan, S.; Zhang, K.; Li, J. Aloe vera: A medicinal plant used in skin wound healing. *Tissue Eng. Part B Rev.* **2021**, *27*, 455–474. [CrossRef]
91. Teplicki, E.; Ma, Q.; Castillo, D.E.; Zarei, M.; Hustad, A.P.; Chen, J.; Li, J. The Effects of Aloe vera on Wound Healing in Cell Proliferation, Migration, and Viability. *Wounds* **2018**, *30*, 263–268. [PubMed]
92. Hamman, J.H. Composition and applications of Aloe vera leaf gel. *Molecules* **2008**, *13*, 1599–1616. [CrossRef]
93. RaesiVanani, A.; Mahdavinia, M.; Kalantari, H.; Khoshnood, S.; Shirani, M. Antifungal effect of the effect of *Securigera securidaca* L. vaginal gel on Candida species. *Curr. Med. Mycol.* **2019**, *5*, 31–35. [CrossRef]
94. Dai, Y.; Chen, S.R.; Chai, L.; Zhao, J.; Wang, Y.; Wang, Y. Overview of pharmacological activities of Andrographis paniculata and its major compound andrographolide. *Crit. Rev. Food Sci. Nutr.* **2019**, *59* (Suppl. 1), S17–S29. [CrossRef]
95. Nagulapalli Venkata, K.C.; Swaroop, A.; Bagchi, D.; Bishayee, A. A small plant with big benefits: Fenugreek (*Trigonella foenum-graecum* Linn.) for disease prevention and health promotion. *Mol. Nutr. Food Res.* **2017**, *61*, 1600950. [CrossRef]
96. Chan, Y.S.; Cheng, L.N.; Wu, J.H.; Chan, E.; Kwan, Y.W.; Lee, S.M.; Leung, G.P.; Yu, P.H.; Chan, S.W. A review of the pharmacological effects of *Arctium lappa* (burdock). *Inflammopharmacology* **2011**, *19*, 245–254. [CrossRef]
97. Wei, W.L.; Zeng, R.; Gu, C.M.; Qu, Y.; Huang, L.F. Angelica sinensis in China-A review of botanical profile, ethnopharmacology, phytochemistry and chemical analysis. *J. Ethnopharmacol.* **2016**, *190*, 116–141. [CrossRef]
98. Givol, O.; Kornhaber, R.; Visentin, D.; Cleary, M.; Haik, J.; Harats, M. A systematic review of Calendula officinalis extract for wound healing. *Wound Repair Regen.* **2019**, *27*, 548–561. [CrossRef]
99. Li, Y.; Dong, M.; Wu, Z.; Huang, Y.; Qian, H.; Huang, C. Activity Screening of the Herb Caesalpinia sappan and an Analysis of Its Antitumor Effects. *Evid. Based Complement. Altern. Med.* **2021**, *2021*, 9939345. [CrossRef]
100. Sugimoto, S.; Matsunami, K. Biological activity of Entada phaseoloides and Entada rheedei. *J. Nat. Med.* **2018**, *72*, 12–19. [CrossRef]
101. Kiefer, D.; Pantuso, T. Panax ginseng. *Am. Fam. Physician.* **2003**, *68*, 1539–1542. [PubMed]
102. Fang, Q.; Yao, Z.; Feng, L.; Liu, T.; Wei, S.; Xu, P.; Guo, R.; Cheng, B.; Wang, X. Antibiotic-loaded chitosan-gelatin scaffolds for infected seawater immersion wound healing. *Int. J. Biol. Macromol.* **2020**, *159*, 1140–1155. [CrossRef] [PubMed]
103. Huang, S.; Lu, G.; Wu, Y.; Jirigala, E.; Xu, Y.; Ma, K.; Fu, X. Mesenchymal stem cells delivered in a microsphere-based engineered skin contribute to cutaneous wound healing and sweat gland repair. *J. Dermatol. Sci.* **2012**, *66*, 29–36. [CrossRef] [PubMed]
104. Li, H.; Wang, F. Core-shell chitosan microsphere with antimicrobial and vascularized functions for promoting skin wound healing. *Mater. Des.* **2021**, *204*, 109683. [CrossRef]
105. Zhang, D.; Ouyang, Q.; Hu, Z.; Lu, S.; Quan, W.; Li, P.; Chen, Y.; Li, S. Catechol functionalized chitosan/active peptide microsphere hydrogel for skin wound healing. *Int. J. Biol. Macromol.* **2021**, *173*, 591–606. [CrossRef]
106. Negut, I.; Grumezescu, V.; Grumezescu, A.M. Treatment Strategies for Infected Wounds. *Molecules* **2018**, *23*, 2392. [CrossRef] [PubMed]
107. MofazzalJahromi, M.; SahandiZangabad, P.; MoosaviBasri, S.M.; SahandiZangabad, K.; Ghamarypour, A.; Aref, A.; Karimi, M.; Hamblin, M.R. Nanomedicine and advanced technologies for burns: Preventing infection and facilitating wound healing. *Adv. Drug Deliv. Rev.* **2018**, *123*, 33–64. [CrossRef] [PubMed]
108. Likus, W.; Bajor, G.; Siemianowicz, K. Nanosilver—Does it have only one face? *Acta Biochim. Pol.* **2013**, *60*, 495–501. [CrossRef] [PubMed]
109. Chen, C.Y.; Yin, H.; Chen, X.; Chen, T.H.; Liu, H.M.; Rao, S.S.; Tan, Y.J.; Qian, Y.X.; Liu, Y.; Hu, X.K. Ångstrom-scale silver particle-embedded carbomer gel promotes wound healing by inhibiting bacterial colonization and inflammation. *Sci. Adv.* **2020**, *6*, eaba0942. [CrossRef]
110. Qiu, L.; Wang, C.; Lan, M.; Guo, Q.; Du, X.; Zhou, S.; Cui, P.; Hong, T.; Jiang, P.; Wang, J.; et al. Antibacterial Photodynamic Gold Nanoparticles for Skin Infection. *ACS Appl. Bio. Mater.* **2021**, *19*, 3124–3132. [CrossRef]
111. García, I.; Henriksen-Lacey, M.; Calvo, J.; de Aberasturi, D.J.; Paz, M.M.; Liz-Marzán, L.M. Size-Dependent Transport and Cytotoxicity of Mitomycin-Gold Nanoparticle Conjugates in 2D and 3D Mammalian Cell Models. *Bioconjug. Chem.* **2019**, *30*, 242–252. [CrossRef]
112. Francesko, A.; Petkova, P.; Tzanov, T. Hydrogel Dressings for Advanced Wound Management. *Curr. Med. Chem.* **2018**, *25*, 5782–5797. [CrossRef]
113. Alven, S.; Aderibigbe, B.A. Chitosan and Cellulose-Based Hydrogels for Wound Management. *Int. J. Mol. Sci.* **2020**, *21*, 9656. [CrossRef]
114. Zhao, X.; Wu, H.; Guo, B.; Dong, R.; Qiu, Y.; Ma, P.X. Antibacterial anti-oxidant electroactive injectable hydrogel as self-healing wound dressing with hemostasis and adhesiveness for cutaneous wound healing. *Biomaterials* **2017**, *122*, 34–47. [CrossRef]
115. Dhaliwal, K.; Lopez, N. Hydrogel dressings and their application in burn wound care. *Br. J. Community Nurs.* **2018**, *23* (Suppl. 9), S24–S27. [CrossRef]

116. Zhang, J.; Zhu, Y.; Zhang, Y.; Lin, W.; Ke, J.; Liu, J.; Zhang, L.; Liu, J. A balanced charged hydrogel with anti-biofouling and antioxidant properties for treatment of irradiation-induced skin injury. *Mater. Sci. Eng. C Mater. Biol. Appl.* **2021**, *131*, 112538. [CrossRef]
117. Tavakoli, S.; Klar, A.S. Advanced Hydrogels as Wound Dressings. *Biomolecules* **2020**, *10*, 1169. [CrossRef]
118. Kharaziha, M.; Baidya, A.; Annabi, N. Rational Design of Immunomodulatory Hydrogels for Chronic Wound Healing. *Adv. Mater.* **2021**, *33*, e2100176. [CrossRef]
119. Niladri, R.; Nabanita, S.; Petr, H.; Petr, S. Permeability and biocompatibility of novel medicated hydrogel wound dressings. *Soft Mater.* **2010**, *8*, 338–357. [CrossRef]
120. Reimer, K.; Fleischer, W.; Brögmann, B.; Schreier, H.; Burkhard, P.; Lanzendörfer, A.; Gümbel, H.; Hoekstra, H.; Behrens-Baumann, W. Povidone-iodine liposomes—An overview. *Dermatology* **1997**, *195* (Suppl. 2), 93–99. [CrossRef]
121. Xu, H.L.; Chen, P.P.; ZhuGe, D.L.; Zhu, Q.Y.; Jin, B.H.; Shen, B.X.; Xiao, J.; Zhao, Y.Z. Liposomes with Silk Fibroin Hydrogel Core to Stabilize bFGF and Promote the Wound Healing of Mice with Deep Second-Degree Scald. *Adv. Healthc. Mater.* **2017**, *6*, 1700344. [CrossRef]
122. Reimer, K.; Vogt, P.M.; Broegmann, B.; Hauser, J.; Rossbach, O.; Kramer, A.; Rudolph, P.; Bosse, B.; Schreier, H.; Fleischer, W. An innovative topical drug formulation for wound healing and infection treatment: In vitro and in vivo investigations of a povidone-iodine liposome hydrogel. *Dermatology* **2000**, *201*, 235–241. [CrossRef]
123. Sağıroğlu, A.A.; Çelik, B.; Güler, E.M.; Koçyiğit, A.; Özer, Ö. Evaluation of wound healing potential of new composite liposomal films containing coenzyme Q10 and d-panthenyl triacetate as combinational treatment. *Pharm. Dev. Technol.* **2021**, *26*, 444–454. [CrossRef]
124. Santos, A.C.; Rodrigues, D.; Sequeira, J.A.D.; Pereira, I.; Simões, A.; Costa, D.; Peixoto, D.; Costa, G.; Veiga, F. Nanotechnological breakthroughs in the development of topical phytocompounds-based formulations. *Int. J. Pharm.* **2019**, *572*, 118787. [CrossRef]
125. Shakeel, F.; Alam, P.; Anwer, M.K.; Alanazi, S.A.; Alsarra, I.A.; Alqarni, M.H. Wound healing evaluation of self-nanoemulsifying drug delivery system containing Piper cubeba essential oil. *3 Biotech* **2019**, *9*, 82. [CrossRef]
126. Koshak, A.E.; Algandaby, M.M.; Mujallid, M.I.; Abdel-Naim, A.B.; Alhakamy, N.A.; Fahmy, U.A.; Alfarsi, A.; Badr-Eldin, S.M.; Neamatallah, T.; Nasrullah, M.Z.; et al. Wound Healing Activity of Opuntia ficus-indica Fixed Oil Formulated in a Self-Nanoemulsifying Formulation. *Int. J. Nanomed.* **2021**, *16*, 3889–3905. [CrossRef]
127. Ponto, T.; Latter, G.; Luna, G.; Leite-Silva, V.R.; Wright, A.; Benson, H.A.E. Novel Self-Nano-Emulsifying Drug Delivery Systems Containing Astaxanthin for Topical Skin Delivery. *Pharmaceutics* **2021**, *13*, 649. [CrossRef]
128. Khan, M.; Nadhman, A.; Sehgal, S.A.; Siraj, S.; Yasinzai, M.M. Formulation and characterization of a Self-Emulsifying Drug Delivery System (SEDDS) of curcumin for the topical application in cutaneous and mucocutaneous leishmaniasis. *Curr. Top. Med. Chem.* **2018**, *18*, 1603–1609. [CrossRef]
129. Anand, S.; Pandey, P.; Begum, M.Y.; Chidambaram, K.; Arya, D.K.; Gupta, R.K.; Sankhwar, R.; Jaiswal, S.; Thakur, S.; Rajinikanth, P.S. Electrospun biomimetic multifunctional nanofibers loaded with ferulic acid for enhanced antimicrobial and wound-healing activities in STZ-Induced Diabetic Rats. *Pharmaceuticals* **2022**, *15*, 302. [CrossRef]
130. Samadian, H.; Zamiri, S.; Ehterami, A. Electrospun cellulose acetate/gelatin nanofibrous wound dressing containing berberine for diabetic foot ulcer healing: In vitro and in vivo studies. *Sci. Rep.* **2020**, *10*, 8312. [CrossRef]
131. Almasian, A.; Najafi, F.; Eftekhari, M.; Shams Ardekani, M.R.; Sharifzadeh, M.; Khanavi, M. Preparation of Polyurethane/Pluronic F127 Nanofibers Containing Peppermint Extract Loaded Gelatin Nanoparticles for Diabetic Wounds Healing: Characterization, in vitro, and in vivo Studies. *Evid. Based Complement. Altern. Med.* **2021**, *2021*, 6646702. [CrossRef]
132. Grip, J.; Engstad, R.; Skjæveland, I.; Škalko-Basnet, N.; Isaksson, J.; Basnet, P.; Holsæter, A.M. Beta-glucan-loaded nanofiber dressing improves wound healing in diabetic mice. *Eur. J. Pharm. Sci.* **2018**, *121*, 269–280. [CrossRef]
133. Xu, X.; Wang, X.; Qin, C.; Khan, A.U.R.; Zhang, W.; Mo, X. Silk fibroin/poly-(L-lactide-co-caprolactone) nanofiber scaffolds loaded with Huangbai Liniment to accelerate diabetic wound healing. *Colloids Surf. B Biointerfaces* **2021**, *199*, 111557. [CrossRef]
134. Alzarea, A.I.; Alruwaili, N.K.; Ahmad, M.M.; Munir, M.U.; Butt, A.M.; Alrowaili, Z.A.; Shahari, M.S.B.; Almalki, Z.S.; Alqahtani, S.S.; Dolzhenko, A.V.; et al. Development and Characterization of Gentamicin-Loaded Arabinoxylan-Sodium Alginate Films as Antibacterial Wound Dressing. *Int. J. Mol. Sci.* **2022**, *23*, 2899. [CrossRef]
135. Lv, F.; Wang, J.; Xu, P.; Han, Y.; Ma, H.; Xu, H.; Chen, S.; Chang, J.; Ke, Q.; Liu, M.; et al. A conducive bioceramic/polymer composite biomaterial for diabetic wound healing. *Acta Biomater.* **2017**, *60*, 128–143. [CrossRef]
136. Li, Y.; Zhang, Z.Z. Sustained curcumin release from PLGA microspheres improves bone formation under diabetic conditions by inhibiting the reactive oxygen species production. *Drug Des. Dev. Ther.* **2018**, *12*, 1453–1466. [CrossRef]
137. Elkomy, M.H.; Eid, H.M.; Elmowafy, M.; Shalaby, K.; Zafar, A.; Abdelgawad, M.A.; Rateb, M.E.; Ali, M.R.A.; Alsalahat, I.; Abou-Taleb, H.A. Bilosomes as a promising nanoplatform for oral delivery of an alkaloid nutraceutical: Improved pharmacokinetic profile and snowballed hypoglycemic effect in diabetic rats. *Drug Deliv.* **2022**, *29*, 2694–2704. [CrossRef]
138. Ternullo, S.; Schulte Werning, L.V.; Holsæter, A.M.; Škalko-Basnet, N. Curcumin-in-Deformable Liposomes-in-Chitosan-Hydrogel as a Novel Wound Dressing. *Pharmaceutics* **2019**, *12*, 8. [CrossRef]
139. Cui, M.D.; Pan, Z.H.; Pan, L.Q. DangguiBuxue Extract-Loaded Liposomes in Thermosensitive Gel Enhance In Vivo Dermal Wound Healing via Activation of the VEGF/PI3K/Akt and TGF-β/SmadsSignaling Pathway. *Evid. Based Complement. Altern. Med.* **2017**, *2017*, 8407249. [CrossRef]

140. Kalantari, K.; Mostafavi, E.; Afifi, A.M.; Izadiyan, Z.; Jahangirian, H.; Rafiee-Moghaddam, R.; Webster, T.J. Wound dressings functionalized with silver nanoparticles: Promises and pitfalls. *Nanoscale.* **2020**, *12*, 2268–2291. [CrossRef]
141. Shalaby, M.A.; Anwar, M.M.; Saeed, H. Nanomaterials for application in wound Healing: Current state-of-the-art and future perspectives. *J. Polym. Res.* **2022**, *29*, 91. [CrossRef]
142. Souriyan-Reyhani pour, H.; Khajavi, R.; Yazdanshenas, M.E.; Zahedi, P.; Mirjalili, M. Cellulose acetate/poly(vinyl alcohol) hybrid fibrous mat containing tetracycline hydrochloride and phenytoin sodium: Morphology, drug release, antibacterial, and cell culture studies. *J. Bioact. Compat. Polym.* **2018**, *33*, 597–611. [CrossRef]
143. Kong, Y.; Xu, R.; Darabi, M.A.; Zhong, W.; Luo, G.; Xing, M.M.Q.; Wu, J. Fast and safe fabrication of a free-standing chitosan/alginate nanomembrane to promote stem cell delivery and wound healing. *Int. J. Nanomed.* **2016**, *11*, 2543–2555. [CrossRef]
144. Lohmann, N.; Schirmer, L.; Atallah, P.; Wandel, E.; Ferrer, R.A.; Werner, C.; Simon, J.C.; Franz, S.; Freudenberg, U. Glycosaminoglycan-based hydrogels capture inflammatory chemokines and rescue defective wound healing in mice. *Sci. Transl. Med.* **2017**, *9*, eaai9044. [CrossRef]
145. Lu, K.J.; Wang, W.; Xu, X.L.; Jin, F.Y.; Qi, J.; Wang, X.J.; Kang, X.Q.; Zhu, M.L.; Huang, Q.L.; Yu, C.H.; et al. A dual deformable liposomal ointment functionalized with retinoic acid and epidermal growth factor for enhanced burn wound healing therapy. *Biomater. Sci.* **2019**, *7*, 2372–2382. [CrossRef]
146. Tao, B.; Lin, C.; Deng, Y.; Yuan, Z.; Shen, X.; Chen, M.; He, Y.; Peng, Z.; Hu, Y.; Cai, K. Copper-nanoparticle-embedded hydrogel for killing bacteria and promoting wound healing with photothermal therapy. *J. Mater. Chem. B* **2019**, *7*, 2534–2548. [CrossRef]
147. Cao, L.; Shao, G.; Ren, F.; Yang, M.; Nie, Y.; Peng, Q.; Zhang, P. Cerium oxide nanoparticle-loaded polyvinyl alcohol nanogels delivery for wound healing care systems on surgery. *Drug Deliv.* **2021**, *28*, 390–399. [CrossRef]
148. Ziv-Polat, O.; Topaz, M.; Brosh, T.; Margel, S. Enhancement of incisional wound healing by thrombin conjugated iron oxide nanoparticles. *Biomaterials* **2010**, *31*, 741–747. [CrossRef]
149. Walker, R.M.; Gillespie, B.M.; Thalib, L.; Higgins, N.S.; Whitty, J.A. Foam dressings for treating pressure ulcers. *Cochrane Database Syst. Rev.* **2017**, *10*, CD011332. [CrossRef]
150. Xue, Y.; Chen, C.; Tan, R.; Zhang, J.; Fang, Q.; Jin, R.; Mi, X.; Sun, D.; Xue, Y.; Wang, Y.; et al. Artificial Intelligence-Assisted Bioinformatics, Microneedle, and Diabetic Wound Healing: A "New Deal" of an Old Drug. *ACS Appl. Mater. Interfaces* **2022**, *14*, 37396–37409. [CrossRef]
151. Yao, S.; Wang, Y.; Chi, J.; Yu, Y.; Zhao, Y.; Luo, Y.; Wang, Y. Porous MOF Microneedle Array Patch with Photothermal Responsive Nitric Oxide Delivery for Wound Healing. *Adv. Sci.* **2022**, *9*, e2103449. [CrossRef]
152. Shan, Y.; Tan, B.; Zhang, M.; Xie, X.; Liao, J. Restorative biodegradable two-layered hybrid microneedles for melanoma photothermal/chemo co-therapy and wound healing. *J. Nanobiotechnol.* **2022**, *20*, 238. [CrossRef]
153. Sillmon, K.; Moran, C.; Shook, L.; Lawson, C.; Burfield, A.H. The Use of Prophylactic Foam Dressings for Prevention of Hospital-Acquired Pressure Injuries: A Systematic Review. *J. Wound Ostomy Cont. Nurs.* **2021**, *48*, 211–218. [CrossRef] [PubMed]
154. Sierra-Sánchez, Á.; Kim, K.H.; Blasco-Morente, G.; Arias-Santiago, S. Cellular human tissue-engineered skin substitutes investigated for deep and difficult to heal injuries. *NPJ Regen. Med.* **2021**, *6*, 35. [CrossRef] [PubMed]
155. Jin, S.; Oh, Y.N.; Son, Y.R.; Kwon, B.; Park, J.H.; Gang, M.J.; Kim, B.W.; Kwon, H.J. Three-Dimensional Skin Tissue Printing with Human Skin Cell Lines and Mouse Skin-Derived Epidermal and Dermal Cells. *J Microbiol. Biotechnol.* **2022**, *32*, 238–247. [CrossRef] [PubMed]
156. Tan, S.H.; Ngo, Z.H.; Sci, D.B.; Leavesley, D.; Liang, K. Recent Advances in the Design of Three-Dimensional and Bioprinted Scaffolds for Full-Thickness Wound Healing. *Tissue Eng. Part B Rev.* **2022**, *28*, 160–181. [CrossRef] [PubMed]
157. Varkey, M.; Visscher, D.O.; van Zuijlen, P.P.M.; Atala, A.; Yoo, J.J. Skin bioprinting: The future of burn wound reconstruction? *Burns. Trauma* **2019**, *7*, 4. [CrossRef]
158. Gupta, P.; Sheikh, A.; Abourehab, M.A.S.; Kesharwani, P. Amelioration of Full-Thickness Wound Using Hesperidin Loaded Dendrimer-Based Hydrogel Bandages. *Biosensors* **2022**, *12*, 462. [CrossRef]
159. Vedhanayagam, M.; Unni Nair, B.; Sreeram, K.J. Collagen-ZnO Scaffolds for Wound Healing Applications: Role of Dendrimer Functionalization and Nanoparticle Morphology. *ACS Appl. Bio. Mater.* **2018**, *1*, 1942–1958. [CrossRef]
160. Patrulea, V.; Borchard, G.; Jordan, O. An Update on Antimicrobial Peptides (AMPs) and Their Delivery Strategies for Wound Infections. *Pharmaceutics* **2020**, *12*, 840. [CrossRef]
161. Jiang, Y.; Zhao, W.; Xu, S.; Wei, J.; Lasaosa, F.L.; He, Y.; Mao, H.; BoleaBailo, R.M.; Kong, D.; Gu, Z. Bioinspired design of mannose-decorated globular lysine dendrimers promotes diabetic wound healing by orchestrating appropriate macrophage polarization. *Biomaterials* **2022**, *280*, 121323. [CrossRef]
162. Wang, Y.; Sun, H. Polymeric Nanomaterials for Efficient Delivery of Antimicrobial Agents. *Pharmaceutics* **2021**, *13*, 2108. [CrossRef]
163. Jiang, G.; Liu, S.; Yu, T.; Wu, R.; Ren, Y.; van der Mei, H.C.; Liu, J.; Busscher, H.J. PAMAM dendrimers with dual-conjugated vancomycin and Ag-nanoparticles do not induce bacterial resistance and kill vancomycin-resistant Staphylococci. *Acta Biomater.* **2021**, *123*, 230–243. [CrossRef] [PubMed]
164. Abdel-Sayed, P.; Kaeppeli, A.; Siriwardena, T.; Darbre, T.; Perron, K.; Jafari, P.; Reymond, J.L.; Pioletti, D.P.; Applegate, L.A. Anti-Microbial Dendrimers against Multidrug-Resistant P. aeruginosa Enhance the Angiogenic Effect of Biological Burn-wound Bandages. *Sci. Rep.* **2016**, *6*, 22020. [CrossRef] [PubMed]
165. Dongargaonkar, A.A.; Bowlin, G.L.; Yang, H. Electrospun blends of gelatin and gelatin-dendrimer conjugates as a wound-dressing and drug-delivery platform. *Biomacromolecules* **2013**, *14*, 4038–4045. [CrossRef] [PubMed]

166. Zhang, Y.; Wang, B.; Meng, X.; Sun, G.; Gao, C. Influences of acid-treated multiwalled carbon nanotubes on fibroblasts: Proliferation, adhesion, migration, and wound healing. *Ann. Biomed. Eng.* **2011**, *39*, 414–426. [CrossRef]
167. Kittana, N.; Assali, M.; Abu-Rass, H.; Lutz, S.; Hindawi, R.; Ghannam, L.; Zakarneh, M.; Mousa, A. Enhancement of wound healing by single-wall/multi-wall carbon nanotubes complexed with chitosan. *Int. J. Nanomed.* **2018**, *13*, 7195–7206. [CrossRef] [PubMed]
168. Liu, S.; Wu, G.; Chen, X.; Zhang, X.; Yu, J.; Liu, M.; Zhang, Y.; Wang, P. Degradation Behavior In Vitro of Carbon Nanotubes (CNTs)/Poly(lactic acid) (PLA) Composite Suture. *Polymers* **2019**, *11*, 1015. [CrossRef]
169. Liu, J.; Ismail, N.A.; Yusoff, M.; Razali, M.H. Physicochemical Properties and Antibacterial Activity of Gellan Gum Incorporating Zinc Oxide/Carbon Nanotubes Bionanocomposite Film for Wound Healing. *Bioinorg. Chem. Appl.* **2022**, *2022*, 3158404. [CrossRef]
170. Khalid, A.; Madni, A.; Raza, B.; Islam, M.U.; Hassan, A.; Ahmad, F.; Ali, H.; Khan, T.; Wahid, F. Multiwalled carbon nanotubes functionalized bacterial cellulose as an efficient healing material for diabetic wounds. *Int. J. Biol. Macromol.* **2022**, *203*, 256–267. [CrossRef]
171. Chen, G.; Wu, Y.; Yu, D.; Li, R.; Luo, W.; Ma, G.; Zhang, C. Isoniazid-loaded chitosan/carbon nanotubes microspheres promote secondary wound healing of bone tuberculosis. *J. Biomater. Appl.* **2019**, *33*, 989–996. [CrossRef]
172. Zhao, D.; Jing, H.; Li, X.; Zhao, W. Application of Nano-Composite Technology for Multi-Empty Carbon Nanotubes in Dressing Change Care. *J. Nanosci. Nanotechnol.* **2021**, *21*, 1300–1306. [CrossRef] [PubMed]
173. Yin, M.; Wu, J.; Deng, M.; Wang, P.; Ji, G.; Wang, M.; Zhou, C.; Blum, N.T.; Zhang, W.; Shi, H.; et al. Multifunctional Magnesium Organic Framework-Based Microneedle Patch for Accelerating Diabetic Wound Healing. *ACS Nano* **2021**, *15*, 17842–17853. [CrossRef]
174. Wang, Y.; Lu, H.; Guo, M.; Chu, J.; Gao, B.; He, B. Personalized and Programmable Microneedle Dressing for Promoting Wound Healing. *Adv. Healthc. Mater.* **2022**, *11*, e2101659. [CrossRef]
175. Chi, J.; Zhang, X.; Chen, C.; Shao, C.; Zhao, Y.; Wang, Y. Antibacterial and angiogenic chitosan microneedle array patch for promoting wound healing. *Bioact. Mater.* **2020**, *5*, 253–259. [CrossRef]
176. Chi, J.; Sun, L.; Cai, L.; Fan, L.; Shao, C.; Shang, L.; Zhao, Y. Chinese herb microneedle patch for wound healing. *Bioact. Mater.* **2021**, *6*, 3507–3514. [CrossRef]
177. Wang, Y.; Gao, B.; He, B. Toward Efficient Wound Management: Bioinspired Microfluidic and Microneedle Patch. *Small* **2022**, *19*, e2206270. [CrossRef]
178. Liu, A.; Wang, Q.; Zhao, Z.; Wu, R.; Wang, M.; Li, J.; Sun, K.; Sun, Z.; Lv, Z.; Xu, J.; et al. Nitric Oxide Nanomotor Driving Exosomes-Loaded Microneedles for Achilles Tendinopathy Healing. *ACS Nano* **2021**, *15*, 13339–13350. [CrossRef]
179. Gao, S.; Zhang, W.; Zhai, X.; Zhao, X.; Wang, J.; Weng, J.; Li, J.; Chen, X. An antibacterial and proangiogenic double-layer drug-loaded microneedle patch for accelerating diabetic wound healing. *Biomater. Sci.* **2023**, *11*, 533–541. [CrossRef]
180. Xu, F.W.; Lv, Y.L.; Zhong, Y.F.; Xue, Y.N.; Wang, Y.; Zhang, L.Y.; Hu, X.; Tan, W.Q. Beneficial Effects of Green Tea EGCG on Skin Wound Healing: A Comprehensive Review. *Molecules* **2021**, *26*, 6123. [CrossRef]
181. Barnum, L.; Samandari, M.; Schmidt, T.A.; Tamayol, A. Microneedle arrays for the treatment of chronic wounds. *Expert Opin. Drug Deliv.* **2020**, *17*, 1767–1780. [CrossRef] [PubMed]
182. Faraji Rad, Z.; Prewett, P.D.; Davies, G.J. An overview of microneedle applications, materials, and fabrication methods. *Beilstein J. Nanotechnol.* **2021**, *12*, 1034–1046. [CrossRef] [PubMed]

pharmaceutics

MDPI

Review

Current Status of Polysaccharides-Based Drug Delivery Systems for Nervous Tissue Injuries Repair

Caterina Valentino, Barbara Vigani, Giuseppina Sandri, Franca Ferrari and Silvia Rossi *

Department of Drug Sciences, University of Pavia, Via Taramelli 12, 27100 Pavia, Italy
* Correspondence: silvia.rossi@unipv.it; Tel.: +39-0382-987357

Abstract: Neurological disorders affecting both CNS and PNS still represent one of the most critical and challenging pathologies, therefore many researchers have been focusing on this field in recent decades. Spinal cord injury (SCI) and peripheral nerve injury (PNI) are severely disabling diseases leading to dramatic and, in most cases, irreversible sensory, motor, and autonomic impairments. The challenging pathophysiologic consequences involved in SCI and PNI are demanding the development of more effective therapeutic strategies since, as yet, a therapeutic strategy that can effectively lead to a complete recovery from such pathologies is not available. Drug delivery systems (DDSs) based on polysaccharides have been receiving more and more attention for a wide range of applications, due to their outstanding physical-chemical properties. This review aims at providing an overview of the most studied polysaccharides used for the development of DDSs intended for the repair and regeneration of a damaged nervous system, with particular attention to spinal cord and peripheral nerve injury treatments. In particular, DDSs based on chitosan and their association with alginate, dextran, agarose, cellulose, and gellan were thoroughly revised.

Keywords: polysaccharides; drug delivery systems; spinal cord injury; peripheral nerve injury

Citation: Valentino, C.; Vigani, B.; Sandri, G.; Ferrari, F.; Rossi, S. Current Status of Polysaccharides-Based Drug Delivery Systems for Nervous Tissue Injuries Repair. *Pharmaceutics* **2023**, *15*, 400. https://doi.org/10.3390/pharmaceutics15020400

Academic Editors: Silvia Tampucci and Daniela Monti

Received: 21 December 2022
Revised: 18 January 2023
Accepted: 20 January 2023
Published: 25 January 2023

1. Introduction

The nervous system is the apparatus of the human body that functions to connect the various structures of the organism, but also to react to external stimuli [1]. It is composed of two major regions: The central nervous system (CNS) and the peripheral nervous system (PNS). The first is the most complex apparatus of the human body and includes the brain and its caudal prolongation, the spinal cord [2,3]. The CNS is connected to the periphery of the body by an extensive network of nerves composing the PNS [4]. The PNS includes the neural tissue outside the CNS, such as paravertebral and neuro-vegetative ganglia, peripheral nerves that extend from the brain and the spinal cord (cranial and spinal nerves, respectively), and specific sensory organs [5]. Neurological disorders affecting both CNS and PNS still represent one of the most critical and challenging pathologies, therefore many researchers have been focusing on such field in the last decades [6].

Spinal cord injury (SCI) and peripheral nerve injury (PNI) are severely disabling diseases leading to dramatic sensory, motor, and autonomic function impairments. Mechanical trauma causing compression of the spinal cord is, generally, the main and most common cause of SCI, which can be the result of motor vehicle accidents, falls, sports-related injuries, or violence [7]. The consequence of such compression results in a cascade of events including cellular, biochemical, and vascular events, exacerbating the damaged area and promoting the formation of a glial scar, responsible for the establishment of physical and chemical barriers to whatever endeavor to stimulate axonal regeneration [6,8]. On the other hand, PNI is defined as the loss of structure and/or function of peripheral nerves, characterized by permanent disablement and severe motor function defects, which could lead to the complete paralysis of the affected limb or the development of intractable neuropathic pain, with a severe impact on a patient's lifestyle [9,10]. Such pathology can be caused by accidents, traumatic conditions, surgery, other events such as ischemia, and chemical or thermal causes [9,11]. Main causes and symptoms of SCI and PNI are represented in Figure 1.

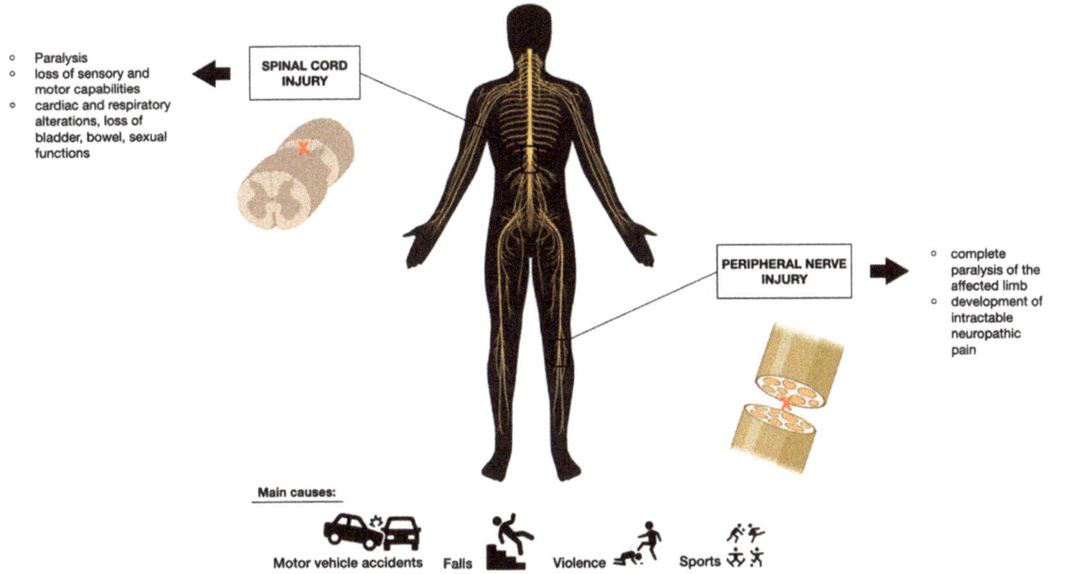

Figure 1. Spinal cord injury and peripheral cord injury: Main causes and symptoms.

Generally, PNS injuries are more common than those of the CNS, due to the lack of protection provided by the blood–brain barrier and the braincase [12]. Notably, in contrast to CNS, peripheral nerve fibers exhibit a remarkable intrinsic potential for self-regeneration. Schwann cells (SCs) are the main cells responsible for the activation of a cell-intrinsic myelin breakdown process through autophagy. Since myelin is the first responsible for storing substances (known as myelin-associated glycoproteins) that prevent the regeneration of damaged axons, this mechanism is acknowledged as the primary mechanism of the regenerative potential after PNI. [13,14]. Nevertheless, the achievement of total regeneration is generally unsatisfactory, providing weak motor recovery and irreversible sensory dysfunction [9]. In fact, an endogenous spontaneous repair of the damaged peripheral nerve is possible only for small gaps (<5 mm); it is partial and, in most cases, accompanied by a reduction in sensory and motor functions [11].

The challenging pathophysiologic consequences involved in SCI and PNI demand the development of more effective therapeutic strategies since, as yet, a therapeutic protocol that can effectively lead to a complete recovery from such pathologies is not available. Nowadays, the therapeutic approach of SCI is mostly limited to providing supportive relief to patients and is focused on the modulation of secondary complications, which hinder SCI treatment after the damage, and the stimulation of functional recovery via rehabilitation [15,16]. Currently, the most common treatments involve neurorrhaphy and allo- and autologous nerve grafting, which represents gold-standard treatments for nerve gaps smaller or higher than 1 cm, respectively [17,18]. These techniques have been, however, overcome due to their various critical issues and limitations, namely, excessive stretching of the nervous tissue and an impairment of nerve vascularization, in the case of neurorrhaphy; on the other hand, nerve grafting is often compromised by the scarce availability of nerve grafts, the morbidity of the donor site, the immunological responses, and the possible neuroma formation [17,19].

The development of tissue-engineered grafts represents a promising strategy for the treatment of lesions of the nervous system; in the case of PNI, such systems are generally named neural guide conduits (NGCs) [19]. The versatility of the geometries of NGCs and the possibility of loading various therapeutic substances (including drugs, cells, and/or growth factors) allow these products to be used in a combined neuroprotective and neuroregenerative approach in order to promote neuronal recovery [20–22].

Polysaccharides are natural polymers consisting of repeated mono- or disaccharide units linked via glycosidic bonds, which are generally isolated from several sources, such as plants, terrestrial and marine animals, or microorganisms [23]. Drug delivery systems (DDSs) composed of polysaccharides have been receiving more and more attention for a wide range of applications, due to their outstanding physical-chemical properties [24]. Due to the possibility of being neutral or presenting positive or negative charges, the linear or branched molecular structure and their wide range of molecular weights (from a few hundred to several thousand Daltons) are some of their attractive physical-chemical properties. Moreover, polysaccharides are recognized for their biocompatibility, bioactivity, biodegradability, and low immunogenicity [23,25–27].

Figure 2 reports a graphical description of the main polysaccharides and their principal sources and properties.

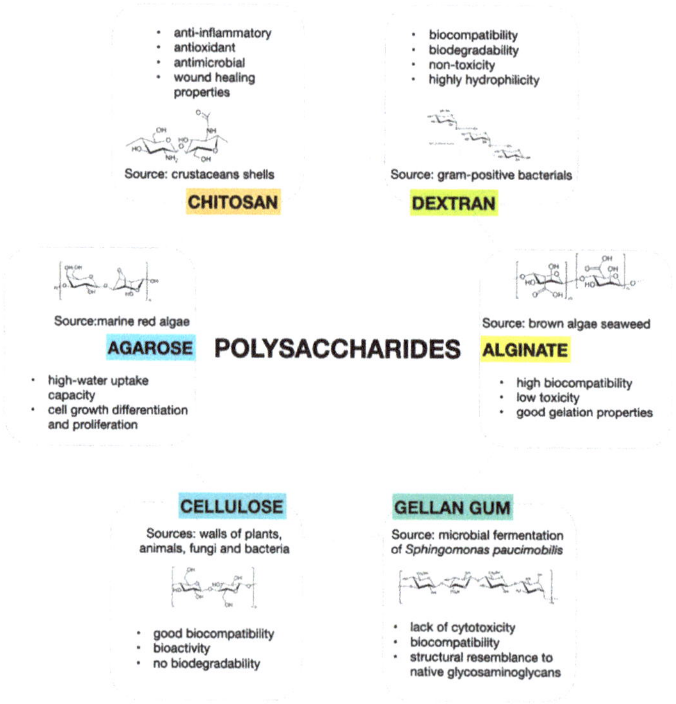

Figure 2. Graphical description of main polysaccharides, their principal sources, and properties.

Due to the above-mentioned characteristics, polysaccharides can be regarded as functional excipients for the setup of effective DDSs for the treatment of nervous tissue injuries [24,28].

This review aims to provide an overview of the most studied polysaccharides used for the development of DDSs intended for the repair and regeneration of a damaged nervous system, with particular attention to spinal cord and peripheral nerve injury treatments. In particular, DDSs based on chitosan and their association with alginate and hyaluronic acid, alginate, dextran, agarose, cellulose, and gellan were thoroughly revised.

2. Chitosan (CS)

Chitosan (CS) is a promising polysaccharide consisting of N-acetyl-D-glucosamine and D-glucosamine units linked by 1-4-β-glycosidic bonds, which is generally obtained from the deacetylation of chitin, derived from crustacean shells [29]. This polysaccharide has the unique properties of being polycationic and is characterized by a multitude of features, such as anti-inflammatory, antioxidant, antimicrobial, and wound-healing properties [30]. CS has been widely used for neural tissue engineering, due to its physical, chemical, and mechanical properties, and especially for its similarity to glycosaminoglycans forming the extracellular matrix [31].

As for SCI treatment, both hydrogels and nano-scale DDSs endowed with antioxidant and/or anti-inflammatory properties able to fill the gap generated by the injury and control the release of bioactive molecules were proposed in the last decade.

The first evidence of CS application for SCI treatment was reported by Skop et al., who developed genipin-crosslinked CS-based microspheres, produced via the coaxial airflow technique and intended for the delivery of cells and growth factors for nervous tissue regeneration. CS was ionically bound with heparin, a well-known anionic glycosaminoglycan with anticoagulant and anti-inflammation properties, as well as a high affinity for growth factors. The system was characterized by high biocompatibility towards the neural stem cell line and easy binding to the fibroblast growth factor, which is an important factor for neural stem cell survival [32].

Wu and coworkers developed nanoparticles (Nps) with neuroprotective potential, based on the chemical interaction between glycol CS, a water-soluble CS derivative, and ferulic acid (FA). Glycol CS and FA can form hydrophobically self-assembled Nps consisting of a hydrophilic shell (glycol CS) and a hydrophobic core (FA). The neuroprotective effect of Nps was assessed by a glutamate-induced excitotoxicity model on primary spinal cord neuron culture. Moreover, Nps were injected in a rat spinal cord contusion injury model to assess the bioavailability, pharmacokinetics, and functional recovery after Nps systemic administration. Axons and neuron cells at the lesion site were greatly recovered by the delivery of FA-glycol CS Nps, and the number of activated astrocytes and macrophages was reduced. These neuroprotective benefits ultimately contributed to SCI functional recovery [33].

Another CS–drug complex was investigated by Gwak et al., who prepared CS Nps via an ionotropic gelation process, based on the amide-coupling method, to provide intracellular delivery of methylprednisolone (MD), a cortico-steroid used for the treatment of acute SCI as an anti-inflammatory agent and to reduce neurological deficits after injury. In addition to drug delivery, the authors aimed at investigating possible gene delivery due to the positive charges of chitosan, which can bind with the negative charges of DNA through electrostatic interaction. Nps determined low cytotoxicity on mouse neural stem cells, and plasmid DNA was efficiently delivered once Nps were injected in vivo in a compressed spinal cord injury model, producing effective protein expression. Inflammation and apoptosis were reduced even at a low MD dose [34].

Ni and coworkers explored the sustained delivery of chondroitinase ABC (ChABC), an enzyme able to decompose glycosaminoglycans chains of chondroitin sulfate proteoglycans (CSPGs), which are recognized for their negative impact on axon regeneration after SCI, and which, therefore, are able to elicit an indirect positive effect on axonal regeneration. For this purpose, ChABC-loaded CS microparticles were prepared via ionotropic gelation, using tripolyphosphate (TPP) as cross-linking agent. Solid microparticles were then mixed with polypropylene carbonate and subjected to electrospinning to obtain microfibers containing CS-ChABC microspheres. The systems were able to assure a stable and prolonged release of ChABC in vitro. Moreover, fibers containing microspheres were implanted in vivo in a hemisected thoracic spinal cord with improved axonal regeneration and animal functional recovery [35].

Wang and collaborators investigated the release of valproic acid from CS Nps. Valproic acid was found to be neuroprotective for microglia and reduce inflammation induced by nervous tissue injury. After intravenous injection in a rat SCI model, Nps enhanced the functional recovery of nervous tissue and inhibited astrocytes activity and, thus, inflammation. Furthermore, the disruption of the blood spinal cord barrier occurring after SCI was recovered, and neuroprotection was successfully achieved [36].

Alizadeh et al. developed a thermo-sensitive hydrogel based on CS for the delivery of the nerve growth factor (NGF). The gel consisted of CS, β-glycerol phosphate disodium salt pentahydrate as a gelling agent and hydroxyethylcellulose as a cross-linking agent; the hydrogel solutions were prepared at 4 °C. The hydrogel was used as a vehicle for the delivery of lentiviral-mediated NGF-overexpressing human adipose-derived mesenchymal stem cells (hADSCs), which are recognized to improve neural growth and neural regeneration due to the secretion of neurotrophic and neurovascular factors and are able to differentiate into Schwann-like cells. Both transduced hADSCs alone and in combination with the CS-based hydrogel were injected in vivo in a contusive spinal cord injury model one week after surgery, and the hydrogel effectiveness was evaluated two months after surgery. The combination of transduced hADSCs and the hydrogel was more effective than transduced hADSCs alone in both repairing SCI and providing the functional recovery of animals [37].

A CS-collagen-based hydrogel loaded with a serine protease inhibitor, serpine (Serp-1), which is an immune-modulating biologic drug, was developed for the treatment of crush-induced SCI by Kwiecien and collaborators. Both low (10 μg) and high doses (100 μg) of Serp-1 were loaded into CS-collagen hydrogels, which were then injected in a dorsal column crush SCI rat model. Locomotor functionality was assessed and histopathologic analysis was performed; the authors proved that hydrogel loaded with the highest Serp-1 dose reduced the damaged area, resulting in better and faster motor recovery and diminished neurological deficits, in comparison with the low-dose-loaded and pristine hydrogel. Furthermore, a reduction of neural injury was observed with the high-dose-loaded hydrogel, which was attributed to a Serp-1-induced reduction of apoptosis [38].

Wang and coworkers designed stearic acid (SA)-CS nanomicelles loaded with sesamol, a polyphenol with strong antioxidant properties but low cellular uptake and biocompatibility and, therefore, requiring a drug carrier system for suitable administration. CS nanomicelles conjugated with SA were developed via centrifugation followed by freeze-drying to obtain a solid product. Nanomicelles proved to be stable in phosphate buffer solution (PBS) for 15 days and able to provide a sustained release of sesamol, reaching almost 100% after 50 h from the beginning of the in vitro release experiment. In vitro tests were performed on NSC-34 cells, a hybrid motoneuron-like cell line; cells were treated with lipopolysaccharide (LPS) before being incubated with the samples (nanomicelles or free sesamol), in order to investigate nanomicelles' protective effect. Other in vitro tests were carried out on the same cell line (MTT, LDH, and intracellular ROS assay). The best results were always observed for loaded nanomicelles, confirming their neuroprotective potential and antioxidant properties. Moreover, loaded nanomicelles were also able to modulate the levels of apoptotic genes and reduce the expression of inflammation-related genes [39].

A CS-based hydrogel was prepared by Javdani et al. by adding a sodium hydroxide solution to a CS solution; the hydrogel was then loaded with Selenium-loaded Nps. Selenium was selected for its relevance as a nutrient for humans and animals and for its fundamental role in many metabolic pathways, namely for its beneficial effect on acute SCI. In vivo studies were performed on rat models: Animals were subjected to aneurysm clamping at the level of the thoracic vertebrae to induce SCI, and the effect of both free and loaded hydrogels was compared with that of the control group (no drug intervention). Histological evaluation of the group treated with the loaded hydrogel highlighted a reduction in the severity of bleeding and the number of inflammatory cells, as well as the occurrence of new nerve fibers. The presence of selenium within the hydrogel was responsible for the inhibition of the cellular pathways related to inflammation and was demonstrated to provide neuronal recovery and protection. Hence, the selenium-loaded CS hydrogel showed potential antioxidant and anti-inflammatory properties for the treatment of SCI [40].

Song and collaborators developed a CS scaffold intended for the SCI treatment. A sandwich-structured composite system for long-lasting controlled release of NGF (2 months, required by the healing process of SCI) was prepared via electrospray and electrospinning techniques. A poly(lactic acid) (PLA) film was used as a sealing layer to impair drug diffusion outside the system and provide mechanical support; the sandwich layer consisted of poly(lactic-co-glycolic acid) PLGA microspheres loaded with NGF, and a CS film constituted the lower layer to host bone marrow mesenchymal stem cells (BMSCs). The PLA and CS films were prepared by electrospinning whilst the NGF-loaded PLGA microspheres were obtained via ultrasonication. In detail, the final composite system was obtained by electrospraying NGF-loaded PLGA microspheres above the PLA fibrous membrane and subsequently electrospinning the CS solution above the previous two layers. A schematic representation of the composite scaffold is displayed in Figure 3.

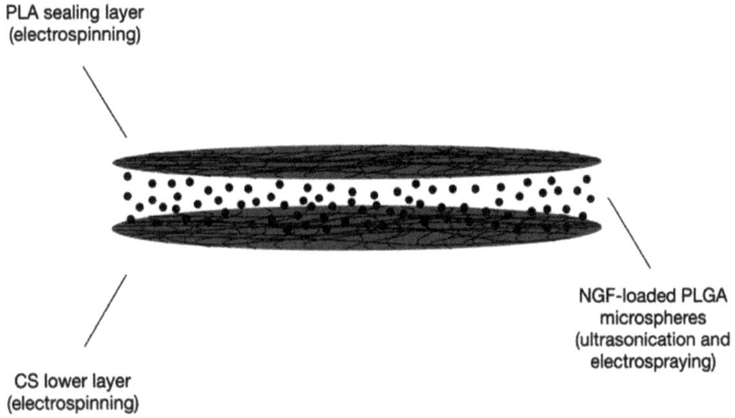

Figure 3. Schematic representation of the sandwich-structured drug delivery composite scaffold developed by Song and coworkers [41].

In vitro drug release measurements demonstrated that the composite system enables a prolonged release of NGF for more than 2 months. The biocompatibility of the system was tested in vitro on a PC-12 cell line for 7 days; neurite outgrowth from PC-12 cells was promoted by the composite system. BMSCs were then seeded onto the CS film. When implanted in a rat model, the final system was found to promote and improve neuroregeneration and locomotor functional recovery of SCI within 8 weeks of evaluation [41].

In addition to the treatment of SCI, researchers have focused their efforts on developing CS-based DDSs for the treatment of PNI. In this context, CS employment as an NGC component seemed particularly promising, due to the physical and chemical structure similarity of these systems with the multi-layer 3D architecture of peripheral nerves. CS is also capable of supporting axonal regrowth, ameliorating functional recovery after injury, healing, and reducing scar formation [31,42].

Regarding PNI, a variety of systems of various types, including NGCs, hydrogels, and Nps loaded with various active molecules with immunosuppressant, neuroprotective, and nerve regeneration enhancement properties, were developed during the last decade.

Li and collaborators developed a CS guide loaded with FK506, an FDA-approved immunosuppressant agent generally used to avoid allograft rejection after transplantation and considered for its neuroprotective and neurotrophic potential. A tubular scaffold was obtained by solvent casting into a specific mold presenting an inner stainless-steel tube and an outer tube, and subsequent soaking in a NaOH solution to obtain a tubular gel. These guides were implanted in vivo on a sciatic nerve injury rat model, and after 6–8 weeks, electrophysiological and histological analyses were performed. The FK506 loaded-CS guide decreased the inflammatory reaction and allowed faster reinnervation, in comparison with the groups treated with a silica guide or CS guide without FK506 [43].

A CS conduit was designed by Farahpour and Ghayour. It was loaded with Acetyl-L-carnitine (ALC), a natural amino acid derivative with both neuroprotective and antinociceptive effects recognized for its potential in improving nerve regeneration after PNI. A CS solution was prepared at 50 °C, and the addition of glycerol was exploited to improve CS mechanical properties, excluding fragility. The CS conduit was prepared by injecting a CS/glycerol solution in a homemade mold with an internal diameter of 1.8 mm and an external one of 2 mm. As for in vivo efficacy studies, animals (rats) were divided into four groups: (i) The transected control group, (ii) the sham-surgery group, (iii) the group treated with CS alone, and (iv) the group treated with ALC-loaded CS conduit. The last group demonstrated improved functional recovery of the transected sciatic nerve, with the promotion of both motor and sensory regeneration and reinnervation of the injury [44].

CS Nps were employed to obtain a sustained gene release for peripheral nerve regeneration after intramuscular administration, which is clinically relevant and not considered invasive. In detail, a thiolated trimethyl CS (TMCSH) was in contact with pDNA encoding for the brain-derived neurotrophic factor (BDNF) for 15 min, which is well-known for its protective action towards neuron survival after injury. BDNF-loaded Nps were prepared according to the method reported in Figure 4; a plasmid encoding for tetanus neurotoxin (HC), which can modulate nanoparticle retrograde transport after peripheral administration, was added as the HC-PEG solution to the TMCSH solution to obtain TMSCH-HC Nps. For in vivo studies, rats were injected with Nps 8 days before nerve crush injury to investigate the effect of a therapeutic agent on the prevention of axon degeneration and/or nerve regeneration. After in vivo administration, Nps proved to stimulate injury recovery, avoid nerve degeneration, and improve nerve regeneration [45].

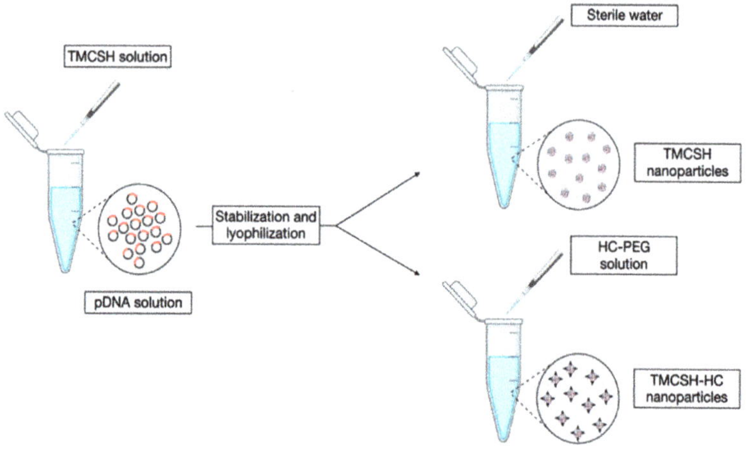

Figure 4. Schematic representation of TMCSH-based nanoparticle preparation, inspired by Lopes and collaborators [45].

A CS-based guide conduit was conceived by Manoukian et al. for the sustained release of 4-Aminopyridine (4AP), a potassium-channel blocker able to accelerate nerve innervation. The nerve guidance conduit was characterized by the presence of aligned microchannels obtained through unidirectional freezing, achieved by pouring a CS solution into a homemade mold that was then submersed in liquid N_2; subsequent freeze-drying of the CS solution allowed the attainment of a highly porous scaffold with a foam-like structure. Halloysite nanotubes (HNTs) loaded with 4AP were also added to the CS solution to improve the strength of the final system and provide a prolonged release of the drug, controlling the initial burst effect. The final porous system was cross-linked using alkaline epichlorohydrin. The authors observed an in vitro drug release from the 4AP-HNTs-loaded CS conduit of almost 98% within 7 days, differently from the 4AP-HNTs and 4AP- load CS matrix, which both showed faster release. The cross-linked 4AP-HNTs-loaded CS conduit displayed the longest sustained release profile (almost 30% at 7 days). In vitro studies carried out on SCs proved the positive effect of 4AP as an enhancer of the upregulation of NGF or BDNF, which are key trophic factors for axon regeneration and nerve remyelination after injury. A sciatic nerve defect rat model was exploited for in vivo preliminary assessment, which corroborated the biocompatibility of the scaffold and the infiltration of SCs within the lumen of the scaffold, following the organized structure of the aligned channels [22].

CS Nps loaded with curcumin and containing SCs cells were developed by Jahromi et al. and loaded into poly-L-lactide acid (PLLA)-based multi-wall carbon nanotube conduits. Curcumin has been demonstrated to decrease SCs apoptosis and enhance the number of myelinated axons inside the injury. Therefore, the authors investigated curcumin association with SCs with the purpose of improving axons regeneration. Specifically, a hollow PLLA/multi-wall carbon nanotube conduit (prepared by electrospinning apparatus equipped with a rotating rod) was filled with a fibrin-based hydrogel containing curcumin-encapsulated CS Nps and SCs. The final scaffold showed good biocompatibility and the ability to promote SCs adhesion, mostly when the PLLA/multi-wall carbon nanotube conduit was at the lowest concentration. A sciatic nerve injury rat model was employed to evaluate the scaffold ability of damage recovery; in particular, the composite system was demonstrated to boost nerve regeneration and improve locomotor functionality when compared to the autograft used as control [46].

Zeng et al. developed CS/PLGA microspheres for the delivery of NGF; microspheres were prepared by a re-emulsification TPP ionic cross-linking method. The release of NGF from microspheres was between 65 and 45% on the 49th day, depending on the TPP concentration used to cross-link the microspheres; upon increasing TPP concentration, a higher reduction in drug release rate in comparison with lower TPP concentration was observed. Microspheres demonstrated their ability to promote neurite formation when tested on PC12 cells thanks to NGF binding to the tyrosine kinase receptor, which activates intracellular pathways that induce neurite extension. In vivo experiments on the sciatic nerve rat model highlighted the ability of axons to regenerate the loaded microspheres [47].

2.1. CS Associations

Chitosan has also been employed in association with other polysaccharides to develop DDSs useful for both SCI and PNI repair and regeneration. Chitosan's association with other polysaccharides is linked to its cation charges, which enable ionic interaction with anionic polysaccharides, such as hyaluronic acid (HA) and alginate (ALG) [48–51]. The association with other polysaccharides can be useful not only to enhance mechanical and biomimetic properties of the systems developed, but also to better control the drug release from these systems [48,52].

2.1.1. CS/HA Association

In the context of SCI regeneration, HA and glycol CS were employed in association to develop a hydrogel loaded with tauroursodeoxycholic acid (TUDCA), a bile acid with cyto-protective activity and anti-neuroinflammatory properties. In detail, oxidized hyaluronate and glycol CS solutions were mixed in different ratios and then cross-linked with TUDCA. The gel obtained was subjected to freeze-drying, and a porous lyophilized product with regular pore size was obtained. The cross-linked gel was tested in vivo in a mechanical SCI model on rats and demonstrated its ability to promote functional recovery by reducing the expression of pro-inflammatory cytokines and thus inhibiting inflammatory pathways involved in SCI [50].

As for PNI treatment, the complementary advantages of CS and HA hydrogel were investigated by Zhang and collaborators, who developed an injectable hydrogel for the prolonged delivery of NGF for peripheral nerve regeneration. An injectable CS/HA 1:1 hydrogel was prepared at 37 °C, using ethyl-3-(3-dimethylaminopropyl) carbodiimide (EDC) and N-hydroxysuccinimide (NHS) to promote an amide reaction between components. The gelation time, measured by a tilt test, was very fast (3 min) at pH 7.4. The freeze-dried hydrogel was characterized by a highly porous interconnected structure with macropores, with the cross-section desirable to enhance axon and nerve cell growth and the absorption of bioactive substances for nerve regeneration. NGF release in PBS shows an initial rapid release due to the diffusion of NGF and the swelling of the hydrogel in an aqueous medium; then a slower release was observed, reaching approximately 80% of the loaded drug in 56 days. The loaded hydrogel demonstrated the highest biocompatibility towards Bone Marrow Mesenchymal Stromal Cells (BMMSCs) when compared to the controls (blank group without NGF, pure components, and NGF-free hydrogel). Moreover, freeze-dried loaded hydrogel determined high cell adhesion and promoted their differentiation [48].

2.1.2. CS/ALG Association

Chitosan has also been used in association with ALG by Vigani and coworkers for the delivery of a sigma 1 receptor agonist, named RC-33, intended for the treatment of SCI. RC-33 is a promising active molecule able to promote neurite outgrowth in PC12 cells induced by the nerve growth factor. Thus, the study was carried out to develop a scaffold characterized by both neuroregenerative potential due to the natural polysaccharide-based scaffold and neuroprotective action due to the presence of the loaded active molecule. RC-33 was incorporated into ALG electrospun nanofibers due to ionic bonds formed between the anionic polysaccharide and the cationic drug candidate; loaded fibers were then subjected to cross-linking with calcium chloride to obtain a water-insoluble product. Cross-linked nanofibers were subsequently loaded into a CS film produced via solvent casting to obtain a flexible and easy-to-handle final product. Good cell biocompatibility of the scaffold was evidenced in human neuroblastoma SH-SY5Y cells [51].

The association of CS with ALG was also investigated by Rahmati and collaborators to develop a hydrogel for the delivery of berberine (Ber) for the treatment of PNI, especially for sciatic nerve regeneration. Ber was employed because of its ability to improve peripheral nerve damage thanks to its antibacterial, immunostimulant, anticancer, and antimotility properties and beneficial effect on neurological disorders. An ALG solution was cross-linked with $CaCl_2$ and added dropwise to a CS solution containing β-glycerol phosphate as a cross-linking agent. Different concentrations (0.1, 1, and 10% *w*/*w*) of Ber were then added to the ALG/CS solution. Loaded freeze-dried hydrogels displayed a highly interconnected porous structure, capable of providing a constant Ber release for 24 days. Hemocompatibility of the hydrogels was demonstrated on human anticoagulated blood collected from volunteers, and their biocompatibility was confirmed on PC12 cells. In vivo studies on a crush-induced sciatic nerve rat model demonstrated a positive effect, especially for the 1% Ber loaded-hydrogel, indicating that Ber has potential activity in nerve regeneration [52].

Table 1 presents a summary of all the CS-based DDSs discussed in the text.

Table 1. Summary of CS- and CS-association-based DDSs for SCI and PNI application. DDS description, production technique, therapeutic agent delivered, and in vitro and in vivo models exploited for efficacy assessment are reported.

DDS	Production Technique	Therapeutic Agent	Application	In Vitro Model	In Vivo Model	Reference
CS microspheres	Genepin-cross-linking and coaxial airflow technique	Heparin	SCI	Neural stem cell line	-	[32]

Table 1. *Cont.*

DDS	Production Technique	Therapeutic Agent	Application	In Vitro Model	In Vivo Model	Reference
Glycol/CS Nps	Self-assembling	Ferulic acid	SCI	Primary spinal cord neurons culture	Rat spinal cord contusion injury model	[33]
CS Nps	Ionotropic gelation	Methylprednisolone	SCI	Mouse neural stem cells	Compressed spinal cord injury rat model	[34]
Microfibers con-taininigChABC-loaded CS microparticles	Ionotropic gelation (with TPP)and electrospinning	Chondroitinase ABC	SCI	-	Hemisected thoracic rad spinal cord model	[35]
CS Nps	conjugating by coupling carboxyl to amino group in the presence of modification reagents and dyalization to isolate the conjugates	Valproic acid	SCI	-	Spinal cord contusion rat injury model	[36]
CS-based thermo-sensitive hydrogel	Hydroxyethylcellulose as cross-linking agent, β-glycerol phosphate disodium salt pentahydrate as gelling agent for CS solution	Lentiviral mediated NGF–overexpressing hADSCs	SCI	-	Contusive rat spinal cord injury model	[37]
CS-collagen based hydrogel loaded with Serp-1	Lyophilization	Serpine (Serp-1)	SCI	-	Dorsal column crush rat model	[38]
CS-stearic acid conjugated nanomicelles loaded with sesamol	Centrifugation followed by freeze-drying	Sesamol	SCI	NSC-34 cell line	-	[39]
CS Hydrogel loaded with Nps	Addition of sodium hydroxide	Selenium	SCI	-	Aneurysm clamping at the level of thoracic vertebrae	[40]
Sandwich system: PLA fibers; NGF-loaded PLGA-microspheres-CS fibers	Electrospinning (PLA fibers and CS fibers), ultrasonication and electrospraying (PLGA microspheres)	NGF	SCI	PC-12 cell line	Allen's SCI models on rats	[41]
CS tubular conduit	Solvent casting in tubular mold and subsequent immersion in NaOH	FK506	PNI	-	Sciatic nerve injury rat model	[43]
CS/glycerol tubular conduit	Home-made tubular mold	Acetyl-L-carnitine	PNI	-	Left sciatic nerve transection on rats	[44]

Table 1. *Cont.*

DDS	Production Technique	Therapeutic Agent	Application	In vitro Model	In Vivo Model	Reference
TMCSH-HC Nps	Mixing of TMCSH and pDNA, lyophilization and addition of HC.	pDNA encoding for BDNF	PNI	-	Injection of nanoparticles before nerve crush injury induction	
CS Nerve guide conduit with aligned microchannels loaded with halloysite nanotubes	Unidirectional freezing in N_2 and freeze-drying, cross-linking with epichlorohydrin	Aminopyridine	PNI	Schwann cell line	Sciatic nerve defect rat model	[22]
PLLA nanotubes containing fibrin hydrogel loaded with curcumin encapsulated CS Nps and SCs	Electrospinning (PLLA nanotubes)	Curcumin	PNI	Schwann cells	Sciatic nerve injury rat model	[46]
CS/PLGA microspheres	Re-emulsification TPP ionic cross-linking method	NGF	PNI	PC12 cells	Sciatic nerve injury rat model	[47]
			Associations			
Oxidized HA/glycol CS hydrogel	Cross-linking and freeze-drying	tauroursodeoxycholic acid	SCI	-	Mechanical SCI rat model	[50]
HA/CS injectable hydrogel	Prepared at 37 °C, using ethyl-3-(3-dimethylaminopropyl) carbodiimide (EDC) and N-hydroxysuccinimide (NHS) and freeze-drying	NGF	PNI	BMMSCs	-	[48]
RC-33 loaded-ALG nanofibers embedded in CS film	Electrospinning (nanofibers), solvent casting (film)	RC-33	SCI	SH-SY5Y cells	-	[51]
Berberine-loaded ALG/CShydrogel	$CaCl_2$ cross-linked ALG added dropwise to CS solution containing β-glycerol phosphate	berberine	PNI	PC12 cells	crush-induced sciatic nerve rat model	[52]

3. ALGINATE

ALG is a natural linear anionic polysaccharide, generally extracted from brown algae seaweed, consisting of repeated units of (1-4)-β-d-mannuronic acid and an α-l-guluronic acid building block [53]. ALG possesses several fruitful features, encompassing high biocompatibility, low toxicity, and good gelation properties, which make it an ideal polymer for the development of a scaffold intended for tissue engineering and drug delivery, due to its interaction with bivalent cations. Namely, ALG has been widely used in the field of nervous tissue engineering to develop various novel DDSs: Injectable or non-injectable hydrogels, microfibers, and more complex composite systems with both neuroprotective and neuroregenerative potential [13,54].

ALG was exploited by Downing and coworkers in 2012 to produce a microfibrous drug delivery system containing rolipram, an anti-inflammatory drug with many positive properties for the treatment of nerve damage and, in particular, SCI. The system developed was a poly(L-lactide)-based microfibrous platform produced via the electrospinning technique and coated with a $CaCl_2$- cross-linked ALG hydrogel layer employed for the controlled and local delivery of rolipram. The developed platform confirmed its ability to control drug delivery, showing a burst release of approximately 40% rolipram within the first 18 h and a controlled release even after 1.5 days. The efficacy of low drug content in the platforms was pointed out after the treatment of rats subjected to a C5 hemisection lesion, with improvements in axon regeneration and functional and anatomical recovery [55].

Ansorena et al. developed an injectable ALG hydrogel embedded with microspheres loaded with the glial-derived neurotrophic factor (GDNF), a growth factor able to stimulate functional recovery by promoting survival and neurite growth. GDNF was encapsulated into polylactic-co-glycolic acid (PLGA) microspheres prepared via solvent extraction/evaporation by means of Total Recirculation One-Machine System (TROMS) technology to obtain a homogeneous product with a high encapsulation efficiency. ALG was mixed with fibrinogen to improve biocompatibility and cross-linked with calcium chloride ($CaCl_2$) to obtain a hydrogel; prior to gelation, GNDF microspheres or free GNFD were added to the ALG:fibrinogen solution. The system was able to control the release of GNDF when encapsulated in PLGA microspheres and when in free form, but the release was slower in the presence of microspheres. A bioactivity assay of the released GNDF, performed on rat pheochromocytoma PC-12 cells, demonstrated the ability to promote cell differentiation in the presence of the growth factor, thanks to the development of neurites from cells. The performance of the ALG:fibrinogen hydrogel containing GNDF microspheres or free GNDF was tested in vivo on a rat model (the hydrogel was formed in situ via the direct injection of the components on hemisected spinal cords of rats); the ability of the GNDF-microsphere-loaded hydrogel to promote neurite formation around the lesion was proven; in contrast, the free GNDF-loaded hydrogel promoted neurite ingrowth at the lesion site. Finally, the functional recovery of the rats was then assessed, which showed the best overall recovery when treated with the free GNDF-loaded hydrogel, in comparison with non-treated animals and the GNDF-microspheres-loaded hydrogel; these results can be conceivably attributed to the different release profile of GNDF at the site of injury [56].

A similar injectable $CaCl_2$ cross-linked ALG:-fibrinogen-based hydrogel was employed by des Rieux and coworkers as a carrier for the vascular endothelial growth factor (VEGF) for its neuroprotective properties in the spinal cord regeneration process. VEGF was used in its free form and encapsulated in CS–dextran sulfate (CS/Dx) Nps or in PLGA microspheres. The free VEGF loaded-hydrogel showed modest proliferation on human neuronal-like cells (SH-SY5Y cells) but not on mouse fibroblast-like cells (NIH-3T3 cells), even when supplemented with fibrinogen. Therefore, the free VEGF-loaded hydrogel stimulated neurite growth in ex vivo dorsal root ganglia cultures, but not when tested in a rat spinal cord hemisection model. VEGF Nps or microspheres were first characterized for the growth factor release profile; both systems provided a slower release when compared to the free VEGF-loaded hydrogel, which was too slow in the case of VEGF microspheres. Both free VEGF and VEGF Nps were then incorporated into the ALG:fibrinogen hydrogel to combine fast and sustained release, and the system was injected into a rat spinal cord hemisection model. VEGF was always demonstrated to stimulate angiogenesis in the lesion site and promote neurite growth in and around the lesion. However, the VEGF-loaded hydrogel was not able to improve the functional recovery of rats [57].

A drug delivery system intended to provide both neuroprotective and neuroregenerative effects was developed by Nazemi and collaborators; minocycline hydrochloride (MH) and paclitaxel (PCX) were chosen for their neuroprotective and neuro-regeneration activity, respectively. A dual system was prepared, composed of PLGA-based microspheres embedded in an ALG hydrogel; ALG sulfate (ALG-S) was also considered for its bioaffinity, due to electrostatic interactions. The synergistic effect of a formulation containing both MH and PCX-loaded PLGA microspheres was investigated. Regarding MH, an antibiotic and anti-inflammatory drug with potent neuroprotective activities, a complex with ALG or ALG-S was prepared, achieved via the formation of an electrostatic interaction and metal-ion chelation, by using $CaCl_2$ or $MgCl_2$ as cross-linking agents. The complex was obtained by mixing equal volumes of the drug solution with ALG or ALG-S solutions (drug and polymer at the same concentrations) and the drug–polymer interaction products were recovered after centrifugation and lyophilization. After UV-Vis spectroscopy analysis of the supernatants deriving from centrifugation, the MH ALG-S complex cross-linked with $MgCl_2$ resulted in being the complex with the highest encapsulation efficiency. Three different hydrogel formulations were prepared by embedding either the free drug directly into the hydrogel or by its indirect incorporation when entrapped into the complex. In detail, two hydrogel formulations were obtained by directly mixing the free drug solution with either an ALG solution or an ALG:ALG-S solution at a ratio of 9:1, and a third hydrogel formulation was prepared via the indirect incorporation of the drug by embedding the MH ALG-S complex within an ALG solution. In all formulations, calcium D-gluconate monohydrate was used as a cross-linking agent to allow the formation of hydrogels. PCX was encapsulated in PLGA microspheres prepared via the single (oil/water) emulsion/solvent evaporation method. Loaded PLGA microspheres were added to the ALG:ALG-S solution and the hydrogel was obtained following the same protocol as for MH. An ALG: ALG-S hydrogel containing either PCX-loaded PLGA microspheres and MH was finally prepared. The final formulation containing both the drugs proved to be able to reduce the activity of inflammatory cells, thanks to the activity of MH, and to decrease the formation of fibrotic scars, as a result of the presence of PCX, when tested on a left lateral hemisection animal (rat) model of SCI. Moreover, an improvement in the functional recovery of animals was observed for the dual-drug delivery system, thus demonstrating its promising regeneration-enhancing properties toward nervous injuries [58].

Recently, in a paper of ours, ALG was cross-linked with Spermidine (SP) to obtain nanogels as innovative tools for peripheral nerve repair. SP is a naturally occurring bioamine endowed with neuroprotective activity and a cationic nature, which allows its interaction with the anionic ALG. ALG at high and medium viscosity was cross-linked via the ionotropic gelation process with SP at different concentrations, and nano/microgels dispersions were obtained (Figure 5). A DoE approach was exploited in order to find the best combination of the two components in terms of the mean hydrodynamic diameter. Viscosity measurements and the solid-state characterization (FT-IR analysis) allowed us to confirm the occurrence of interactions between ALG and SP. The addition of trehalose as a cryoprotecting agent was also considered for the freeze-drying process, which was exploited to obtain a stable solid product. In vitro studies on SCs demonstrated the high biocompatibility of nanogels; antioxidant and anti-inflammatory properties of SP remained even after cross-linking with ALG, meaning that its incorporation in nanogels did not impair its bioactivity [59].

A summary of all the ALG-based DDSs is reported in Table 2.

Table 2. Summary of ALG-based DDSs for SCI and PNI application. DDS description, production technique, therapeutic agent delivered, in vitro and in vivo models exploited for efficacy assessment are reported.

DDS	Production Technique	Therapeutic Agent	Application	In Vitro Model	In Vivo Model	Reference
PLA-based microfibers coated with a $CaCl_2$-cross-linked ALG hydrogel layer	Electrospinning (microfibers), cross-linking (hydrogel)	Rolipram	SCI	-	Rats subjected to C5 hemisection lesion	[55]

Table 2. *Cont.*

DDS	Production Technique	Therapeutic Agent	Application	In Vitro Model	In Vivo Model	Reference
$CaCl_2$ cross-linked ALG-fibrinogen hydrogel embedded with PLGA microspheres	Solvent extraction/evaporation (microspheres)	GDNF	SCI	PC-12 cells	Rat spinal cord hemisection model	[56]
$CaCl_2$ cross-linked ALG: fibrinogen-based hydrogel containing loaded CS-dextran sulfate Nps or PLGA microspheres	$CaCl_2$ cross-linking (hydrogel); not reported for microspheres and Nps	VEGF	SCI	SH-SY5Y and NIH-3T3 cells	Rat spinal cord hemisection model	[57]
MH ALG or ALG-S complex; PLGA-based microspheres embedded in an ALG or ALG-S hydrogel	Lyophilization for Hydrogel; single (oil/water) emulsion/solvent evaporation method for microspheres	MH and PCX	SCI	-	Left lateral hemisection animal (rats) model	[58]
ALG-spermidine cross-linked hydrogel	Ionotropic gelation and freeze-drying	Spermidine	PNI	Schwann cells	-	[59]

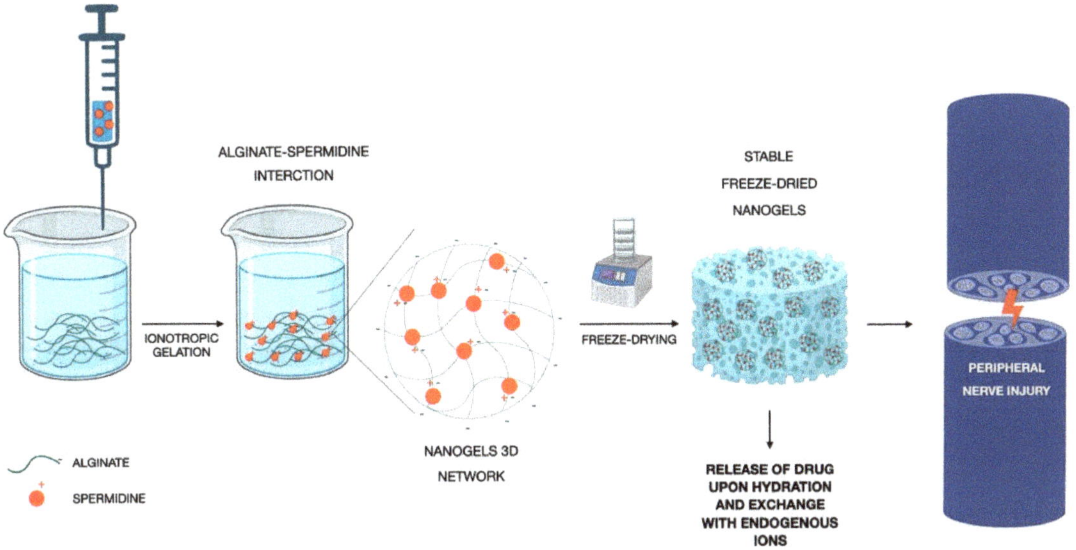

Figure 5. Graphical description of the development of SP-ALG cross-linked nanogels modified by Valentino et al. [59].

4. DEXTRAN

Dextran (Dx) is a complex branched glucan characterized by α-1,6 glycosidic bonds between glucose monomers; it is generally obtained by several Gram-positive, facultatively anaerobe cocci such as *Leuconostoc* and *Streptococcus* strains [60]. It is a biocompatible, biodegradable, nontoxic, and highly hydrophilic polysaccharide that is extensively employed in medicinal products for its peripheral blood-flow-enhancing properties, namely the reduction of blood viscosity and the avoidance of blood clot formation, which prompt Dx use as antithrombolytic agent. Moreover, Dx has been widely used in nanomedicine and for drug delivery; its neutral charge makes it an ideal candidate for Nps synthesis, as it improves nonspecific cellular uptake [61].

As for the employment in nervous system injuries, Dx was used for the synthesis of DDSs intended for spinal cord damage repair by intravenous injection; Nps consisted of an association of ibuprofen and Dx for the delivery of methylprednisolone (MP). Ibuprofen and Dx were combined via esterification between the hydroxyl groups of Dx and the carboxylic acid groups of ibuprofen, activated with N, N-carbonyldiimidazole. System biocompatibility was demonstrated in vitro on BV-12 microglial cells. The blood plasma concentration of MP was further evaluated after intraperitoneal injection in an SCI rat model comparing MP-loaded NPs with a reference solution of MP; for all the times considered, the MP concentration from loaded NPs was higher than that of the reference solution, indicating NPs potential for longer drug circulation and better drug bioavailability. Locomotor functionality recovery and the rehabilitation of neurological deficits and nerve functions after SCI were also improved by ibuprofen-loaded NPs. These Nps were, moreover, found to be successful in the prevention of and the slowdown of neuronal regeneration, due to their anti-inflammatory effect [62].

Dx-loaded Nps were synthesized by Liu and collaborators for the delivery of PCX. Acetalated-Dx Nps were prepared for their reported neuroprotective activity by the microprecipitation method, finding a 1:5 PCX: acetalated-Dx ratio as the best for encapsulation efficiency, loading degree, and PCX release profile. Loaded Nps provided continuous PCX release for 7 days after injection at the site of injury, promoting neural regeneration, neuroprotection, and enhanced locomotor recovery in rats [63].

Acetalated-Dx was also employed by Li and coworkers to obtain microspheres loading Nps, intended for MP sustained release in the case of SCI. Nano-in-micro structured microspheres were prepared by means of a microfluidic flow-focusing device, to exploit a controlled in-droplet precipitation and thus ensure a high drug loading. The nano-in-micro particles obtained allowed a gradual release and higher stability of the loaded drug. Interestingly, the authors succeeded in preparing a system with a high mass fraction of MP, which is particularly desired after intrathecal administration in rats, thanks to the limited volume that can be administered. Such microspheres were injected in vivo after weight drop injury of the spinal cord and provided constant drug release, reduction of the damaged area, and the recovery of motor functionality 28 days after injury [64].

The three Dx DDSs are grouped in Table 3.

Table 3. Summary of Dx-based DDSs for SCI and PNI application. DDS description, production technique, therapeutic agent delivered, and in vitro and in vivo models exploited for efficacy assessment are reported.

DDS	Production Technique	Therapeutic Agent	Application	In Vitro Model	In Vivo Model	Reference
Ibuprofen-Dx Nps	Esterification between the hydroxyl groups of Dx and the carboxylic acid groups of ibuprofen, activated with N, N-carbonyldiimidazole.	Methylprednisolone	SCI	BV-12 microglial cells	Intraperitoneal injection in an SCI rat model	[62]
Acetalated-Dx Nps	Microprecipitation method	PCX	SCI	-	Mechanical SCI rat model	[63]
Acetalated-Dx Nps nano-in-micro structured microspheres	Microfluidic flow-focusing device	Methylprednisolone	SCI	-	Mechanical SCI rat model	[64]

5. AGAROSE (AG)

AG is a water-soluble natural biocompatible polysaccharide extracted from marine red algae, consisting of repeating units of the disaccharide agarobiose, which, in turn, is composed of D-galactose and 3,6-anhydro-L-galactopyranose [65]. AG is characterized by the unique property of self-gelation at 37 °C, without the need for cross-linking agents, thus avoiding their eventual toxicity; such a polymer is also recognized for its high water-uptake capacity, which can promote cell growth, differentiation, and proliferation [66]. In addition, this marine polysaccharide exhibits switchable chemical reactivity for functionalization, strong bioactivity, remarkable mechanical properties, and strict similarity with the natural extracellular matrix (ECM). For these reasons, AG has gained special attention as a biomaterial for the setup of complex carriers for controlled DDSs. In particular, AG's ease of cross-linking through physical interactions has encouraged its application in DDSs [67].

Hence, AG has been widely used for tissue engineering and drug delivery systems intended for several applications, such as cartilage and bone regeneration, brain and nervous defects, cardiovascular diseases, and skin wounds [65]. In the context of SCI and PNI, different systems have been investigated, especially in situ gelling systems or hydrogels with complex geometries, used as platforms to hold and host various other systems such as Nps and microtubes loaded with an active ingredient.

In situ gelling hydrogels for the repair of SCI and local delivery of BDNF were developed by Jain et al. in 2006. The ability of AG regarding in situ gelling and adapting to the shape of the nervous tissue injury was investigated. Since the optimal gelling temperature of AG is 17 °C, the authors developed a cooling system to cool AG in a few seconds when applied at the site of injury and assured its maintenance within the lesion without leaking the defect. BDNF was selected due to the need for neurotrophic factors for neuronal survival and axonal growth after trauma. Lipid microtubes loaded with BDNF were embedded within the AG scaffold. The formulation was able to determine a diffusive-based sustained release of the growth factors in 5–7 weeks and improve the ability of restored nervous fibers to enter the hydrogel thanks to BDNF chemo-attractive action. The scaffold was implanted in vivo in a dorsal over-hemisection rat model and proved to reduce the reactivity of the astrocytes and the production of chondroitin sulfate proteoglycans (CSPGs) due to the presence of BDNF. Moreover, 6 weeks after implantation, a minimal inflammatory response was observed [68].

In 2011, the same authors investigated the release of different bioactive molecules from the AG hydrogel scaffold embedded with lipid microtubes. The scaffold was loaded with, in addition to BDNF, the constitutively active (CA) cell division control protein 42 homolog (CA-Cdc42) and CA Rac Family Small GTPase 1 (CA-Rac1), namely Rho GTPases, which are responsible for the filopodial and lamellipodial extension of axonal growth cones after SCI. The study proved the sustained release of the loaded molecules for at least 2 weeks and the efficacy in vivo of the scaffold in terms of a reduction of astrocytes and CSPG deposition [69].

An AG hydrogel was also developed by Chvatal and coworkers to achieve a localized release of MD. In detail, MD was encapsulated in PLGA Nps produced via the double-emulsion method, and the loaded Nps were mixed with the AG hydrogel. The hydrogel–nanoparticle system was able to determine a slow release of the drug over 6 days. Moreover, such a system, once topically delivered in vivo in a spinal cord injury rat model, proved to promote the reduction of the injury volume and the decrease in secondary injury-related inflammation events within 7 days [70].

Lee and colleagues exploited AG for the preparation of lipid microtubes embedded in an AG-based hydrogel for the delivery of chondroitinase ABC (chABC), an enzyme able to digest CSPGs, which is responsible for the inhibition of axon growth after SCI. Before incorporation into lipid microtubes, chABC was stabilized with trehalose due to its thermal instability at 37 °C. The system was then injected in vivo in a spinal cord lesion rat model, where it determined a sustained delivery of chABC, which, in turn, maintained its ability to digest CSPGs for 2 weeks after injury. Axonal growth and functional recovery of the rat model employed, due to the action of thermally stabilized chABC, was also observed [71].

A complex AG scaffold consisting of microchannels organized in a honeycomb arrangement was developed by Gao and coworkers. In detail, the scaffold consisted of an AG platform characterized by inner micro (166 μm) multi-channel guides, deriving from fiber bundles consisting of hexagonally packed (honeycomb architecture) polystyrene fibers in a linear orientation. The system was loaded with syngeneic marrow stromal cells expressing BDNF, produced via a retroviral vector coding for full-length human BNDF. The authors proved that their bioengineered scaffold could support and guide axonal growth in a clinically relevant in vivo model, namely, the complete transection of a severe rat spinal cord injury [72].

Cox and coworkers developed PLGA Nps loaded with estrogen (E2), which is known for its anti-inflammatory, antioxidant, anti-apoptotic, and neurotrophic properties and its potential for neuroprotection in SCI. E2-loaded PLGA Nps, obtained via the nanoprecipitation method, were dispersed in an AG hydrogel and the efficacy of the scaffold was assessed in a moderate to severe SCI rat model. The E2 concentration in plasma was found to be twice that of the physiological one, and a wide cytokine profiling range was observed [73].

Shultz and collaborators developed an AG hydrogel for the local controlled delivery of physiological doses of thyroid hormone 3,3′,5-triiodothyronine (T3). The rationale is that the loss of oligodendrocytes occurring after SCI results in demyelination and a spared axons formation, which impair healing. Various molecules were compared in vitro, and the authors found that T3 was the most effective to promote oligodendrocytes differentiation. T3 must be necessarily delivered in situ due to its potential systemic side effects, therefore it was loaded onto an AG hydrogel as insoluble particles. A clinically appropriate unilateral cervical spinal cord contusion injury rat model was exploited and exhibited the ability of the scaffold to release a T3 dose consistent with safe doses for humans, the enhancement of oligodendrocytes differentiation, and remyelination after injury [74].

Wang et al. studied the neuroprotective activity of MH, which can target the secondary mechanism after injury, for the treatment of SCI. The drug was complexed with Dx sulfate by the metal ion-assisted interaction to achieve a sustained release; the complex was then loaded onto an injectable AG hydrogel in order to assure the localization of the system in the intrathecal space of the injury. The efficacy of the scaffold was assessed using a clinically relevant unilateral cervical contusion injury model in which the spinal cords of adult rats were impacted at the C5 level. The authors demonstrated that a high concentration of MH requested for SCI repair, not achievable through systemic administration but, however, safe for humans, can be employed if locally delivered [75].

Another research group explored the local delivery of BDNF from an AG hydrogel with the aim of solving the diaphragmatic respiratory impairment attributed to cervical SCI. AG hydrogel was used as a host scaffold for Dx-CS-BDNF particles, self-assembled by electrostatic interactions. The final complex, once injected into the intrathecal space at the injury site in a unilateral cervical contusion rat model, demonstrated its potential efficacy for the effective restoration of respiratory functions [76].

Similarly, Gao and coworkers developed a multi-channel guidance scaffold for the regeneration of PNI. The developed system loaded with BDNF was the same as the previous work reported [72], but in this case, the microchannels had a diameter of 200 μm. After implantation in 15 mm long rat sciatic nerve gaps, the scaffold successfully guided linear axons regeneration over long distances, due to the system architecture and BDNF release at the distal and proximal nerve ending [77].

Table 4 reports all the DDSs based on AG for SCI applications.

Table 4. Summary of Dx-based DDSs for SCI. DDS description, production technique, therapeutic agent delivered, and in vitro and in vivo models exploited for efficacy assessment are reported.

DDS	Production Technique	Therapeutic Agent	Application	In Vitro Model	In Vivo Model	Reference
Loaded-lipid microtubes embedded within AG injectable hydrogel	Self-assembling of lipid microtubes, then added to AG solution.	BNDF	SCI	-	Dorsal over-hemisection rat model	[68]

Table 4. *Cont.*

DDS	Production Technique	Therapeutic Agent	Application	In Vitro Model	In Vivo Model	Reference
Loaded-lipid microtubes embedded within AG injectable hydrogel	Self-assembling of lipid microtubes, then added to AG solution	BNDF, CA-Cdc42 and CA-Rac1	SCI	-	Modified dorsal-over hemisection rat model	[69]
AG hydrogel containing loaded-PLGA Nps	Double emulsion method (Nps)	PCX	SCI	-	Mechanical SCI rat model	[70]
Loaded-lipid microtubes embedded in an AG-based hydrogel	Thermal stabilization of chondroitinase ABC with trehalose	Chondroitinase ABC	SCI	-	Dorsal-over-hemisection injury	[71]
AG scaffold with hexagonally packed multi-channel guides	Multi-component fiber bundle templates (channels diameter: 166 μm)	Syngeneic marrow stromal cells expressing BDNF	SCI	-	Complete transection of rat severe injury model	[72]
AG hydrogel dispersed with loaded PLGA Nps	Nanoprecipitation method and lyophilization (Nps)	Estrogen (E2)	SCI	-	Moderate to severe SCI rat model	[73]
Loaded-AG hydrogel	AG prepared in artificial cerebrospinal fluid and then added with T3 as insoluble particles (obtained neutralizing)	Thyroid hormone 3,3′,5-triiodothyronine (T3).	SCI	-	Unilateral cervical spinal cord contusion injury rat model	[74]
AG containing loaded-Dx sulfate complex	Metal ion-assisted interaction	MH	SCI	-	Unilateral cervical spinal cord contusion injury rat model	[75]
AG embedded with loaded- Dx-CS-particles	Self-assembling by electrostatic interactions (particles)	BNDF	SCI	-	Unilateral cervical spinal cord contusion injury rat model	[76]
AG scaffold with hexagonally packed multi-channel guides	Multi-component fiber bundle templates (channels diameter of 200 μm)	BNDF	SCI	-	Sciatic nerve gaps rat model	[77]

6. CELLULOSE (CL)

Cellulose (CL) is the most abundant biopolymer found in the cell walls of plants or produced by animals, fungi, and bacteria. Structurally, it is primarily composed of D-glucopyranose ring units linked by β-1, 4-glycosidic linkages and organized in chains forming fibrillary units, which are, in turn, assembled in microfibrils [78]. CL has been widely employed for many pharmaceutical purposes due to its tunable properties, in terms of chemical, physical, and mechanical cues; moreover, it is characterized by biocompatibility and bioactivity, although it is not biodegradable [79]. This last feature could be useful for long-term applications, such as the case of nerve tissue recovery after injuries. Bacterial CL has been widely used for nervous tissue damage regeneration, as reported by Jabbari et al., 2022 [80]. For instance, biosynthesized cellulose (BC) obtained from *Gluconacetobacter hansenii* was employed by Stumpf and coworkers to prepare nerve guides able to enhance nerve regeneration after SCI. Authors prepared BC tubes (BCTs) containing various BC amounts depending on *G. hansenii* cells' cultivation at different times (3, 6, 9, 12, 16, 18, and 22 days). BCTs were characterized by an inner diameter of 3.53 mm, a length of 5 cm, and mechanical properties similar to those of native neuronal tissue. BCTs were loaded with NGF as a neurite stimulation agent. NGF-loaded BCTs demonstrated good mechanical properties; in particular, BCTs after 22 days of cultivation (BCTs22) showed Young's Modulus similar to that of the spinal cord. BCTs22 enabled an NGF-controlled release for 7 days, with an initial (8 h) burst release, due to the release of NGF molecules from the outer tube surface. Furthermore, NGF maintained its bioactivity on PC12 cells after release from BC tubes [81].

CL in association with a soy protein isolate was employed by Luo et al. for the development of tubes with an inner diameter of 1.5 mm and an outer diameter of 1.8 mm, intended for the repair of PNI. The soy protein isolate was used for its high biocompatibility, biodegradability, and processability. Tubes were seeded with SCs and pyrroloquinolinequinone (PQQ), a neurotrophic factor able to enhance SCs proliferation and migration. In vivo studies on a rat model were carried out. An autograft nerve, retained as the 'gold standard', was employed as a positive control. It highlighted the ability of the scaffold, especially when seeded with both cells and PQQ, to enhance nerve regeneration, even if to a minor extent in comparison with the control [82].

Table 5 summarizes all the DDSs based on CL described here.

Table 5. Summary of CL-based DDSs for SCI. DDS description, production technique, therapeutic agent delivered, and in vitro and in vivo models exploited for efficacy assessment are reported.

DDS	Production Technique	Therapeutic Agent	Application	In Vitro Model	In Vivo Model	Reference
Biosynthesized cellulose (BC) tubes	Custom-designed bioreactor with silicon tube as mold	NGF	SCI	PC12 cells	-	[81]
CL-soy protein tubes seeded with SC cells	Tubular mold	Pyrroloquinolinequinone (PQQ)	SCI	-	Sciatic nerve injury rat model	[82]

7. GELLAN GUM (GG)

Gellan gum (GG) is a linear anionic exopolysaccharide, produced by microbial fermentation of *Sphingomonas paucimobilis*, composed of tetrasaccharide (1,3-β-D-glucose, 1,4-β-D-glucuronic acid, 1,4-β-D-glucose, and 1,4-α-l-rhamnose (Rha) repeating units with one carboxyl lateral group [83]. More specifically, the structure of GG is characterized by repeating tetramers of L-rhamnose, D-glucuronic acid, and two D-glucose subunits. GG is ductile, thermo-responsive, and can withstand acid and heat stress [84]. The most appealing properties of GG that qualify it as a suitable material for tissue engineering and DDS development are the lack of cytotoxicity, biocompatibility, structural resemblance to native glycosaminoglycans, and mechanical properties comparable to the elastic moduli of common tissues [85]. These properties make it an interesting material for the setup of various DDSs, including particles, films, fibers, and hydrogels. The pH-dependent swelling and stability of GG are desirable properties for DDS formulation, as well as its anionic nature that enables its gelation when combined with monovalent or divalent cations [86].

GG has been widely exploited for nerve tissue regeneration and, as reported in this section, both nanofibers and hydrogels were prepared and loaded with active molecules with positive charges for the treatment of both SCI and PNI.

As for SCI treatment, GG was employed by Vigani and coworkers to develop a composite system for the delivery of RC-33, a drug candidate already used by the same authors, which cross-links with anionic polymers due to its positive charge, as illustrated in Figure 6 [51]. $CaCl_2$-cross-linked GG nanofibers were prepared by the electrospinning process via the addition of two grades of poly (ethylene oxide) (PEO) (high (h-PEO) and low (l-PEO) molecular weight) and poloxamer (P407) to enhance GG electrospinnability.

Electrospun nanofibers were embedded within an outer freeze-dried matrix, composed of an RC-33/GG interaction product and glycine as a cryoprotectant agent, where different concentrations of the drug candidate were loaded. The RC-33/GG interaction product was investigated by dialysis equilibrium, and the maximum binding capacity of GG for RC-33 was found. Four different loaded matrices, which provided the outer structure of the composite system, were obtained varying the concentration of RC-33, based on the results of dialysis equilibrium analysis. It was found that the matrix loaded with the highest amount of RC-33 (interacting with 40% of GG binding sites) was the optimal candidate in terms of mechanical and hydration properties. Finally, cross-linked nanofibers were immersed in a 40%RC-33/GG solution and freeze-dried to obtain the final composite system, characterized by improved mechanical properties [87].

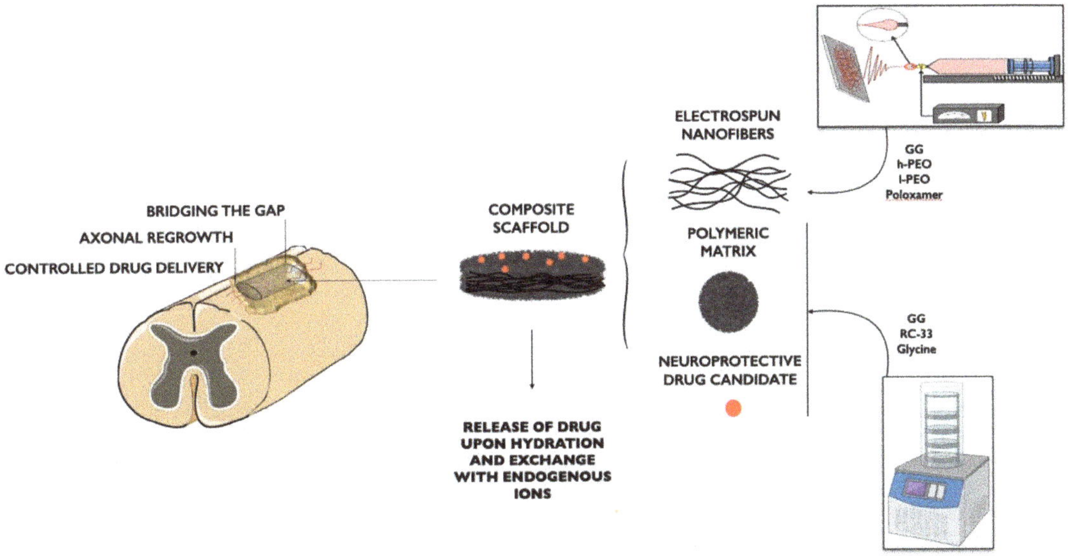

Figure 6. Illustration of the composite scaffold composed of GG electrospun fibers embedded within a GG porous matrix containing RC-33 drug candidate [87].

The same authors used GG to prepare nanofibers loaded with SP and gelatin (GL), intended for both SCI and PNI treatment. Due to its positive charges, SP was used both as a cross-linking agent and as the active component to be released with a positive effect in the treatment of peripheral nerve regeneration. Mixtures containing GG and increasing SP concentrations (0–0.125% *w*/*w*) were prepared to investigate GG/SP interaction. The addition of GL was found to enhance system biomimetic properties. The best mixture of GG/SP/GL was electrospun to obtain cross-linked nanofibers (Figure 7). These fibers were characterized for mechanical properties and morphology after soaking in water for 24 h, thus confirming the formation of an interaction product since nanofibers were insoluble in an aqueous medium. Nanofiber biocompatibility was evaluated on SCs, showing that the presence of GL was mandatory to enhance nanofibers' compatibility with cells [88].

Regarding PNI treatment, porous neurodurable hydrogel conduits made of GG and xanthan gum intercalated with polymethyl methacrylate (PMMA) particles were developed by Ramburrun and collaborators, for the controlled release of two model compounds: Bovine serum albumin (BSA) and diclofenac sodium. Hydrogel conduits were synthesized via a thermal-ionic cross-linking mechanism with direct addition via the intercalation of PMMA. Fifteen different formulations, chosen on the basis of a Box–Behnken experimental design, were evaluated for drug release, swelling, degradation, and textural properties. All the formulations gave an almost zero-order release of BSA and diclofenac sodium over the course of 20 and 30 days, respectively. This result was obtained due to the combination of a pH-responsive (pH 7.4) dissolution of the PMMA particles and the distinct gelling and degradation properties due to the presence of xanthan gum. The variation of the gellan–xanthan ratio and the pore-inducing properties of intercalated PMMA enabled fine-tuning of the mechanical properties of the hydrogel matrices, particularly the matrix rigidity and flexibility [89].

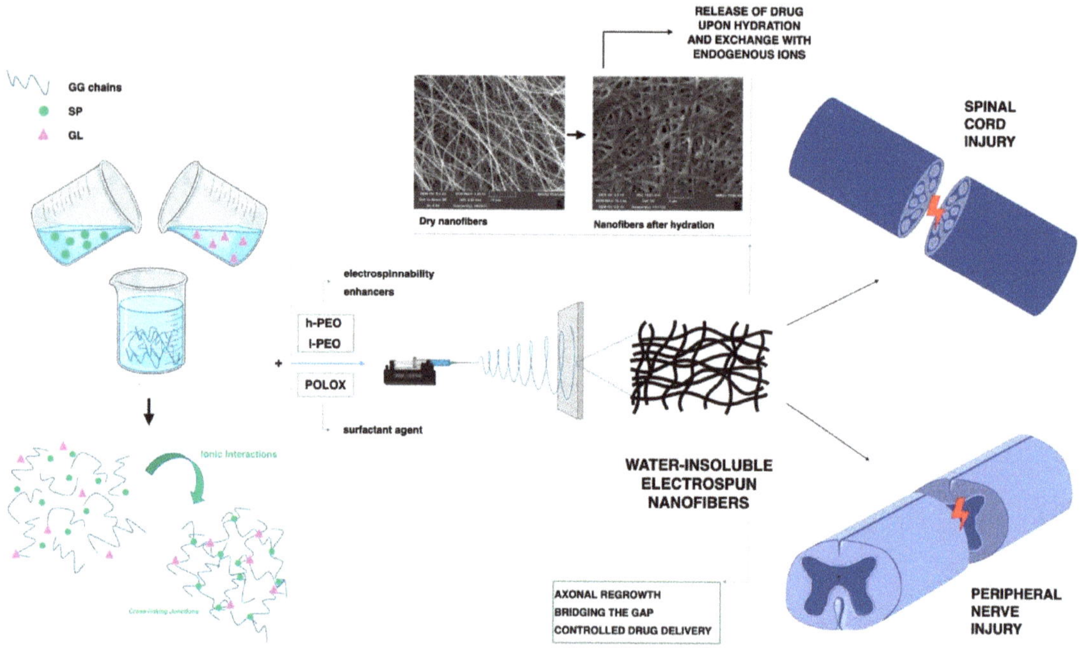

Figure 7. Representative scheme of preparation of GG-based nanofibers containing gelatin and SP developed by Vigani and coworkers for nervous tissue injuries [88].

Li and coworkers developed a GG hydrogel loaded with laminin and NGF to promote neuronal stem cells' proliferation and differentiation. Laminin is recognized as one of the most important elements of ECM of peripheral nerve cells and is responsible for the enhancement of neurite outgrowth. In detail, thiolated GG hydrogel was obtained by dissolving GG at 60 °C and, and after cooling at room temperature, the gel was mixed with a cell suspension containing laminin and NGF. NGF release was studied by immersing the hydrogel in PBS at 37 °C and analyzing the collected solution with a Sandwich ELISA test. The release was rather fast (approximately 60% in 20 h) and the release profile of NGF corresponded to a Fickian diffusion-controlled mechanism. In vitro experiments were carried out with neural stem cells harvested by the cerebral cortex of rats; cell viability after contact with the loaded GG hydrogel was manually determined after 72 h via a cell-counting kit (CCK-8), and cell proliferation was measured by fluorescence microscopy. Neural stem cells were found to be highly interconnected to the hydrogel, creating a whole network with an oval or triangular architecture [90].

Table 6 summarizes all the GG DDSs developed during recent years of research.

Table 6. Summary of GG-based DDSs for SCI. DDS description, production technique, therapeutic agent delivered, and in vitro and in vivo models exploited for efficacy assessment are reported.

DDS	Production Technique	Therapeutic Agent	Application	In Vitro Model	In Vivo Model	Reference
$CaCl_2$-cross-linked GG nanofibers embedded with loaded-GG freeze-dried matrix,	Electrospinning (nanofibers), freeze-drying (porous matrix)	RC-33	SCI	-	-	[87]
hydrogel conduits of GG and xanthan gum intercalated with polymethyl methacrylate particles	Thermal-ionic cross-linking mechanism	Bovine serum albumin (BSA) and diclofenac sodium	PNI	-	-	[89]
thiolated GG hydrogel	Prepared at 60 °C and then poured into a mold to form in situ gel	Laminin and NGF	PNI	Rat neural stem cell	-	[90]
GG/GL nanofibers	Electrospinning	Spermidine	SCI/PNI	Schwann cells		[88]

8. Conclusions

The success of DDSs in SCI and PNI treatment depends on different factors: The bioactive properties of the constituent polymers; the architecture and mechanical properties of the DDS that should mimic the natural extracellular matrix to support cell growth and assembly; and the efficacy of the loaded drug and the kinetics of drug release from the DDS.

Polysaccharides are the first choice of polymers since they are characterized by excellent biocompatibility with intrinsic biological cues and excellent intrinsic properties that facilitate their use in the fabrication of 3D structures able to mimic ECM and control the release of the loaded drug.

Moreover, since these polymers are generally derived from renewable sources or are byproducts of industrial processes, their use is to be considered collaborative in waste management and promising for an improved sustainable circular economy.

The revision of the literature reported here demonstrated how active the research in this field is. In recent years, many polysaccharide-based DDSs have been developed and have been proven effective in the treatment of nerve tissue injuries in animal models. Most of them are complex systems obtained by the assembly of different architectures (such as nanoparticles, fibers, films, and hydrogels) that require the employment of different techniques and, thus, a multistep process for their fabrication. Making the manufacturing process as simple as feasible, removing the crucial processes, and favoring its scalability are issues that will need to be addressed in the upcoming years.

Therefore, future research should be focused on the development of bioactive polysaccharide-based DDSs able to control the release of the loaded drug according to the therapeutic needs and characterized by a simple and scalable manufacturing process. Among the formulations reported, micro/nanofibers and micro/nanoparticles obtained by electrospinning and spray-drying, respectively, that are simple, versatile, highly efficient, and easily scalable using continuous manufacturing technologies are those that fulfill this requirement. Another issue of concern is the sterilization process employed. In many of the papers reviewed, no mention of this aspect is reported. The complexity of the systems developed could make it difficult to find a suitable method. Moreover, it must be underlined that no information about the translatability in humans of the DDSs reviewed here is present in the literature. The lack of in vivo tests on humans could arise as a problem, as mentioned above.

Author Contributions: Conceptualization, C.V., S.R., B.V.; writing—original draft preparation, C.V.; writing—review and editing, C.V., B.V., F.F., S.R.; visualization, C.V., B.V.; supervision, S.R., G.S. All authors have read and agreed to the published version of the manuscript.

Funding: This research received no external funding.

Institutional Review Board Statement: Not applicable.

Informed Consent Statement: Not applicable.

Data Availability Statement: Not applicable.

Conflicts of Interest: The authors declare no conflict of interest.

References

1. Birch, R.; Birch, R.; Birch, R.; Birch, R. The Peripheral Nervous System: Anatomy and Function. In *Peripheral Nerve Injuries: A Clinical Guide*; Springer: London, UK, 2013; pp. 1–67.
2. Thau, L.; Reddy, V.; Singh, P. *Anatomy, Central Nervous System*; StatPearls Publishing: Treasure Island, FL, USA, 2022.
3. Gautam, A. Afferent and Efferent Impulses. In *Encyclopedia of Animal Cognition and Behavior*; Springer International Publishing: Cham, Switzerland, 2017; pp. 1–2.
4. Vijayavenkataraman, S. Nerve guide conduits for peripheral nerve injury repair: A review on design, materials and fabrication methods. *Acta Biomater.* **2020**, *106*, 54–69. [CrossRef] [PubMed]
5. Catala, M.; Kubis, N. Gross anatomy and development of the peripheral nervous system. In *Handbook of Clinical Neurology*; Elsevier: Amsterdam, The Netherland, 2013; Volume 115, pp. 29–41. [CrossRef]
6. Rossi, S.; Vigani, B.; Sandri, G.; Bonferoni, M.C.; Ferrari, F. Design and criteria of electrospun fibrous scaffolds for the treatment of spinal cord injury. *Neural Regen. Res.* **2017**, *12*, 1786–1790. [CrossRef] [PubMed]
7. Ahuja, C.S.; Wilson, J.R.; Nori, S.; Kotter, M.R.N.; Druschel, C.; Curt, A.; Fehlings, M.G. Traumatic Spinal Cord Injury. *Nat. Rev. Dis. Primers* **2017**, *3*, 17018. [CrossRef] [PubMed]
8. Silva, N.A.; Sousa, N.; Reis, R.L.; Salgado, A.J. From basics to clinical: A comprehensive review on spinal cord injury. *Prog. Neurobiol.* **2014**, *114*, 25–57. [CrossRef]
9. Li, R.; Li, D.-H.; Zhang, H.-Y.; Wang, J.; Li, X.-K.; Xiao, J. Growth factors-based therapeutic strategies and their underlying signaling mechanisms for peripheral nerve regeneration. *Acta Pharmacol. Sin.* **2020**, *41*, 1289–1300. [CrossRef]
10. Grinsell, D.; Keating, C.P. Peripheral Nerve Reconstruction after Injury: A Review of Clinical and Experimental Therapies. *BioMed Res. Int.* **2014**, *2014*, 698256. [CrossRef]
11. Yousefi, F.; Arab, F.L.; Nikkhah, K.; Amiri, H.; Mahmoudi, M. Novel approaches using mesenchymal stem cells for curing peripheral nerve injuries. *Life Sci.* **2019**, *221*, 99–108. [CrossRef]
12. Carvalho, C.R.; Silva-Correia, J.; Oliveira, J.M.; Reis, R.L. Nanotechnology in peripheral nerve repair and reconstruction. *Adv. Drug Deliv. Rev.* **2019**, *148*, 308–343. [CrossRef]
13. Carvalho, C.R.; Reis, R.L.; Oliveira, J.M. Fundamentals and Current Strategies for Peripheral Nerve Repair and Regeneration. In *Advances in Experimental Medicine and Biology*; Springer: Singapore, 2020; Volume 1249, pp. 173–201. [CrossRef]
14. Sullivan, R.; Dailey, T.; Duncan, K.; Abel, N.; Borlongan, C.V. Peripheral Nerve Injury: Stem Cell Therapy and Peripheral Nerve Transfer. *Int. J. Mol. Sci.* **2016**, *17*, 2101. [CrossRef]
15. Oyinbo, C.A. Secondary injury mechanisms in traumatic spinal cord injury: A nugget of this multiply cascade. *Acta Neurobiol. Exp. (Wars)* **2011**, *71*, 281–299.
16. Ramer, L.M.; Ramer, M.S.; Bradbury, E.J. Restoring function after spinal cord injury: Towards clinical translation of experimental strategies. *Lancet Neurol.* **2014**, *13*, 1241–1256. [CrossRef]
17. Boni, R.; Ali, A.; Shavandi, A.; Clarkson, A.N. Current and novel polymeric biomaterials for neural tissue engineering. *J. Biomed. Sci.* **2018**, *25*, 90. [CrossRef]
18. Wieringa, P.; Pinho, A.R.; Micera, S.; A Van Wezel, R.J.; Moroni, L. Biomimetic Architectures for Peripheral Nerve Repair: A Review of Biofabrication Strategies. *Adv. Health Mater.* **2018**, *7*, e1701164. [CrossRef]
19. Houshyar, S.; Bhattacharyya, A.; Shanks, R. Peripheral Nerve Conduit: Materials and Structures. *ACS Chem. Neurosci.* **2019**, *10*, 3349–3365. [CrossRef]
20. Sarker; Naghieh, S.; McInnes, A.D.; Schreyer, D.J.; Chen, X. Strategic Design and Fabrication of Nerve Guidance Conduits for Peripheral Nerve Regeneration. *Biotechnol. J.* **2018**, *13*, e1700635. [CrossRef]
21. Anjum, A.; Yazid, M.; Daud, M.F.; Idris, J.; Ng, A.; Naicker, A.S.; Ismail, O.; Kumar, R.A.; Lokanathan, Y. Spinal Cord Injury: Pathophysiology, Multimolecular Interactions, and Underlying Recovery Mechanisms. *Int. J. Mol. Sci.* **2020**, *21*, 7533. [CrossRef]
22. Manoukian, O.S.; Arul, M.R.; Rudraiah, S.; Kalajzic, I.; Kumbar, S.G. Aligned microchannel polymer-nanotube composites for peripheral nerve regeneration: Small molecule drug delivery. *J. Control. Release* **2019**, *296*, 54–67. [CrossRef]
23. Sun, Y.; Ma, X.; Hu, H. Marine Polysaccharides as a Versatile Biomass for the Construction of Nano Drug De-livery Systems. *Mar. Drugs* **2021**, *19*, 345. [CrossRef]

24. Rial-Hermida, M.I.; Rey-Rico, A.; Blanco-Fernandez, B.; Carballo-Pedrares, N.; Byrne, E.M.; Mano, J.F. Recent Progress on Polysaccharide-Based Hydrogels for Controlled Delivery of Therapeutic Biomolecules. *ACS Biomater. Sci. Eng.* **2021**, *7*, 4102–4127. [CrossRef]
25. Shelke, N.B.; James, R.; Laurencin, C.T.; Kumbar, S.G. Polysaccharide biomaterials for drug delivery and regenerative engineering. *Polym. Adv. Technol.* **2014**, *25*, 448–460. [CrossRef]
26. Prasher, P.; Sharma, M.; Mehta, M.; Satija, S.; Aljabali, A.A.; Tambuwala, M.M.; Anand, K.; Sharma, N.; Dureja, H.; Jha, N.K.; et al. Current-status and applications of polysaccharides in drug delivery systems. *Colloid Interface Sci. Commun.* **2021**, *42*, 100418. [CrossRef]
27. Chandrika, K.P.; Prasad, R.; Godbole, V. Development of chitosan-PEG blended films using Trichoderma: Enhancement of antimicrobial activity and seed quality. *Int. J. Biol. Macromol.* **2018**, *126*, 282–290. [CrossRef]
28. Sood, A.; Gupta, A.; Agrawal, G. Recent advances in polysaccharides based biomaterials for drug delivery and tissue engineering applications. *Carbohydr. Polym. Technol. Appl.* **2021**, *2*, 100067. [CrossRef]
29. Ways, T.M.M.; Lau, W.M.; Khutoryanskiy, V.V. Chitosan and Its Derivatives for Application in Mucoadhesive Drug Delivery Systems. *Polymers* **2018**, *10*, 267. [CrossRef] [PubMed]
30. Aranaz, I.; Alcántara, A.R.; Civera, M.C.; Arias, C.; Elorza, B.; Caballero, A.H.; Acosta, N. Chitosan: An Overview of Its Properties and Applications. *Polymers* **2021**, *13*, 3256. [CrossRef] [PubMed]
31. Boecker, A.; Daeschler, S.C.; Kneser, U.; Harhaus, L. Relevance and Recent Developments of Chitosan in Peripheral Nerve Surgery. *Front Cell Neurosci.* **2019**, *13*, 104. [CrossRef]
32. Skop, N.B.; Calderon, F.; Levison, S.; Gandhi, C.D.; Cho, C.H. Heparin crosslinked chitosan microspheres for the delivery of neural stem cells and growth factors for central nervous system repair. *Acta Biomater.* **2013**, *9*, 6834–6843. [CrossRef]
33. Wu, W.; Lee, S.-Y.; Wu, X.; Tyler, J.Y.; Wang, H.; Ouyang, Z.; Park, K.; Xu, X.-M.; Cheng, J.-X. Neuroprotective ferulic acid (FA)–glycol chitosan (GC) nanoparticles for functional restoration of traumatically injured spinal cord. *Biomaterials* **2014**, *35*, 2355–2364. [CrossRef]
34. Gwak, S.-J.; Koo, H.; Yun, Y.; Yhee, J.Y.; Lee, H.Y.; Yoon, D.H.; Kim, K.; Ha, Y. Multifunctional nanoparticles for gene delivery and spinal cord injury. *J. Biomed. Mater. Res. Part A* **2015**, *103*, 3474–3482. [CrossRef]
35. Ni, S.; Xia, T.; Li, X.; Zhu, X.; Qi, H.; Huang, S.; Wang, J. Sustained delivery of chondroitinase ABC by poly(propylene carbonate)-chitosan micron fibers promotes axon regeneration and functional recovery after spinal cord hemisection. *Brain Res.* **2015**, *1624*, 469–478. [CrossRef]
36. Wang, D.; Wang, K.; Liu, Z.; Wang, Z.; Wu, H. Valproic acid-labeled chitosan nanoparticles promote recovery of neuronal injury after spinal cord injury. *Aging (Albany NY)* **2020**, *12*, 8953–8967. [CrossRef]
37. Alizadeh, A.; Moradi, L.; Katebi, M.; Ai, J.; Azami, M.; Moradveisi, B.; Ostad, S.N. Delivery of injectable thermo-sensitive hydrogel releasing nerve growth factor for spinal cord regeneration in rat animal model. *J. Tissue Viability* **2020**, *29*, 359–366. [CrossRef]
38. Kwiecien, J.M.; Zhang, L.; Yaron, J.R.; Schutz, L.N.; Kwiecien-Delaney, C.J.; Awo, E.A.; Burgin, M.; Dabrowski, W.; Lucas, A.R. Local Serpin Treatment via Chitosan-Collagen Hydrogel after Spinal Cord Injury Reduces Tissue Damage and Improves Neurologic Function. *J. Clin. Med.* **2020**, *9*, 1221. [CrossRef]
39. Wang, N.; Yu, H.; Song, Q.; Mao, P.; Li, K.; Bao, G. Sesamol-loaded stearic acid-chitosan nanomicelles mitigate the oxidative stress-stimulated apoptosis and induction of pro-inflammatory cytokines in motor neuronal of the spinal cord through NF-kB signaling pathway. *Int. J. Biol. Macromol.* **2021**, *186*, 23–32. [CrossRef]
40. Javdani, M.; Ghorbani, R.; Hashemnia, M. Histopathological Evaluation of Spinal Cord with Experimental Traumatic Injury Following Implantation of a Controlled Released Drug Delivery System of Chitosan Hy-drogel Loaded with Selenium Nanoparticle. *Biol. Trace Elem. Res.* **2021**, *199*, 2677–2686. [CrossRef]
41. Song, X.; Xu, Y.; Wu, J.; Shao, H.; Gao, J.; Feng, X.; Gu, J. A sandwich structured drug delivery composite membrane for improved recovery after spinal cord injury under longtime controlled release. *Colloids Surf. B Biointerfaces* **2020**, *199*, 111529. [CrossRef]
42. Stenberg, L.; Kodama, A.; Lindwall-Blom, C.; Dahlin, L.B. Nerve regeneration in chitosan conduits and in autologous nerve grafts in healthy and in type 2 diabetic Goto-Kakizaki rats. *Eur. J. Neurosci.* **2015**, *43*, 463–473. [CrossRef]
43. Li, X.; Wang, W.; Wei, G.; Wang, G.; Zhang, W.; Ma, X. Immunophilin FK506 loaded in chitosan guide promotes peripheral nerve regeneration. *Biotechnol. Lett.* **2010**, *32*, 1333–1337. [CrossRef]
44. Farahpour, M.R.; Ghayour, S.J. Effect of in situ delivery of acetyl-L-carnitine on peripheral nerve regeneration and functional recovery in transected sciatic nerve in rat. *Int. J. Surg.* **2014**, *12*, 1409–1415. [CrossRef]
45. Lopes, C.D.; Gonçalves, N.P.; Gomes, C.P.; Saraiva, M.J.; Pêgo, A.P. BDNF gene delivery mediated by neuron-targeted nanoparticles is neuroprotective in peripheral nerve injury. *Biomaterials* **2017**, *121*, 83–96. [CrossRef]
46. Jahromi, M.; Razavi, S.; Seyedebrahimi, R.; Reisi, P.; Kazemi, M. Regeneration of Rat Sciatic Nerve Using PLGA Conduit Containing Rat ADSCs with Controlled Release of BDNF and Gold Nanoparticles. *J. Mol. Neurosci.* **2020**, *71*, 746–760. [CrossRef] [PubMed]
47. Zeng, W.; Hui, H.; Liu, Z.; Chang, Z.; Wang, M.; He, B.; Hao, D. TPP ionically cross-linked chitosan/PLGA microspheres for the delivery of NGF for peripheral nerve system repair. *Carbohydr. Polym.* **2021**, *258*, 117684. [CrossRef] [PubMed]
48. Zhang, L.; Chen, Y.; Xu, H.; Bao, Y.; Yan, X.; Li, Y.; Li, Y.; Yin, Y.; Wang, X.; Qiu, T.; et al. Preparation and evaluation of an injectable chitosan-hyaluronic acid hydrogel for peripheral nerve regeneration. *J. Wuhan Univ. Technol. Sci. Ed.* **2016**, *31*, 1401–1407. [CrossRef]

49. Xu, H.; Zhang, L.; Bao, Y.; Yan, X.; Yin, Y.; Li, Y.; Wang, X.; Huang, Z.; Xu, P. Preparation and characterization of injectable chitosan–hyaluronic acid hydrogels for nerve growth factor sustained release. *J. Bioact. Compat. Polym.* **2016**, *32*, 146–162. [CrossRef]
50. Han, G.H.; Kim, S.J.; Ko, W.-K.; Lee, D.; Lee, J.S.; Nah, H.; Han, I.-B.; Sohn, S. Injectable Hydrogel Containing Tauroursodeoxycholic Acid for Anti-neuroinflammatory Therapy After Spinal Cord Injury in Rats. *Mol. Neurobiol.* **2020**, *57*, 4007–4017. [CrossRef]
51. Vigani, B.; Rossi, S.; Sandri, G.; Bonferoni, M.C.; Rui, M.; Collina, S.; Fagiani, F.; Lanni, C.; Ferrari, F. Dual-Functioning Scaffolds for the Treatment of Spinal Cord Injury: Alginate Nanofibers Loaded with the Sigma 1 Receptor (S1R) Agonist RC-33 in Chitosan Films. *Mar. Drugs* **2019**, *18*, 21. [CrossRef]
52. Rahmati, M.; Ehterami, A.; Saberani, R.; Abbaszadeh-Goudarzi, G.; Kolarijani, N.R.; Khastar, H.; Garmabi, B.; Salehi, M. Improving sciatic nerve regeneration by using alginate/chitosan hydrogel containing berberine. *Drug Deliv. Transl. Res.* **2020**, *11*, 1983–1993. [CrossRef]
53. Ashraf, R.; Sofi, H.S.; Beigh, M.A.; Majeed, S.; Arjamand, S.; Sheikh, F.A. Prospects of Natural Polymeric Scaffolds in Peripheral Nerve Tissue-Regeneration. In *Advances in Experimental Medicine and Biology*; Springer: New York, NY, USA, 2018; Volume 1077, pp. 501–525. [CrossRef]
54. Grijalvo, S.; Nieto-Díaz, M.; Maza, R.M.; Eritja, R.; Díaz, D.D. Alginate Hydrogels as Scaffolds and Delivery Systems to Repair the Damaged Spinal Cord. *Biotechnol. J.* **2019**, *14*, 1900275. [CrossRef]
55. Downing, T.L.; Wang, A.; Yan, Z.-Q.; Nout, Y.; Lee, A.L.; Beattie, M.S.; Bresnahan, J.C.; Farmer, D.L.; Li, S. Drug-eluting microfibrous patches for the local delivery of rolipram in spinal cord repair. *J. Control. Release* **2012**, *161*, 910–917. [CrossRef]
56. Ansorena, E.; De Berdt, P.; Ucakar, B.; Simón-Yarza, T.; Jacobs, D.; Schakman, O.; Jankovski, A.; Deumens, R.; Blanco-Prieto, M.J.; Préat, V.; et al. Injectable alginate hydrogel loaded with GDNF promotes functional recovery in a hemisection model of spinal cord injury. *Int. J. Pharm.* **2013**, *455*, 148–158. [CrossRef]
57. Des Rieux, A.; De Berdt, P.; Ansorena, E.; Ucakar, B.; Damien, J.; Schakman, O.; Audouard, E.; Bouzin, C.; Auhl, D.; Simón-Yarza, T.; et al. Vascular endothelial growth factor-loaded injectable hydrogel enhances plasticity in the injured spinal cord: Vegf-Loaded Injectable Hydrogel. *J. Biomed. Mater. Res. Part A* **2014**, *102*, 2345–2355. [CrossRef]
58. Nazemi, Z.; Nourbakhsh, M.S.; Kiani, S.; Heydari, Y.; Ashtiani, M.K.; Daemi, H.; Baharvand, H. Co-delivery of minocycline and paclitaxel from injectable hydrogel for treatment of spinal cord injury. *J. Control. Release* **2020**, *321*, 145–158. [CrossRef]
59. Caterina, V.; Barbara, V.; Ilaria, F.; Dalila, M.; Giorgio, M.; Lorenzo, M.; Ferrari, F.; Giuseppina, S.; Silvia, R. Development of alginate-spermidine micro/nanogels as potential antioxidant and anti-inflammatory tool in peripheral nerve injuries. *Formul. Stud. Phys. Chem. Characterization. Int. J. Pharm.* **2022**, *626*, 122168. [CrossRef]
60. Hu, Q.; Lu, Y.; Luo, Y. Recent advances in dextran-based drug delivery systems: From fabrication strategies to applications. *Carbohydr. Polym.* **2021**, *264*, 117999. [CrossRef]
61. Wasiak, I.; Kulikowska, A.; Janczewska, M.; Michalak, M.; Cymerman, I.A.; Nagalski, A.; Kallinger, P.; Szymanski, W.W.; Ciach, T. Dextran Nanoparticle Synthesis and Properties. *PLoS ONE* **2016**, *11*, e0146237. [CrossRef]
62. Qi, L.; Jiang, H.; Cui, X.; Liang, G.; Gao, M.; Huang, Z.; Xi, Q. Synthesis of methylprednisolone loaded ibuprofen modified dextran based nanoparticles and their application for drug delivery in acute spinal cord injury. *Oncotarget* **2017**, *8*, 99666–99680. [CrossRef]
63. Liu, W.; Quan, P.; Li, Q.; Tang, P.; Chen, J.; Jiang, T.; Cai, W. Dextran-based biodegradable nanoparticles: An alternative and convenient strategy for treatment of traumatic spinal cord injury. *Int. J. Nanomed.* **2018**, *13*, 4121–4132. [CrossRef]
64. Li, W.; Chen, J.; Zhao, S.; Huang, T.; Ying, H.; Trujillo, C.; Molinaro, G.; Zhou, Z.; Jiang, T.; Li, L.; et al. High drug-loaded microspheres enabled by controlled in-droplet precipitation promote functional recovery after spinal cord injury. *Nat. Commun.* **2022**, *13*, 1262. [CrossRef]
65. Zarrintaj, P.; Manouchehri, S.; Ahmadi, Z.; Saeb, M.R.; Urbanska, A.M.; Kaplan, D.L.; Mozafari, M. Aga-rose-Based Biomaterials for Tissue Engineering. *Carbohydr. Polym.* **2018**, *187*, 66–84. [CrossRef]
66. Azarfam, M.Y.; Nasirinezhad, M.; Naeim, H.; Zarrintaj, P.; Saeb, M. A Green Composite Based on Gelatin/Agarose/Zeolite as a Potential Scaffold for Tissue Engineering Applications. *J. Compos. Sci.* **2021**, *5*, 125. [CrossRef]
67. Yazdi, M.K.; Taghizadeh, A.; Taghizadeh, M.; Stadler, F.J.; Farokhi, M.; Mottaghitalab, F.; Zarrintaj, P.; Ramsey, J.D.; Seidi, F.; Saeb, M.R.; et al. Agarose-based biomaterials for advanced drug delivery. *J. Control. Release* **2020**, *326*, 523–543. [CrossRef] [PubMed]
68. Jain, A.; Kim, Y.-T.; McKeon, R.J.; Bellamkonda, R.V. In situ gelling hydrogels for conformal repair of spinal cord defects, and local delivery of BDNF after spinal cord injury. *Biomaterials* **2006**, *27*, 497–504. [CrossRef] [PubMed]
69. Jain, A.; McKeon, R.J.; Brady-Kalnay, S.M.; Bellamkonda, R.V. Sustained Delivery of Activated Rho GTPases and BDNF Promotes Axon Growth in CSPG-Rich Regions Following Spinal Cord Injury. *PLoS ONE* **2011**, *6*, e16135. [CrossRef]
70. Chvatal, S.A.; Kim, Y.-T.; Bratt-Leal, A.M.; Lee, H.; Bellamkonda, R.V. Spatial distribution and acute anti-inflammatory effects of Methylprednisolone after sustained local delivery to the contused spinal cord. *Biomaterials* **2008**, *29*, 1967–1975. [CrossRef] [PubMed]
71. Lee, H.; McKeon, R.J.; Bellamkonda, R.V. Sustained delivery of thermostabilized chABC enhances axonal sprouting and functional recovery after spinal cord injury. *Proc. Natl. Acad. Sci. USA* **2009**, *107*, 3340–3345. [CrossRef]
72. Gao, M.; Lu, P.; Bednark, B.; Lynam, D.; Conner, J.M.; Sakamoto, J.; Tuszynski, M.H. Templated agarose scaffolds for the support of motor axon regeneration into sites of complete spinal cord transection. *Biomaterials* **2013**, *34*, 1529–1536. [CrossRef]

73. Cox, A.; Varma, A.; Barry, J.; Vertegel, A.; Banik, N. Nanoparticle Estrogen in Rat Spinal Cord Injury Elicits Rapid Anti-Inflammatory Effects in Plasma, Cerebrospinal Fluid, and Tissue. *J. Neurotrauma* **2015**, *32*, 1413–1421. [CrossRef]
74. Shultz, R.B.; Wang, Z.; Nong, J.; Zhang, Z.; Zhong, Y. Local delivery of thyroid hormone enhances oligodendrogenesis and myelination after spinal cord injury. *J. Neural Eng.* **2017**, *14*, 036014. [CrossRef]
75. Wang, Z.; Nong, J.; Shultz, R.B.; Zhang, Z.; Kim, T.; Tom, V.J.; Ponnappan, R.K.; Zhong, Y. Local delivery of minocycline from metal ion-assisted self-assembled complexes promotes neuroprotection and functional recovery after spinal cord injury. *Biomaterials* **2017**, *112*, 62–71. [CrossRef]
76. Ghosh, B.; Wang, Z.; Nong, J.; Urban, M.W.; Zhang, Z.; Trovillion, V.A.; Wright, M.C.; Zhong, Y.; Lepore, A.C. Local BDNF delivery to the injured cervical spinal cord using an engineered hydrogel enhances diaphragmatic respiratory function. *J. Neurosci.* **2018**, *38*, 5982–5995. [CrossRef]
77. Gao, M.; Lu, P.; Lynam, D.; Bednark, B.; Campana, W.M.; Sakamoto, J.; Tuszynski, M. BDNF gene delivery within and beyond templated agarose multi-channel guidance scaffolds enhances peripheral nerve regeneration. *J. Neural Eng.* **2016**, *13*, 066011. [CrossRef]
78. Dutta, S.D.; Patel, D.K.; Lim, K.-T. Functional cellulose-based hydrogels as extracellular matrices for tissue engineering. *J. Biol. Eng.* **2019**, *13*, 55. [CrossRef]
79. Hickey, R.J.; Pelling, A.E. Cellulose Biomaterials for Tissue Engineering. *Front. Bioeng. Biotechnol.* **2019**, *7*, 45. [CrossRef]
80. Jabbari, F.; Babaeipour, V.; Bakhtiari, S. Bacterial cellulose-based composites for nerve tissue engineering. *Int. J. Biol. Macromol.* **2022**, *217*, 120–130. [CrossRef]
81. Stumpf, T.R.; Tang, L.; Kirkwood, K.; Yang, X.; Zhang, J.; Cao, X. Production and evaluation of biosynthesized cellulose tubes as promising nerve guides for spinal cord injury treatment. *J. Biomed. Mater. Res. Part A* **2020**, *108*, 1380–1389. [CrossRef]
82. Luo, L.; Gan, L.; Liu, Y.; Tian, W.; Tong, Z.; Wang, X.; Huselstein, C.; Chen, Y. Construction of nerve guide conduits from cellulose/soy protein composite membranes combined with Schwann cells and pyrroloquinoline quinone for the repair of peripheral nerve defect. *Biochem. Biophys. Res. Commun.* **2015**, *457*, 507–513. [CrossRef]
83. Muthukumar, T.; Song, J.E.; Khang, G. Biological Role of Gellan Gum in Improving Scaffold Drug Delivery, Cell Adhesion Properties for Tissue Engineering Applications. *Molecules* **2019**, *24*, 4514. [CrossRef]
84. Kirchmajer, D.M.; Steinhoff, B.; Warren, H.; Clark, R.; Panhuis, M.I.H. Enhanced gelation properties of purified gellan gum. *Carbohydr. Res.* **2014**, *388*, 125–129. [CrossRef]
85. Stevens, L.R.; Gilmore, K.J.; Wallace, G.G.; Panhuis, M.I.H. Tissue engineering with gellan gum. *Biomater. Sci.* **2016**, *4*, 1276–1290. [CrossRef]
86. Sweety, J.P.; Selvasudha, N.; Dhanalekshmi, U.M.; Sridurgadevi, N. Gellan Gum and Its Composites: Suitable Candidate for Efficient Nanodrug Delivery. In *Nanoengineering of Biomaterials*; Wiley: Hoboken, NJ, USA, 2021; pp. 33–61. [CrossRef]
87. Vigani, B.; Valentino, C.; Cavalloro, V.; Catenacci, L.; Sorrenti, M.; Sandri, G.; Bonferoni, M.; Bozzi, C.; Collina, S.; Rossi, S.; et al. Gellan-based composite system as a potential tool for the treatment of nervous tissue injuries: Cross-linked electrospun nanofibers embedded in a rc-33-loaded freeze-dried matrix. *Pharmaceutics* **2021**, *13*, 164. [CrossRef]
88. Vigani, B.; Valentino, C.; Sandri, G.; Caramella, C.M.; Ferrari, F.; Rossi, S. Spermidine crosslinked gellan gum-based "hydrogel nanofibers" as potential tool for the treatment of nervous tissue injuries: A formulation study. *Int. J. Nanomed.* **2022**, *17*, 3421–3439. [CrossRef] [PubMed]
89. Ramburrun, P.; Kumar, P.; Choonara, Y.E.; du Toit, L.C.; Pillay, V. Design and characterization of neurodurable gellan-xanthan pH-responsive hydrogels for controlled drug delivery. *Expert Opin. Drug Deliv.* **2016**, *14*, 291–306. [CrossRef] [PubMed]
90. Li, W.; Huang, A.; Zhong, Y.; Huang, L.; Yang, J.; Zhou, C.; Zhou, L.; Zhang, Y.; Fu, G. Laminin-modified gellan gum hydrogels loaded with the nerve growth factor to enhance the proliferation and differentiation of neuronal stem cells. *RSC Adv.* **2020**, *10*, 17114–17122. [CrossRef] [PubMed]

pharmaceutics

MDPI

Review

Recent Advances in Targeted Nanocarriers for the Management of Triple Negative Breast Cancer

Rajesh Pradhan [1], Anuradha Dey [2], Rajeev Taliyan [1,*], Anu Puri [3], Sanskruti Kharavtekar [1] and Sunil Kumar Dubey [1,2,†,*]

1 Department of Pharmacy, Birla Institute of Technology and Science, Pilani 333031, India
2 Medical Research, R&D Healthcare Division, Emami Ltd., Kolkata 700056, India
3 RNA Structure and Design Section, RNA Biology Laboratory (RBL), Center for Cancer Research, National Cancer Institute—Frederick, Frederick, MD 21702, USA
* Correspondence: rajeev.taliyan@pilani.bits-pilani.ac.in (R.T.); sunilbit2014@gmail.com (S.K.D.); Tel.: +91-6378-364-745 (R.T.); +91-8239-703-734 (S.K.D.)
† Present Affiliation.

Abstract: Triple-negative breast cancer (TNBC) is a life-threatening form of breast cancer which has been found to account for 15% of all the subtypes of breast cancer. Currently available treatments are significantly less effective in TNBC management because of several factors such as poor bioavailability, low specificity, multidrug resistance, poor cellular uptake, and unwanted side effects being the major ones. As a rapidly growing field, nano-therapeutics offers promising alternatives for breast cancer treatment. This platform provides a suitable pathway for crossing biological barriers and allowing sustained systemic circulation time and an improved pharmacokinetic profile of the drug. Apart from this, it also provides an optimized target-specific drug delivery system and improves drug accumulation in tumor cells. This review provides insights into the molecular mechanisms associated with the pathogenesis of TNBC, along with summarizing the conventional therapy and recent advances of different nano-carriers for the management of TNBC.

Keywords: triple-negative breast cancer; biological barriers; tumor microenvironment; nano-therapeutics; inorganic nanoparticles; organic nanoparticles; organic-inorganic hybrid nanoparticle

Citation: Pradhan, R.; Dey, A.; Taliyan, R.; Puri, A.; Kharavtekar, S.; Dubey, S.K. Recent Advances in Targeted Nanocarriers for the Management of Triple Negative Breast Cancer. *Pharmaceutics* **2023**, *15*, 246. https://doi.org/10.3390/pharmaceutics15010246

Academic Editors: Silvia Tampucci and Daniela Monti

Received: 22 November 2022
Revised: 1 January 2023
Accepted: 4 January 2023
Published: 11 January 2023

1. Introduction

Triple-negative breast cancer (TNBC) is an aggressive type of heterogenous breast cancer that tests negative for the expression of estrogen, progesterone, and human epidermal growth factor 2 receptors (HER2). These tumors have an aggressive nature, with a tendency for early relapse and metastatic spreading toward the lungs, liver, and central nervous system, as well as a poor prognosis [1,2]. The risk of TNBC is higher in certain ethnic groups, such as Latin, African, and African-American women, as well as women with breast cancer 1 (*BRCA1*) gene mutations. This subtype is most prevalent in young women, and it accounts for 15–20% of all breast cancers diagnosed [3,4]. According to current molecular and morphological studies, invasive ductal carcinomas account for around 90% of TNBC cases, while lobular, apocrine, adenoid cystic, and metaplastic carcinomas account for the rest. Despite sharing the triple negative phenotype, the prognosis of every class is distinct [5]. Further, TNBC is classified into six subtypes (Basal-like 1 & 2, Immunomodulatory, Luminal androgen receptor, Mesenchymal, and Mesenchymal stem-like) based on gene expression profiles [6]. Similarly, Burstein et al. proposed a classification for TNBC that included four subtypes: basal-like immune-suppressed, basal-like immune-activated, luminal androgen receptor, and mesenchymal [7]. In the same study, the basal-like immune-activated subtype was found to be associated with a better prognosis, which is consistent with the findings of other studies showing that TNBC with lymphocytic infiltration had a better prognosis [7]. Recently, TNBC's diversity in terms of gene expression and the variety of genetic events has been studied. The diverse nature of TNBC, which usually results in distinct responses to therapy, implies that it should further be subdivided from a therapeutic perspective. The discovery of specific molecular markers for TNBC subtypes would surely enhance the diagnosis process, as well as aid the development of predictive biomarkers and targeted therapeutics [8,9].

As there is still no ideal treatment option available for TNBC, this review focuses on discussing the pathophysiological aspects of TNBC and the receptors involved, which helps in achieving target-specific drug delivery using a variety of nanotechnologies. The current treatment approach for TNBC is either individual use of anticancer drugs or along with surgery or radiotherapy. The chemotherapeutic approach involves the use of anticancer agents belonging to classes such as anthracyclines, platinum compounds, and taxanes. Although there is an unmet need in the therapeutic field against TNBC, developing a novel drug is a relatively expensive process. Hence, the scientific community has recently become more interested in studying new drug delivery systems in order to enhance the therapeutic potential of currently available drug molecules [10]. Thus, various nanotechnological approaches involving the use of various organic and inorganic nanoparticles are being explored with essential modifications in order to be effective against TNBC. Herein, we give a comprehensive review focusing on novel drug delivery approaches for TNBC. Furthermore, this review also discusses several distinct approaches for treating TNBC, including gene therapy, photodynamic (PDT), and photothermal therapies (PTT). As it is important to understand the pathophysiology of TNBC to develop new therapeutic approaches, this review also includes a detailed discussion of the molecular mechanisms involved in TNBC pathogenesis.

2. Pathogenesis of Triple Negative Breast Cancer

TNBC is the most fatal type amongst all breast cancer. It shows the lack of expression on estrogen receptors (ER), progesterone receptors (PR), and HER2 receptor amplification. ER and PR present on the myoepithelial cells and HER2 on the epithelial cells of the breasts are absent in TNBC. The definition of absence here is "less than 1% stain of ER or PR detected by immunohistochemistry and the IHC (immunohistochemistry) score of HER2 to be 0/+1 or +2 with FISH negative" [11,12]. One of the significant causes of the development of ER, PR, and HER2 negative cells is the *BRCA* gene mutation. The *BRCA1* gene regulates the DNA damage checkpoints and has a significant role in the DNA repair process. It can convert ER-negative cells to ER-positive cells and is required for the normal development of mammary glands. *BRCA1* and *BRCA2* repair the double-stranded break in DNA by a process known as homologous recombination. Homologous recombination is the process, which is governed by ATM, *BRCA1*, and *BRCA2*. When DNA is damaged, ATM is activated, which phosphorylates *BRCA1*, and the homologous recombination process begins with the help of *BRCA2* and RAD51 [13]. The *BRCA* gene mutation causes the repair of DNA double-strand breaks by mutagenic mechanisms rather than homologous recombination, resulting in genetic instability. This genetic instability leads to the development of cancer [12]. Another reason for the occurrence of TNBC is the transformation of the mammalian target of rapamycin (mTOR) pathway. This transformation makes the disease aggressive and invasive. In TNBC, the phosphoinositide 3 kinase (PI3K)/Akt/mTOR (PAM) is activated, which causes cancer angiogenesis, cell growth, and cell proliferation. Activation of mTOR results in the generation of two complexes, namely mTORC1 and mTORC2. mTORC1 is responsible for stimulating the growth of cells, and mTORC2 mediates AKT phosphorylation [14,15]. Additionally, in tumor cells, there is no control over the cell cycle progression, which permits the cells having faults in DNA to move forward to the subsequent step of the cycle, or alternatively, it causes more genetic diversity. TNBC cells undergo repeated genetic alterations in the tumor suppressor protein (p53) and Rb pathway inactivation, while c-Myc pathways mutations are infrequent. In the cell cycle control process, the Rb pathway in the G1/S checkpoint is triggered by phosphorylation of retinoblastoma through cyclin/CDK complex activation, which leads to the instigation of transcription factor E2F via Myc activation. In contrast, inhibition of CDK2 activity via up-regulation of CDNK2 encourages the inactivation of suppressor gene p53, which blocks the senescence in apoptosis pathways. The amplified expression of CDNK2A and related CDNK2′s complex permits evading the cell-cycle regulatory checks in the Rb pathways and allows progression into the G1/S phase. Hence, it leads to high expression of cellular proliferation and growth. This recurrent genetic alteration causes TNBC [16,17]. The representative pathology has shown in Figure 1.

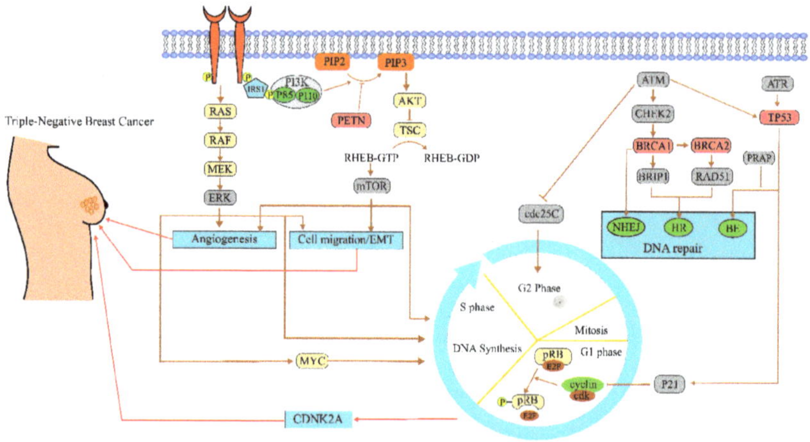

Figure 1. Pathogenesis of TNBC.

Earlier, scientists considered basal-like breast cancer and TNBC as the same because they have similar pathological features. With the advancement in genetics, it was discovered that most basal-like cancers are triple-negative, but not all of them possess a triple-negative phenotype. Breast cancers are basal-like if the tumor cells express the genes that are generally expressed by myoepithelial or basal cells, such as CK, CK 5/6, CK14, CK17, fascin, p-cadherin, vimentin, αβ crystallin, and caveolins 1 and 2. Basal-like breast cancers show negative ER, PR, and HER2 receptor expression and express epidermal growth factor receptor (EGFR). According to this concept, all basal-like breast cancers should be categorized as triple-negative. However, TNBCs differ from basal-like cancers with respect to their prognostic characteristics. TNBCs show short disease-free survival as compared to triple-negative non-basal-like cancers. The metastatic spread of triple-negative basal-like cancer appears to be different than that of non-basal-like phenotype. The basal-like phenotype of TNBC metastasizes to the brain and lungs and is less likely to spread to bones and axillary nodes [18,19]. Some of the basal cancers have shown the presence of ER and HER2 receptors. Some TNBCs were tested to be negative for basal-like markers such as CK5/6, CK14, and EGFR. Hence, we can conclude that not all basal-like cancers are triple-negative, nor all TNBCs are basal-like [20].

3. Conventional Approaches for TNBC

In the current era, the medical field is engrossed with TNBC due to its aggressiveness [21]. The treatment of tumors has progressed from surgery to the use of X-rays. The selection of the treatment for TNBC and its advancement depends on the origin of cancer development, cancer type, and molecular site, as well as the stage of progression [9]. In the late twentieth century, for the management of TNBC, various conventional therapeutic options, including cytotoxic agents and chemotherapy emerged. Also, research work on targeted cancer therapy, which is based on specific molecular targets in neoplastic processes for the betterment of TNBC, is going on. On the other hand, the biotechnology field and

PDT have also offered their contributions as innovative developments in clinical oncology [3]. The most commonly used treatment approaches such as chemotherapy, surgery, and radiotherapy are discussed in detail below.

3.1. Chemotherapy

Chemotherapeutic drugs are well-established therapeutic approaches for various cancer treatments. It works biologically by interfering with cellular pathways, stopping tumor progression by impairing tumor cells' ability to proliferate, resulting in apoptosis. This therapy includes alkylating agents, antimetabolites, mitotic inhibitors, and topoisomerase inhibitors [22,23]. Surprisingly, it has been found that TNBC had an excellent response to chemotherapy as compared to other breast cancer subtypes. It was reported that only one-third of patients who had undergone anthracycline or a combination of anthracyclines and taxane chemotherapy showed a complete pathological response due to chemoresistance [24]. Anthracycline and taxanes were found to be substrates of the multidrug resistance 1 transporter, which is the Pgp transporter. This transporter is majorly involved in the efflux of these drugs, decreasing their effectiveness [22]. Further, non-specific targeting, toxicity, poor solubility, and low bioavailability of anti-cancer drugs restrict the use of conventional therapies. Chemotherapy causes several unwanted side effects, such as nausea, alopecia, vomiting, myelosuppression, and fatigue [25]. Chemotherapy causes a number of other side effects that are both stressful and can be fatal for patients. Furthermore, there are high chances of the development of oral and or gastrointestinal mucositis, leading to ulcerations pain anorexia, weight loss, anemia, fatigue, and chances of sepsis formation. In another retrospective study, individual patient data from four different randomized clinical trials were studied for the presence of hematological non-hematological side effects of anthracycline involving chemotherapy. It was observed that hematological toxicities were common, but the incidence increased with age. It was also observed that taxane involving chemotherapy showed increased incidences of sensory neuropathy of grades 3–4. Mucositis grades 3–4 was more common in the case of elderly patients. The study concluded that the dose-intensified regimes showcased increased toxicities and the incidence of hematological toxicities was high in elderly patients after the use of neoadjuvant chemotherapy regimes against breast cancer.

Additionally, chemotherapy is primarily utilized in combination with surgery to treat different cancer types. Chemotherapy-based drugs are often administered to patients before or after surgery to improve the therapeutic effectiveness. Based on the therapeutic utilization, it is classified into two types i.e., Adjuvant therapy, after surgery and Neoadjuvant therapy i.e., before surgery [26]. A brief discussion related to adjuvant and neoadjuvant chemotherapy has been mentioned below.

3.1.1. Adjuvant Chemotherapy

Adjuvant therapies enhance both overall and disease-free survival. It prevents the risk of metastasis and tumor recurrence activity. Several research findings have reported that taxanes at various doses have shown good efficacy against metastatic breast cancer (MBC). On the other hand, treatment with taxane has not shown significant results for TNBC. Thus, combination based therapy has gained attention towards the management of TNBC. But, it has also found that women with TNBC mostly do not show any improvement after undergoing this therapy as combination approaches of anthracyclines and taxanes show drug resistance [27]. Further, based on combining anthracyclines and taxane, many clinical trials have been conducted with various chemotherapeutic drugs, such as capecitabine plus taxotere, ixabepilone plus capecitabine [15]. It was observed that the efficacy of the ixabepilone and capecitabine combination showed improvement in progression-free survival. Similarly, a pivotal trial of capecitabine plus taxotere showed that capecitabine improves survival when combined with docetaxel, but there is no significant evidence against TNBC [15,27]. Another trial report on anthracyclines showed the benefit of adjuvant anthracyclines. Patients with more than nine affected lymph nodes were assigned either dose-dense conventional chemotherapy or a rapidly cycled tandem high-dose regimen. The results indicated that young TNBC patients benefited from a quickly cycled tandem approach. It resulted in a 5-year event-free survival rate of 71% in TNBC patients, compared to 26% in TNBC patients receiving dose-dense conventional therapy [22]. Thus, it is essential to conduct more clinical trials on different chemotherapy regimens for the advancement of TNBC treatment.

3.1.2. Neoadjuvant Chemotherapy

Neoadjuvant therapy is cytotoxic chemotherapy that is given before surgery and offers a higher pathological complete response rate in TNBC compared to other types of breast cancer. It is useful in the treatment of tumors $\geq$2 cm [28]. Individuals who have TNBC responded initially to neoadjuvant therapy. It was estimated that the possibility of relapse was greater in the first 5 years compared to other types of cancer. Its individualization is needed as *BRCA1* mutations are basal, but not all basal cancers exhibit *BRCA1* mutations [15]. It was found that neoadjuvant therapy with taxanes and anthracyclines exhibited a pathologic complete response (pCR) rate of 30%. Several limitations have also been reported, including changes in the stage of cancer, remains of a residual intraductal component after breast-conserving surgery, and overtreatment [29,30]. Furthermore, Telli et al. conducted a phase II study of neoadjuvant therapy of gemcitabine, carboplatin, and iniparib for tumor management, and this approach was found to be effective against TNBC-affected patients [29]. Another study conducted by Gerber et al. reported that the addition of bevacizumab to anthracyclines and taxane enhanced the complete pathological response from 27.9% to 39.3% in TNBC patients [31]. Wu et al. evaluated the response and prognosis of taxanes and anthracyclines as neoadjuvant therapies in TNBC patients. The results indicated that patients suffering from TNBC were more sensitive to neoadjuvant docetaxel and epirubicin chemotherapy [32]. Neoadjuvant therapy with anthracycline and cyclophosphamide was found to show efficacy, but drug resistance remained to be a problem [15]. Another study conducted by Fisher et al. compared neoadjuvant and adjuvant chemotherapy, they selected 385 patients suffering from stage I-III TNBC. Patients were segregated as neoadjuvant with pCR, neoadjuvant without pCR, and adjuvant settings. Fisher's exact test and analysis of variance were used to compare the data. 17% of patients receiving neoadjuvant therapy reported a pathological complete response. The study suggested neoadjuvant therapy to be more beneficial as compared to adjuvant therapy [33].

3.2. Surgery

Surgery is a well-established traditional technique used to treat cancer without any preventable tissue damage. Surgery offers minimal damage to the surrounding tissues over radiation therapy and chemotherapy [21]. There are two types of surgical processes involved in TNBC, i.e., lumpectomy and mastectomy. In the case of lumpectomy, only the part of the breast having cancer along with the surrounding normal tissue was removed. While in mastectomy the entire breast is removed. Studies have shown lumpectomy to give better loco-regional recurrence-free survival (LRRFS), disease-free survival (DFS), and overall survival (OS) compared to mastectomy [34]. It has been reported that the combination of breast-conserving surgery with postoperative radiotherapy improves the overall survival and breast cancer-specific survival (BCSS) compared to total mastectomy TNBC patients [35]. Although surgery is a better option, it also has certain limitations, including post-operative pain, side effects due to heavy medications, and nosocomial infections along with influencing the lifestyle of the patient [21].

3.3. Radiation Therapy

Generally, after mastectomy or conservation breast surgery (CBS), radiotherapy is given as it can improve loco-regional control in breast cancer [36,37]. In addition to this, reports exhibit *BRCA1* mutation leading to high sensitivity to radiation. It has been reported that removal of occult *BRCA1* deficient tumor foci from the breast and surrounding tissue can occur if CBS is followed by radiotherapy, leading to a decreased locoregional recurrence [22,38]. The associated toxicities and the lack of treatment guides for radiotherapy in TNBC have restricted its use [39,40]. One major limitation of this radiotherapy is the cost involved due to the use of complex machinery and technology. Furthermore, it is also associated with some major off-target side effects.

4. Challenges towards Management of TNBC

4.1. Biological Barrier

Despite the several advantages offered by nanocarriers, overcoming the physicochemical and biological barriers has remained a concern for researchers. In a meta-analysis study conducted, it was estimated that less than 1% of the nanocarriers are targeted to the high enhanced permeability and retention (EPR) region of the tumor. This is attributed to physiological barriers such as endothelial barriers, cellular barriers, clearance by the mononuclear phagocyte system (MPS), and endosomal escape [41]. These barriers are part of the defense system of our body, which offers protection against foreign substances [42]. The electrostatic interactions, steric interactions, hydrophobic interactions, biological interactions, and solvent interactions with the cell membrane depend on the physicochemical properties of nanoparticles, including size, elasticity, surface charge, shape, pKa, and chemistry of ligand. This, in turn, affects the accumulation of nanocarriers in tumor cells [41,43]. Scale-up, approval by regulatory agencies, and efflux by transporters are other hurdles for nanoparticulate drug delivery. The MPS consists of tissue macrophages, bone marrow progenitors, blood monocytes, and dendritic cells. They play a role in the clearance of the nanocarriers and hence affect their pharmacokinetics and tissue distribution. It affects the efficacy and biological half-life of the nanocarrier. After entering the blood, plasma proteins, apolipoproteins, and immunoglobulins get adsorbed on the nanocarrier, forming protein corona. The process is termed as opsonization. The phagocytes take up these nanocarriers attached to the protein corona, thereby clearing them from the body. It mainly occurs in the liver, lymph nodes, and spleen [44]. PEGylation has been found to be one of the approaches for avoiding the nanocarrier's uptake by phagocytes. It induces the formation of a hydrating layer composed of ethylene glycol units and water molecules. This prevents its recognition by phagocytes and thereby its uptake. It enhances the circulating life of nanocarriers [45]. In order to reach the tumor site, the nanocarrier has to cross the endothelial layer of blood vessels. The endothelial layer is composed of a glycocalyx and proteoglycan layer, which guides the entry of solutes and macromolecules. The continuous, fenestrated, and discontinuous endothelium acts as a barrier for nanocarrier [42,46].

The endothelial gaps present in tumor blood vessels are not uniform throughout the tissue. This results in the uneven distribution of the nanocarrier in tumor tissue. Perfusion also guides the accumulation of nanocarriers in the tumor tissue. Heterogeneous perfusion also results in the uneven distribution of nanocarriers in tumor tissue. The interstitial tumor matrix forms a barrier for the nanocarriers that do not extravasate into tumor tissue. Decreasing the particle size can be one of the approaches to enhance penetration in the interstitial matrix [41]. Extravasation has also been found to be dependent on the hemodynamics of nanocarriers. However, its effect needs to be investigated further. The next barrier is the tumor microenvironment (TME). Collagen and elastic fibers of proteins and glycosaminoglycans form the interstitial space. In tumor tissues, the collagen content is more as compared to the normal tissues. This makes the ECM extremely rigid and limits the entry of nanocarriers in tumor cells [42]. Interstitial fluid pressure and hypoxic core are some other factors that affect the distribution of nanocarriers in tumor tissue. Further, for the nanocarrier to enter the nucleus and exert their effect, they have to cross the cell membrane. Nanoparticle internalization takes place through the process of endocytosis. Nanocarriers reach the lysosomes if taken up by endosomes and phagosomes related to the phagocytotic and clathrin-mediated endocytotic pathway. Caveolin mediated endocytosis has been found to bypass lysosomes in some instances. Several approaches have been investigated to escape the endosomes, including the coating of polymers or membrane-stabilizing peptides. Caveolin mediated endocytosis is seen in nanocarriers when their surface is functionalized by cholesterol, folic acid, or albumin [45,46].

Several studies have revealed that particles of size less than 20 nm significantly bind to lipoproteins, whereas particles above 100 nm were found to have an affinity for complement-related proteins. Particles having a size of more than 200 nm were found to be taken up by the liver and spleen because of complement activation. Further, smaller particles were found to exhibit longer circulation half-life. Further, the shape of the nanocarrier determines cycling time, uptake by macrophages, ability to overcome biological barriers, and targeting effects. Solubility, stability, cytotoxicity, and cellular uptake are determined by surface charge. Elasticity has been found to affect tumor uptake and blood circulation.

4.2. Tumor Microenvironment

The tumor cells have several other cells in the neighborhood including immune cells, fibroblasts, lymphocytes, tumor cells, signaling factors, dendritic cells, adipocytes, macrophages, growth factors, tumor vasculature, and proteins that are secreted by these cells. These cells form the tumor microenvironment (TME). The tumor cells alter the components of the TME in such a way that it promotes the growth of the tumor. The cross-talk between the stromal cells and the tumor cells also contribute to the growth and sustenance of the tumor [47]. The TME has been found to govern carcinogenesis, impact the growth and the malignant behavior of mammary cancer cells, and also re-program the surrounding cells [48].

Several mechanisms of tumorigenesis and disease progression are being investigated based on the TME. Biomarkers in the stromal environment have been identified and studied to develop new therapies for the treatment of TNBC. The quick progression and poor prognosis of TNBC can be explained by the unique features observed in its TME [48]. Cancer associated fibroblasts (CAF) secrete cytokine and inflammatory mediators which promote the growth and invasiveness of tumor cells. They are also involved in cancer pathogenesis as they promote angiogenesis, extracellular matrix (ECM) remodeling, cause inflammation, and govern differential events in epithelial cells. It has been found that CAF help in TNBC progression by activating TGF-β [47]. It is believed that they originate from bone marrow derived cells in TME, normal fibroblasts which respond to cancer cell signals, and TME of epithelial cells. It is also found to be involved in brain metastases [49]. They produce metalloproteinases which are involved in metastasis and invasion [47]. Tumor-infiltrating lymphocytes mainly consist of T-cells. It includes CD8+ cytotoxic T cells, CD4+ cytotoxic helper cells, and CD4+ regulatory T cells. It is observed that TNBC is characterized by a large amount of tumor-infiltrating lymphocytes (TIL) due to somatic mutations. High TIL has been considered as an indication of complete pathological response in TNBC patients. Cancers exhibiting high TILs are generally ER negative. T-cells regulate the immune response in tumors in the initial stages of cancer. However, their interaction with tumor cells forming Tregs promotes cancer progression, aggressiveness, and development [47,48]. These Treg cells secret TGF-β, and IL-10, suppressing immune function. They were found to suppress contact with cells by expressing cytotoxic T lymphocyte associated antigen 4(CTLA-4), preventing its recognition by tumor cells. Tumor macrophages are involved in metastases, invasion, migration, and poor prognosis. They secrete TGF-β and IL-10 which suppress the immune response. Cancer associated adipocytes have the capability to produce cytokines, hormones, growth factors, and adipokines. It was found that they are involved in promoting tumorigenesis and enhancing the aggressiveness of cancer cells. The extracellular matrix is mainly composed of proteins such as structural proteins, glycoproteins, and proteoglycans. It has been found to be involved in the migration, invasion, and growth of tumor tissue [48]. The major constituent of ECM, that is collagen IV plays an important role in cancer invasion by its degradation [49]. Given the characteristic changes that are associated with the TME of TNBC, exploring its nuances and developing strategies for modulating the same makes TME a promising therapeutic target for treating TNBC whilst ensuring that the challenges associated with are addressed as well.

5. Newly Developed Therapeutic Approaches against TNBC

The management of TNBC is based on its severity and its molecular site. Although various conventional therapeutic approaches are already established for TNBC, there is still an unmet need pertaining to the treatment to be taken as they have several limitations, including non-selective targeting, side effects, and chronic toxicities such as mucositis, thrombocytopenia, and alopecia. Further, the efflux transporters pose a hurdle for anti-cancer drugs. Mainly the ATP binding cassette transporter glycoprotein, which is a P-gp efflux transporter, acts as a significant barrier to several drugs, including taxanes and anthracyclines [50]. The poor prognosis and poor targeting and drug resistance of conventional medicine against TNBC have promoted the interest of scientists in developing new targets and therapies. Since then, newer therapies, such as PDT, vascular endothelial growth factor (VEGF) inhibitors, Gene therapy, PTT, EGRF, etc., have been developed and utilized for effective treatment. These therapies are briefly detailed below, and a pictorial representation of different novel therapies is shown in Figure 2.

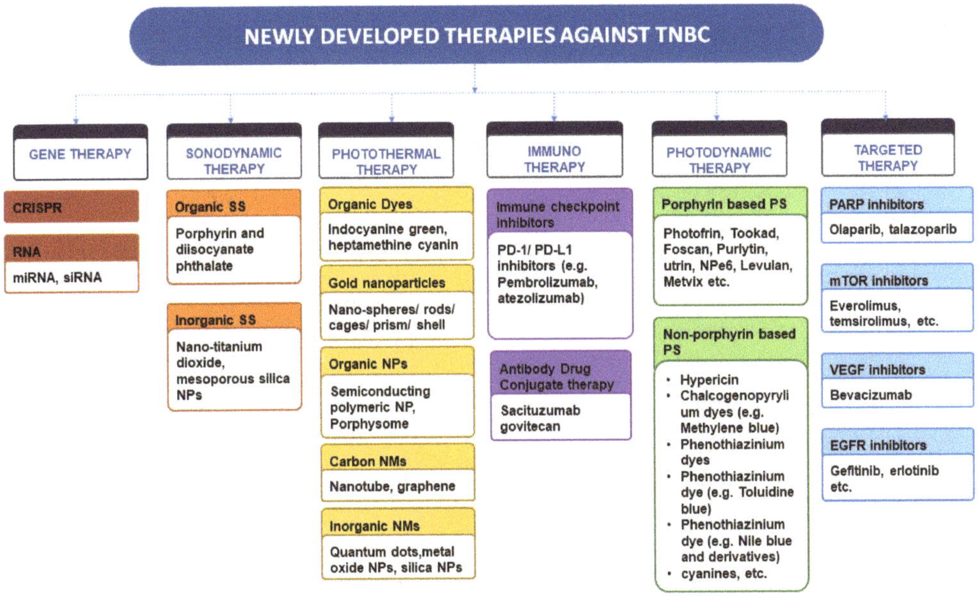

Figure 2. Newly developed therapeutic approaches for TNBC. CRISPR: Clustered Regularly Interspaced Short Palindromic Repeats; EGFR: Epidermal growth factor receptor; miRNA: Micro RNA; mTOR: Mammalian target of rapamycin; NM: Nanomaterials; NPs: Nanoparticles; PARP: Poly (ADP-ribose) polymerases; PD-1/PD-L1: Programmed death ligand 1; PS: Photosensitizer; RNA: Ribonucleic acid; siRNA: Small interfering RNA; SS: Sonodynamic sensitizers; VEGF: Vascular endothelial growth factor.

5.1. VEGF Inhibitor

Vascular endothelial growth factor-A (VEGF) promotes vascular development and angiogenesis in 30–60% of triple-negative breast cancers after binding to VEGF receptor family member 2 (VEGFR) [51,52]. One study reported that mutant p53 along with SWI/SNF, stimulates the expression of VEGFR-2. Thus, mutant p53 can be targeted for breast cancer treatment. Also, JAK2/STAT3 can be targeted, which is recruited by VEGFR-2 to inactivate MYC and SOX2 in breast cancer stem cells [51]. In a recent study, a biocompatible copolymer nanocomplex was developed to deliver VEGF siRNA for TNBC, which inhibited serum degradation, exhibited negligible toxicity, high tumor penetration, high cellular uptake, and high transfection efficiency. In addition, it has significantly inhibited the migration and invasion of TNBC cells [53]. From a recent study, it is found that DEAE (Diethylamino ethyl cellulose)—Dextran coated paclitaxel nanoparticles developed by the solvent evaporation provided the maximum cellular uptake by MDA-MB-231 cells after 1 h. Further, the siRNA prepared by the self-assembly method enhanced the cellular uptake and endosomal escape of nanoparticles and downregulated the intratumoral VEGF protein levels. Also, the strong anti-tumor activity in A-549 xenografts, significant rise in receptor gene expression, and efficiently targeted gene silencing were observed [54]. A study demonstrated that using Bevacizumab monoclonal antibody as a neoadjuvant increased (pCR) while having no effect on DFS or OS in the adjuvant scenario was seen [52].

5.2. EGFR Inhibitors

The epidermal growth factor receptor (EGFR) is a tyrosine kinase receptor belonging to the ErbB family that plays a major role in cell proliferation, differentiation, angiogenesis, metastasis, and protection against cell death pathways as well as inhibition of apoptosis [52,55]. Tyrosine kinase inhibitors such as gefitinib and monoclonal antibodies such as cetuximab have been found to target EGFR. The synergistic effect was observed when a combination of gefitinib, docetaxel, and carboplatin was used. Several monoclonal antibodies target these EGFRs by different mechanisms, including ligand-receptor blockade, inhibiting dimerization, and cell survival signaling pathways [56]. The anti-EGFR antibody-drug conjugate for TNBC is more advantageous than the antibody-only and chemotherapy-only treatments. It targets tumor cell proliferation and inhibits DNA repair via cell membrane or nuclear membrane modulation. It is desirable as it does not affect normal tissues, reduces undesirable effects, and delivers small molecules [57]. Another approach is PEGylated nanomedicine like PEG-liposomal doxorubicin, which displays a conditional internalization by PEG engager as it remains in surface contact with TNBC and shows 100-fold more anti-proliferative activity to EGFR+ TNBC [58]. Furthermore, docetaxel-loaded PEG-poly(epsilon-caprolactone) (PCL) nanoparticles exhibit prolonged drug release for up to 30 days, improved dispersibility, and strong anti-proliferative activity against animal models of TNBC and MDA-MB-231 [59].

5.3. PARP Inhibitors

PARP is an initiation and damage recognition repair protein for single-strand breaks (SSB) in DNA. Its inhibition results in the accumulation of SSBs, leading to double-strand break (DSB) formation. PARP inhibitors are novel oral anticancer drugs that show promising results in TNBC treatment in *BRCA*-mutated breast cancer [60]. There are many PARP inhibitors available for use, like Olaparib, veliparib, niraparib, rucaparib, and talizumab [61]. Studies have shown that PARP inhibitors' benefits are transient in patients with *BRCA*-associated TNBC. Further, several research studies have reported that the high-dimensional single-cell profiling of human TNBC showed that macrophages are the most prevalent immune cell type invading *BRCA* susceptibility associated TNBC. PARP inhibitors have been shown to improve both anti- and pro-tumor features of macrophages by reprogramming glucose and lipid metabolism via the sterol regulatory element-binding protein 1 (SREBF1, SREBP1) pathway. Thus, PARP inhibitors combined with antibodies that target the colony-stimulating factor 1 receptor (CSF1R) significantly increased innate and adaptive immune responses [62].

5.4. mTOR Inhibitors

As explained in the pathophysiology of TNBC, activation of mTOR is responsible for several functions that promote carcinogenesis. This pathway involves targeting PI3K, AKT, or both PI3K and mTOR [63]. Targeting both mTOR and PI3K results in increased efficacy as well as toxicity. Significant functions of PAM pathways include cell proliferation, survival, metabolism, and migration. Its role in malignant cell transformation has also been investigated. Ipatasertib is a pan-AKT inhibitor targeting a phosphorylated form of AKT. In the preclinical studies, synergy is observed when ipatasertib and paclitaxel were combined. Another small molecular AKT inhibitor capivasertib's preclinical evaluation has shown activity even in models with alterations in PIK3CA, AKT, and PTEN [64]. Further, Cretella et al. proved the enhanced anti-tumor efficacy of CDK4/6 inhibitors by combining mTOR inhibitors with impaired glucose metabolism. The results indicated superior efficacy in combining palbociclib and mTOR inhibitors. However, preclinical and clinical studies have to be performed to confirm these results [65]. Kim et al. studied the effects of combining ipatasertib and paclitaxel. They conducted a randomized, double-blind, placebo-controlled, phase 2 trial. The results indicated that the combination showed enhanced progression-free survival. The adverse effects observed were manageable. Further clinical trials have to be performed to confirm its safe and effective use in patients with TNBC [66].

5.5. Immunotherapy

The aggressive nature of TNBC has led to the development of targeted immunotherapies. Unlike other breast cancer subtypes, TNBC was found to be immunogenic. Hence, researchers started investigating approaches to boost the host's immune system. The characteristics that make TNBC immunogenic include higher mutational burden, higher quantities of tumor-infiltrating lymphocytes, and higher programmed death-ligand 1 (PD-L1) expression. Higher immunogenic mutations cause tumor cells to produce new antigens. Present immunotherapy approaches for the destruction of tumor cells include cancer vaccines, immune checkpoint blockades such as PD-1/PD-L1 inhibitors, CTLA-4 inhibitors, induction of cytotoxic T-lymphocytes, adoptive cell transfer-based therapy, and modulation of the TME to increase CTL activity [67,68].

A novel class of bispecific antibodies targeting CD3-Trop2 (trophoblast cell-surface antigen 2) or CD3-CEACAM5 (carcinoembryonic antigen-related cell adhesion molecule 5) significantly inhibited TNBC cell growth when combined with human PBMCs (peripheral blood mononuclear cells). TNBC cells express bispecific antibodies-Fc fusion proteins that target EGFR, human epidermal growth factor receptor 3 (HER3), receptor tyrosine kinase, human epidermal growth factor receptor 3 (HER3), and Notch. Several targets, such as Trop2, CEACAM5, EphA10, P-cadherin, EpCAM, EGFR, and mesothelin, have also been included in immune cell-redirecting bispecific antibodies, and combination therapy with immune checkpoint inhibitors has been studied. Also, TNBC cells overexpress two receptor tyrosine kinases, Axl and cMet, and blocking both of them with bispecific antibodies helps to target TNBC cells [69]. A study was published that used nanomicelles to target the CD44 mediated apoptosis pathway using a hyaluronic acid (HA) coated biocompatible oligomer containing vitamin E and Styrene Maleic Anhydride (SMA) (HA-SMA-TPGS). The HA-SMA-TPGS was self-assembled and encapsulated with a poorly soluble and potent curcumin analogue (CDF) to produce nanomicelles (NM) with excellent parenteral delivery properties [70].

Moreover, Immune Checkpoint Inhibitors (ICI) is another promising immunotherapy. These are the cell surface membrane proteins such as the programmed cell death-1 (PD-1) receptor that are expressed on T-cells. It belongs to the B7 family of checkpoints. Tumor cells express PD-L1, which attaches to the PD-1 receptor present on T-cells. This inactivates T-cells and hinders tumor destruction caused by T-cells. Researchers have developed several anti-PD-1 antibodies and anti-PDL-1 therapeutic antibodies that disturb the immune regulatory checkpoints and activate anti-tumor immune responses by blocking these receptors and avoiding the inactivation of T-cells [71,72]. The ICIs have provided a new horizon in TNBC treatment. Various ICIs are under clinical trials. For example, anti-CTLA-4 mAbs (Ipilimumab), anti-PD-1 mAbs (atezolizumab, durvalumab, avelumab) have shown excellent initial results. But larger clinical trials are required to establish response rate and determine long term efficiency of ICIs [73]. J.A. Kagihara et al. successfully developed Nab-paclitaxel and atezolizumab for the treatment of PDL-1 positive metastatic TNBC. Food and Drug Administration (FDA) has given approval for this combination, which enhanced the progression-free survival in TNBC patients. However, it was observed that it causes some adverse events, including hypothyroidism and rash. Further investigations confirmed that these adverse events were infrequent and consistent as compared to other atezolizumab trials [74].

CTLA-4, also known as CD152, is widely expressed in CD4+, CD8+, NK, and FOXP3+ cells and regulates the T-cell mediated immune response. B7-1 (CD80) and B7-2 (CD-86) ligands are present on antigen-presenting cells with which CTLA-4 associates and negatively regulates T-cell activation. This suppresses the T-cell dependent immune response. Hence, scientists have developed agents that block CTLA-4 and activate the T-cell-dependent immune response. But it was reported that CTLA-4 inhibitors have more side effects as compared to PD-1 inhibitors. For melanoma, Ipilimumab was the first CTLA-4 inhibitor used, whereas, for breast cancer, Tremelimumab and Ipilimumab are being investigated. Other immune checkpoint targets have also been investigated by researchers, including BTLA, VISTA, TIM3, LAG3, and CD47 [72,75].

5.6. Gene Therapy

Silencing of genes using microRNAs (miRNAs) and small interfering RNAs (siRNAs) is a rapidly growing approach in cancer treatment [76]. This therapy is used for both diagnostic and therapeutic purposes (theranostics). The gene silencing by siRNAs regulates gene expression through RNA interference, which can be used to treat cancer. The siRNA delivery system can be labeled with imaging agents (for example, dextran-coated superparamagnetic nanosized particles) for non-invasive real-time imaging of siRNA delivery to the tumor using magnetic resonance imaging (MRI). The siRNA-labelled mediated delivery may also help in monitoring and predicting the therapeutic outcome [77]. The delivery of siRNA is a major problem since it is easily broken down by nucleases and its negative charge makes it difficult to localize in cells [76]. Liu et al. formulated siRNA in cationic lipid-assisted PEG-PLA nanoparticles, targeting cyclin-dependent kinase 1 for TNBC treatment. They showed that these nanoparticles promote cell death in c-myc overexpressed TNBC cells by inhibiting CDK-1 expression [78]. Alshaer et al. formulated aptamer guided siRNA nanoparticles targeting CD-44 in triple-negative breast cancer. The core was made up of a siRNA-protamine complex, and the shell had an aptamer ligand for targeting CD-44 cells. The results indicated that the formulation exhibited anti-tumor activity [79]. Tang et al. loaded siRNA into lipid-coated calcium phosphate nanoparticles, which increased the delivery of siRNA to TNBC cells. It was observed that the nanoparticles accumulated in cancer cells because of enhanced permeability and retention effect and the ability of the formulation to target cancer cells [80].

MicroRNAs (miRNAs) play a key role in the genesis and progression of TNBC. Hence, they have the potential to serve as diagnostic biomarkers [81]. Generally, there is a downregulation of miRNAs in tumor cells, but some miRNAs are upregulated [82]. The miRNA558 is overexpressed and the cluster miR-17/92, miR-106b, miR-200 family (miR-200a, miR-200b, and miR-200c), miR-155 and miR-21 are also the highly expressed ones. TNBC with the metastasis to lymph node revealed that six miRNAs, namely iR-125a-5P, miR-579, miR-627, miR-424, miR-101, and let-7g, were expressed in lymph node tissues [15]. One reported study demonstrated that combining Orlistat-loaded nanoparticles with doxorubicin or antisense-miR-21-loaded NPs significantly increased the apoptotic impact when compared to single doxorubicin, antisense-miR-21-loaded NPs, orlistat-loaded NPs, or free orlistat treatment [83].

5.7. Photodynamic Therapy

This therapy involves photosensitizers, light activatable molecules, the light of wavelength (near IR light), and oxygen. The therapy consists of two stages: photosensitizer administration and illumination with light. Between these two stages lies the incubation period, which decides the location where the photosensitizer will be released, thereby affecting the efficacy of therapy [84,85] After the administration of the photosensitizer, it is distributed in healthy cells as well as tumor cells. Normal cells eliminate the photosensitizer, whereas the tumor cells accumulate it due to morphological differences between the cells since the tumor cells have impaired vasculature and lymphatic drainage [86]. As a result, it does not allow photosensitizers' elimination. This therapy causes the destruction of tumor cells by the excited-state photosensitizer's reaction with triplet-state molecular oxygen present in the body to produce reactive oxygen species (ROS) [87,88]. As the near IR illuminates the photosensitizer, the photon gets absorbed, leading to the formation of an excited singlet stage which is unstable. This unstable state is again stabilized when the photosensitizer returns to the ground state by emitting florescence or through internal conversion by radiating heat. This helps to study the pharmacokinetic distribution of photosensitizers in the body. The excited singlet state of the photosensitizer can also undergo intersystem crossing and destroy tumor tissue by type 1 or type 2 reaction. In a type 1 reaction, electrons or hydrogen are directly abstracted from amino acids or guanine in nucleic acid or NADPH. This leads to the formation of a radical anion of a photosensitizer, which donates its electron to oxygen and produces a superoxide anion radical, restoring the photosensitizer. Type 2 reactions include the transfer of energy to molecular oxygen, resulting in the formation of singlet oxygen, a ROS. The type of reaction that will take place depends on oxygen, substrate concentration, and the type of photosensitizer. However, type 2 reaction is more prevalent as energy transfer occurs at a higher rate than electron abstraction. The superoxide anion radical produced by the type 1 reaction is less reactive than the singlet oxygen produced by the type 2 reaction [89,90].

PDT is a promising therapy for TNBC because of its minimally invasive nature, high accuracy, and precise controllability. Cancers that can be easily assessed by light are treated with photodynamic therapy. They include breast cancer, head and neck cancer, lung cancer, oesophageal, oral, and laryngeal cancer. PDT has been proven to kill tumor cells through three mechanisms. The mechanism that will be followed depends on the localization of the photosensitizer, PS type, and concentration. Direct cell death mechanisms include apoptosis, necrosis, and autophagy, and the other two mechanisms include targeting vascular effects and immune reactions. Apoptosis occurs when the photosensitizer targets the mitochondria. Light-induced damage to mitochondria increases the permeability of the mitochondrial membrane; this, in turn, releases cytochrome c in the cytoplasm, which is necessary to activate the caspase-mediated apoptotic pathway. Cathepsins released after PDT damage to mitochondria causes cleavage of proapoptotic protein Bid to tBid. This then interacts with mitochondria, releasing cytochrome c in the cytoplasm and activating intrinsic apoptosis. This mechanism has been found to be exhibited by NPe6. When the damage caused to the cell is quite high that the apoptotic pathways are destroyed, and the photosensitizer is localized in the plasma membrane, necrosis predominates. This results in the release of intracellular materials outside the cell, inducing inflammation [89,91–93].

Autophagy is induced with low PDT damage to the cell. Autophagy is normally a protective mechanism, however when the lysosome is damaged, autophagy's protective ability is exceeded, and cell death occurs. Negatively charged porphyrins such as phenothiazinium methylene blue and NPe6 were found to localize in the lysosome. Lysosome damage was found to activate apoptosis as well as necrosis. Necroptosis is also induced in some PDT damaged cells due to the imbalance of extracellular and intracellular homeostasis, which depends on the mixed lineage kinase domain like-protein and kinase activities of RIPK 3 and RIPK 1. Blood vessel formation is essential for providing nutrition to cancer cells. Hematoporphyrin has been found to obstruct blood flow. Damage to the tumor vasculature damages the endothelial and subendothelial cells. This causes them to round up, increasing the inter-endothelial cell junctions. These damaged cells release clotting factors such as the von Willebrand factor, resulting in the activation of platelets. Thrombus formation, platelet aggregation, and vessel occlusion are induced by platelet interaction and the released sub-endothelium. They also cause vasoconstriction, decreasing blood flow. This reduced blood flow resulted in hypoxia and decreased nutrition, thereby resulting in tumor destruction. PDT causes damage to cells and induces apoptosis and necrosis; these damaged cells release DAMPs (damage associated membrane proteins). DAMPs include HSP70, calreticulin, ATP, and arachidonic acid. PDT-induced damage also causes the release of inflammatory factors, transcription factors, and heat shock proteins. Tumor antigens bind to the heat shock proteins and interact with toll-like receptors, thereby activating antigen-presenting cells. These antigen-receding cells present the antigens to CD4 helper T-cells, activating cytotoxic CD8+ cells and causing the destruction of tumor cells. The main advantage of PDT induced tumor destruction is that it bypasses the resistance mechanism displayed by the tumor cells [90,93].

Shemesh et al. studied the potential of PDT to kill TNBC cells by using thermosensitive liposomes of indocyanine green (ICG). ICG produces ROS on exposure to a near-IR laser of 808 nm. They concluded that it successfully accumulated in tumor cells. Formulating ICG in thermosensitive liposomes offered several advantages, including stability at physiological temperatures and enhanced permeation [94]. Sun et al. designed a multifunctional cationic porphyrin grafted microbubble loaded with HIF 1 alpha siRNA and delivered by ultrasound targeted PDT to treat TNBC. The siHIF cationic porphyrins were converted to nanoparticles inside the body, which resulted in the targeted accumulation of siRNA and porphyrin in cancer tissue. They concluded that this therapy is effective in TNBC treatment [95].

5.8. Photothermal Therapies

PTT is an emerging and highly effective non-invasive treatment therapy for cancer that utilizes the photothermal effect of photothermal agents (PTAs) to destroy cancer cells without causing harm to normal cells. This therapy causes thermal burns or thermal ablations on the tumor upon exposure as it produces heat from absorbed light [96]. The thermal damage caused disrupts the cell membrane and denatures the protein, thereby killing the cancer cells. PTT needs a bio-compatible photothermal agent capable of absorbing near-infrared light that is a NIR source. Several nanoparticles with the ability to absorb NIR light are used in photothermal therapy [97]. PTT has high research value as it exhibits rapid recovery in short treatment time without including a complex operation [96].

In the study reported by Valcourt et al. formulated Polylactic-co-glycolic acid (PLGA) nanoparticles loaded with IR820 for the treatment of TNBC using photothermal therapy which developed biodegradable and polymeric nano particles inducing cell death primarily through apoptosis [98]. A study by Zhao et al. formulated HA-coated gold nanobipyramids to treat TNBC through PTT. They coated the nanoparticles with high and low molecular weight hyaluronic acid, 380 kDa and 102 kDa, respectively. Upon irradiation with 808 nm, it was found that high molecular weight hyaluronic acid showed superior efficacy in targeting overexpressed CD44 cells in TNBC cells [70,99]. Zhang et al. formulated gold nanoparticles targeting the epidermal growth factor receptor for enhancing autophagic cell death in TNBC patients on treatment with photothermal therapy. On treatment with this therapy, there was an increase in autophagic proteins such as beclin-1, p62, and autophagic vesicles [100]. Another study by Wang et al. selectively sensitized malignant cancer cells to PTT by targeting CD44 and reducing HSP72 (heat shock protein 72). They formulated gold nanostar and siRNA to reduce HSP72. This selectively sensitized TNBC cells to hyperthermia. Some of the other advantages shown by the formulation were endosome escape, robust siRNA loading capacity, high hemocompatibility, easy RNA synthesis, and high biocompatibility [101]. Tian et al. developed polydopamine nanoparticles that are loaded with JQ1 to inhibit c-myc and PD-L1 to increase photothermal therapy for the treatment of TNBC. JQ1 was found to decrease the expression of PD-L1 and inhibit the BRD4-c-MYC axis. The synergistic treatment induces the activation of cytotoxic T-lymphocytes, enhancing the immune response [102].

5.9. Sonodynamic Therapy

Sonodynamic therapy (SDT) represents use of ultrasound waves to enhance the activity of chemotherapeutic agents [103]. Ultrasound is a mechanical wave which has frequency equal to or higher than 20 kHz [104]. These high frequency waves are used for improving the efficacy of chemotherapeutic agents. Anticancer drugs like nitrogen mustard, cyclophosphamide, bleomycin, adriamycin, etc. can serve as sonosensitizer. SDT is a method based on the synergistic interaction between ultrasound and a substance known as a "sonosensitizer". Traditional sonosensitizers have very limited clinical applications due to their low efficiency, low retention in cancer cells, and low tumor selectivity. XiaoHan et al. have developed PEG-IR780@Ce6 for SDT, a new sonosensitizer that inhibits cancer cells as well as suppressing their migration and invasion [105]. The increased efficacy of chemotherapeutic agent by SDT is due to its ability to porate (sonoporation) cell membrane for effective penetration of chemotherapeutic agents [106] or its ability to disperse chemotherapeutic drugs in poorly vascularised tissues in solid tumors [107]. This therapy has been widely used to combat TNBC as the tumor cells of this subtype multiply and progress very fast. Xiaolan Feng et al., proposed a deep-penetrating sonochemistry nanoplatform (Pp18-lipos@SRA737&DOX, PSDL) composed of Pp18 liposomes (Plipo), SRA737 (a CHK1 inhibitor), and doxorubicin (DOX) for the controlled production of reactive oxygen species (ROS) upon ultrasound activation and the resulting intercalation of DOX into the nucleus DNA and induction of cell death [108]. By enclosing manganese-protoporphyrin (MnP) in folic acid-liposomes, Huaqing Chen et al. developed a multifunctional nanosonosensitizer system (FA-MnPs). The nanoparticles of FA-MnPs are characterized by strong depth-responsive SDT as well as a strong immunological response that is mediated by SDT. In the presence of ultrasound radiation, FA-MnPs show great acoustic intensity in mimic tissue up to 8 cm deep, and they also produce significant amounts of singlet oxygen ($^1O_2$). The good depth-responsive SDT of FA-MnPs efficiently slackens the growth of both the surface tumors and the deep lesion in TNBC mouse models [109].

6. Emerging Nanotechnology Based Delivery Approaches toward TNBC Treatment

Nanotechnology is one of the finest tools that has emerged in the past couple of decades, having the potential to fight against severe, difficult-to-manage diseases like cancer. Biomedical sciences have seen the evolution of nanotechnology in targeting cancer, via various approaches in order to have a robust and targeted delivery of diagnostics and therapeutics [110]. Researchers are considering the option of nanomedicine for combinational therapy or as a multidrug delivering system, in case of cancer so as to employ several advantages such as improving the pharmacokinetic profile, decreasing the free drug toxicity, and having a synergistic pharmacological effect of the drugs employed [111]. Other advantages offered by the nano systems are improved therapeutic index, with an increase in localization of the drug in the blood, decreased off-target secondary pharmacological effects, with increased localized targeted action in tumor cells [112,113]. The employment of nanoparticles with primary purpose targeted delivery is based on the characteristic features of the nanoparticles including, average nanoparticulate size, uniformity, surface potential as well as drug loading capacity [112]. The specific advantages offered by nanoparticles can be attributed to the increased large surface area to volume ratio of nanocarriers. Furthermore, the entrapment or loading of drugs in or onto the nanocarriers, has increased the chances of efficient targeting of the tumor cells via the use of single drug molecules or via combination therapy. The nanomedicines are required to be explored more, in targeting the TNBC, as it's already known that TNBC lacks ER, PR, and HER2 receptors on their membrane. The absence of these receptors makes it difficult for conventional drug systems to target the tumor cells. As a consequence, nanotherapeutics is looked upon as an emerging non-conventional approach in targeting TNBC [114]. One of the basic requirements in the formulation of any nanoparticles is its size, which is desired to be in the range of 1 to 200 nm, modifications in this may alter the dynamical pathway of the particle [115]. The adoption of nano systems in targeting cancer is increased, as it increases the permeation and localization of the drug. Various nanoparticle systems consisting of Liposomes, Micelles, Dendrimers, Solid-lipid Nanoparticles, Polymeric Nanoparticles, Gold Nanoparticles etc. and their details are shown Figure 3.

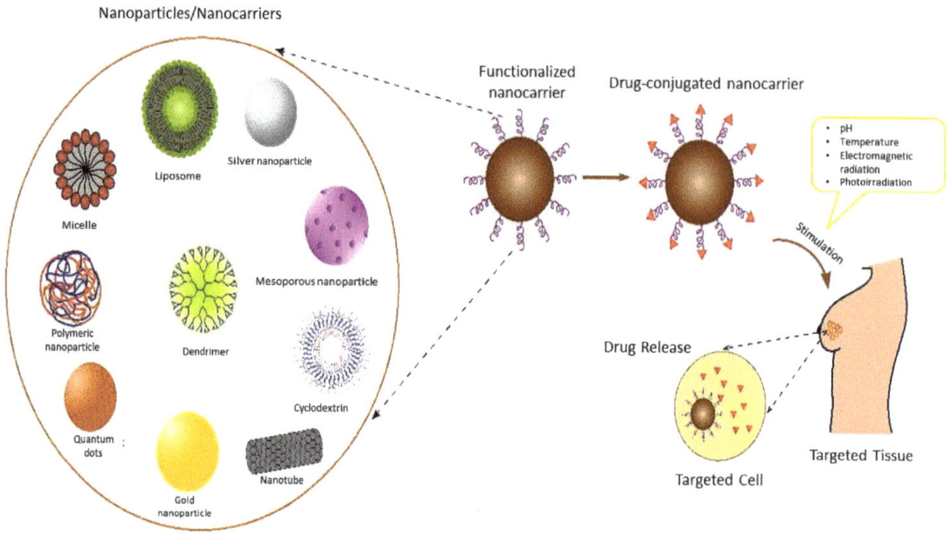

Figure 3. Nanotechnology-based approaches for TNBC treatment.

6.1. Organic Nanoparticles

Organic nanoparticles (NPs) are obtained from natural or synthetic organic molecules. The advantage of this nanoparticle includes tailorable synthesis, facile processability, excellent biocompatibility, and low cytotoxicity, revealing outstanding potential in drug delivery, bioimaging, phototherapy, and biomedical application [116,117]. These nanoparticles also offer better advancement including improved drug encapsulation, triggered release, and site-specific targeting. Also, These nanoparticles deliver distinct benefits for toxic drugs such as chemotherapeutics, where off-target toxicity is a key hurdle [118]. These organic nanoparticles are Lipidic nanocarriers, Biological nanocarriers, Polymeric nanocarriers, etc.

6.1.1. Lipidic Nanocarriers

Lipidic nanocarriers have pronounced advanatages such as they deliver greater availibilty of drug in TME, enhance the biomimetic delivery, achieve reduced side effects, and avoid the development of multidrug resistance (MDR). These nanocarriers can be fabricated from numerous biomaterials including lipid moieties such as fatty acid, phospholipids, lipids etc. These materials can provide several benefits such as being biomimetic, being biodegradable, being more biocompatible, achieving targeted action and improved penetration. These materials are utilised for the development of nanoplatforms such as liposomes, nanostructured lipid carriers (NLCs), Solid lipid nanoparticle (SLNs), and self-micro/nano emulsified drug delivery systems (SMEDDS/SNEDDS), etc. [119].

- Liposomes

Liposomes are the nanoconstructs that can deliver both hydrophobic and hydrophilic drugs. Liposomes were the very first class of nanotherapeutics that got approval for cancer treatment. Liposomes still cover a larger share of nanoparticles in clinical stages [120,121]. These liposomes provide the benefit of biodegradability, biocompatibility, and non-toxic and non-immunogenic nature. On the other hand, the properties of liposomes can be altered considerably by altering the lipid composition, size, and charge on the surface. Liposomes are small spherical-shaped vesicles, which are formulated using cholesterol and natural non-toxic phospholipids. In the case of chemotherapeutics, liposomal formulations enable enhanced efficacy as well as site-specificity in tumor tissues [122,123]. Due to the presence of a phospholipid bilayer, which is amphiphilic in nature, it resembles the mammalian cell membrane, this in return increases the cellular uptake of the same [124]. The liposomal formulations are comparatively low in toxicity parameters as that of drugs alone, and thus improve the drug delivery to the site effectively [125]. Liposomes are being utilized to conjugate many ligands in order to target tumor cells, such as peptides, oligonucleotides, monoclonal antibodies in the case of immunotherapy, and antigen-binding fragments [126]. mAB liposomes which are established and synthesized with optimized procedures offer advantages like cancer specific targeting, high packaging capacity, and prolonged half-life with attached PEG. Y. Si et al. proved that combined standard Gemcitabine and Mertansine which is a polymerization inhibitor led to the formation of mAb-Liposomes against TNBC cells. The inhibition of TNBC cell growth by this combination was observed in cell line xenograft models as well as the patient [127].

- Self-emulsified drug delivery system (SNEDDS/SMEDDS based nanocarrier)

Self-emulsified drug delivery systems serve to make provision for loading a high payload to highly lipophilic drugs via improving its solubility by multiple folds compared to free drug [128,129]. For instance, Guru et al. developed a lipid based self emulsifying system of docetaxol for the management of solid tumors. It has been observed that the carrier provides better retention of drug in biological environment. Also, this study showed greater oral availibilty against marketed formulation (taxotere) [130]. Likewise, Nupur et al. fabricated co-loaded tamoxifen and resveratrol in a SNEDD based nanocarrier for breast cancer treatment. They found that the developed nanocarrier showed enhanced oral bioavailability and improved cytotoxicvity against MCF-7 cell line [131]. In another research report, co-delivery of DOX and LyP-1 in SMEDDS was prepared which reduced the tumor growth and metastasis as well. Also, in-vitro cytotoxicity assays were conducted in p32-expressing BC cells, 4T1 and MDA-MB-231 (TNBC) cell lines and resulted in profound cell killing. LyP-1 is a specific peptide towards the p32 receptor, highly expressed in malignant BC cells [132]. Therefore, self emulsified delivery system is a promising therapeutic approach for the delivery of drug for the management of cancer.

- Solid Lipid Nanoparticle (SLN) and Nanostructured Lipid Carrier (NLC)

Solid lipid nanoparticles (SLNs) and Nanostructured liquid carrier (NLC) are colloidal nanoparticles consisting of an mixture of both liquid lipids and solid lipids which are stabilized by aqueous solution of surfactants or a mixture of surfactants. The lipids utilised for the fabrication of these nanocarriers are many, including triglycerides, fatty acids, steroids, and waxes [119].

Several findings suggest that SLNs can not only uniformly solubilize the hydrophobic drug in lipid matrix system but also provide a drug-enriched shell surrounding the lipidic core. Furthermore, the depostion of drug in the matrix i.e., core type or shell type also plays a crucial role in drug release. For example, drug-enriched SLNs reveal a biphasic drug-release profile; intially nanocarrier show a burst release from the outer shell followed by slow type release pattern release from the lipidic core On the other hand, drug enriched core type provide a long acting sustained release pattern due to enhanced drug diffusional distance from the lipidic core [133].

Kothari et al., developed SLN for the delivery of docetaxel –alpha-lipoic acid using various lipids such as GMS, SA, and Compritol ATO 888 for the treatment of TNBC. The finding suggested that the developed carrier exihibited improved cytotoxicity against DTX-SLNs, ALA-SLNs, and free drugs. Additionally, this carrier revealed enhanced apoptosis i.e., 32% with free drug [134]. In another study that was conducted by Pindiprolu et al., they fabricated SLN for the delivery of niclosamine to treat TNBC As per the results, it was found that the drug loaded carrier displayed improved cytotoxicity with enhanced cellular internalization against free drug Furthermore, these SLNs also exhibited increased cellular transfection due to their ability to avoid the efflux pump and hence improve the bioavailability of drugs within the tumor cells [135]. Although, SLN have several advantages over conventional approaches, nonetheless, they have some limitaions as well chances of leakage, less-loading capacity etc. Thus, second generation lipidic nanoparticle i.e., NLC came into existence which offer high drug loading capacity, have decreased risk of gelation, and restricted leakage of the drug upon storage in ccomparison to SLN. Moreover, NLCs also provide a long term exposure of drug towards cancer cell and ultimately, they can enhance the therapeutic efficacy of drug at the tumor site [136]. Sun et al. fabricated NLC based formulation of qurecetin to enhance the oral bioavailability by using soy lecithin, glyceryl tridecanate, glyceryl tripalmitate, Kolliphor HS15. Also, it was shown the improved entrapment efficiency with the prolong -release pattern of quercetin which enhanced cell killing potential against MCF-7 and MDA-MB-231 cells [137]. In another research work by Andey et al., wherein the authors developed lipidic nanocarrier such as liposome, SLN and NLC for the delivery of estrogenic derivative (ESC8). It was observed that lipidic carrier showed good cytotoxicity against standard drug cisplatin. Moreover, SLN showed greater oral bioavailability of ESC8 in comparision with free ESC8 and also showed reduced tumor growth on the xenograft TNBC model [138]. These lipidic nanocarrier showed good pharmacokinetic profile, better therapeutic effectiveness and improved site specific targeting. Therefore, these lipid should fall under GRAS (Generally Recognized as Safe) regulations during the the formrulation development [119].

6.1.2. Micelles

Polymeric micelles are considered an advanced drug delivery system for various therapeutic compounds. Polymeric micelles are advantageous specifically in cases, wherein a poorly water-soluble drug with increased potency as well as toxicity is to be delivered to its targeted site. Polymeric micelles drug delivery technology provides various opportunities for its appropriate use in delivery of the therapeutics. One of the applications is the use of amphiphilic block copolymers as carriers for poorly soluble drugs with the intention of improving the specific targeting as well as the therapeutic potential of the drug. Because of all such advantages, micelles are considered a potential drug delivery system. In one such study, wherein Cetuximab was conjugated with vitamin E forming micelles were used for targeted delivery of docetaxel against TNBC. The formulation of these micelles enhanced the therapeutics effects of docetaxel with the employment of the formulation of Cetuximab conjugated micelles [139]. Iruthayapandi et al. improved the loading efficiency, sustained release, and biocompatibility via the use of protein-based polymeric micelles. Furthermore, in the treatment of TNBC, gelatin-based polymeric micelles hold much importance. Thus, to increase the bioavailability, biodegradability, cellular uptake, increased drug loading, and targeted delivery of poorly soluble drugs, micelles are considered the ideal choice of nano-carriers [140]. Not just the micelles alone, but micelles when combined with the quantum dots along with anti-EGFR and aminoflavone were also employed against TNBC. The conjugated aminoflavone demonstrated enhanced accumulation of the drug at targeted sites of the tumor as compared to encapsulated but non-conjugated aminoflavone. Thus, this resulted in the regression of tumor size in the TNBC xenograft mouse model. The use of such conjugation facilitated hardly any form of systemic toxicity during the treatment period [141]. On the other hand, a few researchers, also designed activated platelets targeting micelles for treating primary and metastatic TNBC. The study included modification of redox-responsive paclitaxel loaded micelles with P-selectin targeting peptide, thus adhering to the surface of the activated platelets and thus targeting circulating tumor cells in the blood, as a result preventing metastasis. Thus, it can also be concluded that the treatment outcomes are improved, as the micelles target tumor-infiltrating platelets against TNBC [142].

6.1.3. Dendrimers

Dendrimers are synthetic, three-dimensional, highly branched macromolecules having one of the dimensions in nanometers. Dendrimers comprise a central core and various functionalities at the periphery [143]. In a detailed manner, the dendrimer has three different parts, namely, a focal core, and building blocks which are repeated units of dendrimers and has multiple functional groups at the peripheral end [144]. Dendrimers are the ideal choice of nanoformulations in case of drugs that have poor solubility in the body's aqueous environment. Dendrimers as drug delivery vehicles provide characteristic advantages of monodispersity and multivalency [145]. The dendrimers are not only considered for the option of gene delivery but also are associated with gene delivery in the case of TNBC [146]. For instance, Ghosh et al. developed carbon quantum dots from the sweet lemon peel, and conjugated the same with varied generations of polymers including polyamidoamine (PAMAM) in order to form carbon quantum dots-PAMAM conjugates. They have been exploiting these multi-utility dendrimers for diagnostic purposes as well as gene delivery purposes [147]. Dendrimers are garnering a lot of interest due to their use against Polo-like kinase (PLK-1) in TNBC. Dendrimers are modified and complexed in order to deliver the PLK1 siRNA. The dendriplexes formed ultimately lead to an increase in cellular uptake of siPLK1 in MCF-7 and MDA-MB-231 cells. Furthermore, these dendriplexes also enhanced the cell cycle arrest in the sub-G1 phase. Thus, concluding the potential use of phosphorous as well as PAMAM dendrimers in delivering the siPLK1 against TNBC [148]. In another attempt to utilize dendrimers, a doxorubicin-loaded dual-functional drug delivery system was developed by conjugating PAMAM dendrimer, EBP-1, and the cell-penetrating peptide derived from trans-activating transcriptional activator (TAT). This modified drug delivery system was successful in enhancing the anti-proliferative effect of anticancer drugs against MDA-MB-23 cells in vitro, compared to that of the free Doxorubicin effect and convention nano-carrier system. This modification of the drug delivery system not only improved the anti-proliferative activity but also enhanced the drug accumulation at the cancer site in vivo [149]. In another approach for gene delivery through dendrimers, formulations of PAMAM dendrimers were functionalized for delivering the siRNA for TWIST1 gene in TNBC. This silencing was proven efficient because of reduced tumor invasion as well as metastasis [150]. Several other studies substantiate the effectiveness of dendrimers for utilization in cancer therapeutics [151].

6.1.4. Polymeric Nanoparticles

Polymeric nanoparticles have gained tremendous attention over the last decades in terms of their applications in the fields of biology and medicine [152]. Polymeric nanoparticles are mainly colloidal drug delivery systems comprising either natural or synthetic polymers. These have additional advantages over other conventional nanocarriers, in terms of scale-up and manufacturing of the same. Furthermore, these provide the required stability to the systems in the biological fluids [153]. Also because of the absolute pharmacokinetic properties of polymeric nanoparticles, they are considered efficient nanocarriers. [154]. To enhance drug delivery, polymeric engineering of the nanoparticles is one of the highly efficient ways of targeting strategies for anticancer drugs. Silica nanoparticles, loaded with doxorubicin were modified with a polymer, polyethyleneimine, this modified drug delivery system, induced endosomal rupture, and the formulation included HA, binding to CD44 receptors on the cells that are largely expressed in cancer cells. These engineered modifications of the drug delivery system, involving a polymer increased the therapeutic outcome of the delivery system, with a lower dose requirement of the chemotherapeutic agent, thus decreasing the chances of systemic toxicity [155]. In an attempt in using polymersomes, in TNBC, Doxorubicin was encapsulated in the polymersomes which were targeted with hypoxia-responsive peptides, which results in the formation of targeted polymersomes. It was observed that the targeted polymersomes decreased the TNBC cell viability, as well as proved the anti-tumor activity in animal models of TNBC [156]. Furthermore, the therapeutic efficacy of quercetin, a plant-derived flavonoid was enhanced by the development of a nanoparticulate system from D-α-tocopherol polyethylene glycol 1000 succinate (TPGS) and PLGA and was further evaluated in vitro and in vivo against TNBC. Thus, the study concluded that the use of Quercetin nanoparticles formed via the use of the polymer improved the antitumor and repressed the metastatic effect via inhibiting uPA, thus enabling a newer approach against targeting TNBC [157]. The Polymeric Nanoparticles can also be exploited not only to deliver drugs but also in delivering miRNA and siRNA in association with the chemotherapeutic drug, which would ultimately lead to a reduction in tumor volume and thus its growth. PLGA-b-PEG polymer Nanoparticles delivered both antisense-miR-10b and antisense-miR-21 with a dose 0.15 mg/kg drug dose but when siRNA and Doxorubicin loaded nanoparticles were delivered, the overall tumor size reduction in terms of growth and volume was 8 fold more [158,159]. To form a nanoparticle, a protein polymer, called as elastin-like polypeptide was employed and its surface was coated with FK506 binding protein 12, a cognate receptor for the drug Rapamycin which is a potent drug but having poor solubility. The modification and delivery of the drug through the nanoparticle formed from the polymer enhanced the anti-tumor effect in the TNBC xenograft mouse model [160].

6.1.5. Biological Nanocarrier

Biological nanocarriers are bicompartible and biomimetic type of carrier which consist of various biological cells and its components such as exsosome, platelet, RBCs, nucleic acids, peptides, aptamers, antibodies genes etc. These carriers provide the benefits of mediating better site specific tumor targeting actions and triggering detrimental effects on the tumor cells [161]. Some biological nanocarriers have been discussed below:

- Erythrocyte based nanoparticles

Erythrocytes nanoparticles have risen as one of efficient biocarriers due to its easy method of preparation and high drug loading, biodegradability, and ability to provide a long circulation half-life in the biological environment [162]. Cheng yet al. developed a biotinyl modified cRGD conjugated-RBCs nanoparticle of doxrubucin to treat cancer, wherein, they found accurate targeting, high drug loading, and controlled drug release pattern against free drug. Also, these carriers exhibit better cytotoxicity against MCF-7 cell in comparison to free doxrubucin [163].

- CRISPR nanoparticles

The major drawback in the treatment of breast cancer therapy is drug resistance. The leading reasons for the development of drug resistance are transporter efflux and specific modification in tumor cell genetic composition. Due to this concern, CRISPR based therapy is rising attention towards cancer treatment [164]. Clustered regularly interspaced short palindromic repeats abbreviated as CRISPR is the latest genetic editing tool which upon combination with nanotechnology stands to be a promising therapeutic approach for treating cancers [165]. Zhang et al. fabricated PEG phospholipid-modified cationic lipid nanoparticles (PLNP) for CRISPR/Cas9 to enhance its transfection into the cellular site. In this work, they loaded the Cas9/sgRNA plasmid to obtain core-shell component (PLNP/DNA) which helps in effective transfection of Cas9/sgPLK-1 plasmid into tumor cells. Also the result have shown a subsequent reduction in tumor growth after administration of Cas9/sgPLK-1 plasmid to mouse [166]. Thus, this approach delivers greater efficacy both in-vivo and in-vitro which could be useful for clinical translation against various cancers.

- Exosomes

Exosomes are surface membrane vesicles; chemically they consist of protein and lipid in the body, which are in association with numerous pathological and biological processes. Now a days, exosomes are utilized for the drug delivery as it is a body component which provides benefits such as non-immunogenicity and enhanced circulation time [161]. Tian et al. developed DOX loaded exosomes for effective targeting and site specific delivery to αv integrin-positive BC cells. After administration of DOX loaded exosomes, it was reported that the carrier demonstrated low toxicity with reduced the tumor growth., moreover, this exosome can further undergo surface modification to develop other delivery systems like polymeric, lipidic nanoparticles etc. to obtain site specific targeting [167]. Singh et al. reported that miR-10b (microRNA) loaded exosome could treat metastatic BC cells (MDA-MB-231). These carriers enhance the uptake of miR-10b which help in the reduction of protein expression of associated target genes such as KLF4 and HOXD10 [168]. Several findings suggest that exosomes can modulate the TME and provide a platform for drug delivery to various tumors.

6.1.6. Carbon Nanotubes

The carbon nanotubes can be used in the delivery and development of therapeutic agents such as peptides, proteins, nucleic acids, genes, vaccines, etc. Carbon Nanotubes can be combined with bioactive peptides, proteins, nucleic acids, and drugs, and thus functionalized for delivering their cargo to cells and organs [169]. Carbon nanotubes are looked upon as the ideal nanocarriers because the functionalization leads to a decrease in systemic toxicity of the nanotubes as well as a decrease in its immunogenic effect. The functionalization also helps in enabling as well as enhancing the dispersion of carbon nanotubes in the biological aqueous based fluids. To enhance the delivering effect of nanotubes, an attempt was made to develop multiwalled carbon nanotubes for carrying and delivering platinum against TNBC. Thus, the results indicated that the functionalized multiwalled carbon nanotubes decreased the cell viability by 40% based on 48 h. exposure. Although, they decreased the caspase-3 and p53 expression, indicating failure to overcome TNBC resistance [170]. Another approach in targeting TNBC is the use of platinum based acridine loaded in carbon nanotubes. With the help of noncovalent πstacking, the platinum acridines are adsorbed over the surface of carbon nanotubes. This bonding and interaction between the carbon nanotubes and platinum acridines ensure the controlled release of high doses of platinum in the TME, thus limiting dose-proportional toxicities. It was observed that these platinum-acridine carbon nanotubes were effective against various models of TNBC, including SUM159, BT20, MDA-MB-468, and MDA-MB-231 [171]. In another study, the researchers combined the multi-walled carbon nanotubes with Hyaluronic acid, and ∝-Tocopherol succinate was loaded with the cytotoxic drug, Doxorubicin against TNBC cells. Specifically, these TNBC cells were overexpressing the CD44 receptors. The study resulted in significant growth repression and increased apoptotic effect when compared with other formulations [172]. Thus, it is observed that carbon nanotubes are being greatly utilized in the chemotherapeutic field.

6.2. *Inorganic Nanoparticles*

Inorganic nanoparticles (NPs) are those nano-carriers that are developed using various metals (e.g., copper, silver, iron), semiconductors, and carbon dots. They are being explored quite extensively recently for theranostics purposes in cancer research [173,174]. Nowadays, it has gained attention due to exclusive features such as good biocompatibility, hydrophilic nature, high stability, and wide surface conjugation chemistry over organic nanoparticles for better imaging and drug delivery [173,174]. Although, it has numerous advantages for cancer treatment a few inorganic NPs have been successfully transformed for clinical use due to photo-induced tissue damage, non-specific toxicity, and immunogenicity related challenges. Thus, complete research on safety, synthesis, pharmacokinetics, and bio distribution is required in the case of inorganic NPs [173,174]. Some brief details pertaining to different inorganic nanoparticles are mentioned below.

6.2.1. Silica Nanoparticles

The silica nanoparticles are generally spherical. The natural silica itself is the crystalline base material in these nanoparticles. The nanoparticles can be modified into varied sizes and could be furthermore surface modified in order to target specifically. The nonporous form of these nanoparticles can be absorbent and abrasive in nature, on the other hand, mesoporous form is the one that is currently being used in drug delivery applications. The two types of silica base materials include Colloidal Silica and Colloidal Mesoporous Silica. Colloidal silicas are used to prepare nano drug carriers. These are synthesized specifically in basic environment (pH 7–10), whereas mesoporous silica is mainly silica precursors arranged into mesophases which are crystalline in nature. Further, these nanoparticles if surface modified using polymers, lipid bilayers are thus designed to load, retain, protect and release the drugs. Periodic mesoporous nanoparticles are synthesized from a sol-gel procedure [175]. Cheng and colleagues designed and applied the silica nanoparticles in a very unique manner against TNBC. The study included the designing of dendritic large pore mesoporous silica nanoparticles which were deposited with copper sulphide nanoparticles, and the immune adjuvant Resiquimod drug was also loaded in a controlled manner. These were further coated with the homogenous cancer cell membrane and thus conjugated with anti-PD1 peptide AUNP-12 by using a PEG linker, forming an acid-labile benzoic-imine bond. The final formulation was evaluated on metastatic TNBC in vitro as well as in vivo. It was observed that the designed nanoparticles were having high potential of targeting TNBC and also inducing photothermal ablation over primary TNBC tumors. This study thus enabled the pathway of using intelligent biomimetic nanoplatform in the treatment of metastatic TNBC [176]. In another study, the large pore mesoporous silica nanoparticles were deposited with a leukocyte/platelet hybrid and were also co-loaded with Doxorubicin. These silica-based nanoparticles were observed to showcase excellent TNBC-targeting ability as well as increased photothermal activity in vitro and in vivo. Thus, even this approach could be employed for the treatment of metastases [177]. The mesoporous silica nanoparticles could further be modified with RGD peptides and deposited with arsenic trioxide (ATO) in the treatment of MDA-MB-231 TNBC in vivo. The modified nanoparticles, mesoporous silica-RGD peptide-Arsenic trioxide, showcased a better therapeutic effect as compared to nonmodified silica nanoparticles, free arsenic trioxide, and arsenic trioxide combined with silica nanoparticles [178].

6.2.2. Silver Nanoparticles

Although silver metal is being used for several years, silver nanoparticles are a recent development and are being considered for delivering therapeutics for the past few decades. Silver nanoparticles are being used in medicine as well as the agricultural field as antibacterial, antifungal, and antioxidant. In a broader sense, silver nanoparticles have varied applications as biomedical devices, nano drug carriers, imaging probes, etc. [179,180]. The silver nanoparticles were coated with albumin and were evaluated on MDA-MB-231, the TNBC cell line. The formulation ultimately demonstrated cell-based apoptosis, and reduction of gland tumor sizes in mice. Furthermore, it was observed from this study that LD50 of albumin coated Silver Nanoparticles was found to be 30 times more compared to that of normal white blood cells [181]. The systematically dosed silver nanoparticles were found to be effective even at non-toxic doses in repressing the growth of TNBC xenografts in mice [182]. It has been observed that the silver nanoparticles showed selective toxicity in TNBC cell lines even at particular concentrations which are non-lethal to non-cancerous cells. Silver nanoparticles hold the potential for inducing selective toxicity in and sensitizing the TNBC cells to ionizing radiation and also have the potential of PTT heat transducer. Thus, the results indicated that the developed silver nanoparticle plates were effective as selective cytotoxic multimodal therapy against TNBC cell lines and thus to decrease the viability of TNBC cell lines, the combination of silver nanoparticles, PTT as well as ionizing radiations as compared to individualized treatment [183].

6.2.3. Gold Nanoparticles

Recent trends have seen an exploration of the therapeutic and cytotoxic potential of gold nanoparticles. Gold nanoparticles can be developed and synthesized in varied shapes and configurations such as Gold nanoshells, gold nanorods, gold nanocages for delivering cytotoxic drugs [110]. One of the studies conducted explored the mechanism behind the cytotoxicity against TNBC cells induced by differentially charged gold nanoparticles. The mechanism involved induction of oxidative stress, modification of WNT signaling pathway as well as epigenetic modifications, specifically histone H3 modifications at serine 10 and lysine 9/14 residues respectively. The presence of the type of charge also modified the duration for induction of cell death. The postive charges induced cell death abruptly whereas the negative charges slowly induced cell death in MDA-MB-231 cells [184]. Another way of harboring the cytotoxic properties of gold nanoparticles is the use of negatively charged gold NPs which bind to the cell-cell junction which is charged positively charged, thus inducing the formation of a leaky effect owing to the charge of the nanoparticles and thus enabling the access to cancer cells [110]. The gold nanoparticles also sensitize the TNBC cell line, MDA-MB-231 to cytotoxic drug 5-Fluorouracil via the mechanism of reducing the expression of thymidylatesynthetase [184]. In order to develop a treatment modality against TNBC, 21 nm gold nanoparticles were functionalized with Chlorine e6 and Epidermal Growth Factor (EGF). This bifunctionalization helped the specific targeting of TNBC cells mediated via PDT as well [185]. Thus, these gold nanoparticles can be explored furthermore as a nano vehicle as well as therapeutic agents against TNBC.

6.3. Organic-Inorganic Hybrid Nanocarriers

Organic/inorganic hybrid nanocarriers are established to offer the combinatorial benefits of both organic and inorganic nanoparticles. This includes good loading capacity, better release profile, biomimetic system, improved selectivity along with providing better therapeutic efficacy over the parent nanocarriers [186]. For example, surface fuctionalisation of Mesoporous Silica Nanoparticle (MSN) with polyethyleneimine (PEI) provides a contemporaneous benefit of improved cellular uptake of MSNs and also offers efficient nucleic acid delivery as PEI is a cationic polymer which imparts cationicity to the nanoparticle [187]. Manisha et al. developed a surface functionalized CD44 conjugated MSN for miRNA delivery using HA-PLGA for treating TNBC. It was reported that these surface coated MSN nanocarriers provide a better stability to the nanoparticle, have improved effective internalistion into the cellular environment, representing a stable and sustained release of microRNAs from the nanoparticle which further helps to kill the cancer cells [188]. In an another study, folic acid decorated metal organic framework nanocarriers were developed by Laha et al. for the treatment of TNBC [189]. This carrier provided greater uptake and sustained intracellular delivery of curcumion and also enhanced the the therapeutic effectiveness at the tumor site.

7. Stimuli Based Targeting Approaches in Nano Drug Delivery System for TNBC

Stimuli-sensitive systems are the systems that are being explored for the benefit they provide, that is controlled and sustained drug delivery. The stimuli-sensitive systems release the drug based on the internal or external stimuli which can be encompassed under environmental stimuli. These stimuli-sensitive systems can be delivered through various routes, including parenteral, ocular, rectal, vaginal, and dermal as well as transdermal delivery systems [190]. The use of these stimuli-based release systems is currently being worked on for specific targeting and release against TNBC [59]. The stimulus for these stimuli-sensitive release systems can be of two forms, either internal (endogenous) or external (exogenous). The internal stimuli involve pH, GSH, and enzymes whereas the external stimuli involve light (laser beams), temperature, magnetic field, and ultrasound [191]. Among these, the utilization of high temperature is considered one of the best approaches as many medical researchers have found the beneficial uses of hyperthermia in the treatment of solid tumors [192]. A pictorial representation of different stimuli-based nanoparticles is shown in Figure 4. The employment of stimuli-sensitive nanocarriers enables the nanocarriers to actively participate in drug and gene delivery compared to the conventional use of nanocarriers just as vehicles. In order to get the desired stimuli-sensitive nanocarriers, the composition of the nanocarrier assemblies could be modified accordingly. Another advantage of using the stimuli-sensitive nanocarriers is that the stimuli which can be harnessed for the release of the drug are specific to that tumor or disease pathology itself. This allows the modified stimuli-sensitive nanocarrier to release the drug over specific pathological triggers only [193]. Furthermore, if the TME is explored and understood in a detailed manner, it will enable the development and clinical use of the stimuli-sensitive systems, responding to abnormal conditions of the tumor tissues.

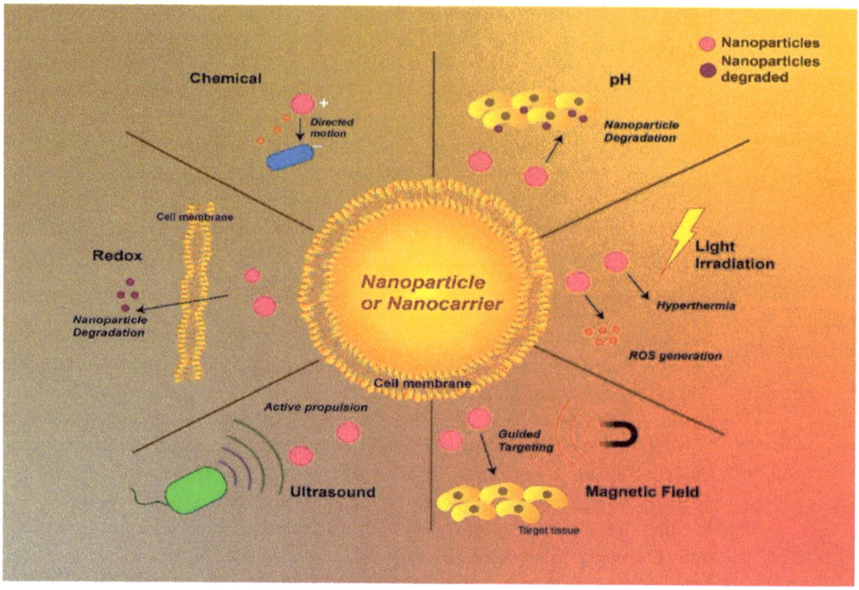

Figure 4. Different stimuli-responsive nanoparticles.

Overexposure to the specific stimulus, would convert these stimuli-responsive systems into controlled drug-releasing systems or would also enhance the uptake or penetration of the drug in the tumor microenvironment. The engineering of the stimuli-sensitive drug delivery systems, mainly relies on the specific biochemical contents of the affected areas and thus leads to specific temporal and spatial drug delivery [194–196]. Some recently reported stimuli-based nanoformulations and their outcome are displayed in Table 1.

Table 1. Stimuli-based targeting approaches in nano drug delivery system for TNBC.

S.No.	Type of Stimulus	Polymer/Nanoplatform	Drug	Significant Outcome	References
1.	pH (internal)	PEG and Poly (β–L–malic acid)	Doxorubicin	Inhibits in vitro cell growth in MDA-MB-468 & MDA-MB-231	[197]
2.	pH (internal)	Aldehyde functionalized PEG polymers	Doxorubicin	Hydrogel inhibited invitro cell growth in MDA-MB-231	[198]
3.	pH (internal)	Dextran-b-poly(histidine)	Doxorubicin	Viability of tumor cells decreased upon pH sensitive controlled release of drug	[199]
4.	pH (internal)	MPEG-poly-(beta-amino ester)	Camptothecin and Tetramethylrho-damine	Invitro significantly inhibited the growth in 5.MDA-M6.B-231 and 7 in vivo 1 time more accumulative compared to non pH sensitive micelles	[200]
5.	Temperature (internal)	poly(*N*-vinylcaprolactam) (PNVCL)-chitosan	Doxorubicin	Decreased off target activity compared to free drug, Invivo reduced tumor volume with decreased systemic side effects	[201]
6.	Reactive Oxygen Species (internal)	Amphiphilic Synthetic Polymers: mPEG and *L*-phenylalanine *N*-carboxyanhydride	Paclitaxel	Increased localization, decreased systemic toxicity in both in vitro and in vivo	[202]
7.	Reactive Oxygen Species (internal)	PLGA	Doxorubicin	Significant increment in ROS levels, increased cytotoxic effect in vitro via expression of cytochrome c, caspase-9, caspase-3, MMP-9	[203]
8.	Enzyme (internal)	Chondroitin sulfate	Doxorubicin hydrochloride	Significant increase in cellular uptake as well improved release profile of the nanoparticles loaded with the drug	[204]
9.	Magnetic field (external)	Chitosan	Rifampicin and Adriamycin	Pulsatile, controlled release of the drug	[205]
10.	Magnetic field (external)	PLGA	Doxorubicin	Controlled release of the cytotoxic drug, Doxorubicin	[206]
11.	Electrical (external)	Polydopamine-Polypyrrole	Doxorubicin	Increased bioavailability, higher drug loading, increased cell adherence	[207]
12.	Ultrasound (external)	PLGA	Mitoxantrone	Increased drug release up to 90%, enhanced controlled release, enhanced blood half-life time	[208]
13.	Ultrasound and thermal (external)	PEG and 4,4-azobis(4-cyanovaleric acid)	Topotecan	Increased circulation time and cellular uptake	[209]

7.1. pH Responsive Delivery System

pH responsive drug delivery system works to protect the drug in blood at pH 7.3–7.4 whereas this system releases the drug or expresses other functions as enhancement of cell penetration potential in the TME with pH 6.8–7.2. This pH-sensitive modified system can also release the loaded drug in response to changes in pH in the cytoplasm (pH greater than 7), endosomes (pH 5–6), and lysosomes (pH 4.5–5.5). If a higher concentration of the drug is localized in the tumor cell cytoplasm, this would enable the increased cytotoxic property of the drug and thus overcome the Pgp efflux capacity and ultimately have a cytotoxic effect on tumor cells [210,211]. Hence, the pH-responsive drug delivery systems exploit the difference in the pH of the cancer tissues and that of the normal tissues, and the acidic pH in endosomal and lysosomal regions for the release of chemotherapeutic drugs.

The strategy used for these pH-responsive drug delivery systems is that upon being triggered by an altered pH (stimuli), they result in modifications in the polymer of nanocarrier in terms of its physical properties like shape, size, or its nature of hydrophobicity [211]. There are two approaches that are being explored in order to achieve pH-responsive release in the TME. One of the approaches is the use of pH labile chemical bonds while conjugation of the drugs with polymers for example use of pH-sensitive polymer-drug conjugates. Examples of such linkages are hydrazone linkage, cis acotinyl linkage as well as acetal linkages. These acid-labile linkages are stable at the physiological pH, neutral pH, or alkaline pH but would break in the acidic pH (of the TME) [193,194,212].

In one such attempt of the application of hydrazone linkages, Patil et al. designed Doxorubicin conjugated PEG nanoparticles via pH-sensitive hydrazone linkages to a nanoconjugate platform of Poly (β–L-malic acid). It was observed that the pH sensitive conjugates were stable in physiological pH whereas released the loaded drug and inhibits in vitro cancer cell growth of MDA-MB-468 and MDA-MB-231 cell lines [197]. Furthermore, another application is the use of self-healable, injectables, which are pH responsive in nature and are the hybrid hydrogels formed via the formation of hydrazone linkages between hydrazide functionalized gelatin (Gel-ADH) and Aldehyde functionalized PEG polymers. The researchers investigated this modified pH-responsive hydrogel against cell lines MDA-MB-231 and observed the potential use of the modified hydrogel in the same [198].

Another approach is the use of copolymers involving titratable groups such as carboxylic acids, amines, which have the potential to get protonated and thus regulate the micelle formation based on the alteration of pH. The change in pH induces micelle formation and thus induces dissociation or alteration in the internal structure of the formulation. As the pH varies, this causes protonation of the pH-sensitive polymers which at normal physiological pH forms a hydrophobic core. As the pH lowers, with lower pKa protonatable groups get charged which result in destabilization and separation of the polymeric chain from the micellar chain. Examples of such protonatable polymers are poly(histidine), poly(acrylicacid), polysulfonamides [211]. Various researchers explore the use of poly (histidine) in the formulation of pH-sensitive systems against the tumor cells. In one such attempt, Hwang et al. explored the use of copolymer dextran-b-poly (histidine) in the formulation of Doxorubicin loaded nanoparticles via the nanoprecipitation method. From the study, it was observed that the nanoparticle releases the drug in the acidic tumor pH, in a pH-dependent controlled release manner [199]. Furthermore, Ko and colleagues designed Camptothecin and Tetramethylrhodamine (TRiTC) in MPEG Poly micelles. This formulation showcased micelle formation and destabilization at pH 6.8, with a pH-dependent drug release. These Camptothecin loaded micelles were evaluated for their cytotoxicity on MDA-MB-231 cell lines. In vivo, the tetramethylrhodamine loaded MPEG micelles were observed to be 11 times more accumulative compared to Tetramethylrhodamine conjugated Camptothecin without pH-sensitive micelles [200].

7.2. Temperature-Sensitive Delivery System

One of the characteristics of the tumor pathological tissues is local hyperthermia. This difference in temperature of normal healthy tissues and the tumor tissues can be harnessed and thus exploited for triggering the release from temperature-sensitive drug delivery system. Furthermore, so as to regulate the temperature for specific locations for the release of the drug, external heat sources could be used. These external heat sources include dielectric heating using microwave induction, ultrasound application, heating using an electrode with higher frequencies, heating using fiberoptics, laser photocoagulation as well as water bath heating [213,214]. The heating of the tumor location results in an increase in the size of the endothelial pore as well as there is an increase in the blood flow to that site, leading to enhanced extravasation of the nanocarriers [196]. The usage of certain lipids with a specific gel-to-liquid transition temperature is one of the approaches used for temperature-sensitive drug delivery. At that specific transition temperature, the liposomal membrane destabilizes, and the drug releases at temperatures of 40–42 °C in clinical hyperthermia protocols. One such good example of the liposomal composition is use of DPPC/MSPC/DSPE-PEG2000 in the ratio of 90:10:4 mole ratio [215,216]. Another most studied lipid for this temperature-sensitive drug delivery systems is dipalmitoylphosphatidylcholine (DPPC), having the transition temperature gel to liquid form of 41 °C [217,218]. In an attempt to use this temperature-sensitive drug delivery system, a solution was developed over the nanoplatform of thermo-sensitive poly (N-vinylcaprolactam)-chitosan nanoparticles. This was altered by the addition of cell-penetrating peptide and loading of the cytotoxic drug Doxorubicin. At the increased temperature of the TME, the base copolymer would lead to phase change, resulting in drug release and thus localization of the cytotoxic drug in the tumor tissues. This modified formulation resulted in decreased off-target cytotoxicity compared to the free drug form of Doxorubicin. It was observed in vivo that this formulation was effective in significantly reducing the tumor volume with decreased systemic side effects [201,219]. In a similar type another study, YC. Ou et al. developed a PTT based gold nanoparticle against the TNBC cells. In this work, a multi-branched gold antenna facilitates elevated temperature via the conversion of NIR light to heat and thus, delivers doxorubicin effectively to the cancer tissue. The modified combination therapy designed in this study, showcased enhanced cytotoxicity in TNBC cell lines compared to the free drug Doxorubicin, with decreased off-target cytotoxicity [220].

7.3. ROS Responsive

ROS responsive nanocarriers are those nanocarriers, which alter their properties based on the presence of ROS. Exposure to in vivo ROS leads to change (physical or chemical) in the ROS-responsive nanocarriers. These ROS-responsive nanocarriers have certain specific advantages compared to conventional nanocarriers. Such ROS-responsive agents can have varied applications including, imaging agents, and site-specific delivery of cytotoxic drugs. These ROS-responsive agents could also modify the tissue microenvironment [221–223]. So as to achieve increased drug localization in tumor tissues, Xi et al. developed redox-responsive diselenide containing dipeptide and further co-precipitated them with amphiphilic co-polymers, which has the potential to selectively target TNBC. This modified formulation exhibited increased localization of the drug in TNBC cells with decreased systemic toxicity in both in vitro and in vivo experiments [202]. In another study, the authors investigated a unique form of nanoparticles, which activate over the Fenton reaction. These nanoparticles contain polymers that are ROS responsive and thus also activate the cascade biological reaction of ROS in tumor cells, The significant increase in the elevated level of ROS resulted in an evident antitumor effect in vitro via expression of cytochrome c, caspase-9, and caspase-3 as well as inhibition of Matrix Metallo Protein–9, thus promoting apoptosis and inhibiting metastasis [203].

7.4. Enzyme-Responsive Drug Delivery System

The enzyme-responsive drug delivery system, in presence of an enzyme, undergoes various structural/physical changes via biocatalytic actions of the specific enzymes. The TME has the presence of certain specific upregulated enzymes, which enhances the potential use of enzymes responsive drug delivery system. The enzyme-responsive drug delivery systems, has unique properties including, sensitivity, selectivity, catalytic efficacy, and biorecognition [224]. The polymeric content of the formulation degrades and releases the drug upon the action of a specific enzyme.

Proteases are the enzymes, which are gaining attention in order to develop the enzyme-responsive drug delivery system. This is because the proteases are often overexpressed in terminal diseases like cancer [225]. In one such attempt of targeting tumor cells, Radhakrishnan et al. designed dual enzyme-responsive nanocapsules loaded with anticancer drugs for their intracellular delivery. The enzyme trypsin or hyaluronidase, causes the capsule wall of polymer to degrade, leading to intracellular delivery of anticancer drug in the tumor tissue [204]. A representative diagram for enzyme responsive drug delivery system has shown in Figure 5. Moreover, in another study that was conducted by Liu et. al., they fabricated an enzyme responsive liposome for co-delivery of tariquidar and doxurubicin to eradicate TNBC using mmp-2 enzyme. This targeted nanocarrier provided potential therapeutics to control tumor growth in drug-resistant TNBC via EGFR targeting and also exhibited controlled release, and prevent transporter efflux which helps the drug to be bioavailable at the tumor site. This study suggested that enzyme stimuli based targeted nanoplatform has the potential to treat drug-resistant TNBC and also enhanced the penetration of drug into the TME [226].

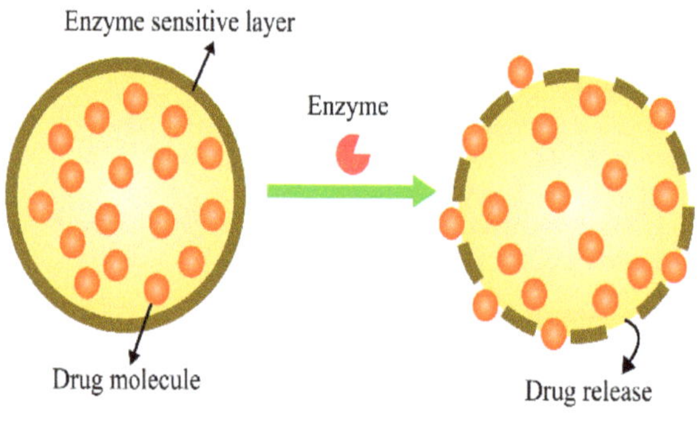

Figure 5. Enzyme responsive drug delivery system.

7.5. Magnetic Responsive Drug Delivery Systems

The magnetic rays have the potential to penetrate into the body and thus are used for body imaging in MRI [227]. These magnetic-responsive drug delivery systems are used for the controlled release of drug targeting the tumor sites. The peculiar characteristics of these systems include biocompatibility, biodegradability, easy synthesis as well as modification [228]. These are potential drug delivery systems as they have a smaller size and thus could easily target the tumor cells. In general, the mechanism involves the generation of heat by the magnetic responsive drug delivery systems, using alternating magnetic frequency. The magnetic drug delivery systems, are being utilized in the therapeutic area, on the basis of two different mechanisms, involving magnetic field-guided drug targeting and magnetic field-induced hyperthermia [206,229]. The hyperthermia-based mechanism has been widely applied to deliver drugs to cancer cells. Furthermore, the resulting hyperthermia via magnetic response also leads to inhibition of the tumor cells as well as provides an opportunity for imaging scans. In an attempt to explore the application of magnetic-responsive drug delivery systems [230]. Wang et al. developed an implantable magnetic chitosan hydrogel which was loaded with the drug Rifampicin, a hydrophobic drug, and Adriamycin, a hydrophilic. The external stimuli applied here was a very low frequency alternating magnetic field which lead to a pulsatile drug release from the delivery system with no involvement of hyperthermia associated with a magnetic field. The authors were successful in developing a magnetic responsive controlled release drug delivery system [205]. Thirunavukkarasu et al. developed superparamagnetic iron oxide nanoparticles, which were loaded with Doxorubicin in a PLGA matrix. This PLGA responds to the application of a magnetic field, leading to the subsequent release of Doxorubicin at specific tissue. The iron oxide nanoparticles undergo a change, leading to a change in the temperature of the solution, which further causes the transition in PLGA, ultimately leading to the release of the drug Doxorubicin [206].

7.6. Electrical Responsive Drug Delivery System

Electro-responsive drug delivery systems act as a delivery system with an external electric field, the drug from the nanoparticle leads to release after the application of a weak electric field. There are various mechanisms, behind the controlled drug release via electrical stimulation, consisting of an oxidation-reduction reaction, degradation of carrier structure, and stimulation of thermo-responsive carriers [231–233]. Xie et al. developed an electroresponsive polydopamine-polypyrrole microcapsules for dexamethasone delivery. The electroresponsive mechanism behind the release was the redox mechanism of polypyrrole. Furthermore, the nanoformulation also was observed to showcase, increased bioavailability, higher drug loading as well as increased cell adherence ability [207]. Neumann et al. used another approach via the application of electrical stimuli. The authors utilized the alterations in the pH change as a result of external electrical stimuli for regulating the release of the drug. The pH-sensitive drug-polymer used was poly(methyl methacrylate-co-methacrylic acid). In addition to this, the authors also observed that pH alterations were reversed after the removal of the external electrical stimuli thus leading to the stopping of the drug release [234].

7.7. Ultrasound-Responsive Drug Delivery Systems

The ultrasound-responsive drug delivery systems have unique advantageous properties, including non-invasiveness, safety, and tissue penetration [235]. The three forces, thermal, mechanical as well as radiation forces are the forces behind the release from ultrasound responsive drug delivery system [236]. In a similar attempt, to use these unique ultrasound responsive drug delivery systems, Xin et al. developed such nanoparticles using PLGA, for inducing ultrasound-responsive vibrations, leading to disruption of the liposomal membrane [208].

Paris et al. developed mesoporous silica nanoparticles which were PEGylated via thermosensitive linker 4,4′-azobis(4-cyanovaleric acid). This thermosensitive linker cleaved upon application of external stimuli of ultrasound waves, leading to drug release protected by PEGylation. This also gave a positive charge to the silica nanoparticles, which in turn increased the cellular uptake [209]. In another study, the researchers developed Ultrasound responsive liposomes using PLGA, which were loaded with the drug Mitoxantrone, which increased the drug release approximately up to 90% compared to the non-ultrasound responsive release of approximately 50%. Furthermore, this ultrasound-responsive liposomal nano-formulation showcased increased blood half-life time, making it a potential drug delivery system [208,237].

7.8. Hypoxia Responsive Drug Delivery Systems

Hypoxia is one of the crucial parameters driving the cellular mechanisms in TME. Primarily, it is involved in tumor angiogenesis, invasion, metastasis and immunosuppression [238]. Due to this evidences, researchers have an increased interest towards developing a target-based nanomaterial for hypoxic TME. Liu et al. developed a hypoxia responsive self-assembled micelle of doxorubicin and ICG using NIDH (nitroimidazole hexylamine) block copolymers. This nanocarrier is capable of detecting hypoxia in the TME to trigger the cytotoxicity and immunogenic responses towards the effective treatment of 4T1 breast cancer. This work suggested that this targeted long-acting nanocarrier was utilized to diagnose the triple negative 4T1 breast tumors via photoacoustic imaging technique, and successfully killing tumors without recurrence [239]. In another research work, Zhang et al. developed a hypoxia responsive drug-drug conjugated nanoparticle using chemical linker Azobenzene for three drugs (combretastatin A-4, irinotecan and cyclopamine) to treat MCF-7 cancer. This carrier could effectively trigger the apoptosis and prevent the proliferation and differentiation cancer. Also it could improve the the cellular uptake and the permeability of the drug [240]. These work demonstrated that this nanocarrier provided a promising therapeutic strategy for tumor management.

8. Regulatory Considerations of Nanoformulations for TNBC

Nowadays, the nanomaterial application in drug delivery has amplified, with many different advanced nano formulation approaches being exploited within the clinical translation. These nanomaterials have numerous unique characteristics which help them in clinical applications. One of its unique properties is narrow range particle size i.e., 10–200 nm which permits the nanoparticle for systemic circulation over a long period of time and also helps to avoid clearance via renal and complement systems [241,242]. This feature of nanoparticles provides a momentous approach to the drug for cancer treatment as it was shown high EPR effect and greater penetration of the drug into the cancer microenvironment. Furthermore, Surface modification of these nanoparticles with various biomarkers including protein, aptamer, folic acid, RGD, etc. delivers site-specific targeting and selectivity characteristics toward the treatment of cancer. One more important characteristic of nanoparticles is electronic and optical characteristics, predominantly in metallic nanoparticles [243]. These characteristics are permitting the nanoparticles to be exploited for numerous applications including bio-imaging, effective targeting, and drug delivery. However, pertaining to nanomaterials, the scientific community has very high expectations regarding disease treatments, still, a lot of obstacles are being observed in the path of development of nano-therapeutics as a standard regulatory guideline for nanomedicine is not available. Despite the absence of regulatory guidelines, there are several nanoparticles already established in the market. Mostly, these are used for cancer treatment; examples of these nanoparticles include Doxil®, AmBisome®, Abraxane®, etc. [21,244]. Additionally, there are some nanomaterials have been under clinical trials for TNBC (Table 2). Although these studies are showing potential effectiveness against TNBC, still their production, scale up and safety is other concerns. Lack of standardization and scientific expertise makes it difficult for the regulatory agency to monitor the safety and effectiveness of these nanomaterials.

Table 2. Key Clinical trials and their outcomes which aimed to explore the nanotechnology-based approaches against TNBC.

Intervention	Nanocarrier	Outcome	Status & Primary Completion Date	Company	NCT & Reference
Docataxel	Nanosomal Liquid Suspension	Proportion of the patients with Objective Response Rate (i.e., CR + PR) as the Best Overall Response Rate (i.e., CR + PR) in the test arm (NDLS) compared to reference arm (Taxotere) Progression free survival (PFS) To evaluate the overall survival (OS) of the patients Incidence of adverse events as assessed by clinical examination, and/or laboratory parameters	Phase III, 31 March 2022	Jina Pharmaceuticals, Intas Pharmaceuticals, Lambda Therapeutic Research Ltd.	NCT03671044
HLX10	In combination with Chemotherapy (as Neoadjuvant Therapy)	Tumor assessment	Phase III, 7 September 2022	Shanghai Henlius Biotech	NCT04301739
Atezolizumab + Nab-Paclitaxel	Nanoparticle albumin bound paclitaxel	Percentage of Participants with treatment-emergent Grade $\geq$ 3 AEs Percentage of Participants with treatment-emergent Grade $\geq$ 2 imAEs Percentage of Participants with all treatment-emergent AEs	Phase III, 30 June 2022	Hoffmann-La Roche	NCT04177108
Nab- Paclitaxel & Bevacizumab	Nanoparticle albumin bound paclitaxel	Progression free survival. Response rate. Overall survival. Safety and toxicity. Exploratory biomarkers will be assessed as potential predictors of response to treatment including: expression of epidermal growth factor receptor (EGFR) and secreted protein acidic and rich in cysteine (SPARC) in the primary tumor and changes in levels of circulating tumor cells (CTCs) and circulating endothelial cells (CECs).	Phase II, 5 July 2017	Sponsor: University of Washington Collaborator: National Cancer Institute (NCI)	NCT00733408
CORT125134 in Combination with Nab-paclitaxel	Nanoparticle albumin bound paclitaxel	Maximum Tolerated Dose of CORT125134 in Combination with nab-paclitaxel, Number of Treatment-Related Adverse Events as Assessed by CTCAE version 4.0 for Patients with Solid Tumors Treated with CORT125134 in combination with nab-paclitaxel	Phase I/II, May 2020	Corcept Therapeutics	NCT02762981

Moreover, most of the nanomedicines are working by interacting with genetic materials or biochemical which is participated in normal genome function and cellular activity. All of these nanoparticles can produce severe toxicity to the body and also can cause cancer via various abnormal pathways i.e., oxidative and nitrosative stress, lipid peroxidation, oxidative DNA damage etc. Therefore, regulatory agencies and the research community is need to work on the development a standard guideline [245]. Currently, USFDA, Health and Consumer Protection Directorate of the European Commission and US-EPA have engaged so as to deal with adverse effect by nano-therapeutics. But there are numerous challenges in the way of the progression of guidelines. One of the major challenges in the regulation of nanomedicine is regulatory bodies i.e., USFDA practice safety data sheet for bulk materials that do not show a similar pharmacokinetic and efficacy profile as in nanomedicine. it can create a severe issue on safety and efficacy of nano-therapeutics. Other issues related to manufacturing, scale-up, and stability, as the nanomedicine is having complex structure and characteristics, it is highly tedious to prepare a consistent batch. Thus, critical Quality Attributes (CQA) is required to monitor the screening of raw material, manufacturing process, scale up and stability that can support understanding nanomedicine. During the clinical translation, the drug mechanism and preclinical safety profile of nanomedicine is needed before approval of clinical trials. Otherwise, it may be compromised the safety and effectiveness of the product. All of the above-discussed challenges are restricting the nanomedicine for their future development.it can directly deter product safety, and quality of products, and could be leading to unproductive control of nanoparticles due to the absence of product-specific safety protocol [21,244,246]. Therefore, to establish the efficient benefits accessed by nano-therapeutics against cancers or other life-threatening diseases, many critical concerns on the regulatory protocol is need to be addressed by the regulatory bodies.

9. Conclusions and Future Perspectives

Triple-negative breast cancer is a heterogeneous metastatic form of breast cancer with unique biological features. Its aggressiveness, poor prognosis, and high rate of recurrence make its management highly challenging. Conventional approaches such as use of cytotoxic chemotherapy (taxanes, anthracyclines, or platinum agents), surgery, and radiotherapy were widely being evaluated by the researchers for their effectiveness in triple-negative breast cancer. However, these therapies involve certain limitations, such as multidrug resistance, poor bioavailability, unwanted side effects, and poor prognosis. Scientists also developed targeted therapies, including EGFR inhibitors, VEGF inhibitors, mTOR inhibitors, and PARP inhibitors. Despite being specific in their action, resistance offered by cancer cells to these inhibitors has hindered their use for TNBC. Nanotherapeutics have surmounted the difficulties of traditional and targeted therapies. They selectively target tumor cells by enhanced permeation and retention effect or by active targeting. Encapsulation of anti-cancer drugs in these nanocarriers protects them from the external environment and enhances their biological half-life, decreasing the dose required for the desired effect. Researchers have developed several nanoparticles for treating as well as monitoring cancer, making them promising delivery systems. Further, various ligands can be attached to the nanocarriers to target them to the desired site. Various organic and inorganic nanoparticles have been developed for TNBC. siRNA and miRNA have also gained the attention of researchers as a therapeutic option for cancer. Delivery of such organic particles is the current need of scientists as TNBC cells lack biomarkers and unifying molecular features. Characteristics of the TME have led to the progress of immunotherapy in TNBC. In another perspective of employing immunotherapy, nanocarriers are employed to carry agents targeting the immune components involved in TNBC progression. However, they have to cross the biological barriers that constitute the natural defense system of our body. Further, their difficult scale-up, efflux by the transporter, and regulatory approval limit their use for successfully treating cancers. There is a need for universities, governments, and pharmaceutical industries to join hands and work on regulatory guidelines for nanotherapeutics.

In the future, precision nanomedicine, which takes into account genetic differences as well as physiological and pathological variabilities amongst patients will be the focus of scientists. Successfully developing novel targeted nanomaterial can be utilized in both treatment and diagnosis of various diseases which will provide a promising future in cancer research. These nanomaterials deliver lot of advantages such as possibility to load/encapsulate drug which can offers stability to the drug from the biological system. As a result, it could enhance biological half-life of anti-cancer drug. Further, these can deliver better targetability to tumor tissue, greater transfection towards cancer cell, as it could facilitate its conjugation with various biomarkers (Protein, ligand, nucleic acid, stimuli-responsive chemical etc.). In contrast, these nanomaterials have shown lot of challenges such as immunogenicity, lack of understanding of nanomaterial on molecular and cellular interaction and lack of clinical translation due to limited availability of TNBC (cell or animal) model. Thus, there is requirement of innovative therapeutic approaches in the future. At the same time, attention has to be given towards developing toxicity-free nanoparticles with the capability of targeting desired tissue and exhibit desired pharmacokinetics selectively. Understanding the complex heterogeneity of TNBC and the identification of new biomarkers that can be targeted will effectively treat TNBC. Further progress in the field of nanotechnology may guarantee success against this life-threatening cancer.

Author Contributions: Writing—original draft, R.P., A.D. and S.K.; Writing—review & editing, R.P., A.D. and S.K.D.; Resources, Supervision, S.K.D., R.T. and A.P.; Conceptualization—S.K.D., R.T. and A.P. All authors have read and agreed to the published version of the manuscript.

Funding: This research received no external funding.

Institutional Review Board Statement: Not applicable.

Informed Consent Statement: Not applicable.

Data Availability Statement: Not applicable.

Conflicts of Interest: The authors declare no conflict of interest.

References

1. Anders, C.; Carey, L.A. Understanding and treating triple-negative breast cancer. *Oncology* **2008**, *22*, 1233–1239. [PubMed]
2. Shekar, N.; Mallya, P.; Gowda, D.V.; Jain, V. Triple-negative breast cancer: Challenges and treatment options. *Int. J. Res. Pharm. Sci.* **2020**, *11*, 1977–1986. [CrossRef]
3. Dietze, E.C.; Carolina, N.; Carolina, N.; Seewaldt, V.L. Triple-negative breast cancer in African-American women: Disparities versus biology. *Nat. Rev. Cancer* **2017**, *15*, 248–254. [CrossRef] [PubMed]
4. Lee, E.; McKean-Cowdin, R.; Ma, H.; Spicer, D.V.; Van Den Berg, D.; Bernstein, L.; Ursin, G. Characteristics of triple-negative breast cancer in patients with a BRCA1 mutation: Results from a population-based study of young women. *J. Clin. Oncol.* **2011**, *29*, 4373–4380. [CrossRef] [PubMed]
5. Neven, P.; Brouckaert; Wildiers; Floris. Update on triple-negative breast cancer: Prognosis and management strategies. *Int. J. Womens Health* **2012**, *4*, 511. [CrossRef]
6. Yin, L.; Duan, J.J.; Bian, X.W.; Yu, S.C. Triple-negative breast cancer molecular subtyping and treatment progress. *Breast Cancer Res.* **2020**, *22*, 1–13. [CrossRef]
7. Burstein, M.D.; Tsimelzon, A.; Poage, G.M.; Covington, K.R.; Contreras, A.; Fuqua, S.A.W.; Savage, M.I.; Osborne, C.K.; Hilsenbeck, S.G.; Chang, J.C.; et al. Comprehensive genomic analysis identifies novel subtypes and targets of triple-negative breast cancer. *Clin. Cancer Res.* **2015**, *21*, 1688–1698. [CrossRef]
8. Mendes, T.F.S.; Kluskens, L.D.; Rodrigues, L.R. Triple Negative Breast Cancer: Nanosolutions for a Big Challenge. *Adv. Sci.* **2015**, *2*, 1–14. [CrossRef]
9. Lehmann, B.D.; Jovanović, B.; Chen, X.; Estrada, M.V.; Johnson, K.N.; Shyr, Y.; Moses, H.L.; Sanders, M.E.; Pietenpol, J.A. Refinement of triple-negative breast cancer molecular subtypes: Implications for neoadjuvant chemotherapy selection. *PLoS ONE* **2016**, *11*, 1–22. [CrossRef]
10. Singhvi, G.; Rapalli, V.K.; Nagpal, S.; Dubey, S.K.; Saha, R.N. Nanocarriers as Potential Targeted Drug Delivery for Cancer Therapy. In *Nanoscience in Medicine*; Daima, H.K., Pn, N., Ranjan, S., Dasgupta, N., Lichtfouse, E., Eds.; Springer International Publishing: Cham, Switzerland, 2020; Volume 1, pp. 51–88.

11. Penault-Llorca, F.; Viale, G. Pathological and molecular diagnosis of triple-negative breast cancer: A clinical perspective. *Ann. Oncol.* **2012**, *23* (Suppl. 6), 22–25. [CrossRef]
12. Bosch, A.; Eroles, P.; Zaragoza, R.; Viña, J.R.; Lluch, A. Triple-negative breast cancer: Molecular features, pathogenesis, treatment and current lines of research. *Cancer Treat. Rev.* **2010**, *36*, 206–215. [CrossRef]
13. Roy, R.; Chun, J.; Powell, S.N. BRCA1 and BRCA2: Different roles in a common pathway of genome protection. *Nat. Rev. Cancer* **2011**, *12*, 68–78. [CrossRef]
14. Khan, M.A.; Jain, V.K.; Rizwanullah, M.; Ahmad, J.; Jain, K. PI3K/AKT/mTOR pathway inhibitors in triple-negative breast cancer: A review on drug discovery and future challenges. *Drug Discov. Today* **2019**, *24*, 2181–2191. [CrossRef]
15. Medina, M.A.; Oza, G.; Sharma, A.; Arriaga, L.G.; Hernández, J.M.H.; Rotello, V.M.; Ramirez, J.T. Triple-negative breast cancer: A review of conventional and advanced therapeutic strategies. *Int. J. Environ. Res. Public Health* **2020**, *17*, 2078. [CrossRef]
16. Engebraaten, O.; Vollan, H.K.M.; Børresen-Dale, A.L. Triple-negative breast cancer and the need for new therapeutic targets. *Am. J. Pathol.* **2013**, *183*, 1064–1074. [CrossRef]
17. Podo, F.; Buydens, L.M.C.; Degani, H.; Hilhorst, R.; Klipp, E.; Gribbestad, I.S.; Van Huffel, S.; van Laarhoven, H.W.M.; Luts, J.; Monleon, D.; et al. Triple-negative breast cancer: Present challenges and new perspectives. *Mol. Oncol.* **2010**, *4*, 209–229. [CrossRef]
18. Borri, F.; Granaglia, A. Pathology of triple negative breast cancer. *Semin. Cancer Biol.* **2021**, *72*, 136–145. [CrossRef]
19. Reis-Filho, J.S.; Tutt, A.N.J. Triple negative tumours: A critical review. *Histopathology* **2008**, *52*, 108–118. [CrossRef]
20. Elias, A.D. Triple-negative breast cancer: A short review. *Am. J. Clin. Oncol. Cancer Clin. Trials* **2010**, *33*, 637–645. [CrossRef]
21. Hejmady, S.; Pradhan, R.; Alexander, A.; Agrawal, M.; Singhvi, G.; Gorain, B.; Tiwari, S.; Kesharwani, P.; Dubey, S.K. Recent advances in targeted nanomedicine as promising antitumor therapeutics. *Drug Discov. Today* **2020**, *25*, 2227–2244. [CrossRef]
22. Wahba, H.A.; El-hadaad, H.A. Current approaches in treatment of triple-negative breast cancer Treatment modalities of TNBC. *Cancer Biol. Med.* **2015**, *12*, 106–116. [PubMed]
23. Nedeljkovi, M. Mechanisms of Chemotherapy Resistance in Triple-Negative Breast Cancer—How We Can Rise to the Challenge. *Cell* **2019**, *8*, 957. [CrossRef] [PubMed]
24. Andreopoulou, E.; Sparano, J.A. Chemotherapy in Patients with Anthracycline- and Taxane-Pretreated Metastatic Breast Cancer: An Overview. *Curr. Breast Cancer Rep.* **2013**, *5*, 42–50. [CrossRef]
25. Anampa, J.; Makower, D.; Sparano, J.A. Progress in adjuvant chemotherapy for breast cancer: An overview. *BMC Med.* **2015**, *13*, 1–13. [CrossRef] [PubMed]
26. Burotto, M.; Wilkerson, J.; Stein, W.D.; Bates, S.E.; Fojo, T. Adjuvant and neoadjuvant cancer therapies: A historical review and a rational approach to understand outcomes. *Semin. Oncol.* **2019**, *46*, 83–99. [CrossRef]
27. Perez, E.A.; Moreno-Aspitia, A.; Thompson, E.A.; Andorfer, C.A. Adjuvant therapy of triple negative breast cancer. *Breast Cancer Res. Treat.* **2010**, *120*, 285–291. [CrossRef]
28. Montemurro, F.; Nuzzolese, I.; Ponzone, R. Neoadjuvant Or Adjuv. Chemother. Early Breast Cancer? *Expert Opin. Pharmacother.* **2020**, *21*, 1071–1082. [CrossRef]
29. Akashi-Tanaka, S.; Watanabe, C.; Takamaru, T.; Kuwayama, T.; Ikeda, M.; Ohyama, H.; Mori, M.; Yoshida, R.; Hashimoto, R.; Terumasa, S.; et al. BRCAness predicts resistance to taxane-containing regimens in triple negative breast cancer during neoadjuvant chemotherapy. *Clin. Breast Cancer* **2015**, *15*, 80–85. [CrossRef]
30. Inaji, H.; Komoike, Y.; Motomura, K.; Kasugai, T.; Koyama, H. The role of neoadjuvant chemotherapy for breast cancer treatment. *Gan. Kagaku Ryoho.* **2002**, *29*, 1113–1119.
31. Gerber, B.; Loibl, S.; Eidtmann, H.; Rezai, M.; Fasching, P.A.; Tesch, H.; Eggemann, H.; Schrader, I.; Kittel, K.; Hanusch, C.; et al. Neoadjuvant bevacizumab and anthracycline-taxane-based chemotherapy in 678 triple-negative primary breast cancers; results from the geparquinto study (GBG 44). *Ann. Oncol.* **2013**, *24*, 2978–2984. [CrossRef]
32. Wu, J.; Li, S.; Jia, W.; Su, F. Response and prognosis of taxanes and anthracyclines neoadjuvant chemotherapy in patients with triple-negative breast cancer. *J. Cancer Res. Clin. Oncol.* **2011**, *137*, 1505–1510. [CrossRef]
33. Fisher, C.S.; Ma, C.X.; Gillanders, W.E.; Aft, R.L.; Eberlein, T.J.; Gao, F.; Margenthaler, J.A. Neoadjuvant chemotherapy is associated with improved survival compared with adjuvant chemotherapy in patients with triple-negative breast cancer only after complete pathologic response. *Ann. Surg. Oncol.* **2012**, *19*, 253–258. [CrossRef]
34. Kim, K.; Park, H.J.; Shin, K.H.; Kim, J.H.; Choi, D.H.; Park, W.; Ahn, S.D.; Kim, S.S.; Kim, D.Y.; Kim, T.H.; et al. Breast conservation therapy versus mastectomy in patients with T1-2N1 triple-negative breast cancer: Pooled analysis of KROG 14-18 and 14-23. *Cancer Res. Treat.* **2018**, *50*, 1316–1323. [CrossRef]
35. Guo, L.; Xie, G.; Wang, R.; Yang, L.; Sun, L.; Xu, M.; Yang, W.; Chung, M.C. Local treatment for triple-negative breast cancer patients undergoing chemotherapy: Breast-conserving surgery or total mastectomy? *BMC Cancer* **2021**, *21*, 1–12. [CrossRef]
36. Yao, Y.; Chu, Y.; Xu, B.; Hu, Q.; Song, Q. Radiotherapy after Surgery Has Significant Survival Benefits for Patients with Triple-Negative Breast Cancer. *Cancer Med.* **2019**, *8*, 554–563. [CrossRef]
37. Baskar, R.; Lee, K.A.; Yeo, R.; Yeoh, K.W. Cancer and radiation therapy: Current advances and future directions. *Int. J. Med. Sci.* **2012**, *9*, 193–199. [CrossRef]
38. Abdulkarim, B.S.; Cuartero, J.; Hanson, J.; Deschênes, J.; Lesniak, D.; Sabri, S. Increased risk of locoregional recurrence for women with T1-2N0 triple-negative breast cancer treated with modified radical mastectomy without adjuvant radiation therapy compared with breast-conserving therapy. *J. Clin. Oncol.* **2011**, *29*, 2852–2858. [CrossRef]

39. Bellon, J.R.; Burstein, H.J.; Frank, E.S.; Mittendorf, E.A.; King, T.A. Multidisciplinary Considerations in the Treatment of Triplegative Breast Cancer. *Can. CA. Cancer J. Clin.* **2020**, *70*, 432–442. [CrossRef]
40. He, M.Y.; Rancoule, C.; Rehailia-Blanchard, A.; Espenel, S.; Trone, J.C.; Bernichon, E.; Guillaume, E.; Vallard, A.; Magné, N. Radiotherapy in triple-negative breast cancer: Current situation and upcoming strategies. *Crit. Rev. Oncol. Hematol.* **2018**, *131*, 96–101. [CrossRef]
41. Rosenblum, D.; Joshi, N.; Tao, W.; Karp, J.M.; Peer, D. Progress and challenges towards targeted delivery of cancer therapeutics. *Nat. Commun.* **2018**, *9*, 1–12. [CrossRef]
42. Barua, S.; Mitragotri, S. Challenges associated with Penetration of Nanoparticles across Cell and Tissue Barriers: A Review of Current Status and Future Prospects. *Nano. Today* **2014**, *9*, 223–243. [CrossRef] [PubMed]
43. Lin, J.; Miao, L.; Zhong, G.; Lin, C.-H.; Dargazangy, R.; Alexander-Katz, A. Understanding the synergistic effect of physicochemical properties of nanoparticles and their cellular entry pathways. *Commun. Biol.* **2020**, *3*, 205. [CrossRef] [PubMed]
44. Zhao, Z.; Ukidve, A.; Krishnan, V.; Mitragotri, S. Effect of physicochemical and surface properties on in vivo fate of drug nanocarriers. *Adv. Drug Deliv. Rev.* **2019**, *143*, 3–21. [CrossRef] [PubMed]
45. Blanco, E.; Shen, H.; Ferrari, M. Principles of nanoparticle design for overcoming biological barriers to drug delivery. *Nat. Biotechnol.* **2015**, *33*, 941–951. [CrossRef] [PubMed]
46. Sriraman, S.K.; Aryasomayajula, B.; Torchilin, V.P. Barriers to drug delivery in solid tumors. *Tissue Barriers* **2014**, *2*, e29528. [CrossRef]
47. Deshmukh, S.K.; Srivastava, S.K.; Tyagi, N.; Ahmad, A.; Singh, A.P.; Ghadhban, A.A.L.; Dyess, D.L.; Carter, J.E.; Dugger, K.; Singh, S. Emerging evidence for the role of differential tumor microenvironment in breast cancer racial disparity: A closer look at the surroundings. *Carcinogenesis* **2017**, *38*, 757–765. [CrossRef]
48. Yu, T.; Di, G. Role of tumor microenvironment in triple-negative breast cancer and its prognostic significance. *Chin. J. Cancer Res.* **2017**, *29*, 237–252. [CrossRef]
49. Soysal, S.D.; Tzankov, A.; Muenst, S.E. Role of the Tumor Microenvironment in Breast Cancer. *Pathobiology* **2015**, *82*, 142–152. [CrossRef]
50. Johnson, R.; Sabnis, N.; McConathy, W.J.; Lacko, A.G. The potential role of nanotechnology in therapeutic approaches for triple negative breast cancer. *Pharmaceutics* **2013**, *5*, 353–370. [CrossRef]
51. Zhu, X.; Zhou, W. The emerging regulation of VEGFR-2 in triple-negative breast cancer. *Front. Endocrinol.* **2015**, *6*, 1–7. [CrossRef]
52. Sukumar, J.; Gast, K.; Quiroga, D.; Lustberg, M.; Williams, N. Triple-negative breast cancer: Promising prognostic biomarkers currently in development. *Expert Rev. Anticancer Ther.* **2021**, *21*, 135–148. [CrossRef]
53. Zhao, Z.; Li, Y.; Shukla, R.; Liu, H.; Jain, A.; Barve, A.; Cheng, K. Development of a Biocompatible Copolymer Nanocomplex to Deliver VEGF siRNA for Triple Negative Breast Cancer. *Theranostics* **2019**, *9*, 4508–4524. [CrossRef]
54. Ghosh, S.; Javia, A.; Shetty, S.; Bardoliwala, D.; Maiti, K.; Banerjee, S.; Khopade, A.; Misra, A.; Sawant, K.; Bhowmick, S. Triple negative breast cancer and non-small cell lung cancer: Clinical challenges and nano-formulation approaches. *J. Control Rel.* **2021**, *337*, 27–58. [CrossRef]
55. Nakai, K.; Hung, M.C.; Yamaguchi, H. A perspective on anti-EGFR therapies targeting triple-negative breast cancer. *Am. J. Cancer Res.* **2016**, *6*, 1609–1623.
56. Wang, Y.; Zhang, T.; Kwiatkowski, N.; Abraham, B.J.; Lee, T.I.; Xie, S.; Yuzugullu, H.; Von, T.; Li, H.; Lin, Z.; et al. CDK7-Dependent Transcriptional Addiction in Triple-Negative Breast Cancer. *Cell* **2015**, *163*, 174–186. [CrossRef]
57. Si, Y.; Xu, Y.; Guan, J.S.; Chen, K.; Kim, S.; Yang, E.S.; Zhou, L.; Liu, X.M. Anti-EGFR antibody-drug conjugate for triple-negative breast cancer therapy. *Eng. Life Sci.* **2021**, *21*, 37–44. [CrossRef]
58. Su, Y.C.; Burnouf, P.A.; Chuang, K.H.; Chen, B.M.; Cheng, T.L.; Roffler, S.R. Conditional internalization of PEGylated nanomedicines by PEG engagers for triple negative breast cancer therapy. *Nat. Commun.* **2017**, *8*, 1–12. [CrossRef]
59. Pawar, A.; Prabhu, P. Biomedicine & Pharmacotherapy Nanosoldiers: A promising strategy to combat triple negative breast cancer. *Biomed. Pharmacother.* **2019**, *110*, 319–341. [CrossRef]
60. Geenen, J.J.J.; Linn, S.C.; Beijnen, J.H.; Schellens, J.H.M. PARP Inhibitors in the Treatment of Triple-Negative Breast Cancer. *Clin Pharmacokinet.* **2018**, *57*, 427–437. [CrossRef]
61. Lyons, T.G. Targeted Therapies for Triple-Negative Breast Cancer. *Curr. Treat. Options Oncol.* **2019**, *20*, 1–13. [CrossRef]
62. Mehta, A.K.; Cheney, E.M.; Hartl, C.A.; Pantelidou, C.; Oliwa, M.; Castrillon, J.A.; Lin, J.R.; Hurst, K.E.; de Oliveira Taveira, M.; Johnson, N.T.; et al. Targeting immunosuppressive macrophages overcomes PARP inhibitor resistance in BRCA1-associated triple-negative breast cancer. *Nat. Cancer* **2021**, *2*, 66–82. [CrossRef] [PubMed]
63. Massihnia, D.; Galvano, A.; Fanale, D.; Perez, A.; Castiglia, M.; Incorvaia, L.; Listì, A.; Rizzo, S.; Cicero, G.; Bazan, V.; et al. Triple negative breast cancer: Shedding light onto the role of pi3k/akt/mtor pathway. *Oncotarget* **2016**, *7*, 60712–60722. [CrossRef] [PubMed]
64. Chan, J.J.; Tan, T.J.; Dent, R. Novel therapeutic avenues in triple-negative breast cancer: PI3K/AKT inhibition, androgen receptor blockade, and beyond. *Ther. Adv. Med. Oncol.* **2019**, *11*, 1758835919880429. [CrossRef] [PubMed]
65. Cretella, D.; Ravelli, A.; Fumarola, C.; La Monica, S.; Digiacomo, G.; Cavazzoni, A.; Alfieri, R.; Biondi, A.; Generali, D.; Bonelli, M.; et al. The anti-tumor efficacy of CDK4/6 inhibition is enhanced by the combination with PI3K/AKT/mTOR inhibitors through impairment of glucose metabolism in TNBC cells. *J. Exp. Clin. Cancer Res.* **2018**, *37*, 1–12. [CrossRef] [PubMed]

66. Kim, K.Y.; Park, K.I.; Kim, S.H.; Yu, S.N.; Park, S.G.; Kim, Y.W.; Seo, Y.K.; Ma, J.Y.; Ahn, S.C. Inhibition of autophagy promotes salinomycin-induced apoptosis via reactive oxygen species-mediated PI3K/AKT/mTOR and ERK/p38 MAPK-dependent signaling in human prostate cancer cells. *Int. J. Mol. Sci.* **2017**, *18*, 1088. [CrossRef]
67. Marra, A.; Viale, G.; Curigliano, G. Recent advances in triple negative breast cancer: The immunotherapy era. *BMC Medicine* **2019**, *17*, 1–9. [CrossRef]
68. Stovgaard, E.S.; Nielsen, D.; Hogdall, E.; Balslev, E. Triple negative breast cancer–prognostic role of immune-related factors: A systematic review. *Acta Oncol.* **2018**, *57*, 74–82. [CrossRef]
69. Dees, S.; Ganesan, R.; Singh, S.; Grewal, I.S. Bispecific Antibodies for Triple Negative Breast Cancer. *Trends Cancer* **2021**, *7*, 162–173. [CrossRef]
70. Wang, Z.; Sau, S.; Alsaab, H.O.; Iyer, A.K. CD44 directed nanomicellar payload delivery platform for selective anticancer effect and tumor specific imaging of triple negative breast cancer. *Nanomed. Nanotechnol. Biol Med.* **2018**, *14*, 1441–1454. [CrossRef]
71. Lee, J.; Kim, D.M.; Lee, A. Prognostic role and clinical association of tumor-infiltrating lymphocyte, programmed death ligand-1 expression with neutrophil-lymphocyte ratio in locally advanced triple-negative breast cancer. *Cancer Res. Treat.* **2019**, *51*, 649–663. [CrossRef]
72. Mina, L.A.; Lim, S.; Bahadur, S.W.; Firoz, A.T. Immunotherapy for the treatment of breast cancer: Emerging new data. *Breast Cancer Targets Ther.* **2019**, *11*, 321–328. [CrossRef]
73. Singh, S.; Numan, A.; Maddiboyina, B.; Arora, S.; Riadi, Y.; Md, S.; Alhakamy, N.A.; Kesharwani, P. The emerging role of immune checkpoint inhibitors in the treatment of triple-negative breast cancer. *Drug Discov. Today* **2021**, *26*, 1721–1727. [CrossRef]
74. Kagihara, J.A.; Andress, M.; Diamond, J.R. Nab-paclitaxel and atezolizumab for the treatment of PD-L1-positive, metastatic triple-negative breast cancer: Review and future directions. *Expert Rev. Precis. Med. Drug Dev.* **2020**, *5*, 59–65. [CrossRef]
75. Du, X.; Tang, F.; Liu, M.; Su, J.; Zhang, Y.; Wu, W.; Devenport, M.; Lazarski, C.A.; Zhang, P.; Wang, X.; et al. A reappraisal of CTLA-4 checkpoint blockade in cancer immunotherapy. *Cell Res.* **2018**, *28*, 416–432. [CrossRef]
76. Ahmadzada, T.; Reid, G.; McKenzie, D.R. Fundamentals of siRNA and miRNA therapeutics and a review of targeted nanoparticle delivery systems in breast cancer. *Biophys. Rev.* **2018**, *10*, 69–86. [CrossRef]
77. Mirza, Z.; Karim, S. Nanoparticles-based drug delivery and gene therapy for breast cancer: Recent advancements and future challenges. *Semin. Cancer Biol.* **2021**, *69*, 226–237. [CrossRef]
78. Liu, Y.; Zhu, Y.H.; Mao, C.Q.; Dou, S.; Shen, S.; Tan, Z.B.; Wang, J. Triple negative breast cancer therapy with CDK1 siRNA delivered by cationic lipid assisted PEG-PLA nanoparticles. *J. Control. Rel.* **2014**, *192*, 114–121. [CrossRef]
79. Alshaer, W.; Hillaireau, H.; Vergnaud, J.; Mura, S.; Deloménie, C.; Sauvage, F.; Ismail, S.; Fattal, E. Aptamer-guided siRNA-loaded nanomedicines for systemic gene silencing in CD-44 expressing murine triple-negative breast cancer model. *J. Control. Rel.* **2018**, *271*, 98–106. [CrossRef]
80. Tang, J.; Howard, C.B.; Mahler, S.M.; Thurecht, K.J.; Huang, L.; Xu, Z.P. Enhanced delivery of siRNA to triple negative breast cancer cells in vitro and in vivo through functionalizing lipid-coated calcium phosphate nanoparticles with dual target ligands. *Nanoscale* **2018**, *10*, 4258–4266. [CrossRef]
81. Qattan, A. Novel mirna targets and therapies in the triple-negative breast cancer microenvironment: An emerging hope for a challenging disease. *Int. J. Mol. Sci.* **2020**, *21*, 8905. [CrossRef]
82. Zhu, H.; Dai, M.; Chen, X.; Chen, X.; Qin, S.; Dai, S. Integrated analysis of the potential roles of miRNA-mRNA networks in triple negative breast cancer. *Mol. Med. Rep.* **2017**, *16*, 1139–1146. [CrossRef] [PubMed]
83. Bhargava-Shah, A.; Foygel, K.; Devulapally, R.; Paulmurugan, R. Orlistat and antisense-miRNA-loaded PLGA-PEG nanoparticles for enhanced triple negative breast cancer therapy. *Nanomedicine* **2016**, *11*, 235–247. [CrossRef] [PubMed]
84. Dobson, J.; de Queiroz, G.F.; Golding, J.P. Photodynamic therapy and diagnosis: Principles and comparative aspects. *Vet. J.* **2018**, *233*, 8–18. [CrossRef] [PubMed]
85. Fernandes, S.R.G.; Fernandes, R.; Sarmento, B.; Pereira, P.M.R.; Tomé, J.P.C. Photoimmunoconjugates: Novel synthetic strategies to target and treat cancer by photodynamic therapy. *Org. Biomol. Chem.* **2019**, *17*, 2579–2593. [CrossRef] [PubMed]
86. Nitheesh, Y.; Pradhan, R.; Hejmady, S.; Taliyan, R.; Singhvi, G.; Alexander, A.; Kesharwani, P.; Dubey, S.K. Surface engineered nanocarriers for the management of breast cancer. *Mater. Sci. Eng. C* **2021**, *130*, 112441. [CrossRef]
87. Sivasubramanian, M.; Chuang, Y.C.; Lo, L.W. Evolution of nanoparticle-mediated photodynamic therapy: From superficial to deep-seated cancers. *Molecules* **2019**, *24*, 520. [CrossRef]
88. Banerjee, S.M.; Macrobert, A.J.; Mosse, C.A.; Periera, B.; Bown, S.G.; Keshtgar, M.R.S. Photodynamic therapy: Inception to application in breast cancer. *Breast* **2017**, *31*, 105–113. [CrossRef]
89. Dos Santos, A.F.; De Almeida, D.R.Q.; Terra, L.F.; Baptista, M.S.; Labriola, L. Photodynamic therapy in cancer treatment–An update review. *J. Cancer Metastasis Treat.* **2019**, *5*, 25. [CrossRef]
90. Van Straten, D.; Mashayekhi, V.; De Bruijn, H.S.; Oliveira, S.; Robinson, D.J. Oncologic Photodynamic Therapy: Basic Principles, Current Clinical Status and Future Directions. *Cancers* **2017**, *9*, 19. [CrossRef]
91. Kwiatkowski, S.; Knap, B.; Przystupski, D.; Saczko, J.; Kędzierska, E.; Knap-Czop, K.; Kotlińska, J.; Michel, O.; Kotowski, K.; Kulbacka, J. Photodynamic therapy—mechanisms, photosensitizers and combinations. *Biomed. Pharmacother.* **2018**, *106*, 1098–1107. [CrossRef]
92. Castano, A.P.; Demidova, T.N.; Hamblin, M.R. Mechanisms in photodynamic therapy: Part one—Photosensitizers, photochemistry and cellular localization. *Photodiagnosis Photodyn. Ther.* **2004**, *1*, 279–293. [CrossRef]

93. Castano, A.P.; Mroz, P.; Hamblin, M.R. Photodynamic therapy and anti-tumour immunity. *Nat. Rev. Cancer* **2006**, *6*, 535–545. [CrossRef]
94. Shemesh, C.S.; Hardy, C.W.; Yu, D.S.; Fernandez, B.; Zhang, H. Indocyanine green loaded liposome nanocarriers for photodynamic therapy using human triple negative breast cancer cells. *Photodiagnosis Photodyn. Ther.* **2014**, *11*, 193–203. [CrossRef]
95. Sun, S.; Xu, Y.; Fu, P.; Chen, M.; Sun, S.; Zhao, R.; Wang, J.; Liang, X.; Wang, S. Ultrasound-targeted photodynamic and gene dual therapy for effectively inhibiting triple negative breast cancer by cationic porphyrin lipid microbubbles loaded with HIF1α-siRNA. *Nanoscale* **2018**, *10*, 19945–19956. [CrossRef]
96. Zhao, L.; Zhang, X.; Wang, X.; Guan, X.; Zhang, W.; Ma, J. Recent advances in selective photothermal therapy of tumor. *J. Nanobiotechnol.* **2021**, *19*, 1–15. [CrossRef]
97. Gao, G.; Jiang, Y.W.; Guo, Y.; Jia, H.R.; Cheng, X.; Deng, Y.; Yu, X.W.; Zhu, Y.X.; Guo, H.Y.; Sun, W.; et al. Enzyme-Mediated Tumor Starvation and Phototherapy Enhance Mild-Temperature Photothermal Therapy. *Adv. Funct. Mater.* **2020**, *30*, 1–13. [CrossRef]
98. Valcourt, D.M.; Dang, M.N.; Day, E.S. IR820-loaded PLGA nanoparticles for photothermal therapy of triple-negative breast cancer. *J. Biomed. Mater. Res. Part A* **2019**, *107*, 1702–1712. [CrossRef]
99. Zhao, S.; Tian, Y.; Liu, W.; Su, Y.; Zhang, Y.; Teng, Z.; Zhao, Y.; Wang, S.; Lu, G.; Yu, Z. High and low molecular weight hyaluronic acid-coated gold nanobipyramids for photothermal therapy. *RSC Adv.* **2018**, *8*, 9023–9030. [CrossRef]
100. Zhang, M.; Kim, H.S.; Jin, T.; Woo, J.; Piao, Y.J.; Moon, W.K. Near-infrared photothermal therapy using anti-EGFR-gold nanorod conjugates for triple negative breast cancer. *Oncotarget* **2017**, *8*, 86566–86575. [CrossRef]
101. Wang, S.; Tian, Y.; Tian, W.; Sun, J.; Zhao, S.; Liu, Y.; Wang, C.; Tang, Y.; Ma, X.; Teng, Z.; et al. Selectively Sensitizing Malignant Cells to Photothermal Therapy Using a CD44-Targeting Heat Shock Protein 72 Depletion Nanosystem. *ACS Nano* **2016**, *10*, 8578–8590. [CrossRef]
102. Tian, Y.; Lei, M. Polydopamine-Based Composite Nanoparticles with Redox-Labile Polymer Shells for Controlled Drug Release and Enhanced Chemo-Photothermal Therapy. *Nanoscale Res. Lett.* **2019**, *14*, 1–14. [CrossRef] [PubMed]
103. Rosenthal, I.; Sostaric, J.Z.; Riesz, P. Sonodynamic therapya review of the synergistic effects of drugs and ultrasound. *Ultrason. Sonochem.* **2004**, *11*, 349–363. [CrossRef] [PubMed]
104. Carovac, A.; Smajlovic, F.; Junuzovic, D. Application of Ultrasound in Medicine. *Acta Inf. Med.* **2011**, *19*, 168. [CrossRef] [PubMed]
105. Han, X.; Song, Z.; Zhou, Y.; Zhang, Y.; Deng, Y.; Qin, J.; Zhang, T.; Jiang, Z. Mitochondria-targeted high-load sound-sensitive micelles for sonodynamic therapy to treat triple-negative breast cancer and inhibit metastasis. *Mater. Sci. Eng. C* **2021**, *124*, 112054. [CrossRef] [PubMed]
106. Li, Y.S.; Reid, C.N.; McHale, A.P. Enhancing ultrasound-mediated cell membrane permeabilisation (sonoporation) using a high frequency pulse regime and implications for ultrasound-aided cancer chemotherapy. *Cancer Lett.* **2008**, *266*, 156–162. [CrossRef] [PubMed]
107. Nomikou, N.; Li, Y.S.; McHale, A.P. Ultrasound-enhanced drug dispersion through solid tumours and its possible role in aiding ultrasound-targeted cancer chemotherapy. *Cancer Lett.* **2010**, *288*, 94–98. [CrossRef]
108. Feng, X.; Wu, C.; Yang, W.; Wu, J.; Wang, P. Mechanism-Based Sonodynamic–Chemo Combinations against Triple-Negative Breast Cancer. *Int. J. Mol. Sci.* **2022**, *23*, 7981. [CrossRef]
109. Chen, H.; Liu, L.; Ma, A.; Yin, T.; Chen, Z.; Liang, R.; Qiu, Y.; Zheng, M.; Cai, L. Noninvasively immunogenic sonodynamic therapy with manganese protoporphyrin liposomes against triple-negative breast cancer. *Biomaterials* **2021**, *269*, 120639. [CrossRef]
110. Thakur, V.; Kutty, R.V. Recent advances in nanotheranostics for triple negative breast cancer treatment. *J. Exp. Clin. Cancer Res.* **2019**, *38*, 1–22. [CrossRef]
111. El-Sahli, S.; Hua, K.; Sulaiman, A.; Chambers, J.; Li, L.; Farah, E.; McGarry, S.; Liu, D.; Zheng, P.; Lee, S.-H.; et al. A triple-drug nanotherapy to target breast cancer cells, cancer stem cells, and tumor vasculature. *Cell Death Dis.* **2021**, *12*, 8. [CrossRef]
112. Mukherjee, A.; Waters, A.K.; Kalyan, P.; Achrol, A.S.; Kesari, S.; Yenugonda, V.M. Lipid-polymer hybrid nanoparticles as a next-generation drug delivery plkatform: State of the art, emerging technologies, and perspectives. *Int. J. Nanomed.* **2019**, *14*, 1937–1952. [CrossRef]
113. Tran, S.; DeGiovanni, P.; Piel, B.; Rai, P. Cancer nanomedicine: A review of recent success in drug delivery. *Clin. Transl. Med.* **2017**, *6*, 1–21. [CrossRef]
114. Teles, R.H.G.; Moralles, H.F.; Cominetti, M.R. Global trends in nanomedicine research on triple negative breast cancer: A bibliometric analysis. *Int. J. Nanomed.* **2018**, *13*, 2321–2336. [CrossRef]
115. Saeed, N.A.; Hamzah, I.H.; Mahmood, S.I. The applications of nano-medicine in the breast cancer therapy. *J. Phys. Conf. Ser.* **2021**, *1853*, 012061. [CrossRef]
116. Fang, F.; Li, M.; Zhang, J.; Lee, C.S. Different Strategies for Organic Nanoparticle Preparation in Biomedicine. *ACS Mater. Lett.* **2020**, *2*, 531–549. [CrossRef]
117. Romero, G.; Moya, S.E. Synthesis of organic nanoparticles. *Front. Nanosci.* **2012**, *4*, 115–141. [CrossRef]
118. Mitragotri, S.; Stayton, P. Organic nanoparticles for drug delivery and imaging. *MRS Bull.* **2014**, *39*, 219–223. [CrossRef]
119. Chaudhuri, A.; Kumar, D.N.; Shaik, R.A.; Eid, B.G.; Abdel-Naim, A.B.; Md, S.; Ahmad, A.; Agrawal, A.K. Lipid-Based Nanoparticles as a Pivotal Delivery Approach in Triple Negative Breast Cancer (TNBC) Therapy. *Int. J. Mol. Sci.* **2022**, *23*, 10068. [CrossRef]
120. Shi, J.; Kantoff, P.W.; Wooster, R.; Farokhzad, O.C. Cancer nanomedicine: Progress, challenges and opportunities. *Nat. Rev. Cancer* **2017**, *17*, 20–37. [CrossRef]

121. Bourquin, J.; Milosevic, A.; Hauser, D.; Lehner, R.; Blank, F.; Petri-Fink, A.; Rothen-Rutishauser, B. Biodistribution, Clearance, and Long-Term Fate of Clinically Relevant Nanomaterials. *Adv. Mater.* **2018**, *30*, 1704307. [CrossRef]
122. Bozzuto, G.; Molinari, A. Liposomes as nanomedical devices. *Int. J. Nanomed.* **2015**, *10*, 975–999. [CrossRef] [PubMed]
123. Zamani, P.; Momtazi-Borojeni, A.A.; Nik, M.E.; Oskuee, R.K.; Sahebkar, A. Nanoliposomes as the adjuvant delivery systems in cancer immunotherapy. *J. Cell Physiol.* **2018**, *233*, 5189–5199. [CrossRef] [PubMed]
124. Gonda, A.; Zhao, N.; Shah, J.V.; Calvelli, H.R.; Kantamneni, H.; Francis, N.L.; Ganapathy, V. Engineering Tumor-Targeting Nanoparticles as Vehicles for Precision Nanomedicine. *Med. One* **2019**, *4*, e190021. [PubMed]
125. Beltrán-Gracia, E.; López-Camacho, A.; Higuera-Ciapara, I.; Velázquez-Fernández, J.B.; Vallejo-Cardona, A.A. Nanomedicine review: Clinical developments in liposomal applications. *Cancer Nanotechnol.* **2019**, *10*, 11. [CrossRef]
126. Immordino, M.L.; Dosio, F.; Cattel, L. Stealth liposomes: Review of the basic science, rationale, and clinical applications, existing and potential. *Int. J. Nanomed.* **2006**, *1*, 297–315.
127. Si, Y.; Zhang, Y.; Ngo, H.G.; Guan, J.-S.; Chen, K.; Wang, Q.; Singh, A.P.; Xu, Y.; Zhou, L.; Yang, E.S.; et al. Targeted Liposomal Chemotherapies to Treat Triple-Negative Breast Cancer. *Cancers* **2021**, *13*, 3749. [CrossRef]
128. Maji, I.; Mahajan, S.; Sriram, A.; Medtiya, P.; Vasave, R.; Khatri, D.K.; Kumar, R.; Singh, S.B.; Madan, J.; Singh, P.K. Solid self emulsifying drug delivery system: Superior mode for oral delivery of hydrophobic cargos. *J. Control. Release* **2021**, *337*, 646–660. [CrossRef]
129. Gursoy, R.N.; Benita, S. Self-emulsifying drug delivery systems (SEDDS) for improved oral delivery of lipophilic drugs. *Biomed Pharmacother.* **2004**, *58*, 173–182. [CrossRef]
130. Valicherla, G.R.; Dave, K.M.; Syed, A.A.; Riyazuddin, M.; Gupta, A.P.; Singh, A.; Wahajuddin; Mitra, K.; Datta, D.; Gayen, J.R. Formulation optimization of Docetaxel loaded self-emulsifying drug delivery system to enhance bioavailability and anti-tumor activity. *Sci. Rep.* **2016**, *6*, 1–11. [CrossRef]
131. Shrivastava, N.; Parikh, A.; Dewangan, R.P.; Biswas, L.; Verma, A.K.; Mittal, S.; Ali, J.; Garg, S.; Baboota, S. Solid Self-Nano Emulsifying Nanoplatform Loaded with Tamoxifen and Resveratrol for Treatment of Breast Cancer. *Pharmaceutics* **2022**, *14*, 1486. [CrossRef]
132. Timur, S.S.; Yöyen-Ermiş, D.; Esendağlı, G.; Yonat, S.; Horzum, U.; Esendağlı, G.; Gürsoy, R.N. Efficacy of a novel LyP-1-containing self-microemulsifying drug delivery system (SMEDDS) for active targeting to breast cancer. *Eur. J. Pharm. Biopharm.* **2019**, *136*, 138–146. [CrossRef]
133. Scioli Montoto, S.; Muraca, G.; Ruiz, M.E. Solid Lipid Nanoparticles for Drug Delivery: Pharmacological and Biopharmaceutical Aspects. *Front. Mol. Biosci.* **2020**, *7*, 1–24. [CrossRef]
134. Kothari, I.R.; Mazumdar, S.; Sharma, S.; Italiya, K.; Mittal, A.; Chitkara, D. Docetaxel and alpha-lipoic acid co-loaded nanoparticles for cancer therapy. *Ther. Deliv.* **2019**, *10*, 227–240. [CrossRef]
135. Ss Pindiprolu, S.K.; Krishnamurthy, P.T.; Ghanta, V.R.; Chintamaneni, P.K. Phenyl boronic acid-modified lipid nanocarriers of niclosamide for targeting triple-negative breast cancer. *Nanomedicine* **2020**, *15*, 1551–1565. [CrossRef]
136. Garg, J.; Pathania, K.; Sah, S.P.; Pawar, S.V. Nanostructured lipid carriers: A promising drug carrier for targeting brain tumours. *Futur. J. Pharm. Sci.* **2022**, *8*, 1–31. [CrossRef]
137. Haider, M.; Abdin, S.M.; Kamal, L.; Orive, G. Nanostructured lipid carriers for delivery of chemotherapeutics: A review. *Pharmaceutics* **2020**, *12*, 288. [CrossRef]
138. Andey, T.; Sudhakar, G.; Marepally, S.; Patel, A.; Banerjee, R.; Singh, M. Lipid nanocarriers of a lipid-conjugated estrogenic derivative inhibit tumor growth and enhance Cisplatin activity against triple-negative breast cancer: Pharmacokinetic and efficacy evaluation. *Mol. Pharm.* **2015**, *12*, 1105–1120. [CrossRef]
139. Kutty, R.V.; Feng, S.-S. Cetuximab conjugated vitamin E TPGS micelles for targeted delivery of docetaxel for treatment of triple negative breast cancers. *Biomaterials* **2013**, *34*, 10160–10171. [CrossRef]
140. Selestin Raja, I.; Thangam, R.; Fathima, N.N. Polymeric Micelle of a Gelatin-Oleylamine Conjugate: A Prominent Drug Delivery Carrier for Treating Triple Negative Breast Cancer Cells. *ACS Appl. Bio. Mater.* **2018**, *1*, 1725–1734. [CrossRef]
141. Wang, Y.; Wang, Y.; Chen, G.; Li, Y.; Xu, W.; Gong, S. Quantum-Dot-Based Theranostic Micelles Conjugated with an Anti-EGFR Nanobody for Triple-Negative Breast Cancer Therapy. *ACS Appl. Mater. Interfaces* **2017**, *9*, 30297–30305. [CrossRef]
142. Zhang, Y.; Zhu, X.; Chen, X.; Chen, Q.; Zhou, W.; Guo, Q.; Lu, Y.; Li, C.; Zhang, Y.; Liang, D.; et al. Activated Platelets-Targeting Micelles with Controlled Drug Release for Effective Treatment of Primary and Metastatic Triple Negative Breast Cancer. *Adv. Funct. Mater.* **2019**, *29*, 1806620. [CrossRef]
143. Dubey, S.K.; Salunkhe, S.; Agrawal, M.; Kali, M.; Singhvi, G.; Tiwari, S.; Saraf, S.; Saraf, S.; Alexander, A. Understanding the Pharmaceutical Aspects of Dendrimers for the Delivery of Anticancer Drugs. *Curr. Drug. Targets* **2020**, *21*, 528–540. [CrossRef]
144. Jang, W.-D.; Selim, K.; Lee, C.-H.; Kang, I.-K. Bioinspired application of dendrimers: From bio-mimicry to biomedical applications. *Prog. Polym. Sci.* **2009**, *34*, 1–23. [CrossRef]
145. Lin, Q.; Jiang, G.; Tong, K. Dendrimers in Drug-Delivery Applications. *Des. Monomers Polym.* **2012**, *13*, 301–324. [CrossRef]
146. Dubey, S.K.; Kali, M.; Hejmady, S.; Saha, R.N.; Alexander, A.; Kesharwani, P. Recent advances of dendrimers as multifunctional nano-carriers to combat breast cancer. *Eur. J. Pharm. Sci. Off J. Eur. Fed. Pharm. Sci.* **2021**, *164*, 105890. [CrossRef] [PubMed]
147. Ghosh, S.; Ghosal, K.; Mohammad, S.A.; Sarkar, K. Dendrimer functionalized carbon quantum dot for selective detection of breast cancer and gene therapy. *Chem. Eng. J.* **2019**, *373*, 468–484. [CrossRef]

148. Jain, A.; Mahira, S.; Majoral, J.-P.; Bryszewska, M.; Khan, W.; Ionov, M. Dendrimer mediated targeting of siRNA against polo-like kinase for the treatment of triple negative breast cancer. *J. Biomed. Mater Res. A* **2019**, *107*, 1933–1944. [CrossRef]
149. Liu, C.; Gao, H.; Zhao, Z.; Rostami, I.; Wang, C.; Zhu, L.; Yang, Y. Improved tumor targeting and penetration by a dual-functional poly(amidoamine) dendrimer for the therapy of triple-negative breast cancer. *J. Mater. Chem. B.* **2019**, *7*, 3724–3736. [CrossRef]
150. Finlay, J.; Roberts, C.M.; Lowe, G.; Loeza, J.; Rossi, J.J.; Glackin, C.A. RNA-based TWIST1 inhibition via dendrimer complex to reduce breast cancer cell metastasis. *BioMed Res Int.* **2015**, *2015*, 382745. [CrossRef]
151. Surekha, B.; Kommana, N.S.; Dubey, S.K.; Kumar, A.V.P.; Shukla, R.; Kesharwani, P. PAMAM dendrimer as a talented multifunctional biomimetic nanocarrier for cancer diagnosis and therapy. *Colloids Surf. B Biointerfaces* **2021**, *204*, 111837. [CrossRef]
152. Mogoşanu, G.D.; Grumezescu, A.M.; Bejenaru, C.; Bejenaru, L.E. Polymeric protective agents for nanoparticles in drug delivery and targeting. *Int. J. Pharm.* **2016**, *510*, 419–429. [CrossRef]
153. van Vlerken, L.E.; Vyas, T.K.; Amiji, M.M. Poly(ethylene glycol)-modified Nanocarriers for Tumor-targeted and Intracellular Delivery. *Pharm. Res.* **2007**, *24*, 1405–1414. [CrossRef]
154. Li, B.; Li, Q.; Mo, J.; Dai, H. Drug-Loaded Polymeric Nanoparticles for Cancer Stem Cell Targeting. *Front. Pharmacol.* **2017**, *8*, 51. [CrossRef]
155. Fortuni, B.; Inose, T.; Ricci, M.; Fujita, Y.; Van Zundert, I.; Masuhara, A.; Fron, E.; Mizuno, H.; Latterini, L.; Rocha, S.; et al. Polymeric Engineering of Nanoparticles for Highly Efficient Multifunctional Drug Delivery Systems. *Sci. Rep.* **2019**, *9*, 2666. [CrossRef]
156. Mamnoon, B.; Loganathan, J.; Confeld, M.I.; De Fonseka, N.; Feng, L.; Froberg, J.; Choi, Y.; Tuvin, D.M.; Sathish, V.; Mallik, S. Targeted polymeric nanoparticles for drug delivery to hypoxic, triple-negative breast tumors. *ACS Appl. Bio. Mater.* **2021**, *4*, 1450–1460. [CrossRef]
157. Zhou, Y.; Chen, D.; Xue, G.; Yu, S.; Yuan, C.; Huang, M.; Jiang, L. Improved therapeutic efficacy of quercetin-loaded polymeric nanoparticles on triple-negative breast cancer by inhibiting uPA. *RSC Adv.* **2020**, *10*, 34517–34526. [CrossRef]
158. Devulapally, R.; Sekar, N.M.; Sekar, T.V.; Foygel, K.; Massoud, T.F.; Willmann, J.K.; Paulmurugan, R. Polymer nanoparticles mediated codelivery of antimiR-10b and antimiR-21 for achieving triple negative breast cancer therapy. *ACS Nano* **2015**, *9*, 2290–2302. [CrossRef]
159. Deng, Z.J.; Morton, S.W.; Ben-Akiva, E.; Dreaden, E.C.; Shopsowitz, K.E.; Hammond, P.T. Layer-by-layer nanoparticles for systemic codelivery of an anticancer drug and siRNA for potential triple-negative breast cancer treatment. *ACS Nano* **2013**, *7*, 9571–9584. [CrossRef]
160. Shi, P.; Aluri, S.; Lin, Y.-A.; Shah, M.; Edman, M.; Dhandhukia, J.; Cui, H.; MacKay, J.A. Elastin-based protein polymer nanoparticles carrying drug at both corona and core suppress tumor growth in vivo. *J. Control Release* **2013**, *171*, 330–338. [CrossRef]
161. Jain, V.; Kumar, H.; Anod, H.V.; Chand, P.; Gupta, N.V.; Dey, S.; Kesharwani, S.S. A review of nanotechnology-based approaches for breast cancer and triple-negative breast cancer. *J. Control Release* **2020**, *326*, 628–647. [CrossRef]
162. Koleva, L.; Bovt, E.; Ataullakhanov, F.; Sinauridze, E. Erythrocytes as carriers: From drug delivery to biosensors. *Pharmaceutics* **2020**, *12*, 276. [CrossRef] [PubMed]
163. Wang, C.; Wang, M.; Zhang, Y.; Jia, H.; Chen, B. Cyclic arginine-glycine-aspartic acid-modified red blood cells for drug delivery: Synthesis and in vitro evaluation. *J. Pharm. Anal.* **2022**, *12*, 324–331. [CrossRef] [PubMed]
164. Chen, Y.; Zhang, Y. Application of the CRISPR/Cas9 System to Drug Resistance in Breast Cancer. *Adv. Sci.* **2018**, *5*, 1700964. [CrossRef] [PubMed]
165. Farheen, J.; Hosmane, N.S.; Zhao, R.; Zhao, Q.; Iqbal, M.Z.; Kong, X. Nanomaterial-assisted CRISPR gene-engineering—A hallmark for triple-negative breast cancer therapeutics advancement. *Mater Today Bio.* **2022**, *16*, 100450. [CrossRef] [PubMed]
166. Zhang, L.; Wang, P.; Feng, Q.; Wang, N.; Chen, Z.; Huang, Y.; Zheng, W.; Jiang, X. Lipid nanoparticle-mediated efficient delivery of CRISPR/Cas9 for tumor therapy. *NPG Asia Mater.* **2017**, *9*, 3–10. [CrossRef]
167. Tian, Y.; Li, S.; Song, J.; Ji, T.; Zhu, M.; Anderson, G.J.; Wei, J.; Nie, G. A doxorubicin delivery platform using engineered natural membrane vesicle exosomes for targeted tumor therapy. *Biomaterials* **2014**, *35*, 2383–2390. [CrossRef]
168. Singh, R.; Pochampally, R.; Watabe, K.; Lu, Z.; Mo, Y.Y. Exosome-mediated transfer of miR-10b promotes cell invasion in breast cancer. *Mol. Cancer* **2014**, *13*, 1–11. [CrossRef]
169. Prodana, M.; Ionita, D.; Ungureanu, C.; Bojin, D.; Demetrescu, I. Enhancing Antibacterial Effect of Multiwalled Carbon Nanotubes using Silver Nanoparticles. *Dig. J. Nanomater Biostruct.* **2011**, *6*, 549–556.
170. Badea, M.A.; Prodana, M.; Dinischiotu, A.; Crihana, C.; Ionita, D.; Balas, M. Cisplatin Loaded Multiwalled Carbon Nanotubes Induce Resistance in Triple Negative Breast Cancer Cells. *Pharmaceutics* **2018**, *10*, 228. [CrossRef]
171. Fahrenholtz, C.; Ding, S.; Bernish, B.; Wright, M.; Bierbach, U.; Singh, R. Abstract B05: Self-assembling platinum-acridine loaded carbon nanotubes for triple-negative breast cancer chemotherapy. *Mol. Cancer Res.* **2016**, *14*, B05. [CrossRef]
172. Singhai, N.J.; Maheshwari, R.; Ramteke, S. CD44 receptor targeted 'smart' multi-walled carbon nanotubes for synergistic therapy of triple-negative breast cancer. *Colloid Interface Sci. Commun.* **2020**, *35*, 100235. [CrossRef]
173. Kim, D.; Kim, J.; Park, Y.I.; Lee, N.; Hyeon, T. Recent Development of Inorganic Nanoparticles for Biomedical Imaging. *ACS Cent Sci.* **2018**, *4*, 324–336. [CrossRef]
174. Anselmo, A.C.; Mitragotri, S. A Review of Clinical Translation of Inorganic Nanoparticles. *AAPS J.* **2015**, *17*, 1041–1054. [CrossRef]

175. Croissant, J.G.; Butler, K.S.; Zink, J.I.; Brinker, C.J. Synthetic amorphous silica nanoparticles: Toxicity, biomedical and environmental implications. *Nat. Rev. Mater.* **2020**, *5*, 886–909. [CrossRef]
176. Cheng, Y.; Chen, Q.; Guo, Z.; Li, M.; Yang, X.; Wan, G.; Chen, H.; Zhang, Q.; Wang, Y. An Intelligent Biomimetic Nanoplatform for Holistic Treatment of Metastatic Triple-Negative Breast Cancer via Photothermal Ablation and Immune Remodeling. *ACS Nano* **2020**, *14*, 15161–15181. [CrossRef]
177. Zhang, T.; Liu, H.; Li, L.; Guo, Z.; Song, J.; Yang, X.; Wan, G.; Li, R.; Wang, Y. Leukocyte/platelet hybrid membrane-camouflaged dendritic large pore mesoporous silica nanoparticles co-loaded with photo/chemotherapeutic agents for triple negative breast cancer combination treatment. *Bioact. Mater.* **2021**, *6*, 3865–3878. [CrossRef]
178. Wu, X.; Han, Z.; Schur, R.M.; Lu, Z.-R. Targeted Mesoporous Silica Nanoparticles Delivering Arsenic Trioxide with Environment Sensitive Drug Release for Effective Treatment of Triple Negative Breast Cancer. *ACS Biomater. Sci. Eng.* **2016**, *2*, 501–507. [CrossRef]
179. Siddiqi, K.S.; Husen, A.; Rao, R.A.K. A review on biosynthesis of silver nanoparticles and their biocidal properties. *J. Nanobiotechnol.* **2018**, *16*, 14. [CrossRef]
180. Lee, S.H.; Jun, B.-H. Silver Nanoparticles: Synthesis and Application for Nanomedicine. *Int. J. Mol. Sci.* **2019**, *20*, 865. [CrossRef]
181. Azizi, M.; Ghourchian, H.; Yazdian, F.; Bagherifam, S.; Bekhradnia, S.; Nyström, B. Anti-cancerous effect of albumin coated silver nanoparticles on MDA-MB 231 human breast cancer cell line. *Sci. Rep.* **2017**, *7*, 5178. [CrossRef]
182. Swanner, J.; Fahrenholtz, C.; Tenvooren, I.; Bernish, B.; Sears, J.; Hooker, A.; Furdui, C.; Alli, E.; Li, W.; Donati, G.; et al. Silver nanoparticles selectively treat triple negative breast cancer cells without affecting non-malignant breast epithelial cells in vitro and in vivo. *FASEB BioAdv.* **2019**, *10*, 639–660. [CrossRef] [PubMed]
183. Sears, J.J. Nanoparticle Based Multi-Modal Therapies Against Triple Negative Breast Cancer. Ph.D. Thesis, Wake Forest University, Winston-Salem, NC, USA, 2018.
184. Surapaneni, S.K.; Bashir, S.; Tikoo, K. Gold nanoparticles-induced cytotoxicity in triple negative breast cancer involves different epigenetic alterations depending upon the surface charge. *Sci. Rep.* **2018**, *8*, 12295. [CrossRef] [PubMed]
185. Castilho, M.L.; Jesus, V.P.S.; Vieira, P.F.A.; Hewitt, K.C.; Raniero, L. Chlorin e6-EGF conjugated gold nanoparticles as a nanomedicine based therapeutic agent for triple negative breast cancer. *Photodiagn. Photodyn Ther.* **2021**, *33*, 102186. [CrossRef] [PubMed]
186. Madni, A.; Tahir, N.; Rehman, M.; Raza, A.; Mahmood, M.A.; Khan, M.I.; Kashif, P.M. Hybrid Nano-carriers for Potential Drug Delivery. *Adv. Technol. Deliv. Ther.* **2017**, *2017*, 53–87.
187. Xia, T.; Kovochich, M.; Liong, M.; Meng, H.; Kabehie, S.; George, J.I.; Zink, S.; Nel, A.E. Polyethyleneimine Coating Enhances the Cellular Uptake of Mesoporous Silica Nanoparticles and Allows Safe Delivery of siRNA and DNA Constructs. *ACS Nano* **2009**, *3*, 3273–3286. [CrossRef]
188. Ahir, M.; Upadhyay, P.; Ghosh, A.; Sarker, S.; Bhattacharya, S.; Gupta, P.; Ghosh, S.; Chattopadhyay, S.; Adhikary, A. Delivery of dual miRNA through CD44-targeted mesoporous silica nanoparticles for enhanced and effective triple-negative breast cancer therapy. *Biomater. Sci.* **2020**, *8*, 2939–2954. [CrossRef]
189. Laha, D.; Pal, K.; Chowdhuri, A.R.; Parida, P.K.; Sahu, S.K.; Jana, K.; Karmakar, P. Fabrication of curcumin-loaded folic acid-tagged metal organic framework for triple negative breast cancer therapy in in vitro and in vivo systems. *New J. Chem.* **2019**, *43*, 217–229. [CrossRef]
190. Bhardwaj, A.; Kumar, L.; Mehta, S.; Mehta, A. Stimuli-sensitive systems—An emerging delivery system for drugs. *Artif. Cells Nanomed. Biotechnol.* **2015**, *43*, 299–310. [CrossRef]
191. Yu, J.; Chu, X.; Hou, Y. Stimuli-responsive cancer therapy based on nanoparticles. *Chem. Commun.* **2014**, *50*, 11614–11630. [CrossRef]
192. Nakayama, M.; Okano, T.; Miyazaki, T.; Kohori, F.; Sakai, K.; Yokoyama, M. Molecular design of biodegradable polymeric micelles for temperature-responsive drug release. *J. Control Release* **2006**, *115*, 46–56. [CrossRef]
193. Ganta, S.; Devalapally, H.; Shahiwala, A.; Amiji, M. A review of stimuli-responsive nanocarriers for drug and gene delivery. *J. Control Release* **2008**, *126*, 187–204. [CrossRef]
194. Fleige, E.; Quadir, M.A.; Haag, R. Stimuli-responsive polymeric nanocarriers for the controlled transport of active compounds: Concepts and applications. *Adv. Drug Deliv. Rev.* **2012**, *64*, 866–884. [CrossRef]
195. Cheng, R.; Meng, F.; Deng, C.; Klok, H.-A.; Zhong, Z. Dual and multi-stimuli responsive polymeric nanoparticles for programmed site-specific drug delivery. *Biomaterials* **2013**, *34*, 3647–3657. [CrossRef]
196. Torchilin, V.P. Multifunctional, stimuli-sensitive nanoparticulate systems for drug delivery. *Nat. Rev. Drug Discov.* **2014**, *13*, 813–827. [CrossRef]
197. Patil, R.; Portilla-Arias, J.; Ding, H.; Konda, B.; Rekechenetskiy, A.; Inoue, S.; Black, K.L.; Holler, E.; Ljubimova, J.Y. Cellular delivery of doxorubicin via pH-controlled hydrazone linkage using multifunctional nano vehicle based on poly(β-l-malic acid). *Int. J. Mol. Sci.* **2012**, *13*, 11681–11693. [CrossRef]
198. Cimen, Z.; Babadag, S.; Odabas, S.; Altuntas, S.; Demirel, G.; Demirel, G.B. Injectable and Self-Healable pH-Responsive Gelatin–PEG/Laponite Hybrid Hydrogels as Long-Acting Implants for Local Cancer Treatment. *ACS Appl. Polym. Mater.* **2021**, *3*, 3504–3518. [CrossRef]

199. Hwang, J.-H.; Choi, C.W.; Kim, H.-W.; Kim, D.H.; Kwak, T.W.; Lee, H.M.; hyun Kim, C.; Chung, C.W.; Jeong, Y.-I.; Kang, D.H. Dextran-b-poly (L-histidine) copolymer nanoparticles for pH-responsive drug delivery to tumor cells. *Int. J. Nanomed.* **2013**, *8*, 3197.
200. Min, K.H.; Kim, J.-H.; Bae, S.M.; Shin, H.; Kim, M.S.; Park, S.; Lee, H.; Park, R.-W.; Kim, I.-S.; Kim, K. Tumoral acidic pH-responsive MPEG-poly (β-amino ester) polymeric micelles for cancer targeting therapy. *J. Control. Release* **2010**, *144*, 259–266. [CrossRef]
201. Niu, S.; Williams, G.R.; Wu, J.; Wu, J.; Zhang, X.; Chen, X.; Li, S.; Jiao, J.; Zhu, L.-M. A chitosan-based cascade-responsive drug delivery system for triple-negative breast cancer therapy. *J. Nanobiotechnol.* **2019**, *17*, 95. [CrossRef]
202. He, X.; Zhang, J.; Li, C.; Zhang, Y.; Lu, Y.; Zhang, Y.; Liu, L.; Ruan, C.; Chen, Q.; Chen, X.; et al. Enhanced bioreduction-responsive diselenide-based dimeric prodrug nanoparticles for triple negative breast cancer therapy. *Theranostics* **2018**, *8*, 4884–4897. [CrossRef]
203. Zhang, J.; Zuo, T.; Liang, X.; Xu, Y.; Yang, Y.; Fang, T.; Li, J.; Chen, D.; Shen, Q. Fenton-reaction-stimulative nanoparticles decorated with a reactive-oxygen-species (ROS)-responsive molecular switch for ROS amplification and triple negative breast cancer therapy. *J. Mater. Chem. B* **2019**, *7*, 7141–7151. [CrossRef]
204. Radhakrishnan, K.; Tripathy, J.; Gnanadhas, D.P.; Chakravortty, D.; Raichur, A.M. Dual enzyme responsive and targeted nanocapsules for intracellular delivery of anticancer agents. *RSC Adv.* **2014**, *4*, 45961–45968. [CrossRef]
205. Wang, Y.; Li, B.; Xu, F.; Han, Z.; Wei, D.; Jia, D.; Zhou, Y. Tough Magnetic Chitosan Hydrogel Nanocomposites for Remotely Stimulated Drug Release. *Biomacromolecules* **2018**, *19*, 3351–3360. [CrossRef] [PubMed]
206. Thirunavukkarasu, G.K.; Cherukula, K.; Lee, H.; Jeong, Y.Y.; Park, I.-K.; Lee, J.Y. Magnetic field-inducible drug-eluting nanoparticles for image-guided thermo-chemotherapy. *Biomaterials* **2018**, *180*, 240–252. [CrossRef] [PubMed]
207. Xie, C.; Li, P.; Han, L.; Wang, Z.; Zhou, T.; Deng, W.; Wang, K.; Lu, X. Electroresponsive and cell-affinitive polydopamine/polypyrrole composite microcapsules with a dual-function of on-demand drug delivery and cell stimulation for electrical therapy. *NPG Asia Mater.* **2017**, *9*, e358. [CrossRef]
208. Xin, Y.; Qi, Q.; Mao, Z.; Zhan, X. PLGA nanoparticles introduction into mitoxantrone-loaded ultrasound-responsive liposomes: In vitro and in vivo investigations. *Int. J. Pharm.* **2017**, *528*, 47–54. [CrossRef] [PubMed]
209. Paris, J.L.; Manzano, M.; Cabañas, M.V.; Vallet-Regí, M. Mesoporous silica nanoparticles engineered for ultrasound-induced uptake by cancer cells. *Nanoscale* **2018**, *10*, 6402–6408. [CrossRef]
210. Schmaljohann, D. Thermo- and pH-responsive polymers in drug delivery. *Adv. Drug Deliv. Rev.* **2006**, *58*, 1655–1670. [CrossRef]
211. Liu, Y.; Wang, W.; Yang, J.; Zhou, C.; Sun, J. pH-sensitive polymeric micelles triggered drug release for extracellular and intracellular drug targeting delivery. *Asian J. Pharm. Sci.* **2013**, *8*, 159–167. [CrossRef]
212. Gao, W.; Chan, J.M.; Farokhzad, O.C. pH-responsive nanoparticles for drug delivery. *Mol. Pharm.* **2010**, *7*, 1913–1920. [CrossRef]
213. Dutz, S.; Hergt, R. Magnetic nanoparticle heating and heat transfer on a microscale: Basic principles, realities and physical limitations of hyperthermia for tumour therapy. *Int. J. Hyperth. Off. J. Eur. Soc. Hyperth. Oncol. N. Am. Hyperth. Gr.* **2013**, *29*, 790–800. [CrossRef]
214. Ortega, D.; Pankhurst, Q.A. Magnetic hyperthermia. In *Nanoscience: Volume 1: Nanostructures through Chemistry*; The Royal Society of Chemistry: London, UK, 2013; pp. 60–88.
215. Anyarambhatla, G.R.; Needham, D. Enhancement of the Phase Transition Permeability of DPPC Liposomes by Incorporation of MPPC: A New Temperature-Sensitive Liposome for use with Mild Hyperthermia. *J. Liposome Res.* **1999**, *9*, 491–506. [CrossRef]
216. Needham, D.; Anyarambhatla, G.; Kong, G.; Dewhirst, M.W. A new temperature-sensitive liposome for use with mild hyperthermia: Characterization and testing in a human tumor xenograft model. *Cancer Res.* **2000**, *60*, 1197–1201.
217. Li, R.; Wu, W.; Liu, Q.; Wu, P.; Xie, L.; Zhu, Z.; Yang, M.; Qian, X.; Ding, Y.; Yu, L.; et al. Intelligently targeted drug delivery and enhanced antitumor effect by gelatinase-responsive nanoparticles. *PLoS ONE* **2013**, *8*, e69643. [CrossRef]
218. Torchilin, V.P. Fundamentals of Stimuli-responsive Drug and Gene Delivery Systems. In *Stimuli-Responsive Drug Delivery Systems*; Royal Society of Chemistry: London, UK, 2018; pp. 1–32.
219. Dubey, S.K.; Bhatt, T.; Agrawal, M.; Saha, R.N.; Saraf, S.; Saraf, S.; Alexander, A. Application of chitosan modified nanocarriers in breast cancer. *Int. J. Biol. Macromol.* **2022**, *194*, 521–538. [CrossRef]
220. Ou, Y.-C.; Webb, J.; Faley, S.; Shae, D.; Talbert, E.; Lin, S.; Cutright, C.; Wilson, J.; Bellan, L.; Bardhan, R. Gold Nanoantenna-Mediated Photothermal Drug Delivery from Thermosensitive Liposomes in Breast Cancer. *ACS Omega* **2016**, *1*, 234–243. [CrossRef]
221. Tapeinos, C.; Pandit, A. Physical, Chemical, and Biological Structures based on ROS-Sensitive Moieties that are Able to Respond to Oxidative Microenvironments. *Adv. Mater.* **2016**, *28*, 5553–5585. [CrossRef]
222. Saravanakumar, G.; Kim, J.; Kim, W.J. Reactive-Oxygen-Species-Responsive Drug Delivery Systems: Promises and Challenges. *Adv. Sci.* **2017**, *4*, 1600124. [CrossRef]
223. Liu, J.; Li, Y.; Chen, S.; Lin, Y.; Lai, H.; Chen, B.; Chen, T. Biomedical Application of Reactive Oxygen Species-Responsive Nanocarriers in Cancer, Inflammation, and Neurodegenerative Diseases. *Front. Chem.* **2020**, *8*, 838. [CrossRef]
224. Rasheed, T.; Bilal, M.; Abu-Thabit, N.Y.; Iqbal, H.M. *Stimuli Responsive Polymeric Nanocarriers for Drug Delivery Applications*; Woodhead Publishing: Cambridge, MA, USA, 2018.
225. Basel, M.T.; Shrestha, T.B.; Troyer, D.L.; Bossmann, S.H. Protease-sensitive, polymer-caged liposomes: A method for making highly targeted liposomes using triggered release. *ACS Nano* **2011**, *5*, 2162–2175. [CrossRef]

226. Liu, C.; Zhao, Z.; Gao, R.; Zhang, X.; Sun, Y.; Wu, J.; Liu, J.; Chen, C. Matrix Metalloproteinase-2-Responsive Surface-Changeable Liposomes Decorated by Multifunctional Peptides to Overcome the Drug Resistance of Triple-Negative Breast Cancer through Enhanced Targeting and Penetrability. *ACS Biomater. Sci. Eng.* **2022**, *8*, 2979–2994. [CrossRef]
227. Wang, Y.; Kohane, D.S. External triggering and triggered targeting strategies for drug delivery. *Nat. Rev. Mater.* **2017**, *2*, 1–14. [CrossRef]
228. Yang, H.Y.; Li, Y.; Lee, D.S. Multifunctional and Stimuli-Responsive Magnetic Nanoparticle-Based Delivery Systems for Biomedical Applications. *Adv. Ther.* **2018**, *1*, 1800011. [CrossRef]
229. Schleich, N.; Danhier, F.; Préat, V. Iron oxide-loaded nanotheranostics: Major obstacles to in vivo studies and clinical translation. *J. Control. Release* **2015**, *198*, 35–54. [CrossRef] [PubMed]
230. Zhou, X.; Wang, L.; Xu, Y.; Du, W.; Cai, X.; Wang, F.; Ling, Y.; Chen, H.; Wang, Z.; Hu, B. A pH and magnetic dual-response hydrogel for synergistic chemo-magnetic hyperthermia tumor therapy. *RSC Adv.* **2018**, *8*, 9812–9821. [CrossRef] [PubMed]
231. Jeon, G.; Yang, S.Y.; Byun, J.; Kim, J.K. Electrically actuatable smart nanoporous membrane for pulsatile drug release. *Nano Lett.* **2011**, *11*, 1284–1288. [CrossRef]
232. Servant, A.; Bussy, C.; Al-Jamal, K.; Kostarelos, K. Design, engineering and structural integrity of electro-responsive carbon nanotube-based hydrogels for pulsatile drug release. *J. Mater Chem. B* **2013**, *1*, 4593–4600. [CrossRef]
233. Ge, J.; Neofytou, E.; Cahill, T.J., 3rd; Beygui, R.E.; Zare, R.N. Drug release from electric-field-responsive nanoparticles. *ACS Nano* **2012**, *6*, 227–233. [CrossRef]
234. Hosseini-Nassab, N.; Samanta, D.; Abdolazimi, Y.; Annes, J.P.; Zare, R.N. Electrically controlled release of insulin using polypyrrole nanoparticles. *Nanoscale* **2017**, *9*, 143–149. [CrossRef]
235. Paris, J.L.; Cabañas, M.V.; Manzano, M.; Vallet-Regí, M. Polymer-Grafted Mesoporous Silica Nanoparticles as Ultrasound-Responsive Drug Carriers. *ACS Nano* **2015**, *9*, 11023–11033. [CrossRef]
236. Luo, Z.; Jin, K.; Pang, Q.; Shen, S.; Yan, Z.; Jiang, T.; Zhu, X.; Yu, L.; Pang, Z.; Jiang, X. On-Demand Drug Release from Dual-Targeting Small Nanoparticles Triggered by High-Intensity Focused Ultrasound Enhanced Glioblastoma-Targeting Therapy. *ACS Appl. Mater. Interfaces* **2017**, *9*, 31612–31625. [CrossRef]
237. Raza, A.; Rasheed, T.; Nabeel, F.; Hayat, U.; Bilal, M.; Iqbal, H.M.N. Endogenous and Exogenous Stimuli-Responsive Drug Delivery Systems for Programmed Site-Specific Release. *Molecules* **2019**, *24*, 1117. [CrossRef]
238. Pham, S.H.; Choi, Y.; Choi, J. Stimuli-responsive nanomaterials for application in antitumor therapy and drug delivery. *Pharmaceutics* **2020**, *12*, 630. [CrossRef]
239. Liu, J.; Ai, X.; Cabral, H.; Liu, J.; Huang, Y.; Mi, P. Tumor hypoxia-activated combinatorial nanomedicine triggers systemic antitumor immunity to effectively eradicate advanced breast cancer. *Biomaterials* **2021**, *273*, 120847. [CrossRef]
240. Zhang, R.; Li, Y.; Zhang, M.; Tang, Q.; Zhang, X. Hypoxia-responsive drug–drug conjugated nanoparticles for breast cancer synergistic therapy. *RSC Adv.* **2016**, *6*, 30268–30276. [CrossRef]
241. Gurpreet, K.; Singh, S.K. Review of nanoemulsion formulation and characterization techniques. *Indian J. Pharm. Sci.* **2018**, *80*, 781–789. [CrossRef]
242. Di, J.; Gao, X.; Du, Y.; Zhang, H.; Gao, J.; Zheng, A. Size, shape, charge and "stealthy" surface: Carrier properties affect the drug circulation time in vivo. *Asian J. Pharm Sci.* **2021**, *16*, 444–458. [CrossRef]
243. Patra, J.K.; Das, G.; Fraceto, L.F.; Vangelie, E.; Campos, R.; Rodriguez, P.; Susana, L.; Torres, A.; Armando, L.; Torres, D.; et al. Nano based drug delivery systems: Recent developments and future prospects. *J. Nanobiotechnol.* **2018**, *16*, 1–33. [CrossRef]
244. Hejmady, S.; Singhvi, G.; Saha, R.N.; Dubey, S.K. Regulatory aspects in process development and scale-up of nanopharmaceuticals. *Ther. Deliv.* **2020**, *11*, 341–343. [CrossRef]
245. Ajdary, M.; Moosavi, M.A.; Rahmati, M.; Falahati, M.; Mahboubi, M.; Mandegary, A.; Jangjoo, S.; Mohammadinejad, R.; Varma, R.S. Health concerns of various nanoparticles: A review of their in vitro and in vivo toxicity. *Nanomaterials* **2018**, *8*, 634. [CrossRef]
246. Hua, S.; de Matos, M.B.C.; Metselaar, J.M.; Storm, G. Current trends and challenges in the clinical translation of nanoparticulate nanomedicines: Pathways for translational development and commercialization. *Front. Pharmacol.* **2018**, *9*, 1–14. [CrossRef]

pharmaceutics

Article

Formulation Development of Fast Dissolving Microneedles Loaded with Cubosomes of Febuxostat: In Vitro and In Vivo Evaluation

Brijesh Patel and Hetal Thakkar *

Centre for Post-Graduate Studies and Research in Pharmaceutical Sciences, Shri G.H. Patel Pharmacy Building, Faculty of Pharmacy, The Maharaja Sayajirao University of Baroda, Vadodara 390002, Gujarat, India
* Correspondence: hetal.thakkar-pharmacy@msubaroda.ac.in

Abstract: Febuxostat is a widely prescribed drug for the treatment of gout, which is a highly prevalent disease worldwide and is a major cause of disability in mankind. Febuxostat suffers from several limitations such as gastrointestinal disturbances and low oral bioavailability. Thus, to improve patient compliance and bioavailability, transdermal drug delivery systems of Febuxostat were developed for obtaining enhanced permeation. Cubosomes of Febuxostat were prepared using a bottom-up approach and loaded into a microneedle using a micromolding technique to achieve better permeation through the skin. Optimization of the process and formulation parameters were achieved using our design of experiments. The optimized cubosomes of Febuxostat were characterized for various parameters such as % entrapment efficiency, vesicle size, Polydispersity index, Transmission electron microscopy, in vitro drug release, Small angle X-ray scattering, etc. After loading it in the microneedle it was characterized for dissolution time, axial fracture force, scanning electron microscopy, in vitro drug release, pore closure kinetics, etc. It was also evaluated for various ex vivo characterizations such as in vitro cell viability, ex vivo permeation, ex vivo fluorescence microscopy and histopathology which indicates its safety and better permeation. In vivo pharmacokinetic studies proved enhanced bioavailability compared with the marketed formulation. Pharmacodynamic study indicated its effectiveness in a disease-induced rat model. The developed formulations were then subjected to the stability study, which proved its stability.

Keywords: gout; Febuxostat; cubosomes; Microneedles; xanthine oxidase inhibitors

Citation: Patel, B.; Thakkar, H. Formulation Development of Fast Dissolving Microneedles Loaded with Cubosomes of Febuxostat: In Vitro and In Vivo Evaluation. *Pharmaceutics* **2023**, *15*, 224. https://doi.org/10.3390/pharmaceutics15010224

Academic Editors: Daniela Monti and Silvia Tampucci

Received: 17 November 2022
Revised: 3 January 2023
Accepted: 4 January 2023
Published: 9 January 2023

1. Introduction

Gout is a systemic disease that results from the deposition of monosodium urate crystals (MSU) in the tissues. Increased serum uric acid (SUA) above a specific threshold is a main reason for the formation of uric acid crystals. MSU crystals can deposit in all the tissues, mainly present in and around the joints forming tophi [1,2]. According to the American College of Rheumatology, medication therapy of gout involves the use of analgesics, NSAIDs, corticosteroids, colchicine, xanthine oxidase inhibitors and uricosurics [3–5].

A xanthine oxidase inhibitor (e.g., Allopurinol and Febuxostat) is a substance that inhibits the activity of xanthine oxidase, an enzyme involved in purine metabolism. In humans, inhibition of xanthine oxidase reduces the production of uric acid, and thus they are indicated for the treatment of hyperuricemia and related medical conditions including gout [1,5]. Febuxostat (FBX) is a novel, potent, non-purine selective xanthine oxidase inhibitor and has been reported to be more effective in lowering and maintaining serum urate levels than allopurinol. It is potent and more selective than allopurinol, and can be used safely in patients with hypersensitivity reactions towards allopurinol. Febuxostat is available as a tablet dosage form in the market for once a day administration in 40, 80 and 120 mg strengths. The tablets of Febuxostat are marketed under the names of Fabulas, Feboxa, Febuget, Febucip etc. The oral bioavailability of Febuxostat is 38% and is affected by the presence of food. The oral bioavailability of FBX is hampered by its low (<15 μg/mL) aqueous solubility and extensive enzymatic degradation in the intestine and liver. Furthermore, FBX's peak plasma concentration (C_{max}) is reduced by 38–49% in the presence of food [6]. FBX has its shortcomings, for example, poor bioavailability, food dependent absorption, gut-wall metabolism and gastrointestinal disturbances (nausea, diarrhea, stomach pain, ulcers, vomiting) [7]. The delivery of the drug through transdermal delivery will avoid all these problems associated with the present marketed formulation.

The transdermal route of drug delivery was selected to overcome the various limitations associated with FBX as cited above. However, the use of this route of drug delivery is limited by the presence of the outermost nonviable layer of stratum corneum [8–10]. Therefore, to overcome this limitation several mechanical, chemical approaches or novel drug delivery systems have been reported [11,12]. Nisomal gel [13], ethosomes [14], nanoemulsion [15] and self-nanoemulsion-loaded transdermal film [7] have been reported for the transdermal delivery of Febuxostat.

In the present investigation, a novel drug delivery system (Cubosomes) of Febuxostat was developed and characterized due to its efficiency of deeper penetration. Moreover, cubosomes also have advantages such as the ease of preparation, high drug loading and entrapment, low cost of raw materials, better skin penetration, and the fact it is bioadhesive and biocompatible etc. Cubosomes are the colloidal dispersion (size ranging from 100 to 300 nm) made by dispersing the bicontinuous cubic liquid crystalline structures in an aqueous medium that have surface active agents [16,17]. Cubosomes have the ability to encapsulate a variety of drug molecules falling into the hydrophilic, lipophilic and amphiphilic classes [18]. However, the use of cubosomes or any other nanocarrier systems does not ensure the complete permeation of therapeutic molecules across the skin which will decrease the bioavailability of drug molecules as reported in many literatures [19–21]. To improve the permeation across the skin, a combination of two or more approaches of the enhancement of transdermal permeation was suggested in many of the literatures, among which the combination of a Microneedles (MNs) patch with nanocarriers is the most prominent way to improve permeation across the skin [19,20,22]. Zhang P. et al., reported the use of MNs combined with chitosan nanoparticles to improve the permeation of insulin across the skin. They obtained a 4.2 fold increase in the permeation of insulin across the skin compared with the chitosan nanoparticles [19].

MN patches are generally made up of small micron sized needles attached to the patch. The application of MN proves to be advantageous as it facilitates the permeation across the toughest barrier, the stratum corneum, without causing any kind of pain [23]. Various materials can be used in the fabrication of (MN) and they can be biodegradable or non-biodegradable [24]. Biodegradable MNs are prepared using materials such as amylopectin, polyvinyl alcohol, poly vinyl pyrrolidone, poly lactic acid, PLGA, etc., [23]. They are preferred over the non-biodegradable ones as the accidental breakage of them later during insertion may lead to complications such as sepsis due to the broken part of the MN [10,23]. Thus, for the present project, MNs were fabricated using hydrophilic biodegradable polymers [23,25–28].

The objective of the present study was to develop, optimize and compare cubosomal formulation of FBX and an MN patch loaded with cubosomes of FBX for the enhancement of bioavailability by circumventing the first pass metabolism of FBX and to avoid gastrointestinal disturbances.

2. Materials and Method

Febuxostat and Glyceryl Monooleate (GMO) were obtained as a gift sample from Ami drugs and specialty chemicals Pvt. Ltd., India and Mohini Organics, Mumbai, India, respectively. DermaStamp DTR-150 and YYR-150 (35 and 80 titanium MNs, respectively) were procured from Guangzhou Junguan Beauty Co. Ltd., Guangzhou, China. Polydimethylsiloxane (Sylgard® 184) was procured from Dow corning, Midland, MI, USA. Polyvinyl alcohol (PVA)-6000 and Lactose were purchased from Acros Organic, New Jersey,

USA and Hi-media, Mumbai, India, respectively. Acetonitrile (HPLC grade), and methanol (HPLC grade) were purchased from Rankem Fine Ltd., Mumbai, India. Formic acid (AR grade), Disodium hydrogen phosphate, potassium dihydrogen phosphate, sodium chloride, sodium hydroxide were obtained from Spectrochem Labs Ltd., Mumbai, India. Thiazolyl Blue Tetrazolium Bromide (MTT) was bought from Sigma-Aldrich, USA. Antibiotic antimycotic solution 100×, Gamma irradiated fetal bovine serum and DMEM (Dulbecco's Modified Eagle Medium) were purchased form Hi-Media, Mumbai, India. Potassium oxonate was obtained from TCI, Chennai, India. Preparation of double distilled water was performed in the laboratory, filtered with 0.2 μ membrane filter (store airtight container) and utilized within a maximum of 7 days.

2.1. Preparation of FBX Cubosomes

Bottom-up approach was utilized for the preparation of Cubosomes of FBX. A QbD approach was employed for the optimization of cubosomes of FBX using particle size and entrapment efficiency as Critical Quality Attributes (CQAs). In this process, concentration of lipid and PVA was selected as Critical Material Attributes (CMAs). Optimized composition for FBX cubosomes was mentioned in Table 1. For the preparation of cubosomes of FBX two solutions were prepared: (A) organic phase and (B) aqueous phase. For the preparation of organic phase (A), X % *w*/*v* of Glyceryl monoolein (GMO) was taken in a 10 mL glass beaker and dissolved in 4 mL of ethanol, then 40 mg of FBX was added to it. For aqueous phase (B), Y % *w*/*v* of PVA was dissolved in 20 mL of water. Both solutions were initially kept at a temperature of 60 °C and were continuously stirred for 5–10 min. Afterward, organic phase was added to aqueous phase in a drop-wise manner with continuous stirring and the addition of organic phase was maintained at a rate of 1 mL/min and 1000 rpm on magnetic stirrer. The resulting medium was cooled down to room temperature and then continuously stirred for 30 min using a magnetic stirrer. This medium was then introduced to the rotary evaporator at a temperature of 50 °C under vacuum for removing ethanol from the dispersion and the volume of the prepared batch was reduced to 10 mL. The resulting Cubosomal dispersion was exposed to centrifugation with process parameters, i.e., for a period of 10 min at 5000 rpm and the temperature was set as 25 °C for facilitating the sedimentation of free drug. Care was taken while separating the supernatant of cubosomal dispersion so as to not disturb the free drug pellet which is deposited at the bottom of the centrifuge tube. Finally, the resulting separated cubosomal dispersion was stored for utilization in future tests in glass vials at room temperature [27,29].

Table 1. Composition of optimized formulation.

For Cubosomes of FBX	
Conc. of GMO * (%*w*/*v*)	**Conc. of PVA** ** (%*w*/*v*)
7.6	1.2
For MN patch of cubosomes of FBX	
Conc. of PVA ** (%*w*/*v*)	**Conc. of lactose** (%*w*/*v*)
33.46	8.7

* GMO: Glyceryl monooleate. ** PVA: Polyvinyl alcohol 6000.

2.2. Preparation of MN Patch

A QbD approach was employed for the optimization of cubosomal FBX MN using axial fracture force and dissolution time as CQAs and concentration of PVA and lactose as CMAs. Optimized composition for cubosomal FBX MN was mentioned in Table 1. For the preparation of MN matrix, polyvinyl alcohol (PVA) was used as a polymer at concentration of X % *w*/*v* and Lactose was used as a filler at a concentration of Y % *w*/*v*. Hot plate magnetic stirrer was employed for solubilizing optimum amounts of PVA and lactose in 10 mL of FBX cubosomal dispersion (3.4 mg/mL FBX). This mixture was gently stirred at a temperature of 50 °C on the hot plate magnetic stirrer. Further, this mixture was cooled and was brought back to ambient room temperature. The resulting 1 mL of PVA/Lactose blend solutions which contain approximately 3.4 mg of drug were shifted into PDMS micromolds which have capacity of 2 mL. In order to fill the cavities of MN of the micromolds, it was centrifuged at 3000 RPM at 25 °C for 10 min. Further, to ensure that the MN structure was properly hardened, micromolds were placed in vacuum desiccators for a period of 24 h to evaporate the water. Then, 1567 high adhesion double coated medical tape (3M™, St. Paul, Minnesota, USA) which had backing film on the opposite adhesive side was employed to extract the MN arrays from micromolds. MN patches prepared from the aforesaid procedure were kept in airtight containers and calcium oxide along with silica gel was employed as desiccants [30,31].

2.3. *Characterization of Cubosomes*

2.3.1. Vesicle Size Analysis

Ref. [32] The dispersions of FBX-loaded cubosomes were diluted up to 10 times with pre-filtered distilled water. Further, the dispersions were taken into disposable sizing cuvette and the vesicle size and poly-dispersity index (PDI) were analyzed with the help of Nano-ZS zetasizer which calculates vesicle size and PDI based on dynamic light scattering (DLS). For the calculation of mean diameter of cubosomes, the instrument examines angular scattering of a laser beam during its passage through the dispersed cubosomal sample and use the Mie theory of light scattering.

2.3.2. Zeta Potential Measurement

Ref. [32] Nano-ZS zetasizer by Malvern Instruments Ltd., Bristol, UK, was used for the analysis of zeta potential of FBX-loaded cubosomes. For this, the dispersion of FBX-loaded cubosomes was taken and was diluted up to 10 times and the dilution was performed using pre-filtered distilled water. Then, the dispersion was taken in disposable folded capillary cells and was evaluated for zeta potential. Nano ZS zetasizer uses Smoluchowski equation for the calculation of zeta potential centered on the amount of doppler shift occurring due to electrophoretic mobility of colloidal particles in response to the electric field applied to the dispersion.

2.3.3. % Entrapment Efficiency

Ref. [33] For the determination of entrapment efficiency, free FBX was separated from entrapped FBX in cubosomes by centrifuging it at 6000 rpm for a period of 15 min at a temperature of 25 °C using Remi Centrifuge. Then, supernatant of the centrifuge tube which contains cubosomal dispersion of FBX was separated carefully without disturbing the hard pallet of free drug, which was formed at the bottom of the centrifuge tube. Cubosomal dispersion of FBX was broken down using ACN:Methanol (9:1) for quantitative analysis of FBX. The absorbance of the prepared sample was then calculated using UV visible spectrophotometer at a wavelength of 315 nm. The % entrapment efficiency was calculated with the help of Equation (1).

$$\%EE = \frac{Amount\ of\ entrapped\ drug}{Total\ drug\ added} \times 100 \quad (1)$$

2.3.4. Total Drug Content (% Assay)

For the determination of total drug content of the prepared formulation, 1 mL of the cubosomal dispersion, which was equivalent to 4 mg of FBX, was accurately withdrawn and was dissolved in 10 mL of ACN. The prepared samples of FBX were then analyzed using UV visible spectrophotometer at a wavelength of 315 nm. The % total drug content of cubosomes of FBX was calculated using Equation (2).

$$\%Total\ drug\ content = \frac{Amount\ of\ total\ drug\ estimated}{Total\ drug\ added} \times 100 \quad (2)$$

2.3.5. Shape and Surface Morphology

Ref. [29] Transmission electron microscopy was employed for the evaluation of shape and surface morphology of the FBX-loaded cubosomes. For performing the test, the dispersion was smeared on a carbon-coated grid, and any extra material was removed and the grid was dried at room temperature for a period of 5 hrs. Transmission electron microscope (CM 200, Philips, Amsterdam, Netherlands) was employed with the following process parameters, i.e., the operating voltage was set in a range of 20–200 kV to visualize cubosomes at suitable magnification with an accelerating voltage of 20 kV.

2.3.6. Small Angle X-rays Scattering

Ref. [29] Bruker Nanostar Xeuss 2.0 model was employed for conducting SAXS experiments furnished with a rotating anode and three-pinhole collimation. The device employs Cu-Kα radiation having a λ_{max} of 1.54 Å and a sample to detect a length of approximately 105 cm. Anode was set at 45 kV and 100 mA current. The samples were transferred in a 2 mm quartz capillary (from Charles-Supper, Westborough, MA, USA) having 10 µm wall thickness. For keeping reference, scattering from glassy carbon film was employed. The temperature of sample holder was maintained by Peltier unit. The obtained data was taken on a HISTAR gas filled multi-wire detector. Further, the 2D data was circularly averaged for the conversion of data to 1D. The scanning of samples was performed for a period enough to obtain at least two million counts. Further, these were normalized with the transmission coefficient of the sample and the acquisition time. The scattering emerging from silver behenate was employed for the calibration of Detector.

2.3.7. Headspace Gas Chromatography (HS-GC) Testing for Residual Solvent

Standard Preparation

In 10 mL volumetric flask, 0.13 mL of ethanol equivalent to 0.1 g was taken and the final volume was made up to mark using DMF (dimethyl formamide) which gave final concentration of 10,000 ppm. In other 10 mL volumetric flask, 1 mL of above obtained solution was taken and final volume was made up to mark using deionized water to obtain final concentration of 1000 ppm [34].

Sample Preparation

A volume of formulation (0.105 mL) equivalent to 0.1 g was taken in 10 mL volumetric flask and the final volume was made up to the mark with DMF. From the above solution 1 mL was taken in 10 mL volumetric flask and volume was made up using deionized water. Sample was injected into column (capillary column: CR-624, Dimensions: 30 m, 0.53 mm, 3.00 µm) at 80 °C using nitrogen as carrier gas. Other parameters such as carrier gas flow rate, H_2 gas flow rate, air flow rate, injection volume, injector temperature and detector temperature were set to 40 mL/min, 30 mL/min, 300 mL/min, 0.2 µL, 260 °C and 260 °C, respectively. Total run time was set at 20 min [34].

2.4. Characterization of MN Patch

2.4.1. Axial Fracture Force

Brookfield CT3 texture analyzer was employed for the measurement of axial needle fracture force. For performing this, double sided adhesive tape was used for placing MN arrays on the texture analyzer's mobile probe. This step was performed carefully in a manner that axis of MN aligns parallel with axis of mobile probe. After this, probe was automated for pressing the MNs on a rigid, flat steel [35]. During testing, the needle strength graph was made with the help of Texture Pro CT data acquisition software. An axial fracture force using following Equation (3) with the help of peak load obtained from the graph generated by Texture pro CT.

$$F = mg \tag{3}$$

where: m = mass applied for breaking of MN. g = gravitational force.

2.4.2. In Vitro Dissolution Study

Preparation of Gelatin Slab

Stratum corneum exhibits a water content of approx. 30 ± 5%. To simulate the conditions of stratum corneum, artificial gelatin skin was designed having a similar hydration level. This was achieved by adding 35% water and 65% gelatin. For preparing the artificial gelatin skin, 6.5 g of gelatin was taken and was transferred in 10 mL water and this solution was kept for hydration for a period of 30 min. Further, the gelatin was solubilized with the help of water bath at a temperature at 60 °C with continuous stirring. The resulting solution was transferred into a glass Petri dish and the water was permitted to vaporize till a weight of 10 g was attained. The resulting film obtained after evaporation was cut out into small square pieces.

In Vitro Dissolution Study

Optimized batch of MN patch was implanted in the gelatin film and removed at various time intervals (15 and 30 s, 1, 1.5, 2, 2.5, 3, 3.5, 4, 4.5, and 5 min) and viewed under microscope [36].

2.4.3. Shape and Surface Morphology

For characterization of MN's shape and their surface morphology, fast dissolving MN patches were affixed on sample stub and observed under JSM-5610LV scanning electron microscope (JEOL, Tokyo, Japan) wherein the accelerating voltage was set at 20 kV [36].

2.4.4. Skin Penetrability

For checking the skin penetrability, double sided adhesive tapes were employed for mounting the MNs on mobile probe of Brookfield CT3 texture analyzer. Care was taken that the axis of MN was aligned parallel to the axis of the mobile probe. Probe was automated for pressing the MNs on the full thickness of pinned pig ear skin on a soft sponge pad under slight tension for simulating in situ mechanical support. The insertion speed of the moving probe was set at a speed of 20 mm/s. Skin area, on which MN was applied, was treated with trypan blue dye for a period of 5 min. A tissue paper was used for wiping off the excess dye from the skin. Further, digital camera was employed for taking the pictures of stained pores [37].

2.4.5. Pore Closure Kinetic

For performing this study, rat skin was pinned with slight tension onto the soft board, for simulating in situ mechanical support. Five pieces of skin were used for the study and one patch was applied on each piece. At different time intervals, i.e., 0, 3, 6, 12 and 24 h, MN patch was detached from one piece of skin. The sections of epidermis were taken using cryo-microtome to observe for pore closure with the help of Eclipse H600L inverted microscope (Nikon, Tokyo, Japan) [37].

2.4.6. Physical Stability of Cubosomes in Fast Dissolving MN Patch

After preparing the fast dissolving MN patch, instantly the physical stability of the cubosomes in MN patch was examined for vesicle size and entrapment efficiency. For this, the MN patch was solubilized in pre-filtered distilled water for obtaining the dispersion of cubosomes. The vesicle size and entrapment efficiency were estimated by the methods described above [37].

2.4.7. Total Drug Content

Double distilled water was taken at a volume of 10 mL and used for solubilizing prepared MN patch of FBX. Then, 1 mL of prepared samples were diluted with acetonitrile:

Methanol (9:1), respectively, and quantitatively analyzed using the UV spectrophotometer at λ_{max} of 315 nm [37].

2.5. In Vitro Drug Release Study

A dialysis membrane having molecular weight cut off in range of 12–14 K dalton in the Franz diffusion cell was employed for conducting in vitro drug release study. In case of Franz diffusion cell, the donor compartment has a volume capacity of 20 mL. To perform in vitro drug release, 30% ethanolic phosphate buffer pH 7.4 served as a diffusion medium [38]. For performing the study, plain drug suspension in water (1 mL), cubosomes of FBX (1 mL), MN patch containing plain FBX, and MN patch containing FBX cubosomes all equivalent to 3.4 mg were placed in the donor compartment. Further, from the receptor compartment, samples (1.0 mL) were removed at steady time intervals (0.5, 1, 2, 3, 4, 5, 6, 8, 12 and 24 h) and the identical volume (1.0 mL) was replaced by a fresh diffusion medium. Samples were evaluated using the UV spectrophotometer (Shimadzu, Kyoto, Japan, model: UV 1800) at λ_{max} of 315 nm. Triplicate readings of all experiments were recorded and further average of these readings was considered [36,39].

2.6. In Vitro Cell Viability Study

2.6.1. Cell Culturing and Sub-Culturing

The cell culture of fibroblast 3T3 was bought from NCCS, Pune. The received flask was kept in an anaerobic incubator for a period of 24 h at a temperature of 37 °C and 5% CO_2 without removing the media. Later, culture medium from the flasks was taken out and the adherent cells were washed with the help of PBS pH 7.4. Freshly prepared Trypsin-EDTA solution was then poured into the flask in order to completely cover the cell monolayer and was kept in the incubator for 2 min at 37 °C for detachment of adherent cells. For neutralizing trypsin's activity in the flask, fresh growth medium was poured into the flask. Further, the cell culture was exposed to centrifugation at 1200 rpm for a period of 5 min. Then, after discarding the supernatant, resulting cells were re-suspended in a fresh growth medium. Cells were counted using neubauer counting chamber and transferred into new flasks at a plating density of 1×10^4 cells/cm^2. These flasks were kept in incubator set at a temperature of 37 °C and 5% CO_2 to facilitate cell growth. The growth media was renewed every third day and passaging was conducted once the culture attained 80–90% confluency [40,41].

2.6.2. MTT Assay

For the determination of safety, viability evaluation of fibroblast 3T3 cells was performed with the help of 3-(4,5-dimethylthiazol-2-yl)-2,5-diphenyltetrazolium bromide (MTT) assay [42]. Principle of this assay is based on the fact that mitochondrial dehydrogenase is responsible for the reduction of yellow-colored tetrazolium MTT whose production is found in viable (metabolically active) cells. The resultant intracellular purple formazan is solubilized and quantified with the help of spectrophotometer. Suspension of 3T3 cells in growth media was prepared from its culture using the same method as described above. A 96 well plate was used for seeding of the cells (5000 cells/well) and then it was kept in the incubator for a period of 24 h to facilitate cell growth and its attachment to the plate surface. After 24 h, growth media was discarded and 200 µL of fresh treatment media was transferred to these wells. Then, cubosomal dispersion of FBX and FBX suspension were diluted in growth media to obtain 1000 µg/mL of FBX. From these prepared cubosomal dispersions in fresh growth medium, 100 µL was added in different wells of 96 well plate. The plate was then incubated for 24 h and then the treatment media present was discarded. Then, 200 µL of growth media and 100 µL of MTT solution were transferred to each well and the plate was kept in the incubator for a period of 4 h. After this, 200 µL of dimethyl sulfoxide was transferred to each well for solubilizing formazan crystals, after removing growth media and MTT solution carefully. A microplate reader 690 XR from Bio-Rad, California, USA was utilized for the measurement of absorbance of the resultant solution. Measurement was performed at 570 nm. Cells viability in wells were treated with phosphate buffer saline pH 7.4, which acted as negative control, and was considered as 100% for the calculation of "% cell viability" [42].

2.7. In Vitro Permeation Study

MN patch of FBX, FBX cubosomes loaded MN, suspension of FBX and optimized cubosomes of FBX were tested for deposition profile and permeation with the help of full thickness rat abdominal skin. An abdominal area of rat was shaved to remove hair before harvesting the skin. A harvested rat skin was stored at −20 °C until it was needed for FBX permeation study. The evaluation was conducted by employing a Franz-type diffusion cell having a 20 mL receptor chamber. For performing this experiment, 30% ethanol solution prepared in distilled water was used for filling the receptor compartment and circular water bath was employed for maintaining its temperature at 37 °C. Before initiating the permeation experiment, the skin sections were thawed at room temperature. The skin sections were kept over a soft sponge pad and 30% v/v ethanol solution prepared in distilled water was used to impregnate the skin for a period of 30 min. This was performed for equilibration. Further, the skin sections were affixed between the receptor and donor compartment. Care was taken that the stratum corneum faces the donor compartment. Diffusion media used in the Franz diffusion cell was stirred at a speed of 100 rpm. After equilibration was achieved, cubosomal dispersion of FBX and FBX suspension (equivalent to 3.41 mg of FBX) was added in donor compartment. MN patch of FBX and cubosomes-loaded MN patch that had an identical amounts of drug were applied over the skin sections. This was achieved by the application of mild pressure using thumb on the skin which was kept under slight tension. Subsequently, it was affixed in place. From the sampling arm of the diffusion cell, samples that had a volume of 0.5 mL were taken at various time points, i.e., 0.5, 1, 1.5, 2, 3, 4, 5, 6, 7, 8, 12 and 24 h. Further, fresh diffusion media of the same volume were replaced in order to maintain the total volume. The skin section was removed from the Franz diffusion cell after 24 h and the skin was washed with 5 mL diffusion media three times. For calculation of the drug adhered to the skin, washings of the skin were saved. A scalpel was used for cutting the washed skin into small pieces. Then, these pieces were suspended in methanol, homogenized in cold conditions for a period of 5 min and then were sonicated using bath sonicator for a period of 15 min. For quantification of the drug accumulated in skin, the drug was removed by centrifuging it at an rpm of 5000 for a period of 10 min. All the samples were filtered with the help of 0.2 μm membrane filter and the quantification of the drug was performed by employing HPLC. The cumulative quantity of drug that permeated through the skin (per cm^2 surface area of skin) was calculated. Finally, a graph was plotted with the concentration of the cumulative amount of drug permeated per cm^2 surface area of skin against time. For transdermal steady flux (JSS; $\mu g/cm^2/h$), a slop of the terminal portion of graph of concentration of FBX permeated across skin against time was found [37,43]. Equation (4) was used for the calculation of permeation enhancement ratio [44].

$$PER = \frac{J_{SS}^{test}}{J_{SS}^{control}} \tag{4}$$

where J_{SS}^{test} is steady state flux via test formulation and $J_{SS}^{control}$ is steady state flux via FBX suspension.

2.8. In Vitro Fluorescence Microscopy Study

With the help of fluorescence microscopy, permeation behavior of the formulations which were developed was illustrated. FITC suspension, its MN patch, optimized cubosomes and cubosomes (FITC loaded)-loaded MN patch were formulated and utilized for the study. FITC-loaded cubosomes and MN patch were prepared by replacing drug by FITC in optimized compositions. The rat skin was thawed at room temperature, equilibrated and fastened on Franz diffusion cell in the same way as explained in ex vivo skin permeation study. FITC-loaded formulations were smeared onto the stratum corneum layer of the skin in a similar way as explained in ex vivo skin permeation study. After a period of 12 h, skin sectioning was performed in dark environment using cryo-microtome, sections were fixed on a glass slide. Confocal laser scanning microscope was utilized for examining the fluorescence on the slide [37,45].

2.9. Histopathological Studies

Ref. [46] The study was conducted using rat abdominal skin, which was obtained from the sacrificed animals via academic protocol approved by the institutional animal ethics committee of the Maharaja Sayajirao University of Baroda (MSU/IAEC/2019–20/1902). Cubosomes-loaded MN patch of FBX, cubosomes of FBX and FBX drug suspension were applied on freshly excised rat abdominal skin. Apart from this, isopropyl alcohol (IPA) and PBS treated abdominal rat skin were used as positive control and negative control, respectively. After four hours, skins were immersed in 10% buffered formalin, dehydrated gradually increasing concentration of ethanol, immersed in xylene and finally embedded in paraffin. The 5-μm thick sections of skin were cut from these paraffin blocks using microtome and placed on glass slides. The paraffin wax was removed by gently warming the slides and washing the molten wax with xylene. Sections were then washed with absolute alcohol and water and stained with haematoxylin and eosin to determine gross histopathology. Commercial glycerol's mounting fluid was used to finally mount the stained sections. Negative control and positive control slides were also prepared by treating rat skin with phosphate buffer solution pH 6.8 and isopropyl alcohol, respectively, using the same method. The slides were analyzed at 10-fold magnification using optical microscope [30,37].

2.10. In Vivo Pharmacokinetic Study

Sprague–Dawley rats weighing 200–270 g were procured from an official CPCSEA breeder. Rats which were obtained were placed in cages present in the animal house wherein the temperature was set at 22 ± 3 °C and light-dark cycle of fixed 12 h was maintained. Handling of the animals was performed with compliance to CPCSEA guidelines, Department of Animal Welfare, Government of India. Rats were kept on a standard chow diet and were given water as desired. In total, 30 rats were allocated to 5 groups randomly as shown in Table 2. Each group had two sets and each set had three animals. All group animals were fasted 12 h before starting the experiment. Marketed FBX tablet in a suspension form was administered to Group 1 animals through the oral route. Transdermal patch of FBX was applied on Group 2 animals. Group 3 animals were applied cubosomes of FBX, and MN patch of FBX was applied on Group 4 animals. Cubosomes-loaded MN patches were applied to Group 5 animals. Diethyl ether was employed as anesthetic agent while collecting the blood samples (not more than 0.5 mL) from retro orbital plexus. The collected blood samples were transferred to microcentrifuge tubes containing heparin at 1, 3, 5, 8, and 24 h from set-1 and 2, 4, 6, and 12 h from set-2 resulting 9 time points (1, 2, 3, 4, 5, 6, 8, 12, 24 h). The rats were replenished with saline solution. These blood samples were exposed to centrifugation at 3500 RPM for a period of 10 min at a temperature of 4 °C. The harvested samples of plasma were analyzed using a developed HPLC method at λ_{max} of 315 nm to estimate pharmacokinetic parameters such as C_{max}, T_{max}, $T_{1/2}$, AUC and MRT [30,37,47,48].

Table 2. Animal grouping for pharmacokinetic study of FBX.

Sr. No.	Groups		
	Treatment	**Set-I**	**Set-II**
1	Marketed oral suspension of FBX (4.07 mg/kg)	3 *	3 *
2	Transdermal patch of FBX (4.07 mg/kg)	3 *	3 *
3	Developed cubosomes of FBX(4.07 mg/kg)	3 *	3 *
4	MN patch of FBX (4.07 mg/kg)	3 *	3 *
5	Cubosomes of FBX loaded MN patch (4.07 mg/kg)	3 *	3 *
	Total	**30**	

* Not sacrificed, rehabilitated and used in pharmacodynamic study after washing period.

2.11. Pharmacodynamic Study

Sprague–Dawley rats weighing 200–270 g were procured from an official CPCSEA breeder. Rats were handled in a similar manner as described in pharmacokinetic study (Section 2.10). Twelve rats were allocated to 4 groups randomly as shown in Table 3. All animals except Normal control (group 1) were sensitized with Potassium oxonate (PO) (250 mg/kg in 0.9% saline solution, intraperitoneally- IP) for induction of gout [49,50]. Initial paw volumes of rats were determined in all groups. Group 3 was treated with FBX (4.07 mg/kg orally) for 28 days as a standard control. Group 4 was treated with Cubosomes-loaded MN patch of FBX for 28 days. After 0 and 28th day, not more than 0.5 mL blood was withdrawn from retro-orbital plexus route. All animals were euthanized humanely using overdose of diethyl ether for assessing biochemical parameter (uric acid) and X-ray of rat paw [48].

Table 3. Animal grouping for pharmacodynamic study of FBX.

Sr. No.	Group	Treatment	No of animals
1	Normal control	Distilled water	3 **
2	Model control	PO * (250 mg/kg in 0.9% saline solution, Intraperitoneally-IP) [49,50]	3 **
3	Standard control	FBX (4.07 mg/kg orally) + PO (250 mg/kg in 0.9% saline solution, Intraperitoneally-IP) [49,50]	3 **
4	Test control I	Cubosomes-loaded MN patch of FBX (4.07 mg/kg transdermally) + PO (250 mg/kg in 0.9% saline solution, Intraperitoneally) [49,50]	3 **
		Total	**12 ****

* PO: Potassium Oxonate. ** Sacrificed to harvest rat paw for X-ray.

2.12. Measurement of Uric Acid (UA)

When gout was induced in rat model, there was an increase in uric acid levels in rat blood which was measured quantitatively using uric acid enzyme kit which was purchased from Coral Clinical Systems, U.S. Nagar, Uttarakhand, India [51]. For measuring the concentration of uric acid, 3 test tubes were taken and labelled as blank, standard and sample. Then, 1.0 mL of working reagent (uricase) was added in each test tube and warmed at 37 °C for 5 min. After pre-warming, 25 μL sample (rat plasma) and standard (uric acid solution) were added in sample and standard test tube, respectively. In blank, only uricase solution was taken. All test tubes were incubated at 37 °C for minimum of 10 min. After 10 min, absorbance of all prepared samples was measured at 520 nm using UV spectrophotometer (Shimadzu, Japan UV-1800) against blank. A concentration of uric acid was measured using Equation (5).

$$Uric\ Acid\ Concentration = \frac{Abs\ (sample)}{Abs\ (Standard)} \times Concentration\ standard\ (\mathrm{mg/dL}) \quad (5)$$

2.13. X-ray

X-ray of rat's paw was taken at Angela Lobo clinic, Vadodara, after harvesting it from the rat after euthanizing humanely to study change in bone shape after completion of pharmacodynamic study. X-rays of all animals from all groups were taken to study efficacy of the developed formulation.

2.14. Stability Study

Ref. [52] Stability study of the optimized Febuxostat-loaded cubosomes and MN patch of FBX cubosomes was performed as per the ICH guidelines of stability study. Three sample from the prepared optimized formulation were stored in airtight vials at room temperature (25–30 °C) and at (40–50 °C) in stability chamber. After time intervals of one, two and three months, cubosomes were analyzed for vesicle size and entrapment efficiency and MN patch of FBX cubosomes were evaluated for in vitro dissolution time and AFF (Axial Fracture Force).

3. Results and Discussion

The cubosomes of FBX were prepared using a bottom-up approach and then they were loaded in an MN patch to bypass the stratum corneum. The process and formulation parameters were optimized using the design of experiments. The mean vesicle size, PDI and zeta potential of the FBX cubosomes were found to be 157.5 nm, 0.165 and −17.2 mv, respectively. The vesicle size was small enough to penetrate the stratum corneum and further permeate to the blood vessels. According to the literature, if nanocarriers have a vesicle size below 300 nm, it can efficiently reach to the deeper layer of skin [53]. Here, cubosomes which have a vesicle size of less than 300 nm were successfully prepared. Due to the smaller size, they can efficiently reach the dermis layer of the skin, and can get absorbed in the systemic circulation to obtain the desired therapeutic concentration in the blood. Zeta potential most commonly indicates the stability of the colloidal formulation. Various components used in the preparation of colloidal dispersion contribute to the development of zeta potential on the vesicles. The optimum zeta potential required for the stability of colloidal dispersion is ± 30 mV according to various literatures [54]. The optimized cubosomes of FBX have zeta potential of −17 mV which is way below that required for the stability of cubosomal dispersion. The negative zeta potential was obtained due to the presence of free oleic acid in the lipid. However, the prepared cubosomal dispersion was found to be stable at room temperature due to the stealthing effect of the stabilizer (PVA) used in the preparation of cubosomes of FBX [55]. Moreover, the optimized cubosomes were not going to be stored in the dispersion form but were loaded into MNs, which are solid structures. Thus, zeta potential of −17 mv did not affect the physical stability of cubosomes in the MNs. The mean % entrapment efficiency of the optimized formulation was found to be 85.2%, while the total drug content of cubosomes was found to be 97.28%, i.e., 1 mL of cubosomal dispersion contained 3.89 mg FBX. In the case of entrapment efficiency, high efficiency and the drug loading of FBX was obtained in cubosomes. The lipophilic property of the entrapped FBX is responsible for the high % of entrapment efficiency of the optimized formulation (Febuxostat- log p value- 3.3). Additionally, cubosomes have a distinct advantage of providing high entrapment efficiency of the encapsulated drug according to the literature [27,56].

The TEM image of the optimized FBX cubosomes is shown in Figure 1. The TEM image of cubosomes of FBX indicate that the cubosomes have a cubical shape with smooth surface [57]. The size of the cubosomes seen in the image was found to be in line with the results of vesicle size data obtained by the Malvern Zetasizer.

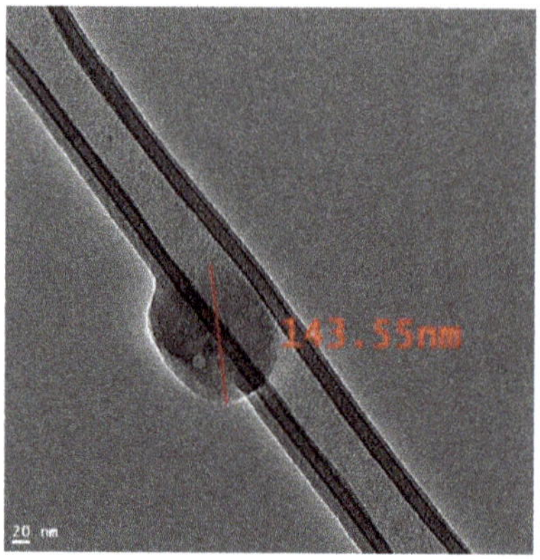

Figure 1. TEM image of Febuxostat-loaded cubosomes. From the TEM image of FBX-loaded cubosomes it can be observed that it has cubical shape.

SAXS was used for the investigation of the liquid crystalline structure of the prepared cubosomes and the results are shown in Figure 2. It showed a sequence of two well-defined scattering patterns and one diffuse diffraction pattern at Q values of 0.12, 1.75 and 1.0–2.25 A^{-1} region with relative positions on a curve, respectively. The peak at the Q value of 0.12 A^{-1} indicates characteristic scattering peaks due to the cubic phase, whereas the peak at 1.75 A^{-1} reveals a scattering pattern due to the whole cubosome liquid crystalline structure. The key characteristic of this X-Ray scattering diagram was a diffuse scattering pattern of low intensity in the region of 1.0–2.25 A^{-1} indicating the presence of water channels inside cubosomes which is a unique feature among all nanocarriers [58].

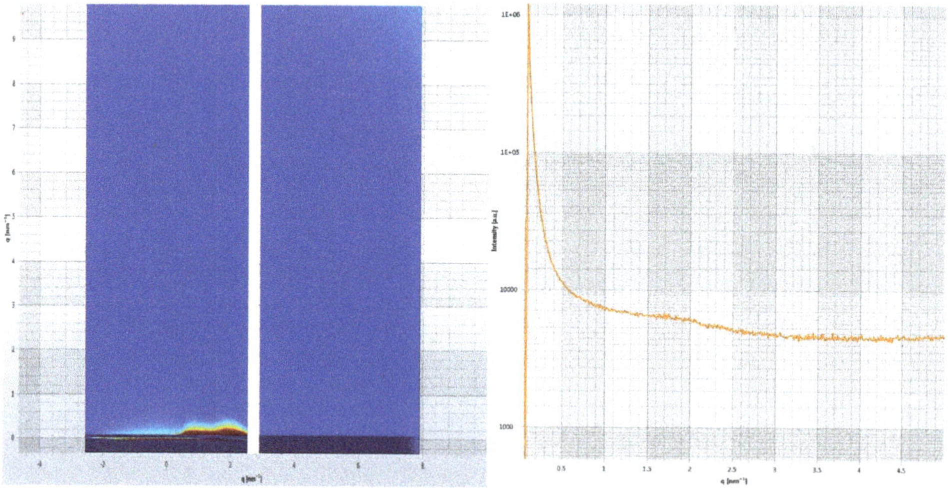

Figure 2. Scattering pattern of optimized cubosomes of FBX. SAXS analysis of prepared formulation was conducted to investigate the liquid crystalline structure of cubosomes. The diffuse scattering pattern of low intensity in the region of 1.0–2.25 A^{-1} indicating presence of water channels inside cubosomes which is a unique feature among all nanocarriers.

Ethanol was used in the preparation of the cubosomal dispersion and hence the final product was evaluated for the presence of ethanol using HS-GC. The results indicated that 167.55 pm of ethanol was present in the final formulation. The presence of organic solvents in the final formulation may pose the risk of toxicity. In order to prepare the cubosomes, ethanol was used as a co-solvent. The limit of ethanol in the final formulation is 1500 ppm according to the ICH guidelines Q3C (R6) for residual solvents [59]. The ethanol content in the prepared cubosomes of FBX was within permitted concentrations as per ICH guideline Q3C (R6) indicating no risk of toxicity due to the presence of ethanol [59].

The optimized cubosomal dispersion was then used to formulate the fast dissolving MN patch. Characterization of the MN patch was performed for various parameters. The Axial fracture force (AFF) is a good indicator of the mechanical strength of MNs and thus it was determined. The MNs should have sufficient mechanical strength, i.e., 0.03 N/MN so that it can breach the stratum corneum without breaking and deliver the loaded carrier systems directly to the dermis layer of skin [60]. The high value of Axial fracture force (1.2 N-calculated from Supplementary Figure S1) indicates the sufficient mechanical strength of the developed MN patch of FBX. The MNs with low mechanical strength may break during application. From the dermis layer, FBX can be absorbed directly in the systemic circulation and inhibit the xanthine oxidase enzyme present in blood pool. This enzyme is responsible for the formation of uric acid [1]. Thus, it can be interpreted that the prepared MN patch of cubosomes of FBX has a required minimum mechanical strength according to the literature [60].

The observation of the in vitro dissolution time of MNs indicates that MNs start to dissolve immediately and complete dissolution is obtained in 1.25 min (Supplementary Figure S2). The intention in the present research work was to develop a fast dissolving MN patch in order to obtain a rapid drug release.

SEM images of a fast dissolving MN patch are presented in Figure 3. These images showed smooth surfaced, conical MNs that have a length of 1.5 mm, and a base diameter of approx. 200 µm. The prepared MNs of cubosomes loaded with FBX and MNs of pure FBX were analyzed for SEM (Scanning Electron Microscopy). From the SEM images it was found that FBX cubosomes-loaded MNs have a smooth surface, while MNs of pure FBX have a rough surface. The rough surface of MNs of FBX was obtained due the insolubilized FBX. While cubosomes of FBX remained in a dispersed form and provided a smooth texture to the prepared MNs.

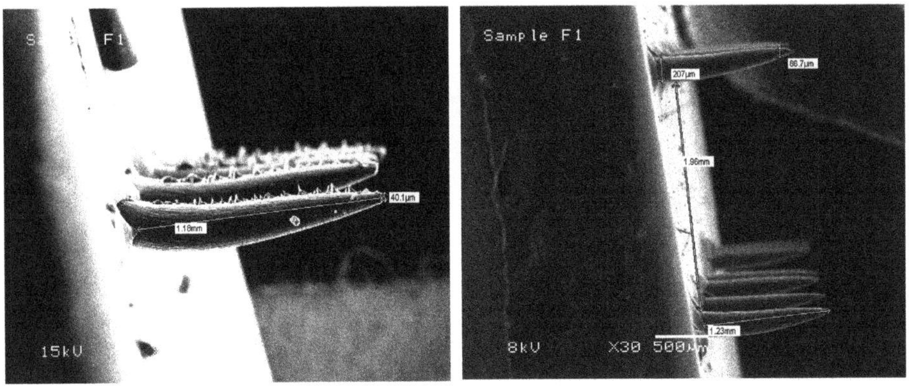

(A) (B)

Figure 3. SEM of MN patch (**A**) MN patch of FBX-plain drug (**B**) MN patch loaded with cubosomes of FBX. SEM analysis was performed to analyse the difference between prepared MN patches.

Skin penetrability of the prepared MN patches containing cubosomes of FBX was performed and the presence of tiny blue stains on the rat skin proves that the prepared MN patches were able to create the pores in the skin and deliver FBX directly to the dermis layer. (Supplementary Figure S3).

The application of an MN patch on rat skin formed pores which were detected using a microscope (4× and 10×) and its average pore diameters were measured as shown in

Figure 4. From pore closure kinetic studies, it was found that the size of the micropores formed in the skin decreased with time and complete pore closure was observed in 24 h. The results are in concurrence with the reported literature [61]. Closure of the pores created due to the MNs is necessary in order to avoid the chances of infection resulting from the open pores.

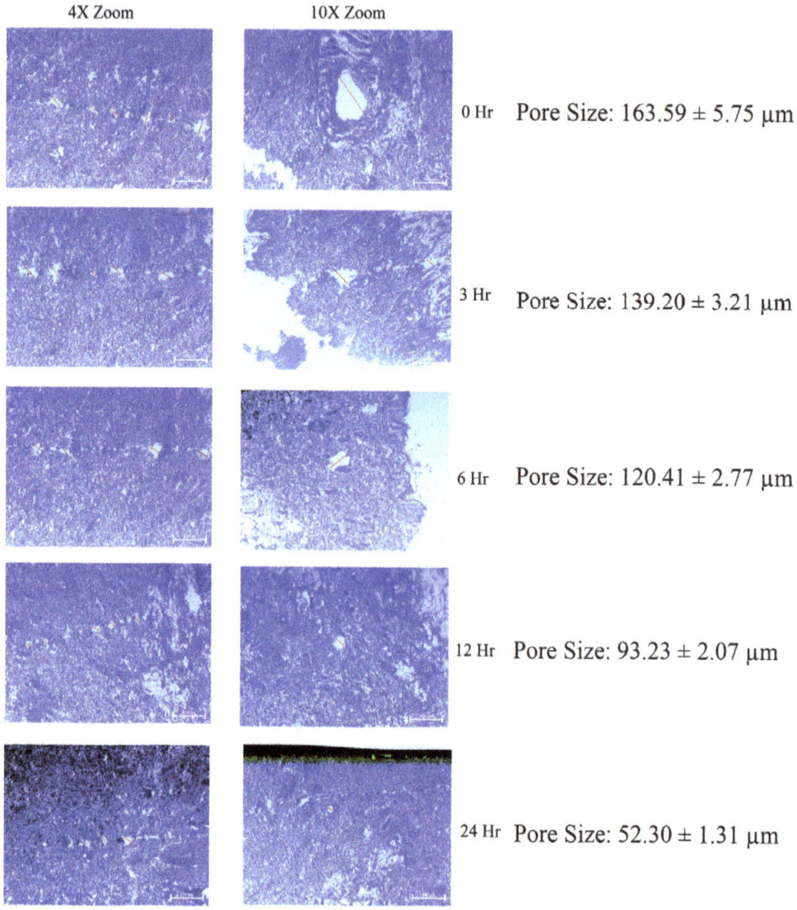

Figure 4. Pore closure kinetics for MN patch loaded with cubosomes of FBX. Pore closure kinetic study was performed to find out that how long the prepared pores will remain open after applying MN patch.

The physical stability of the cubosomes in the MN patch was ensured by the determination of the particle size and entrapment efficiency of cubosomes after a complete dissolution of MN patch in water. The initial vesicle size and entrapment efficiency of the cubosomes of FBX were found to be 157.5 ± 4.16 nm and 85.2 ± 2.68%, respectively, while the vesicle size and entrapment efficiency after the dissolution of the MN patch were found to be 154 ± 3.94 nm and 84.9 ± 1.88%, respectively. No significant change in the vesicle size and entrapment efficiency of cubosomes of FBX was observed after loading them in the MN patch, suggesting its stability in the MN patch. The total drug content of the MN patch containing FBX cubosomes was found to be 98.14% which means each MN patch of FBX cubosomes contains 3.336 mg of FBX.

3.1. In Vitro Drug Release Study

The cumulative percent release of the drug from all four formulations at various time intervals are shown in Figure 5. Various mathematical models were applied to the data of drug release from all formulations to find out the release behavior of FBX and the results are listed in Table 4. The results of the in vitro release study showed that more than 60% of FBX was released from cubosomes in 24 h, which was significantly greater than the release from plain drug suspension (39.97%) probably due to its low solubility in phosphate buffer pH 7.4. FBX is practically insoluble in water. Thus, when its suspension was prepared in phosphate buffer pH 7.4 and filled in a diffusion bag, it was not able to solubilize. In order to permeate across the dialysis bag, the drug molecules must be present in a solubilized state. Thus, due to the poor solubility of FBX, only a limited amount of the drug was able to permeate through the dialysis bag. On the other hand, cubosomes have the advantage of improving the surface area which is in contact with the phosphate buffer pH 7.4. Therefore, the more the amount of the drug dissolved in a phosphate buffer pH 7.4 diffuses to the donor compartment. Due to this reason, a higher amount of FBX was able to permeate the diffusion membrane and a higher in vitro release of FBX was obtained in the case of cubosomes of FBX [62,63]. The highest drug release was obtained from MN loaded either with the plain drug or with cubosomes of FBX. The micropores formed in the dialysis membrane seem to be responsible for allowing higher amounts of the drug to permeate through the membrane. Various mathematical models were applied for the in vitro drug release study. In the case of cubosomes, the highest R^2 value was obtained for the Korsmeyer–Peppas model suggesting a diffusion-controlled system, where the release rate was dependent on the drug concentration remaining within the cubosomes. Moreover, the n value of 0.607 for the Korsmeyer–Peppas model suggests non-Fickian diffusion of the drug from cubosomes [64]. Non-Fickian diffusion means that the diffusion of the drug from the cubosomes does not follow Fick's laws of diffusion. However, in the case of MN loaded with FBX cubosomes, the highest R^2 value was obtained for the first order indicating that the drug release rate depends on the concentration gradient across the membrane.

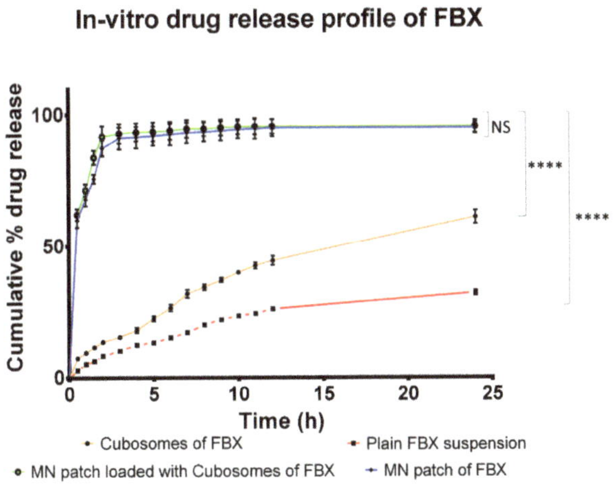

Figure 5. In vitro drug release from various developed formulations of FBX. In vitro drug release study was performed to find out the release time of FBX from various formulations. Each release study was performed in triplicate and SD was also included in figure in horizontal bar. NS indicates Not significant while **** indicates significance between data with p value < 0.0001.

Table 4. Various mathematical models and their correlation coefficient values.

Mathematical Models		MN Patch of FBX	MN Patch of FBX Cubosomes	Cubosomes of FBX	FBX Suspension
Zero order	R^2	0.6601	0.5432	0.7267	0.3063
First order		0.9719	0.9801	0.9498	0.5742
Higuchi		0.7631	0.7074	0.9568	0.9608
Hixon–Crowell		0.8492	0.7902	0.9114	0.4905
Korsmeyer–Peppas		0.8778	0.8339	0.9813	0.9653
	N	0.104	0.088	0.607	0.465

An in vitro release study of MNs of FBX and MNs loaded with cubosomes of FBX shows that more than 90% of the drug released in 2 h from both. The results showed a significant increase in drug release when compared with the drug release profile of FBX cubosomes. The optimized MN patches were able to penetrate the diffusion bag due to the sharp MN structure. After penetration in the membrane, MN structures rapidly dissolved due to the use of fast dissolving polymers leading to the rapid release of FBX in the diffusion medium. Whereas in the case of cubosomes, the drug has to diffuse through the intact diffusion bag and thus release is slow. Such observations suggest that due to the penetrating ability of MNs, a highly significant increase in drug release was possible. A Bonferroni's multiple comparison test of two-way ANOVA was applied in between the column and relative symbology was marked in Figure 5. Based on this, it can be said that there is a significant difference between the release of FBX from cubosomes and FBX suspension compared with the MN patch loaded with cubosomes of FBX while there is no significant difference between the MN patch of FBX and the MN patch loaded with cubosomes of FBX. Various mathematical models were applied for the in vitro release study of both MN patches and, according to the R^2 value for first order model, it was found to be highest which suggest that the release rate is dependent on the drug concentration in the carrier.

3.2. *In Vitro Cell Viability Study*

The cell viability data for FBX formulations are summarized graphically in Figure 6. In the cell viability study, a significantly lower viability of cells treated with Triton X 100 (6.82 ± 3.65%) indicated the validity of the positive control. The viability of cells treated with FBX-loaded cubosomes (90.32 ± 9.93%) was found to be significantly higher than the positive control and near to the negative control. A Dunnett's multiple comparison test of two-way ANOVA between the column was applied to prove the significance difference between the columns. There is significance difference among FBX suspension and triton X compared with cubosomes of FBX while there is no significant difference among PBS 7.4 and the placebo compared to cubosomes of FBX. Moreover, the viability of cells treated with FBX suspension was 67.48 ± 12.36% which is significantly less than the cubosomes of FBX. This indicated a less toxic nature of developed formulations compared with the FBX suspension and triton X 100.

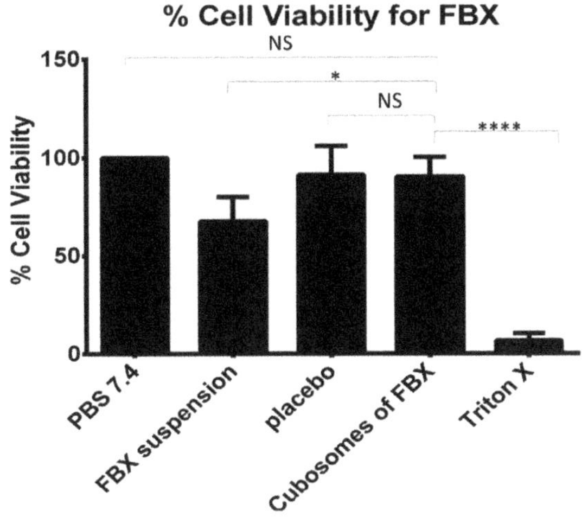

NS indicates Not Significant, "*" level of significance

Figure 6. % viability of fibroblast 3T3 cells after treatment with cubosomes of Febuxostat. Cell viability study was performed to investigate the toxicity of prepared formulation on fibroblast 3T3 cells. This study was performed in triplicate. NS indicates Not significant while **** indicates significance between data with p value < 0.0001.

3.3. *In Vitro Permeation Study*

The results of the in vitro permeation studies through the full-thickness rat skin are shown in Table 5 and Figure 7. A Bonferroni's multiple comparison test of two-way ANOVA was applied in between the column and the relative symbology was marked in Figure 7. Based on this, it can be said that there is a significant difference between the permeation of the drug across the skin from the cubosomes of FBX, the MN patch of FBX and FBX suspension compared with the MN patch loaded with cubosomes of FBX. A slight improvement in FBX permeation was observed with the MN patch of FBX (J_{ss}—6.45 μg/cm^2/h) as compared with FBX suspension (J_{ss}—4.20 μg/cm^2/h) due to the MN's ability to permeate the skin barrier. However, in the case of the cubosomes of FBX-loaded MN patch, an 8.34-fold increase in permeation (J_{ss}—35.06 μg/cm^2/h, PER—8.34) was observed due to the microporation of the skin. The results reflected that cubosomes of FBX (J_{ss}—18.43 μg/cm^2/h) can permeate the skin barrier less significantly than the cubosomes of FBX-loaded MN patch. The amount of the drug retained on the surface and deposited within the skin was determined after the completion of release experiments and the data are presented in Figure 8. On the basis of the results of permeability from various formulations, they are organized in order of increasing permeability: FBX Suspension < FBX MN patch < FBX cubosomes < cubosomes of FBX-loaded MN Patch. In the case of the MN patch, permeation obtained through rat skin was less than through FBX cubosomes and MN patch loaded with cubosomes of FBX due to the very poor solubility of FBX in water. The synergistic effect of microporation on cubosome's permeability was established as a significant enhancement was observed in the permeability through the cubosomes of FBX-loaded MN patch [37,65]. From Figure 8, it has been noted that that FBX MN exhibits the maximum deposition of FBX. The reason for this finding may be the lipophilic nature of FBX, which may not have been able to permeate to the systemic circulation in sufficient concentrations due to the deposition in the hydrophilic dermis layer. It can be concluded that permeation was better with the cubosomes of FBX-loaded MN when compared with cubosomes of FBX. In case of cubosomes of FBX, the retention of FBX on the skin was more compared with the cubosomes of FBX-loaded MN. Moreover, in the case of drug suspension, the maximum retention of the drug on the skin suggests that the drug alone is unable to cross the skin barrier efficiently.

Table 5. Amount of FBX permeated across Rat skin.

Time	Amount of Drug Permeated Per Unit Area of Rat Skin (μg/cm²)			
	FBX Suspension	**FBX Cubosomes**	**FBX MNP**	**Cubosomes Loaded MNP**
0.5	7.10 ± 0.42	34.20 ± 0.90	16.41 ± 0.76	61.49 ± 1.07
1.0	7.86 ± 0.31	39.68 ± 0.37	26.54 ± 0.63	78.60 ± 1.38
1.5	9.74 ± 0.66	48.13 ± 0.69	38.12 ± 1.41	91.56 ± 1.92
2.0	13.50 ± 0.88	64.21 ± 1.10	47.66 ± 5.22	113.67 ± 0.89
3.0	17.89 ± 0.30	87.16 ± 2.79	56.23 ± 2.17	163.46 ± 2.72
4.0	20.24 ± 0.41	101.08 ± 3.39	60.57 ± 1.98	196.81 ± 4.42
5.0	24.98 ± 0.39	124.46 ± 4.76	71.68 ± 2.29	234.77 ± 6.28
6.0	29.07 ± 0.68	145.33 ± 5.98	78.16 ± 2.32	251.92 ± 5.72
7.0	33.84 ± 0.79	161.77 ± 6.98	89.20 ± 4.67	266.54 ± 9.39
8.0	38.49 ± 0.93	179.10 ± 8.56	100.11 ± 1.69	293.75 ± 8.84
9.0	40.67 ± 1.12	189.07 ± 6.93	115.08 ± 5.78	326.34 ± 10.70
10.0	43.63 ± 1.58	208.46 ± 9.93	125.48 ± 2.384	354.80 ± 13.05
11.0	45.26 ± 1.24	223.01 ± 12.80	143.79 ± 5.70	378.48 ± 13.94
12.0	49.78 ± 1.99	247.49 ± 13.65	153.26 ± 3.38	410.72 ± 14.54
24.0	69.78 ± 2.75	345.10 ± 19.12	204.94 ± 6.44	503.84 ± 21.21
J_{ss}	**4.20**	**18.43**	**6.45**	**35.06**
PER	1	4.38	1.54	8.34

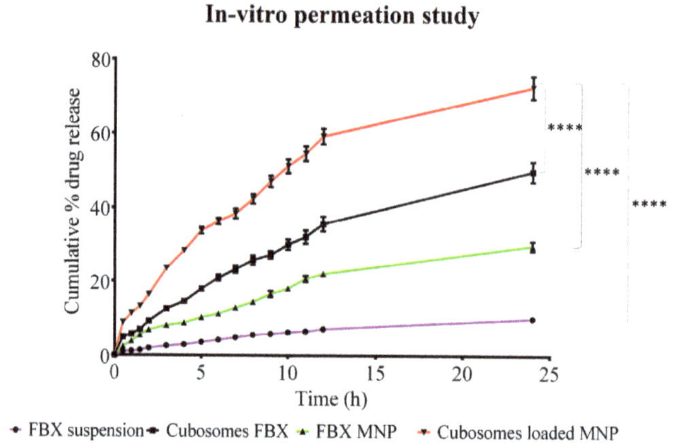

Figure 7. In vitro permeation of FBX from various developed formulations using rat skin. In vitro permeation was performed to investigate that the prepared formulations were able to cross the stratum corneum or not and release the drug to the release media. NS indicates Not significant while **** indicates significance between data with p value < 0.0001.

3.4. *Ex Vivo Fluorescence Microscopy Study*

Sections of the rat skin were exposed to FITC-loaded formulations for a period of 12 h. Further, after exposure, fluorescence microscopic images of rat skin sections were taken and the images are presented in Figure 9. Based on the results of the ex vivo fluorescence microscopy study, formulations are organized in increasing order of the fluorescence: FITC Suspension < FITC MN patch < FITC cubosomes < MN patch loaded with cubosomes of FITC. It was noted that the data collected from the fluorescence microscope experiment complied with the ex vivo permeation and deposition data wherein maximum fluorescence was reported in sections of skin which were exposed to the MN patch loaded with FITC cubosomes. Therefore, it can be concluded that there is enhanced permeation through developed nanocarriers-loaded fast dissolving MN patches [37].

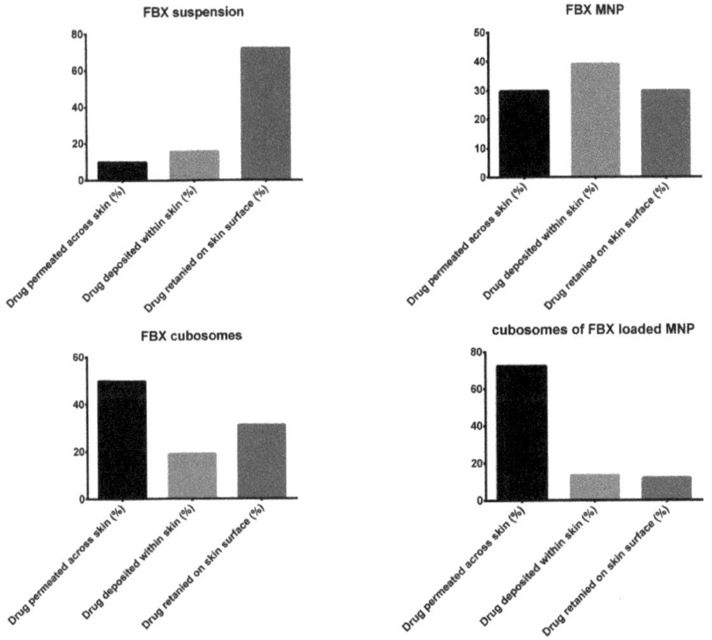

Figure 8. FBX distribution profile after 24 h of permeation study. During in vitro permeation study, the fraction of drug retained on skin surface, drug deposited within skin and drug permeated across the skin could be understood from this figure.

3.5. Histopathology Studies

The haematoxylin and eosin-stained sections of rat abdominal skins treated with developed cubosomes of FBX and cubosomes-loaded MNs of FBX were examined under a microscope for any pathological changes and compared with negative (PBS 7.4) and positive controls (IPA) to study the safety aspects of using an MN patch. The microscopic images have been shown in Figure 10. The sections of skins treated with developed cubosomes of FBX and cubosomes-loaded MNs of FBX showed almost similar cellular integrity as compared to skin treated with phosphate buffer saline (pH 7.4) as a negative control with no sign of inflammation. The section of skin treated with isopropyl alcohol as a positive control showed considerable damage to skin layers as an indication of irritation and toxicity. Therefore, it can be concluded that there is no evidence of damage to the skin samples after treatment with FBX suspension, cubosomes of FBX and MN loaded with cubosomes of FBX. This finding proves that the prepared formulations and drug samples are non-toxic and non-irritant to the skin samples. Moreover, in the case of skin sections treated with MN, microchannels were observed, which were created due to the insertion of the MN. This proves that MNs can effectively bypass the stratum corneum and deliver the drug directly to the dermis layer of skin for systemic delivery of the drug.

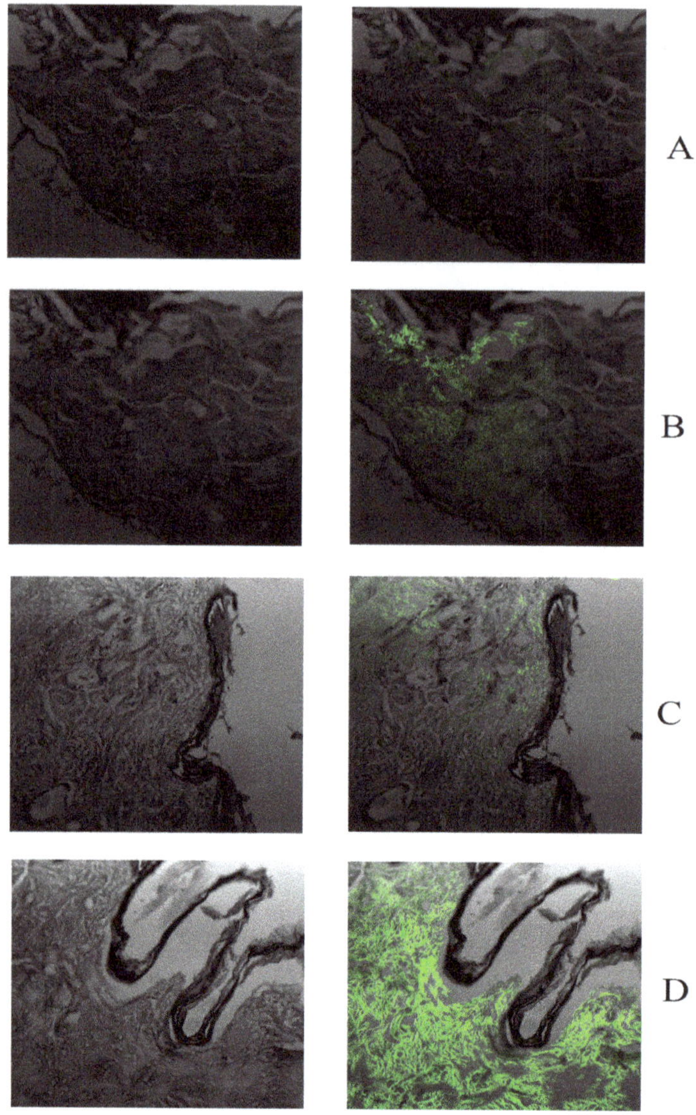

Figure 9. Fluorescence microscopic images of rat skin sections after 12 h of treatment with (**A**) FITC suspension, (**B**) MN patch of FITC, (**C**) cubosomes loaded with FITC, (**D**) MN patch with FITC cubosomes. From this figure, the permeation of drug across the skin from various formulations was well understood.

3.6. Pharmacokinetic Study

HPLC was employed for the determination of concentrations of FBX in the blood plasma of rats and the data are represented graphically in Figure 11. Thermo Scientific™ Kinetica Software was utilized for the calculation of various PK parameters from the collected data and is summarized in Table 6.

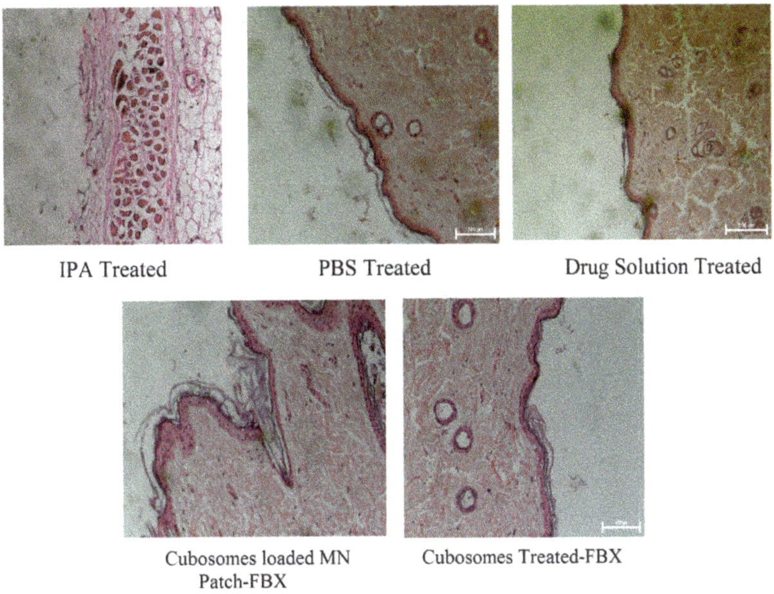

Figure 10. Histopathology study of developed formulation. Histopathology study was performed to understand the toxicity various formulation on the skin using positive and negative control.

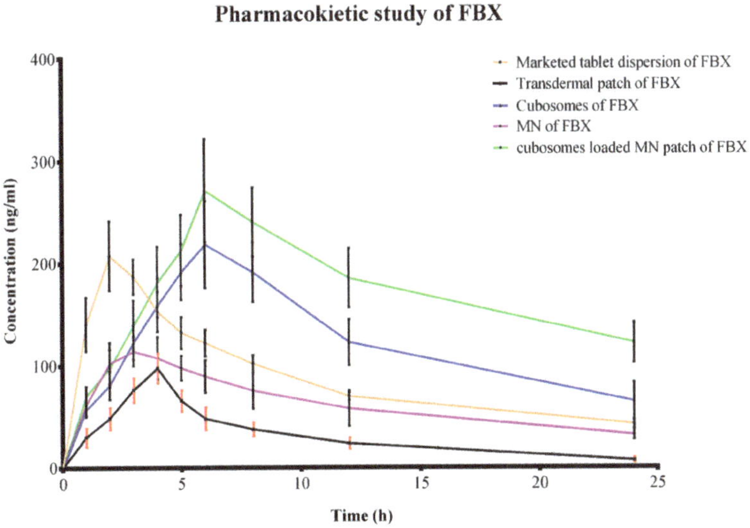

Figure 11. Plasma FBX concentration vs. Time profile of various dosage forms in Sprague–Dawley rats. Pharmacokinetic study was performed to understand the absorption of FBX from various formulations. Drug absorption profiles of various formulation were also compared using this figure.

Table 6. Pharmacokinetic parameters (FBX) computed using Kinetica Software.

Parameters	Marketed Oral Tablet	Transdermal Film	Cubosomal Gel	MN Patch	Cubosomal Loaded MN Patch
C_{max} (ng/mL)	207.53	97.58	218.44	112.48	271.03
T_{max}	2.00	4.00	6.00	5.00	6.00
AUC_{0-t} (ng × h/mL)	2784.06	798.84	3825.48	1957.89	7328.70
$T_{1/2}$ (h)	11.72	6.32	11.28	11.15	18.57
MRT (h)	16.91	9.79	18.43	17.72	29.05
F_{rel}	1	0.29	1.37	0.70	2.63

The results shown in Table 6 indicate that in the case of marketed oral tablets in suspension form, high C_{max} and low t_{max} values are obtained which indicate the rapid and good absorption of the drug through the oral route. Moreover, the value of $T_{1/2}$ and MRT indicate the slow elimination from the body leading to BID (twice a day) administration. In the case of simple transdermal film, very low values of C_{max} are obtained indicating much less absorption of the drug through the transdermal route in the absence of any penetration enhancement. On the other hand, cubosomal gel, FBX MN and FBX cubosomal MN show significantly higher values of C_{max} indicating an enhanced permeation of FBX compared with the transdermal film of FBX. Apart from this, cubosomal gel and FBX cubosomal MNs achieved C_{max} comparable to the marketed oral tablet and FBX cubosomal MNs obtained the highest C_{max} (271.03 ng/mL). The comparison of the AUC_{0-t} values indicate a highly significant difference between the different groups and the maximum value was obtained from FBX cubosome MNs. This is due to the synergistic effect of the cubosomes and the MN in enhancing the skin permeation. Both the approaches, viz., cubosomes and MNs individually led to an increase in transdermal permeation. However, FBX cubosomal MNs show a maximum AUC_{0-t} and MRT suggesting the maximum absorption of FBX and FBX is available in systemic circulation for a maximum period of time compared with other all formulations. However, a combination of both approaches results in a synergistic effect and an almost three-fold enhancement in bioavailability was observed. The results are in concurrence with the earlier reports [37]. An increase in the bioavailability can lead to a reduction in the dose and the associated side effects making the treatment safer and more patient compliant. The higher values of $T_{1/2}$ and MRT also indicate the possibility of a reduction in the dosage frequency. Upon the release of the FBX cubosomes after the MNs dissolution in the skin, the drug is released in a sustained manner from the cubosomes. The results are in concurrence with the in vitro release studies which indicate a sustained release of the drug from cubosomes.

3.7. Pharmacodynamic Study

As shown in Figure 12, there was an increase in uric acid levels in rat blood in the animals in which an induction of gout was performed. As shown in Figure 12A, the standard and test control animal groups have lower blood uric acid than the model control group indicating the efficacy of the developed formulation.

The bone X-rays of rats from all groups are depicted in Figure 12B. From Figure 12B, it was observed that there was no tophi formation in any set of the animal groups. Moreover, no structural change in bones of rat was observed during the study.

In the pharmacodynamic study, gout was induced in the Wistar rat using potassium oxonate. In the rat, uricase enzymes are responsible for the metabolism of uric acid. Thus, to increase uric acid levels in rat blood, uricase enzymes must be inhibited. Potassium oxonate inhibits this uricase enzyme and increases uric acid levels in rat blood [66]. When the blood levels of uric acid increases above 6 mg/dL, it results in the formation of Monosodium urate (MSU) crystals. These MSU crystals then deposit in various joints and form a tophus-like structure. However, these elevated levels of uric acid do not result in the formation of MSU crystals every time. Thus, in many cases the hyperuricemia patients do not develop this tophus-like structure. However, these patients are prone to the formation of tophi around various bone joints if uric acid levels in blood are not controlled [1].

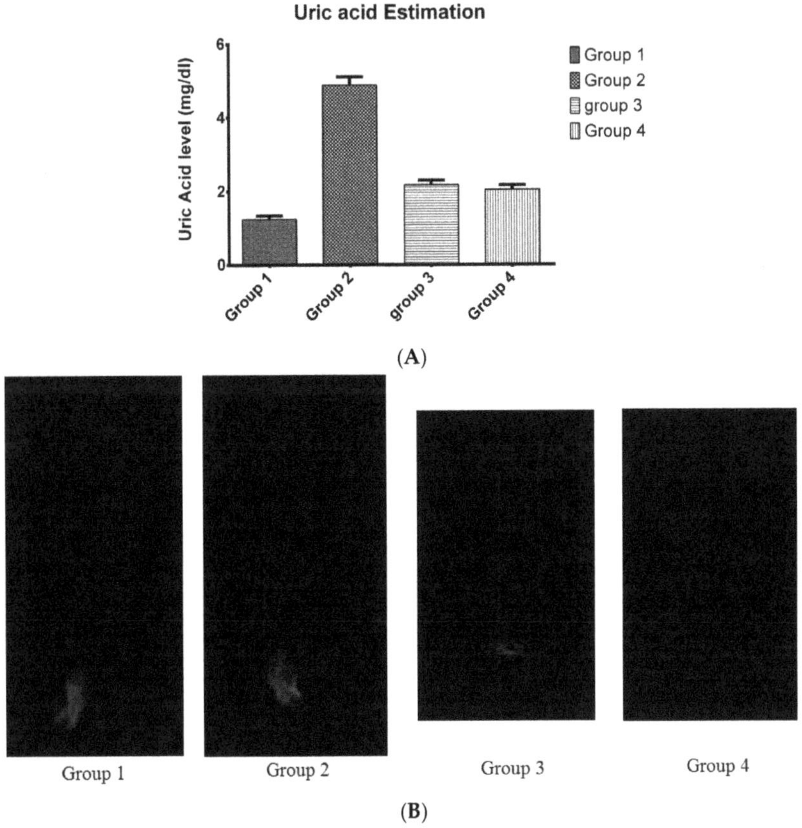

Figure 12. Pharmacodynamic study (**A**) Measurement of serum uric acid level in rat serum, (**B**) X-ray of rat's paw to study efficacy of the developed formulation of FBX. The efficacy of the prepared formulation was compared to the marketed formulation by its ability to reduce the serum uric acid level.

Due to this reason, bone X-rays of rats did not confirm the formation of any tophi during the induction of gout. However, increased levels of uric acid suggests the induction of hyperuricemia, and this increased level of uric acid in the blood is responsible for the induction of gout according to the literature [1,67,68]. The uric acid level was controlled using a developed and marketed formulation and proved the efficacy of the MN patch loaded with cubosomes of FBX and cubosomal gel of FBX.

3.8. Stability Study

The results of AFF (Axial Fracture Force) and the in vitro dissolution time of MNP and vesicle size, PDI and the percent of drug entrapment of cubosomes stored for up to three months are summarized in Figure 13. In the stability study it was found out that, under storage conditions, AFF of the MN patch was slightly decreased while there is no effect on the in vitro dissolution time of developed MNP. Similarly, a slight increase in vesicle size and PDI, as well as a slight decrease in drug entrapment, was evident in the storage in cubosomal formulations. However, the values observed after three months were found to be within the desirable limits required for formulations to perform effectively. Such observations at intermediate temperatures and a high relative humidity could interpret that solid MN patches should be stored in airtight containers with silica bags to absorb moisture content while cubosomal dispersion should be stored in airtight containers at room temperature. Bonferroni's multiple comparison test of two-way ANOVA was applied in between the column and relative symbology was marked in Figure 13. From this, it can be concluded that there is no significant difference between the obtained data sets meaning the developed formulations are stable in test conditions.

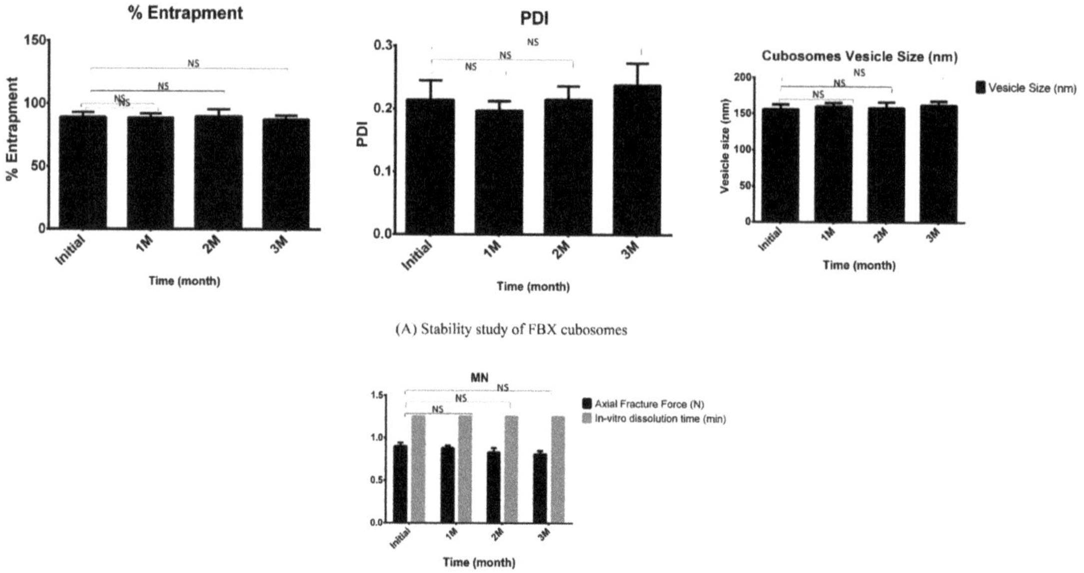

(A) Stability study of FBX cubosomes

(B) Stability study of MN patch loaded with FBX cubosome

Figure 13. Stability Study of developed formulations. Short term stability study of the prepared formulations was conducted and they were characterized for various tests such as vesicle size, PDI, % entrapment efficiency for FBX cubosomes and AFF (Axial Fracture Force) and in vitro dissolution time for FBX cubosomes of MN patch. NS indicates Not significant.

During the development of cubosomal FBX MN, processes that ease up the lab process to scale up to commercial scale is necessary. For instance, cubosomes preparation using a bottom-up approach, MN patch preparation by a micromold casting method, etc. However, the scale up for cubosomes and MN preparation still needs to be investigated. To ensure patient concern due to deposition of polymers used to prepare polymeric MN needs to be investigated. Like other polymeric MNs, cubosomal FBX MNs were also anticipated to deposit small amounts of polymers with every application which may further enhance upon several uses over an extended period of time. However, the molecular weights of these polymers play a significant role in their elimination. For example, a report on the safety assessment of PVP suggests that for its complete excretion, the polymer size must be below the glomerular threshold [69,70]. Hence, while selecting the polymers, consideration was given to low molecular weight polymers, viz., poly vinyl pyrollidone (MW, 3.5 kD) and poly vinyl alcohol (MW, 6 kD). Additionally, rotating the site of application may also be helpful in avoiding local accumulation. Thus, polymer accumulation during multiple uses of MNs still needs to be investigated. Skin often repairs itself against minor mechanical trauma associated with injuries such as scrapes and scratches without developing infection. Hence, despite the breaching of the stratum corneum to deliver the payload to viable epidermis and dermis, MNP poses less risk of infection via microbial influx through MN-induced microchannels as experienced during their appropriate clinical usage [71]. Further, an in vitro study conducted to evaluate the ability of microorganisms to cause infection when allowed to ingress via MN-induced microchannels revealed that no microorganisms were able to penetrate viable epidermis and dermis making it unlikely to develop infections in immunecompetent patients [72]. Therefore, it still remains for the present study to prove that microorganisms are not able to cross the microchannels that are created by MNs. However, in order to assure safe use in every patient, the need for sterility in MNPs has been widely discussed and seems to be critical for regulatory approval [73]. Aseptic manufacturing as well as terminal sterilization via moist heat, dry heat and gamma radiation methods have already been investigated [74]. Aseptic manufacturing is inconvenient and fairly expensive while terminal sterilization often adversely affects the physicochemical properties of polymeric MNPs or their cargo. Hence, a suitable sterilization method must be considered carefully to avoid such issues where the use of antimicrobial agents has also been suggested to achieve a self-sterilization effect at the injection site [73,75]. Furthermore, being fast dissolving in nature, cubosomal FBX MNs will not produce any sharp waste and, hence, will not require safe disposal.

4. Conclusions

The present investigation was aimed to overcome the poor oral bioavailability and gastrointestinal-related disturbances associated with the FBX via the development of a transdermal formulation which can deliver FBX in a controlled manner to achieve therapeutic goals with better patient compliance. In the present study, in the first phase, cubosomes of FBX were prepared with GMO as a lipid and PVA as a stabilizer using a bottom-up approach and were evaluated for various characterization tests. Then, in the next phase, the developed cubosomes of FBX were loaded in the MN patch and evaluated for various characterization tests. The QbD methodology was applied for the optimization of both formulations. The particle size and entrapment efficiency were selected as CQAs for cubosomes of FBX while the axial fracture force and dissolution time of MNs were selected as CQAs for FBX cubosomal MNs. The cubosomal FBX and FBX cubosomal MNs were subjected to the stability study and were found to be stable during the stability period. However, the MN patch required special airtight container closure systems due to the hydroscopic nature of a polymer used in preparation of the MN patch. Form the ex vivo study, it can be concluded that the MN patch loaded with cubosomes of FBX has the highest transdermal flux followed by cubosomes of FBX. From the histopathological evolution of the prepared formulations, it can be established that the prepared formulations are nontoxic to the rat skin and both MN patches are capable of breaching the stratum corneum successfully. From the pharmacokinetic study, it can be concluded that the MN patch loaded with cubosomes of FBX can permeate the stratum corneum and deliver the FBX to systemic circulation more successfully than any other formulation, followed by the cubosomal gel of FBX. In the case of the pharmacodynamic study, it was observed that the MN patch loaded with cubosomes of FBX was able to control the uric levels in rat blood more successfully than other formulations, followed by the cubosomal gel of FBX.

Supplementary Materials: The following supporting information can be downloaded at: https://www.mdpi.com/article/10.3390/pharmaceutics15010224/s1, Figure S1. Load vs. Time curve for MN patch loaded with cubosomes of FBX; Figure S2. In-vitro dissolution study of MN patch containing FBX loaded cubosomes; Figure S3. Skin penetrability of Cubosomes of FBX loaded MN Patch.

Author Contributions: B.P.: Conceptualization, Methodology, Investigation, Analysis, Data Curation, Writing—Original Draft Preparation. H.T.: Conceptualization, Writing—Review and Editing, Visualization, Supervision. All authors have read and agreed to the published version of the manuscript.

Funding: The present research work was funded by the ICMR (Indian Council of Medical Research)-Government of India.

Institutional Review Board Statement: An animal study was performed as per animal protocol number MSU/IAEC/2019-20/1921 approved by the Institutional Animal Ethics Committee of The Maharaja Sayajirao University of Baroda, Vadodara, Gujarat-390002, India on 22 February 2020.

Informed Consent Statement: No human study was performed in this study.

Data Availability Statement: A data is not available due to privacy reason.

Acknowledgments: Authors are thankful to Ami drugs and specialty chemicals Pvt. Ltd. and Mohini Organics for providing Febuxostat and GMO, respectively, as gift samples. Authors are also thankful to ICMR for research funding.

Conflicts of Interest: The authors declare no conflict of interest.

References

1. Ragab, G.; Elshahaly, M.; Bardin, T. Gout: An old disease in new perspective—A review. *J. Adv. Res.* **2017**, *8*, 495–511. [CrossRef] [PubMed]
2. Dalbeth, N.; Choi, H.K.; Joosten, L.A.; Khanna, P.P.; Matsuo, H.; Perez-Ruiz, F.; Stamp, L.K. Gout (Primer). *Nat. Rev. Dis. Prim.* **2019**, *5*, 69. [CrossRef] [PubMed]
3. Engel, B.; Just, J.; Bleckwenn, M.; Weckbecker, K. Treatment options for gout. *Dtsch. Ärzteblatt Int.* **2017**, *114*, 215. [CrossRef] [PubMed]
4. Sahai, R.; Sharma, P.K.; Misra, A.; Dutta, S. Pharmacology of the Therapeutic Approaches of Gout. In *Recent Advances in Gout*; IntechOpen: London, UK, 2019.
5. Tripathi, K. *Essentials of Medical Pharmacology*; JP Medical Ltd.: London, UK, 2013.
6. Vohra, A.M.; Patel, C.V.; Kumar, P.; Thakkar, H.P. Development of dual drug loaded solid self microemulsifying drug delivery system: Exploring interfacial interactions using QbD coupled risk based approach. *J. Mol. Liq.* **2017**, *242*, 1156–1168. [CrossRef]
7. Alhakamy, N.A.; Fahmy, U.A.; Ahmed, O.A.; Almohammadi, E.A.; Alotaibi, S.A.; Aljohani, R.A.; Alharbi, W.S.; Alfaleh, M.A.; Alfaifi, M.Y. Development of an optimized febuxostat self-nanoemulsified loaded transdermal film: In-vitro, ex-vivo and in-vivo evaluation. *Pharm. Dev. Technol.* **2020**, *25*, 326–331. [CrossRef]
8. Bariya, S.H.; Gohel, M.C.; Mehta, T.A.; Sharma, O.P. Microneedles: An emerging transdermal drug delivery system. *J. Pharm. Pharmacol.* **2012**, *64*, 11–29. [CrossRef]
9. Desale, R.; Wagh, K.; Akarte, A.; Baviskar, D.; Jain, D. Microneedle technology for advanced drug delivery: A review. *Int. J. PharmTech Res.* **2012**, *4*, 181–189.
10. Larrañeta, E.; Lutton, R.E.; Woolfson, A.D.; Donnelly, R.F. Microneedle arrays as transdermal and intradermal drug delivery systems: Materials science, manufacture and commercial development. *Mater. Sci. Eng. R Rep.* **2016**, *104*, 1–32. [CrossRef]
11. Tuan-Mahmood, T.-M.; McCrudden, M.T.; Torrisi, B.M.; McAlister, E.; Garland, M.J.; Singh, T.R.R.; Donnelly, R.F. Microneedles for intradermal and transdermal drug delivery. *Eur. J. Pharm. Sci.* **2013**, *50*, 623–637. [CrossRef]
12. van der Maaden, K.; Jiskoot, W.; Bouwstra, J. Microneedle technologies for (trans) dermal drug and vaccine delivery. *J. Control. Release* **2012**, *161*, 645–655. [CrossRef]
13. Singh, S.; Parashar, P.; Kanoujia, J.; Singh, I.; Saha, S.; Saraf, S.A. Transdermal potential and anti-gout efficacy of Febuxostat from niosomal gel. *J. Drug Deliv. Sci. Technol.* **2017**, *39*, 348–361. [CrossRef]
14. El-Shenawy, A.A.; Abdelhafez, W.A.; Ismail, A.; Kassem, A.A. Formulation and Characterization of Nanosized Ethosomal Formulations of Antigout Model Drug (Febuxostat) Prepared by Cold Method: In Vitro/Ex Vivo and In Vivo Assessment. *AAPS PharmSciTech* **2020**, *21*, 31. [CrossRef] [PubMed]
15. Kanke, P.K.; Pathan, I.B.; Jadhav, A.; Usman, M.R.M. Formulation and evaluation of febuxostat nanoemulsion for transdermal drug delivery. *J. Pharm. BioSciences* **2019**, *7*, 1–7.
16. Karami, Z.; Hamidi, M. Cubosomes: Remarkable drug delivery potential. *Drug Discov. Today* **2016**, *21*, 789–801. [CrossRef] [PubMed]
17. Garg, G.; Saraf, S.; Saraf, S. Cubosomes: An overview. *Biol. Pharm. Bull.* **2007**, *30*, 350–353. [CrossRef] [PubMed]
18. Chong, J.Y.; Mulet, X.; Boyd, B.J.; Drummond, C.J. Steric stabilizers for cubic phase lyotropic liquid crystal nanodispersions (cubosomes). In *Advances Planar Lipid Bilayers Liposomes*; Elsevier: Amsterdam, The Netherlands, 2015; Volume 21, pp. 131–187.
19. Zhang, P.; Zhang, Y.; Liu, C.-G. Polymeric nanoparticles based on carboxymethyl chitosan in combination with painless microneedle therapy systems for enhancing transdermal insulin delivery. *RSC Adv.* **2020**, *10*, 24319–24329. [CrossRef] [PubMed]
20. Alimardani, V.; Abolmaali, S.S.; Yousefi, G.; Rahiminezhad, Z.; Abedi, M.; Tamaddon, A.; Ahadian, S. Microneedle arrays combined with nanomedicine approaches for transdermal delivery of therapeutics. *J. Clin. Med.* **2021**, *10*, 181. [CrossRef]
21. Larrañeta, E.; McCrudden, M.T.; Courtenay, A.J.; Donnelly, R.F. Microneedles: A new frontier in nanomedicine delivery. *Pharm. Res.* **2016**, *33*, 1055–1073. [CrossRef]
22. Park, J.-H.; Allen, M.G.; Prausnitz, M.R. Biodegradable polymer microneedles: Fabrication, mechanics and transdermal drug delivery. *J. Control. Release* **2005**, *104*, 51–66. [CrossRef]
23. Prausnitz, M.R.; Langer, R. Transdermal drug delivery. *Nat. Biotechnol.* **2008**, *26*, 1261–1268. [CrossRef]

24. Lee, J.W.; Park, J.-H.; Prausnitz, M.R. Dissolving microneedles for transdermal drug delivery. *Biomaterials* **2008**, *29*, 2113–2124. [CrossRef] [PubMed]
25. Donnelly, R.F.; Singh, T.R.R.; Woolfson, A.D. Microneedle-based drug delivery systems: Microfabrication, drug delivery, and safety. *Drug Deliv.* **2010**, *17*, 187–207. [CrossRef] [PubMed]
26. Hong, X.; Wei, L.; Wu, F.; Wu, Z.; Chen, L.; Liu, Z.; Yuan, W. Dissolving and biodegradable microneedle technologies for transdermal sustained delivery of drug and vaccine. *Drug Des. Dev. Ther.* **2013**, *7*, 945.
27. Patel, B.; Thakkar, H.P. Cubosomes: Novel Nanocarriers for Drug Delivery. In *Nanocarriers: Drug Delivery System*; Springer: Berlin/Heidelberg, Germany, 2021; pp. 227–254.
28. Parinaz, S.; Renata, I.; Ben, J. Impact of Preparation Method and Variables on the Internal Structure, Morphology, and Presence of Liposomes in Phytantriol-Pluronic (r) F127 Cubosomes. *Colloids Surf. B Biointerfaces* **2016**, *145*, 845–853.
29. Thakkar, H.; Pandya, K.; Patel, B. Microneedle-mediated transdermal delivery of tizanidine hydrochloride. In *Drug Delivery Systems*; Springer: Berlin/Heidelberg, Germany, 2020; pp. 239–258.
30. Li, Z.; He, Y.; Deng, L.; Zhang, Z.-R.; Lin, Y. A fast-dissolving microneedle array loaded with chitosan nanoparticles to evoke systemic immune responses in mice. *J. Mater. Chem. B* **2020**, *8*, 216–225. [CrossRef]
31. Mansouri, S.; Cuie, Y.; Winnik, F.; Shi, Q.; Lavigne, P.; Benderdour, M.; Beaumont, E.; Fernandes, J.C. Characterization of folate-chitosan-DNA nanoparticles for gene therapy. *Biomaterials* **2006**, *27*, 2060–2065. [CrossRef]
32. Qi, L.; Xu, Z.; Jiang, X.; Hu, C.; Zou, X. Preparation and antibacterial activity of chitosan nanoparticles. *Carbohydr. Res.* **2004**, *339*, 2693–2700. [CrossRef]
33. Amidi, M.; Romeijn, S.G.; Borchard, G.; Junginger, H.E.; Hennink, W.E.; Jiskoot, W. Preparation and characterization of protein-loaded N-trimethyl chitosan nanoparticles as nasal delivery system. *J. Control. Release* **2006**, *111*, 107–116. [CrossRef]
34. Pan, Y.; Li, Y.-j.; Zhao, H.-y.; Zheng, J.-m.; Xu, H.; Wei, G.; Hao, J.-s. Bioadhesive polysaccharide in protein delivery system: Chitosan nanoparticles improve the intestinal absorption of insulin in vivo. *Int. J. Pharm.* **2002**, *249*, 139–147. [CrossRef]
35. Deshpande, S.; Venugopal, E.; Ramagiri, S.; Bellare, J.R.; Kumaraswamy, G.; Singh, N. Enhancing cubosome functionality by coating with a single layer of poly-ε-lysine. *ACS Appl. Mater. Interfaces* **2014**, *6*, 17126–17133. [CrossRef]
36. Rizwan, S.; Dong, Y.-D.; Boyd, B.J.; Rades, T.; Hook, S. Characterisation of bicontinuous cubic liquid crystalline systems of phytantriol and water using cryo field emission scanning electron microscopy (cryo FESEM). *Micron* **2007**, *38*, 478–485. [CrossRef] [PubMed]
37. Witschi, C.; Doelker, E. Residual solvents in pharmaceutical products: Acceptable limits, influences on physicochemical properties, analytical methods and documented values. *Eur. J. Pharm. Biopharm.* **1997**, *43*, 215–242. [CrossRef]
38. Srivastava, P.K.; Thakkar, H.P. Vinpocetine loaded ultradeformable liposomes as fast dissolving microneedle patch: Tackling treatment challenges of dementia. *Eur. J. Pharm. Biopharm.* **2020**, *156*, 176–190. [CrossRef] [PubMed]
39. Sheshala, R.; Anuar, N.K.; Samah, N.H.A.; Wong, T.W. In vitro drug dissolution/permeation testing of nanocarriers for skin application: A comprehensive review. *AAPS PharmSciTech* **2019**, *20*, 1–28. [CrossRef]
40. Avachat, A.M.; Parpani, S.S. Formulation and development of bicontinuous nanostructured liquid crystalline particles of efavirenz. *Colloids Surf. B Biointerfaces* **2015**, *126*, 87–97. [CrossRef]
41. Segeritz, C.-P.; Vallier, L. Cell culture: Growing cells as model systems in vitro. In *Basic Science Methods for Clinical Researchers*; Elsevier: Amsterdam, The Netherlands, 2017; pp. 151–172.
42. Helgason, C.D.; Miller, C.L. *Basic Cell Culture Protocols*; Humana Press: Totowa, NJ, USA, 2005.
43. Basak, V.; Bahar, T.E.; Emine, K.; Yelda, K.; Mine, K.; Figen, S.; Rustem, N. Evaluation of cytotoxicity and gelatinases activity in 3T3 fibroblast cell by root repair materials. *Biotechnol. Biotechnol. Equip.* **2016**, *30*, 984–990. [CrossRef]
44. Indermun, S.; Choonara, Y.E.; Kumar, P.; du Toit, L.C.; Modi, G.; van Vuuren, S.; Luttge, R.; Pillay, V. Ex vivo evaluation of a microneedle array device for transdermal application. *Int. J. Pharm.* **2015**, *496*, 351–359. [CrossRef]
45. Salah, S.; Mahmoud, A.A.; Kamel, A.O. Etodolac transdermal cubosomes for the treatment of rheumatoid arthritis: Ex vivo permeation and in vivo pharmacokinetic studies. *Drug Deliv.* **2017**, *24*, 846–856. [CrossRef]
46. Li, Y.; Wang, C.; Wang, J.; Chu, T.; Zhao, L.; Zhao, L. Permeation-enhancing effects and mechanisms of O-acylterpineol on isosorbide dinitrate: Mechanistic insights based on ATR-FTIR spectroscopy, molecular modeling, and CLSM images. *Drug Deliv.* **2019**, *26*, 107–119. [CrossRef]
47. Chen, C.-H.; Shyu, V.B.-H.; Chen, C.-T. Dissolving microneedle patches for transdermal insulin delivery in diabetic mice: Potential for clinical applications. *Materials* **2018**, *11*, 1625. [CrossRef]
48. Bancroft, J.D.; Gamble, M. *Theory and Practice of Histological Techniques*; Elsevier Health Sciences: Amsterdam, The Netherlands, 2008.
49. Thakkar, H.P.; Savsani, H.; Kumar, P. Ethosomal hydrogel of raloxifene HCl: Statistical optimization & ex vivo permeability evaluation across microporated Pig ear skin. *Curr. Drug Deliv.* **2016**, *13*, 1111–1122.
50. Slaoui, M.; Fiette, L. Histopathology procedures: From tissue sampling to histopathological evaluation. In *Drug Safety Evaluation*; Springer: Berlin/Heidelberg, Germany, 2011; pp. 69–82.
51. Wang, Z.Z.; Liu, F.; Gong, Y.F.; Huang, T.Y.; Zhang, X.M.; Huang, X.Y. Antiarthritic effects of sorafenib in rats with adjuvantbiticval arthritis. *Anat. Rec.* **2018**, *301*, 1519–1526. [CrossRef] [PubMed]

52. Li, L.; Teng, M.; Liu, Y.; Qu, Y.; Zhang, Y.; Lin, F.; Wang, D. Anti-gouty arthritis and antihyperuricemia effects of sunflower (*Helianthus annuus*) head extract in gouty and hyperuricemia animal models. *BioMed Res. Int.* **2017**, *2017*, 5852076. [CrossRef] [PubMed]
53. De Souza, M.R.; de Paula, C.A.; de Resende, M.L.P.; Grabe-Guimarães, A.; de Souza Filho, J.D.; Saúde-Guimarães, D.A. Pharmacological basis for use of Lychnophora trichocarpha in gouty arthritis: Anti-hyperuricemic and anti-inflammatory effects of its extract, fraction and constituents. *J. Ethnopharmacol.* **2012**, *142*, 845–850. [CrossRef] [PubMed]
54. Zhang, G.-B.; Ren, S.-S.; Wang, B.-Y.; Tian, L.-Q.; Bing, F.-H. Hypouricemic effect of flaccidoside II in rodents. *J. Nat. Med.* **2017**, *71*, 329–333. [CrossRef] [PubMed]
55. Zhao, Y.; Yang, X.; Lu, W.; Liao, H.; Liao, F. Uricase based methods for determination of uric acid in serum. *Microchim. Acta* **2009**, *164*, 1–6. [CrossRef]
56. Guideline, I.H.T. Stability testing of new drug substances and products. *Q1A (R2) Curr. Step* **2003**, *4*, 1–24.
57. Danaei, M.; Dehghankhold, M.; Ataei, S.; Hasanzadeh Davarani, F.; Javanmard, R.; Dokhani, A.; Khorasani, S.; Mozafari, M. Impact of particle size and polydispersity index on the clinical applications of lipidic nanocarrier systems. *Pharmaceutics* **2018**, *10*, 57. [CrossRef]
58. Samimi, S.; Maghsoudnia, N.; Eftekhari, R.B.; Dorkoosh, F. Lipid-based nanoparticles for drug delivery systems. In *Characterization and Biology of Nanomaterials for Drug Delivery*; Elsevier: Amsterdam, The Netherlands, 2019; pp. 47–76.
59. Freitas, C.; Müller, R.H. Effect of light and temperature on zeta potential and physical stability in solid lipid nanoparticle (SLN™) dispersions. *Int. J. Pharm.* **1998**, *168*, 221–229. [CrossRef]
60. Jain, D.; Athawale, R.; Bajaj, A.; Shrikhande, S.; Goel, P.N.; Gude, R.P. Studies on stabilization mechanism and stealth effect of poloxamer 188 onto PLGA nanoparticles. *Colloids Surf. B Biointerfaces* **2013**, *109*, 59–67. [CrossRef]
61. European Medicines Agency. ICH Guideline Q3C (R5) on Impurities: Guideline for Residual Solvents. Available online: https://www.tga.gov.au/sites/default/files/ichq3cr5.pdf (accessed on 3 January 2023).
62. Olatunji, O.; Das, D.B.; Garland, M.J.; Belaid, L.; Donnelly, R.F. Influence of array interspacing on the force required for successful microneedle skin penetration: Theoretical and practical approaches. *J. Pharm. Sci.* **2013**, *102*, 1209–1221. [CrossRef] [PubMed]
63. Kalluri, H.; Kolli, C.S.; Banga, A.K. Characterization of microchannels created by metal microneedles: Formation and closure. *AAPS J.* **2011**, *13*, 473–481. [CrossRef] [PubMed]
64. Ahirrao, M.; Shrotriya, S. In vitro and in vivo evaluation of cubosomal in situ nasal gel containing resveratrol for brain targeting. *Drug Dev. Ind. Pharm.* **2017**, *43*, 1686–1693. [CrossRef] [PubMed]
65. Peng, X.; Zhou, Y.; Han, K.; Qin, L.; Dian, L.; Li, G.; Pan, X.; Wu, C. Characterization of cubosomes as a targeted and sustained transdermal delivery system for capsaicin. *Drug Des. Dev. Ther.* **2015**, *9*, 4209. [CrossRef] [PubMed]
66. Wu, I.Y.; Bala, S.; Škalko-Basnet, N.; Di Cagno, M.P. Interpreting non-linear drug diffusion data: Utilizing Korsmeyer-Peppas model to study drug release from liposomes. *Eur. J. Pharm. Sci.* **2019**, *138*, 105026. [CrossRef] [PubMed]
67. Qiu, Y.; Gao, Y.; Hu, K.; Li, F. Enhancement of skin permeation of docetaxel: A novel approach combining microneedle and elastic liposomes. *J. Control. Release* **2008**, *129*, 144–150. [CrossRef] [PubMed]
68. Nuki, G.; Simkin, P.A. A concise history of gout and hyperuricemia and their treatment. *Arthritis Res. Ther.* **2006**, *8*, 1–5. [CrossRef] [PubMed]
69. Dehlin, M.; Jacobsson, L.; Roddy, E. Global epidemiology of gout: Prevalence, incidence, treatment patterns and risk factors. *Nat. Rev. Rheumatol.* **2020**, *16*, 380–390. [CrossRef]
70. Nair, B. Final report on the safety assessment of polyvinylpyrrolidone (PVP). *Int. J. Toxicol.* **1998**, *17*, 95–130. [CrossRef]
71. Quinn, H.L.; Larrañeta, E.; Donnelly, R.F. Dissolving microneedles: Safety considerations and future perspectives. *Ther. Deliv.* **2016**, *7*, 283–285. [CrossRef]
72. Donnelly, R.F.; Singh, T.R.R.; Tunney, M.M.; Morrow, D.I.; McCarron, P.A.; O'Mahony, C.; Woolfson, A.D. Microneedle arrays allow lower microbial penetration than hypodermic needles in vitro. *Pharm. Res.* **2009**, *26*, 2513–2522. [CrossRef] [PubMed]
73. Lee, K.J.; Jeong, S.S.; Roh, D.H.; Kim, D.Y.; Choi, H.K.; Lee, E.H. A practical guide to the development of microneedle systems–In clinical trials or on the market. *Int. J. Pharm.* **2020**, *573*, 118778. [CrossRef] [PubMed]
74. McCrudden, M.T.; Alkilani, A.Z.; Courtenay, A.J.; McCrudden, C.M.; McCloskey, B.; Walker, C.; Alshraiedeh, N.; Lutton, R.E.; Gilmore, B.F.; Woolfson, A.D.; et al. Considerations in the sterile manufacture of polymeric microneedle arrays. *Drug Deliv. Transl. Res.* **2015**, *5*, 3–14. [CrossRef] [PubMed]
75. Amodwala, S.; Kumar, P.; Thakkar, H.P. Statistically optimized fast dissolving microneedle transdermal patch of meloxicam: A patient friendly approach to manage arthritis. *Eur. J. Pharm. Sci.* **2017**, *104*, 114–123. [CrossRef] [PubMed]

pharmaceutics

Review

Nanoemulgel: A Novel Nano Carrier as a Tool for Topical Drug Delivery

Mahipal Reddy Donthi [1,†], Siva Ram Munnangi [1,2,†,‡], Kowthavarapu Venkata Krishna [1,3,§], Ranendra Narayan Saha [1], Gautam Singhvi [1] and Sunil Kumar Dubey [1,4,*,||]

1 Department of Pharmacy, Birla Institute of Technology and Science, Pilani (BITS-PILANI), Pilani Campus, Pilani 333031, India
2 Department of Pharmaceutics and Drug Delivery, School of Pharmacy, The University of Mississippi, Oxford, MS 38677, USA
3 Center for Pharmacometrics and Systems Pharmacology, Department of Pharmaceutics, College of Pharmacy, University of Florida, Orlando, FL 32827, USA
4 R&D Healthcare Division Emami Ltd., 13, BT Road, Kolkata 700056, India
* Correspondence: sunilbit2014@gmail.com; Tel.: +91-8239703734
† These authors contributed equally to this work.
‡ Current Affiliation: Department of Pharmaceutics and Drug Delivery, School of Pharmacy, The University of Mississippi, Oxford, MS 38677, USA.
§ Current Affiliation: Center for Pharmacometrics and Systems Pharmacology, Department of Pharmaceutics, College of Pharmacy, University of Florida, Orlando, FL 32827, USA.
|| Current Affiliation: R&D Healthcare Division Emami Ltd., 13, BT Road, Kolkata 700056, India.

Abstract: Nano-emulgel is an emerging drug delivery system intended to enhance the therapeutic profile of lipophilic drugs. Lipophilic formulations have a variety of limitations, which includes poor solubility, unpredictable absorption, and low oral bioavailability. Nano-emulgel, an amalgamated preparation of different systems aims to deal with these limitations. The novel system prepared by the incorporation of nano-emulsion into gel improves stability and enables drug delivery for both immediate and controlled release. The focus on nano-emulgel has also increased due to its ability to achieve targeted delivery, ease of application, absence of gastrointestinal degradation or the first pass metabolism, and safety profile. This review focuses on the formulation components of nano-emulgel for topical drug delivery, pharmacokinetics and safety profiles.

Keywords: nano emulgel; topical delivery; permeation; surfactant; bioavailability

Citation: Donthi, M.R.; Munnangi, S.R.; Krishna, K.V.; Saha, R.N.; Singhvi, G.; Dubey, S.K. Nanoemulgel: A Novel Nano Carrier as a Tool for Topical Drug Delivery. *Pharmaceutics* **2023**, *15*, 164. https://doi.org/10.3390/pharmaceutics15010164

Academic Editors: Silvia Tampucci and Daniela Monti

Received: 20 November 2022
Revised: 24 December 2022
Accepted: 27 December 2022
Published: 3 January 2023

1. Introduction

The recent progress in drug synthesis and high throughput screening have steered drug discovery and development toward lipophilic drug moieties. Currently, 90% of drugs in the discovery pipeline and more than 40% of the drugs present in the market are of lipophilic nature [1]. The lipophilic nature of the drugs leads to problems like poor solubility, unpredictable absorption, and inter and intra-subject variability concerning pharmacokinetics. Various techniques have been employed to increase the solubility of active moieties. These techniques include physical and chemical modification of API along with formulation strategies, which include particle size reduction, complexation, amorphization, and nano-carrier drug delivery systems as represented in Figure 1 [2–4].

Despite of employing various technologies for enhancing the solubility, delivering the drugs via the oral route is not always feasible owing to their low bioavailability associated with poor absorption, first-pass metabolism, chemical and enzymatic degradation [5,6]. In addition, clinical complications and low concentrations of the drug at the site of action hinder drug delivery through the oral route. For example, the oral administration of Disease-modifying anti-rheumatic drugs (DMARDs) used in the treatment of arthritis are associated with various side effects like carcinogenicity, hepatotoxicity, and hematologic toxicity [7,8]. These clinical complications can be mitigated by delivering the drug through the topical route [9].

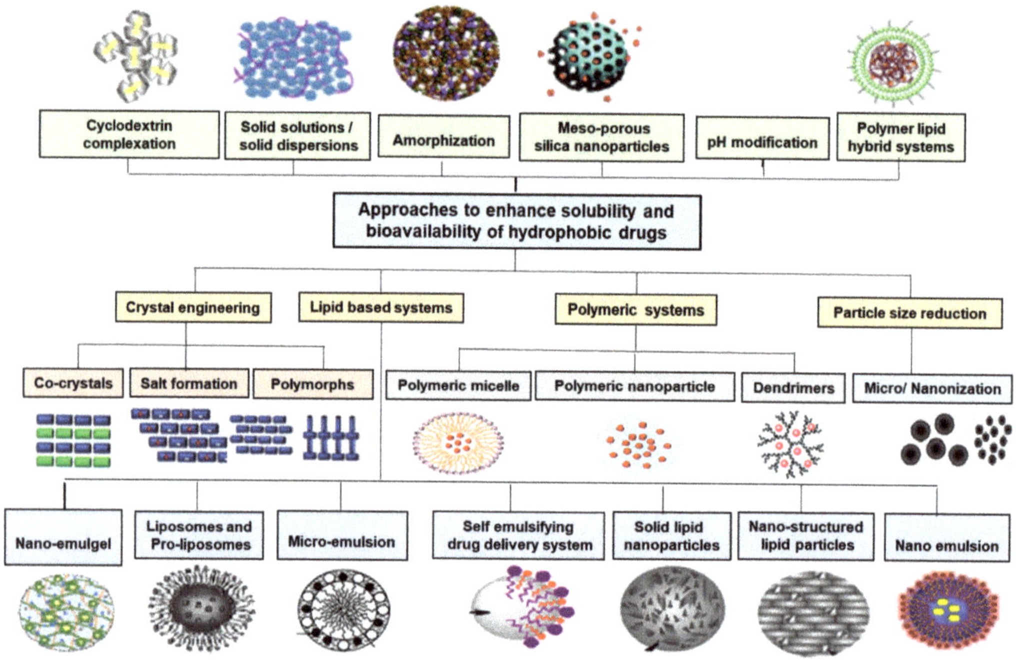

Figure 1. Strategies to improve solubility and bioavailability of lipophilic drugs.

In topical delivery, skin being a fundamental defense layer, considers the API's as external components and restricts their entry into the body. The outer most layer of epidermis called stratum corneum is the first and firm layer to overcome for drug penetration into the skin [10]. Various mechanisms have been explored to enhance the drug permeation. One such mechanism involves disruption of skin layer structure, which can be achieved using techniques such as chemical penetration enhancers, ultrasound, iontophoresis, sonophoresis, electroporation and microneedles [11]. In contrary, the use of nanocarriers was observed to be an effective strategy for circumventing the SC barrier without exacerbating skin damage and achieving efficient drug penetration. They facilitate the drug delivery through the skin utilizing intra and inter cellular transport mechanisms, interacting with skin components to mediate transport or to create depots of the drug for sustained or stimuli-induced release. These novel carrier for topical administration includes but not limited to emulsions (nano/micro), micelles, dendrimers, liposomes, solid lipid nanoparticles and nano-structured lipid carriers [12–14]. Among these, nano-emulsions are found to be a potential drug delivery system because of their high drug-loading capacities, solubilizing capacities, ease of manufacturability, stability and controlled release patterns. These nano-emulsions owing to their lipophilic core allow the movement of more lipophilic molecules across the topical membranes compared to the liposomes [15]. In addition, liposomes stability has always been an issue, as they disintegrate during the penetration process. Likewise, the low drug loading capacity and uncontrolled release hinders the application of solid-lipid nanoparticles in dermal drug delivery. Similarly, micelles exhibit poor stability and encapsulation efficiency. In the same way, the toxicity and poor controlled release behavior of dendrimers limits its topical application [16].

Nano-emulsions are heterogeneous colloidal mixtures of oil and water, with one component as a dispersed phase and the other as a continuous phase. A surfactant known as an emulsifier is adsorbed at the interface between the dispersed and continuous phases, lowering the surface tension and thus stabilizing the system. These systems possess high thermodynamic stability leading to longer shelf life compared to simple emulsions, micelles or suspensions, etc. Despite having various advantages, nano-emulsions are limited by their low viscosity leading to low retention time and spreadability [17]. These problems can be resolved by modifying the nano-emulsion into a nano-emulgel by using a suitable gelling agent [18].

The nano-emulgel acts as a colloidal system consisting of a mixture of emulsion and gel. The emulsion part protects the drug from enzymatic degradation, and hydrolysis and improves the permeation like other nano-carriers. Besides enhancing the penetration of the drug through the skin, it is equally important to retain the therapeutic concentrations of the drug for a sufficient period of time. The gel part improves the viscosity and spreadability resulting in improved retention time, and also reduces the surface and interfacial tension, thus improving the thermodynamic stability. Nano-emulgel possesses various advantages having high drug loading capacity, better penetration, diffusion, and low skin irritation compared to other nano-carriers [19,20].

This article aims to provide insight into the selection of formulation ingredients of a nano-emulgel, characteristics and formulation aspects, advantages, pharmacokinetics and pharmacodynamics, and safety of the same. The objective here is to give an overview of the future and rationale behind the nano-emulgel drug delivery system.

2. Drug Delivery through a Topical Route

The characteristics of any ideal formulation are patient compliance, self-administration, non-invasiveness, fewer side effects, and better pharmacological action. The topical route administering formulations possess most of the aforementioned characteristics [21]. The benefits of the topical route of administration comprise of avoiding the hepatic first-pass effect, decreased side effects due to the local site of action, enhancement in percutaneous absorption and topical usage may even increase bioavailability with a sustained deposition [22]. Further, the reduced drug loss due to metabolism or decomposition, and the ability to specifically target the drug at the desired site are also some of the advantages. Minimization of drug breakdown coupled with constant delivery of drug for a prolonged period results in prominent movement of the drug across the barrier of stratum corneum, leading to improved bioavailability [23,24].

An increase in the bioavailability of drugs via the topical route has been proved in various research works. For example, Flurbiprofen nano-emulsion showed 4.4 times increment in bioavailability upon topical administration compared to oral delivery [25]. Zhou et al. prepared the nano-emulsion of nile red dye, which displayed 10-fold increase in the penetration of dye across the skin compared to an emulsion formulation [26]. Gannu et al. reported a 3.5-fold increase in Lacidipine bioavailability via the transdermal route of administration using microemulsions. The group reported that this improvement could be due to the avoidance of the first-pass effect on the drug upon topical application [24]. Further, an enhancement in the therapeutic and pharmacological effect of therapeutically active agents has been demonstrated with topical formulations.

Conventional topical formulations that are used are solutions, ointments, lotions, creams, patches, gels, etc. [27]. But these topical formulations have to traverse the remarkably effective and competent stratum corneum barrier along with viable epidermis of the skin as shown in Figure 2. The SC is a 10–20 μm thick lipid-interspersed matrix of terminally formed keratinocytes, which causes a huge challenge for the delivery of therapeutically active agents via the topical or transdermal route of administration [28–30]. Thus, reducing the amount of drug reaching the target site. This is reflected in the marketed topical preparations possessing low permeation leading to poor therapeutic effect [31–33]. Therefore, research in this area mainly focuses on the development of topical formulations with appropriate permeability and ensuring delivery by numerous mechanisms. The direction of research work in recent years has shifted towards novel carrier systems with the intent to alter the permeability of hydrophobic drugs through the skin. New formulation development techniques and strategies are emerging in recent years but the main drawback of the recent strategies is the usage of chemicals and non-green solvents for enhancing the permeation. The usage of these preparations for a long period would lead to various skin complications [34,35]. Besides, various limitations posed by the skin, there are certain characteristics that an active moiety should possess in order to be suitable for the topical route of administration as represented in Table 1 [36,37].

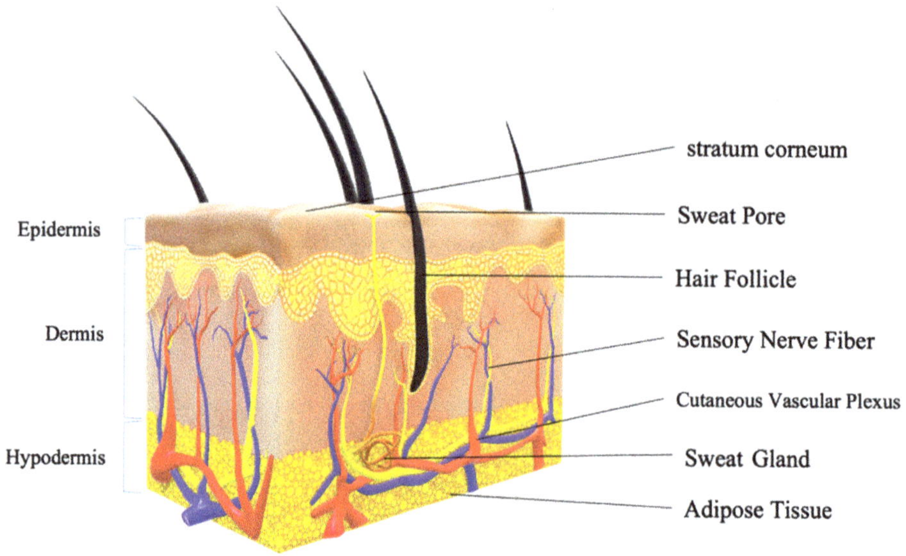

Figure 2. Skin morphology.

Table 1. Primary requirement of active moiety for topical delivery.

Properties	Conditions
$t_{1/2}$	$\leq$10 h
Molecular mass	$\leq$500 Daltons The limit can be exceeded by altering the permeability of skin
Molecular size	Small
Polarity	Non-polar is desirable
Log P	0.8–5
pKa	Higher
Irritation on skin	Non-irritating
Skin Permeability coefficient	$\geq 0.5 \times 10^{-3}$ cm/h

3. Nano Emulsions in Topical Delivery

An upgrade and innovation of topical and transdermal drug delivery systems led to the development of lipid-based nano-formulations. Though there are various formulations, research has deepened pertaining to nano-emulsions due to the aforementioned advantages and their ability to deliver hydrophobic drugs non-invasively and without the need for a penetration enhancer [38]. Nano-emulsions are an isotropic biphasic mixture consisting of two portions: water and oil, where one phase is dispersed in the other as nanosized droplets. The system is stabilized by the utilization of an interfacial layer of surfactants [39]. The difference between nano-emulsions and traditional emulsions is that the former has decreased propensity to undergo phase separation [40,41]. A prominent number of in-vivo studies have been carried out demonstrating the applications and feasibility of topical micro and nano-emulsions. In-vitro works have also supported the use of these topical lipidic formulations [42,43]. These nano-emulsion systems possess a translucent or transparent appearance. The thermodynamic stability of nano-emulsions is greater than other lipid carriers. Nano-emulsions exhibit an increased solubilization capacity as compared to solutions of simple micelles [29,44]. These formulations can solubilize and incorporate large amounts of active drug substances due to the increased surface area because of the nano-size of oil droplets [45]. The phenomena of creaming or sedimentation are the general issues faced in an emulsion. The improved stability in a nano-emulsion is due to Brownian motion and less gravitational force acting on the particles because of their nano-size, thus preventing the stability issues like sedimentation and creaming [46]. Numerous studies have demonstrated the enhanced permeation of drugs upon administered as nano-emulsion systems in comparison to other formulations like emulsions, creams, and ointment gels [47–49]. The enhanced permeation is because of the ability of nano-emulsion to overcome the firmly bonded lipid bi-layers, thus able to penetrate deep into the skin and deliver the drug to systemic circulation because of smaller sized dispersed droplets, which facilitate transcellular in addition to paracellular transport [18].

4. Nano-Emulgel Drug Delivery System

Despite possessing many advantages, nano-emulsions lack spreadability because of their low viscosity resulting in poor retention of formulation over the skin [50]. This limitation hampers the clinical applications of nano-emulsions [51]. This issue has been resolved by incorporating a gelling agent into the nano-emulsion, thus forming a nano-emulgel [52]. Huge quantities of aqueous or hydroalcoholic bases are employed in a colloidal particulate system to prepare gels [53]. Nano-emulgel is formed by incorporating the nano-emulsion into a hydrogel matrix, which reduces the thermodynamic instability of the emulsion. The improved thermodynamic stability is due to the reduction in the portability of the non-aqueous phase because of the increased consistency of the external medium. The increased retention time and thermodynamic stability enable the formulation to release the drug over a period, making nano-emulgel a controlled release dosage form for topical administration benefiting the drugs with a short half-life [19,54].

The incorporation of nano-emulsion into a gelling system helps to annihilate the disadvantages of both individual systems. The combined nano-emulgel enjoys the properties of a gel with the refined characteristics of a nano-emulsion. The Table 2 discloses the advantages of nano-emulgel over conventional emulgel owing to its particle size and thermodynamic stability. The variety of benefits offered by nano-emulgels is enhanced skin permeation, greater loading of an active moiety, less irritation, and greater spreadability. This is apparent in comparison with different nano-carriers such as solid lipid nanoparticles and liposomes. The nano-emulsion is made suitable for topical use due to the increased viscosity of the gel. To achieve the same, various gelling agents compatible with skin like xanthan gum, carbomer 980, Pluronic's, carrageenan, and carbomer 934 are used for topical application [55]. Acceptable localization and drug dispersion through adequate percutaneous absorption across the skin is achieved in nano-emulsions. This helps to increase the efficacy locally and also systematically via the skin. This system can also be used to deliver drugs to the central nervous system (CNS) due to its ability to cross the blood-brain barrier when applied through the nasal route [56,57]. Non-irritant and non-greasy nature of nano-emulgel facilitates better patient compliance [53] In addition, pharmacokinetic properties like enhanced bioavailability and decreased side effects are added advantages for these systems [58]. The hydrogel matrix, consistency, and homogeneity have added to the growing focus on nano-emulgels. Furthermore, various studies have shown that nanoemulgel has increased stability due to less Oswalt ripening caused by decreased mobility of oil globules in gel matrix [59]. For instance, Kaur et al. developed a topical nanoemulgel loaded with TPGS containing mefenamic acid. In the pharmacodynamic investigation, the optimized nanoemulgel inhibited inflammation and enhanced percent reaction time with improved analgesic efficacy. The formulated nanoemulgel outperformed other traditional topical formulations in terms of long-term stability and drug penetration [60].

Table 2. Comparison between conventional emulgel and nano-emulgel.

Parameter	Conventional Emulgel	Nano-Emulgel
Thermodynamic stability	Not stable because of natural tendence of coalescence leading to sedimentation or creaming [61]	Stable–because of their smaller particle size, Brownian motion provides enough stability against gravity, preventing sedimentation or creaming [54]
Particle size	Greater than >500 nm [18]	Less than 100 nm [62]
Bioavailability	Comparatively less bioavailable than Nano-emulgel [63]	Enhanced bioavailability, attributed to small size and large surface area [64]
Permeation	Comparatively lower permeation [65]	High permeation owing to its lower particle size [54,65]
Preparation	Require high energy techniques [66]	It can be prepared either by using high or low energy techniques [20]
Systemic absorption	Very minimal	Higher compared to conventional emulgel due to the small particle size and large surface area [54]
Ability to cross BBB	Cannot cross BBB [67]	Can Cross BBB because of its small particle size [68]

Besides these, nano-emulgel is devoid of other formulation stability limitations like the problem of destabilization faced with conventional emulgels, the problem of moisture entrapment faced with powders, the problem of cake formation faced with suspensions, the problem of coalescence of oil globules, formation of agglomerates in case of suspensions, along with the problem of poor adherence and excessive spreadability that is faced with nano-emulsions [53]. Due to these factors, nano-emulgel is often thought of as an improved and different topical drug delivery approach over the standard marketed dosage forms. This novel formulation is welcome for research targeting various skin diseases and disorders. Nano-emulgel will soon be capturing the market in the topical delivery segment as a favorable substitute over conventional forms and some are currently being marketed as in Table 3. Many preclinical (Table 4) and clinical studies (Table 5) are being conducted to evaluate the efficacy of nanoemulgel.

Table 3. Examples of marketed emulgels for topical application.

Marketed Product	Active Pharmaceutical Ingredient	Manufacturing Company
Voltaren Emulgel	Diclofenac diethylamine	GlaxoSmithKline
Isofen Emulgel	Ibuprofen	Beit Jala Pharmaceutical Co.
Benzolait Emulgel	Benzoyl peroxide & Biguanide	Roydermal
Miconaz-H Emulgel	Miconazole nitrate & Hydrocartisone	Medical Union Pharmaceuticals
Derma Feet	Urea	Herbitas
Adwiflam Emulgel	Diclofenac diethylamine, Methyl Salicylate & Menthol	Saja Pharmaceuticals
Nucoxia Emulgel	Etoricoxib	Zydus Cadila Healthcare LTD

Table 4. Pre-clinical Studies on the nano-emulgel dosage form.

Active Ingredient	Composition	In Vivo Model	Route of Administration	Therapeutic Outcome	Reference
Curcumin	Oil: Labrofac PG + transcutol HP Surfactant mixture: Tween 20 + solutol HS15 Gelling agent: Carbopol 934	BALB/c mice	Topical	Psoriatic mice treated with the curcumin nano-emulgel showed faster and earlier healing than those treated with curcumin plus betamethasone-17-valerate gel	[69]
Thymoquinone	Oil: Black seed oil Surfactant mixture: Kolliphor EL + transcutol HP Gelling agent: Carbopol 940	Wistar rat	Topical	Nano-emulgel administration of thymoquinone improves its therapeutic efficiency in wound healing studies in Wistar rats	[70]
Curcumin and Resveratrol	Oil: Labrofac PG Surfactant mixture: Tween 80 Gelling agent: Carbopol	Wistar rat	Topical	Curcumin and resveratrol nano-emulgel technology revealed drastically increased curcumin and resveratrol deposition in skin layers. The in-vivo investigation revealed that the NEG formulation resulted in improved burn healing, with histological findings comparable to standard control skin. Thymoquinone nano-emulgel delivery method improves thymoquinone therapeutic effectiveness in wound healing studies in Wistar rats.	[71]
Brucine	Oil: Myrrh oil Surfactant mixture: Tween 80 + PEG 400 Gelling agent: Carboxymethylcellulose sodium	BALB/c mice and Wistar rats	Topical	Brucine-loaded nanoemulgel has shown improved anti-inflammatory and anti-nociceptive efficacy.	[72]
Curcumin	Oil: Labrofac PG Surfactant mixture: Tween 80 + PEG 400 Gelling agent: Carbopol 940	Albino rats	Topical	Curcumin nanoemulgel improved the wound-healing efficacy of curcumin compared to the conventional gel formulation.	[69]
Raloxifene hydrochloride	Oil: Peceol Surfactant mixture: Tween 20 + transcutol HP Gelling agent: Chitosan	Wistar rats	Topical	Raloxifene hydrochloride (RH) loaded nanoemulgel formulation for enhanced bioavailability and anti-anti-osteoporotic efficacy of RH. The bioavailability improved by 26-fold compared oral marketed product.	[73]
Eprinomectin	Oil: Castor oil Surfactant mixture: Tween 80 + Labrasol Gelling agent: Carbomer 940-1	Wistar rats	Topical	Naoemulgel formulation showed improved skin permeability of 1.45-fold compared to emulgel and had no skin-irritating property	[74]
Amisulpride	Oil: Maisine CC Surfactant mixture: Labrosol + transcutol HP Gelling agent: Poloxamer 407, Gellan gum	Wistar rats	Intranasal	Improved pharmacokinetic profile. The C_{max} of API in brain after administering through in-situ nano-emulgel improved by 3.39-fold compared to intravenous administration of nano-emulsion.	[75]
Disulfiram	Oil: Ethyl oleate Surfactant mixture: Tween 80 + transcutol HP Gelling agent: Deacetylated gellan gum	Sprague Dawley rats	Intranasal	Improved survival rate of rats and reduced tumor progression (Glioblastoma). The survival time of in-situ nano-emulgel treated group is 1.6 times higher than control group	[76]

Table 5. Clinical studies on emulgel dosage form.

Identifier No	Active Constituent	Titile of the Study	Conditions	Referance
NCT05536193	Metformin and salicylic acid	Topical Metformin Emulgel VS Salicylic Acid Peeling in Treatment of Acne Vulgaris	Acne Vulgaris	[77]
NCT03074162	Diclofenac sodium & Capsaicin	Comparison of the Bioavailability of Diclofenac in a Combination Product (Diclofenac 2% + Capsaicin 0.075% Topical Gel) With Two Diclofenac Only Products, Diclofenac Mono Gel 2% and Voltarol® 12 Hour Emulgel 2.32% Gel, in Healthy Volunteers	Inflammatory	[78]
NCT04579991	Visnadin, ethyl ximeninate, coleus barbatus	Effects of Visnadin, Ethyl Ximeninate, Coleus Barbatus and Millet in Emulgel on Sexual Function in Postmenopausal Women	Female Sexual FunctionVulvovaginal AtrophyPost-menopausal Atrophic Vaginitis	[79]
NCT04110860	Voriconazole	Clinical Assessment of Voriconazole Self Nano Emulsifying Drug Delivery System Intermediate Gel	Tinea Versicolor	[80]
NCT04110834	Itraconazole	Clinical Assessment of Itraconazole Self Nano Emulsifying Drug Delivery System Intermediate Gel	Tinea Versicolor	[81]
NCT03492541	Propylene glycol-based eye drops	Evaluation of the Clinical Efficacy and Tolerability of SYSTANE Complete in Adult Patients With Dry Eye Disease Following Topical Ocular Use for 4 Weeks: A Multicenter Trial	Dry eye disease	[82]
NCT05641246	Carbamide diclofenac	Effect of Topical Diclofenac on Clinical Outcome in Breast Cancer Patients Treated With Capecitabine: A Randomized Controlled Trial.	Hand and Foot Syndrome	[83]

5. Formulation Components

Nano-emulgels are made up of two individual systems; the gelling agent and the nano-emulsion i.e., emulsion consisting of nano droplets which are of o/w or w/o type. Both emulsion types possess an aqueous and an oily phase. The gel base consists of polymers that can swell on the absorption of a liquid. The various components in the nano-emulgel formulation are provided in Table 6 [53,84]. The overview of the selection criteria of the essential components in a nano-emulgel have been discussed below.

Table 6. Details of commonly used excipients in nano-emulgel formulations.

S.No	Disease/ Disorder	Active Pharmaceutical Ingredient	Composition				References
			Oil	Surfactant	Co-Surfactant	Gelling Agent	
1	Anti-inflammatory	Curcumin	Emu oil	Cremophor RH40	Labrafil M2125CS	Carbopol	[85]
2	Anti-inflammatory	Diclofenac sodium	Isopropyl myristate	Tween 20	Labrafil M2125CS	Carbopol 980	[86]
3	Anti-inflammatory	Meloxicam	Almond and peppermint oil (1:2)	Tween 80	Ethanol	Carbopol 940	[87]
4	Antimicrobial and Anti-Inflammatory	Quercetin	Cinnamon oil	Tween 80	Carbitol	Poloxamer	[88]
5	Antifungal	Itraconazole	Eugenol	Labrasol	TranscutolP, Lecithin	Carbolpol	[89]
6	Antifungal	Fluconazole	Capmul MCM	Tween 80	Transcutol P	Carbopol 934	[90]
7	Anti-hyperglycemic	Glibenclamide	Labrafac: Triacetin (1:1)	Tween 80	Diethylene glycol monoethyl ether	Carbopol 934	[91]
8	Antihypertensive	Carvedilol	Oleic acid: IPM (3:1)	Tween 20	Carbitol	Carbopol-934	[92]
9	Immunosuppressive agent	Cyclosporine	Oleic acid	Tween 80	Transcutol P	Guar gum.	[93]
10	Anti-cancer	Chrysin	Capryol 90	Tween 80	Transcutol HP	Pluronic F127	[94]
11	Wound Healing	Atorvastatin Calcium	Liquid Paraffin	Tween 80	Propylene glycol	Sodium carboxymethyl cellulose	[95]
12	Anti-inflammatory	Curcumin	Myrrh Oil	Tween 80	Ethanol	Sodium carboxymethyl cellulose	[96]
13	Wound Healing	Curcumin	Labrofac PG	Tween 80	Propylene glycol 400	Carbopol 940	[69]
14	Anti-fungal	Terbinafine HCl	Peceol oil	Tween 80	Propanol	Carbopol 940	[97]
15	Anti-fungal	Ebselen	Captex	Kolliphor ELP	Dimethylacetamide	Soluphus (10% *w/v*) & HPMC K4M (2.5% *w/v*)	[98]

5.1. Oil Phase

The selection of oil and its quantity depends on the application and utility of the nano-emulgel. The permeability, stability, and viscosity of the prepared nano-emulsion depends on the type and quantity of chosen lipid component, i.e., oil phase. Primarily in case of pharmaceutical and cosmetic applications, the oil phase is made up of either naturally or synthetically originated lipids, unless the oil phase itself is an active ingredient. The consistency of the lipids may vary from liquid to high molecular solids. The hydrophobicity of an oil plays a crucial role in forming a stable emulsion, wherein poor hydrophobicity of the oil is shown to increase the emulsification, concurrently affecting the solubility of lipophilic moieties [99]. Thus, choosing an oil is an essential prerequisite for nano-emulgel development as a novel drug delivery system [100].

Natural oils exhibit an additional medicinal significance leading to an increase in the researcher's interest to use these additive properties supporting the pharmacological action of the active moiety. For example, oleic acid is frequently used oil in nano-emulgel formulations and is obtained from vegetable and animal sources. It is a biodegradable and biocompatible omega-nine fatty acid and has elevated solubilization characteristics along with improving percutaneous absorption [101]. Antioxidants present in oleic acid contribute to cellular membrane integrity. It also repairs cell damage and showcases formulation stabilization [55,102]. Arora et al. confirmed that an increase in oleic acid content in the preparation increases the rate of permeation. In their study, using 6% oleic acid instead of 3% in the preparation nanoemulgel of drastically improved the permeability of ketoprofen [55].

Another natural oil called Emu oil is being appreciated for its analgesic, antipruritic, and antioxidant characteristics. Jeengar and group prepared nano-emulgel of curcumin with emu oil to treat the disease of joint synovium, the formulation demonstrated enhanced permeability and better pharmacological activity compared to pure curcumin [85,103]. The use of emu oil has been encouraged in the cosmetic field as well [85]. It moisturizes the skin and has high amounts of unsaturated fatty acids like oleic acid, thus improving the penetration of the drug [104].

The therapeutically active agent may also be used as the oil component in nano-emulgel preparation. Active moieties from Swietenia macrophylla have anti-inflammatory action and are self-employed as an oily phase in nano-emulgel. The therapeutic effect was found to be better in this nano-carrier preparation as opposed to the parent form [44]. Further, the edible oils considered to be the preferred lipid excipient of choice for the development of emulsions, are not frequently chosen due to their poor ability to dissolve large amounts of lipophilic drugs. Therefore, these oils are chemical modification or hydrolyzed to form an appropriate oil, which upon combining with a suitable surfactant enhances the solubility of hydrophobic compounds for nano-emulgel formulation [104].

5.2. Surfactant System

Surfactants are an essential ingredient in nano-emulsion, which are utilized in the stabilization of the unstable mix of two immiscible phases. This is achieved by a decreasing the interfacial tension amongst the two phases and alteration of dispersion entropy. The surfactant should show quick adsorption along the interface of the liquids. The final result is a decrease of interfacial tension and inhibition of coalescence of the individual nano-sized droplets [105].

The HLB value of the surfactant is an important variable for selecting the proper surfactant. The surfactants are either w/o type (HLB of 3–8) or o/w type (HLB of 8–16). In w/o emulsions, low HLB value surfactants i.e., less than 8 are utilized. Alternately Spans and Tweens are used for o/w emulsion as their HLB value is more than 8. A mixture of Span and Tween provides better stability to an emulsion system compared to pure Span or Tween containing preparations. Thus, using a proper mixture of surface-active agents is essential to formulate an ideal nano-emulsion. Based on the charge, the surfactants are of four main categories i.e., cationic, non-ionic, anionic, and zwitterionic nature. Examples of cationic surfactants are hexadecyl trimethyl ammonium bromide, cetyl trimethyl ammonium bromide, quaternary ammonium compounds, and dodecyl dimethyl ammonium bromide [106,107]. Poloxamer 124 and 188, Tween 20 and Caproyl 90 are some of the non-ionic surfactants [108,109]. Anionic surfactants are sodium dodecyl sulphate and sodium bis-2-ethylhexylsulfosuccinate [110]. Phospholipids such as phosphatidylcholine are part of zwitterion surfactants [111]. Toxicity should be considered while selecting the surfactant as it may lead to irritation of the gastrointestinal tract or skin based on the route of administration. Ionic surfactants are usually not preferred due to their toxicity and non-biocompatibility. The safety, biocompatibility and being unaffected by pH or ionic strength alteration make non-ionic surfactants an appropriate choice [112].

The surfactants derived from natural sources such as bacteria, fungi, and animals are being considered as a potential option, due to their safety, biodegradability, and biocompatibility. Bio-surfactants show a similar mechanism in decreasing surface tension along the interface due to amphiphilic properties. This is mainly due to the presence of non-polar short fatty acids and polar functionalities as the tail and head respectively [113]. They are more bio-compatible and safer than synthetic surfactants.

5.3. Co-Surfactant System

Co-surfactants support surfactants during the emulsification of oil in the water phase. Co-surfactants are required for decreasing the interfacial tension and improving the emulsification [114]. Flexibility is added to the interfacial film along with attaining transient negative interfacial tension due to co-surfactants. The association between the surfactant and co-surfactant along with the partitioning of the drug in immiscible phases decides the drug release from the nano-emulgel. Hence co-surfactant selection is equally important as surfactant. The commonly used co-surfactants are PEG- 400, transcutol® HP, absolute ethyl alcohol, and carbitol [115]. Alcohol based co-surfactants are most preferred because of their ability to partition between the oil and water phase thereby improving their miscibility.

The concentration of co-surfactant being used has to be chosen cautiously, since it may affect the emulsification by surfactant. Also, a combination of surfactant and co-surfactant with closer HLB values does not produce a stable emulsion as produced by non-ionic surfactants with different HLB values. The reason may be due to the solubilization of higher HLB value surfactants in the aqueous phase. Whereas, lower HLB value surfactants solubilize in the non-aqueous phase, enabling more intense association with the mixture of surfactant and co-surfactant [116]. Therefore, the choice of various formulation components and the rationale behind them is a very demanding and stimulating exercise.

5.4. Gelling Agents

Gelling agents upon addition to the appropriate media as a colloidal mixture forms a weakly cohesive three-dimensional structural network with a high degree of cross-linking either physically or chemically providing consistency to nano-emulgel [117–119]. In topical applications, these agents are used to stabilize the formulation, to attain optimum delivery of the drug across the skin. They play an important role in determining various parameters of the formulation like consistency, rheological properties, bio-adhesive properties, pharmacokinetics, spreadability, and extrudability. Based on the origin, these gelling agents are divided into natural, synthetic, and semi-synthetic. The Table 7 gives information on the concentration and pharmaceutical adaptability of various gelling agents used to prepare nano-emulgel. Natural gelling agents are bio-polysaccharides or their derivatives and proteins. The pectin, carrageenan, alginic acid, locust bean gum, and gelatine, etc., are bio-polysaccharides, while xanthan gum, starch, dextran, and acacia gum, etc., are derivatives of bio-polysaccharides. Though they provide excellent biocompatibility and biodegradability, the major limitation of natural gelling agents is microbial degradation [119,120]. Like natural gelling agents, semisynthetic gelling agents also offer good biocompatibility and biodegradability [121]. These agents are usually the derivatives of cellulose like hydroxypropyl cellulose, ethyl cellulose, sodium alginate, etc. The semisynthetic agents are comparatively more stable than natural gelling agents and are more responsive to chemical, biological and environmental changes like pH and temperature [122]. Synthetic gelling agents are prepared by chemical synthesis, some of them are FDA-approved e.g., carbomers and poloxamers [123,124]. Carbomers are polymerized acrylic acids, while poloxamers are triblock non-iconic copolymers comprising two hydrophilic units of polyoxyethylene attached to a central hydrophobic chain of polypropylene [124,125]. The FDA-approved synthetic agents are non-toxic and offer a wide range of rheological properties based on the molecular weight of the polymer, thus suitable for a wide range of applications.

Table 7. Various gelling agents and their pharmaceutical adaptability for use in topical emulgel.

Gelling Agent	Concentration Range (%*w*/*w*)	Pharmaceutical Adaptability	Reference
HPMC	2–6%	• Forms neutral gels • Can provide good stability • Resists microbial growth	[126,127]
Carbomer (Carbopol) Grades–ETD 2020, 171, 910, 934, 934P, 940, 1342 NF, 1971P	0.1–1.5%	• Forms high viscous gel • Forms gel at very low concentration • Provides controlled releasep • H dependent gelling	[126,128]
NaCMC	3–6%	• It withstands autoclaving. Therefore, can be used in sterile gels • Stable between pH 2 to 10	[129,130]
Poloxamer Grades–124, 182, 188, 407	20–30%	• Possess better solubility in cold water • Thermoreverisble gelation–gel at room temperature and liquid at refrigerated conditions	[131,132]
Combination of HPMC & Carbopol	1.2%	• Combination can improve stability of emulsion compared to individual components	[133,134]

6. Preparation of Nano-Emulgel

Nano-emulgel is a non-equilibrium formulation of structured liquids requiring energy, surfactant, or both for its preparation. They are spontaneously formulated by mixing the components. This is undertaken by introducing energy in the biphasic system or decreasing the interfacial tension between the interfaces of the two immiscible phases [135].

There are various nano-emulgel preparation methods reported based on the order of mixing of oil and aqueous phase [136]. Lupi et al. (2014) as illustrated in Figure 3A solubilized the drug in the oil phase and gelling agent in the water phase separately. The oil phase is added to the aqueous gel phase under stirring followed by homogenization to form an emulsion. The sol form of gelling agent in the emulsion is converted to gel by various mechanisms like adding a complexing agent or adjusting to the required pH [137]. Dong et al. (2015) as illustrated in Figure 3B divided the total quantity of water required for the preparation into two parts. One part of the divided quantity is used to prepare pre-emulsion and the other part is used for the preparation of gel. Later, these two components are mixed together under stirring [138]. Jeengar et al. (2016) prepared the emulsion and gel separately, followed by mixing them together in a 1:1 *w*/*w* ratio [85].

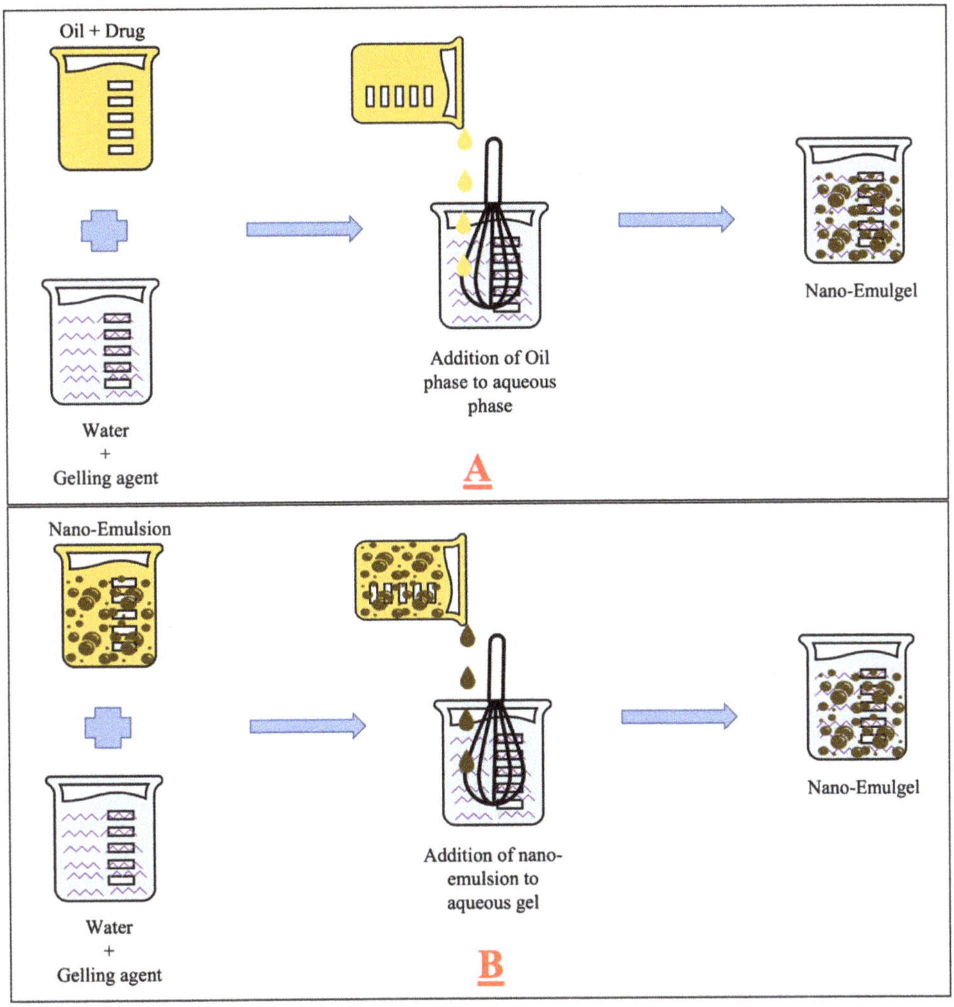

Figure 3. Schematic representation for the preparation of nano-emulgel by (**A**) adding Oil (oil + drug) phase to aqueous (water + gelling agent) phase (**B**) adding nano-emulsion to aqueous (water + gelling agent) phase.

Nano-emulgel formulation preparation can be further divided into two types based on the implementation of high-energy and low-energy emulsification techniques. High energy method involves the use of mechanical devices to produce a highly disruptive force in which both phases undergo size reduction. Hence this method may lead to the heating up of components in the formulation causing thermodynamic instability of the formulation and making it not suitable for thermo-labile drugs. Microfluidizers, high-pressure homogenizers, and ultrasonicated are high-energy methods employed to obtain a nanosized emulsion. This method is used for preparing nano-formulation of sizes of about 1 nm.

Phase inversion, self-emulsification, temperature, and phase transition are techniques of low energy approach. These methods provide the required thermodynamic stability to the nano-emulsion. The spontaneous method involves mixing oil, surfactant, and water in the best ratio possible and is most applicable for thermolabile compounds. The emulsification process is based on the surfactant and co-surfactant characteristics and their order of addition. Temperature-based alterations in HLB are utilized for non-ionic surfactants like Tween 20 Tween 60, Tween 80, Labrasol [139]. This method is mostly utilized for phase transition during phase inversion. Application of cooling with constant stirring will lead to a reversal of emulsion prepared at inversion temperature. Reduction in phase inversion temperature facilitates the inclusion of thermolabile components using this technique [140]. The second step incorporates gelling agent to change the liquid state to gel in the nano-emulsion. The thixotropic nature of the gelling agent facilitates the gel-solution conversion when shear stress is applied to the preparation keeping the volume constant. This leads to thickening in o/w nano-emulsion because of the creation of a gelled structure.

7. Permeability of Nano-Emulgel

In the preparation of emulsion-based gels, it is necessary to examine the important process parameters that have a significant effect on the size and formulation stability. In order to accomplish this, we must select the proper preparation process at the early stages. Emulsions are developed using various techniques, such as mechanical (or rotor-stator), high-pressure, microfluidization, and ultrasonic methods. The mechanical system comprises a colloid mill, that has a complex geometry, and the droplets of an emulsion generated by this system are several microns in size, making it the least desirable approach for manufacturing nanoemulsions [141]. Achieving an optimum droplet size is highly challenging. However, a droplet size of less than a micron can be achieved using high-pressure homogenization and sonication techniques, which in turn helps extend the shelf life of emulsions by lowering the creaming rate. For this reason, homogenization and sonication are considered to be efficient methods for the development of nanoemulsion [142,143]. In addition, increasing the homogenization speed or duration by itself is not enough to decrease the size of the globules, however, the use of the optimum concentration of an emulsifier is necessary to maintain control over the re-coalescence of the emulsion. For instance, Sabna Kotta et al. made a nanoemulsion utilizing the phase inversion and homogenization methods. In this formulation, gelucire 44/14 was used as a surfactant and transcutol-HP as a co-surfactant. They employed both the proposed techniques to produce nano-sized emulsion globules. In the case of homogenization, the large globule size was observed, despite increased pressure, and increased cycles at lower concentrations of an emulsifier. This demonstrates that the globule size of the formulation could not be decreased by homogenization alone. When the optimum concentration of an emulsifier is combined with increasing homogenization pressure and cycles, the size of the globules decreases. Because homogenization alone can break down globule size to nano, but with a lower concentration of surfactant, the newly formed globule surface would be improperly covered with a surfactant, resulting in re-coalescence. With the optimum concentration of an emulsifier and increased homogenization pressure and cycles, a smaller globule size with a good polydispersity index could be achieved. As a result, the author came to the conclusion that, throughout the preparation process, the desired particle size was obtained with a lower PDI by the combination of the surfactant, homogenization pressure, and cycle duration [142].

Mohammed S. et al. used ultrasonication to develop a thymoquinone-loaded topical nanoemulgel for wound healing. They used black seed oil (oil vehicle), Kolliphor El (surfactant), and Transcutol HP (co-surfactant). Nanoemulgel was prepared using different time intervals (3, 5, and 10 min) of ultrasonication at a 40% amplitude. When the concentration of surfactant decreased with 10 min of ultrasonication, the globular size increased. Meanwhile, increasing the concentration of surfactant with 10 min of sonication time resulted in a smaller globular size. The authors concluded that sonication is more effective when the appropriate concentration is used [70]. Monitoring the process control parameters and taking into account the composition of the excipients is both necessary steps in the process of optimizing the formulation.

8. Permeability of Nano-Emulgel

Skin shows an inherent property of acting as a protective barrier against external agents. Therefore, penetration through the skin is a major complication associated with topical delivery systems. The outermost layer of skin is the stratum corneum, which is followed by stratum granulosum and stratum lucidum. The stratum corneum is loosely composed of keratinized cells, waxy lipids, fatty acids, and cholesterol. All these constituents of stratum corneum help in retaining moisture and provide a hydrophobic barrier over the skin [18]. After the stratum corneum, there is the epidermis which is followed by dermis and subcutaneous layer. After crossing the subcutaneous layer, the active moiety will finally reach the systemic circulation. The primary hurdle for the drug moiety after reaching out from gel matrix is crossing the stratum corneum, from here the nano sized droplet due to the virtue of small diameter traverses basically through two different pathways as shown in Figure 4. One is cell to cell transfer involving concentration gradient-based movement called transcellular transport or intracellar transport, while the other is a passage through intercellular spaces or paracellular transport [118]. Whereas there is a third pathways called transappendageal transport, its influence on drug penetration is limited because hair follicles and glandular ducts make up negligible portion of the total surface area of the skin [16].

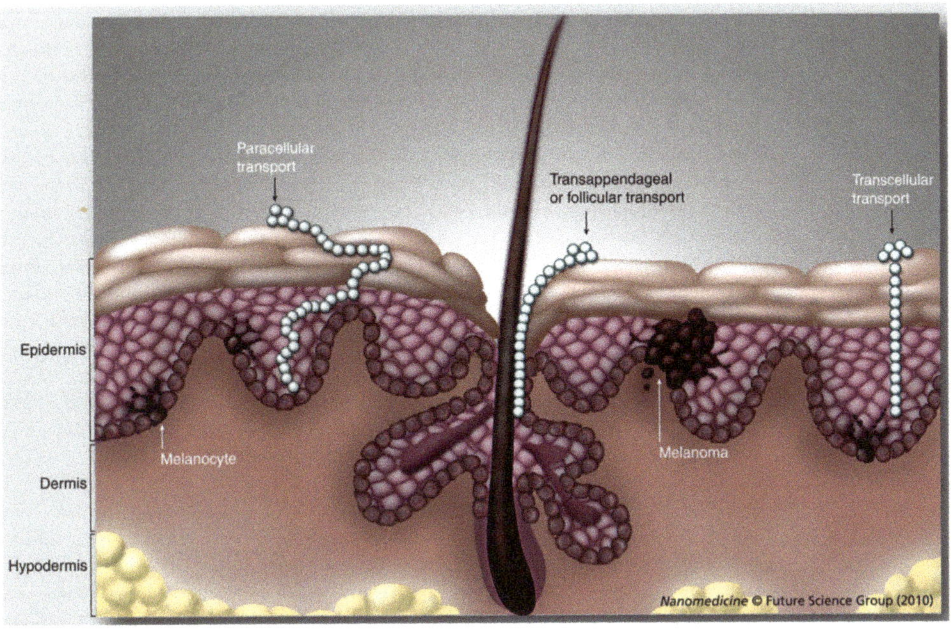

Figure 4. Graphical representation of entry of nano-emulgel into skin [144]. Adapted from Nanomedicine, 3 December 2010; 5(9): 1385–1399. Copyright (2010) Future Science Group.

Generally, ex-vivo permeation studies involve the examination of nano-emulgel formulation on isolated tissue in a simulated biological medium. Ex-vivo studies give a comparative analysis of penetration with different types of topical dosage forms and an idea about the flux rate of the drug inside the skin. Jeengar et al. made a nano-emulsion with emu oil as the oil phase. Optimized nanoemulsion was amalgamated with Carbopol gel to form nano emulgel and used for topical delivery of curcumin as an anti-inflammatory agent in rheumatoid arthritis. Ex-vivo permeation studies showed that permeation through the skin was higher for nano-emulgel as the retention of the formulation was higher compared to the nano-emulsion [145,146]. Elmateeshy and group formulated nano-emulgel by incorporating terbinafine HCl (TB) nano-emulsion formed from peceol as oil phase and (TWEEN 80/propanolol) as surfactant mixture and Carbopol as a gelling agent. The enhanced permeation of peceol oil-based nano-emulgel was observed in ex-vivo studies compared to available marketed products [97]. Similarly, Mulia et al. developed nano-emulgel for mangosteen extract composed of o/w nano-emulsion with virgin coconut oil as the oil phase and Tween 80/SPAN 80 as surfactant mixtures. The gel base was made with xanthan gum and phenoxyethanol was supplemented as a preservative. In vitro permeation studies demonstrated elevated penetration compared to nano-emulsion [29,147,148]. In addition, Bhattacharya et al., formulated celecoxib nano-emulgel with carbopol-940 hydrogel base, while Tween 80 and Acconon MC8-2EP as surfactants. Both in-vitro drug release and ex-vivo studies showed positive results. After the twelfth hour of diffusion, the optimized formulation displayed 95.5% cumulative release of the drug, whereas the commercially available formulation showed only 56.90% release. A higher penetration coefficient is displayed by nano-emulgel compared to commercial formulation [149]. In the same way, Chin et al. also developed a nano-emulgel of telmisartan for intranasal delivery using different molecular weight chitosan polymers. The ex-vivo penetration studies showed an improved permeation profile. The group also demonstrated the improvement in permeation is attributed to the molecular weight of the polymer, where the medium molecular weight chitosan provides higher permeation [150].

A nano-emulgel preparation for delivery through the transdermal route of tacrolimus was formulated by Begur et al. using almond gum as gel and oleic acid as the lipophilic phase. Cremophor was used as a surfactant to improve penetration. Examinations on rat abdominal skin showed a substantial increase in penetration [34]. Similarly, butenafine an antifungal agent available as a cream in the market, Syamala et al. prepared a nano-emulgel formulation for the same drug and found considerable results. Ex-vivo penetration studies showed a substantial increase in permeation over marketed creams [97]. In the same way, an increment in permeation of ketoconazole by about 53% was observed by delivering the drug in nano-emulgel formulation compared to normal marketed cream. The quality of life of the patient could be improved by implementing these types of dosage forms [151]. These studies showcase the ability of nano-emulgel in enhancing the permeation of the active moiety compared to nano-emulsion and conventional topical dosage forms. The permeation of the nano-emulgel is affected by various factors like gelling agents, surfactants, and permeation enhancers, etc. The gelling agents improve the permeation by improving the adherence of formulation upon the skin. While the surfactant alone or in combination with a co-surfactant will improve the permeation by disrupting the lipid bilayer. All these components can improve the permeation of active moiety.

9. Characterization Studies of Nano-Emulgel

The pharmaceutical product must be evaluated to ensure quality and consistency between different batches. These tests help in understanding the product's behavior and stability. According to USP, there are few universal tests for any given dosage form e.g., description, identification, assay, and impurities. A topical dosage form should undergo a few specific tests set by USP on a case-by-case basis: uniformity of dosage units, water content, microbial limits, antimicrobial and antioxidant content, pH, particle size, sterility, and API's polymorphic nature. Apart from the tests required for topical dosage form, nanoemulgel consists of nanosized globules, which need to be evaluated for zeta potential, droplet size and polydispersity index (PDI). Along with these physiochemical tests, a dosage form needs to be evaluated for its in-vitro release, spreadability, bio-adhesive tests, skin-irritation, ex-vivo permeability and in-vivo bioavailability can be performed to understand the behavior of nanoemulgel. Methods and techniques for analyzing significant properties of a nanoemulgel are briefly described below:

9.1. Zeta Potential

The particles in a solution usually possess a layer of ions on their surface, referred to as the stern layer. Adjacent to the stern layer, there exists a diffuse layer of loosely bounded ions, which along with the stern layer collectively called an electrical double layer. There is a boundary between the ions in the diffuse layer that move with the particle and the ions that remain with the bulk dispersant. The zeta potential is the electrostatic potential at this "slipping plane" boundary [152]. Zeta potential measurement provides an indirect measure of the net charge and is a tool to compare batch-to-batch consistency. The higher the zeta potential, the greater the repulsion resulting in increased stability of the formulation. For example, the high zeta potential of emulsion globules prevents them from coalescing. A surface charge modifier may also be used to adjust the surface charge. For instance, if a negatively charged surface modifier is used, the zeta-potential value becomes negative, and vice-versa [153,154]. Surface active ingredients (such as anionic or cationic surfactants) thus play an important role in emulsion stability, and the zeta potential can be measured using various instruments such as the ZC-2000 (Zeecom-2000, Microtec Co. Ltd., Chiba, Japan), Malvern Nanosizer/Zetasizer® nano-ZS ZEN 3600 (Malvern Instruments, Westborough, MA, USA), and others.

9.2. Droplet Size Measurement and Polydispersity Index (PDI)

The size of globule in nanoemulgel is referred as its hydrodynamic diameter, which is a diameter of equivalent hard sphere that diffuses at the same rate as the active moiety [155]. The PDI determines the distribution of droplet size and is defined as the standard deviation of droplet size divided by mean droplet size. The droplet size and the polydispersity index are closely connected to the stability and drug release, as well as the ex-vivo and in-vivo performance of the dosage form. In addition, it is important to measure consistency between different batches. The globule size and PDI of the formulation can be measured using a zeta sizer or master sizer. The globule size of the emulsion can be determined using the principle of dynamic light scattering, in which the transitional diffusion coefficient is measured by monitoring the interaction between the laser beam and dispersion, as well as the Polydispersity index [156,157].

9.3. Rheological Characterizations

Rheology is the study of the deformation and flow of materials. The rheological characterization of materials reveals the influence of excipient concentrations like oils, surfactants, and gelling agents on the formulation's viscoelastic flow behavior. If a formulation's viscosity and flow characteristics vary, this may influence its stability, drug release, and other in-vivo parameters. In this instance, the formulation's shear thinning tendency generates a thin layer on the skin surface, improving permeability, whereas a thicker formulation decreases permeation. Therefore, the rheological behavior is an extremely important factor in the formulation of nanoemulgel and several unique types of viscometers can be used to determine the rheological behavior [20]. FDA recommends the evaluation of complete flow curves whenever possible, plotted as both heat stress versus shear rate and viscosity versus shear rate across multiple shear rates until low or high plateaus are observed. If a formulation exhibits plastic flow, yield stress values should be evaluated.

9.4. Spreadability Testing

The spreadability property of the topical dosage form ensures the evenly spreading of the dosage form, thus delivering a stranded dose subsequently affecting the efficacy. The viscosity of the nanoemulgel greatly affects the spreadability property [158]. To date, no standard method has been established for measuring the spreadability of the dosage form. A few tests, that are commonly used for a good approximation of spreadability are a parallel-plate method and human subject assessment etc. The parallel-plate method (slip and drag method) is a widely employed technique because of its simplicity and relatively economic [158]. The instrumental setup consists of two glass slides of the same length, one of which is stationarily attached to the wooden block, and the other glass slide is mobile attached to a pulley at one end to measure spreadability. Spreadability is determined by the emulgel's 'Slip' and 'Drag' qualities. The nanoemulgel dosage form will be placed on a stationary glass slide, which is then squeezed in between stationary and mobile glass slides. The formulation is squeezed firmly for uniformly spreading formulation between two slides and to remove any air bubbles. The known weights are added to the pulley until the upper slide slips off from the lower slide. The time required for slipping off is recorded, which is used to calculate spreadability using the following equation [159].

$$S = M * L/T$$

where, S, M, L and T respectively represent the spreadability, weight bounded to the upper slide, Length of the slide, and Time taken to detach the slides.

9.5. In-Vitro Release Test (IVRT)

The efficacy and safety of the API are associated with drug release from the dosage form. The IVRT serves as a tool for assessing the quality of the drug product [160]. According to FDA, the IVRT studies for semi-solid dosage forms are conducted using either the vertical diffusion cell or an immersion cell. The vertical diffusion cell consists of receptor and donor chambers, separated by a receptor membrane. The donor chamber holds the sample of dosage form, while the receptor chamber holds the receptor media. The receptor media can be a buffer or hydro-alcoholic solution, selected based on the solubility, sink condition, and stability of the API. The skin-like receptor membrane is selected based on the effective pore size, high permeability and expected inertness towards the API. If necessary, the receptor membrane should be saturated with release media. The temperature of the media should be maintained around 32 °C ± 1 °C for topical administering products, for products intended for mucosal membrane the temperature should be 37 °C ± 1 °C. A Teflon-coated magnetic stirrer is used for stirring the receptor media. While the immersion cell model has a cell body, which acts as a reservoir [161]. The cell body is covered with a membrane and closed using a leakproof seal (retaining ring cap) that ensures no leakage of the dosage form. The retaining ring cap possesses an opening on the top, and it should be adjusted in such a way that the membrane is in contact with the dosage form on the bottom and release media on the top. The whole setup is used along with the USP-2 apparatus, wherein the immersion cell is placed in flat bottomed dissolution vessel with a usual volume of 150–200 mL. A mini spin-paddle is used for stirring or agitating the media [162].

9.6. Bio-Adhesive Property

Bio-adhesive strength is used to determine the force required to detach the drug carrier system from a biological surface. This property is important for a topical dosage form if prolonged contact is required [163]. This test is usually performed using rat or pig skin, the latter is preferred because of its resemblance to human skin. There are various techniques to measure this property but none of them is approved by FDA. The texture analyzer is one such technique, where the upper mobile probe and stationary lower base plate will be covered with skin. The dosage form is placed on the skin of the base plate. The upper probe is lowered to contact the lower base plate and the contact is maintained for at least a minute. The upper probe is lifted slowly until the separation of skin sheets. The force required to separate the two skin sheets will be measured by the instrument and represented as the area under the force-distance curve [164].

10. Safety Issues

One of the crucial concerns while developing a skin-related formulation is toxicity and skin irritation [165]. Impairment of enzyme activity, disturbance in normal physiological functions, and sometimes carcinogenic effects (For e.g., being caused by Sodium do-decyl benzene sulfonate) are some common toxicity issues related to surfactants [166]. Smith and the group analyzed the effect of two surface active agents' sodium dodecyl sulfate and dodecyl trimethyl ammonium bromide on penetration and skin perturbation. They concluded that disruption in the layer of skin is primarily caused by elevated concentrations of micelle agglomerate and monomers [167].

Irritation caused by topical nano-emulgel can be examined by applying it on the shaven skin of a rat, then observation of redness and other signs of inflammation on the skin were made and then graded based on the number of eurythmic spots to assess their clinical implication as give in Table 8 [167]. In general, a grade scale up to 2 is safe. Azeem et al. prepared ropinirole nano-emulgel using caproyl 90, tween 20 and carbital. The skin irritation studies showed a grade 2 erythema index, which is safe [168]. Gannu et al. also performed skin irritation studies of nano-emulgel prepared using non-ionic surfactant the Tween 80 and co-surfactant labrasol. They observed no signs of skin irritation as the used surfactants are generally considered safe [25]. Usually, major toxic effects are observed with cationic surfactants so they are avoided in preparations associated with topical delivery. While nonionic surfactants are mostly preferred as they cause minimum perturbation of the skin layer [151,166].

Table 8. Skin irritation grading scale and their clinical implications.

Clinical Portrayal	Grade
No erythema	0
Slight erythema that is barely perceptible	1
Moderate erythema that is visible	2
Erythema and papules	3
Severe Edema	4
Erythema, edema, and papules	5
Vesicular eruption	6
Strong reaction spreading beyond the application sight	7

11. Challenges

The impartment of large drug entities with molecular weight exceeding 400 Dalton is hindered in this dosage form, as they show difficulty during size reduction and are found to leach out of the gel mesh network. A limited number of safe surfactants and co-surfactants are available for emulgel preparation. Not much maneuvering can be done with the selection of surfactant as it can have hazardous consequences. The abundance of surfactant in emulgel can lead to skin problems like contact dermatitis, erythema, redness of skin, skin layer perturbation [169]. High susceptibility of the gelling agent toward variations in pH and temperature can lead to the breaking of gel structure and the leaching of chemicals [170].

Capriciousness in nano-emulsion is caused due to Ostwald ripening, which is associated with nano size of oil droplets, preferably nano-emulsion is prepared shortly before its application. Optimizing the speed of the stirrer in the homogenizer (as required to produce an inflexible and non-cracking gel), mixing appropriate quantities of surface-active agents, and selecting a reliable packing material are very pivotal tasks associated with the stability of nano-emulgel [171]. Highly specialized instruments are required for size reduction to nanoscale, which requires handling by skilled labour. Expensive sustenance of high energy homogenizers and production cost is one of the critical limitations associated with scale up of nano-emulgel formulation. Besides these disadvantages, the comforting prospect of nano-emulgel is increased adherent property and elevated embranglement of the drug in the gel mesh [53]. Also, prevalent drawbacks associated with conventional topical dosage forms i.e., emulsion, ointment, lotions etc. such as creaming, phase disruption, oxidation induced degradation of ointments are overcome by forming emulgel [44].

12. Current and Future Prospects of Nanoemulgel

Delivering hydrophobic drugs to the biological systems has been a major challenge in formulation development owing to their low solubility, leading to poor bioavailability. Some of the topical formulations include creams, ointments, and lotions. They possess good emollient characteristics, however, has slow drug release kinetics due to the presence of hydrophobic oleaginous bases such as petrolatum, beeswax, and vegetable oils, which inhibit the incorporation of water or aqueous phase. On contrary, topical aqueous-based formulations like gels enhance the drug release from the medication since it provides an aqueous environment for medicament. Therefore, hydrophobic APIs are blended with oily bases to form an emulgel, which further undergoes nanonization to form a nanoemulgel with enhanced properties. The superior properties of a nanoemulgel like thermodynamic stability, permeation enhancement, and sustained release make it an excellent dosage form. There are several marketed emulgels and patents being filed (Table 9) for the same, demonstrating its tremendous progress in this field. By making advancements in the ongoing research, nanoemulgel, as a delivery system would outshine, in formulating the drugs that are being eliminated from the development pipeline owing to their poor bioavailability, therapeutic non-efficacy, etc. Despite these advantages, the manufacturing of nano-emulsion limits its commercialization. However, with the progressing technology, commercially feasible and profitable manufacturing techniques could be possible in the future. With the advantages of nano-emulgel over other formulations, a tremendous increase in the production of nano-emulgel can be foreseen.

Table 9. Recent patents on nano-emulgel.

Patent Number	API	Title	Disease Indication	Current Assignee/Inventors	Granted/Publication Year	Reference
US11185504B2	Aromatase inhibitors	Transdermal non-aqueous nanoemulgels for systemic delivery of aromatase inhibitor	breast cancer	Qatar University	2021	[172]
CA3050535C	Anti-inflammatory nutraceuticals e.g., resveratrol, cinnamaldehyde, green tea polyphenols, lipoic acid etc.	Methods of treating inflammatory disorders and global inflammation with compositions comprising phospholipid nanoparticle encapsulations of anti-inflammatory nutraceuticals	Inflammatory Disorders	Nanosphere Health Sciences Inc	2021	[173]
CN107303263B	Tripterygium glycosides	Tripterygium glycosides nanoemulsion gel and preparation method thereof	Immune diseases e.g., clinical rheumatoid arthritis and psoriasis etc.	Second Military Medical University SMMU	2020	[174]
EP3099301B1	Besifloxacin	Besifloxacin for the treatment of resistant acne	Acne vulgaris	Vyome Therapeutics Ltd.	2019	[175]
WO2020240451A1	Brinzolamide	In-situ gelling nanoemulsion of brinzolamide	glaucoma	Hemant Hanumant BHALERAO, Sajeev Chandran	2020	[176]
WO2020121329A1	Minoxidil and castor oil	Minoxidil and castor oil nanoemulgel for alopecia	androgenic alopecia	Sudha suresh Dr. Rathodsoniya ramesh devasani	2020	[177]
BR102019014044A2	Ketoconazole	Nanoemulgel based on ucúuba fat (*Virola surinamensis*) for transungual administration of antimicotics	Onychomycosis	Rayanne Rocha Pereira et al.	2021	[178]

13. Conclusions

The selection of ingredients and their appropriate ratios play a vital role in deciding the properties of a nano-emulgel. Deviation from this could affect the conversion of a nano-emulsion to a nano-emulgel and its thermodynamic stability. The nano-emulgel is more stable compared to that of a nano-emulsion mainly due to its less mobile dispersed phase and the decreased interfacial tension. Thus, the former is a better alternative in delivering lipophilic moieties mainly due to improved permeation, and better pharmacokinetics, which subsequently improves the pharmacological effect. Patient compliance is also elevated due to its non-greasy and improved spreading properties on topical administration. Despite of its advantages, nano-emulgel is still at its infancy in the prospect of the pharmaceutical industry. However, various emulgels are being marketed e.g., Voltron emulgel, which holds out hope for the commercialization of nano-emulgel in near future. Hence it has the potential to become a center of attention due to its safety, efficacy, and user-friendly nature for topical drug delivery. Despite some disadvantages, nano-emulgel is a tool for the future which may be an alternative to traditional formulations.

Author Contributions: Writing—original draft, M.R.D.; Writing—review & editing, S.R.M.; Writing, review & editing, K.V.K.; Resources, Supervision, R.N.S.; editing, Resources, Supervision, G.S.; Conceptualization, Review & editing, Resources, Supervision S.K.D. All authors have read and agreed to the published version of the manuscript.

Funding: This review received no external funding.

Conflicts of Interest: The authors declare that they have no known competing financial interest or personal relationships that could have appeared to influence the work reported in this paper.

References

1. Kalepu, S.; Nekkanti, V. Insoluble Drug Delivery Strategies: Review of Recent Advances and Business Prospects. *Acta Pharm. Sin. B* **2015**, *5*, 442–453. [CrossRef] [PubMed]
2. Donthi, M.R.; Munnangi, S.R.; Krishna, K.V.; Marathe, S.A.; Saha, R.N.; Singhvi, G.; Dubey, S.K. Formulating Ternary Inclusion Complex of Sorafenib Tosylate Using β-Cyclodextrin and Hydrophilic Polymers: Physicochemical Characterization and In Vitro Assessment. *AAPS PharmSciTech* **2022**, *23*, 1–15. [CrossRef] [PubMed]
3. Singh, G.; Singh, D.; Choudhari, M.; Kaur, S.D.; Dubey, S.K.; Arora, S.; Bedi, N. Exemestane Encapsulated Copolymers L121/F127/GL44 Based Mixed Micelles: Solubility Enhancement and in Vitro Cytotoxicity Evaluation Using MCF-7 Breast Cancer Cells. *J. Pharm. Investig.* **2021**, *51*, 701–714. [CrossRef]
4. Alekya, T.; Narendar, D.; Mahipal, D.; Arjun, N.; Nagaraj, B. Design and Evaluation of Chronomodulated Drug Delivery of Tramadol Hydrochloride. *Drug Res.* **2018**, *68*, 174–180. [CrossRef] [PubMed]
5. Homayun, B.; Lin, X.; Choi, H.J. Challenges and Recent Progress in Oral Drug Delivery Systems for Biopharmaceuticals. *Pharmaceutics* **2019**, *11*, 129. [CrossRef]
6. Donthi, M.R.; Dudhipala, N.R.; Komalla, D.R.; Suram, D.; Banala, N. Preparation and Evaluation of Fixed Combination of Ketoprofen Enteric Coated and Famotidine Floating Mini Tablets by Single Unit Encapsulation System. *J. Bioequiv. Availab.* **2015**, *7*, 1–5. [CrossRef]
7. Wang, W.; Zhou, H.; Liu, L. Side Effects of Methotrexate Therapy for Rheumatoid Arthritis: A Systematic Review. *Eur. J. Med. Chem.* **2018**, *158*, 502–516. [CrossRef]
8. Rajitha, R.; Narendar, D.; Arjun, N.; Nagaraj, B. Colon Delivery of Naproxen: Preparation, Characterization and Clinical Evaluation in Healthy Volunteers. *Int. J. Pharm. Sci. Nanotechnol.* **2016**, *9*, 3383–3389. [CrossRef]
9. Garg, N.K.; Singh, B.; Tyagi, R.K.; Sharma, G.; Katare, O.P. Effective Transdermal Delivery of Methotrexate through Nanostructured Lipid Carriers in an Experimentally Induced Arthritis Model. *Colloids Surf. B Biointerfaces* **2016**, *147*, 17–24. [CrossRef]
10. Szumała, P.; Macierzanka, A. Topical Delivery of Pharmaceutical and Cosmetic Macromolecules Using Microemulsion Systems. *Int. J. Pharm.* **2022**, *615*, 121488. [CrossRef]
11. Gupta, R.; Dwadasi, B.S.; Rai, B.; Mitragotri, S. Effect of Chemical Permeation Enhancers on Skin Permeability: In Silico Screening Using Molecular Dynamics Simulations. *Sci. Rep.* **2019**, *9*, 1456. [CrossRef] [PubMed]
12. Saka, R.; Jain, H.; Kommineni, N.; Chella, N.; Khan, W. Enhanced Penetration and Improved Therapeutic Efficacy of Bexarotene via Topical Liposomal Gel in Imiquimod Induced Psoriatic Plaque Model in BALB/c Mice. *J. Drug Deliv. Sci. Technol.* **2020**, *58*, 101691. [CrossRef]
13. Pandi, P.; Jain, A.; Kommineni, N.; Ionov, M.; Bryszewska, M.; Khan, W. Dendrimer as a New Potential Carrier for Topical Delivery of SiRNA: A Comparative Study of Dendriplex vs. Lipoplex for Delivery of TNF-$α$ SiRNA. *Int. J. Pharm.* **2018**, *550*, 240–250. [CrossRef] [PubMed]

14. Sarathlal, K.C.S.; Kakoty, V.; Krishna, K.V.; Dubey, S.K.; Chitkara, D.; Taliyan, R. Neuroprotective Efficacy of Co-Encapsulated Rosiglitazone and Vorinostat Nanoparticle on Streptozotocin Induced Mice Model of Alzheimer Disease. *ACS Chem. Neurosci.* **2021**, *12*, 1528–1541. [CrossRef]
15. Nafisi, S.; Maibach, H.I. Nanotechnology in Cosmetics. In *Cosmetic Science and Technology: Theoretical Principles and Applications*; Elsevier: Amsterdam, The Netherlands, 2017; pp. 337–369. [CrossRef]
16. Tiwari, N.; Osorio-Blanco, E.R.; Sonzogni, A.; Esporr\'\in-Ubieto, D.; Wang, H.; Calderon, M. Nanocarriers for Skin Applications: Where Do We Stand? *Angew. Chem. Int. Ed.* **2022**, *61*, e202107960. [CrossRef]
17. Shukla, T.; Upmanyu, N.; Agrawal, M.; Saraf, S.; Saraf, S.; Alexander, A. Biomedical Applications of Microemulsion through Dermal and Transdermal Route. *Biomed. Pharmacother.* **2018**, *108*, 1477–1494. [CrossRef]
18. Nastiti, C.M.R.R.; Ponto, T.; Abd, E.; Grice, J.E.; Benson, H.A.E.; Roberts, M.S. Topical Nano and Microemulsions for Skin Delivery. *Pharmaceutics* **2017**, *9*, 37. [CrossRef]
19. Aithal, G.C.; Narayan, R.; Nayak, U.Y. Nanoemulgel: A Promising Phase in Drug Delivery. *Curr. Pharm. Des.* **2020**, *26*, 279–291. [CrossRef]
20. Anand, K.; Ray, S.; Rahman, M.; Shaharyar, A.; Bhowmik, R.; Bera, R.; Karmakar, S. Nano-Emulgel: Emerging as a Smarter Topical Lipidic Emulsion-Based Nanocarrier for Skin Healthcare Applications. *Recent Pat. Antiinfect. Drug Discov.* **2019**, *14*, 16–35. [CrossRef]
21. Murthy, S.N. Approaches for Delivery of Drugs Topically. *AAPS PharmSciTech* **2019**, *21*, 1–2. [CrossRef]
22. Iqbal, M.A.; Md, S.; Sahni, J.K.; Baboota, S.; Dang, S.; Ali, J. Nanostructured Lipid Carriers System: Recent Advances in Drug Delivery. *J. Drug Target.* **2012**, *20*, 813–830. [CrossRef] [PubMed]
23. Bhowmik, D.; Gopinath, H.; Kumar, B.P.; Duraivel, S.; Kumar, K.P.S. Recent Advances In Novel Topical Drug Delivery System. *Pharma Innov. J.* **2012**, *1*, 12–31.
24. Gannu, R.; Palem, C.R.; Yamsani, V.V.; Yamsani, S.K.; Yamsani, M.R. Enhanced Bioavailability of Lacidipine via Microemulsion Based Transdermal Gels: Formulation Optimization, Ex Vivo and in Vivo Characterization. *Int. J. Pharm.* **2010**, *388*, 231–241. [CrossRef] [PubMed]
25. Bhaskar, K.; Anbu, J.; Ravichandiran, V.; Venkateswarlu, V.; Rao, Y.M. Lipid Nanoparticles for Transdermal Delivery of Flurbiprofen: Formulation, in Vitro, Ex Vivo and in Vivo Studies. *Lipids Health Dis.* **2009**, *8*, 1–15. [CrossRef]
26. Zhou, J.; Zhou, M.; Yang, F.F.; Liu, C.Y.; Pan, R.L.; Chang, Q.; Liu, X.M.; Liao, Y.H. Involvement of the Inhibition of Intestinal Glucuronidation in Enhancing the Oral Bioavailability of Resveratrol by Labrasol Containing Nanoemulsions. *Mol. Pharm.* **2015**, *12*, 1084–1095. [CrossRef]
27. Chang, R.K.; Raw, A.; Lionberger, R.; Yu, L. Generic Development of Topical Dermatologic Products: Formulation Development, Process Development, and Testing of Topical Dermatologic Products. *AAPS J.* **2013**, *15*, 41–52. [CrossRef]
28. Cevc, G. Lipid Vesicles and Other Colloids as Drug Carriers on the Skin. *Adv. Drug Deliv. Rev.* **2004**, *56*, 675–711. [CrossRef]
29. Chellapa, P.; Mohamed, A.T.; Keleb, E.I.; Elmahgoubi, A.; Eid, A.M.; Issa, Y.S.; Elmarzugi, N.A. Nanoemulsion and Nanoemulgel as a Topical Formulation. *IOSR J. Pharm.* **2015**, *5*, 43–47.
30. Marto, J.; Baltazar, D.; Duarte, A.; Fernandes, A.; Gouveia, L.; Militão, M.; Salgado, A.; Simões, S.; Oliveira, E.; Ribeiro, H.M. Topical Gels of Etofenamate: In Vitro and in Vivo Evaluation. *Pharm. Dev. Technol.* **2015**, *20*, 710–715. [CrossRef]
31. Lau, W.; White, A.; Gallagher, S.; Donaldson, M.; McNaughton, G.; Heard, C. Scope and Limitations of the Co-Drug Approach to Topical Drug Delivery. *Curr. Pharm. Des.* **2008**, *14*, 794–802. [CrossRef]
32. Raza, K.; Kumar, M.; Kumar, P.; Malik, R.; Sharma, G.; Kaur, M.; Katare, O.P. Topical Delivery of Aceclofenac: Challenges and Promises of Novel Drug Delivery Systems. *Biomed Res. Int.* **2014**, *2014*, 406731. [CrossRef] [PubMed]
33. Somagoni, J.; Boakye, C.H.A.; Godugu, C.; Patel, A.R.; Faria, H.A.M.; Zucolotto, V.; Singh, M. Nanomiemgel–a Novel Drug Delivery System for Topical Application–in Vitro and in Vivo Evaluation. *PLoS ONE* **2014**, *9*, e115952. [CrossRef] [PubMed]
34. Begur, M.; Vasantakumar Pai, K.; Gowda, D.V.; Srivastava, A.; Raghundan, H.V. Development and Characterization of Nanoemulgel Based Transdermal Delivery System for Enhancing Permeability of Tacrolimus. *Adv. Sci. Eng. Med.* **2016**, *8*, 324–332. [CrossRef]
35. Ngawhirunpat, T.; Worachun, N.; Opanasopit, P.; Rojanarata, T.; Panomsuk, S. Cremophor RH40-PEG 400 Microemulsions as Transdermal Drug Delivery Carrier for Ketoprofen. *Pharm. Dev. Technol.* **2013**, *18*, 798–803. [CrossRef]
36. Bos, J.D.; Meinardi, M.M.H.M. The 500 Dalton Rule for the Skin Penetration of Chemical Compounds and Drugs. *Exp. Dermatol.* **2000**, *9*, 165–169. [CrossRef]
37. Phad, A.R.; Dilip, N.T.; Sundara Ganapathy, R. Emulgel: A Comprehensive Review for Topical Delivery of Hydrophobic Drugs. *Asian J. Pharm.* **2018**, *12*, 382.
38. Azeem, A.; Rizwan, M.; Ahmad, F.J.; Iqbal, Z.; Khar, R.K.; Aqil, M.; Talegaonkar, S. Nanoemulsion Components Screening and Selection: A Technical Note. *AAPS PharmSciTech* **2009**, *10*, 69. [CrossRef]
39. Gao, F.; Zhang, Z.; Bu, H.; Huang, Y.; Gao, Z.; Shen, J.; Zhao, C.; Li, Y. Nanoemulsion Improves the Oral Absorption of Candesartan Cilexetil in Rats: Performance and Mechanism. *J. Control. Release* **2011**, *149*, 168–174. [CrossRef]
40. Kim, B.S.; Won, M.; Lee, K.M.; Kim, C.S. In Vitro Permeation Studies of Nanoemulsions Containing Ketoprofen as a Model Drug. *Drug Deliv.* **2008**, *15*, 465–469. [CrossRef]
41. Akhter, S.; Jain, G.; Ahmad, F.; Khar, R.; Jain, N.; Khan, Z.; Talegaonkar, S. Investigation of Nanoemulsion System for Transdermal Delivery of Domperidone: Ex-Vivo and in Vivo Studies. *Curr. Nanosci.* **2008**, *4*, 381–390. [CrossRef]

42. El Maghraby, G.M. Transdermal Delivery of Hydrocortisone from Eucalyptus Oil Microemulsion: Effects of Cosurfactants. *Int. J. Pharm.* **2008**, *355*, 285–292. [CrossRef] [PubMed]
43. Huang, Y.B.; Lin, Y.H.; Lu, T.M.; Wang, R.J.; Tsai, Y.H.; Wu, P.C. Transdermal Delivery of Capsaicin Derivative-Sodium Nonivamide Acetate Using Microemulsions as Vehicles. *Int. J. Pharm.* **2008**, *349*, 206–211. [CrossRef] [PubMed]
44. Eid, A.M.; El-Enshasy, H.A.; Aziz, R.; Elmarzugi, N.A. Preparation, Characterization and Anti-Inflammatory Activity of Swietenia Macrophylla Nanoemulgel. *J. Nanomed. Nanotechnol.* **2014**, *5*, 1–10. [CrossRef]
45. Algahtani, M.S.; Ahmad, M.Z.; Ahmad, J. Nanoemulgel for Improved Topical Delivery of Retinyl Palmitate: Formulation Design and Stability Evaluation. *Nanomaterials* **2020**, *10*, 848. [CrossRef]
46. Bernardi, D.S.; Pereira, T.A.; Maciel, N.R.; Bortoloto, J.; Viera, G.S.; Oliveira, G.C.; Rocha-Filho, P.A. Formation and Stability of Oil-in-Water Nanoemulsions Containing Rice Bran Oil: In Vitro and in Vivo Assessments. *J. Nanobiotechnol.* **2011**, *9*, 44. [CrossRef]
47. Bolzinger, M.A.; Briançon, S.; Pelletier, J.; Fessi, H.; Chevalier, Y. Percutaneous Release of Caffeine from Microemulsion, Emulsion and Gel Dosage Forms. *Eur. J. Pharm. Biopharm.* **2008**, *68*, 446–451. [CrossRef]
48. Fini, A.; Bergamante, V.; Ceschel, G.C.; Ronchi, C.; Moraes, C.A.F. Control of Transdermal Permeation of Hydrocortisone Acetate from Hydrophilic and Lipophilic Formulations. *AAPS PharmSciTech* **2008**, *9*, 762. [CrossRef]
49. Teichmann, A.; Heuschkel, S.; Jacobi, U.; Presse, G.; Neubert, R.H.H.; Sterry, W.; Lademann, J. Comparison of Stratum Corneum Penetration and Localization of a Lipophilic Model Drug Applied in an o/w Microemulsion and an Amphiphilic Cream. *Eur. J. Pharm. Biopharm.* **2007**, *67*, 699–706. [CrossRef]
50. Khurana, S.; Jain, N.K.; Bedi, P.M.S. Nanoemulsion Based Gel for Transdermal Delivery of Meloxicam: Physico-Chemical, Mechanistic Investigation. *Life Sci.* **2013**, *92*, 383–392. [CrossRef]
51. Mou, D.; Chen, H.; Du, D.; Mao, C.; Wan, J.; Xu, H.; Yang, X. Hydrogel-Thickened Nanoemulsion System for Topical Delivery of Lipophilic Drugs. *Int. J. Pharm.* **2008**, *353*, 270–276. [CrossRef]
52. Pund, S.; Pawar, S.; Gangurde, S.; Divate, D. Transcutaneous Delivery of Leflunomide Nanoemulgel: Mechanistic Investigation into Physicomechanical Characteristics, in Vitro Anti-Psoriatic and Anti-Melanoma Activity. *Int. J. Pharm.* **2015**, *487*, 148–156. [CrossRef] [PubMed]
53. Dev, A.; Chodankar, R.; Shelke, O. Emulgels: A Novel Topical Drug Delivery System. *Pharm. Biol. Eval.* **2015**, *2*, 64–75.
54. Sengupta, P.; Chatterjee, B. Potential and Future Scope of Nanoemulgel Formulation for Topical Delivery of Lipophilic Drugs. *Int. J. Pharm.* **2017**, *526*, 353–365. [CrossRef] [PubMed]
55. Arora, R.; Aggarwal, G.; Harikumar, S.L.; Kaur, K. Nanoemulsion Based Hydrogel for Enhanced Transdermal Delivery of Ketoprofen. *Adv. Pharm.* **2014**, *2014*, 468456. [CrossRef]
56. Gorain, B.; Choudhury, H.; Tekade, R.K.; Karan, S.; Jaisankar, P.; Pal, T.K. Comparative Biodistribution and Safety Profiling of Olmesartan Medoxomil Oil-in-Water Oral Nanoemulsion. *Regul. Toxicol. Pharmacol.* **2016**, *82*, 20–31. [CrossRef]
57. Dubey, S.K.; Ram, M.S.; Krishna, K.V.; Saha, R.N.; Singhvi, G.; Agrawal, M.; Ajazuddin; Saraf, S.; Saraf, S.; Alexander, A. Recent Expansions on Cellular Models to Uncover the Scientific Barriers Towards Drug Development for Alzheimer's Disease. *Cell. Mol. Neurobiol.* **2019**, *39*, 181–209. [CrossRef]
58. Formariz, T.P.; Sarmento, V.H.V.; Silva-Junior, A.A.; Scarpa, M.V.; Santilli, C.V.; Oliveira, A.G. Doxorubicin Biocompatible O/W Microemulsion Stabilized by Mixed Surfactant Containing Soya Phosphatidylcholine. *Colloids Surf. B Biointerfaces* **2006**, *51*, 54–61. [CrossRef]
59. Ahmad, M.Z.; Ahmad, J.; Alasmary, M.Y.; Akhter, S.; Aslam, M.; Pathak, K.; Jamil, P.; Abdullah, M.M. Nanoemulgel as an Approach to Improve the Biopharmaceutical Performance of Lipophilic Drugs: Contemporary Research and Application. *J. Drug Deliv. Sci. Technol.* **2022**, *72*, 103420. [CrossRef]
60. Kaur, G. TPGS Loaded Topical Nanoemulgel of Mefenamic Acid for the Treatment of Rheumatoid Arthritis. *Int. J. Pharm. Pharm. Res.* **2019**, *15*, 64–107.
61. Tesch, S.; Gerhards, C.; Schubert, H. Stabilization of Emulsions by OSA Starches. *J. Food Eng.* **2002**, *54*, 167–174. [CrossRef]
62. Ingle, A.P.; Shende, S.; Gupta, I.; Rai, M. Recent Trends in the Development of Nano-Bioactive Compounds and Delivery Systems. In *Biotechnological Production of Bioactive Compounds*; Elsevier: Amsterdam, The Netherlands, 2020; pp. 409–431.
63. Ashara, K.C. Microemulgel: An overwhelming approach to improve therapeutic action of drug moiety. *Saudi Pharm. J.* **2014**, *24*, 452–457. [CrossRef] [PubMed]
64. Sultana, N.; Akhtar, J.; Khan, M.I.; Ahmad, U.; Arif, M.; Ahmad, M.; Upadhyay, T. Nanoemulgel: For Promising Topical and Systemic Delivery. In *Drug Development Life Cycle*; IntechOpen: London, UK, 2022.
65. Zhou, H.; Yue, Y.; Liu, G.; Li, Y.; Zhang, J.; Gong, Q.; Yan, Z.; Duan, M. Preparation and Characterization of a Lecithin Nanoemulsion as a Topical Delivery System. *Nanoscale Res. Lett.* **2010**, *5*, 224–230. [CrossRef] [PubMed]
66. Li, P.; Nielsen, H.M.; Müllertz, A. Oral Delivery of Peptides and Proteins Using Lipid-Based Drug Delivery Systems. *Expert Opin. Drug Deliv.* **2012**, *9*, 1289–1304. [CrossRef] [PubMed]
67. Bahadur, S.; Pardhi, D.M.; Rautio, J.; Rosenholm, J.M.; Pathak, K. Intranasal Nanoemulsions for Direct Nose-to-Brain Delivery of Actives for Cns Disorders. *Pharmaceutics* **2020**, *12*, 1230. [CrossRef] [PubMed]
68. Chatterjee, B.; Gorain, B.; Mohananaidu, K.; Sengupta, P.; Mandal, U.K.; Choudhury, H. Targeted Drug Delivery to the Brain via Intranasal Nanoemulsion: Available Proof of Concept and Existing Challenges. *Int. J. Pharm.* **2019**, *565*, 258–268. [CrossRef] [PubMed]

69. Algahtani, M.S.; Ahmad, M.Z.; Nourein, I.H.; Albarqi, H.A.; Alyami, H.S.; Alyami, M.H.; Alqahtani, A.A.; Alasiri, A.; Algahtani, T.S.; Mohammed, A.A.; et al. Preparation and Characterization of Curcumin Nanoemulgel Utilizing Ultrasonication Technique for Wound Healing: In Vitro, Ex Vivo, and in Vivo Evaluation. *Gels* **2021**, *7*, 213. [CrossRef]
70. Algahtani, M.S.; Ahmad, M.Z.; Shaikh, I.A.; Abdel-Wahab, B.A.; Nourein, I.H.; Ahmad, J. Thymoquinone Loaded Topical Nanoemulgel for Wound Healing: Formulation Design and In-Vivo Evaluation. *Molecules* **2021**, *26*, 3863. [CrossRef] [PubMed]
71. Alyoussef, A.; El-Gogary, R.I.; Ahmed, R.F.; Ahmed Farid, O.A.; Bakeer, R.M.; Nasr, M. The Beneficial Activity of Curcumin and Resveratrol Loaded in Nanoemulgel for Healing of Burn-Induced Wounds. *J. Drug Deliv. Sci. Technol.* **2021**, *62*, 102360. [CrossRef]
72. Abdallah, M.; Lila, A.; Unissa, R.; Elsewedy, H.S.; Elghamry, H.A.; Soliman, M.S. Preparation, Characterization and Evaluation of Anti-Inflammatory and Anti-Nociceptive Effects of Brucine-Loaded Nanoemulgel. *Colloids Surf. B Biointerfaces* **2021**, *205*, 111868. [CrossRef]
73. Zakir, F.; Ahmad, A.; Mirza, M.A.; Kohli, K.; Ahmad, F.J. Exploration of a Transdermal Nanoemulgel as an Alternative Therapy for Postmenopausal Osteoporosis. *J. Drug Deliv. Sci. Technol.* **2021**, *65*, 102745. [CrossRef]
74. Mao, Y.; Chen, X.; Xu, B.; Shen, Y.; Ye, Z.; Chaurasiya, B.; Liu, L.; Li, Y.; Xing, X.; Chen, D. Eprinomectin Nanoemulgel for Transdermal Delivery against Endoparasites and Ectoparasites: Preparation, in Vitro and in Vivo Evaluation. *Taylor Fr.* **2019**, *26*, 1104–1114. [CrossRef]
75. Gadhave, D.; Tupe, S.; Tagalpallewar, A.; Gorain, B.; Choudhury, H.; Kokare, C. Nose-to-Brain Delivery of Amisulpride-Loaded Lipid-Based Poloxamer-Gellan Gum Nanoemulgel: In Vitro and in Vivo Pharmacological Studies. *Int. J. Pharm.* **2021**, *607*, 121050. [CrossRef]
76. Qu, Y.; Li, A.; Ma, L.; Iqbal, S.; Sun, X.; Ma, W.; Li, C. Nose-to-Brain Delivery of Disulfiram Nanoemulsion in Situ Gel Formulation for Glioblastoma Targeting Therapy. *Int. J. Pharm.* **2021**, *597*, 120250. [CrossRef]
77. Topical Metformin Emulgel VS Salicylic Acid Peeling in Treatment of Acne Vulgaris—Full Text View—ClinicalTrials.Gov. Available online: https://clinicaltrials.gov/ct2/show/NCT05536193 (accessed on 1 December 2022).
78. Comparison of the Bioavailability of Diclofenac in a Combination Product (Diclofenac 2% + Capsaicin 0.075% Topical Gel) with Two Diclofenac Only Products, Diclofenac Mono Gel 2% and Voltarol®12 Hour Emulgel 2.32% Gel, in Healthy Volunteers—Full Text View—ClinicalTrials.Gov. Available online: https://clinicaltrials.gov/ct2/show/NCT03074162 (accessed on 1 December 2022).
79. Effects of Visnadin, Ethyl Ximeninate, Coleus Barbatus and Millet in Emulgel on Sexual Function in Postmenopausal Women—Full Text View—ClinicalTrials.Gov. Available online: https://clinicaltrials.gov/ct2/show/NCT04579991 (accessed on 1 December 2022).
80. Clinical Assessment of Voriconazole Self Nano Emulsifying Drug Delivery System Intermediate Gel—Full Text View—ClinicalTrials.Gov. Available online: https://www.clinicaltrials.gov/ct2/show/NCT04110860 (accessed on 1 December 2022).
81. Clinical Assessment of Itraconazole Self Nano Emulsifying Drug Delivery System Intermediate Gel—Full Text View—ClinicalTrials.Gov. Available online: https://www.clinicaltrials.gov/ct2/show/NCT04110834 (accessed on 1 December 2022).
82. Study of Efficacy and Tolerability of SYSTANE Complete in Patients with Dry Eye Disease—Full Text View—ClinicalTrials.Gov. Available online: https://clinicaltrials.gov/ct2/show/NCT03492541 (accessed on 1 December 2022).
83. Santhosh, A.; Kumar, A.; Pramanik, R.; Gogia, A.; Prasad, C.P.; Gupta, I.; Gupta, N.; Cheung, W.Y.; Pandey, R.M.; Sharma, A.; et al. Randomized Double-Blind, Placebo-Controlled Study of Topical Diclofenac in the Prevention of Hand-Foot Syndrome in Patients Receiving Capecitabine (the D-TORCH Study). *Trials* **2022**, *23*, 420. [CrossRef]
84. Eswaraiah, S.; Swetha, K. Emulgel: Review on Novel Approach to Topical Drug Delivery. Available online: https://asianjpr.com/AbstractView.aspx?PID=2014-4-1-2 (accessed on 1 December 2022).
85. Jeengar, M.K.; Rompicharla, S.V.K.; Shrivastava, S.; Chella, N.; Shastri, N.R.; Naidu, V.G.M.; Sistla, R. Emu Oil Based Nano-Emulgel for Topical Delivery of Curcumin. *Int. J. Pharm.* **2016**, *506*, 222–236. [CrossRef]
86. Md, S.; Alhakamy, N.A.; Aldawsari, H.M.; Kotta, S.; Ahmad, J.; Akhter, S.; Alam, M.S.; Khan, M.A.; Awan, Z.; Sivakumar, P.M. Improved Analgesic and Anti-Inflammatory Effect of Diclofenac Sodium by Topical Nanoemulgel: Formulation Development—In Vitro and in Vivo Studies. *J. Chem.* **2020**, *2020*, 4071818. [CrossRef]
87. Drais, H.K.; Hussein, A.A. Formulation Characterization and Evaluation of Meloxicam Nanoemulgel to Be Used Topically. *Iraqi J. Pharm. Sci.* **2017**, *26*, 9–16.
88. Aithal, G.C.; Nayak, U.Y.; Mehta, C.; Narayan, R.; Gopalkrishna, P.; Pandiyan, S.; Garg, S. Localized In Situ Nanoemulgel Drug Delivery System of Quercetin for Periodontitis: Development and Computational Simulations. *Molecules* **2018**, *23*, 1363. [CrossRef]
89. Wankar, J.; Ajimera, T. Design, Development and Evaluation of Nanoemulsion and Nanogel of Itraconazole for Transdermal Delivery. *J. Sci. Res. Pharm.* **2014**, *3*, 6–11.
90. Pathak, M.K.; Chhabra, G.; Pathak, K. Design and Development of a Novel PH Triggered Nanoemulsified In-Situ Ophthalmic Gel of Fluconazole: Ex-Vivo Transcorneal Permeation, Corneal Toxicity and Irritation Testing. *Drug Dev. Ind. Pharm.* **2013**, *39*, 780–790. [CrossRef]
91. Wais, M.; Samad, A.; Nazish, I.; Khale, A.; Aqil, M.; Khan, M. Formulation development ex-vivo and in-vivo evaluation of nanoemulsion for transdermal delivery of glibenclamide. *Int. J. Pharm. Pharm. Sci.* **2013**, *5*, 747–754.
92. Pratap, S.B.; Brajesh, K.; Jain, S.K.; Kausar, S. Development and Characterization of A Nanoemulsion Gel for Transdermal Delivery of Carvedilol. *Int. J. Drug Dev. Res.* **2012**, *4*, 151–161.

93. Begur, M.; Pai, V.; Gowda, D.V.; AtulSrivastava; Raghundan, H.V.; Shinde, C.G.; Manusri, N. Enhanced Permeability of Cyclosporine from a Transdermally Applied Nanoemulgel. *Der Pharm. Sin.* **2015**. Available online: https://hal.archives-ouvertes.fr/hal-03627803 (accessed on 1 December 2022).
94. Nagaraja, S.; Basavarajappa, G.M.; Attimarad, M.; Pund, S. Topical Nanoemulgel for the Treatment of Skin Cancer: Proof-of-Technology. *Pharmaceutics* **2021**, *13*, 902. [CrossRef]
95. Morsy, M.A.; Abdel-Latif, R.G.; Nair, A.B.; Venugopala, K.N.; Ahmed, A.F.; Elsewedy, H.S.; Shehata, T.M. Preparation and Evaluation of Atorvastatin-Loaded Nanoemulgel on Wound-Healing Efficacy. *Pharmaceutics* **2019**, *11*, 609. [CrossRef]
96. Soliman, W.E.; Shehata, T.M.; Mohamed, M.E.; Younis, N.S.; Elsewedy, H.S. Enhancement of Curcumin Anti-Inflammatory Effect via Formulation into Myrrh Oil-Based Nanoemulgel. *Polymers* **2021**, *13*, 577. [CrossRef]
97. Elmataeeshy, M.E.; Sokar, M.S.; Bahey-El-Din, M.; Shaker, D.S. Enhanced Transdermal Permeability of Terbinafine through Novel Nanoemulgel Formulation; Development, in Vitro and in Vivo Characterization. *Future J. Pharm. Sci.* **2018**, *4*, 18–28. [CrossRef]
98. Vartak, R.; Menon, S.; Patki, M.; Billack, B.; Patel, K. Ebselen Nanoemulgel for the Treatment of Topical Fungal Infection. *Eur. J. Pharm. Sci.* **2020**, *148*, 105323. [CrossRef]
99. Vandamme, T.F. Microemulsions as Ocular Drug Delivery Systems: Recent Developments and Future Challenges. *Prog. Retin. Eye Res.* **2002**, *21*, 15–34. [CrossRef]
100. Bashir, M.; Ahmad, J.; Asif, M.; Khan SU, D.; Irfan, M.; Ibrahim, A.Y.; Asghar, S.; Khan, I.U.; Iqbal, M.S.; Haseeb, A.; et al. Undefined Nanoemulgel, an Innovative Carrier for Diflunisal Topical Delivery with Profound Anti-Inflammatory Effect: In Vitro and in Vivo Evaluation. *Int. J. Nanomed.* **2021**, *16*, 1457. [CrossRef]
101. Williams, A.C.; Barry, B.W. Penetration Enhancers. *Adv. Drug Deliv. Rev.* **2012**, *64*, 128–137. [CrossRef]
102. Aggarwal, G.; Dhawan, B.; Harikumar, S. Enhanced Transdermal Permeability of Piroxicam through Novel Nanoemulgel Formulation. *Int. J. Pharm. Investig.* **2014**, *4*, 65. [CrossRef] [PubMed]
103. Jeengar, M.K.; Sravan Kumar, P.; Thummuri, D.; Shrivastava, S.; Guntuku, L.; Sistla, R.; Naidu, V.G.M. Review on Emu Products for Use as Complementary and Alternative Medicine. *Nutrition* **2015**, *31*, 21–27. [CrossRef]
104. Tayel, S.A.; El-Nabarawi, M.A.; Tadros, M.I.; Abd-Elsalam, W.H. Positively Charged Polymeric Nanoparticle Reservoirs of Terbinafine Hydrochloride: Preclinical Implications for Controlled Drug Delivery in the Aqueous Humor of Rabbits. *AAPS PharmSciTech* **2013**, *14*, 782–793. [CrossRef] [PubMed]
105. Silva, H.D.; Cerqueira, M.A.; Vicente, A.A. Influence of Surfactant and Processing Conditions in the Stability of Oil-in-Water Nanoemulsions. *J. Food Eng.* **2015**, *167*, 89–98. [CrossRef]
106. Rajpoot, K.; Tekade, R.K. Microemulsion as Drug and Gene Delivery Vehicle: An inside Story. In *Drug Delivery Systems*; Academic Press: Cambridge, MA, USA, 2019; pp. 455–520. [CrossRef]
107. Rousseau, D.; Rafanan, R.R.; Yada, R. Microemulsions as Nanoscale Delivery Systems. *Compr. Biotechnol. Second Ed.* **2011**, *4*, 675–682. [CrossRef]
108. Khachane, P.V.; Jain, A.S.; Dhawan, V.V.; Joshi, G.V.; Date, A.A.; Mulherkar, R.; Nagarsenker, M.S. Cationic Nanoemulsions as Potential Carriers for Intracellular Delivery. *Saudi Pharm. J.* **2015**, *23*, 188–194. [CrossRef]
109. Shakeel, F.; Haq, N.; Alanazi, F.K.; Alsarra, I.A. Impact of Various Nonionic Surfactants on Self-Nanoemulsification Efficiency of Two Grades of Capryol (Capryol-90 and Capryol-PGMC). *J. Mol. Liq.* **2013**, *182*, 57–63. [CrossRef]
110. Mantzaridis, C.; Mountrichas, G.; Pispas, S. Complexes between High Charge Density Cationic Polyelectrolytes and Anionic Single- and Double-Tail Surfactants. *J. Phys. Chem. B* **2009**, *113*, 7064–7070. [CrossRef]
111. Zakharova, L.Y.; Pashirova, T.N.; Fernandes, A.R.; Doktorovova, S.; Martins-Gomes, C.; Silva, A.M.; Souto, E.B. Self-Assembled Quaternary Ammonium Surfactants for Pharmaceuticals and Biotechnology. In *Organic Materials as Smart Nanocarriers for Drug Delivery*; William Andrew Publishing: Norwich, NY, USA, 2018; pp. 601–618. [CrossRef]
112. Bali, V.; Ali, M.; Ali, J. Study of Surfactant Combinations and Development of a Novel Nanoemulsion for Minimising Variations in Bioavailability of Ezetimibe. *Colloids Surf. B. Biointerfaces* **2010**, *76*, 410–420. [CrossRef]
113. Hu, J.; Chen, D.; Jiang, R.; Tan, Q.; Zhu, B.; Zhang, J. Improved Absorption and in Vivo Kinetic Characteristics of Nanoemulsions Containing Evodiamine–Phospholipid Nanocomplex. *Int. J. Nanomed.* **2014**, *9*, 4411–4420. [CrossRef]
114. Poré, J. *Emulsions, Micro-Émulsions, Émulsions Multiples*; Editions Techniques des Industries des Corps Gras: Neuilly sur Seine, France, 1992; ISBN 9782950724106.
115. Wang, Z.; Mu, H.-J.; Zhang, X.-M.; Ma, P.-K.; Lian, S.-N.; Zhang, F.-P.; Chu, S.-Y.; Zhang, W.-W.; Wang, A.-P.; Wang, W.-Y.; et al. Lower Irritation Microemulsion-Based Rotigotine Gel: Formulation Optimization and in Vitro and in Vivo Studies. *Int. J. Nanomed.* **2015**, *10*, 633–644.
116. Syed, H.K.; Peh, K.K. Identification of Phases of Various Oil, Surfactant/ Co-Surfactants and Water System By Ternary Phase Diagram. *Acta Pol. Pharm.* **2014**, *71*, 301–309. [PubMed]
117. Shah, H.; Jain, A.; Laghate, G.; Prabhudesai, D. Pharmaceutical Excipients. In *Remington*; Academic Press: Cambridge, MA, USA, 2021; pp. 633–643. [CrossRef]
118. Ojha, B.; Jain, V.K.; Gupta, S.; Talegaonkar, S.; Jain, K. Nanoemulgel: A Promising Novel Formulation for Treatment of Skin Ailments. *Polym. Bull.* **2021**, *79*, 1–25. [CrossRef]
119. Dubey, S.K.; Alexander, A.; Sivaram, M.; Agrawal, M.; Singhvi, G.; Sharma, S.; Dayaramani, R. Uncovering the Diversification of Tissue Engineering on the Emergent Areas of Stem Cells, Nanotechnology and Biomaterials. *Curr. Stem Cell Res. Ther.* **2020**, *15*, 187–201. [CrossRef]

120. Ajazuddin; Alexander, A.; Khichariya, A.; Gupta, S.; Patel, R.J.; Giri, T.K.; Tripathi, D.K. Recent Expansions in an Emergent Novel Drug Delivery Technology: Emulgel. *J. Control. Release* **2013**, *171*, 122–132. [CrossRef]
121. Deshmukh, K.; Basheer Ahamed, M.; Deshmukh, R.R.; Khadheer Pasha, S.K.; Bhagat, P.R.; Chidambaram, K. Biopolymer Composites with High Dielectric Performance: Interface Engineering. In *Biopolymer Composites in Electronics*; Elsevier: Amsterdam, The Netherlands, 2017; pp. 27–128. [CrossRef]
122. Vlaia, L.; Coneac, G.; Olariu, I.; Vlaia, V.; Lupuleasa, D. Cellulose-Derivatives-Based Hydrogels as Vehicles for Dermal and Transdermal Drug Delivery. In *Emerging Concepts in Analysis and Applications of Hydrogels*; IntechOpen: London, UK, 2016. [CrossRef]
123. Hashemnejad, S.M.; Badruddoza, A.Z.M.; Zarket, B.; Ricardo Castaneda, C.; Doyle, P.S. Thermoresponsive Nanoemulsion-Based Gel Synthesized through a Low-Energy Process. *Nat. Commun.* **2019**, *10*, 2749. [CrossRef]
124. Perale, G.; Veglianese, P.; Rossi, F.; Peviani, M.; Santoro, M.; Llupi, D.; Micotti, E.; Forloni, G.; Masi, M. In Situ Agar–Carbomer Hydrogel Polycondensation: A Chemical Approach to Regenerative Medicine. *Mater. Lett.* **2011**, *65*, 1688–1692. [CrossRef]
125. Braun, S. Encapsulation of Cells (Cellular Delivery) Using Sol–Gel Systems. *Compr. Biomater.* **2011**, *4*, 529–543. [CrossRef]
126. Daood, N.M.; Jassim, Z.E.; Ghareeb, M.M.; Zeki, H. Studying the Effect of Different Gelling Agent on The Preparation and Characterization of Metronidazole as Topical Emulgel. *Asian J. Pharm. Clin. Res.* **2019**, *12*, 571–577. [CrossRef]
127. Kathe, K.; Kathpalia, H. Film Forming Systems for Topical and Transdermal Drug Delivery. *Asian J. Pharm. Sci.* **2017**, *12*, 487–497. [CrossRef] [PubMed]
128. Rapalli, V.K.; Mahmood, A.; Waghule, T.; Gorantla, S.; Kumar Dubey, S.; Alexander, A.; Singhvi, G. Revisiting techniques to evaluate drug permeation through skin. *Expert Opin. Drug Deliv.* **2021**, *18*, 1829–1842. [CrossRef] [PubMed]
129. Ibrahim, M.M.; Shehata, T.M. The Enhancement of Transdermal Permeability of Water Soluble Drug by Niosome-Emulgel Combination. *J. Drug Deliv. Sci. Technol.* **2012**, *22*, 353–359. [CrossRef]
130. Dixit, A.S.; Charyulu, N.; Nayari, H. Design and Evaluation of Novel Emulgel Containing Acyclovir for Herpes Simplex Keratitis. *Lat. Am. J. Pharm.* **2011**, *30*, 844–852.
131. Salem, H.F.; Kharshoum, R.M.; Abou-Taleb, H.A.; Naguib, D.M. Nanosized Nasal Emulgel of Resveratrol: Preparation, Optimization, in Vitro Evaluation and in Vivo Pharmacokinetic Study. *Drug Dev. Ind. Pharm.* **2019**, *45*, 1624–1634. [CrossRef] [PubMed]
132. De Souza Ferreira, S.B.; Bruschi, M.L. Investigation of the Physicochemical Stability of Emulgels Composed of Poloxamer 407 and Different Oil Phases Using the Quality by Design Approach. *J. Mol. Liq.* **2021**, *332*, 115856. [CrossRef]
133. Shahin, M.; Abdel Hady, S.; Hammad, M.; Mortada, N. Novel Jojoba Oil-Based Emulsion Gel Formulations for Clotrimazole Delivery. *AAPS PharmSciTech* **2011**, *12*, 239–247. [CrossRef] [PubMed]
134. El-Setouhy, D.A.; Ahmed El-Ashmony, S.M. Ketorolac Trometamol Topical Formulations: Release Behaviour, Physical Characterization, Skin Permeation, Efficacy and Gastric Safety. *J. Pharm. Pharmacol.* **2010**, *62*, 25–34. [CrossRef]
135. Anton, N.; Vandamme, T.F. The Universality of Low-Energy Nano-Emulsification. *Int. J. Pharm.* **2009**, *377*, 142–147. [CrossRef]
136. Sharma, V.; Nayak, S.K.; Paul, S.R.; Choudhary, B.; Ray, S.S.; Pal, K. Emulgels. In *Polymeric Gels*; Woodhead Publishing: Sawston, UK, 2018; pp. 251–264. [CrossRef]
137. Lupi, F.R.; Gabriele, D.; Seta, L.; Baldino, N.; de Cindio, B.; Marino, R. Rheological Investigation of Pectin-Based Emulsion Gels for Pharmaceutical and Cosmetic Uses. *Rheol. Acta* **2015**, *54*, 41–52. [CrossRef]
138. Dong, L.; Liu, C.; Cun, D.; Fang, L. The Effect of Rheological Behavior and Microstructure of the Emulgels on the Release and Permeation Profiles of Terpinen-4-Ol. *Eur. J. Pharm. Sci.* **2015**, *78*, 140–150. [CrossRef]
139. Solè, I.; Pey, C.M.; Maestro, A.; González, C.; Porras, M.; Solans, C.; Gutiérrez, J.M. Nano-Emulsions Prepared by the Phase Inversion Composition Method: Preparation Variables and Scale Up. *J. Colloid Interface Sci.* **2010**, *344*, 417–423. [CrossRef] [PubMed]
140. Lovelyn, C.; Attama, A.A.; Lovelyn, C.; Attama, A.A. Current State of Nanoemulsions in Drug Delivery. *J. Biomater. Nanobiotechnol.* **2011**, *2*, 626–639. [CrossRef]
141. Van der Schaaf, U.S.; Nanoemulsions, H.K. Fabrication of Nanoemulsions by Rotor-Stator Emulsification. In *Nanoemulsions*; Academic Press: Cambridge, MA, USA, 2018; pp. 141–174. [CrossRef]
142. Kotta, S.; Khan, A.W.; Ansari, S.H.; Sharma, R.K.; Ali, J. Formulation of Nanoemulsion: A Comparison between Phase Inversion Composition Method and High-Pressure Homogenization Method. *Drug Deliv.* **2015**, *22*, 455–466. [CrossRef] [PubMed]
143. Juttulapa, M.; Piriyaprasarth, S.; Takeuchi, H.; Sriamornsak, P. Effect of High-Pressure Homogenization on Stability of Emulsions Containing Zein and Pectin. *Asian J. Pharm. Sci.* **2017**, *12*, 21–27. [CrossRef]
144. Bei, D.; Meng, J.; Youan, B.B.C. Engineering Nanomedicines for Improved Melanoma Therapy: Progress and Promises. *Nanomedicine* **2010**, *5*, 1385–1399. [CrossRef]
145. Gorantla, S.; Singhvi, G.; Rapalli, V.K.; Waghule, T.; Dubey, S.K.; Saha, R.N. Targeted Drug-Delivery Systems in the Treatment of Rheumatoid Arthritis: Recent Advancement and Clinical Status. *Ther. Deliv.* **2020**, *11*, 269–284. [CrossRef]
146. Prathyusha, E.; Prabakaran, A.; Ahmed, H.; Dethe, M.R.; Agrawal, M.; Gangipangi, V.; Sudhagar, S.; Krishna, K.V.; Dubey, S.K.; Pemmaraju, D.B.; et al. Investigation of ROS Generating Capacity of Curcumin-Loaded Liposomes and Its in Vitro Cytotoxicity on MCF-7 Cell Lines Using Photodynamic Therapy. *Photodiagnosis Photodyn. Ther.* **2022**, *40*, 103091. [CrossRef]
147. Mulia, K.; Ramadhan, R.M.A.; Krisanti, E.A. Formulation and Characterization of Nanoemulgel Mangosteen Extract in Virgin Coconut Oil for Topical Formulation. *MATEC Web Conf.* **2018**, *156*, 01013. [CrossRef]

148. Chellapa, P.; Eid, A.M.; Elmarzugi, N.A. Preparation and Characterization of Virgin Coconut Oil Nanoemulgel. *J. Chem. Pharm. Res.* **2015**, *7*, 787–793. Available online: https://www.jocpr.com (accessed on 1 December 2022).
149. Bhattacharya, S.; Prajapati, B.G. Formulation and Optimization of Celecoxib Nanoemulgel. *Asian J. Pharm. Clin. Res.* **2017**, *10*, 353–365. [CrossRef]
150. Chin, L.Y.; Tan, J.Y.P.; Choudhury, H.; Pandey, M.; Sisinthy, S.P.; Gorain, B. Development and Optimization of Chitosan Coated Nanoemulgel of Telmisartan for Intranasal Delivery: A Comparative Study. *J. Drug Deliv. Sci. Technol.* **2021**, *62*, 102341. [CrossRef]
151. Khullar, R.; Kumar, D.; Seth, N.; Saini, S. Formulation and Evaluation of Mefenamic Acid Emulgel for Topical Delivery. *Saudi Pharm. J.* **2012**, *20*, 63–67. [CrossRef] [PubMed]
152. Clogston, J.D.; Patri, A.K. Zeta Potential Measurement. *Methods Mol. Biol.* **2011**, *697*, 63–70. [CrossRef] [PubMed]
153. Krishna, K.V.; Saha, R.N.; Dubey, S.K. Biophysical, Biochemical, and Behavioral Implications of ApoE3 Conjugated Donepezil Nanomedicine in a Aβ1-42Induced Alzheimer's Disease Rat Model. *ACS Chem. Neurosci.* **2020**, *11*, 4139–4151. [CrossRef]
154. Khosa, A.; Krishna, K.V.; Saha, R.N.; Dubey, S.K.; Reddi, S. A Simplified and Sensitive Validated RP-HPLC Method for Determination of Temozolomide in Rat Plasma and Its Application to a Pharmacokinetic Study. *J. Liq. Chromatogr. Relat. Technol.* **2018**, *41*, 692–697. [CrossRef]
155. Manus Maguire, C.; Rösslein, M.; Wick, P.; Prina-Mello, A. Characterisation of Particles in Solution–a Perspective on Light Scattering and Comparative Technologies. *Taylor Fr.* **2018**, *19*, 732–745. [CrossRef]
156. Sneha, K.; Kumar, A. Nanoemulsions: Techniques for the Preparation and the Recent Advances in Their Food Applications. *Innov. Food Sci. Emerg. Technol.* **2022**, *76*, 102914. [CrossRef]
157. Khosa, A.; Krishna, K.V.; Dubey, S.K.; Saha, R.N. Lipid Nanocarriers for Enhanced Delivery of Temozolomide to the Brain. *Methods Mol. Biol.* **2020**, *2059*, 285–298. [CrossRef]
158. Garg, A.; Aggarwal, D.; Garg, S.; America, A.S.-T.N. Spreading of Semisolid Formulations: An Update. *Pharm. Technol. N. Am.* **2002**, *26*, 84.
159. Nikumbh, K.V.; Sevankar, S.G.; Patil, M.P. Formulation Development, in Vitro and in Vivo Evaluation of Microemulsion-Based Gel Loaded with Ketoprofen. *Drug Deliv.* **2015**, *22*, 509–515. [CrossRef] [PubMed]
160. Shah, V.P.; Simona Miron, D.; Ștefan Rădulescu, F.; Cardot, J.M.; Maibach, H.I. In Vitro Release Test (IVRT): Principles and Applications. *Int. J. Pharm.* **2022**, *626*, 122159. [CrossRef]
161. Sheshala, R.; Anuar, N.K.; Abu Samah, N.H.; Wong, T.W. In Vitro Drug Dissolution/Permeation Testing of Nanocarriers for Skin Application: A Comprehensive Review. *AAPS PharmSciTech* **2019**, *20*, 164. [CrossRef] [PubMed]
162. Kanfer, I.; Rath, S.; Purazi, P.; Mudyahoto, N.A. In Vitro Release Testing of Semi-Solid Dosage Forms. *Dissolut. Technol.* **2017**, *24*, 52–60. [CrossRef]
163. Shaikh, R.; Raj Singh, T.; Garland, M.; Woolfson, A.; Donnelly, R. Mucoadhesive Drug Delivery Systems. *J. Pharm. Bioallied Sci.* **2011**, *3*, 89–100. [CrossRef] [PubMed]
164. Amorós-Galicia, L.; Nardi-Ricart, A.; Verdugo-González, C.; Arroyo-García, C.M.; García-Montoya, E.; Pérez-Lozano, P.; Suñé-Negre, J.M.; Suñé-Pou, M. Development of a Standardized Method for Measuring Bioadhesion and Mucoadhesion That Is Applicable to Various Pharmaceutical Dosage Forms. *Pharmaceutics* **2022**, *14*, 1995. [CrossRef]
165. Yuan, C.; Xu, Z.Z.; Fan, M.; Liu, H.; Xie, Y.; Zhu, T. Study on Characteristics and Harm of Surfactants. *J. Chem. Pharm. Res.* **2014**, *6*, 2233–2237.
166. Lewis, M.A. Chronic Toxicities of Surfactants and Detergent Builders to Algae: A Review and Risk Assessment. *Ecotoxicol. Environ. Saf.* **1990**, *20*, 123–140. [CrossRef]
167. James-Smith, M.A.; Hellner, B.; Annunziato, N.; Mitragotri, S. Effect of Surfactant Mixtures on Skin Structure and Barrier Properties. *Ann. Biomed. Eng.* **2011**, *39*, 1215–1223. [CrossRef]
168. Azeem, A.; Ahmad, F.J.; Khar, R.K.; Talegaonkar, S. Nanocarrier for the Transdermal Delivery of an Antiparkinsonian Drug. *AAPS PharmSciTech* **2009**, *10*, 1093–1103. [CrossRef]
169. Patel, B.B.; Patel, J.K.; Chakraborty, S.; Shukla, D. Revealing Facts behind Spray Dried Solid Dispersion Technology Used for Solubility Enhancement. *Saudi Pharm. J.* **2015**, *23*, 352–365. [CrossRef] [PubMed]
170. Wang, W.; Hui, P.C.L.; Kan, C.W. Functionalized Textile Based Therapy for the Treatment of Atopic Dermatitis. *Coatings* **2017**, *7*, 82. [CrossRef]
171. Gutiérrez, J.M.; González, C.; Maestro, A.; Solè, I.; Pey, C.M.; Nolla, J. Nano-Emulsions: New Applications and Optimization of Their Preparation. *Curr. Opin. Colloid Interface Sci.* **2008**, *13*, 245–251. [CrossRef]
172. Sallam, A.A.N.; Younes, H.M. Transdermal Non-Aqueous Nanoemulgels for Systemic Delivery of Aromatase Inhibitor. EP3727363A1, 28 October 2020.
173. Kaufman, R.C. Methods of Treating Inflammatory Disorders and Global Inflammation with Compositions Comprising Phospholipid Nanoparticle Encapsulations of NSAIDS 2018. CA2970917C, 17 September 2019.
174. Xiaoling, F.; Xiao, W.; Xianyi, S. Tripterygium Glycoside-Containing Micro-Emulsified Gel Transdermal Preparation and Preparation Method Thereof. CN107303263B, 31 July 2020.
175. Sengupta, S.; Chawrai, S.R.; Ghosh, S.; Ghosh, S.; Jain, N.; Sadhasivam, S.; Buchta, R.; Bhattacharyya, A. Besifloxacin for the Treatment of Resistant Acne. EP3099301B1, 18 December 2019.
176. Bhalerao, H.; Koteshwara, K.; Chandran, S. Design, Optimisation and Evaluation of in Situ Gelling Nanoemulsion Formulations of Brinzolamide. *Drug Deliv. Transl. Res.* **2020**, *10*, 529–547. [CrossRef] [PubMed]

177. Suresh, S.; Rathod, S.; Devasani, R. Minoxidil and Castor Oil Nanoemulgel for Alopecia. WO2020121329A1, 16 June 2020.
178. Somvico, R.R.P.O.C.S.J.M.R.C.P. Nanoemulgel Based on Ucúuba Fat (Virola Surinamensis) for Transungual Administration of Antimicotics. BR102019014044A2, 13 October 2021.

pharmaceutics

Article

Effect of Ciprofloxacin-Loaded Niosomes on *Escherichia coli* and *Staphylococcus aureus* Biofilm Formation

Linda Maurizi [1,†], Jacopo Forte [2,†], Maria Grazia Ammendolia [3], Patrizia Nadia Hanieh [2], Antonietta Lucia Conte [1], Michela Relucenti [4], Orlando Donfrancesco [4], Caterina Ricci [5], Federica Rinaldi [2,*], Carlotta Marianecci [2], Maria Carafa [2] and Catia Longhi [1]

1 Dipartimento di Sanità Pubblica e Malattie Infettive, Sapienza Università di Roma, Piazzale Aldo Moro, 5-00185 Roma, Italy
2 Dipartimento di Chimica e Tecnologie del Farmaco, Sapienza Università di Roma, Piazzale Aldo Moro, 5-00185 Roma, Italy
3 Centro Nazionale Tecnologie Innovative in Sanità Pubblica, Istituto Superiore di Sanità, Viale Regina Elena, 299-00161 Roma, Italy
4 Dipartimento di Scienze Anatomiche, Istologiche, Medico-Legali e dell'Apparato locomotore, Sapienza Università di Roma, Via Alfonso Borelli, 50-00161 Roma, Italy
5 Dipartimento di Biotecnologie Mediche e Medicina Traslazionale, Università di Milano, Via Fratelli Cervi, 93-20090 Milano, Italy
* Correspondence: federica.rinaldi@uniroma1.it; Tel.: +39-06-4991-3970
† These authors contributed equally to this work.

Citation: Maurizi, L.; Forte, J.; Ammendolia, M.G.; Hanieh, P.N.; Conte, A.L.; Relucenti, M.; Donfrancesco, O.; Ricci, C.; Rinaldi, F.; Marianecci, C.; et al. Effect of Ciprofloxacin-Loaded Niosomes on *Escherichia coli* and *Staphylococcus aureus* Biofilm Formation. *Pharmaceutics* **2022**, *14*, 2662. https://doi.org/10.3390/pharmaceutics14122662

Academic Editor: Udo Bakowsky

Received: 9 November 2022
Accepted: 27 November 2022
Published: 30 November 2022

Abstract: Infections caused by bacterial biofilms represent a global health problem, causing considerable patient morbidity and mortality in addition to an economic burden. *Escherichia coli*, *Staphylococcus aureus*, and other medically relevant bacterial strains colonize clinical surfaces and medical devices via biofilm in which bacterial cells are protected from the action of the immune system, disinfectants, and antibiotics. Several approaches have been investigated to inhibit and disperse bacterial biofilms, and the use of drug delivery could represent a fascinating strategy. Ciprofloxacin (CIP), which belongs to the class of fluoroquinolones, has been extensively used against various bacterial infections, and its loading in nanocarriers, such as niosomes, could support the CIP antibiofilm activity. Niosomes, composed of two surfactants (Tween 85 and Span 80) without the presence of cholesterol, are prepared and characterized considering the following features: hydrodynamic diameter, ζ-potential, morphology, vesicle bilayer characteristics, physical-chemical stability, and biological efficacy. The obtained results suggest that: (i) niosomes by surfactants in the absence of cholesterol are formed, can entrap CIP, and are stable over time and in artificial biological media; (ii) the CIP inclusion in nanocarriers increase its stability, with respect to free drug; (iii) niosomes preparations were able to induce a relevant inhibition of biofilm formation.

Keywords: niosomes; drug delivery; ciprofloxacin; anti biofilm activity; bladder cells

1. Introduction

Biofilms, communities of microorganisms that live in a self-produced extracellular matrix, are an important virulence factor that causes severe problems to public health. Biofilms allow pathogens to escape host defenses and resist antimicrobial treatment [1]. Of particular concern is that biofilm formation on indwelling medical devices can lead to serious, recalcitrant infections [2].

Uropathogenic *Escherichia coli* (UPEC) and *Staphylococcus aureus* are frequently detected in patients with indwelling urinary tract devices [3–6]. Urinary tract infections (UTIs) are considered to be the most common bacterial infections, affecting around 150–250 million people each year worldwide. These infections account for 75% of infections in community settings and 50–65% of those in healthcare settings [3]. Almost all healthcare-associated

UTIs are caused by instrumentation of the urinary tract, whose permanence can lead to complications such as prostatitis, epididymitis, and orchitis in males, and cystitis, pyelonephritis, endocarditis, and meningitis in patients [7,8]. Ciprofloxacin (CIP) is one of the most commonly prescribed fluoroquinolone antibiotics for UTIs to which both *E. coli* and *S. aureus* have become resistant [9–11].

Progresses in the field of nanotechnologies applied to therapy and diagnosis have led to the birth of a new branch of science defined as "nanomedicine", within which a prominent space is occupied by a particular class of unconventional pharmaceutical forms known as pharmaceutical nanocarrier [12].

Nanotechnology has led to the discovery of various types of nanocarriers, which bring several features: favorable physical-chemical features, protection of the loaded active compound, and especially controlled and targeted drug delivery towards the active site decreasing, therefore, unwanted effects of the drug on surrounding tissues, and improving drug half-life in the body [13]. Moreover, the use of nanocarriers as a drug-delivery system thus allows the drug to reach the active site in higher concentration, reducing the needed dosage amount and increasing patient compliance. So, the design of an efficient drug delivery system is fundamental and crucial.

Many delivery modalities find clinical practicality in the field of urology, specifically in the treatment of UTIs, and offer advantages over conventional methods.

Intravesical therapy is a local drug administration, and it is now considered highly promising, providing a high concentration of drugs with the great advantage of minimal systemic side effects. This administration route could be useful to improve the treatment of pathological bladder conditions such as interstitial cystitis, bladder pain syndrome, or urinary infections in which both *E. coli* and *S. aureus* could be involved. Unfortunately, many drugs are not stable in the hostile urine environment, so also, in this case, drug delivery systems, able to protect the loaded drug from degradation, could represent an efficient strategy [14].

Among the various nanocarriers, niosomes are successfully used in different pharmaceutical applications. They are stable, non-immunogenic vesicles suitable for hydrophilic and lipophilic drug loading and delivery. Moreover, niosomes are vesicles produced by the self-assembly of surfactants, which are more stable and less expensive compared to the phospholipids used for liposomes [15].

The most used surfactants for niosomal preparation are the non-ionic ones (which do not possess a charged group in their hydrophilic heads) because they are biocompatible, more stable, and less toxic with respect to the other types of surfactants such as amphoteric, anionic, or cationic ones [16]. Non-ionic surfactants possess a hydrophilic head and a hydrophobic tail, which, when in contact with an aqueous environment, arrange to form a vesicle with a hydrophilic inner core surrounded by one or more concentric lipophilic bilayers.

Recently, the advances in pharmaceutical technology and, in particular, in nanocarrier preparation and characterization provided a promising tool for enhancing the activity and safety of available antimicrobial agents.

Various studies described the antimicrobial and antibiofilm activity of peculiar CIP-loaded niosome formulations against Gram-negative and Gram-positive bacteria [17–19] but according to our literature review, only few authors investigated Span 80-Tween 85 based niosomes (from Scopus.com accessed on 20 October 2022).

This study aims to prepare and characterize specific surfactant based nanocarriers entrapping CIP and testing their potential antimicrobial effect against strong biofilm producer strains. Niosomal vesicles composed of Tween 85 and Span 80 (in an equimolar mixture) have been prepared. Both selected surfactants were chosen for the presence, in their chemical structure, of oleic acid (3 molecules in Tween 85 and 1 in Span 80) moiety able to reduce expression of inflammatory molecules. Moreover, the lipophilic bilayer structure of niosomes (administered by intravesical route) could potentially facilitate their adherence to the apical membrane of the vesical surface and enhance CIP efficacy [20]. In addition, Hayashi et al. demonstrate that Span 80 niosomes are able to perturb phospholipid membranes. This could be useful in the interaction of niosomes with the microbial membranes and biofilm environment and in achieving a more efficient internalization of the loaded drug [21,22].

Here, the proposed nanocarriers have been deeply characterized considering several physical-chemical features such as hydrodynamic diameter, ζ-potential, morphological and bilayer characteristics, and biological effectiveness. The antibiofilm activity of CIP-loaded niosomes towards strong biofilm producer bacterial strains was also studied.

2. Materials and Methods

2.1. Materials

Tween 85, Span 80, ciprofloxacin (CIP), diphenylhexatriene (DPH), pyrene, Hepes, ethanol F.U., methanol, chloroform, were purchased by Sigma-Aldrich (St. Louis, MO, USA).

2.2. Preparation of Niosomes and Drug-Loaded Niosomes

The niosomal vesicles were obtained through the film layer preparation technique, better known as Thin Layer Evaporation (TLF) [23]. The components of the vesicles are weighed on the balance according to data reported in Table 1. An organic mixture of chloroform-ethanol 3:1 (v/v) was useful to solubilize the lipophilic compounds. Subsequently, it is removed under reduced pressure Rotavapor® R-210 (Büchi-Italia S.r.l., Assago (MI), Italy) at room temperature for one hour, leading to a thin layer of film in the test tube. Finally, an oil pump is applied for another hour to remove any residue of the organic solvent.

Table 1. Sample compositions.

Sample	Tween 85 (mM)	Span 80 (mM)	Ciprofloxacin (mg/mL)
A	22.5	22.5	-
B	22.5	22.5	2

Then the sample is hydrated; in the case of "empty" niosomes, the hydration is carried out by 5 mL of Hepes buffer (pH = 7.4, M = 0.01), while in the case of "loaded" niosomes, 5 mL of a CIP solution (2 mg/mL) are used. A small amount of HCl is used to prepare the CIP solution, to facilitate the solubilization of the drug in Hepes buffer; HCl will then be removed by dialysis (dialysis tube cut-off 10.000 Da) for a period of 3 h.

Subsequently, the vortex action on the sample allows the detachment of the film layer from the wall of the test tube, and a suspension of multilamellar vesicles is obtained. This suspension is then sonicated with an ultrasonic disruptor sonicator (Vibracell-VCX 500, Sonics, Taunton, MA, USA) to obtain unilamellar vesicles (5 min, 65 °C, and 25% amplitude).

In order to remove any impurities or substances not taking part of the vesicular structure, the sample was purified by size exclusion chromatography on glass column of Sephadex G75. Finally, filtration is performed (MF-Millipore®, Ireland, E.U. 0.22 μm) in order to retain impurities and to sterilize the sample in accordance with Ph. Eur.

2.3. Small Angle X-ray Scattering

Small Angle X-ray Scattering (SAXS) experiments were carried out at the ID02 SAXS beamline of ESRF (Grenoble, France) DOI:10.15151/ESRF-ES-624938971. Purified suspensions were put in Kapton capillaries at room temperature, irradiated with a monochromatic beam, λ = 0.1 nm, for short exposure times (0.5–1 s) to avoid radiation damage. The intensity spectra were acquired at two different sample-to-detector distances, namely 1 m and 10 m, and joined after careful background subtraction to obtain the intensity spectra in $0.006 < q < 5\ nm^{-1}$ momentum transfer range, where $q = 4\pi sen(\vartheta/2)/\lambda$, being ϑ the scattering angle. The intensity decay gave information on the internal structure of nanosized particles [24]. To this end, the profiles were fitted with a core-multishell model that describes the scattering from vesicles and niosomes, with an aqueous spherical core and a layered shell composed of a hydrophobic stratum inserted between two hydrophilic layers [25].

2.4. Dynamic Light Scattering (DLS) and ζ-Potential Measurements

The prepared samples were characterized by evaluating: hydrodynamic diameter, ζ-potential and PDI (polydispersity index that gives information on size distribution), employing a Malvern Nano ZS90 apparatus (Malvern Instruments, Worcestershire, UK), equipped with a 5 mW HeNe laser, λ = 632.8 nm.

The scattering angle was 90°, and the analysis of the intensity autocorrelation function was carried out using the Contin algorithm and analyzed by using the cumulant method to get the values of the particle dimensions and size distribution (PDI) [26].

The calculated mean hydrodynamic diameter corresponds to the intensity-weighted average [27]. Electrophoretic mobility of the vesicles was measured by laser Doppler anemometry using the Malvern Zetasizer Nano ZS90 apparatus (Malvern Instruments, Worcestershire, UK). The ζ-potential was obtained by converting the mobility (u) using the Smoluchowski relation $\zeta = u\eta/\eth$, where η is the viscosity and the permittivity of the solvent phase [28].

2.5. Transmission Electron Microscopy (TEM)

Morphology of niosomes was obtained by visualizing the samples by TEM analyses. One drop of empty and loaded nanocarriers was placed into a formvar carbon-coated grid. After 2 min adsorption, niosomes were negatively stained with 2% (v/v) filtered aqueous sodium phosphotungstate acid (PTA) and examined by a FEI 208S transmission electron microscope (FEI Company, Hillsboro, OR, USA) with an accelerating voltage of 100 kV. To optimize image editing, Adobe Photoshop software was used.

2.6. Fluorometric Measurements

Information about the lipophilic bilayer of niosomes and CIP-loaded niosomes (samples A and B) was obtained by measuring the DPH fluorescence anisotropy, which is a parameter correlated to membrane rigidity or fluidity [29]. The samples were prepared to dissolve in the organic mixture of the probe (2×10^{-4} M), together with the other components, following the same preparation method described in paragraph 2.2. DPH fluorescent measurements were performed using a luminescence spectrometer (LS5013, PerkinElmer, Waltham, MA, USA) with excitation λ ex = 350 nm and detecting the fluorescence intensity at λ em = 428 nm [30]. In employing Equation (1), the fluorescence anisotropy (r) was determined:

$$\text{Fluorescence anisotropy (r)} = \frac{(I_{vv} - I_{vh}) \times G}{(I_{vv} + 2I_{vh}) \times G} \quad (1)$$

where I_{vv}, I_{vh}, I_{hv}, and I_{hh} are fluorescent intensities, subscript v (vertical) and h (horizontal) represent the orientation of polarized light, and $G = I_{hv}/I_{hh}$ factor is the ratio of sensitivity of the detection system for vertically and horizontally polarized light.

Additionally, bilayer characterization studies were also performed utilizing a different fluorescent probe: the pyrene. This probe is useful to evaluate the polarity and microviscosity of the vesicular bilayer and was used both in empty niosomes and in CIP-loaded ones. Samples A and B were prepared by adding Pyrene (4 mM) to the components (following the same preparation method described above). By fluorescence measurements, it is possible to investigate the lateral distribution and the mobility of the membrane compounds. Pyrene is a fluorescence probe with a spectrum characterized by five emission peaks as the monomer (from I1 to I5) and one as excimer (IE). In particular, the ratio I1/I3, corresponding to the first and third vibration bands of the Pyrene spectrum, is related to the polarity of the probe environment. Pyrene can form intramolecular excimer based on the viscosity of the probe microenvironment [31].

2.7. Drugs Entrapment Efficiency (EE%)

Utilizing the UV-vis spectrophotometer (Lambda 25, PerkinElmer, Waltham, MA, USA), the entrapment efficiency (E.E.) of CIP inside niosomes was determined. In particular, the CIP entrapped amount was calculated by using the calibration curve previously defined. Loaded niosomes were diluted in Hepes buffer, and the absorbance of drugs at $\lambda = 271$ nm was measured [32].

E. E. % was calculated as (2):

$$E.E.\ (\%) = \frac{\text{Entrapped drug (mg)}}{\text{Total drug used (mg)}} \times 100 \quad (2)$$

2.8. Physicochemical Stability

Both unloaded and loaded niosomes were stored at two different temperatures: room temperature/4 °C for a period of 90 days. The data concerning nanocarrier stability were collected employing DLS (Malvern Instruments, Worcestershire, UK).

These experiments consist of monitoring over time the dimension and ζ-potential variations of the samples by DLS measurements.

Empty and loaded niosomes were also subject to stability studies performed in simulated biological fluids (Artificial Urine, pH 6.6) in order to evaluate the niosomal stability. Artificial urine was prepared according to Monika Pietrzyńska et al., 2017 [33], and the composition is reported in Table 2.

Table 2. Artificial body urine (pH 6.6): chemical composition.

Reagent	Dosage (g)
Urea	25.0
NaCl	9.0
NH4CL	3.0
Creatinine	2.0
Na2HPO4	2.5
KH2PO4	2.5
Na2SO3	3.0
Distilled water	Total 1.0 L

These tests were executed by carrying out DLS measurements to assess that the size, the PDI, and the ζ-potential of the vesicular suspensions remained constant.

In order to carry out these experiments, 1 mL of the sample was added to 1 mL of the artificial body urine and put into a test tube, subject to a magnetic stirrer at 37 °C to mimic the body temperature. This experiment lasted 24 h.

Moreover, free CIP and CIP-loaded niosomes were evaluated over time, for a period of 90 days, at two different storage temperatures (25 °C and 4 °C).

This experiment consists of monitoring over time the drug stability by means of a UV-vis spectrophotometer, observing the intensity and shape of the peak at 271 nm [32].

2.9. In Vitro Release Studies

In vitro drug release experiments were carried out by inserting in a dialysis tube (molecular weight cut-off: 8000 MW by Spectra/Por®) the drug-loaded niosomal suspension. The dialysis tube was immersed in the release medium (Hepes Buffer 10 mM, pH 7.4 or Artificial Urine, pH 6.6) at 37 °C and gently magnetically stirred during the experiment.

The drug concentration amount in the release medium was detected by UV spectrophotometer (Lambda 25, PerkinElmer, Waltham, MA, USA) at different time points until up 48 h. In order to perform UV analysis, 1ml di external medium was withdrawn and immediately analyzed to the spectrophotometer and then re-inserted back.

Reported values represent the mean values over three repeated independent experiments, and errors are the standard deviation.

2.10. Bacterial Strains

Uropathogenic *Escherichia coli* ATCC 700928 (CFT073) and *Staphylococcus aureus* ATCC 6538P were biofilm-forming reference strains obtained from the American Type Culture Collection (ATCC, Manassas, VA, USA). *E. coli* K-12 MG1655 was a weak biofilm producer strain. The microorganisms were grown in Brain Heart Infusion broth (BHI) (Oxoid) and stored in 15% glycerol-BHI at −80 °C.

2.11. Determination of Minimum Inhibitory Concentration (MIC) of CIP-Loaded Niosomes

The MIC determination of CIP-loaded niosomes was performed by the microdilution method and carried out in triplicate. Exponentially growing bacterial cultures were diluted to cell density 0.5 McFarland, and 10 µL of bacterial suspension was added to 190 µL of BHI (Oxoid) containing CIP-loaded in niosomes at concentrations from 125 µg/mL to 7.8 µg/mL. Empty niosomes were diluted and tested similarly. After the incubation at 37 °C for 24 h, the bacterial growth was evaluated by measuring the optical density at 595 nm. All experiments were conducted in triplicate.

2.12. Effect of CIP-Loaded Niosomes on Bacterial Biofilm Production

A volume of 20 µL of each bacterial strain (1–2 × 10^8 CFU/mL) was inoculated into wells of a 96- well polystyrene plate containing 180 µL of Tryptic Soy Broth (TSB) and incubated for a period of 24 h at 37 °C. Then, after washing with phosphate-buffered saline, the plates were allowed to dry. The wells were stained for 15 min with crystal violet (Sigma-Aldrich, 1% *w*/*v*), a basic dye that binds negatively charged molecules. The dye was solubilized with 95% (*v*/*v*) ethanol for 30 min. The optical density (OD) at 570 nm of each well was measured, and biofilm production was classified as described by Stepanovic et al., 2004 [34]. Based on the cut-off OD, defined as three standard deviations above the mean OD of the negative control (ODc), strains were classified as follows: OD ≤ ODc = no biofilm producers, ODc < OD ≤ (2 × ODc) = weak biofilm producers, (2 × ODc) < OD ≤ (4 × ODc) = moderate biofilm producers, and (4 × ODc) < OD = strong biofilm producers.

In order to measure the biofilm inhibition induced by CIP-loaded niosomes, the growth medium was supplemented with these substances at a concentration of 3.9 µg/mL. The inhibition of cell attachment was evaluated after 24 h incubation at 37 °C. The percentage of biofilm inhibition by sub-MIC niosomes preparation has been calculated using the following formula [35]:

$$Biofilm\ inhibition\ (\%) = 100 - (OD570\ sample/OD570\ control \times 100)$$

Values higher than 40% were considered relevant in biofilm inhibition. Uninoculated TSB broth was used as a negative control. Experiments were run in sextuplicate.

2.13. SEM Analysis

Samples of biofilms grown on aluminum stubs were washed and fixed in 2.5% glutaraldehyde in 0.1 M phosphate buffer pH 7.4 for at least 48 h. Samples were washed overnight in phosphate buffer pH 7.4, and the day after, they were postfixed with OsO_4 1.33 % in H_2O for 1 h at room temperature. Samples were then washed for 20 min with H_2O, and then they were treated for 30 min with tannic acid 1% in H_2O. Samples were then washed for 20 min with H_2O. The excess water was dried carefully with filter paper, and then the samples were mounted on the specimen holder and observed in a Hitachi SU3500 microscope (Hitachi, Japan) at variable pressure conditions of 5 kV and 30 Pa [36,37]. Three-dimensional reconstruction of images was undertaken by Hitachi Map 3D Software (v.8.2., Digital surf, Besançon, France). For each image, a selected area was extracted for the 3D image reconstruction procedure. The surface topography of the extracted area is shown in false colors [38], and it was processed by the particle count procedure to evaluate the size of ECM granules.

2.14. Evaluation of Intracellular Uptake

In order to visualize niosomes intracellular uptake, T24 cells were seeded on 8-well chamber-slides (Falcon) for 24 h at 37 °C and exposed to niosomes loaded with Nile Red dye/CIP-Nile Red co-loaded niosomes [39], for different times. Cells treated with free Nile Red, prepared as 1 mg/mL stock solution in acetone and used at a final concentration of 100 ng/mL, were used as control. After the incubation times, cells were washed with phosphate-buffered saline solution (PBS) at pH 7.4 and fixed in methanol/acetone (1:1) for 5 min at −20 °C. Slides, extensively washed with PBS, were mounted with 0.1% (*w*/*v*) p-phenylenediamine in 10% (*v*/*v*) PBS, 90% (*v*/*v*) glycerol, pH 8.0 and observed by fluorescence microscopy using a Leica DM4000 (Leica Microsystem, Wetzlar, Germany) fluorescence microscope, equipped with an FX 340 digital camera.

2.15. Cytotoxicity Studies

T24 cells (concentration of 1×10^4/well) were seeded in 96-well plates and cultured at 37 °C with 5% CO_2 for 24 h. Different concentrations of CIP-niosomes (from 250 μg/mL to 31.2 μg/mL) were added to cell monolayers and incubated for 24 h. Then, 100 μL of 0.5 mg/mL of 3-(4,5-dimethylthiazol-2-yl)-2,5-diphenyltetrazolium bromide (MTT) reagent was added to each well for an additional 4 h. Afterwards the dye was eluted with 200 μL of DMSO for 10 min at room temperature and, finally the OD at 568 nm was measured using a microplate reader (PerkinElmer, Boston, MA, USA).

2.16. Statistical Analysis

For the statistical analysis of niosomes characterization studies, two-way ANOVA was performed. Multiple comparisons were performed according to Tukey's test for ζ-potential, hydrodynamic diameter, and polydispersity index (PDI), respectively, and Dunnett's test for the Turbidimetric assay. Any p-value < 0.05 was considered statistically significant. For the cytotoxicity assay and antimicrobial studies, all values were reported as mean ± standard deviation (SD). Statistical analyses were performed by one-way ANOVA followed by Tukey's post hoc pairwise tests (Graph Pad Prism, Version 5.0). A p-value of less than 0.05 (* $p < 0.05$) was considered significant.

3. Results and Discussion

3.1. Characterization of Empty and Loaded Niosomes

Empty and loaded niosomes were deeply characterized. First of all, preliminary SAXS analyses have been useful in evaluating the ability of surfactants to form niosomes, even if in the absence of cholesterol. SAXS techniques represent a non-invasive tool to provide ensemble-averaged and, thus, statistically relevant information on the structure and conformation of materials. Even though the samples are in a very dilute solution, the high brilliance of the synchrotron radiation allows for obtaining good statistics of the data in a wide range of q, with the uncertainty on the measure intrinsically represented by the fluctuation of the scattered data. Figure 1 reports the intensity spectrum measured for the suspension composed by Tween 85: Span 80 at a 1:1 molar ratio after purification. The intensity profile is characteristic for quite monodisperse nano-sized particles that have been modeled with a spherical aqueous core (size ≥ 100 nm) surrounded by a single surfactant bilayer (with a hydrophobic core between two hydrophilic regions) of about 5 nm thickness, demonstrating the ability of the selected surfactants to form unilamellar niosomes.

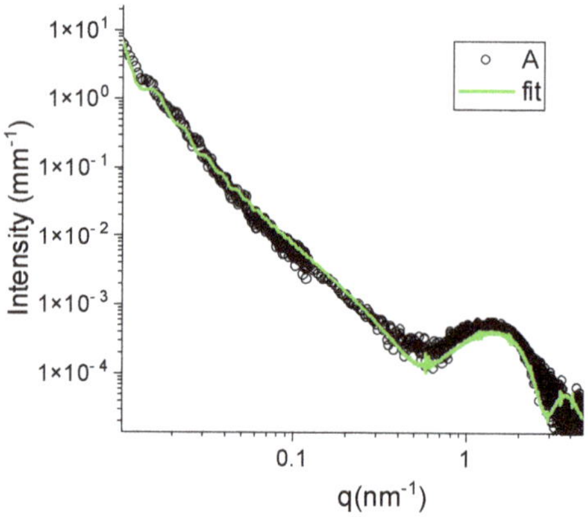

Figure 1. SAXS spectrum of empty niosomes, sample A (22.5 mM, room temperature, open dots) reported in log-log scale. The green line is the best fit.

Moreover, empty and loaded niosomes were characterized by evaluating hydrodynamic diameter, ζ-potential, PDI, and CIP entrapment efficiency (EE%,) and the results have been reported in Table 3. In particular, dynamic light scattering analysis showed that CIP-niosomes are characterized by an increased size with respect to empty ones (from 113.10 nm to 260 nm) that confirm the drug inclusion inside the vesicles [40], while the ζ-potential remains constant and enough negative to assure colloidal stability. As previously described, the polydispersity index (PDI) is an important parameter to be taken into account. Both formulations, loaded and unloaded, showed a PDI of 0.2, which indicates monodisperse samples. Probably, niosomes characterized by a low PDI value could be stable over time, and, in terms of drug pharmacokinetics, they could be characterized by similar behavior after in vivo administration [41].

Table 3. Niosomal formulations characterization: summary of physicochemical features.

Sample	Hydrodynamic Diameter (nm) ± SD	ζ-Potential (mV) ± SD	PDI ± SD	Ciprofloxacin (mg/mL)	Ciprofloxacin E.E. %
A	113.10 ± 3.17	−34.80 ± 1.91	0.20 ± 0.01	-	-
B	260.41 ± 4.03	−37.50 ± 2.12	0.20 ± 0.01	0.40	20

Electronic visualization of empty and CIP-loaded niosomes showed spherical vesicles with size corresponding approximately to dimensions revealed by DLS. The increased size of CIP-loaded niosomes reported by DLS was confirmed by electronic observations that also revealed no changes in a spherical shape (Figure 2).

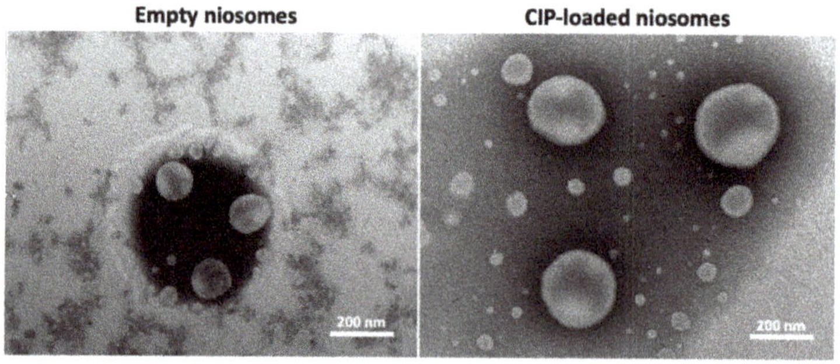

Figure 2. Transmission electron microscopy images of empty and CIP-loaded niosomes.

From Table 3, it is possible to observe that the CIP EE%, obtained by UV analysis, is almost 20% (0.4 mg/mL), which is a concentration useful for biofilm treatment, as demonstrated by the results.

In order to characterize the lipophilic bilayer of empty and loaded niosomes, the microviscosity, polarity, and anisotropy were studied. Following the method previously described, the nanocarriers have been prepared to load the molecular probe (pyrene) inside the vesicles. The values of polarity and microviscosity have been collected to analyze the obtained fluorescence spectrum, and the values of polarity and microviscosity have been collected (Table 4). In particular, the polarity values were quite similar for niosomes (A) and CIP-niosomes (B), but microviscosity values decreased when CIP was loaded in the nanocarrier. The decreased microviscosity could be explained by a partial CIP localization in the niosomal lipophilic compartment. In fact, in agreement with the drug's chemical structure, the CIP lipophilic portion could be located inside the bilayer, while the hydrophilic portion could be inside the aqueous core. According to these results, the anisotropy values increase with drug inclusion. The anisotropy value suggests a more rigid bilayer of the loaded niosomes with respect to empty ones to confirm the partial drug localization in the bilayer [42].

Table 4. Bilayer characterization. (SD values are all in the range ± 0.01–0.02).

Sample	I_1/I_3 (Polarity)	I_E/I_3 (Microviscosity)	Anisotropy A.U. (Fluidity)
A	0.96	1.01	0.22
B	0.97	0.46	0.35

3.2. *Stability Studies*

3.2.1. Physical Stability of Niosomes

Studies on physical stability were also performed according to the method previously described, and the obtained data are reported in Figure 3. In particular, both samples (empty and loaded niosomes) appeared to be stable for 90 days when stored at 4 °C (no significant dimensional change is observed but only minor variations), while the hydrodynamic diameter of sample A decreased when it was stored at room temperature (Figure 3, Panel a). So, it is possible to conclude that the samples have been characterized by significant colloidal stability when maintained at 4 °C [43].

3.2.2. Stability of Niosomes in Artificial Urine

The stability of samples A and B was also studied in artificial urine (Figure 3, Panel b) to mimic the effect of the media on the vesicle stability after intravesical administration. The experiments were carried out at 37 °C evaluating the vesicle size by DLS measurements for 24 h as described in the Material and Methods section (Section 2). During the experiment, no significant changes in hydrodynamic diameter were observed for both samples (empty and loaded vesicles). It is possible to conclude that the artificial media doesn't affect the niosome integrity.

3.2.3. Stability Studies over Time of Free CIP and CIP-Loaded into Niosomes

In order to evaluate CIP stability in terms of decomposition or degradation, the unloaded and loaded drug peak was observed by a UV spectrophotometer. The UV spectra were recorded immediately after sample preparation and after 30, 60, and 90 days at room temperature and 4 °C. The CIP concentration values obtained by these experiments were reported in Figure 3, Panel c. Drug stability (at room temperature and 4 °C) has been enhanced by its inclusion in the vesicles with respect to free drug dissolved in Hepes solution.

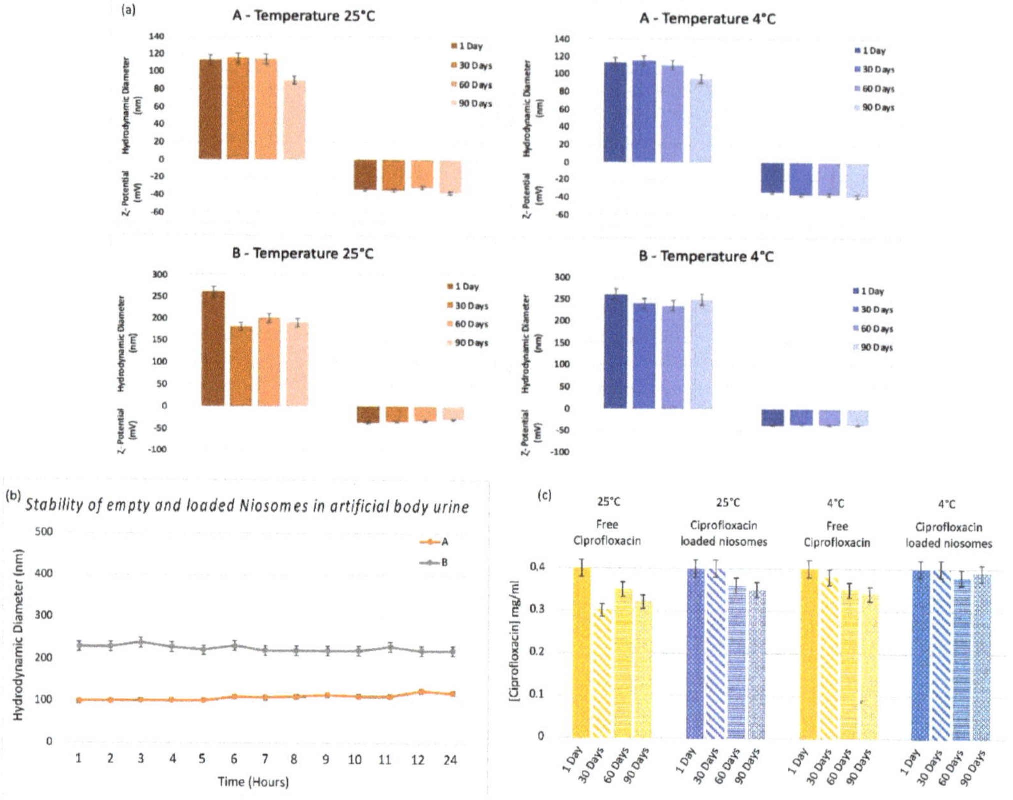

Figure 3. (**a**) Result of investigation on physicochemical stability of empty niosomes (A) and CIP-loaded niosomes (B) in terms of hydrodynamic diameter and ζ-potential until up 90 days at 4 °C and room temperature. (**b**) Stability studies in the presence of artificial body urine, following variation of hydrodynamic diameter and ζ-potential values of CIP-loaded niosomes; Pre-exp values refer to CIP-loaded niosomes before the presence of artificial body urine. (**c**) Stability studies over time of free CIP and CIP-loaded into niosomes at two different storage temperatures over a 90-day period.

3.3. CIP Release Studies

In vitro release studies of niosomal loaded sample in Hepes buffer and artificial urine have been carried out, and data obtained are shown in Figure 4. The total amount of released drug by niosomes is around 50% in Hepes Buffer and around 40% in simulated biological fluid. From these data, it is possible to observe that Sample B is characterized by the same release profiles in both media. Probably, these results are related to physical-chemical features of sample B. In fact, the anisotropy value (A = 0.35) of loaded niosomes could suggest that the CIP is retained by the bilayer vesicles due to their rigidity. Moreover, according to Uhljar et al., 2021 and Volpe 2004, CIP is characterized by a slow drug permeation coefficient (Pbl) that could prevent complete drug diffusion through the vesicle bilayer [44,45]. Concerning the integrity of niosomes, it is possible to affirm (by DLS measurements) that the niosomal formulations are characterized by no vesicle degradation at the experimental condition.

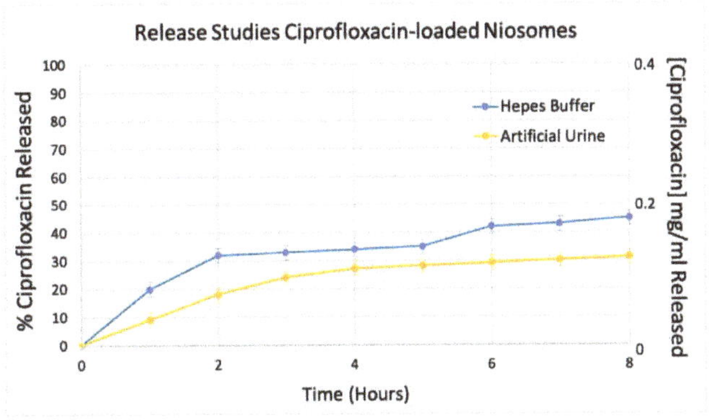

Figure 4. CIP release profile until up 24 h. Data were obtained as the mean of three independent experiments.

3.4. *In Vitro Antibacterial Activity of CIP-Loaded Niosomes*

In our study, the blank niosomal preparations showed no or lower antimicrobial activity compared to CIP-loaded niosomes (Figure 5A).

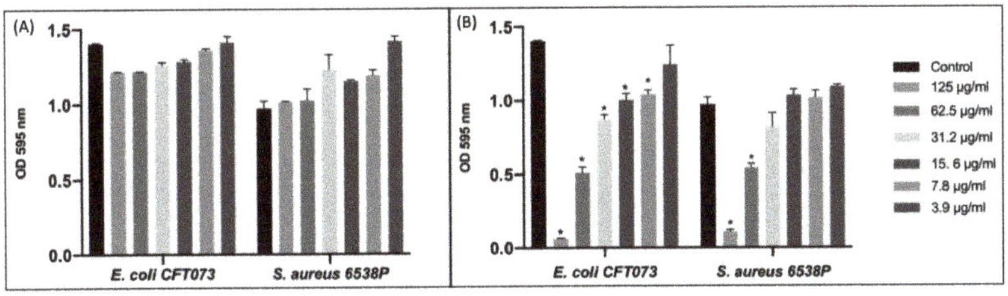

Figure 5. Susceptibility test with empty niosomes (**A**) and CIP-loaded niosomes (**B**). Data were expressed as mean ± SD. All considered conditions were compared to untreated control. * p value ≤ 0.05.

CIP-loaded niosomes significantly inhibited the growth of *E. coli* more than the *S. aureus* strain (Figure 5B). Probably, niosomal vesicles, thanks to specific surfactants employed, are able to perturb microbial phospholipid membranes interacting efficiently with Gram-negative bacteria. Furthermore, since fluoroquinolones can cross the cytoplasmic membrane by simple diffusion, any condition which creates a concentration gradient towards the bacterial cell outer membrane could improve drug permeation [46–48].

3.5. *Biofilm Production and Anti-Biofilm Activity of the Sub-MIC of Formulated Niosomes*

The biofilm production ability of bacterial strains was classified into four groups: no biofilm, weak, moderate, and strong. In our experimental conditions, *E. coli* CFT073 and *S. aureus* 6538P were confirmed as strong biofilm producers [49,50].

In order to verify the effect of CIP-loaded niosomes on biofilm production, we treated bacterial strains with encapsulated CIP for 24 h. The results obtained showed that the sub-MIC concentration of niosome-encapsulated CIP significantly inhibited biofilm formation when compared with empty niosomes (Table 5).

Table 5. Percentage of biofilm inhibition by sub-MIC niosomes. Values higher than 40% were considered relevant in biofilm inhibition.

Strains	**A**	**B**	**Free CIP 1/2 MIC**	**Free CIP 1/4 MIC**
E. coli CFT073	12.1	64.6	3.8	0.4
S. aureus ATCC 6538P	26.3	75.0	9.7	9.7

The antibiofilm activity of antibiotic-loaded niosome preparations against a wide variety of pathogens was previously reported [51–54].

A significant decrease in *S. aureus* biofilm formation by niosome-encapsulated CIP was described by some authors [19]. They suggested the adsorption of the nanocarriers on the biofilm surface with subsequent delivery of the encapsulated drug to the bacterial cells. Moreover, their study, using Real Time PCR, revealed the down-regulation of *ica*B biofilm gene expression compared to the free CIP. They postulated that the reduction of the *ica*B gene expression could be associated with the inhibition of transcription of bacterial genes by reactive oxygen species (ROS) and/or direct interaction of the niosome-encapsulated CIP with the transcription factors involved with the expression of *ica*B.

Dong et al., 2019 revealed that treatment with sub-MICs of CIP for 24 h inhibited biofilm formation and reduced the expression of virulence genes and biofilm formation genes in *E. coli* [55]. Our results, demonstrating the ability to efficiently release the antibiotic during the biofilm formation process, confirmed that CIP-loaded niosome formulation represents a suitable drug delivery system.

3.6. *Ultrastructural Morphology*

As previously described by other authors for the CIP [55], SEM analyses confirmed that the biofilm structure of the *E. coli* changed significantly after treatment with sub-MICs of CIP-loaded niosomes.

The observation of untreated samples (Figure 6A–C) showed that the sample is characterized by the presence of an abundant extracellular matrix (ECM), whose surface appears irregular, being in some areas smooth and compact (c), while in others it appears spongy and rough (s). At higher magnification (B), it was possible to appreciate the ECM ultrastructure in detail. It consists of a 3D network of trabeculae showing a globular aspect (insert). A labyrinthic system of narrow channels develops in the ECM, giving it a spongy appearance. In Figure 6C, the 3D reconstruction of the sample surface topography is represented. White and red areas are ECM trabeculae, formed by globular structures, and the channels perforating the ECM are represented with color shades from green to blue. When the sample is treated with niosome preparation, as illustrated in Figure 6D–F, ultrastructural modifications appear. The picture at low magnification (D) shows the sample presenting a spongy appearance (s) with also compact and smooth areas (c). However, the image at higher magnification (E) clarifies that the spongy areas have a different meshes structure concerning the untreated sample. The treatment affects the trabecular structure disassembling the globular structures into fine filamentous structures. The change is visible in Figure 6F, where the 3D reconstruction of sample surface topography is presented. A lot of channel openings are visible (blue-green areas), which means that the ECM is more perforated. White and red areas are ECM trabeculae, thinner than those of control samples, due to the disassembling action of the treatment.

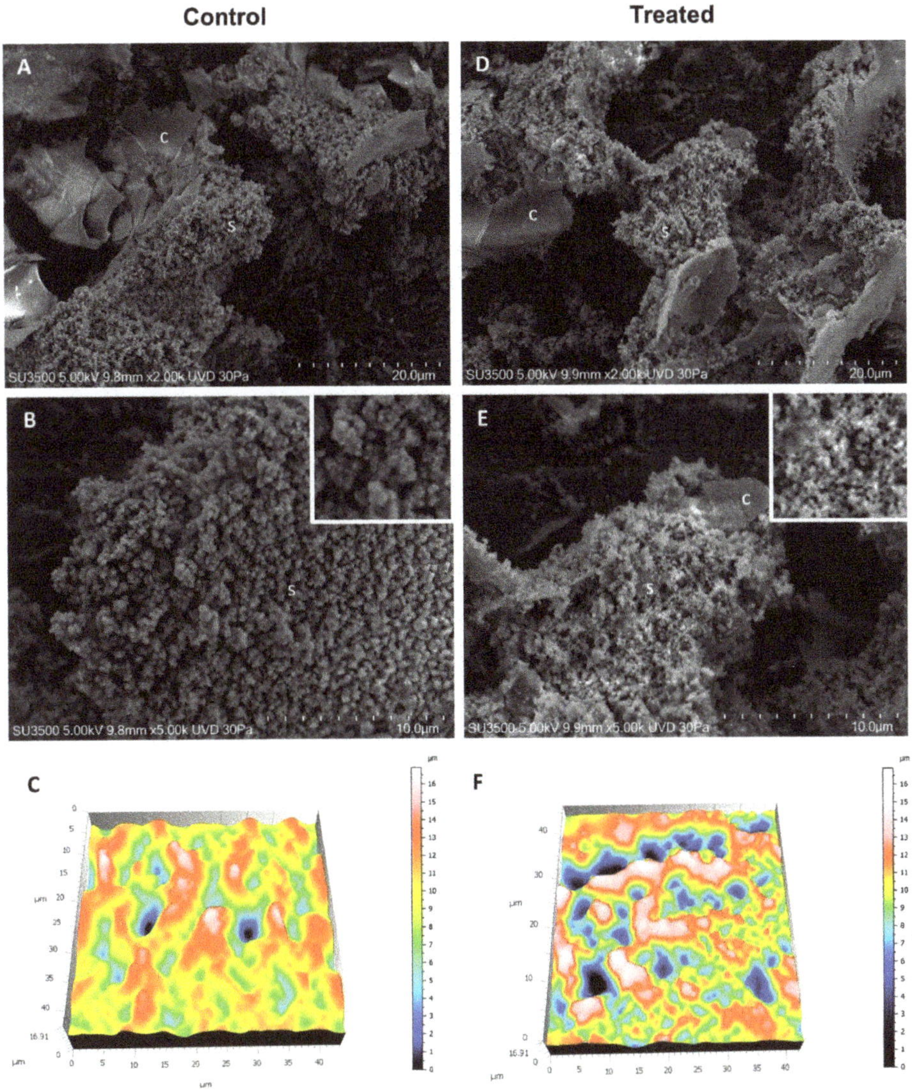

Figure 6. VP-SEM images of *E. coli* (**A–C**) and *E. coli* treated with CIP-loaded niosomes (**D–F**). (**A**) Low magnification (2.00K), ECM shows compact and smooth ECM areas (c), as well as spongy and rough areas (c). (**B**) Higher magnification (5.00K) of the ECM spongy area, ECM trabeculae show a globular structure (inset). (**C**) 3D reconstruction of sample surface topography, ECM trabeculae are represented in white and red areas, and the channels that perforate the ECM are represented with color shades from green to blue. (**D**) Low magnification (2.00K), ECM shows both compact and smooth ECM areas (c), both spongy and rough areas. (**E**) Higher magnification (5.00K) of the ECM spongy area, ECM trabeculae show a fine filamentous network structure (inset). (**F**) 3D reconstruction of sample surface topography, ECM trabeculae are represented in white and red areas, and the channels that perforate the ECM are represented with color shades from green to blue. Note that ECM presents more channels than the control, and trabeculae are thinner than the control sample.

To further quantitatively analyze the effect of niosomes on *E. coli* ECM ultrastructure, we used the Hitachi Map 3D Software (v.8.2., Digital surf, Besançon, France) particle analysis tool to detect globular structures that characterize the trabecular system of the control sample and observe if any difference exists in the treated sample. As shown in Figure 7, the control sample has a number of globular structures, sized 0.05–0.5 µm, which is about double that of the treated sample. This can be explained as a disassembling effect of the treatment. In the control, the ECM filaments are tightly coiled, and the treatment breaks down this compact arrangement producing an ECM with a loose filament arrangement.

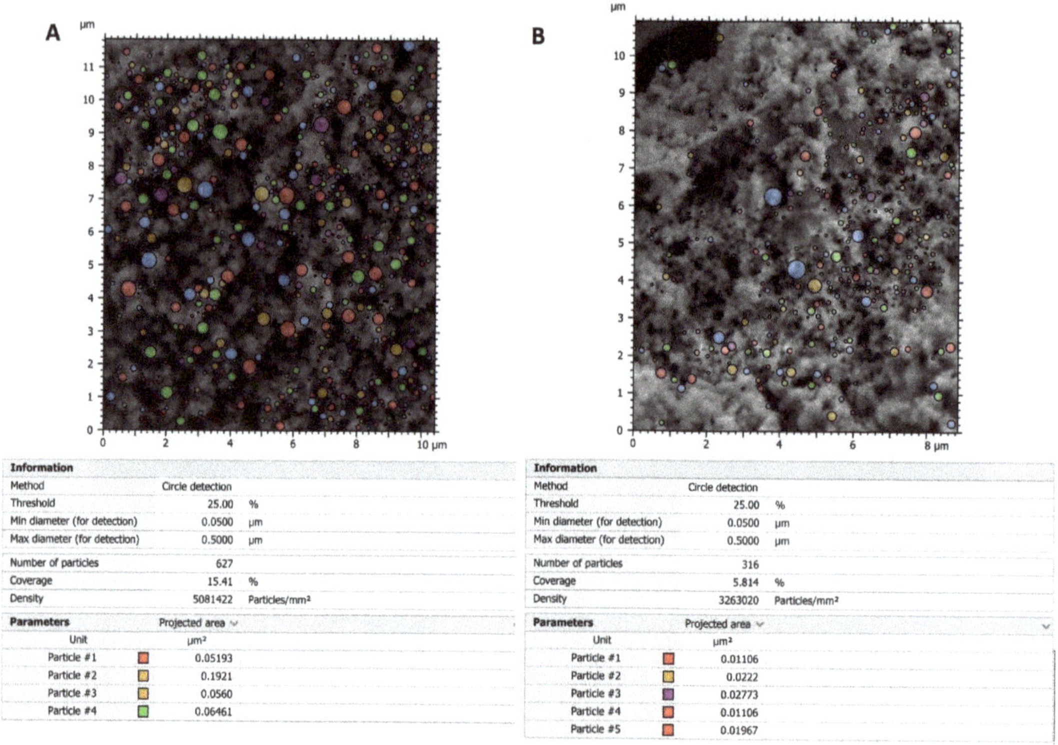

Figure 7. Analysis of globular structures in control (**A**) vs. treated samples (**B**). The particle analysis tool of Hitachi 3D map software revealed that the globular structures in the control sample are about twice those of the treated sample. This may be explained by thinking of a disassembling effect of the treatment on the filament that forms the ECM trabeculae.

3.7. Evaluation of Intracellular Uptake

In order to verify the ability to enter epithelial cell monolayers, the CIP-loaded niosomes were probed by following the aggregation of the Nile Red. A fluorescence microscope was used to assess intracellular uptake efficiency. As shown in Figure 8, at 7 h post-treatment, an increased fluorescence in the cells exposed to the CIP-loaded niosomes, with respect to Nile Red niosomes, was detected.

Figure 8. Untreated T-24 cells (**A**), cells treated for 7h with Nile Red loaded niosomes (**B**) and Nile Red and CIP co-loaded niosomes (**C**).

3.8. Cytotoxicity Studies

In order to evaluate the cytotoxicity activity of these formulations, an MTT assay was performed, incubating niosomal preparations for 24 h with cell monolayers (Figure 9). The treatment with niosomes triggers a cytotoxic effect compared to the untreated control, only at the highest concentration used: 250 µg/mL (for both empty and CIP-niosomes). Moreover, a slight increase in cytotoxicity was observed for CIP-niosomes at the concentration of 125 µg/mL, suggesting the ability of niosomes to target and release the drug. Results obtained confirmed CIP cytotoxicity on T24 cells as well as for other authors [56,57].

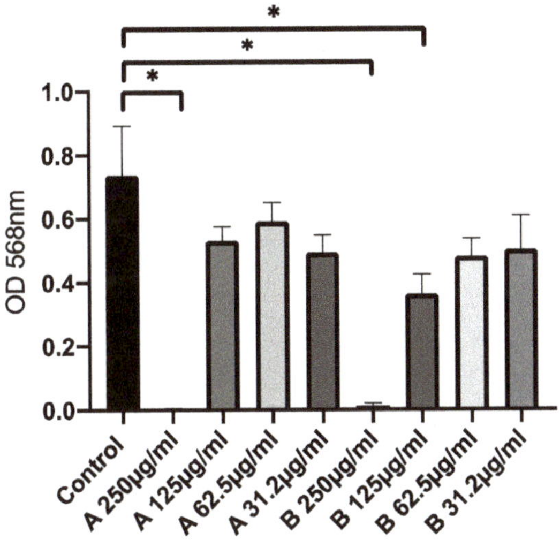

Figure 9. Cytotoxic activity of CIP-loaded niosomes (OD values). MTT assay on human bladder cancer cells (T24). Data were expressed as mean ± SD. All considered conditions were compared to untreated control. * p value ≤ 0.05.

4. Conclusions

Biofilm-related infections remain a serious concern in clinical services. Considering the impact of infectious diseases on human health, the development of advanced delivery systems able to target drugs directly to the site of interest appears to be a priority in the pharmaceutical area. For this purpose, empty and loaded niosomal formulations (composed of surfactant without cholesterol) were prepared and deeply characterized. Results obtained by physical-chemical characterization and by in vitro functional assays demonstrated that CIP-niosomes might be good candidates for the proposed application.

In fact, niosomal formulations exhibited: (i) good stability over time and in a simulated biological fluid, (ii) the capability to protect the entrapped drug by degradation phenomena, and (iii) controlled CIP release and antibiofilm effects against both Gram-positive and Gram-negative strains. In addition, the niosomes showed reduced toxicity on cell monolayers. The results of this study will contribute to developing niosomal formulations that, employed at the appropriate concentration, could represent a promising drug delivery system, stable and inexpensive (due to the low cost of surfactants) to contrast the biofilm development in bacterial infections.

Further investigations will be necessary to assess the safety and efficacy of the niosomes for a potential application in clinical trials.

Author Contributions: Conceptualization, F.R., C.L., C.M., M.G.A. and M.C.; Methodology, L.M., A.L.C., J.F., P.N.H., M.G.A., M.R., O.D. and C.R.; Validation, F.R.; Investigation, L.M., A.L.C., J.F., P.N.H., M.G.A., O.D., and C.R.; Data curation, L.M., C.L., J.F., F.R. and M.R.; writing—original draft preparation, L.M., C.L., J.F., P.N.H., M.R. and M.G.A.; writing—review and editing, L.M., C.L., J.F., and F.R.; Visualization, J.F. and P.N.H.; supervision, C.L., C.M., M.C. and F.R.; Funding Acquisition, C.L. All authors have read and agreed to the published version of the manuscript.

Funding: This research was funded by Ricerca Scientifica Ateneo 2019 "Sapienza" University of Rome to C.L.

Data Availability Statement: Not applicable.

Conflicts of Interest: The authors declare no conflict of interest.

References

1. Flemming, H.C.; Wingender, J.; Szewzyk, U.; Steinberg, P.; Rice, S.A.; Kjelleberg, S. Biofilms: An emergent form of bacterial life. *Nat. Rev. Microbiol.* **2016**, *14*, 563–575. [CrossRef] [PubMed]
2. Wi, Y.M.; Patel, R. Understanding Biofilms and Novel Approaches to the Diagnosis, Prevention, and Treatment of Medical Device-Associated Infections. *Infect. Dis. Clin.* **2018**, *32*, 915–929. [CrossRef] [PubMed]
3. Flores-Mireles, A.L.; Walker, J.N.; Caparon, M.; Hultgren, S.J. Urinary Tract Infections: Epidemiology, Mechanisms of Infection and Treatment Options. *Nat. Rev. Microbiol.* **2015**, *13*, 269–284. [CrossRef] [PubMed]
4. Walker, J.N.; Flores-Mireles, A.L.; Pinkner, C.L.; Schreiber, H.L.; Joens, M.S.; Park, A.M.; Potretzke, A.M.; Bauman, T.M.; Pinkner, J.S.; Fitzpatrick, J.A.J.; et al. Catheterization Alters Bladder Ecology to Potentiate Staphylococcus Aureus Infection of the Urinary Tract. *Proc. Natl. Acad. Sci. USA* **2017**, *114*, E8721–E8730. [CrossRef]
5. Terlizzi, M.E.; Gribaudo, G.; Maffei, M.E. UroPathogenic Escherichia Coli (UPEC) Infections: Virulence Factors, Bladder Responses, Antibiotic, and Non-Antibiotic Antimicrobial Strategies. *Front. Microbiol.* **2017**, *8*, 1566. [CrossRef]
6. Ullah, H.; Bashir, K.; Idrees, M.; Ullah, A.; Hassan, N.; Khan, S.; Nasir, B.; Nadeem, T.; Ahsan, H.; Khan, M.I.; et al. Phylogenetic Analysis and Antimicrobial Susceptibility Profile of Uropathogens. *PLoS ONE* **2022**, *17*, e0262952. [CrossRef]
7. Urinary Tract Infections: Virus-PMC. Available online: https://www.ncbi.nlm.nih.gov/pmc/articles/PMC8357242/ (accessed on 3 October 2022).
8. Lipsky, B.A.; Lipsky, B.A.; Hoey, C.T. Treatment of Bacterial Prostatitis. *Clin. Infect. Dis.* **2010**, *50*, 1641–1652. [CrossRef]
9. Fasugba, O.; Gardner, A.; Mitchell, B.G.; Mnatzaganian, G. Ciprofloxacin Resistance in Community- and Hospital-Acquired Escherichia Coli Urinary Tract Infections: A Systematic Review and Meta-Analysis of Observational Studies. *BMC Infect. Dis.* **2015**, *15*, 545. [CrossRef]
10. Daum, T.E.; Schaberg, D.R.; Terpenning, M.S.; Sottile, W.S.; Kauffman, C.A. Increasing Resistance of Staphylococcus Aureus to Ciprofloxacin. *Antimicrob. Agents Chemother.* **1990**, *34*, 1862–1863. [CrossRef]
11. Soto, S.M. Importance of Biofilms in Urinary Tract Infections: New Therapeutic Approaches. *Adv. Biol.* **2014**, *2014*, 1–13. [CrossRef]
12. Moghassemi, S.; Hadjizadeh, A. Nano-Niosomes as Nanoscale Drug Delivery Systems: An Illustrated Review. *J. Control. Release* **2014**, *185*, 22–36. [CrossRef] [PubMed]
13. Marianecci, C.; Petralito, S.; Rinaldi, F.; Hanieh, P.N.; Carafa, M. Some Recent Advances on Liposomal and Niosomal Vesicular Carriers. *J. Drug Deliv. Sci. Technol.* **2016**, *32*, 256–269. [CrossRef]
14. Sarfraz, M.; Qamar, S.; Rehman, M.U.; Tahir, M.A.; Ijaz, M.; Ahsan, A.; Asim, M.H.; Nazir, I. Nano-Formulation Based Intravesical Drug Delivery Systems: An Overview of Versatile Approaches to Improve Urinary Bladder Diseases. *Pharmaceutics* **2022**, *14*, 1909. [CrossRef] [PubMed]
15. Karim, S.S.A.; Karim, Q.A. Omicron SARS-CoV-2 Variant: A New Chapter in the COVID-19 Pandemic. *Lancet* **2021**, *398*, 2126–2128. [CrossRef] [PubMed]
16. Ag Seleci, D.; Seleci, M.; Walter, J.G.; Stahl, F.; Scheper, T. Niosomes as Nanoparticular Drug Carriers: Fundamentals and Recent Applications. *J. Nanomater.* **2016**, *2016*, 7372306. [CrossRef]
17. Akbari, V.; Abedi, D.; Pardakhty, A.; Sadeghi-Aliabadi, H. Release Studies on Ciprofloxacin Loaded Non-Ionic Surfactant Vesicles. *Avicenna J. Med. Biotechnol.* **2015**, *7*, 69.

18. Kashef, M.T.; Saleh, N.M.; Assar, N.H.; Ramadan, M.A. The Antimicrobial Activity of Ciprofloxacin-Loaded Niosomes against Ciprofloxacin-Resistant and Biofilm-Forming Staphylococcus Aureus. *Infect. Drug Resist.* **2020**, *13*, 1619. [CrossRef]
19. Mirzaie, A.; Peirovi, N.; Akbarzadeh, I.; Moghtaderi, M.; Heidari, F.; Yeganeh, F.E.; Noorbazargan, H.; Mirzazadeh, S.; Bakhtiari, R. Preparation and Optimization of Ciprofloxacin Encapsulated Niosomes: A New Approach for Enhanced Antibacterial Activity, Biofilm Inhibition and Reduced Antibiotic Resistance in Ciprofloxacin-Resistant Methicillin-Resistance Staphylococcus Aureus. *Bioorganic Chem.* **2020**, *103*, 104231. [CrossRef]
20. Tyagi, P.; Kashyap, M.; Majima, T.; Kawamorita, N.; Yoshizawa, T.; Yoshimura, N. Intravesical Liposome Therapy for Interstitial Cystitis. *Int. J. Urol.* **2017**, *24*, 262–271. [CrossRef]
21. Bartelds, R.; Nematollahi, M.H.; Pols, T.; Stuart, M.C.A.; Pardakhty, A.; Asadikaram, G.; Poolman, B. Niosomes, an Alternative for Liposomal Delivery. *PLoS ONE* **2018**, *13*, e0194179. [CrossRef]
22. Nielsen, C.K.; Kjems, J.; Mygind, T.; Snabe, T.; Meyer, R.L. Effects of Tween 80 on Growth and Biofilm Formation in Laboratory Media. *Front. Microbiol.* **2016**, *7*, 1878. [CrossRef] [PubMed]
23. Rinaldi, F.; Seguella, L.; Gigli, S.; Hanieh, P.N.; Del Favero, E.; Cantù, L.; Pesce, M.; Sarnelli, G.; Marianecci, C.; Esposito, G.; et al. InPentasomes: An Innovative Nose-to-Brain Pentamidine Delivery Blunts MPTP Parkinsonism in Mice. *J. Control. Release* **2019**, *294*, 17–26. [CrossRef] [PubMed]
24. Camara, C.I.; Bertocchi, L.; Ricci, C.; Bassi, R.; Bianchera, A.; Cantu', L.; Bettini, R.; Del Favero, E. Hyaluronic Acid—Dexamethasone Nanoparticles for Local Adjunct Therapy of Lung Inflammation. *Int. J. Mol. Sci.* **2021**, *22*, 10480. [CrossRef]
25. Doucet, M.; Cho, J.H.; Alina, G.; Bakker, J.; Bouwman, W.; Butler, P.; Campbell, K.; Gonzales, M.; Heenan, R.; Jackson, A. SasView for Small Angle Scattering Analysis. Zenodo. SasView Version 4.1.org. 2017. Available online: http://www.Sasview.org (accessed on 3 October 2022).
26. Koppel, D.E. Analysis of Macromolecular Polydispersity in Intensity Correlation Spectroscopy: The Method of Cumulants. *J. Chem. Phys.* **1972**, *57*, 4814–4820. [CrossRef]
27. De Vos, C.; Deriemaeker, L.; Finsy, R. Quantitative Assessment of the Conditioning of the Inversion of Quasi-Elastic and Static Light Scattering Data for Particle Size Distributions. *Langmuir* **1996**, *2*, 2630–2636. [CrossRef]
28. Sennato, S.; Bordi, F.; Cametti, C.; Marianecci, C.; Carafa, M.; Cametti, M. Hybrid Niosome Complexation in the Presence of Oppositely Charged Polyions. *J. Phys. Chem. B* **2008**, *112*, 3720–3727. [CrossRef] [PubMed]
29. Lentz, B.R. Membrane "Fluidity" as Detected by Diphenylhexatriene Probes. *Chem. Phys. Lipids* **1989**, *50*, 171–190. [CrossRef]
30. Sciolla, F.; Truzzolillo, D.; Chauveau, E.; Trabalzini, S.; Di Marzio, L.; Carafa, M.; Marianecci, C.; Sarra, A.; Bordi, F.; Sennato, S. Influence of Drug/Lipid Interaction on the Entrapment Efficiency of Isoniazid in Liposomes for Antitubercular Therapy: A Multi-Faced Investigation. *Colloids Surf. B Biointerfaces* **2021**, *208*, 112054. [CrossRef]
31. Zachariasse, K.A. Intramolecular Excimer Formation with Diarylalkanes as a Microfluidity Probe for Sodium Dodecyl Sulphate Micelles. *Chem. Phys. Lett.* **1978**, *57*, 429–432. [CrossRef]
32. Gummadi, S.; Thota, D.; Varri, S.V.; Vaddi, P.; Rao, V.L.N.S. Development and Validation of UV Spectroscopic Methods for Simultaneous Estimation of Ciprofloxacin and Tinidazole in Tablet Formulation. *Int. Curr. Pharm. J.* **2012**, *1*, 317–321. [CrossRef]
33. Pietrzyńska, M.; Voelkel, A. Stability of Simulated Body Fluids Such as Blood Plasma, Artificial Urine and Artificial Saliva. *Microchem. J.* **2017**, *134*, 197–201. [CrossRef]
34. Stepanović, S.; Ćirković, I.; Ranin, L.; Svabić-Vlahović, M. Biofilm Formation by *Salmonella* spp. and Listeria Monocytogenes on Plastic Surface. *Lett. Appl. Microbiol.* **2004**, *38*, 428–432. [CrossRef]
35. Lagha, M.; Bothma, J.P.; Levine, M. Mechanisms of Transcriptional Precision in Animal Development. *Trends Genet.* **2012**, *28*, 409–416. [CrossRef]
36. Bossù, M.; Selan, L.; Artini, M.; Relucenti, M.; Familiari, G.; Papa, R.; Vrenna, G.; Spigaglia, P.; Barbanti, F.; Salucci, A.; et al. Characterization of Scardovia Wiggsiae Biofilm by Original Scanning Electron Microscopy Protocol. *Microorganisms* **2020**, *8*, 807. [CrossRef] [PubMed]
37. Relucenti, M.; Familiari, G.; Donfrancesco, O.; Taurino, M.; Li, X.; Chen, R.; Artini, M.; Papa, R.; Selan, L. Microscopy Methods for Biofilm Imaging: Focus on SEM and VP-SEM Pros and Cons. *Biology* **2021**, *10*, 51. [CrossRef]
38. Relucenti, M.; Miglietta, S.; Bove, G.; Donfrancesco, O.; Battaglione, E.; Familiari, P.; Barbaranelli, C.; Covelli, E.; Barbara, M.; Familiari, G. SEM BSE 3D Image Analysis of Human Incus Bone Affected by Cholesteatoma Ascribes to Osteoclasts the Bone Erosion and VpSEM DEDX Analysis Reveals New Bone Formation. *Scanning* **2020**, *2020*, 9371516. [CrossRef] [PubMed]
39. Rinaldi, F.; Forte, J.; Pontecorvi, G.; Hanieh, P.N.; Carè, A.; Bellenghi, M.; Tirelli, V.; Ammendolia, M.G.; Mattia, G.; Marianecci, C.; et al. PH-Responsive Oleic Acid Based Nanocarriers: Melanoma Treatment Strategies. *Int. J. Pharm.* **2022**, *613*, 121391. [CrossRef]
40. Khatib, I.; Ke, W.-R.; Cipolla, D.; Chan, H.-K. Storage Stability of Inhalable, Controlled-Release Powder Formulations of Ciprofloxacin Nanocrystal-Containing Liposomes. *Int. J. Pharm.* **2021**, *605*, 120809. [CrossRef]
41. Masarudin, M.J.; Cutts, S.M.; Evison, B.J.; Phillips, D.R.; Pigram, P.J. Factors Determining the Stability, Size Distribution, and Cellular Accumulation of Small, Monodisperse Chitosan Nanoparticles as Candidate Vectors for Anticancer Drug Delivery: Application to the Passive Encapsulation of [14C]-Doxorubicin. *Nanotechnol. Sci. Appl.* **2015**, *8*, 67–80. [CrossRef]
42. Rinaldi, F.; Del Favero, E.; Rondelli, V.; Pieretti, S.; Bogni, A.; Ponti, J.; Rossi, F.; Di Marzio, L.; Paolino, D.; Marianecci, C.; et al. PH-Sensitive Niosomes: Effects on Cytotoxicity and on Inflammation and Pain in Murine Models. *J. Enzym. Inhib. Med. Chem.* **2017**, *32*, 538–546. [CrossRef]

43. Khan, D.H.; Bashir, S.; Khan, M.I.; Figueiredo, P.; Santos, H.A.; Peltonen, L. Formulation Optimization and in Vitro Characterization of Rifampicin and Ceftriaxone Dual Drug Loaded Niosomes with High Energy Probe Sonication Technique. *J. Drug Deliv. Sci. Technol.* **2020**, *58*, 101763. [CrossRef]
44. Uhljar, L.É.; Kan, S.Y.; Radacsi, N.; Koutsos, V.; Szabó-Révész, P.; Ambrus, R. In Vitro Drug Release, Permeability, and Structural Test of Ciprofloxacin-Loaded Nanofibers. *Pharmaceutics* **2021**, *13*, 556. [CrossRef] [PubMed]
45. Volpe, D.A. Permeability Classification of Representative Fluoroquinolones by a Cell Culture Method. *AAPS J* **2004**, *6*, 1–6. [CrossRef] [PubMed]
46. Akbarzadeh, I.; Shayan, M.; Bourbour, M.; Moghtaderi, M.; Noorbazargan, H.; Eshrati Yeganeh, F.; Saffar, S.; Tahriri, M. Preparation, Optimization and In-Vitro Evaluation of Curcumin-Loaded Niosome@calcium Alginate Nanocarrier as a New Approach for Breast Cancer Treatment. *Biology* **2021**, *10*, 173. [CrossRef]
47. Satish, J.; Amusa, A.S.; Gopalakrishna, P. In Vitro Activities of Fluoroquinolones Entrapped in Non-Ionic Surfactant Vesicles against Ciprofloxacin-Resistant Bacteria Strains. *J. Pharm. Technol. Drug Res.* **2012**, *1*, 5. [CrossRef]
48. Akbarzadeh, I.; Keramati, M.; Azadi, A.; Afzali, E.; Shahbazi, R.; Chiani, M.; Norouzian, D.; Bakhshandeh, H. Optimization, Physicochemical Characterization, and Antimicrobial Activity of a Novel Simvastatin Nano-Niosomal Gel against E. Coli and S. Aureus. *Chem. Phys. Lipids* **2021**, *234*, 105019. [CrossRef]
49. Gupta, S.; Kumar, P.; Rathi, B.; Verma, V.; Dhanda, R.S.; Devi, P.; Yadav, M. Targeting of Uropathogenic Escherichia Coli PapG Gene Using CRISPR-Dot Nanocomplex Reduced Virulence of UPEC. *Sci. Rep.* **2021**, *11*, 17801. [CrossRef]
50. Chen, Q.; Xie, S.; Lou, X.; Cheng, S.; Liu, X.; Zheng, W.; Zheng, Z.; Wang, H. Biofilm formation and prevalence of adhesion genes among Staphylococcus aureus isolates from different food sources. *Microbiologyopen* **2020**, *9*, e00946. [CrossRef]
51. Barakat, H.S.; Kassem, M.A.; El-Khordagui, L.K.; Khalafallah, N.M. Vancomycin-Eluting Niosomes: A New Approach to the Inhibition of Staphylococcal Biofilm on Abiotic Surfaces. *AAPS PharmSciTech* **2014**, *15*, 1263–1274. [CrossRef]
52. Manosroi, A.; Khanrin, P.; Lohcharoenkal, W.; Werner, R.G.; Götz, F.; Manosroi, W.; Manosroi, J. Transdermal Absorption Enhancement through Rat Skin of Gallidermin Loaded in Niosomes. *Int. J. Pharm.* **2010**, *392*, 304–310. [CrossRef]
53. Piri-Gharaghie, T.; Jegargoshe-Shirin, N.; Saremi-Nouri, S.; Khademhosseini, S.; Hoseinnezhad-Lazarjani, E.; Mousavi, A.; Kabiri, H.; Rajaei, N.; Riahi, A.; Farhadi-Biregani, A. Effects of Imipenem-Containing Niosome Nanoparticles against High Prevalence Methicillin-Resistant Staphylococcus Epidermidis Biofilm Formed. *Sci. Rep.* **2022**, *12*, 5140. [CrossRef] [PubMed]
54. Abdelaziz, A.A.; Elbanna, T.E.; Sonbol, F.I.; Gamaleldin, N.M.; El Maghraby, G.M. Optimization of Niosomes for Enhanced Antibacterial Activity and Reduced Bacterial Resistance: In Vitro and in Vivo Evaluation. *Expert Opin. Drug Deliv.* **2015**, *12*, 163–180. [CrossRef] [PubMed]
55. Dong, G.; Li, J.; Chen, L.; Bi, W.; Zhang, X.; Liu, H.; Zhi, X.; Zhou, T.; Cao, J. Effects of Sub-Minimum Inhibitory Concentrations of Ciprofloxacin on Biofilm Formation and Virulence Factors of *Escherichia coli*. *Braz. J. Infect. Dis.* **2019**, *23*, 15–21. [CrossRef] [PubMed]
56. Kamat, A.M.; Lamm, D.L. Antitumor Activity of Common Antibiotics against Superficial Bladder Cancer. *Urology* **2004**, *63*, 457–460. [CrossRef] [PubMed]
57. Gurtowska, N.; Kloskowski, T.; Drewa, T. Ciprofloxacin Criteria in Antimicrobial Prophylaxis and Bladder Cancer Recurrence. *Med. Sci. Monit.* **2010**, *16*, RA218-23.

pharmaceutics

MDPI

Article

Design of a Transdermal Sustained Release Formulation Based on Water-Soluble Ointment Incorporating Tulobuterol Nanoparticles

Noriaki Nagai *, Fumihiko Ogata, Saori Deguchi, Aoi Fushiki, Saki Daimyo, Hiroko Otake and Naohito Kawasaki

Faculty of Pharmacy, Kindai University, 3-4-1 Kowakae, Higashiosaka 577-8502, Japan
* Correspondence: nagai_n@phar.kindai.ac.jp

Abstract: We aimed to investigate which base was suitable for preparing transdermal formulations incorporating tulobuterol (TUL) nanoparticles (30–180 nm) in this study. Three bases (water-soluble, absorptive, and aqueous ionic cream) were selected to prepare the transdermal formulations, and TUL nanoparticles were prepared with a bead-milling treatment. In the drug release study, the TUL release from the water-soluble ointment was higher than that from the other two ointments. Moreover, the addition of *l*-menthol enhanced TUL nanoparticle release from the ointment, and the rat skin penetration of the TUL water-soluble ointment was also significantly higher than that of the other two ointments. In addition, the drug penetration of the TUL water-soluble ointment with *l*-menthol sustained zero-order release over 24 h, and the skin permeability of TUL increased with TUL content in the ointment. On the other hand, this penetration was significantly inhibited by treatment with a caveolae-mediated endocytosis inhibitor (nystatin). In conclusion, we found that the water-soluble base incorporating TUL nanoparticles and *l*-menthol was the best among those assessed in this study. Furthermore, the pathway using caveolae-mediated endocytosis was related to the skin penetration of TUL nanoparticles in the TUL water-soluble ointment with *l*-menthol. These findings are useful for the design of a transdermal sustained-release formulation based on TUL nanoparticles.

Keywords: tulobuterol; nanoparticle; ointment; transdermal delivery system; endocytosis

Citation: Nagai, N.; Ogata, F.; Deguchi, S.; Fushiki, A.; Daimyo, S.; Otake, H.; Kawasaki, N. Design of a Transdermal Sustained Release Formulation Based on Water-Soluble Ointment Incorporating Tulobuterol Nanoparticles. *Pharmaceutics* **2022**, *14*, 2431. https://doi.org/10.3390/pharmaceutics14112431

Academic Editors: Daniela Monti and Silvia Tampucci

Received: 14 October 2022
Accepted: 8 November 2022
Published: 10 November 2022

1. Introduction

Tulobuterol (TUL), or 2-tert-butylamino-1-(2-chloro-phenyl)-ethanol, is a β2-adrenergic agonist, and transdermal formulations incorporating TUL are used in the treatment of chronic obstructive pulmonary disease (COPD), emphysema, bronchitis, and asthma [1]. Transdermal formulations of TUL can easily maintain plasma drug concentrations at required levels and avoid difficulties associated with oral administration and drug degradation in the gastrointestinal tract; moreover, they provide more consistent serum drug levels, have reduced side effects, and eliminate hepatic first-pass metabolism [2–4]. In addition, the administration of transdermal formulations is simple to discontinue, and medication is easily confirmed [5,6]. Although TUL is dissolved in acrylate and rubber in generic transdermal formulations, the original form of commercially available TUL (CA-TUL tape, Hokunalin® Tape, Hisamitsu Pharmaceutical Co., Inc., Tokyo, Japan) is a transdermal patch preparation with a crystal reservoir system, which has been patented. CA-TUL tape contains molecular and crystallized forms of TUL, provides a favorable pharmacokinetic profile, has β2-agonist activity that can be sustained for 24 h, and is widely used by children [1,7,8]. Therefore, CA-TUL tape can be expected to improve adverse drug reactions, making it a useful, long-acting β2-agonist with good adherence that can be applied to children and the elderly [9,10].

On the other hand, recent studies have showed that drug nanoparticles can penetrate deep into the skin, depending on their size [11], and that the transcellular route, intercellular lipid space route, and transappendageal route (hair follicles and sweat glands)

are related to drug nanoparticle penetration into the skin [12]. Thus, nanotechnology is a modern and rapidly evolving trend that is being applied in transdermal therapeutic systems (TTS), which include nanocrystals, polymeric nanoparticles, nanovesicles, dendrimers, liposomes, nanomicelles, nanoemulsions, and lipid nanocarriers. Nanovesicles and polymeric nanoparticles possess an advantage over other methods in that they promote transdermal permeation without affecting the skin's structure [13–15]. Moreover, the small size (40 nm–800 nm) of lipid nanocarriers allows them to adhere to the lipid film of the stratum corneum and to increase the number of drug molecules that penetrate into deeper layers of the skin [14,15]. We previously prepared a gel formulation based on *l*-menthol (a skin penetration enhancer) and solid NPs of a drug, demonstrating that gels incorporating drug nanoparticles (NPs) could enhance skin penetration in rat and pig skins [16,17]. Moreover, the reduction in particle size and the utilization of chemical penetration enhancers causes a dramatic increase in the cellular uptake of NPs [16–18]. Thus, it is expected that transdermal formulations combined with solid NPs and skin penetration enhancers may be useful as a TTS; however, information on the relationship between bases and NPs in transdermal formulations is not sufficient, and more studies are required.

In this study, we select TUL as a model drug since TUL transdermal formulations are widely used in clinics, and we aim to develop a good transdermal formulation incorporating TUL nanoparticles (TUL-NPs). Ointments are either nongreasy or greasy preparations, depending on the type of base used. Oil-in-water creams are water-soluble and are more cosmetically and aesthetically acceptable than water-in-oil creams. Creams are semisolid emulsions and are usually applied topically as medicated or unmedicated products. We prepare transdermal formulations incorporating TUL-NPs using water-soluble (WS), absorptive (AB), and aqueous ionic cream (AC) ointment bases with and without *l*-menthol (WS, AB, and AC bases). In addition, we investigate the drug release and skin penetration properties of these transdermal formulations.

2. Materials and Methods

2.1. Animals

Seven-week-old male Wistar rats were purchased from Kiwa Laboratory Animals Co., Ltd. (Wakayama, Japan). They were fed a CE-2 formulation diet (Clea Japan Inc., Tokyo, Japan), and water was provided freely. The rats were housed under normal conditions (light during 7:00 am–7:00 pm; 25 °C). The experiments using rats were approved by the animal care and use committee of Kindai University and were carried out in accordance with the Pharmacy Committee Guidelines.

2.2. Chemicals

Pentobarbital was purchased from Sumitomo Dainippon Pharma Co., Ltd. (Toyo, Japan). TUL, cytochalasin D, methyl p-hydroxybenzoate, polyethylene glycol (PEG) 4000, cetyl alcohol, l-menthol, mineral oil, white wax, sodium tetraborate, propylene glycol, beeswax, and sodium dodecyl sulfate were provided by Wako Pure Chemical Industries, Ltd. (Osaka, Japan). Commercially available 0.5 mg TUL tape (CA-TUL, Hokunalin® Tapes 0.5 mg) was obtained from Mylan EPD G.K (Tokyo, Japan). Rottlerin and dynasore were purchased from Nacalai Tesque (Kyoto, Japan), and methylcellulose (MC, SM-4) was obtained from Shin-Etsu Chemical Co., Ltd. (Tokyo, Japan). 2-hydroxypropyl-β-cyclodextrin (HPβCD) was supplied by Nihon Shokuhin Kako Co., Ltd. (Tokyo, Japan), and PEG 400 was provided by Maruishi Pharmaceutical Co., Ltd. (Osaka, Japan). Nystatin and 450 nm pore size MF™ membrane filters were purchased from Sigma-Aldrich Japan (Tokyo, Japan) and Merck Millipore (Tokyo, Japan), respectively. All other chemicals used were of the highest purity commercially available.

2.3. Preparation of Ointments Incorporating TUL-NPs

The TUL-NPs were prepared following methods reported in previous studies [18–21]. Briefly, TUL and methylcellulose (MC) were mixed and milled using a Bead Smash 12

(3000 rpm, 30 s, 4 °C; Wakenyaku Co., Ltd., Kyoto, Japan). Then, 2-hydroxypropyl-β-cyclodextrin (HPβCD) was added and dispersed using distilled water. Thereafter, the dispersions were milled using a Bead Smash 12 (5500 rpm, 60 s × 30 times, 4 °C), providing dispersions containing TUL-NPs. In this study, three bases (water-soluble (WS), absorptive (AB), and aqueous ionic cream (AC) bases) were selected, and the TUL-NP dispersions, as well as the mixtures of TUL, MC, and HPβCD (TUL-MP dispersions), were gelled using the three bases. Table 1 shows the composition of each ointment incorporating TUL. For the WS base, polyethylene glycol (PEG) 4000 and cetyl alcohol were dissolved at 60 °C, and PEG 400 was then added and mixed. The WS/TUL ointments were prepared by the addition of TUL-MPs or TUL-NPs to the WS base. The AB base was prepared as follows: cetyl alcohol, white wax, and mineral oil were dissolved at 60 °C, and subsequently, sodium tetraborate was added. Thereafter, the AB base and TUL (TUL-MPs or TUL-NPs) were mixed (AB/TUL ointments). The AC base was prepared by dissolving cetyl alcohol, beeswax, propylene glycol, and sodium dodecyl sulfate at 60 °C, and the AC/TUL ointments were prepared by the addition of TUL-MPs or TUL-NPs to the AC base. In this study, *l*-menthol was added into these ointments (WS/TUL, AB/TUL, and AC/TUL), and the resulting formulations were denoted as Men-WS/TUL, Men-AB/TUL, and Men-AC/TUL.

Table 1. Compositions of WS/TUL, AB/TUL, and AC/TUL ointments with and without *l*-menthol.

Ointment	Content (%*w*/*w*)					
	WS/TUL	Men-WS/TUL	AB/TUL	Men-AB/TUL	AC/TUL	Men-AC/TUL
TUL	0.2	0.2	0.2	0.2	0.2	0.2
MC	0.05	0.05	0.05	0.05	0.05	0.05
HPβCD	0.5	0.5	0.5	0.5	0.5	0.5
PEG 4000	43.8	43.8	-	-	-	-
PEG 400	43.8	43.8		-	-	-
Cetyl alcohol	0.4	0.4	12.5	12.5	15	15
Mineral oil	-	-	56	56	-	-
White wax	-	-	12	12	-	-
Sodium tetraborate	-	-	0.5	0.5	-	-
Propylene glycol	-	-	-	-	10	10
Beeswax	-	-	-	-	1	1
Sodium dodecyl sulfate	-	-	-	-	2	2
l-menthol	-	2	-	2	-	2
Distilled water ad.	100	100	100	100	100	100

2.4. Measurement of TUL Particles

The size of the TUL-MPs was measured using a SALD-7100 laser diffraction particle size analyzer (Shimadzu Corp., Kyoto, Japan), and the size of the TUL-NPs was determined using both a SALD-7100 and a NANOSIGHT LM10 dynamic light-scattering analyzer (QuantumDesign Japan, Tokyo, Japan). The measurement conditions for the SALD-7100 analyzer were a maximum value of scattered light intensity in the range of 40–60% and a refractive index of 1.60 ± 0.10i. The measurement conditions for the NANOSIGHT LM10 analyzer were a wavelength of 405 nm (blue), a time of 60 s, and a viscosity of 1.27 mPa·s. In addition, the number of TUL-NPs was also detected using a NANOSIGHT LM10 analyzer. Images of ointments incorporating TUL were captured using a scanning probe microscope (SPM). The SPM images were captured with a SPM-9700 (Shimadzu Corp., Kyoto, Japan) instrument. During SPM measurements, the ointments were washed with distilled water, and the isolated TUL particles were measured.

2.5. Drug Solubility of TUL Ointments

Dissolved TUL and TUL-NPs in the ointments were isolated using centrifugation at 100,000× *g* (Beckman Optima^{TM} MAX-XP Ultracentrifuge, Beckman Coulter, Osaka, Japan). Isolated TUL was dissolved using methanol, and the content was measured in order to calculate the solubility of TUL in the ointments. The TUL contents were measured using the HPLC method described below.

2.6. Viscosity of the Ointments

A Brookfield digital viscometer was used to measure the viscosity of the TUL ointments, as presented in Table 1 (Brookfield Engineering Laboratories, Inc., Middleboro, MA, USA).

2.7. Stability of the TUL Ointments

As presented in Table 1, an amount of 0.3 g of each TUL ointment was placed in a beaker, and the ointments were kept in a refrigerator (4 °C) with a lid to keep them from drying out. After 1 month, ointments were removed from beakers, and the changes in particle size and number, as well as content, of TUL were measured with a NANOSIGHT LM10 instrument and the HPLC method described below. When measuring particle characteristics of the ointments, the ointments were distributed in 100 mL of water since accurate particle size could not be measured in the ointment state.

2.8. Drug Release from TUL Ointments

Franz diffusion cells were used to measure the release of TUL from the ointments as previously reported [16–18]. Briefly, 12.2 mL of phosphate-buffered solution (pH 7.2) was transferred into a reservoir chamber and thermoregulated at 37 °C. MF™ membrane filters (450 nm pore size) were set in a Franz diffusion cell. Then, O-ring flanges (1.6 cm i.d.) were placed on the filters, and 0.3 g of each prepared ointment or CA-TUL tape was spread uniformly over the filters. In the experiment, 100 µL of sample solution was withdrawn from the reservoir chamber at 0.5, 1, 3, 6, and 24 h and supplemented with the same volume of pH 7.2 phosphate-buffered solution. The collected samples were used to measure particle size and number, as well as content, of TUL with a NANOSIGHT LM10 analyzer and the HPLC method described below. The area under the TUL concentration–time curve ($AUC_{Release}$) was analyzed according to the trapezoidal rule to the last TUL measurement point (24 h).

2.9. In Vitro Transdermal Penetration of Ointments Incorporating TUL

The in vitro transdermal penetration of the ointments was evaluated following the methods described in previous studies [16–18]. The hair on the abdominal areas of 7-week-old Wistar rats was removed on the day prior to the experiment, and the rats were euthanized by injection with a lethal dose of pentobarbital on the day of the experiment. Thereafter, the abdominal area was collected and set in a Franz diffusion cell. A total of 12.2 mL of phosphate-buffered solution (pH 7.2) was transferred into a reservoir chamber and thermoregulated at 37 °C. Then, O-ring flanges (1.6 cm i.d.) were placed on filters, and 0.3 g of each prepared ointment or CA-TUL tape was spread uniformly over the filters. In the experiment, 100 µL of sample solution was withdrawn from the reservoir chamber at 0.5, 1, 3, 6, and 24 h and supplemented with the same volume of pH 7.2 phosphate-buffered solution. The collected samples were used to measure the particle size and number, as well as content, of TUL with a NANOSIGHT LM10 instrument and the HPLC method described below. The trapezoidal rule to the last TUL measurement point (24 h) was used to analyze the area under the penetrated TUL concentration–time curves (AUC_{Skin}), and pharmacokinetic parameters were analyzed according to Equations (1) and (2) using a nonlinear least-squares computer program (MULTI) [22].

$$D = \frac{\delta^2}{6\tau} \quad (1)$$

$$J_c = \frac{Q}{A \cdot (t - \tau)} = \frac{D \cdot K_m \cdot C_c}{\delta} = K_p \cdot C_c \quad (2)$$

where J_c, K_m, K_p, D, τ, A, δ, Q_t, and C_c are the penetration rate, skin coefficient, preparation partition coefficient, diffusion constant within the skin, penetration coefficient through the skin, lag time, effective area of the skin (2 cm^2), thickness of the skin (0.071 cm, n = 5), and amount of TUL in the reservoir solution at time t, respectively.

Nystatin (54 μM) [23], dynasore (40 μM) [24], rottlerin (2 μM) [16,25], and cytochalasin D (10 μM) [23] were used to inhibit caveolae-mediated endocytosis (CavME), clathrin-mediated endocytosis (CME), micropinocytosis (MP), and phagocytosis, respectively. These pharmacological inhibitors were dissolved in 0.5% DMSO and applied to the removed skin 1 h prior to the application of the transdermal formulations.

2.10. Measurement of TUL by HPLC Method

TUL was measured using an HPLC method. A Shimadzu LC-20AT system equipped with an SIL-20AC auto-injector and a CTO-20 A column oven (Shimadzu Corp.) was used to measure the TUL content. TUL was dissolved in methanol containing internal standard (1 μg/mL methyl p-oxybenzoate), and 10 μL was injected into the HPLC. The column used was an Inertsil® ODS-3 column (3 μm, column size: 2.1 mm × 50 mm; GL Science Co., Inc., Tokyo, Japan), and 0.02 M potassium dihydrogen phosphate:acetonitrile (87:13, v/v) was utilized as the mobile phase at 0.25 mL/min and 35 °C. TUL was detected at 211 nm.

2.11. Statistical Analysis

Statistical analyses were performed using Dunnett's multiple comparisons (one-way analysis of variance), and $p < 0.05$ was chosen as the significance level. The data are expressed as means ± standard error (SE) of the mean.

3. Results

3.1. Evaluation of Ointments Incorporating TUL-NPs

Figure 1 shows the particle size distribution of TUL particles with or without bead-milling treatment. The mean TUL particle size without bead-milling treatment was 23.6 ± 0.28 μm (Figure 1A). Alternately, the size distribution of TUL particles with the bead-milling treatment was 30–180 nm (Figure 1B,C), and AFM imaging of TUL using SPM showed that the TUL particles with bead-milling treatment were uniformly dispersed with no large agglomerates. Figure 2 shows images of the ointments incorporating TUL-MPs and TUL-NPs. Although the TUL particles were observed in the ointments incorporating TUL-MPs, the TUL particles could not be visually confirmed in the ointments incorporating TUL-NPs. The ointments incorporating TUL-NPs appeared whitish in comparison with the ointments incorporating TUL-MPs. Table 2 shows the mean particle size, the solubility, and the viscosity of ointments incorporating TUL-MPs and TUL-NPs. The particle sizes in the ointments incorporating TUL-NPs remained on the nanoscale. The amounts of dissolved TUL in the WS/TUL-NP, AB/TUL-NP, and AC/TUL-NP ointments were higher than those in the corresponding ointments incorporating TUL-MPs. On the other hand, significant differences were not observed between the solubility levels in the WS/TUL, AB/TUL, and AC/TUL ointments with and without *l*-menthol. The viscosities were also similar in the ointments incorporating TUL-MPs and TUL-NPs, with the following order of viscosity: WS/TUL > AC/TUL > AB/TUL. Although the viscosities were similar in the AB/TUL and AC/TUL ointments with and without *l*-menthol, the addition of *l*-menthol decreased the viscosity in the WS/TUL ointments. In this study, we measured concentration and particle size 1 month after preparation. The TUL concentrations in the WS/TUL-NP, AB/TUL-NP, and AC/TUL-NP ointments were not changed and remained nanosized. The mean particle sizes in the WS/TUL-NP, Men-WS/TUL-NP, AB/TUL-NP, Men-AB/TUL-NP, AC/TUL-NP, and Men-AC/TUL-NP ointments were 121, 118, 126, 120, 113, and 119 nm, respectively.

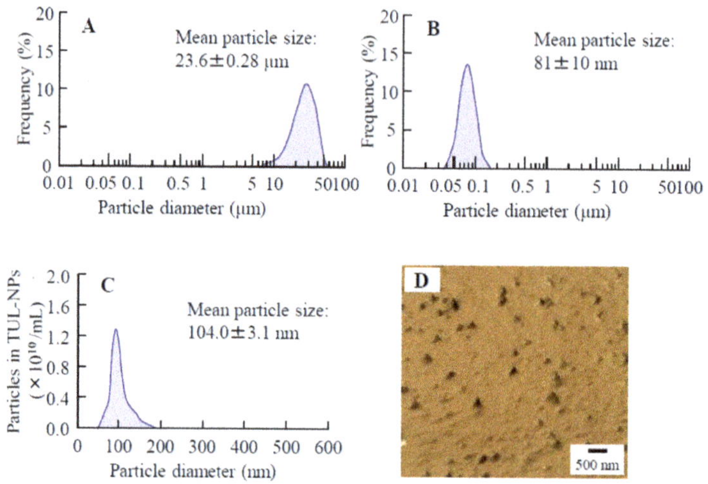

Figure 1. Particle size frequencies and SPM image of TUL in TUL-MP and TUL-NP dispersions. Particle distribution of TUL-MP (**A**) and TUL-NP (**B**) dispersions using laser diffraction particle size analyzer. Particle distribution (**C**) and SPM image (**D**) of TUL-NPs in TUL-NP dispersions using dynamic light-scattering analyzer and SPM-9700 instrument, respectively. Particle sizes of bead-mill-treated TUL particles were on the nanoscale, i.e., 30–180 nm.

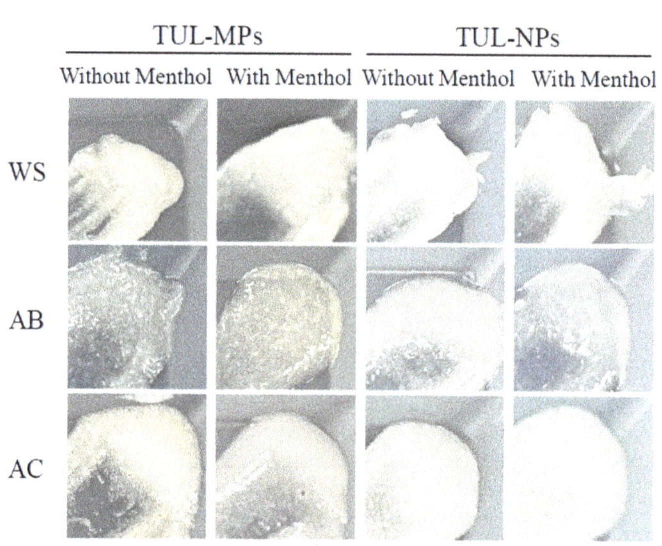

Figure 2. Digital images of ointments incorporating TUL-MPs and TUL-NPs. The compositions of ointments incorporating TUL-MPs and TUL-NPs are presented in Table 1.

Table 2. Changes in particle size, solubility, and viscosity of WS/TUL, AB/TUL, and AC/TUL ointments with and without *l*-menthol.

Ointment		Mean Particle Size (μm)	Solubility (μM)	Viscosity (mPa·s)
WS/TUL	WS/TUL-MP	28.0 ± 0.31	139 ± 7.9	1476 ± 83
	WS/TUL-NP	0.113 ± 0.0038	192 ± 9.3 *,#	710 ± 47 *,#
	Men-WS/TUL-MP	27.5 ± 0.29	139 ± 8.5	1466 ± 86
	Men-WS/TUL-NP	0.113 ± 0.0033	193 ± 9.0 *,#	747 ± 45 *,#
AB/TUL	AB/TUL-MP	28.3 ± 0.32	112 ± 6.9	79.6 ± 7.8
	AB/TUL-NP	0.119 ± 0.0038	189 ± 9.2 *,#	76.3 ± 7.9
	Men-AB/TUL-MP	27.7 ± 0.29	113 ± 8.1	77.5 ± 7.9
	Men-AB/TUL-NP	0.116 ± 0.0035	185 ± 8.8 *,#	70.2 ± 8.4
AC/TUL	AC/TUL-MP	29.2 ± 0.36	116 ± 6.8	599 ± 37
	AC/TUL-NP	0.110 ± 0.0039	186 ± 7.9 *,#	597 ± 35
	Men-AC/TUL-MP	28.1 ± 0.28	120 ± 7.0	606 ± 39
	Men-AC/TUL-NP	0.112 ± 0.0036	188 ± 9.1 *,#	605 ± 38

The compositions of ointments containing TUL-MPs and TUL-NPs are presented in Table 1. n = 6–10. * $p < 0.05$ vs. MP formulation for each category. # $p < 0.05$ vs. MP formulation with *l*-menthol for each category.

3.2. Drug Release of TUL in WS/TUL, AB/TUL, and AC/TUL Ointments with and without l-Menthol

Figure 3 shows the release profiles of TUL from the WS/TUL, AB/TUL, and AC/TUL ointments with and without *l*-menthol. The drug release values from the WS/TUL ointments were higher than those of the ointments with AB and AC bases. The AUC_{Release} values were similar for the WS/TUL and AC/TUL ointments incorporating both MPs and NPs. On the other hand, the TUL release from the AB/TUL-NP ointment was higher than that of the AB/TUL-MP ointment. The addition of *l*-menthol decreased the TUL release from the AB and AC bases. In contrast to the results for the AB and AC bases, the TUL release from the WA base was enhanced by the addition of *l*-menthol. In the AB/TUL-NP and AC/TUL-NP ointments with and without *l*-menthol, the numbers of TUL-NPs were low (AB/TUL-NP, 5.14 ± 0.38 × 10^5; Men-AB/TUL-NP, 1.10 ± 0.21 × 10^6; AC/TUL-Ns, 2.94 ± 0.58 × 10^5; Men-AC/TUL-NP, 2.79 ± 0.85 × 10^6 (particles/mL)). However, more TUL-NPs were released into the reservoir chamber by the WS/TUL-NP and Men-WS/TUL-NP ointments. The particle size frequencies (particle numbers) of the released TUL-NPs were 220.8 ± 11.5 nm (2.22 ± 0.15 × 10^8 particles/mL) and 200.2 ± 10.2 nm (5.39 ± 0.93 × 10^8 particles/mL), as detected in the reservoir chamber 24 h after the application of the WS/TUL-NP and Men-WS/TUL-NP ointments, respectively.

3.3. Transdermal Delivery of TUL in WS/TUL, AB/TUL, and AC/TUL Ointments with and without l-Menthol

Figure 4 shows the skin penetration profiles of the WS/TUL, AB/TUL, and AC/TUL ointments with and without *l*-menthol, and Table 3 shows the pharmacokinetic parameters estimated from the data in Figure 4. The AUC_{Skin} values for the WS base were higher than those of the AB and AC bases. In the AB/TUL and AC/TUL ointments, the skin penetration of TUL in the ointment was decreased by the addition of *l*-menthol, although *l*-menthol enhanced the skin penetration of TUL in the WS/TUL ointment. Although the AUC_{Skin} values in the AB/TUL ointments incorporating TUL-NPs and TUL-MPs were not significantly different, the AUC_{Skin} values for the Men-WS/TUL-NP and Men-AC/TUL-NP ointments were significantly higher than those of the corresponding ointments containing TUL-MPs. In particular, the skin penetration of the Men-WS/TUL-NP ointment was greater than that of the Men-WS/TUL-MP ointment, and the AUC_{Skin} value of the Men-WS/TUL-NP ointment was 2.87-fold that of the Men-WS/TUL-MP ointment. In addition, the J_c, K_m, and K_p values of the Men-WS/TUL-NP ointment were also significantly enhanced in comparison with the Men-WS/TUL-MP ointment. Figure 5 shows the relationship between TUL content and skin penetration in the Men-WS/TUL-NP ointment. The skin penetration of TUL in the Men-WS/TUL-NP ointment linearly increased with TUL content for 24 h, and sustained penetration was observed in comparison with CA-TUL tape since the penetration rate of TUL in CA-TUL tape decreased after 6 h. The skin penetration of CA-TUL tape was higher than that of the Men-WS/TUL-NP ointment at the same content, although the AUC_{Skin} values in ointments incorporating 1.0–1.5% TUL-NPs were similar to that of CA-TUL tape. In addition, we investigated whether the particles penetrated through the skin tissue by the measurement of particles in the receiving phase of the permeation study. In contrast to the results of the drug release experiments using membrane filters, the number of TUL-NPs was not quantified in the reservoir chamber for the in vitro studies using rat skin treated with Men-WS/TUL-NP ointment.

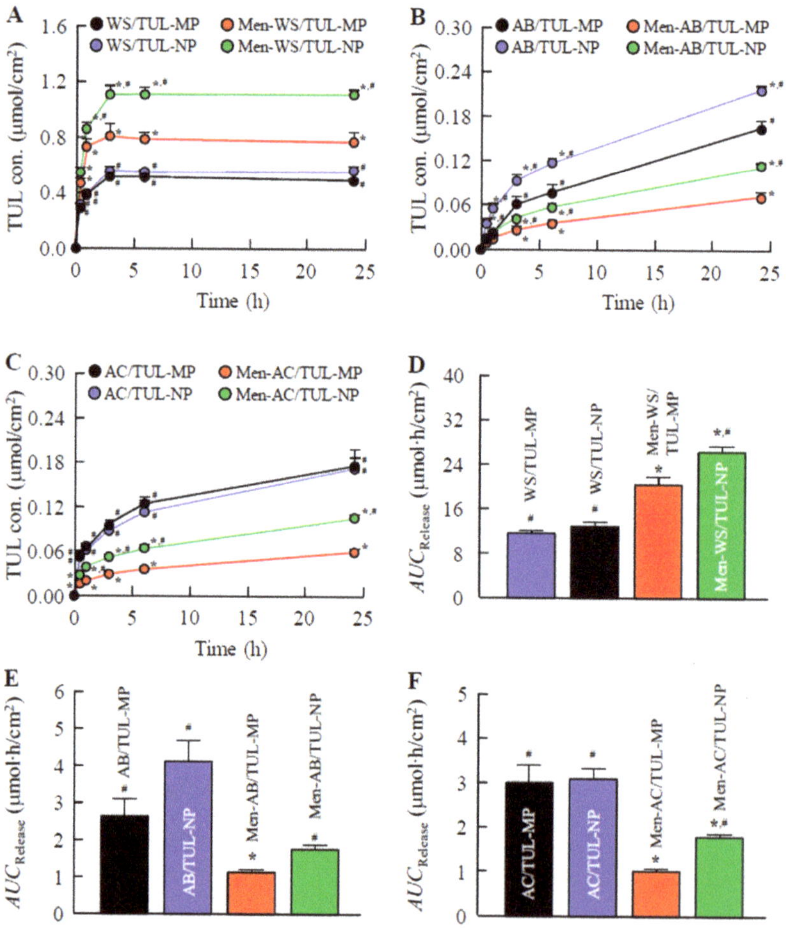

Figure 3. TUL release from ointments incorporating TUL-MPs and TUL-NPs through 450 nm pore membranes. Release of TUL from WS/TUL (**A**); AB/TUL (**B**); and AC/TUL (**C**) ointments with and without *l*-menthol through membranes. $AUC_{Release}$ of WS/TUL (**D**); AB/TUL (**E**); and AC/TUL (**F**) ointments with and without *l*-menthol through 450 nm pore membranes. The compositions of ointments incorporating TUL-MPs and TUL-NPs are presented in Table 1. Results are means ± SE; n = 6–10. * $p < 0.05$ vs. MP formulation for each category. # $p < 0.05$ vs. MP formulation with *l*-menthol for each category. The TUL release values from WS/TUL ointments were higher than those from AB/TUL and AC/TUL ointments, and the combination of NPs and *l*-menthol enhanced drug release from the WS base.

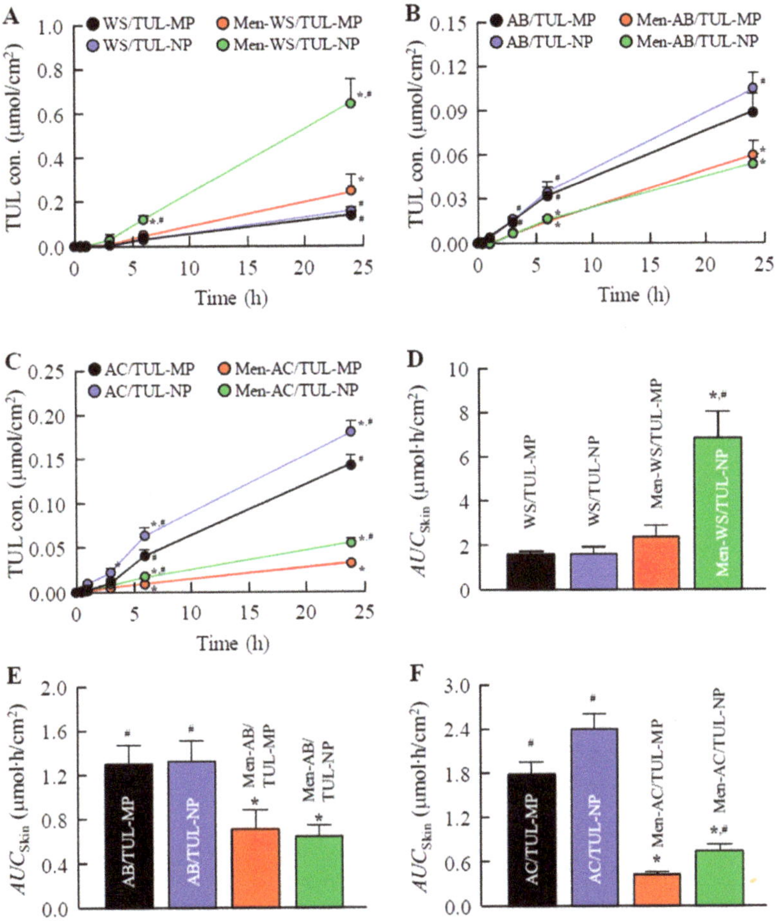

Figure 4. Transdermal penetration of TUL in ointments incorporating TUL-MPs and TUL-NPs. Penetration of WS/TUL (**A**); AB/TUL (**B**); and AC/TUL (**C**) ointments with and without *l*-menthol through rat skin. AUC_{Release} values of WS/TUL (**D**); AB/TUL (**E**); and AC/TUL (**F**) ointments with and without *l*-menthol through rat skin. The compositions of ointments incorporating TUL-MPs and TUL-NPs are presented in Table 1. Results are means ± SE; n = 6–8. * $p < 0.05$ vs. MP formulation for each category. # $p < 0.05$ vs. MP formulation with *l*-menthol for each category. TUL penetration amounts of the WS/TUL ointments were higher in comparison with AB/TUL and AC/TUL ointments, and the combination of NPs and *l*-menthol in the WS base significantly increased skin penetration.

Table 3. Pharmacokinetic analysis of TUL transdermal formulations in the in vitro penetration of rat skin.

	Ointment	J_c ($\times 10^{-2}$ μmol/cm²/h)	K_p ($\times 10^{-3}$ cm/h)	K_m	τ (h)	D ($\times 10^{-3}$ cm²/h)
WS/TUL	WS/TUL-MP	0.09 ± 0.01 #	0.13 ± 0.01 #	0.01 ± 0.01 #	0.09 ± 0.03 #	0.25 ± 0.17 #
	WS/TUL-NP	0.10 ± 0.02 #	0.14 ± 0.02 #	0.01 ± 0.01 #	1.17 ± 0.01	0.14 ± 0.01 #
	Men-WS/TUL-MP	0.62 ± 0.10 *	0.85 ± 0.14 *	0.07 ± 0.01	1.01 ± 0.07	0.83 ± 0.05 *
	Men-WS/TUL-NP	2.79 ± 0.49 *,#	3.82 ± 0.68 *,#	0.36 ± 0.07 *,#	1.13 ± 0.02	0.74 ± 0.02 *
AB/TUL	AB/TUL-MP	0.07 ± 0.05	0.10 ± 0.07	0.03 ± 0.02	0.90 ± 0.43	3.70 ± 2.11
	AB/TUL-NP	0.11 ± 0.08	0.16 ± 0.11	0.02 ± 0.01	0.29 ± 0.14	3.65 ± 2.08
	Men-AB/TUL-MP	0.05 ± 0.02	0.07 ± 0.05	0.01 ± 0.01	0.43 ± 0.15	5.86 ± 3.15
	Men-AB/TUL-NP	0.06 ± 0.03	0.08 ± 0.05	0.01 ± 0.01	0.39 ± 0.18	9.36 ± 4.62
AC/TUL	AC/TUL-MP	0.12 ± 0.06	0.16 ± 0.09	0.01 ± 0.01	0.17 ± 0.08 #	3.66 ± 1.59 #
	AC/TUL-NP	0.05 ± 0.03	0.07 ± 0.04	0.02 ± 0.01	0.28 ± 0.19 #	2.24 ± 1.58 #
	Men-AC/TUL-MP	0.02 ± 0.01	0.03 ± 0.01 *	0.01 ± 0.01	0.83 ± 0.29 *	13.49 ± 5.54 *,
	Men-AC/TUL-NP	0.03 ± 0.01	0.02 ± 0.01 *	0.01 ± 0.01	0.75 ± 0.14 *	11.19 ± 5.15

The compositions of ointment containing TUL-MPs and TUL-NPs are presented in Table 1. n = 6–9. * $p < 0.05$ vs. MP formulation for each category. # $p < 0.05$ vs. MP formulation with *l*-menthol for each category.

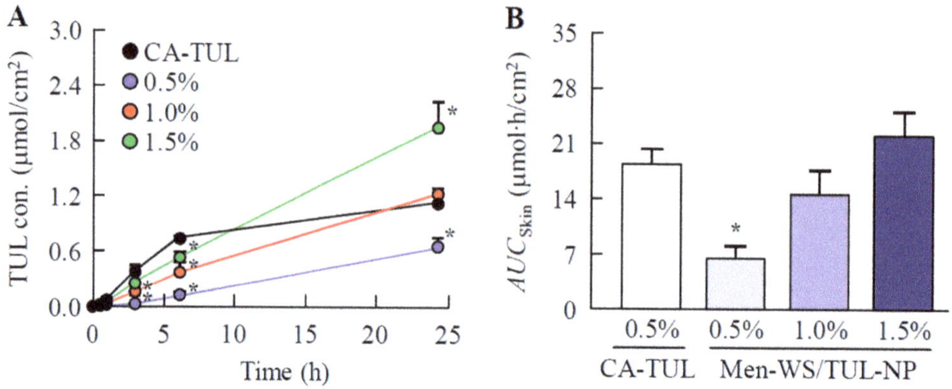

Figure 5. Changes in the transdermal penetration of TUL in Men-WS/TUL ointments containing 0.5–1.5% TUL-NPs. Changes in TUL profiles (**A**) and AUC_{Skin} values (**B**) of Men-WS/TUL-NP ointments (0.5–1.5% TUL). Results are means ± SE; n = 6–8. * $p < 0.05$ vs. CA-TUL tape for each category. The transdermal penetration of TUL in Men-WS/TUL-NP ointment was enhanced with TUL content, and the penetration profile linearly increased.

3.4. Effect of Energy-Dependent Endocytosis on Transdermal Pathway of TUL-NPs in Men-WS/TUL-NP Ointment

Figure 6 shows the skin penetration of TUL-NPs from the Men-WS/TUL-NP ointment into skin treated with endocytosis inhibitors. No significant difference was observed in the transdermal penetration of skin treated with or without rottlerin. On the other hand, transdermal penetration in skin treated with dynasore tended to decrease in comparison with the control group. In addition, nystatin treatment significantly inhibited the transdermal penetration of the Men-WS/TUL-NP ointment. The AUC_{Skin} value in skin treated with nystatin was 60.7% that of the control group.

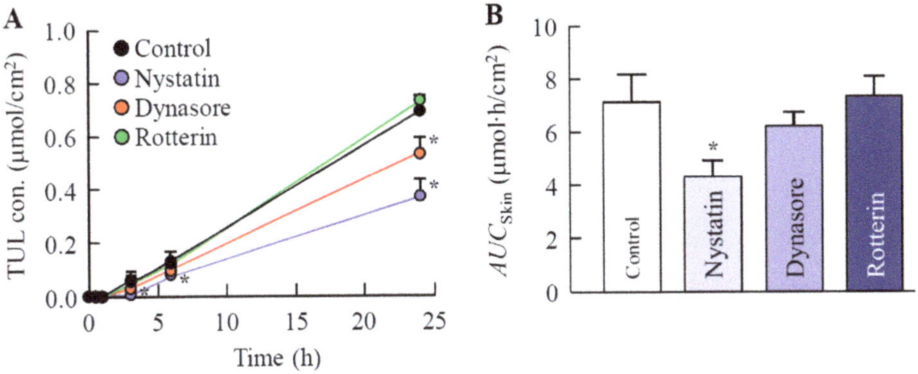

Figure 6. Effect of energy-dependent endocytosis on the transdermal penetration of TUL in the Men-WS/TUL-NP ointment. Changes in the TUL profiles (**A**) and AUC_{Skin} values (**B**) of the Men-WS/TUL-MP ointment after treatment with endocytosis inhibitors (nystatin, dynasore, and rottlerin). The composition of the Men-WS/TUL-NP ointment is presented in Table 1. Results are means ± SE; n = 6–8. * $p < 0.05$ vs. control for each category. The transdermal penetration of TUL-NPs was significantly inhibited by treatment with CavME inhibitor (nystatin).

4. Discussion

The objective of the present study was to evaluate the suitability of WS, AB, and AC bases in the preparation of transdermal formulations incorporating TUL-NPs. We found that the skin penetration of TUL in the Men-WS/TUL-NP ointment linearly increased with TUL content for 24 h, and the sustained penetration was higher than that of CA-TUL tape. In addition, we showed that the CavME pathway was related to the high level of skin penetration for the Men-WS/TUL-NP ointment.

First, we attempted to prepare TUL-NPs. Previous studies have shown that the selection of additives is important for effective milling and that MC enhances the milling efficiency in bead-milling treatment. In addition, HPβCD prevents the aggregation of drug NPs [16–21]. According to these reports, we prepared TUL-NPs using additives (MC and HPβCD) and bead-milling treatment and succeeded in providing dispersions containing TUL-NPs (Figure 1). Following that, we investigated which types of bases were suitable for the development of transdermal formulations incorporating TUL-NPs since the selection of an appropriate formulation base is essential for the ease of application of nanoparticles, as well as the enhancement of dermal and transdermal delivery [26]. Ointments are denoted as oil-based, emulsion-based, and water-soluble-based, and it is suggested that the emulsion-based or water-soluble-based ointments are suitable for maintaining drugs in ointments for solid NPs since hydrophobic drugs are dissolved by oil-based ointments. Ahmed et al. [27] reported that WS, AB, and AC bases provided the best release of a hydrophobic drug (mefenamic acid) among cream and ointment bases. In addition, our previous studies using Carbopol and MC gel bases have shown that drug release decreases with higher polymer concentrations [28,29]. In summary, we selected WS, AB, and AC bases to prepare transdermal formulations incorporating TUL-NPs in this study and determined the lowest concentration of each polymer needed to form gels (Table 1). In any of the bases, the TUL size particle in ointments incorporating TUL-NPs was on the nanoscale, i.e., 30 nm–200 nm (Table 2). In addition, the addition of *l*-menthol did not affect the drug solubility of the TUL ointments (Table 2).

Thereafter, we investigated the drug release and skin penetration of these ointments. The TUL release values from the WS base were higher than those of the AB and AC bases (Figure 3). Moreover, the release of TUL from ointments incorporating TUL-NPs tended to be higher than those incorporating TUL-MPs (Figure 3). *l*-menthol decreased TUL release in the creams (the AB and AC bases); however, the TUL release from hydrophilic ointments (the WA base) was increased by the addition of *l*-menthol. *l*-menthol is a hydrophobic drug; therefore, it was hypothesized that the affinity of creams and *l*-menthol would be higher than that of hydrophilic ointments and that *l*-menthol had a high affinity for TUL. A high affinity between *l*-menthol, TUL, and the bases (AB and AC bases) could cause a decrease in TUL release from the ointment. In addition, it was suggested that the enhanced TUL release from the WC base was due to the low affinity between *l*-menthol and the WC base in comparison with that of the AB and AC bases. Moreover, the viscosity of the WS base was significantly higher than that of the AC and AB bases, and the addition of *l*-menthol decreased the viscosity of the WS base incorporating TUL (Table 2). Our previous studies have shown that enhanced viscosity attenuates the drug release of NPs from ointments [28,29]. From these results, the decreased viscosity could also be related to the enhanced TUL release from the WC base as a result of the addition of *l*-menthol.

It is essential to understand the factors influencing the skin permeability of nanoparticles to provide safe and efficient therapeutic applications. A previous study reported that skin penetration was influenced by the physicochemical characteristics of nanocarriers, such as composition, size, shape, and surface chemistry, as well as skin features [30]. In this study, the skin permeability levels of ointments incorporating TUL were also similar to the drug release tendencies of corresponding ointments, although the skin penetration of the Men-WS/TUL-NP ointment was significantly higher in comparison with the other ointments (Figure 4). It is known that particles over 100 nm in size cannot penetrate skin tissue [18,31], although the addition of *l*-menthol alters the barrier properties of the stratum corneum and causes reversible disruption of the lipid domains since menthol preferentially distributes into the intercellular spaces of the stratum corneum. Thus, the combination of *l*-menthol increases the skin absorption of NPs [18,32,33]. In addition, the release amounts of TUL-NPs from the WS/TUL-NP ointments with and without *l*-menthol were higher than those of the AB and AC bases. In summary, the high drug release from the base and the decrease in barrier function of the stratum corneum resulting from *l*-menthol may cause the high skin permeability found for the Men-WS/TUL-NP ointment. Moreover, the release of NPs from bases incorporating TUL-NPs was observed in the drug release examination; however, TUL-NPs were not detected in the in vitro studies using rat skin treated with the Men-WS/TUL-NP ointment. These results suggest that the released TUL-NPs from the Men-WS/TUL-NP ointment were dissolved in the skin permeation process.

It is important to clarify the levels of absorption and sustained drug release compared with those of CA-TUL tape. We have previously reported that the percutaneous absorption of a transdermal formulation based on drug NPs is higher than that in transdermal formulations based on dissolved and MP drugs [16–18]. In contrast to the data from our previous studies, the transdermal penetration of the Men-WS/TUL-NP ointment was less than that of CA-TUL tape at the same content. CA-TUL tape is a patch preparation, and the differences between patches and ointments may relate to the transdermal penetration of TUL. On the other hand, the skin penetration of TUL in the Men-WS/TUL-NP ointment linearly increased with TUL content, and the sustained efficiency was higher than that of CA-TUL tape (Figure 5). In addition, the skin penetration of the Men-WS/TUL-NP ointment containing 1.5% TUL-NPs was significantly higher than that of CA-TUL tape at 24 h in the in vitro transdermal penetration test using rat skin (Figure 5). Considering rheological properties, drug release, and skin penetration, the WS base incorporating TUL-NPs and *l*-menthol (the Men-WS/TUL-NP ointment) was the best among the studied formulations.

Furthermore, we demonstrated the pathway for the transdermal absorption of TUL-NPs in the Men-WS/TUL-NP ointment. The endocytosis of nanomedicine has drawn tremendous interest in the last decade, and endocytosis is one of the functions for breaching tissue barriers, resulting in the efficient delivery of nanoparticles [34,35]. In particular, it has been reported that energy-dependent endocytosis is the major route by which nanomedicines are transported across membranes [36–39]. Energy-dependent endocytosis processes related to the uptake of NPs are classified as being CavME, CME, or MP, and the sizes of the vesicles vary with the specific pathway of the endocytic processes, i.e., the sizes for the CavME, CME, and MP pathways are <80 nm, <120 nm, and 100 nm–5 μm, respectively [40]. The particle size of the TUL-NPs in the Men-WS/TUL-NP ointment was 30 nm–200 nm (Table 2), and we showed that nystatin attenuated the skin penetration of TUL in the Men-WS/TUL-NP ointment (Figure 6). In summary, we hypothesized that the CavME pathway was the main skin-related pathway of the TUL-NPs in the Men-WS/TUL-NP ointment. In addition, the released TUL-NPs from the Men-WS/TUL-NP ointment could be dissolved in this process, since the TUL-NPs were not detected in the in vitro studies using rat skin treated with the Men-WS/TUL-NP ointment.

Further studies are needed to clarify the CavME pathway and to obtain more pharmacokinetic data on the in vivo percutaneous absorption of TUL-NPs in the Men-WS/TUL-NP ointment in animal models. In addition, it is important to elucidate the usefulness of the Men-WS/TUL-NP ointment. Therefore, we plan to investigate the therapeutic effects of Men-WS/TUL-NP ointments (0.5–1.5% TUL) on the long-term management of asthma and COPD using an animal model.

5. Conclusions

We produced transdermal formulations based on WS, AB, and AC bases incorporating TUL-NPs and evaluated which bases were suitable for preparing transdermal formulations incorporating TUL-NPs. We found that the WS base incorporating TUL-NPs and *l*-menthol (the Men-WS/TUL-NP ointment) was the best among those assessed in this study. In addition, we showed that the penetration of TUL from the Men-WS/TUL-NP ointment sustained zero-order release over 24 h, and its skin permeability increased with TUL content in the Men-WS/TUL-NP ointment. Furthermore, the CavME pathway was related to the skin penetration of TUL-NPs in the Men-WS/TUL-NP ointment (Figure 7). These findings may be utilized to improve the transdermal delivery of TUL in future studies.

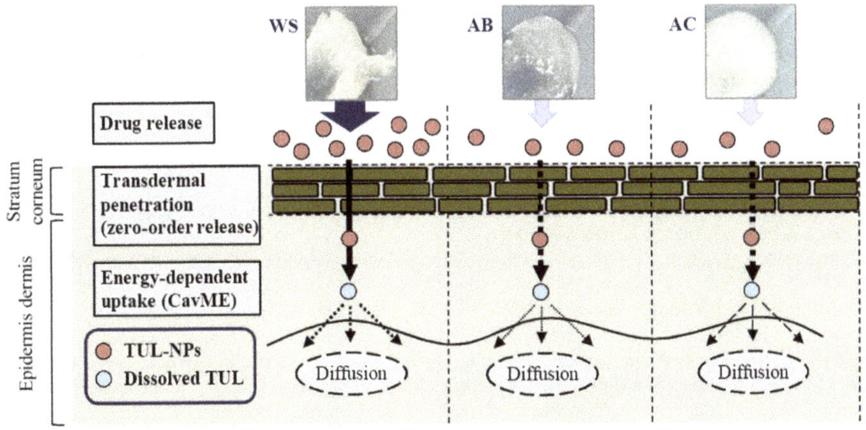

Figure 7. Suggested mechanism for the transdermal penetration process of TUL from ointment bases incorporating TUL-NPs and *l*-menthol.

Author Contributions: Conceptualization, N.N.; data curation, F.O. and S.D. (Saori Deguchi), A.F., S.D. (Saki Daimyo) and H.O.; formal analysis, F.O., S.D. (Saori Deguchi), A.F., S.D. (Saki Daimyo), H.O. and N.K.; funding acquisition, N.N.; investigation, F.O., S.D. (Saori Deguchi) and N.K.; methodology, N.N., F.O. and N.K.; supervision, N.N.; visualization, N.N.; writing—original draft, N.N.; writing—review and editing, N.N. All authors have read and agreed to the published version of the manuscript.

Funding: We gratefully acknowledge the funding received from the Ministry of Education, Culture, Sports, Science, and Technology of Japan (Grant No. 21K06656).

Institutional Review Board Statement: This study was conducted according to the guidelines of the Japanese Pharmacological Society and Kindai University and approved by Kindai University (KAPS-31-001, 1 April 2019).

Informed Consent Statement: Not applicable.

Data Availability Statement: Not applicable.

Conflicts of Interest: The authors declare no conflict of interest. The funders had no role in the design of the study; in the collection, analyses, or interpretation of data; in the writing of the manuscript; or in the decision to publish the results.

References

1. Park, S.I.; Kim, B.H. Bioequivalence assessment of tulobuterol transdermal delivery system in healthy subjects. *Int. J. Clin. Pharmacol. Ther.* **2018**, *56*, 381–386. [CrossRef] [PubMed]
2. Lee, H.; Song, C.; Baik, S.; Kim, D.; Hyeon, T.; Kim, D.H. Device-assisted transdermal drug delivery. *Adv. Drug Deliv. Rev.* **2018**, *127*, 35–45. [CrossRef]
3. Gavin, P.D.; El-Tamimy, M.; Keah, H.H.; Boyd, B.J. Tocopheryl phosphate mixture (TPM) as a novel lipid-based transdermal drug delivery carrier: Formulation and evaluation. *Drug Deliv. Transl. Res.* **2017**, *7*, 53–65. [CrossRef] [PubMed]
4. Alkilani, A.Z.; McCrudden, M.T.C.; Donnelly, R.F. Transdermal Drug Delivery: Innovative Pharmaceutical Developments Based on Disruption of the Barrier Properties of the stratum corneum. *Pharmaceutics* **2015**, *7*, 438–470. [CrossRef] [PubMed]
5. Mofidfar, M.; O'Farrell, L.; Prausnitz, M.R. Pharmaceutical jewelry: Earring patch for transdermal delivery of contraceptive hormone. *J. Control. Release* **2019**, *301*, 140–145. [CrossRef]
6. Marwah, H.; Garg, T.; Goyal, A.K.; Rath, G. Permeation enhancer strategies in transdermal drug delivery. *Drug Deliv.* **2016**, *23*, 564–578. [CrossRef]
7. Horiguchi, T.; Kondo, R.; Miyazaki, J.; Fukumokto, K.; Torigoe, H. Clinical evaluation of a transdermal therapeutic system of the beta2-agonist tulobuterol in patients with mild or moderate persistent bronchial asthma. *Arzneimittelforschung* **2004**, *54*, 280–285. [CrossRef]
8. Katsunuma, T.; Fujisawa, T.; Nagao, M.; Akasawa, A.; Nomura, I.; Yamaoka, A.; Kondo, H.; Masuda, K.; Yamaguchi, K.; Terada, A.; et al. Effects of transdermal tulobuterol in pediatric asthma patients on long-term leukotriene receptor antagonist therapy: Results of a randomized, open-label, multicenter clinical trial in japanease children aged 4–12 years. *Allergol. Int.* **2013**, *62*, 37–43. [CrossRef]
9. Murakami, Y.; Sekijima, H.; Fujisawa, Y.; Ooi, K. Adjustment of Conditions for Combining Oxybutynin Transdermal Patch with Heparinoid Cream in Mice by Analyzing Blood Concentrations of Oxybutynin Hydrochloride. *Biol. Pharm. Bull.* **2019**, *42*, 586–593. [CrossRef]
10. McConville, J. Special Focus Issue: Transdermal, Topical and Folicular Drug Delivery Systems. *Drug Dev. Ind. Pharm.* **2016**, *42*, 845. [CrossRef]
11. Palmer, B.C.; DeLouise, L.A. Nanoparticle-Enabled Transdermal Drug Delivery Systems for Enhanced Dose Control and Tissue Targeting. *Molecules* **2016**, *21*, 1719. [CrossRef] [PubMed]
12. Pegoraro, C.; MacNeil, S.; Battaglia, G. Transdermal drug delivery: From micro to nano. *Nanoscale* **2012**, *4*, 1881–1894. [CrossRef] [PubMed]
13. Williams, A.C.; Barry, B.W. Penetration enhancers. *Adv. Drug Deliv. Rev.* **2004**, *56*, 603–618. [CrossRef]
14. Soma, D.; Attari, Z.; Reddy, M.S.; Damodaram, A.; Koteshwara, K.B.G. Solid lipid nanoparticles of irbesartan: Preparation, characterization, optimization and pharmacokinetic studies. *Braz. J. Pharm. Sci.* **2017**, *53*, e15012. [CrossRef]
15. Montenegro, L.; Lai, F.; Offera, A.; Sarpietro, M.G.; Micicchè, L.; Maccioni, A.M.; Valenti, D.; Fadda, A.M. From nanoemulsions to nanostructured lipid carriers: A relevant development in dermal delivery of drugs and cosmetics. *J. Drug Deliv. Sci. Technol.* **2016**, *32*, 100–112. [CrossRef]
16. Otake, H.; Yamaguchi, M.; Ogata, F.; Deguchi, S.; Yamamoto, N.; Sasaki, H.; Kawasaki, N.; Nagai, N. Energy-Dependent Endocytosis Is Responsible for Skin Penetration of Formulations Based on a Combination of Indomethacin Nanoparticles and l-Menthol in Rat and Göttingen Minipig. *Int. J. Mol. Sci.* **2021**, *22*, 5137. [CrossRef] [PubMed]
17. Nagai, N.; Ogata, F.; Yamaguchi, M.; Fukuoka, Y.; Otake, H.; Nakazawa, Y.; Kawasaki, N. Combination with l-Menthol Enhances Transdermal Penetration of Indomethacin Solid Nanoparticles. *Int. J. Mol. Sci.* **2019**, *20*, 3644. [CrossRef]
18. Nagai, N.; Ogata, F.; Otake, H.; Nakazawa, Y.; Kawasaki, N. Design of a transdermal formulation containing raloxifene nanoparticles for osteoporosis treatment. *Int. J. Nanomedicine* **2018**, *13*, 5215–5229. [CrossRef]
19. Otake, H.; Goto, R.; Ogata, F.; Isaka, T.; Kawasaki, N.; Kobayakawa, S.; Matsunaga, T.; Nagai, N. Fixed-Combination Eye Drops Based on Fluorometholone Nanoparticles and Bromfenac/Levofloxacin Solution Improve Drug Corneal Penetration. *Int. J. Nanomedicine* **2021**, *16*, 5343–5356. [CrossRef]
20. Nagai, N.; Ogata, F.; Otake, H.; Nakazawa, Y.; Kawasaki, N. Energy-dependent endocytosis is responsible for drug transcorneal penetration following the instillation of ophthalmic formulations containing indomethacin nanoparticles. *Int. J. Nanomedicine* **2019**, *14*, 1213–1227. [CrossRef]

21. Nagai, N.; Ito, Y.; Okamoto, N.; Shimomura, Y. A nanoparticle formulation reduces the corneal toxicity of indomethacin eye drops and enhances its corneal permeability. *Toxicology* **2014**, *319*, 53–62. [CrossRef] [PubMed]
22. Nagai, N.; Iwamae, A.; Tanimoto, S.; Yoshioka, C.; Ito, Y. Pharmacokinetics and Antiinflammatory Effect of a Novel Gel System Containing Ketoprofen Solid Nanoparticles. *Biol. Pharm. Bull.* **2015**, *38*, 1918–1924. [CrossRef] [PubMed]
23. Mäger, I.; Langel, K.; Lehto, T.; Eiríksdóttir, E.; Langel, U. The role of endocytosis on the uptake kinetics of luciferin-conjugated cell-penetrating peptides. *Biochim. Biophys. Acta* **2012**, *1818*, 502–511. [CrossRef]
24. Malomouzh, A.I.; Mukhitov, A.R.; Proskurina, S.E.; Vyskocil, F.; Nikolsky, E.E. The effect of dynasore, a blocker of dynamin-dependent endocytosis, on spontaneous quantal and non-quantal release of acetylcholine in murine neuromuscular junctions. *Dokl. Biol. Sci.* **2014**, *459*, 330–333. [CrossRef]
25. Kang, Y.L.; Oh, C.; Ahn, S.H.; Choi, J.C.; Choi, H.Y.; Lee, S.W.; Choi, I.S.; Song, C.S.; Lee, J.B.; Park, S.Y. Inhibition of endocytosis of porcine reproductive and respiratory syndrome virus by rottlerin and its potential prophylactic administration in piglets. *Antivir. Res.* **2021**, *195*, 105191. [CrossRef]
26. Liu, M.; Wen, J.; Sharma, M. Solid Lipid Nanoparticles for Topical Drug Delivery: Mechanisms, Dosage Form Perspectives, and Translational Status. *Curr. Pharm. Des.* **2020**, *26*, 3203–3217. [CrossRef] [PubMed]
27. Ahmed, T.A.; Ibrahim, H.M.; Ibrahim, F.; Samy, A.M.; Fetoh, E.; Nutan, M.T. In vitro release, rheological, and stability studies of mefenamic acid coprecipitates in topical formulations. *Pharm. Dev. Technol.* **2011**, *16*, 497–510. [CrossRef] [PubMed]
28. Minami, M.; Otake, H.; Nakazawa, Y.; Okamoto, N.; Yamamoto, N.; Sasaki, H.; Nagai, N. Balance of Drug Residence and Diffusion in Lacrimal Fluid Determine Ocular Bioavailability in In Situ Gels Incorporating Tranilast Nanoparticles. *Pharmaceutics* **2021**, *13*, 1425. [CrossRef]
29. Nagai, N.; Isaka, T.; Deguchi, S.; Minami, M.; Yamaguchi, M.; Otake, H.; Okamoto, N.; Nakazawa, Y. In Situ Gelling Systems Using Pluronic F127 Enhance Corneal Permeability of Indomethacin Nanocrystals. *Int. J. Mol. Sci.* **2020**, *21*, 7083. [CrossRef]
30. Farjami, A.; Salatin, S.; Jafari, S.; Mahmoudian, M.; Jelvehgari, M. The Factors Determining the Skin Penetration and Cellular Uptake of Nanocarriers: New Hope for Clinical Development. *Curr. Pharm. Des.* **2021**, *27*, 4315–4329. [CrossRef]
31. Nagai, N.; Ito, Y. Therapeutic effects of gel ointments containing tranilast nanoparticles on paw edema in adjuvant-induced arthritis rats. *Biol. Pharm. Bull.* **2014**, *37*, 96–104. [CrossRef] [PubMed]
32. Kaplun-Frischoff, Y.; Touitou, E. Testosterone skin permeation enhancement by menthol through formation of eutectic with drug and interaction with skin lipids. *J. Pharm. Sci.* **1997**, *86*, 1394–1399. [CrossRef] [PubMed]
33. Kunta, J.R.; Goskonda, V.R.; Brotherton, H.O.; Khan, M.A.; Reddy, I.K. Effect of menthol and related terpenes on the percutaneous absorption of propranolol across excised hairless mouse skin. *J. Pharm. Sci.* **1997**, *86*, 1369–1373. [CrossRef] [PubMed]
34. Patel, S.; Kim, J.; Herrera, M.; Mukherjee, A.; Kabanov, A.V.; Sahay, G. Brief update on endocytosis of nanomedicines. *Adv. Drug Deliv. Rev.* **2019**, *144*, 90–111. [CrossRef] [PubMed]
35. Gimeno-Benito, I.; Giusti, A.; Dekkers, S.; Haase, A.; Janer, G. A review to support the derivation of a worst-case dermal penetration value for nanoparticles. *Regul. Toxicol. Pharmacol.* **2021**, *119*, 104836. [CrossRef] [PubMed]
36. Aderem, A.; Underhill, D.M. Mechanisms of phagocytosis in macrophages. *Annu. Rev. Immunol.* **1999**, *17*, 593–623. [CrossRef] [PubMed]
37. Kou, L.; Sun, J.; Zhai, Y.; He, Z. The endocytosis and intracellular fate of nanomedicines: Implication for rational design. *Asian J. Pharm. Sci.* **2013**, *8*, 1–10. [CrossRef]
38. Rappoport, J. Focusing on clathrin-mediated endocytosis. *Biochem. J.* **2008**, *412*, 415–423. [CrossRef]
39. Wang, J.; Byrne, J.D.; Napier, M.E.; DeSimone, J.M. More effective nanomedicines through particle design. *Small* **2011**, *7*, 1919–1931. [CrossRef]
40. Zhang, S.; Li, J.; Lykotrafitis, G.; Bao, G.; Suresh, S. Size-Dependent Endocytosis of Nanoparticles. *Adv. Mater.* **2008**, *21*, 419–424. [CrossRef]

pharmaceutics

MDPI

Article

Novel Pullulan/Gellan Gum Bilayer Film as a Vehicle for Silibinin-Loaded Nanocapsules in the Topical Treatment of Atopic Dermatitis

Mailine Gehrcke [1], Carolina Cristóvão Martins [2], Taíne de Bastos Brum [1], Lucas Saldanha da Rosa [3], Cristiane Luchese [2], Ethel Antunes Wilhelm [2], Fabio Zovico Maxnuck Soares [3] and Letícia Cruz [1,*]

[1] Laboratório de Tecnologia Farmacêutica, Programa de Pós-Graduação em Ciências Farmacêuticas, Centro de Ciências da Saúde, Universidade Federal de Santa Maria, Santa Maria 97105-900, RS, Brazil
[2] Laboratório de Pesquisa em Farmacologia Bioquímica—Centro de Ciências Químicas, Farmacêuticas e de Alimentos, Universidade Federal de Pelotas, Pelotas 96010-900, RS, Brazil
[3] Laboratório de Biomateriais, Centro de Ciências da Saúde, Departamento de Odontologia Restauradora, Universidade Federal de Santa Maria, Santa Maria 97015-372, RS, Brazil
* Correspondence: leticia.cruz@ufsm.br; Tel.: +55-55-3220-9373

Abstract: In this study a novel gellan gum/pullulan bilayer film containing silibinin-loaded nanocapsules was developed for topical treatment of atopic dermatitis (AD). The bilayer films were produced by applying a pullulan layer on a gellan gum layer incorporated with silibinin nanocapsules by two-step solvent casting method. The bilayer formation was confirmed by microscopic analysis. In vitro studies showed that pullulan imparts bioadhesitvity for the films and the presence of nanocapsules increased their occlusion factor almost 2-fold. Besides, the nano-based film presented a slow silibinin release and high affinity for cutaneous tissue. Moreover, this film presented high scavenger capacity and non-hemolytic property. In the in vivo study, interestingly, the treatments with vehicle film attenuated the scratching behavior and the ear edema in mice induced by 2,4-dinitrochlorobenzene (DNCB). However, the nano-based film containing silibinin modulated the inflammatory and oxidative parameters in a similar or more pronounced way than silibinin solution and vehicle film, as well as than hydrocortisone, a classical treatment of AD. In conclusion, these data suggest that itself gellan gum/pullulan bilayer film might attenuate the effects induced by DNCB, acting together with silibinin-loaded nanocapsules, which protected the skin from oxidative damage, improving the therapeutic effect in this AD-model.

Keywords: nanocapsules; films; silibinin; pullulan; gellan gum; atopic dermatitis

Citation: Gehrcke, M.; Martins, C.C.; de Bastos Brum, T.; da Rosa, L.S.; Luchese, C.; Wilhelm, E.A.; Soares, F.Z.M.; Cruz, L. Novel Pullulan/Gellan Gum Bilayer Film as a Vehicle for Silibinin-Loaded Nanocapsules in the Topical Treatment of Atopic Dermatitis. *Pharmaceutics* **2022**, *14*, 2352. https://doi.org/10.3390/pharmaceutics14112352

Academic Editors: Daniela Monti and Silvia Tampucci

Received: 10 October 2022
Accepted: 27 October 2022
Published: 31 October 2022

1. Introduction

Atopic dermatitis (AD), a chronic inflammatory skin condition, results from a complex interaction among the genetic predisposition, the environmental factors, as well as the dysfunctions in the skin barrier and in the permeability of allergens and pathogens agents [1]. These pathological mechanisms could play synergistically to maintain the clinical symptoms of AD, including pruritus, eczematous lesions, remodeling of the skin surface and generalized skin dryness due to the constant inflammation [2]. The current treatments for AD are based on skin barrier recovery and the use of corticosteroids and topical and systemic immunosuppressant, which are used as unique or combined therapy. However, these pharmacological treatments present limited effectiveness and relevant side effects [3].

In this context, the attempts to planning and developing a new, safe, and effective therapeutic strategy for AD have been receiving notable priority. Several studies have been demonstrated the safety and the effectiveness of naturally occurring substances in the AD treatment and thereby, their arise as an interesting option to overcome conventional issues of the available pharmacological treatment [4]. Silibinin (SB), the main biologically active flavonoid extracted from *Silybum marianum* seeds, is well recognized for its potent antioxidant and anti-inflammatory properties [5]. This flavonoid has been widely studied to prevent and to treat inflammatory skin disorders, including dermatoses [6–8]. However, SB is poor soluble in water and its solubilization in others solvents is limited [9]. Few studies have explored the development of formulations that allow the application of this flavonoid on the skin. The existent studies involve the use of nanotechnology to improve the SB solubility and performance against skin diseases [7,8,10].

Moreover, drug delivery systems based on nanotechnology have shown a promising strategy for AD management due to the nanostructures' properties to achieve enhanced skin penetration and retention, release control, and reduced side effects [11–13]. Among these nanostructured systems stand out the nanocapsules (NC) which have a core/shell structure constituted by oil and polymer, respectively [14]. This structural organization favors the encapsulation of lipophilic substances into NC, increasing the solubility and therapeutic efficacy of these substances [7,10,15].

In this same way, polymeric films present many advantages to treat inflammatory skin diseases, such as AD, due to their potential to adhere to the skin providing protection and hydration for this tissue [16–18]. For the films development aiming cutaneous use, natural polysaccharides have shown to be promising candidates due to low toxicity, biocompatibility and biodegradability [19]. Among such polysaccharides is gellan gum, which has been shown to be suitable for the formation of ultrathin films with physical and mechanical resistance and with swelling capacity, being promising for protecting the area and absorbing exudate of lesions [20,21]. Another natural polysaccharide that has attracted the interest of researchers is pullulan, which forms elastic, bioadhesive and fast-dissolving films [18,22]. In addition, it was suggested by Jeong and co-workers that pullulan may present a physical action against AD [16].

Studies have reported the bilayer films as an advantageous alternative to polymeric blends for cutaneous delivery. These films are able to preserve the intrinsic characteristics of each film-forming agent in it respective polymeric layer, improving the film properties [23,24]. Besides, combining nanostructures into polymeric films for cutaneous delivery renders a dosage form that can be administered over the affected skin, promoting a physical barrier for injured skin, and allowing a lower frequency of administration, which improves the therapeutic efficacy [25,26].

Also, NC have shown a promising alternative to incorporate lipophilic substances into hydrophilic films [18]. In a previous study, we demonstrate the promising advantages of SB encapsulation into NCs followed by their incorporation into gellan gum films for cutaneous delivery. The developed films presented physical and mechanical resistance and fluid absorption capacity. Besides, due to the high SB lipophilicity, the nanoencapsulation was fundamental to guarantee it dosage homogeneity into gellan gum film. Also, the NC association into the gellan gum film provided increased SB stability, as well as controlled release and improved skin permeation to the dermis [27]. These results plus to the attractive pullulan films characteristics stimulated our curiosity about combining them in a novel bilayer film to treat AD.

The objective of the present study was to develop a bilayer film composed by gellan gum layer incorporated with SB-loaded NC as the bottom side and with a pullulan layer as the top side, as well as to evaluate the film potential against AD-like skin lesions. This novel nano-based bilayer film was engineered such that the distinct layers provide their respective beneficial features for AD treatment.

2. Materials and Methods

2.1. Materials

SB (98% purity), Span® 80 (sorbitan monooleate) were obtained from Sigma Aldrich (São Paulo, SP, Brazil). Ethylcelullose was donated by Colorcon (Cotia, SP, Brazil). Medium chain triglycerides and Tween® 80 (polysorbate 80) were bought from Delaware (Porto

Alegre, RS, Brazil). Gellan gum (Kelcogel®) and pullulan were donated by CP Kelco (Limeira, RS, Brazil) and Hayashibara, respectively. 2,4-dinitrochlorobenzene (DNCB) was purchased from Sigma (St. Louis, MO, USA) and it was used as an inductor of atopic dermatitis (AD)-like lesions. The hydrocortisone ointment (HC) was obtained commercially and it was used as a reference drug. All other solvents and reagents were analytical grade and used as received.

2.2. Methods

2.2.1. Preparation of Nanocapsule Suspensions and Bilayer Films

SB-loaded NC were prepared by the interfacial deposition of preformed polymer method [28], at a concentration of 2.5 mg/mL as described previously [27]. The organic phase was composed by SB (0.025 g), ethylcellulose (0.25 g), Span® 80 (0.1925 g), medium chain triglycerides (0.75 g) and acetone (68 mL). The aqueous phase was composed of water (132 mL) and Tween® 80 (0.1925 g). Then, the organic phase was added under magnetic stirring into aqueous phase followed by solvent evaporation under reduced pressure to achieve a 10 mL final volume. Unloaded NC suspensions were prepared using this same protocol, omitting SB. After preparation, NC suspensions were characterized in terms of particle size and polydispersity index by photon correlation spectroscopy (1:500 dilution in ultrapure water), and zeta potential by the microelectrophoresis technique (1:500 dilution in 10 mM NaCl). These analyzes were conducted using the Zetasizer Nanoseries® equipment (Malvern Instruments, Malvern, Worcestershire, UK). Besides, the total SB content in the suspension was assessed by its extraction from the nanocapsules using methanol and sonication (5 min), followed by high performance liquid chromatography analysis.

The bilayer films were produced by two-step solvent casting method. For this, firstly, the gellan gum dispersion was prepared by dispersing 0.25 g of this gum in 15 mL distilled water while heating at 80 °C, under magnetic stirring for 2 h. After, an amount of glycerol (1 g) was added to this dispersion. Subsequently, the mixture was removed from the heating and 10 mL of water or NC suspension were added to gellan gum dispersion to produce the first layer of the vehicle film or nano-based films, respectively. After, this mixture was instantly poured into a Petri dish (90 × 13 mm) and was partially dried at 40 °C for 15 h. Then, a water solution containing 3% (*w*/*v*) of pullulan and 0.5% (*w*/*v*) of glycerol was prepared at room temperature and under magnetic stirring for 30 min. After complete pullulan solubilization, this solution was poured on the surface of the first layer and dried at 40 °C for 24 h. The bilayer films were named BF NC SB, BF NC B and BF vehicle for films produced with nanoencapsulated SB, unloaded NC and control film, respectively.

2.2.2. Scanning Electron Microscopy

The structure of the bilayer films was evaluated by scanning electron microscopy (SEM) (JEOL JSM 6360, Akishima, Japan). To visualize the layers, the films were cryofractured in order to analyze the sides sections after fracture. To carry out the analysis, the samples were previously placed on a double–sided adhesive carbon tape, mounted on the sample slab and coated with gold (Denton Vaccum II, 100 Å, Moorestown, EUA) under reduced pressure. The samples were subsequently analyzed using an accelerating voltage of 10 kV. This analysis was also performed for a monolayer vehicle film (MF vehicle) which was produced containing only the layer of gellan gum and served as a control.

2.2.3. Bilayer Films Characterization

The bilayer films were characterized by homogeneity of SB content, thickness, moisture, nanometric size maintenance and swelling index. For thickness measurement, films (n = 3) were prepared for each formulation and then five measurements were performed on each film (four in the corner and one in the middle). Mean thickness values were calculated and expressed in μm. For homogeneity of SB content, the films (n = 3) were cut into three fragments of 1 cm × 1 cm each. The SB content in each fragment was quantified by extracting the phytochemical in methanol, subjecting it to stirring for 20 min followed by sonication for another 20 min. Samples were filtered (0.45 μm) and analyzed by high performance liquid chromatography (HPLC), using a guard column and a Kinetex C_{18} Phenomenex column (250 mm × 4.60 mm, 5 μm; 110 Å) at room temperature. The mobile phase consisted of water pH 3.5 and acetonitrile (60:40, *v*/*v*) at isocratic flow rate (1.0 mL/min) and the detection wavelength used for SB was 288 nm, as described previously [27]. The mean values of the SB content were calculated and expressed in $\mu g/cm^2$, and the content (%) was calculated in relation to the theoretical amount of SB present in the film.

The particle size and the polydispersity index of the NC after their incorporation into films were evaluated by photon correlation spectroscopy (PCS) (ZetaSizer, Malvern, Worcestershire, UK). For this, film fragments (0.1 g) were dispersed in 50 mL of ultrapure water (500× dilution). The nanostructures were extracted from the films under magnetic stirring for 2 h before analysis. For moisture assessment, the films were cut into 2 cm × 2 cm fragments and later placed in an oven at 60 °C [26]. These fragments were weighed after regular time intervals until the weight became constant. The residual water content in the films was determined following Equation (1).

$$\text{Moisture content} = [(\text{Wd} - \text{Wi})/\text{Wi}] \times 100 \quad (1)$$

where: Wd is the weight of the films after drying and Wi is the initial weight of the films.

To evaluate the swelling index, the films were cut into 2 cm × 2 cm pieces and weighed (Wd). Then, these fragments were placed in beakers containing 50 mL of pH 7.4 phosphate buffer at 37 °C for 24 h [26]. Afterwards, the films were removed from contact with the buffer and dried with absorbent paper, and the hydrated fragment was weighed (Ws). The swelling index was calculated following Equation (2).

$$\text{Swelling index} = [(\text{Ws} - \text{Wd})/\text{Wd}] \times 100 \quad (2)$$

where: Ws is the weight of the film after swelling and Wd is the weight of the dried film.

2.2.4. Folding Endurance and Mechanical Properties

Folding endurance was determined by repeatedly folding the films in the same place up to 300 times (n = 3). Then, the films were evaluated for groove formation or breakage. The mechanical properties in terms of tensile strength, deformation and Young's modulus was determined using the universal testing machine (Emic, São José dos Pinhais, Brazil), according to ASTM-D882-02 standards [29]. For this, film samples measuring 60 mm × 45 mm and with about 40 μm thick were individually fixed on the machine probe and a tensile load was applied at an initial separation of 4 cm and 50 mm/min of the cross-head speed. The maximum deformation suffered by the film was determined by the percentage change in the length of the sample in relation to its original size. The tensile strength was determined by the ratio of the force needed to rupture the film and the cross-sectional area of the strip, whereas the young's modulus was calculated by the ratio of stress and strain values.

2.2.5. In vitro Studies

Occlusion Test

The in vitro occlusion test was carried out according to our previous protocol [27]. For this, a 100 mL capacity beaker containing 50 mL of water was sealed with a cellulose acetate filter (90 mm, Sterlitech, Auburn, USA), which was subsequently covered with the bilayer films (n = 3). The films were applied with the pullulan layer in contact with the filter paper. Then, the beakers were stored at 32 °C and at predetermined times (0, 6, 24 and 48 h) they were weighed for the water loss determination. A film-free beaker was used as a negative control and the occlusion factor was calculated according to Equation (3).

$$\text{Occlusion factor} = [(\text{A} - \text{B})/\text{A}] \times 100 \quad (3)$$

where: A is the water loss of the negative control and B is the water loss in film presence.

Skin Preparation

Human skin fragments were obtained from healthy female patients undergoing abdominoplasty surgery. Subcutaneous fat was removed and the skin was stored at −4 °C (freezer) until use. Two different skin conditions were obtained. The first condition was the whole skin (intact cutaneous tissue), with the presence of *stratum corneum*, epidermis and dermis. The second condition was skin with an impaired barrier in which the *stratum corneum* was removed by a successive tape-stripping procedure [30] using 18 pieces of adhesive tape. The protocol was approved by the research committee with humans from the Federal University of Santa Maria—RS without identifying data (CAEE: 27168719.4.0000.5346).

Bioadhesive Strength

To carry out the experiment, an adapted apparatus was used composed of two balanced arms [31]. A plastic frame was connected to one of these arms under which the films were fixed. The skin (intact or injured) was fixed on a glass plate under the frame. The contact between the films and the skin fragment occurred by applying a weight of 1 N for 60 s. Afterwards, water was added at a constant speed of 1 drop/s in an opposite side plastic tube until the separation between skin and film occurred.

All analyzes were performed in triplicate and the volume of water used was measured in a graduated cylinder. Both the top layer (gellan gum layer) and the bottom layer (pullulan layer) of the bilayer films were analyzed. The bioadhesive strength was calculated using Equation (4) and the result was expressed in dyne/cm^2.

$$\text{Bioadhesive strength (dyne/cm}^2\text{)} = (V \times G)/A \quad (4)$$

where: V = amount of water (g) required for the detachment between the sample and the tissue; G = acceleration of gravity (980 cm/s^2); A = area of exposed tissue (cm^2).

SB Release and Skin Permeation/Penetration Study

The in vitro SB release and skin permeation/penetration from films was conducted through vertical Franz diffusion cells with diffusion area of 3.14 cm^2. The receptor medium used in the assays was pH 7.4 phosphate buffer at 32 ± 0.5 °C. The films (1 cm × 1 cm) corresponding to 440 µg of SB were placed on a dialysis membrane (10,000 Da, Sigma Aldrich, São Paulo, SP, Brazil) or on the skin surface with the pullulan layer in contact with the donator medium.

For the release study, at predetermined periods (1, 2, 3, 4, 5, 6, 7, 8, 12 and 24 h), a volume of 300 µL of the receptor medium was removed and replaced by the same volume of medium fresh. The amount of SB released was determined using HPLC method described in the Section 2.2.3. For the skin permeation study, skin samples were obtained and treated as described in the Section "Skin Preparation". This experiment was performed using intact (separating the skin layers only at the end of 24 h) and injured skin (without *stratum corneum*). The circular skin fragments were placed between the donor and recipient compartments with the dermis in contact with the recipient medium. After 24 h, the films were gently removed from the skin surface, the skin was carefully removed and the receptor medium was collected. For intact skin, the *stratum corneum* was removed using 18 pieces of adhesive tape. For both intact and injured skin, the epidermis was separated from the dermis by heating the skin in ultrapure water at 60 °C for 45 s, followed by removing the epidermis with a spatula. The strips containing the *stratum corneum*, and the epidermis and dermis fragments were placed in different test tubes containing methanol and vortexed for 2 min followed by an ultrasound bath for another 30 min. The SB content in the different skin layers and in the receptor medium was determined by HPLC (Section 2.2.3).

Free Radical Scavenging Activity

The antioxidant effect of the film containing nanoencapsulated SB was evaluated through the ability to scavenge the synthetic radical 2,2′-azinobis (3-ethylbenzothiazoline-6-sulfonic acid) ($ABTS^+$), as previously described by Yang et al. [32], with some modifications. First, an $ABTS^+$ solution was prepared by reacting the ABTS stock solution (7 mM) with sodium persulfate (140 mM), 12 h before the assay (final $ABTS^+$ concentration 42.7 Mm). Bilayer films were cut into 0.5 cm × 0.5 cm pieces and added to tubes containing 1 mL of $ABTS^+$ solution. The tubes were mixed by inversion and incubated at room temperature for 30 min. An $ABTS^+$ solution was kept under the same reaction conditions and was used as a negative control. Blank samples containing the film fragments and water were also prepared. After incubation, the films were removed and the absorbance of solution was measured at 734 nm (UV-1800 Spectrophotometer, Shimadzu, Kyoto, Japan). The experiment was carried out in triplicate. Percentage radical scavenging capacity was calculated using the Equation (5).

$$SC\% = 100\ ((AbsA - AbsB) \times 100)/AbsC \quad (5)$$

where: SC%: Scavenging capacity in percentage; AbsA: sample absorbance; AbsB: blank absorbance; AbsC: negative control absorbance.

Hemocompatibility Evaluation of Formulations

The films hemocompatibility was evaluated by direct contact test according to Standard Practice for Assessment of Hemolytic Properties of Materials [33], with some modifications. For this purpose, anti-coagulated blood (9 parts of blood to 1 part of citrate) was collected from a healthy human volunteer. Then, 2 mL of the anticoagulated blood was centrifuged at 2000 rpm for 5 min, followed by discarding the plasma. The resulting pellet was washed with saline solution 3 times to completely remove plasma and obtain only erythrocytes. Afterwards, the erythrocytes were resuspended in saline and at a final concentration of 10% (*v*/*v*). Film fragments measuring 0.5 cm × 0.5 cm were inserted into microtubes containing 700 µL of saline solution and allowed to equilibrate for about 1 h. Afterwards, 100 µL of resuspended erythrocytes were added to the tubes. Positive and negative hemolysis controls were prepared with water and saline, respectively. In addition, blanks were prepared containing the film fragments and water. The tubes were then incubated for 1 h at 37 °C. After incubation, all samples were centrifuged at 2000 rpm for 5 min and the absorbance of the supernatant was measured spectrophotometrically at 540 nm (UV-1800 Spectrophotometer, Shimadzu, Kyoto, Japan). The percentage of hemolysis was calculated according to equation 6. This protocol was approved by research committee with humans from the Federal University of Santa Maria—RS (CAEE: 27168719.4.0000.5346).

$$\%\ of\ hemolysis = (AbA - AbB/AbC) \times 100 \quad (6)$$

where: AbA: sample absorbance; AbB: blank absorbance; AbC: positive control absorbance.

2.2.6. In Vivo Study

Animals

Female BALB-c mice (6–8 weeks old) were housed in a separate animal room at controlled temperature (22 ± 2 °C), under a 12/12-h light/dark cycle (the lights were turned on at 07:00 AM), with free access to standard diet and water. The experimental study was conducted according to the Ethical Research Committee of the Federal University of Pelotas, Rio Grande do Sul, Brazil and registered under the number CEEA 23357-2018/40-2019. The number of animals used was the minimum necessary to evaluate the consistent effects of the treatments and every effort was made to minimize their suffering.

Experimental Design

The experimental design of this study is illustrated in supplementary material (Figure S1). The allergen sensitization and challenge induced by DNCB lead to the development of skin lesions similar to those of AD, as previously described by Chan et al. [34]. The dorsal skin of each mouse was shaved to remove all hair from the area. In the sensitization phase, it was applied 200 μL of 0.5 *v*/*v*-% DNCB dissolved in acetone/olive oil (3:1 ratio) on the shaved area in the first three days of the experimental protocol. These animals were also challenged by applying 20 μL and 200 μL of 1.0 *v*/*v*-% DNCB on the right ear and the dorsal skin, respectively, on days 14, 17, 20, 23, 26 and 29 of the experimental protocol.

In order to evaluate the effects of free SB or bilayer films treatments on the AD-like skin lesions, mice were randomly divided into seven experimental groups ($n = 7$ animals/group): normal control mice were exposed to the vehicle containing acetone/olive oil (3:1) and AD-induced mice were sensitized and challenged with DNCB. All other experimental groups were sensitized and challenged with DNCB, as well as received the following treatments: the Free SB (500 μL); the bilayer vehicle film (BF vehicle) (2.5 H × 2.5 L); the bilayer film containing NC without SB (BF NC B) (2.5 H × 2.5 L), the bilayer film containing nanoencapsulated SB (BF NC SB) (2.5 H × 2.5 L) or 1% of hydrocortisone (HC) (0.5 g), as a comparative drug commonly prescribed for the AD treatment. The Free SB solution was prepared by dissolving the SB in 10 mL of acetone/olive oil (3:1).

The treatments mentioned above were applied in the dorsal region of mice and secured with a bandage starting on day 14 of the experimental protocol. At the same days of mice were challenged with DNCB (14, 17, 20, 23, 26 and 29), the films were changed. The animals were monitored in order to ensure that the films were not removed from the application site. The treatment schedule was based on previous studies that used the same animal model and assessed the pharmacological action of films formulations [35,36].

Followed the last treatment, on day 30 of the experimental protocol, the scratching behavior, one of the hallmark AD-like behaviors, was evaluated in the animals. Twenty-four hours later (day 31), the clinical skin severity scores were determined to assess the manifestations of the AD-like signs in mice. After that, the animals were euthanized by inhalation of isoflurane anesthetic. The samples of dorsal skin of each mouse were rapidly dissected, weighted and frozen at −20 °C to further biochemical analyses. In addition, both ears and the spleen were collected to determine the ear edema and the spleen index, respectively.

2.2.7. AD-like Clinical Signs

Scratching Behavior

The scratching behavior, one of the classic AD-like signs, was evaluated on day 30 of the experimental protocol. The time that mice spent rubbing the dorsal skin, ears and nose with their hind paws was measured and recorded for 30 min [37]. The data was expressed in seconds (s).

Clinical Skin Severity Scores

On day 31 of the experimental protocol, the dorsal skin of mice was photographed and the skin severity scores were assessed according to the method described by Park and Oh [38]. The characteristic signs of skin lesions were classified as: (1) pruritus/itching, (2) erythema/hemorrhage, (3) edema, (4) excoriation/erosion and (5) scaling/dryness. The above-mentioned signs were ranked as: 0 (no signs), 1 (mild), 2 (moderate) and 3 (severe).

2.2.8. Evaluation of the Inflammation Markers

Ear Swelling

On day 31 of the experimental protocol, the animals were euthanized and both ears were cut at the base and weighted on the analytical balance. The ear swelling was measured by the difference between the samples of the DNCB-treated ear (right) and the control ear (left). The results were expressed in mg.

NOx Content

The samples of dorsal skin were homogenized in $ZnSO_4$ (200 mM) and acetonitrile (96%). The homogenates were then centrifuged at 14,000 rpm for 30 min at 4 °C, and the supernatant was collected for the NOx assay. The accumulation of nitrite in the supernatant, an indicator of NO oxidation, was assessed by Griess reaction [39]. Briefly, the NOx content was estimated in a medium containing 2% vanadium chloride (in 5% HCl), 0.1% N-(1-naphthyl) ethylenediamine dihydrochloride and 2% sulfanilamide (in 5% HCl). After incubation at 37 °C for 1h, the color reaction was measured spectrophotometrically at λ = 540 nm. The concentration of nitrite/nitrate in the supernatant was determined from a sodium nitrite standard curve and expressed as nmol NOx/g of tissue.

2.2.9. Evaluation of the Immune Function

Spleen Index

On day 31, mice were sacrificed and the spleen were harvested and weighted on the analytical balance to calculate its relative weight trough the formula: Spleen (g)/Body weight (g). The results were expressed as spleen index.

2.2.10. Evaluation of the Oxidative Stress Markers

Tissue Preparation

To elucidate the involvement of oxidative stress, the samples of dorsal skin were homogenized (1:10, *w/v*) in 50 mM Tris-HCl at pH 7.4. The homogenates were centrifuged at 2500 rpm for 10 min and the low-speed supernatant (S1) was used to determine the thiobarbituric acid reactive species (TBARS) and non-protein thiol (NPSH) levels as well as the catalase (CAT) activity.

Protein Concentration

The protein concentration was estimated according to the method described by Bradford [40], using a bovine serum albumin (1 mg/mL) as a standard. The color reaction was measured spectrophotometrically at λ = 595 nm.

TBARS Levels

TBARS assay was performed to indirectly determine the malondialdehyde (MDA) levels, an important lipid peroxidation marker. As previously described by OHKAWA et al. [41], MDA reacts with 2-thiobarbituric acid (TBA) under acidic conditions and high temperatures to yield the chromogen. The S1 aliquots were incubated with 0.8% TBA, acetic acid buffer (pH 3.4) and 8.1% sodium dodecyl sulfate (SDS) for 2 h at 95 °C. The color reaction was measured at λ = 532 nm and the results were expressed as nmol of MDA/mg of protein, respectively.

NPSH Levels

The NPSH, a non-enzymatic antioxidant defense, was determined by Ellman's method [42]. Briefly, S1 was mixed (1:1) with 10% trichloroacetic acid. After centrifugation (3000 rpm for 10 min), an aliquot of supernatant containing free SH-groups was added in 1 M potassium phosphate buffer pH 7.4 and 10 mM 5,5′-dithiobis-2-nitrobenzoic acid (DTNB). The color reaction was measured at λ = 412 nm and NPSH levels were expressed as nmol of NPSH/g tissue.

CAT Activity

CAT activity was spectrophotometrically determined by monitoring the H_2O_2 consumption at λ = 240 nm, as previously described [43]. In a medium containing an aliquot of S1 and 50 mM potassium phosphate buffer pH 7.0, the enzymatic reaction was started by adding the substrate H_2O_2 (0.3 mM). The enzymatic activity was expressed as Units (1U decomposes 1 mmol of H_2O_2 per minute at pH 7.0 and 25 °C)/mg protein.

2.2.11. Statistical Analysis

Formulations were prepared and analyzed in triplicate and the results were expressed as mean ± standard deviation (SD) or standard error of the mean (SEM). A Gaussian distribution was tested using D'Agostino normality test. For data considered parametric, an one or two-way analysis of variance (ANOVA) followed by post-hoc Tukey's test was performed to compare the significant difference among the experimental groups. All statistical analyses were performed using GraphPad Prism statistical software (version 8.0, San Diego, CA, USA. Values of $p < 0.05$ were considered statistically significant.

3. Results

3.1. The Bilayer Films Presented Suitable Physicochemical Characteristics

Firstly, NC suspensions were produced and characterized. These NC presented particle diameter of 115 ± 3 and 134 ± 5 nm for SB-loaded NC and unloaded NC, respectively. The polydispersity index was below 0.2 and the zeta potential was around −10 mV for both NC suspensions. Besides, the content of SB was 98.9%, being suitable for incorporation into the films. Then, bilayer films were prepared using gellan gum as the first polymeric layer forming agent and pullulan as the second layer. The prepared films presented a macroscopically homogeneous appearance. Besides, the BF NC SB was slightly whitish due the nanostructure presence (Figure 1). Figure 1A shows the images obtained from the side sections of the films at a magnification of 300×. The bilayer formation was clearly observed, without visualization of detachment between the different layers, confirming their adhesion. Besides, the interface between the layers of vehicle film was slightly delimited, which was probably due to the high affinity existing between the gellan gum and pullulan. Whereas, for the films containing the NC, it was observed an irregular interface, probably due to the partial nanoparticles migration from the gellan gum layer to the pullulan layer during the process of the second layer formation. However, it was not possible to observe the presence of these nanostructures in the different polymeric layers at higher magnification (Figure 1B,C), indicating that the nanostructures are intimately embedded between the chains of the film-forming polymers.

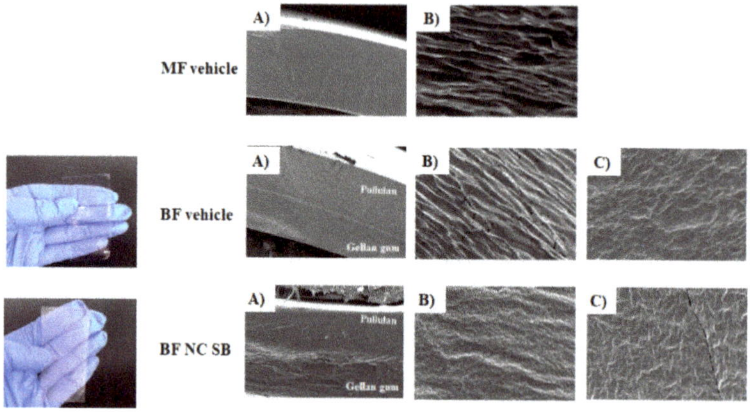

Figure 1. Macroscopic appearance and comparative SEM images of the side sections of monolayer vehicle film (MF vehicle), bilayer vehicle film (BF vehicle) and bilayer film containing silibinin-loaded nanocapsules (BF NC SB) with magnification of 300× (**A**) and magnification of 5000× on the gellan layer (**B**) and pullulan layer (**C**).

Then, bilayer films were examined for thickness, moisture, content homogeneity, swelling index and mechanical properties (Table 1). Analysis of the homogeneity of dosage and thickness were performed to ensure the consistency of dosage and of film formation along the extended area. We found that the films had a homogeneous thickness with values close to 40 μm, as well as the film fragments presented around 440 μg/cm^2 of SB, which represents about 92.25% of theoretical value (477 μg/cm^2).

Table 1. Results of bilayer films characterization.

	BF NC SB	BF NC B	BF Vehicle
Drug content homogeneity (μg/cm^2)	440.61 ± 5.21	-	-
Thickness (μm)	43 ± 7	42 ± 5	40 ± 5
Size (nm)	117 ± 11	137 ± 15	514 ± 143
Polydispersity index	0.28 ± 0.05	0.24 ± 0.05	0.47 ± 0.11
Swelling index (%)	106.95 ± 2.56 *	101.02 ± 6.45 *	134.42 ± 4.78
Moisture (%)	13.89 ± 2.01	12.83 ± 1.07	14.83 ± 2.64
Tensile strength (MPa)	1.09 ± 0.03 *	1.13 ± 0.09 *	0.51 ± 0.05
Elongation (%)	4.93 ± 0.45 *	4.61 ± 0.18 *	7.69 ± 0.15
Young's modulus (MPa)	27.25± 5.95 *	24.25 ± 6.57 *	7.25 ± 5.95

The results are expressed by mean with SD of triplicate. Asterisks denote the significant difference (*) $p < 0.05$ by paired Student's t test between BF vehicle and BF NC SB or BF NC B.

After NC incorporation into bilayer film, it was observed that their nanometric size was maintained. In addition, this inclusion reduced the swelling index and deformation values of the film, as well as increased the tensile strength and Young's Modulus values ($p < 0.05$). The folding endurance test was manually measured in which no film showed formation of cracks or breaks after being folded in the same place for 300 times, suggesting adequate flexibility.

3.2. The Nanocapsules Improved the Occlusive Potential of Bilayer Film

The result of the occlusion test is presented in Table 2. The occlusion factor of nano-based films was significantly higher than BF vehicle at all analyzed time points ($p < 0.001$), indicating that the incorporation of NC into the polymeric matrix increased the capacity of water retention by the film. Besides, no significant difference ($p > 0.05$) was observed between films containing unloaded and the corresponding SB-loaded NC, suggesting that flavonoid encapsulation did not change the occlusion factor.

Table 2. Occlusion factor of bilayer films.

Formulation	Occlusion Factor (%)		
	6 h	24 h	48 h
BF Vehicle	20.18 ± 3.66 **	28.18 ± 2.05 **	29.45 ± 1.47 **
BF NC SB	51.90 ± 2.01	50.16 ± 1.84	55.78 ± 2.35
BF NC B	49.50 ± 4.31	52.30 ± 6.09	50.26 ± 2.45

The results are expressed by mean with SD of triplicate. Asterisks denote the significant difference by One-way ANOVA followed by the Tukey's test. (**) $p < 0.01$ between BF vehicle and BF NC SB or BF NC B.

3.3. The Pullulan Layer Confers Bioadhesion to the Film

Figure 2 shows the results obtained after evaluating the bioadhesive strength of bilayer films using two skin conditions: full thickness and superficial lesion. In both skin conditions evaluated, the films showed higher values of bioadhesive strength when evaluated with the pullulan layer in contact with the skin surface ($p < 0.001$). Bioadhesive strength values were reduced after the NC incorporation into films ($p < 0.001$). The skin condition studied also influenced the films bioadhesion, presenting a reduction in the values of 17,554 ± 1399 to 11,753 ± 528 dyne/cm^2 and of 10934 ± 699 to 7230 ± 863 dyne/cm^2 for BF vehicle and BF NC SB, respectively.

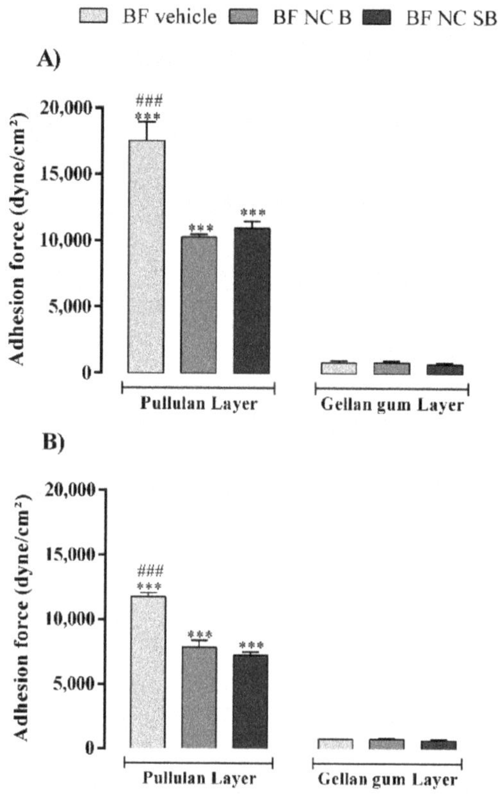

Figure 2. In vitro bioadhesion evaluation using the non-injured (**A**) and injured (**B**) skin of bilayer vehicle film (BF vehicle) and films containing nanocapsules with silibinin (BF NC SB) or without (BF NC B). Each column represents the mean ± SD. The films were evaluated on both sides (pullulan layer and gellan gum layer). (***) $p < 0.001$ significant differences between pullulan layer and gellan gum layer, (###) $p < 0.001$ significant differences between BF vehicle and BF NC SB or BF NC B (Two-way ANOVA followed by Tukey's test).

3.4. The BF NC SB Slowly Releases SB and Retains it in the Skin

The SB release from the films is represented by the cumulative release of substance per area as a function of time in hours (Figure 3A). The BF NC SB film presented a slow release over the period of 24 h, reaching 7.14 ± 1.34% of SB released.

In relation to the skin permeation of SB from the bilayer film, the total SB retained (Figure 3B) on the uninjured was less than injured skin ($p < 0.01$). Besides, regardless the skin condition (uninjured and injured) the SB was not detected in the receptor medium. The Figure 3C,D show the distribution percentage of SB in different layers of skin after 24 h of experiment for uninjured and injured human skin, respectively. In both skin conditions, SB penetrated the cutaneous tissue in quantifiable drug amounts. In uninjured skin, the flavonoid accumulated preferably in the *stratum corneum*. For the injured skin a higher SB amount in the epidermis in comparison to dermis was observed ($p < 0.001$).

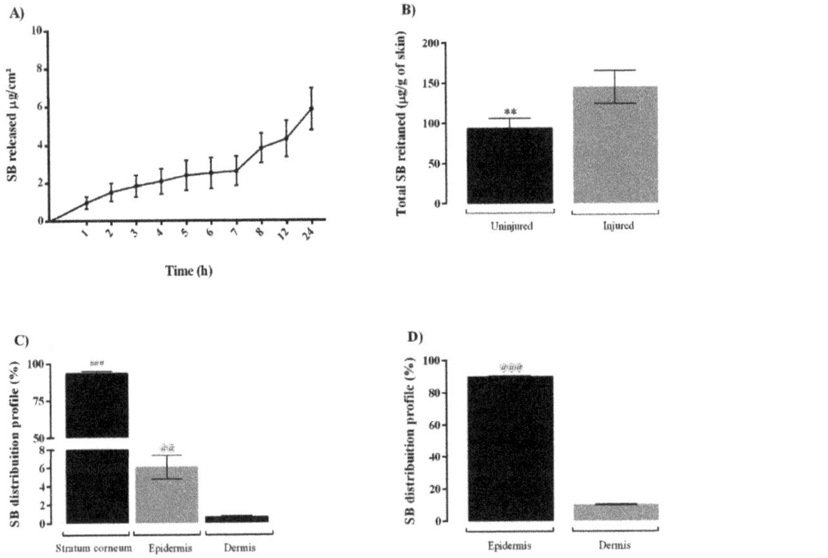

Figure 3. In vitro release profile and skin permeation of silibinin from nano-based bilayer film. (**A**) Cumulative amount of released silibinin from film in phosphate buffer receptor medium pH 7.4 at 32.0 °C (n = 3); (**B**) Cumulative amount of retained silibinin in the uninjured and injured skin after 24 h of incubation (n = 6); (**C**) Percentage of silibinin distribution in the different non-injured skin layers; (**D**) Percentage of silibinin distribution in the different injured skin layers. All the results are expressed as mean ± SEM. (**) $p < 0.01$ significant differences between cumulative amount of SB in the non-injured and injured skin, (###) $p < 0.001$ significant differences between SB quantified in the *stratum corneum* and epidermis in the non-injured skin, (@@) $p < 0.01$ and (@@@) $p < 0.001$ significant differences between compound quantified in the epidermis and dermis in the non-injured or injured skin.

3.5. The BF NC SB Neutralized the ABTS+ Radical

Figure 4 shows radical scavenging activity of SB-loaded NC bilayer film in comparison to vehicle and bilayer film containing unloaded NC. Corroborating the antioxidant activity of SB, the BF NC SB presented high radical scavenging activity (about 100%). Both BF vehicle and BF NC B had a low influence on the neutralization of the ABTS+ radical, demonstrating that there is no false positive in the test performed.

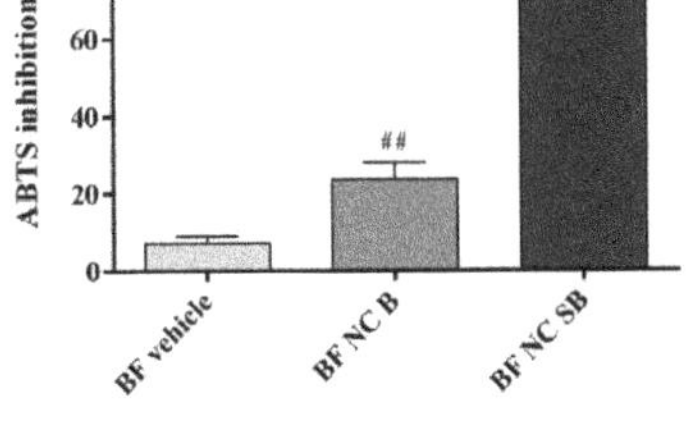

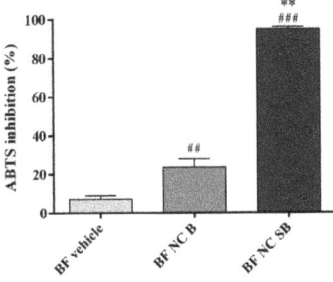

Figure 4. Percentage of ABTS radical inhibition by films. Each column represents the mean ± SD. (**) $p < 0.01$ significant differences between bilayer film containing nanocapsules with (BF NC SB) or without (BF NC B) silibinin, (##) $p < 0.01$ and (###) $p < 0.001$ significant differences between BF NC SB or BF NC B and BF vehicle (One-way ANOVA followed by the Tukey's test).

3.6. The Bilayer Films Are Hemocompatible

The bilayer films were evaluated for their potential to cause lysis in human erythrocytes. The percentage of hemolysis was 0.73 ± 0.05%, 0.56 ± 0.23% and 0.61 ± 0.11% for vehicle film, film containing unloaded NC and film containing SB-loaded NC, respectively. All these values are similar to the hemolytic percentage of saline solution (0.70 ± 0.09%), indicating a good blood compatibility of produced films.

3.7. The BF NC SB Treatment Attenuated the AD-like Clinical Signs Induced by DNCB in Mice

The Figure 5 depicts the effect of free SB or BF NC SB treatments on the severity of the skin lesions and the scratching behavior in mice. Our data demonstrated that all the animals exposed to DNCB exhibited AD-like clinical signs, represented by an increase in the clinical skin severity score when compared with the control group ($p < 0.0001$). The free SB or the BF NC SB treatments markedly decreased the characteristics AD-like signs in the DNCB-exposed mice. In turn, the BF vehicle, the BF NC B and the HC did not alter the severity of lesions induced by DNCB.

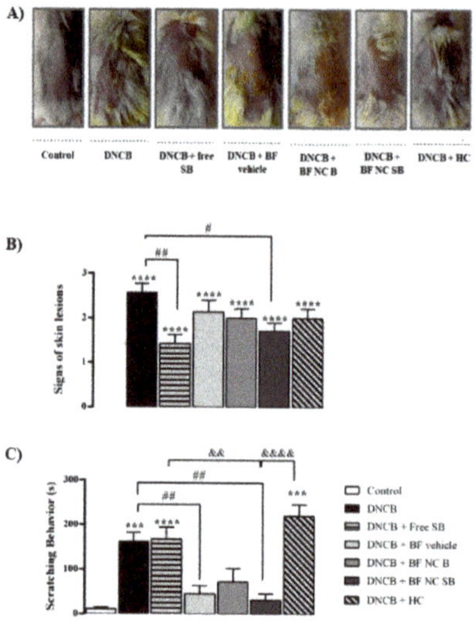

Figure 5. Effect of the different treatments on AD-like clinical signs induced by DNCB in mice. (**A**) Images of the skin lesions from the groups of mice taken on the last day of the experiment (day 31). (**B**) Score of the skin lesions. (**C**) Scratching time evaluated on day 30 of the experimental protocol. Each column represents the mean ± SEM of 7 mice per group. (****) $p < 0.0001$ and (***) $p < 0.001$ compared with the control group, (##) $p < 0.01$ and (#) $p < 0.05$ compared with the DNCB group, (&&&&) $p < 0.0001$ and (&&) $p < 0.01$ compared with the BF NC SB group (One-way ANOVA followed by the Tukey's test).

The results demonstrated that the DNCB-exposed mice exhibited an increase in the scratching time when compared with the control group ($p < 0.001$). In contrast, both free SB and HC treatments did not alter the scratching behavior induced by DNCB in mice ($p > 0.05$). The treatment with BF NC B did not show statistical difference in the scratching time when compared to the DNCB- or vehicle-treated mice ($p > 0.05$). However, repeated appli-cations of the BF vehicle or the BF NC SB reduced this typical AD-like behavior in the DNCB group ($p < 0.01$). In fact, there is a statistically significant difference among the BF NC SB, HC, and free SB treatments, suggesting that the topical application of BF NC SB was more effective to reduce the scratching behavior in DNCB treated mice than the free SB ($p < 0.01$) or HC ($p < 0.0001$).

3.8. The BF NC SB Treatment Suppressed the Ear Swelling Induced by DNCB in Mice

The Figure 6 illustrates the effect of free SB as well as all formulations containing or not SB on the development of ear swelling induced by DNCB in mice. The results evidenced that the DNCB substantially increased the ear swelling when compared with the control group ($p < 0.0001$), whereas the topical application of BF vehicle, BF NC B and BF NC SB reduced the ear swelling induced by DNCB in mice ($p < 0.01$). On the other side, the treatments with free SB or HC had no statistical difference on ear swelling when compared to vehicle or DNCB exposed mice ($p > 0.05$).

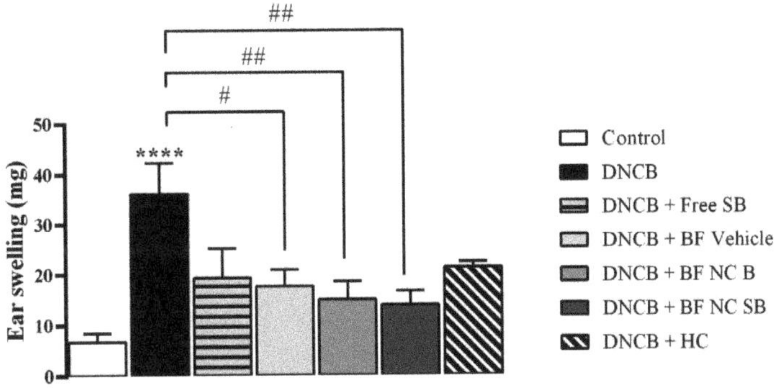

Figure 6. Effect of the different treatments on the ear swelling induced by DNCB in mice. The ear swelling was assessed on day 31 of the experimental protocol. Each column represents the mean ± SEM of 7 mice per group. (****) $p < 0.0001$ compared with the control group, (##) $p < 0.01$ and (#) $p < 0.05$ compared with the DNCB group (One-way ANOVA followed by the Tukey's test).

3.9. The BF NC SB Treatment Did Not Alter the Spleen Index after DNCB Exposure in Mice

As shown in Figure 7, the DNCB-exposed mice significantly increased the spleen index in comparison with the control group ($p < 0.001$). Only the HC treatment attenuated the splenomegaly induced by DNCB in mice ($p < 0.0001$). No statistically significant difference was evidenced in the spleen index after the topical applications of BF vehicle, the BF NC B, and the BF NC SB in the dorsal skin of mice when compared to DNCB or control groups ($p > 0.05$)

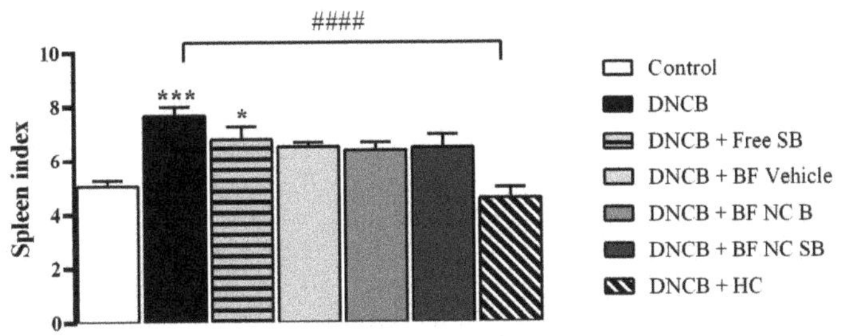

Figure 7. Effect of the different treatments on the spleen index. This parameter was assessed on day 31 of the experimental protocol. Each column represents the mean ± SEM of 7 mice per group. (***) $p < 0.001$ and (*) $p < 0.05$ compared with the control group, (####) $p < 0.0001$ compared with the DNCB group (One-way ANOVA followed by the Tukey's test).

3.10. The BF NC SB Treatment Modulated Some Markers of Oxidative Stress and Inflammation in the Dorsal Skin of Mice Exposed to DNCB

The dorsal skin of DNCB-treated mice exhibited a significant excessive production of NOx levels when compared with the control group ($p < 0.0001$), as shown in Figure 8.

The topical applications of free SB, BF NC SB and HC reduced the NOx levels in the dorsal skin of DNCB exposed mice ($p < 0.01$) whereas the treatments with BF vehicle or BF NC B had no statistical difference on the NOx levels in the dorsal skin of mice when compared to vehicle or DNCB groups ($p > 0.05$).

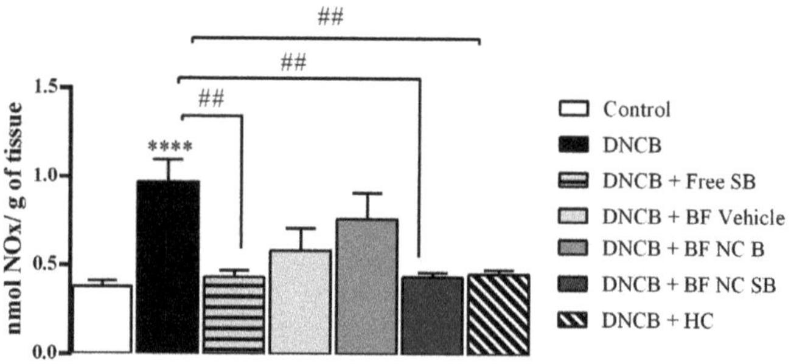

Figure 8. Effect of the different treatments on the NOx levels in the dorsal skin of mice exposed to DNCB. Each column represents the mean ± SEM of 7 mice per group. (****) $p < 0.0001$ compared with the control group and (##) $p < 0.01$ compared with the DNCB group (One-way ANOVA followed by the Tukey's test).

The Figure 9 summarizes the results regarding the effect of the film formulations in some oxidative stress parameters in the dorsal skin of mice exposed to DNCB. In relation to the control group, the DNCB exposure led to an enhancement of TBARS levels in the dorsal skin of mice ($p < 0.05$). In turn, the treatments with BF vehicle, BF NC B, as well as BF NC SB reduced the TBARS levels in animals exposed to DNCB ($p < 0.0001$). The topical applications of free SB or HC did not alter the TBARS levels in the dorsal skin when compared to vehicle or DNCB groups ($p > 0.05$). In this line, the BF NC SB treatment was statistically more effective in reducing the TBARS levels in the dorsal skin of mice exposed to DNCB than the free SB or HC ($p < 0.01$) (Figure 9A).

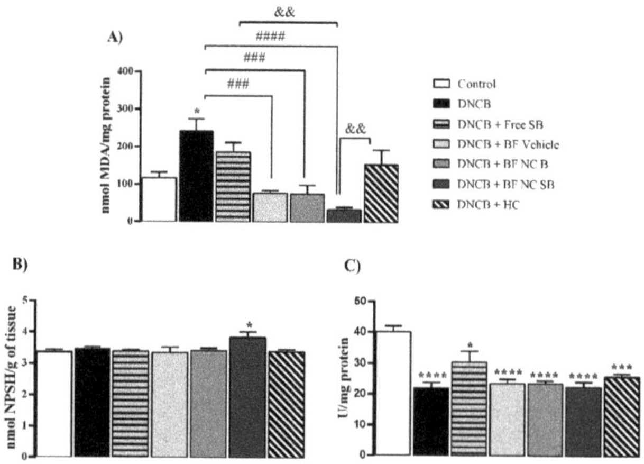

Figure 9. Effect of the different treatments on the levels of TBARS (**A**) and of NPSH (**B**) as well as on the CAT activity (**C**) in the dorsal skin of mice exposed to DNCB. Each column represents the mean ± SEM of 7 mice per group. (****) $p < 0.0001$, (***) $p < 0.001$ and (*) $p < 0.05$ compared with the control group, (####) $p < 0.001$ and (###) $p < 0.01$ compared with the DNCB group, (&&) $p < 0.01$ compared with the BF NC SB group (One-way ANOVA followed by the Tukey's test).

Regarding the non-enzymatic antioxidant defenses, the repeated applications of BF NC SB increased the levels of NPSH in the dorsal skin of mice in relation to the control group ($p < 0.05$) whereas the statistical analysis also revealed similar NPSH levels in the dorsal skin of mice among all other experimental groups ($p > 0.05$) (Figure 9B). Moreover, the animals exposed to DNCB presented an inhibition of CAT activity in the dorsal skin when compared with the control group ($p < 0.0001$). All the formulations tested, as well as the positive control (HC) were unable to restore the CAT activity at the control levels in the dorsal skin of DNCB-treated mice ($p > 0.05$) (Figure 9C).

4. Discussion

Aiming to suppress the clinical AD symptoms, this study was designed such that the immediate layer composed by pullulan could provide an adhesion to the cutaneous tissue, whilst a gellan gum layer containing SB-loaded NC provides a therapeutic effect in sustained manner. For this, firstly, the bilayer films were produced by applying a second pullulan layer on top of the first layer composed by gellan gum incorporated with SB-loaded NC using two-step solvent casting method. Micrographs of side sections of produced films exhibited clearly the formation of two polymeric layers, which seem to be adhered to each other, without gap signals.

After drying, is important to assure the homogeneity of the final formulation, since during this process may be observed agglomeration or sedimentation of solid particles or air bubbles on the surface, leading to homogeneity problems [19]. The bilayer films produced presented a thin and homogeneous thickness, which is required for cutaneous application. In addition, the SB content was in the range of 85 to 115% recommended for polymeric films [44], as well as this flavonoid was evenly distributed throughout the film as demonstrated by the content uniformity test. The residual water content in the films was less than 15% for all films, corroborating the previous reports for films produced by the same technique [26,27]. All these evaluations indicate that the casting solvent method employed to produce the bilayer films was successful.

In the SEM analysis, although a difference was observed between the microstructure of vehicle film and of nano-based film, it was not possible to visualize the NC. However, PCS analysis indicated the nanometric size maintenance of particles in the final dosage form, suggesting that the NC are intimately inserted between the polymeric chains. The nanoparticles presence in the polymeric film network has already been suggested in others researches involving polymeric films containing nanostructured systems [18,45]. This insertion may restrict the free movement of polymer chains [46]. In fact, our results demonstrated that the films containing NC showed higher young's modulus values and lower swelling index than vehicle film, indicating a reduction of both mobility and relaxation of the polymeric chains, making the film stiffer and less susceptible to penetration by fluids. However, all the produced bilayer films showed fluid absorption capacity, as well as remained intact after 24 h in contact with the buffer, suggesting that they are a resistant material for the exudative lesions treatment. This result may be due to the presence of the gellan gum layer in the film, which forms swellable and resistant films, being replaced less frequently in the lesion site [20,21]. Moreover, the young's modulus obtained for the developed films have values within of the young's modulus range of skin, which can vary between 0.02 and 57 MPa [47].

The formation of two distinct layers in the film was also confirmed through the bioadhesion test, in which the layer composed by pullulan presented values of bioadhesive strength greater than the layer composed by gellan gum. In fact, pullulan-based formulations have already been reported in the literature with bioadhesive properties [22,48]. The bioadhesion can contribute to an intimate contact of the pharmaceutical dosage form with the skin, increasing the residence time at the action site and decreasing the administration frequency. Besides, the use of bioadhesive films may reduce the use of adhesive tapes for their application, which are constantly associated with greater skin irritation and pain [49]. Considering that AD is characterized by a loss of skin barrier function, mainly due to a damage to the lipids of the *stratum corneum* [1], the bioadhesion was also evaluated using an injured skin model (without *stratum corneum*). The films bioadhesive potential was lower when in contact with the injured skin. Similar results were observed for hydrogels containing β-caryophyllene nanoemulsions [50]. Besides, the bioadhesion values of nano-based films were lower than vehicle film in both uninjured and injured skin conditions. In fact, it was observed an irregular interface in the SEM images for film containing NC. This result reinforces our argument that NC migrate from the gellan gum layer to the pullulan layer during the film formation.

Distinct theories are used to describe the bioadhesion. Among these theories is adsorption, in which the bioadhesiveness between a substance and a tissue results from van der Waals, hydrophobic or hydrogen bond interactions [51]. The hydrophilic groups of pullulan may be preferentially oriented towards the inside of the film interacting with glycerol used as plasticizers, as showed in others studies involving hydrophilic polymers-based films [49,52]. Thus, the film surface may become slightly more hydrophobic, favoring its interaction with lipophilic surface of *stratum corneum* in uninjured skin. However, simulating a skin damage condition through *stratum corneum* removal occurs viable epidermis exposure, which is less lipophilic than *stratum corneum*, reducing the pullulan interaction. This greater bioadhesion in the intact skin than in the injured skin can be beneficial, since during the film peeling from skin, the lesion area will not be more damaged. In addition, the NC inclusion in polymeric chains may be masking the chemical groups in pullulan that interact with the skin, resulting in lower bioadhesion. In fact, in others studies were also observed a bioadhesive reduction of hydrophilic polymers after inclusion of solid substances into films [49,53].

The skin barrier impairment in patients with AD leads to the transepidermal water loss, as well as the drug permeability barrier is diminished in these patients, increasing the risk of systemic drug absorption [1]. The in vivo skin hydration can be correlated with in vitro occlusion factor in a linear form. In other words, the greater the occlusion factor, the greater the cutaneous hydration observed in vivo [54]. Our results confirm the higher occlusive effect provided by nanostructures, as previously reported in other studies [18,27]. This result points out that nano-based films are promising in the design of novel treatments intended for improving skin hydration.

Regarding permeation study, in the intact skin, the BF NC SB providing a SB deposition in higher amounts in the *stratum corneum* followed by its delivery on epidermis and dermis. This find is in line with the cutaneous permeation mechanism observed for polymeric nanoparticles reported in the literature [55]. As expected, after removing the main barrier to substances penetration across the skin, an increase in the SB accumulation in the epidermis and dermis was observed. However, despite the SB amount increased in the injured skin, no amount of this flavonoid was detected in the receptor medium. This affinity for cutaneous tissue is advantageous for AD treatment because favors a site-specific therapeutic response, without systemic absorption [56]. In addition, in vitro release test evidenced a sustained SB release from the nano-based film. This controlled release may favor the SB delivery locally in the skin in a well-controlled manner, allowing a less frequent film replacement and avoiding greater local irritation and the risk of injured area contamination [26,56].

Previously published data showed that gellan gum films had young's modulus about three times greater and occlusion factor about twice lower than bilayer films produced [27]. Besides, this same gellan gum film presented SB amount released from nano-based film around 5% in 24 h under the same conditions, as well as similar SB distribution profile in the different skin layers to that found for bilayer films. The improvement of young's modulus and occlusive properties observed here are in line with described in the literature for bilayer films. This type of film presents heterogeneous structures, taking advantage of the best characteristics of each individual polymer, and thus, improving the physical and mechanical properties of the final film [24]. Besides, pullulan is highly hydrophilic, forming a fast-dissolving film [22]. Thus, it is possible to infer that the pullulan layer addition confers an increase in bioadhesion and occlusion values, without altering the release and skin permeation profile of SB.

Flavonoids have antioxidant and anti-inflammatory effects which may ameliorate signs of allergic diseases, including AD [4]. Besides, studies describe that the beneficial effects of SB in skin pathologies are mainly due to its action in preventing the generation of oxidative stress. Thus, a preliminary in vitro evaluation of the SB antioxidant effect from the nano-based bilayer films produced was performed. The results showed a high capacity of BF NC SB to neutralize the ABTS+ radical, corroborating others studies that evaluated the antioxidant effect of this flavonoid using this same synthetic radical [32,57]. Next, the hemocompatibility of the developed bilayer films was determined in order to assess their safety when in contact with red blood cells, since AD lesions may bleed. Our results showed a hemolytic index less than 1% for all the bilayer films. According to the Standard Practice for Assessment of Hemolytic Properties of Materials, materials that cause 0–2% hemolysis are considered non-hemolytic and safe [33].

Given the in vitro results obtained, it was performed a pre-clinical study to evidence the therapeutic efficacy of bilayer films against AD using hapten-induced mice. AD is characterized by itchy, red and swollen skin due to chronic inflammation and immune dysregulation. Multiple challenges with haptens, such as DNCB, into the skin of mice mimics the ongoing maintenance and the progression of AD. Thereby, the epicutaneous sensitizer DNCB is commonly used to induce a chemical animal model of dermatitis [58]. The present study reinforces that DNCB model reproduced the typical AD-like signs, as evidenced by an increase in the severity of skin lesions and the scratching behavior. Considered a typical AD symptom, scratching could be responsible for trigger a physical injury to epidermis, as well as aggravate the inflammatory processes. In this line, it is recognized as an essential and a skin specific behavior related to itching [59].

Immunologically, it has been reported that haptens exposure for an extend period evoke primarily the release of T helper 1 (Th1) cytokines that shift to a delayed chronic Th2-dominated inflammatory response, that it is similar to human AD. In this context, the immunological and inflammatory process is closely implicated in the pathogenesis of AD [60]. Some studies established that the local inflammation and the skin irritation induced by the external application of DNCB could also result in ear edema, probably due to an increase in the vascular permeability [36,37]. As expected, our results showed that repeated DNCB exposure promoted a marked ear swelling, as well as an enhancement in the spleen index and the NOx levels in the dorsal skin of mice. Interestingly, the spleen index, known as a peripheral immune organ, may be reflected the stimulation of immunological responses by activating T-lymphocytes [59]. Besides, it is well known that cytokines may lead to an excessive production of NO which aggravate the inflammatory response and consequently sustain the local tissue injury [61].

Previous studies have shown that melanocytes and keratinocytes, some types of skin cells, generates RS [62,63]. Similarly, the immune cells response and the release of inflammatory mediators are often associated with an increase in the production of oxidative molecules and reactive free radicals [64]. Indeed, an overproduction of NO levels lead to the development of oxidative stress that is also involved as a pathological aspect of AD [65,66]. In agreement with those findings, our results showed that repeated DNCB challenges inhibited the CAT activity and increased the TBARS levels, indicating the establishment of oxidative stress in the dorsal skin of mice. Indeed, these data reaffirm that the multifaceted interactions among the inflammatory mediators, the immune cells and the oxidative stress favor the maintenance of skin lesion like-AD.

Although topical corticosteroids are usually recommended as the standard anti-inflammatory treatment to alleviate the severe sings related to AD, their long-term use results in the manifestation of innumerous adverse effects, such as skin atrophy, purpura, dyspigmentation and declined immune function [67]. Consistent with these findings, our results revealed that mice treated with HC exhibited skin lesions like-AD, scratching behavior, a reduction in the spleen index and in the NOx levels on the local injury in mice.

This data agrees with the fact that the long-term steroids use lead to the development of skin thinning, cracking or bleeding, as well as, immunosuppression [68].

All the bilayer films presented a significant reduction of scratching behavior, ear edema and TBARS levels whereas the other treatments did not reduce these parameters. It is known that the skin hydration is related to reduction of pruritus and skin barrier recovery, reducing the AD aggravation [69]. Consistent with this information, the in vitro results showed the bioadhesive and occlusive potential of films, suggesting that they may adhere to skin, improving it hydration. Similar result was obtained by Jeong and co-workers in which pullulan films with and without *Rhus verniciflua* extract ameliorated the AD-like sings as epidermal thickness and reduced the cell infiltration, suggesting that pullulan itself suppress AD development [16]. In fact, the pullulan layer was applied in contact with mice skin surface to act with bioadhesive function, while the gellan gum layer could act as external barrier against aggression and as vehicle for SB-loaded NC. In addition, both pullulan and gellan gum layers were plasticized with glycerol, a humectant used in dermatological formulations. Beyond to act as humectant, glycerol appears to accelerate the skin barrier repair after its disruption [9,69]. Thus, we speculate that bilayer films components may be promoting a physical barrier against scratches and other sensitizing agents, improving the hydration and restoring the skin's barrier function.

Moreover, the current study demonstrated that free SB, BF NC SB and HC treatment reduced the NOx levels in the dorsal skin of mice after DNCB exposure. Considering the involvement of NO in inflammatory conditions, we speculated that a decrease in the NOx levels might trigger, at least in part, a decrease in vascular permeability as well as a lower production of cytokines and prostaglandins, thus decreasing the severity of the inflammatory process on the local injury [70]. NO, a highly reactive free radical, also contribute to oxidative damage that occur by lipid peroxidation and oxidation of proteins and thiols [66].

Although CAT activity was not altered by treatment with SB formulations, the BF NC SB treatment increased the levels of NPSH in the dorsal skin. Tripeptide glutathione (GSH), the major non-protein thiol quantified in the NPSH assay, plays an essential role in the cellular redox homeostasis, protecting organelles from oxidative damage and inflammatory cascade [71]. In this sense, our findings suggest that the BF NC SB enhanced the GSH production which led to an increase in the NPSH levels in an attempt to counteract the lipid peroxidation and the oxidative damage in the dorsal skin of DNCB exposed mice.

Our in vivo results confirmed the benefits of SB nanoencapsulation since topical applications of free SB alleviated the AD-like signs and NOx levels in mice whereas the BF NC SB decreased the incidence of the skin lesions score, the scratching behavior, the ear swelling and oxidative parameters. We believe that NC could increase the residence time of SB in the cutaneous tissue, and thus, exert a better antioxidant performance. In line with these findings, previous studies reported the promising anti-inflammatory effect of hydrogels containing nanoencapsulated SB in a model of contact dermatitis induced by croton oil [7] or DNCB [8]. In addition, specifically for AD, the better performance of polymeric nanoparticles in attenuating the inflammatory and immunological effects of the disease is well documented in the literature when compared to non-nanoencapsulated substances [56,72]. Recently our research group demonstrated that pullulan films incorporated with pomegranate seed oil NC presented better biological effects than this free vegetable oil or associated into nanoemulsions against the same mice model of AD used here, highlighting the superiority of NC formulations [18].

Collectively the results obtained in in vitro and in vivo evaluations suggest that bilayer films containing SB-loaded NC may exert a physical barrier action in AD treatment, as cutaneous bioadhesion, prevention of skin dryness, protection of lesions and ability to absorb exudates, followed by therapeutic action, with sustained and localized SB release and improved antioxidant effect. In this sense, the results obtained here encourage further investigations involving pullulan/gellan gum bilayer film as vehicle for AD control, as well as the combination of NC into films to improve the therapeutic performance. Furthermore, it is important to mention that, to our knowledge, this is the first study where a bilayer film based on gellan gum and pullulan containing nanostructures was produced, characterized and assessed against the AD injuries in a pre-clinical study.

5. Conclusions

This study demonstrated a facile prepare of gellan gum/pullulan bilayer, where the top pullulan layer was able to ensure a good bioadhesion to the skin while the bottom gellan gum layer has swelling capacity and acts as a vehicle for the release of SB-loaded NC. The in vitro studies showed that nano-based film presented higher occlusion factor. Besides, this film released SB in a slow and gradual manner and allowed the retention of this flavonoid in the skin tissue. Also, SB-loaded NC incorporated into bilayer films presented high scavenger capacity and did not present hemolytic degree. The in vivo study provided evidence that topical applications of BF NC SB attenuated the AD-like skin lesions, the scratching behavior and the ear edema by modulating some markers related to redox signaling and the inflammatory process. Interestingly, the bilayer film without SB presence attenuated the scratching behavior, the ear edema and TBARS levels, suggesting that the novel bilayer vehicle film might also provide benefits to treat atopic skin. It should be noted that the efficacy of BF NC SB treatment on behavioral and biochemical parameters was similar or better than the free SB solution, the classical treatment (HC) or bilayer films without SB. In this scenario, our data reinforce that the combination of NC and films could be applied to enhance the performance of antioxidant substances in skin disorders treatment. Particularly, we highlighted that the gellan gum/pullulan bilayer film containing SB-loaded NC might combine skin protection and hydration effects with improved anti-inflammatory and antioxidant effects, being a promising and potential therapeutic alternative for AD treatment.

Supplementary Materials: The following supporting information can be downloaded at: https://www.mdpi.com/article/10.3390/pharmaceutics14112352/s1, Figure S1: Schematic representation of the experimental design of in vivo study.

Author Contributions: Conceptualization, M.G., C.C.M. and L.C.; data curation, M.G., C.C.M., T.d.B.B., C.L., E.A.W. and L.C.; formal analysis, M.G. and L.C.; funding acquisition, L.C.; methodology, M.G., C.C.M., T.d.B.B., L.S.d.R., F.Z.M.S. and L.C.; investigation, M.G., T.d.B.B., C.C.M. and L.S.d.R.; project administration, M.G. and L.C.; writing—original draft preparation, M.G. and L.C.; writing—review and editing, M.G., C.C.M., T.d.B.B., L.S.d.R., F.Z.M.S., C.L., E.A.W. and L.C.; validation, M.G. and L.C.; visualization, M.G. and L.C.; supervision, L.C. All authors have read and agreed to the published version of the manuscript.

Funding: This research was funded by Conselho Nacional de Desenvolvimento Científico e Tecnológico (CNPq—Brazil), grant number 456863/2014-1.

Institutional Review Board Statement: The animal study protocol was approved (approval date November 13, 2019) by the Ethical Research Committee of the Federal University of Pelotas, Rio Grande do Sul, Brazil (CEEA 23357-2018/140-2019).

Informed Consent Statement: Informed consent was obtained from all subjects involved in the study.

Data Availability Statement: Not applicable.

Acknowledgments: The authors express their gratitude to the Cristiane Bona da Silva for Zetasizer access and Charlene Menezes for Zetasizer analysis. The authors acknowledge also the Natália de Freitas Daudt for their assistance with SEM analysis. Letícia Cruz thank CNPq for PQ fellowship (process number: 315612/2020-7). Mailine Gehrcke gratefully acknowledges Coordenação de Aperfeiçoamento de Pessoal de nível Superior (CAPES- BR) for the financial support (88887.463649/2019-00).

Conflicts of Interest: The authors declare no conflict of interest.

References

1. Malik, K.; Heitmiller, K.D. An Update on the Pathophysiology of Atopic Dermatitis. *Dermatol. Clin.* **2017**, *35*, 317–326. [CrossRef] [PubMed]
2. Rerknimitr, P.; Otsuka, A.; Nakashima, C.; Kabashima, K. The Etiopathogenesis of Atopic Dermatitis: Barrier Disruption, Immunological Derangement, and Pruritus. *Inflamm. Regen.* **2017**, *37*, 14. [CrossRef] [PubMed]
3. Mancuso, J.B.; Lee, S.S.; Paller, A.S.; Ohya, Y. Management of Severe Atopic Dermatitis in Pediatric Patients. *J. Allergy Clin. Immunol. Pract.* **2021**, *9*, 1462–1471. [CrossRef] [PubMed]
4. Wu, S.; Pang, Y.; He, Y.; Zhang, X.; Peng, L.; Guo, J.; Zeng, J. A Comprehensive Review of Natural Products against Atopic Dermatitis: Flavonoids, Alkaloids, Terpenes, Glycosides and Other Compounds. *Biomed. Pharmacother.* **2021**, *140*, 111741. [CrossRef]
5. Song, X.; Zhou, B.; Cui, L.; Lei, D.; Zhang, P.; Yao, G.; Xia, M.; Hayashi, T.; Hattori, S.; Ushiki-Kaku, Y.; et al. Silibinin Ameliorates Aβ25-35-Induced Memory Deficits in Rats by Modulating Autophagy and Attenuating Neuroinflammation as Well as Oxidative Stress. *Neurochem. Res.* **2017**, *42*, 1073–1083. [CrossRef] [PubMed]
6. Samanta, R.; Pattnaik, A.; Pradhan, K.; Mehta, B.; Pattanayak, S.; Banerjee, S. Wound Healing Activity of Silibinin in Mice. *Pharmacogn. Res.* **2016**, *8*, 298. [CrossRef]
7. Rigon, C.; Crivellaro, M.; Marchiori, L.; Jardim, S.; Schopf, N.; Chaves, S.; Callegaro, M.; Carlos, R.; Beck, R.; Ferreira, A.; et al. Hydrogel Containing Silibinin Nanocapsules Presents Effective Anti- Inflammatory Action in a Model of Irritant Contact Dermatitis in Mice. *Eur. J. Pharm. Sci.* **2019**, *137*, 104969. [CrossRef]
8. Shrotriya, S.N.; Vidhate, B.V.; Shukla, M.S. Formulation and Development of Silybin Loaded Solid Lipid Nanoparticle Enriched Gel for Irritant Contact Dermatitis. *J. Drug Deliv. Sci. Technol.* **2017**, *41*, 164–173. [CrossRef]
9. Atrux-Tallau, N.; Romagny, C.; Padois, K.; Denis, A.; Haftek, M.; Falson, F.; Pirot, F.; Maibach, H.I. Effects of Glycerol on Human Skin Damaged by Acute Sodium Lauryl Sulphate Treatment. *Arch. Dermatol. Res.* **2010**, *302*, 435–441. [CrossRef]
10. Marchiori, M.C.L.; Rigon, C.; Camponogara, C.; Oliveira, S.M.; Cruz, L. Hydrogel Containing Silibinin-Loaded Pomegranate Oil Based Nanocapsules Exhibits Anti-Inflammatory Effects on Skin Damage UVB Radiation-Induced in Mice. *J. Photochem. Photobiol. B Biol.* **2017**, *170*, 25–32. [CrossRef]
11. Li, G.; Fan, C.; Li, X.; Fan, Y.; Wang, X.; Li, M.; Liu, Y. Preparation and In Vitro Evaluation of Tacrolimus-Loaded Ethosomes. *Sci. World J.* **2012**, *2012*, 874053. [CrossRef]
12. Wang, Y.; Cao, S.; Yu, K.; Yang, F.; Yu, X.; Zhai, Y.; Wu, C.; Xu, Y. Integrating Tacrolimus into Eutectic Oil-Based Microemulsion for Atopic Dermatitis: Simultaneously Enhancing Percutaneous Delivery and Treatment Efficacy with Relieving Side Effects. *Int. J. Nanomed.* **2019**, *14*, 5849–5863. [CrossRef]
13. Badihi, A.; Fru, M.; Soroka, Y.; Benhamron, S.; Tzur, T.; Nassar, T.; Benita, S. Topical Nano-Encapsulated Cyclosporine Formulation for Atopic Dermatitis Treatment. *Nanomed. Nanotechnol. Biol. Med.* **2020**, *24*, 102140. [CrossRef]
14. Mora-Huertas, C.E.; Fessi, H.; Elaissari, A. Polymer-Based Nanocapsules for Drug Delivery. *Int. J. Pharm.* **2010**, *385*, 113–142. [CrossRef]
15. Ferreira, L.M.; Sari, M.H.M.; Azambuja, J.H.; da Silveira, E.F.; Cervi, V.F.; Marchiori, M.C.L.; Maria-Engler, S.S.; Wink, M.R.; Azevedo, J.G.; Nogueira, C.W.; et al. Xanthan Gum-Based Hydrogel Containing Nanocapsules for Cutaneous Diphenyl Diselenide Delivery in Melanoma Therapy. *Investig. New Drugs* **2019**, *38*, 662–674. [CrossRef]
16. Jeong, J.H.; Back, S.K.; An, J.H.; Lee, N.; Kim, D.; Na, C.S.; Jeong, Y.; Han, S.Y. Topical Film Prepared with Rhus Verniciflua Extract-Loaded Pullulan Hydrogel for Atopic Dermatitis Treatment. *J. Biomed. Mater. Res.* **2019**, *107*, 1–10. [CrossRef]
17. Alves, N.O.; Gabriela, T.; Weber, D.M.; Luchese, C.; Wilhelm, E.A.; Fajardo, A.R. Chitosan/Poly (Vinyl Alcohol)/Bovine Bone Powder Biocomposites: A Potential Biomaterial for the Treatment of Atopic Dermatitis-like Skin Lesions. *Carbohydr. Polym.* **2016**, *148*, 115–124. [CrossRef]
18. Cervi, V.F.; Saccol, C.P.; Sari, M.H.M.; Martins, C.C.; Da Rosa, L.S.; Ilha, B.D.; Soares, F.Z.; Luchese, C.; Wilhelm, E.A.; Cruz, L. Pullulan Film Incorporated with Nanocapsules Improves Pomegranate Seed Oil Anti-Inflammatory and Antioxidant Effects in the Treatment of Atopic Dermatitis in Mice. *Int. J. Pharm.* **2021**, *609*, 121144. [CrossRef]
19. Karki, S.; Kim, H.; Na, S.; Shin, D.; Jo, K.; Lee, J. Thin Films as an Emerging Platform for Drug Delivery. *Asian J. Pharm. Sci.* **2016**, *11*, 559–574. [CrossRef]
20. Ismail, N.A.; Amin, K.A.M.; Majid, F.A.A.; Razali, M.H. Gellan Gum Incorporating Titanium Dioxide Nanoparticles Biofilm as Wound Dressing: Physicochemical, Mechanical, Antibacterial Properties and Wound Healing Studies. *Mater. Sci. Eng. C* **2019**, *103*, 109770. [CrossRef]
21. Razali, M.H.; Ismail, N.A.; Amin, K.A.M. Nanotubos de dióxido de titânio incorporaram filme bio-nanocompósito de goma gelana para cicatrização de feridas: Efeito da concentração de nanotubos de TiO_2. *Int. J. Biol. Macromol.* **2020**, *153*, 1117–1135. [CrossRef] [PubMed]
22. Cervi, V.F.; Saccol, C.P.; Da Rosa Pinheiro, T.; Santos, R.C.V.; Cruz, L.; Sari, M.H.M. A Novel Nanotechnological Mucoadhesive and Fast-Dissolving Film for Vaginal Delivery of Clotrimazole: Design, Characterization, and In Vitro Antifungal Action. *Drug Deliv. Transl. Res.* **2022**, *5*, 1–13. [CrossRef]
23. Contardi, M.; Russo, D.; Suarato, G.; Heredia-Guerrero, J.A.; Ceseracciu, L.; Penna, I.; Margaroli, N.; Summa, M.; Spanò, R.; Tassistro, G.; et al. Polyvinylpyrrolidone/Hyaluronic Acid-Based Bilayer Constructs for Sequential Delivery of Cutaneous Antiseptic and Antibiotic. *Chem. Eng. J.* **2019**, *358*, 912–923. [CrossRef]

24. Neto, R.J.G.; Maria, G.; De Almeida, L.; Santos, P.; Agostini, M.; Moraes, D.; Masumi, M. Characterization and in Vitro Evaluation of Chitosan/Konjac Glucomannan Bilayer Film as a Wound Dressing. *Carbohydr. Polym.* **2019**, *212*, 59–66. [CrossRef] [PubMed]
25. Dhal, C.; Mishra, R. In Vitro and in Vivo Evaluation of Gentamicin Sulphate-Loaded PLGA Nanoparticle-Based Film for the Treatment of Surgical Site Infection. *Drug Deliv. Transl. Res.* **2020**, *10*, 1032–1043. [CrossRef]
26. Shahzad, A.; Khan, A.; Afzal, Z.; Farooq, M.; Khan, J.; Majid, G. Formulation Development and Characterization of Cefazolin Nanoparticles-Loaded Cross-Linked Films of Sodium Alginate and Pectin as Wound Dressings. *Int. J. Biol. Macromol.* **2019**, *124*, 255–269. [CrossRef]
27. Gehrcke, M.; Brum, T.B.; Rosa, L.S.; Ilha, B.D.; Soares, F.Z.M.; Cruz, L. Incorporation of Nanocapsules into Gellan Gum Films: A Strategy to Improve the Stability and Prolong the Cutaneous Release of Silibinin. *Mater. Sci. Eng. C* **2021**, *119*, 111624. [CrossRef]
28. Fessi, H.; Puisieux, F.; Devissaguet, J.P.; Ammoury, N.; Benita, S. Nanocapsule Formation by Interfacial Polymer Deposition Following Solvent Displacement. *Int. J. Pharm.* **1989**, *55*, R1–R4. [CrossRef]
29. ASTM. *Standard Test Method for Tensile Properties of Thin Plastic Sheeting-D882-02*; ASTM International: West Conshohocken, PA, USA, 2002.
30. Schlupp, P.; Weber, M.; Schmidts, T.; Geiger, K.; Runkel, F. Development and Validation of an Alternative Disturbed Skin Model by Mechanical Abrasion to Study Drug Penetration. *Results Pharma. Sci.* **2014**, *4*, 26–33. [CrossRef]
31. Osmari, B.F.; Giuliani, L.M.; Reolon, J.B.; Rigo, G.V.; Tasca, T.; Cruz, L. Gellan Gum-Based Hydrogel Containing Nanocapsules for Vaginal Indole-3- Carbinol Delivery in Trichomoniasis Treatment. *Eur. J. Pharm. Sci.* **2020**, *151*, 105379. [CrossRef]
32. Yang, N.; Jia, X.; Wang, D.; Wei, C.; He, Y.; Chen, L.; Zhao, Y. Silibinin as a Natural Antioxidant for Modifying Polysulfone Membranes to Suppress Hemodialysis-Induced Oxidative Stress. *J. Memb. Sci.* **2019**, *574*, 86–99. [CrossRef]
33. ASTM. *Standard Practice for Assessment of Hemolytic Properties of Materials-F 756-00*; ASTM International: West Conshohocken, PA, USA, 2020.
34. Chan, C.C.; Liou, C.J.; Xu, P.Y.; Shen, J.J.; Kuo, M.L.; Len, W.B.; Chang, L.E.; Huang, W.C. Effect of Dehydroepiandrosterone on Atopic Dermatitis-like Skin Lesions Induced by 1-Chloro-2,4-Dinitrobenzene in Mouse. *J. Dermatol. Sci.* **2013**, *72*, 149–157. [CrossRef]
35. Voss, G.T.; Oliveira, R.L.; de Souza, J.F.; Duarte, L.F.B.; Fajardo, A.R.; Alves, D.; Luchese, C.; Wilhelm, E.A. Therapeutic and Technological Potential of 7-Chloro-4-Phenylselanyl Quinoline for the Treatment of Atopic Dermatitis-like Skin Lesions in Mice. *Mater. Sci. Eng. C* **2018**, *84*, 90–98. [CrossRef]
36. Weber, D.M.; Voss, G.T.; De Oliveira, R.L.; da Fonseca, C.A.R.; Paltian, J.; Rodrigues, K.C.; Ianiski, F.R.; Vaucher, R.A.; Luchese, C.; Wilhelma, E.A. Topic Application of Meloxicam-Loaded Polymeric Nanocapsules as a Technological Alternative for Treatment of the Atopic Dermatitis in Mice. *J. Appl. Biomed. J.* **2018**, *16*, 337–343. [CrossRef]
37. Kim, H.; Kim, J.R.; Kang, H.; Choi, J.; Yang, H.; Lee, P.; Kim, J.; Lee, K.W. 7,8,4′-Trihydroxyisoflavone Attenuates DNCB-Induced Atopic Dermatitis-like Symptoms in NC/Nga Mice. *PLoS ONE* **2014**, *9*, e104938. [CrossRef]
38. Park, G.; Oh, M.S. Inhibitory Effects of Juglans Mandshurica Leaf on Allergic Dermatitis-like Skin Lesions-Induced by 2,4-Dinitrochlorobenzene in Mice. *Exp. Toxicol. Pathol.* **2014**, *66*, 97–101. [CrossRef]
39. Green, L.C.; Wagner, D.A.; Glogowski, J.; Skipper, P.L.; Wishnok, J.S.; Tannenbaum, S.R. Analysis of Nitrate, Nitrite, and [15N]Nitrate in Biological Fluids. *Anal. Biochem.* **1982**, *126*, 131–138. [CrossRef]
40. Bradford, M.M. A Rapid and Sensitive Method for the Quantitation of Microgram Quantities of Protein Utilizing the Principle of Protein-Dye Binding. *Anal. Biochem.* **1976**, *72*, 248–254. [CrossRef]
41. Ohkawa, H.; Ohishi, N.; Yagi, K. Assay for Lipid Peroxides in Animal Tissues Thiobarbituric Acid Reaction. *Anal. Biochem.* **1979**, *358*, 351–358. [CrossRef]
42. Ellman, G.L. Tissue sulfhydryl groups. *Am. J. Anal. Chem.* **1959**, *82*, 70–77. [CrossRef]
43. Aebi, H. Catalase in vitro. *Methods Enzymol.* **1984**, *105*, 121–126. [PubMed]
44. Dixit, R.P.; Puthli, S.P. Oral Strip Technology: Overview and Future Potential. *J. Control. Release* **2009**, *139*, 94–107. [CrossRef] [PubMed]
45. Machado, A.; Cunha-reis, C.; Araújo, F.; Nunes, R.; Seabra, V.; Ferreira, D.; Sarmento, B. Development and in Vivo Safety Assessment of Tenofovir-Loaded Nanoparticles-in-Film as a Novel Vaginal Microbicide Delivery System. *Acta Biomater.* **2016**, *44*, 332–340. [CrossRef] [PubMed]
46. Ghorbani, M.; Hassani, N.; Raeisi, M. Development of Gelatin Thin Film Reinforced by Modified Gellan Gum and Naringenin-Loaded Zein Nanoparticle as a Wound Dressing. *Macromol. Res.* **2022**, *30*, 397–405. [CrossRef]
47. Evans, N.D.; Oreffo, R.O.C.; Healy, E.; Thurner, P.J.; Hin, Y. Epithelial Mechanobiology, Skin Wound Healing, and the Stem Cell Niche. *J. Mech. Behav. Biomed. Mater.* **2013**, *28*, 397–409. [CrossRef]
48. De Lima, J.A.; Paines, T.C.; Motta, M.H.; Weber, W.B.; Santos, S.S.; Cruz, L.; Silva, C.D.B. Novel Pemulen/Pullulan Blended Hydrogel Containing Clotrimazole-Loaded Cationic Nanocapsules: Evaluation of Mucoadhesion and Vaginal Permeation. *Mater. Sci. Eng. C* **2017**, *79*, 886–893. [CrossRef]
49. Pagano, C.; Rachele, M.; Calarco, P.; Scuota, S.; Conte, C.; Primavilla, S.; Ricci, M.; Perioli, L. Bioadhesive Polymeric Films Based on Usnic Acid for Burn Wound Treatment: Antibacterial and Cytotoxicity Studies. *Colloids Surf. B Biointerfaces* **2019**, *178*, 488–499. [CrossRef]

50. Parisotto-peterle, J.; Bidone, J.; Lucca, L.G.; de Moraes Soares Araújo, G.; Falkembach, M.C.; da Silva Marques, M.; Horn, A.P.; dos Santos, M.K.; da Veiga, V.F.; Limberger, R.P.; et al. Healing Activity of Hydrogel Containing Nanoemulsified β-Caryophyllene. *Eur. J. Pharm. Sci.* **2020**, *148*, 105318. [CrossRef]
51. Mansuri, S.; Kesharwani, P.; Jain, K.; Tekade, R.K.; Jain, N.K. Mucoadhesion: A Promising Approach in Drug Delivery System. *React. Funct. Polym.* **2016**, *100*, 151–172. [CrossRef]
52. Pagano, C.; Latterini, L.; Di Michele, A.; Luzi, F.; Puglia, D.; Ricci, M.; Viseras, A.; Perioli, L. Polymeric Bioadhesive Patch Based on Ketoprofen-Hydrotalcite Hybrid for Local Treatments. *Pharmaceutics* **2020**, *12*, 733. [CrossRef]
53. Timur, S.S.; Yüksel, S.; Akca, G.; Şenel, S. Localized Drug Delivery with Mono and Bilayered Mucoadhesive Films and Wafers for Oral Mucosal Infections. *Int. J. Pharm.* **2019**, *559*, 102–112. [CrossRef]
54. Montenegro, L.; Parenti, C.; Turnaturi, R.; Pasquinucci, L. Resveratrol-Loaded Lipid Nanocarriers: Correlation between in Vitro Occlusion Factor and in Vivo Skin Hydrating Effect. *Pharmaceutics* **2017**, *9*, 58. [CrossRef]
55. Venturini, C.G.; Bruinsmann, F.A.; Oliveira, C.P.; Contri, R.V.; Pohlmann, A.R.; Guterres, S.S. Vegetable Oil-Loaded Nanocapsules: Innovative Alternative for Incorporating Drugs for Parenteral Administration. *J. Nanosci. Nanotechnol.* **2016**, *16*, 1310–1320. [CrossRef]
56. Hussain, Z.; Katas, H.; Cairul, M.; Mohd, I.; Kumulosasi, E.; Sahudin, S. Antidermatitic Perspective of Hydrocortisone as Chitosan Nanocarriers: An Ex Vivo and In Vivo Assessment Using an NC / Nga Mouse Model. *J. Pharm. Sci.* **2013**, *102*, 1063–1075. [CrossRef]
57. Khelifi, I.; Hayouni, E.A.; Cazaux, S.; Ksouri, R.; Bouajila, J. Evaluation of in Vitro Biological Activities: Antioxidant; Anti-Inflammatory; Anti- Cholinesterase; Anti- Xanthine Oxidase, Anti-Superoxyde Dismutase, Anti-α-Glucosidase and Cytotoxic of 19 Bioflavonoids. *Cell. Mol. Biol.* **2020**, *66*, 9–19. [CrossRef]
58. Kabashima, K.; Nomura, T. Revisiting Murine Models for Atopic Dermatitis and Psoriasis with Multipolar Cytokine Axes. *Curr. Opin. Immunol.* **2017**, *48*, 99–107. [CrossRef]
59. Tang, L.; Li, X.L.; Deng, Z.X.; Xiao, Y.; Cheng, Y.H.; Li, J.; Ding, H. Conjugated Linoleic Acid Attenuates 2,4-Dinitrofluorobenzene-Induced Atopic Dermatitis in Mice through Dual Inhibition of COX-2/5-LOX and TLR4/NF-KB Signaling. *J. Nutr. Biochem.* **2020**, *81*, 108379. [CrossRef]
60. Saini, S.; Pansare, M. New Insights and Treatments in Atopic Dermatitis. *Immunol. Allergy Clin. N. Am.* **2021**, *41*, 653–665. [CrossRef]
61. Fan, P.; Yang, Y.; Liu, T.; Lu, X.; Huang, H.; Chen, L.; Kuang, Y. Anti-Atopic Effect of Viola Yedoensis Ethanol Extract against 2,4-Dinitrochlorobenzene-Induced Atopic Dermatitis-like Skin Dysfunction. *J. Ethnopharmacol.* **2021**, *280*, 114474. [CrossRef]
62. Pelle, E.; Mammone, T.; Maes, D.; Frenkel, K. Keratinocytes Act as a Source of Reactive Oxygen Species by Transferring Hydrogen Peroxide to Melanocytes. *J. Investig. Dermatol.* **2005**, *124*, 793–797. [CrossRef]
63. Rodriguez, K.J.; Wong, H.K.; Oddos, T.; Southall, M.; Frei, B.; Kaur, S. A Purified Feverfew Extract Protects from Oxidative Damage by Inducing DNA Repair in Skin Cells via a PI3-Kinase-Dependent Nrf2/ARE Pathway. *J. Dermatol. Sci.* **2013**, *72*, 304–310. [CrossRef] [PubMed]
64. Steullet, P.; Cabungcal, J.H.; Monin, A.; Dwir, D.; O'Donnell, P.; Cuenod, M.; Do, K.Q. Redox Dysregulation, Neuroinflammation, and NMDA Receptor Hypofunction: A "Central Hub" in Schizophrenia Pathophysiology? *Schizophr. Res.* **2016**, *176*, 41–51. [CrossRef] [PubMed]
65. Li, W.; Wu, X.; Xu, X.; Wang, W.; Song, S.; Liang, K.; Yang, M.; Guo, L.; Zhao, Y.; Li, R. Coenzyme Q10 Suppresses TNF-α-Induced Inflammatory Reaction In Vitro and Attenuates Severity of Dermatitis in Mice. *Inflammation* **2016**, *39*, 281–289. [CrossRef] [PubMed]
66. Wink, D.A.; Hines, H.B.; Cheng, R.Y.S.; Switzer, C.H.; Flores-Santana, W.; Vitek, M.P.; Ridnour, L.A.; Colton, C.A. Nitric Oxide and Redox Mechanisms in the Immune Response. *J. Leukoc. Biol.* **2011**, *89*, 873–891. [CrossRef] [PubMed]
67. Barnes, L.; Kaya, G.; Rollason, V. Topical Corticosteroid-Induced Skin Atrophy: A Comprehensive Review. *Drug Saf.* **2015**, *38*, 493–509. [CrossRef] [PubMed]
68. Slater, N.A.; Morrell, D.S. Systemic Therapy of Childhood Atopic Dermatitis. *Clin. Dermatol.* **2015**, *33*, 289–299. [CrossRef]
69. Danby, S.G.; Draelos, Z.D.; Gold, L.F.S.; Cha, A.; Aikman, L.; Sanders, P.; Wu-linhares, D.; Cork, M.J.; Danby, S.G.; Draelos, Z.D.; et al. Vehicles for Atopic Dermatitis Therapies: More than Just a Placebo. *J. Dermatolog. Treat.* **2022**, *33*, 685–698. [CrossRef]
70. Mcdaniel, M.L.; Kwon, G.; Hill, J.R.; Marshall, C.A.; Corbett, J.A. Cytokines and Nitric Oxide in Islet Inflammation and Diabetes (43950D). *Exp. Biol. Med.* **1996**, *211*, 24–32. [CrossRef]
71. Winterbourn, C.C. Revisiting the Reactions of Superoxide with Glutathione and Other Thiols. *Arch. Biochem. Biophys.* **2016**, *595*, 68–71. [CrossRef]
72. Hussain, Z.; Katas, H.; Cairul, M.; Mohd, I.; Kumolosasi, E.; Buang, F.; Sahudin, S. Self-Assembled Polymeric Nanoparticles for Percutaneous Co-Delivery of Hydrocortisone/Hydroxytyrosol: An Ex Vivo and In Vivo Study Using an NC/Nga Mouse Model. *Int. J. Pharm.* **2013**, *444*, 109–119. [CrossRef]

pharmaceutics

Review

Application of Intranasal Administration in the Delivery of Antidepressant Active Ingredients

Zhiyu Jin, Yu Han, Danshen Zhang, Zhongqiu Li, Yongshuai Jing, Beibei Hu * and Shiguo Sun *

College of Chemistry and Pharmaceutical Engineering, Hebei University of Science and Technology, Shijiazhuang 050018, China
* Correspondence: hub2018@hebust.edu.cn (B.H.); sunsg@nwsuaf.edu.cn (S.S.)

Abstract: As a mental disease in modern society, depression shows an increasing occurrence, with low cure rate and high recurrence rate. It has become the most disabling disease in the world. At present, the treatment of depression is mainly based on drug therapy combined with psychological therapy, physical therapy, and other adjuvant therapy methods. Antidepressants are primarily administered peripherally (oral and intravenous) and have a slow onset of action. Antidepressant active ingredients, such as neuropeptides, natural active ingredients, and some chemical agents, are limited by factors such as the blood–brain barrier (BBB), first-pass metabolism, and extensive adverse effects caused by systemic administration. The potential anatomical link between the non-invasive nose–brain pathway and the lesion site of depression may provide a more attractive option for the delivery of antidepressant active ingredients. The purpose of this article is to describe the specific link between intranasal administration and depression, the challenges of intranasal administration, as well as studies of intranasal administration of antidepressant active ingredients.

Keywords: depression; blood–brain barrier; brain targeting; antidepressant active ingredients; intranasal administration; challenges of delivery

Citation: Jin, Z.; Han, Y.; Zhang, D.; Li, Z.; Jing, Y.; Hu, B.; Sun, S. Application of Intranasal Administration in the Delivery of Antidepressant Active Ingredients. *Pharmaceutics* **2022**, *14*, 2070. https://doi.org/10.3390/pharmaceutics14102070

Academic Editor: Piroska Szabó-Révész

Received: 23 August 2022
Accepted: 24 September 2022
Published: 28 September 2022

1. Introduction

Depression is a very common central nervous system (CNS) disorder disease worldwide. Clinical symptoms include chronic depression, apathy, loss of appetite, and loss of interest. According to the World Health Organization, more than 350 million people worldwide suffer from depression, hundreds of thousands commit suicide each year, and the number is rising rapidly. More than 75% of patients in countries with a shortage of medical resources and healthcare personnel go untreated [1]. The clinical treatment of depression is still characterized by a low cure rate, high recurrence rate, residual symptoms, dysfunction, and high risk of self-injury and suicide, which imposes a serious burden on physical and mental health and economic levels around the world [2,3]. After decades of research, the most widely accepted treatment strategy for depression is a combination of medication, psychotherapy, and physical therapy. Due to the social environment, psychological burden, and other reasons, the use of antidepressants is generally accepted by patients. There are several antidepressants on the market, most of which are administered orally in tablets or capsules (Selegiline in the transdermal patch, Esketamine in the nasal spray, and Brexanolone in the intravenous infusion). Antidepressants are absorbed by different untargeted tissues and organs in the gastrointestinal tract or after absorption into the blood circulatory system, resulting in systemic clearance (in oral and parenteral route) and widespread adverse reactions such as drowsiness, weight gain, constipation, dry mouth, and dizziness [4–7]. Invasive drug administration of the brain can cause great discomfort to patients. Some depressive patients have marked relief or disappearance of depressive symptoms after taking antidepressants for some time. However, there is also a subset of patients (major depressive disorder, MDD) who do not respond to two or more first-line antidepressants and have acute or severe suicidal thoughts. Drug therapy is limited because of the complexity of the pathogenesis of depression and the variability of individual patients [8]. The current clinical situation is that depressive symptoms are relieved after 2 to 3 weeks or even longer after oral first-line antidepressants. The early stage after taking antidepressants is also the stage where adverse reactions are prominent. During this incubation period, patients are at increased risk of disability or suicide (especially in patients $\leq$ 24 years of age), aggravated disease, and decreased medication compliance. Therefore, there is an urgent need to find new antidepressant active ingredients and routes of administration with a fast onset of action and fewer side effects.

Due to BBB, extensive metabolism, high protein binding rate, and systemic side effects, many ingredients with antidepressant activity cannot exert effective therapeutic effects by oral or injection administration. The BBB is mainly composed of microvascular endothelial cells, mural cells, and glial cell astrocytes (Figure 1). In contrast to inner cells in other tissues, brain endothelial cells are connected by tight junctions (TJs) and have very little vesicle-mediated transcellular transport. The endothelium of the brain contains a variety of enzymes that inactivate certain neurotransmitters, antidepressants, and toxins, preventing them from entering the brain. In addition, efflux transporters, which are located on the blood side of the endothelial cells, use energy to transport passively diffused lipophilic molecules back into the blood, especially the P-glycoprotein (P-gp) [9–11]. Preclinical studies have shown that many antidepressants are substrates of P-gp, which may affect the distribution of antidepressants to their target of action [12,13]. The effects of ABCB1 polymorphisms on P-gp expression and antidepressant transport were significantly different between individuals, which may lead to treatment variability [14].

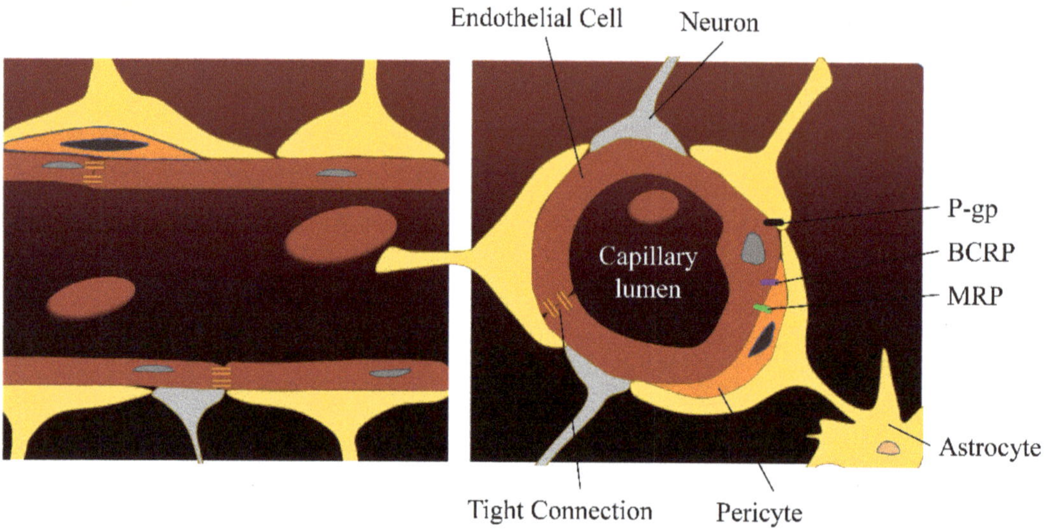

Figure 1. The blood–brain barrier. P-gp, P-glycoprotein; BCRP, breast cancer resistance protein; MRP, multidrug resistance-associated protein.

Some first-line antidepressants, off-label drugs, peptides, natural active ingredients, and other substances with antidepressant activity are limited by the choice of route of administration because of their properties. Even when administered intravenously, some drugs still enter the liver for degradation and are restricted by the BBB. Intracerebral or spinal injections are the most direct methods of delivering drugs to the brain but are invasive and cause great discomfort to patients. Invasive drug delivery is not suitable for depression, which requires long-term treatment. Subsequent increasing studies have shown that intranasal administration can treat brain diseases through the bypassing of BBB, olfactory nerve pathways, trigeminal nerve pathway, and mucosal epithelial pathways, to get the drugs to the CNS [15–18]. Compared with traditional antidepressant administration routes (oral administration, intravenous administration, intramuscular injection, etc.), intranasal has the advantages of avoiding first-pass effect, improving bioavailability, short onset time, small dose of administration, less toxic and adverse reactions on the body, and good patient compliance [17]. The direct non-invasive pathway between the nose and the brain is undoubtedly one of the best options for the limited antidepressant active ingredients to enter the CNS to exert antidepressant effects. The purpose of this review is to present the unique anatomical and physiological link between the nose–brain pathway and the lesion site of depression. For example, the olfactory system has a high degree of overlap with areas that process emotions and memory functions [19,20]. Furthermore, this review also summarizes studies of antidepressant active ingredients (off-label drugs, peptides and natural active ingredients, etc.), in addition to first-line antidepressants and the delivery carriers in intranasal administration.

2. Unique Advantages of Intranasal Administration in the Treatment of Depression

The nose is the starting part of the respiratory pathway and olfactory organs. The nose is divided into two different compartments by the nasal septum. The surface area of the human nasal mucosa is 150 cm^2~160 cm^2, the thickness of nasal mucosa is about 2~5 mm, and the average pH value of nasal mucus is 5.5~6.5 [18,21]. The nasal cavity can be divided into three areas according to their structure and function: the nasal vestibule, olfactory region, and respiratory region.

2.1. Direct Pathway

The nasal cavity is the only non-invasive direct route between the CNS and the external environment. The nasal vestibular area is located in the front of the nasal cavity with a very small surface area, only about 0.6 cm^2. Parts of the nasal vestibular area are skin tissue. There are nasal hairs and mucus that block and filter particulate matter; so, the absorption capacity of antidepressants is limited in the nasal vestibule [22]. The olfactory area is distributed on the medial surface of the superior turbinate, part of the middle turbinate, and the corresponding part of the septum [8]. The surface area is relatively small, only about 10 cm^2. The cells of the olfactory region are composed of olfactory cells, supporting cells, basal cells, and trigeminal nerve cells. The lamina propria (LP)—the relatively loose connective tissue layer beneath the epithelial cells—contains nerves, blood vessels, and lymphatic vessels. Olfactory sensory neurons (OSNs) are bipolar neurons. Axons from OSNs expressing the same odorant receptors aggregate in LP to form axon bundles that pass through the ethmoid plate and terminate in glomeruli formed within the olfactory bulb (Figure 2) [23].

The glomerulus of the olfactory bulb is the only transit point between the peripheral and central olfactory systems. OSNs synapse with mitral cells and tuft cells to project secondary and tertiary olfactory structures. Olfactory information is transmitted to secondary olfactory structures, particularly the piriform cortex and amygdala cortex. Tertiary olfactory structures include the thalamus, hypothalamus, amygdala, hippocampus, orbitofrontal cortex, and insular cortex. These regions are closely associated with the expression of mood and memory function in depression [19,20]. Depressed patients are often associated with reduced olfactory bulb volume and olfactory dysfunction, and people with olfactory dysfunction are at higher risk for depression [25,26]. The causal relationship between the olfactory system and depression is unclear, but the correlation between the severity of olfactory disturbance and the severity of depression have been established. Melatonin MT_1 and MT_2 receptors are expressed in the glomerular layer of the olfactory bulb and coupled to Gi protein. Studies have shown that modulation of melatonin receptors expressed in the olfactory bulb can ameliorate 6-hydroxydopamine-induced depression-like behaviors [27,28]. By intranasal administration, the olfactory bulb can serve as a target for the antidepressant effects of melatonin and melatonin analogs.

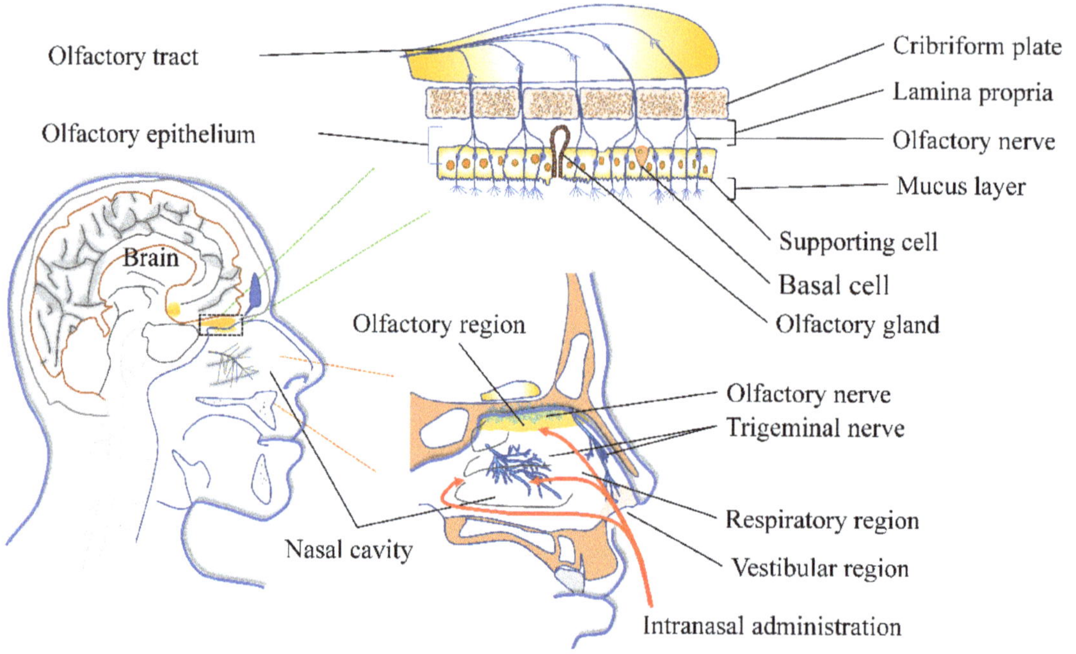

Figure 2. Anatomical schematic of the non-invasive pathway between nose and brain. Adapted with permission from Ref. [24]. Copyright 2018, Elsevier.

Olfactory nerve cilia, located on the surface of the olfactory mucosa, can internalize antidepressants, which are then slowly transported in the axoplasm to synapses within the olfactory bulb that connect with secondary olfactory neurons. Afterward, antidepressants are delivered to the prefrontal cortex, hippocampus, and other parts along the olfactory conduction pathway, possibly by repeating this process. Antidepressants need to be transported through the axons of neurons in the olfactory system at least through the tertiary neurons to reach the hippocampus. There are complex possibilities for interneuronal transfer and axoplasmic transport. Further studies are needed on drug delivery in the perineuronal and intraneuronal spaces and transfer between neurons.

The axons of olfactory neurons are surrounded by a series of olfactory ensheathing cells, and their outer layers are covered by additional neuro fibroblasts (ONFs). The ONFs layer is continuous to the meninges that cover the brain, which means that the perineural space formed between the olfactory sheath cell layer and the neuro fibroblasts are continuous to the subarachnoid space [29]. Antidepressants can enter LP through transcellular or paracellular pathways of the olfactory epithelium, and then along the perineural space into the olfactory bulb and cerebrospinal fluid [30]. Antidepressants need to pass through TJs to reach LP, where OSNs undergo apoptosis and replacement in about 30 to 60 days to protect the brain from airborne contaminants, bacteria, and so on. During this period, a potential delay between lysis and regrowth of OSNs leaves gaps between the tightly connected nasal epithelial cells, resulting in increased permeability [29,31]. In addition, increased permeability of drug absorption is associated with phosphorylation of closed-protein, such as protein kinase signaling or certain substances [32,33]. The characteristic of this pathway is that its effect is faster than that of the olfactory nerve pathway. As the olfactory bulb is located below and anterior to the orbital surface of the frontal lobe of the cerebral hemisphere, antidepressants can be rapidly distributed to the close prefrontal cortex.

The trigeminal nerve is another nerve pathway that connects the nose to the brain, innervating the olfactory and respiratory mucosa [7,29,34]. The trigeminal nerve originates from the pons, where the ocular and maxillary branches supply the nasal cavity. Unlike the axonal bundles of the OSNs, the trigeminal nerve is mainly composed of the myelin sheath made of Schwann cells. The non-myelinated branches of the trigeminal nerve supply olfactory mucosal vessels and regulate blood flow. Trigeminal nerve endings (and their associated arteries) are located below LP and nasal mucosa TJs and do not penetrate the epithelial surface like OSNs [35,36]. Insulin, which has antidepressant activity, was administered intranasally, and fluorescently labeled insulin was found to reach the brain through the extracellular space around the trigeminal nerve [37]. Drugs entering LP may also reach other brain areas, such as the brainstem, through the trigeminal branch [35].

The locus coeruleus and raphe nuclei located in the brainstem are the major sites for the synthesis of norepinephrine (NE) and serotonin (5-HT) in the brain. The locus coeruleus and raphe nuclei have a wide range of neuronal projections and play an important role in regulating neuronal activity in areas such as the prefrontal cortex, amygdala, hippocampus, lateral habenula, and anterior cingulate gyrus related to emotion and memory. Neuronal damage in the locus coeruleus and raphe nucleus is strongly associated with depression, especially MDD. The study found that after intranasal administration of radiolabeled immunoglobulin ([^{125}I]-IgG), the highest signal in the olfactory bulb and brain stem was observed by radiography [38]. Pang et al. evaluated the pharmacokinetics of intranasal insulin in the rat brain and showed that the highest levels were in the brainstem of the brain, followed by the olfactory bulb, cerebellum, hippocampus, hypothalamus, and striatum [39]. Exogenous active ingredients may be directly distributed to the brainstem through the trigeminal nerve and its surrounding space to repair damaged neurons.

According to previous studies, the convection in the brain perivascular space driven by arterial pulsation is thought to be the main reason for the rapid and widespread distribution of antidepressant active ingredients after entering the brain from the nasal cavity [40,41]. As the olfactory bulb is located below and anterior to the orbital surface of the frontal lobe, antidepressants can be rapidly distributed to the adjacent prefrontal cortex through the olfactory bulb. The cerebrospinal fluid and the subarachnoid space have many vascular punctures deep into the brain, separated from the brain parenchyma by the pia mater. The tiny spaces between these blood vessel punctures and the pia mater allow for the circulation of cerebrospinal fluid, known as the cerebrospinal fluid microcirculation. Antidepressants are better distributed in the brain through perineural, perivascular, and cerebrospinal fluid flow.

2.2. Indirect Pathway

The olfactory area, about 130 square centimeters, acts to warm and humidify the inhaled air as well as filter particles and pathogenic microorganisms. The mucosa is the stratified or pseudostratified columnar ciliated epithelium, and the cilia mainly move from the front to the back of the nasopharynx. The mucosa is rich in secretory glands and goblet cells, producing a large number of secretions. The mucosal surface is covered with a layer of mucus blanket, which moves backward with the movement of cilia [42]. The abundant capillaries in the respiratory area make this area a nasal cavity, and when administered, antidepressants are absorbed into the system rather than directly into the brain.

The nasal mucosa is in direct contact with the external environment, where many pathogens exist. Nasal-associated lymphoid tissue (NALT) plays a vital role in maintaining the body's immunity [43]. NALT is located in the LP and submucosal region of the nasal epithelium and connects to the cervical lymph nodes. This is also a potential pathway for the nose-to-brain antidepressant delivery.

Antidepressants can be absorbed by capillaries and lymphatics through the continuous porous endothelium, or enter the blood circulation in the olfactory area through the LP, avoiding first-pass metabolism [44]. Small lipophilic molecules pass more easily. However, the indirectly transported substances still need to pass through the BBB to enter the CNS or brain tissue; so, the substances that enter the brain through the circulation are usually small molecular weight and strong lipophilic substances [45,46].

From the perspective of physiological anatomy and pathological correlation, intranasal administration may have a more direct antidepressant effect than other routes of administration. Results from different studies suggest that intranasal drug delivery may be through a single route or a combination of different routes. Compared with the traditional administration route, intranasal administration can effectively avoid the extensive metabolism and low permeability of the blood–brain barrier caused by the gastrointestinal route. but also avoid the antidepressants in the circulation of the blood as a result of higher plasma protein binding and distribution in the other route, targeting the organization and improve the concentration of antidepressants in the CNS [2,47]. Non-invasive intranasal administration can effectively shorten the onset time and improve compliance in patients with depression requiring long-term treatment [2,48].

3. Challenges of Intranasal Administration

Intranasal administration offers an excellent strategy to overcome the challenge of the complex pathophysiology of brain diseases and drug penetration into the brain. However, the physiological characteristics of the nose, the physicochemical properties of antidepressant active ingredients, and even intranasal drug delivery devices can influence the nose-to-brain delivery of antidepressants [18,49].

TJs of the olfactory and respiratory epithelium and their protective mucus lining act as selective filters, reducing antidepressants diffusion and permeability [50]. Although most lipophilic compounds are more permeable to the nasal mucosa, peptides, macromolecules, and small hydrophilic molecule compounds are generally less permeable [34,51]. Permeation enhancers have been shown to improve the absorption of high molecular weight antidepressants by facilitating the production of hydrophilic pores, and increasing membrane fluidity and permeability of TJs [34,52]. Common Permeation enhancers include cyclodextrin, chitosan (CN), surfactant, Cremophor RH40, saponins, etc. [51,53,54]. Gavini E. et al. prepared solid microparticles based on CN or methyl-beta-cyclodextrin to enhance the nose-to-brain delivery of deferoxamine mesylate [55]. Cell-penetrating peptides and penetration accelerating sequences significantly facilitate the delivery of polypeptide antidepressants, such as insulin [56]. Permeation enhancers may cause irritation and toxicity to the nasal mucosa while increasing the permeability of the compounds. The selection of materials and safety experiments is the premise to ensure the safety and effectiveness of prescription, especially for depression requiring long-term administration. Nasal mucus is mainly composed of 90~95% mucin and 2~3% water and tends to dissolve hydrophilic substances [57]. Mucin fibers are usually composed of a proline–threonine–serine backbone with intermitting cysteine-rich domains. Different amino acids in the PTS region are highly glycosylated via O-linked bonding. The degree of mucin glycosylation affects the permeability and viscosity of the mucus [58,59]. There are many cilia on the surface of the olfactory and respiratory areas, which block particles from the external environment from entering the nasal cavity. The olfactory region cilia do not move, and the respiratory region cilia generally toward the pharynx at an average velocity of 5~8 mm per minute; so, the nasal administration of drug particles tends to be cleared in an average time of 20 to 30 min [60]. Inhibitory substances and mucoadhesive materials such as hyaluronan, poloxamer, carbopol, gellan gum, polycarbophil, and other polymers can effectively delay mucociliary clearance and increase the retention time of antidepressants in the nasal cavity, thus increasing drug intake in the brain by interacting with mucin or reversibly/irreversibly inhibiting cilia movement [51].

Although there are fewer enzymes in the nasal cavity, they still may affect antidepressant absorption, such as cytochrome P-450 enzyme system, glutathione S-transferase, proteolytic enzyme, and other "Pseudo-first-pass effect" on foreign substances. Enzyme inhibitors or some surfactants such as bestatin, amastatin, boroleucine, fusidic acids, and phospholipids can be used to avoid the metabolism of the compound, improve stability, and thus increase the absorption of the original drug [34].

Like the BBB, P-gp act as multidrug resistance pumps, expressed in nasal epithelial cells, and P-gp effusion restricts the entry of most substances into the brain. This effect can be modulated by the use of P-gp inhibitors [61]. For example, cyclosporine A and rifampin as P-gp inhibitors improve the permeability of verapamil after intranasal administration [62].

As the breathing area is much larger than the olfactory area, some of the antidepressants that are administered intranasally will still enter the brain via indirect routes. Therefore, studies have shown that the use of direct access can be increased by combining vasoconstrictors to reduce the absorption of antidepressants through the nasal vessels and to increase the retention time of antidepressants in the nasal mucosa [51,63]. Dhuria S. V. et al. administered an intranasal combination of phenylephrine and neuropeptides and measured the concentration of the drug in CNS tissues and blood. The addition of 1% phenyl epinephrine significantly reduced the absorption of HC and D-KTP in the blood and increased the delivery volume of the olfactory bulb [64]. Vasoconstrictors reduce the peripheral side effects of CNS antidepressants by restricting the absorption of nasal blood vessels into systemic circulation.

In addition to the physiological factors mentioned above, the physicochemical properties and formulation composition of antidepressant active ingredients also have an effect on the nose-to-brain delivery. The molecular weight of the drug is inversely proportional to the percentage absorbed. Lipophilic ingredients are more readily absorbed than hydrophilic ones. Lipophilic small molecule ingredients with a molecular weight of less than 1 kDa can be transported quickly. The absorption of ingredients of less than 300 Da was almost unaffected by molecular weight, while the absorption of antidepressants with molecular weight between 300 Da and 1 kDa was inversely correlated with molecular weight [65]. Currently, the most widely used and marketed antidepressants, such as Escitalopram, Fluoxetine, Duloxetine, and Venlafaxine, have molecular weights ranging from 200 to 400 Da [66,67]. The pH value of the formulation not only ensures the stability of the drug itself but also ensures the stability of the physiological conditions of the nasal cavity. The nasal pH range is 5.0~6.8, and intranasal administration should be close to this to avoid irritation of the nasal mucosa. The drug is always absorbed in a non-ionized state; so, the pKa of the drug should also be taken into account in determining the pH of the formulation [68]. However, changes in temperature, humidity, and some pathological conditions can cause changes in nasal pH [69]. In addition, the prescribed osmotic pressure should also be adapted to the physiological conditions of the nasal cavity; otherwise, it will affect the normal nasal mucosal cell morphology and ciliary movement and further affect the absorption of antidepressants [31,70].

The choice of drug delivery equipment also plays a very important role in whether the drug formulation can be utilized to the maximum extent and play its due therapeutic effect [34,49,71]. Treatment of CNS diseases such as depression requires antidepressants to be delivered to the olfactory region. Traditional pump sprays usually deposit antidepressants in the anterior nasal cavity, which are removed quickly by clearance of the nasal mucosa. The droppers may deposit the olfactory area above the nasal cavity better than nasal pump sprays, but this often requires the patient to lie on his or her back in a head-down, forward–forward position and even professional administration techniques, which can be very inconvenient [72]. Vianase™ device is an electronic atomizer developed by Kurve Technology® that consists of an atomizer and a vortex chamber [73]. The atomizer causes the preparation to produce atomized particles, which, under the action of the vortex chamber, form a vortex, and are ejected from the equipment in this form. Vianase™ device can precisely maximize the delivery of drug to the olfactory region through electronic control. Opt-Poeder device is a bidirectional delivery device that uses the patient's exhalation as the power to deliver the drug to the targeted site. The device is designed so that when a person exhales, the soft palate closes, preventing antidepressants from entering the respiratory tract and depositing in the nose. The Precision Olfactory Delivery device, designed and manufactured by Impl Neuropharma, is a single-dose delivery device that utilizes inert gas as an impeller to deliver drug to the superior nasal cavity [34,74]. Experimental data indicate that 45% of the dose of the Precision Olfactory Delivery device can be deposited in the upper nasal cavity compared with the traditional pump [75]. Other devices such as the Aero Pump System are also well used in intranasal delivery. However, the proportion of drug deposition in the olfactory region is still relatively low due to these devices; thus, the research and development of more convenient and efficient devices is one of the focuses of current research.

4. The Delivery Carriers and Nanocarriers for Intranasal Administration

4.1. Polymer-Based Carriers

The physiological characteristics of the nose–brain pathway and the physicochemical properties of ingredients lead to a certain degree of limitation for intranasal administration [58,59]. So far, many scholars have conducted studies on the nose-to-brain delivery system to improve the safety and effectiveness of ingredients, alongside the development in polymer technology and pharmaceutical technology. The delivery carriers and nanocarriers have made significant contributions to protecting antidepressants from protein degradation, enhancing olfactory mucosal uptake and CNS utilization and prolonging half-life (Figure 3).

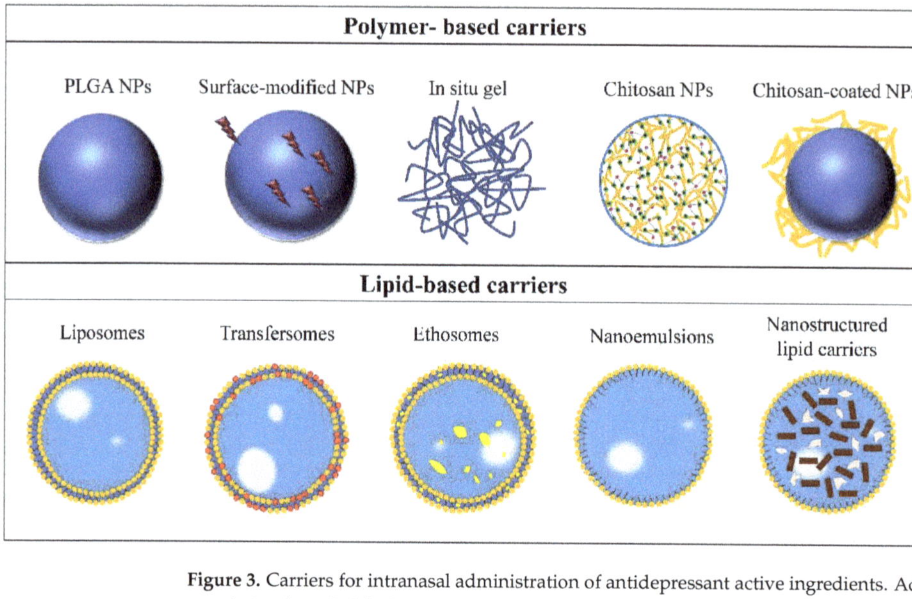

Figure 3. Carriers for intranasal administration of antidepressant active ingredients. Adapted with permission from Ref. [24]. Copyright 2018, Elsevier.

In situ gel refers to a kind of preparation that is transformed from liquid to non-chemical cross-linked semi-solid gel immediately after administration in solution state due to physiological conditions of the administration site or some stimulus factors in the external environment (pH, temperature, ionic strength, etc.) [76,77]. Compared with the traditional drug delivery system, in situ gels prepared with a variety of different polymers or containing different stimulus-induced release factors show good biocompatibility when exposed to the site of administration for a longer time to prolong the retention time of formulation and improve the bioavailability of antidepressants [78–80]. The combination of nanocarriers with in situ gels is also a promising strategy.

Poly (lactic-co-glycolic acid) (PLGA), which is approved for therapeutic applications by the U.S. Food and Drug Administration (FDA), is considered one of the most promising synthesized polymers as a drug delivery system due to its biodegradability, biocompatibility, controllable properties, well-defined formulation techniques, and great potential for targeting [31,81,82]. PLGA is a polymer synthesized by the interaction of glycolic acid and lactic acid monomer.

Chitosan nanoparticles (CN-NPs) not only can open TJs between cells transiently but also reduce the mucociliary clearance, prolonging the retention time of the compound as a biological adhesive material, which enhances the delivery of antidepressants [83,84]. CN is a natural linear polysaccharide cationic and hydrophilic polymer by alkaline hydrolysis of chitin, which is the second most abundant biopolymer in nature. It is an important component of the shells of many lower animals, especially arthropods such as shrimp, crabs, and insects. It also exists in the cell walls of lower plants such as bacteria, algae, and fungi. CN consists of randomly distributed β-(1,4)-linked d-glucosamine (deacetylated) and N-acetyl-d-glucosamine (acetylated) units. Due to its properties of adhesivity, biocompatibility, and biodegradability, CN has been well used in the field of medical engineering [85].

Alginate nanoparticles have also been studied and reported by many researchers in nose-to-brain drug delivery [86,87]. Alginate is extracted from brown Marine algae and then processed several times before it can be used as a polymer in pharmaceutical preparations. Alginates are mainly composed of β-D-mannuronic acid and α-L-guluronic acid, and have properties such as biocompatibility, biodegradability, low toxicity, and pH sensitivity [88,89]. Alginate contains a large number of carboxyl groups and is a hydrophilic anionic polymer, showing a certain adhesion. Divalent cations, such as Ca^{2+}, exchange ions with cations on α-L-guluronic to form a cross-linked network structure, thus forming alginate hydrogels [90].

Nanoemulsion (NE) is a liquid nano-dispersion system formed by two kinds of insoluble liquid stabilized by surfactant and co-surfactant (O/W; W/O). NEs have been widely used in nasal and brain delivery of insoluble antidepressants due to excellent solubility, thermodynamic stability, and easy preparation [91–93]. CN is often added to NEs to reduce clarity and prolong the retention time of antidepressants in the nasal cavity [91,94].

4.2. Lipid-Based Carriers

Compared with polymer carriers, lipid carriers have better biocompatibility than synthetic polymers because the materials of lipid carriers are basically derived from natural materials [95]. The degradation of polymer carriers in vivo is often accompanied by an increase in toxicity, while the degradation products of lipid carriers have low immunogenicity. Lipid carriers can effectively avoid being cleared by the immune system in the body and achieve long cycles. Liposome is a kind of structure similar to the cell membrane, mainly composed of phospholipids' double-layer synthetic membrane. In water-soluble solvents, the hydrophobic tails are clustered close to each other inside the bilayer phospholipids, while the hydrophilic heads are exposed outward to the aqueous phase, forming vesicles with bilayer molecular structure [96]. Liposomes can effectively protect the stability of encapsulated antidepressants, reduce drug toxicity, and play a slow release, prolonging the action time of antidepressants [97]. In addition, some studies have used altered phospholipid vesicles, such as transferosomes, ethosomes, and phospholipid magnesomes, for intranasal administration. Adding glycerol and ethanol makes the phospholipid vesicles softer, enhancing the permeability to the nose–brain pathway [98–100].

It is noteworthy that due to the traditional lipid carrier containing unsaturated chains, it is easy to fuse, oxidize, and hydrolyze as a solution form; so, the half-life of classic liposomes is short and the stability is poor. In order to overcome the defects of liposome preparations, new lipid-based forms for drug delivery have been gradually developed. For instance, solid lipid nanoparticles (SLNs) and nanostructured lipid carriers (NLCs) have been widely used in drug delivery [101,102]. SLNs are nanoparticles made using one or more lipids—such as triglycerides, lecithin, fatty acids, etc.—as carrier materials and combined with surfactants as stabilizers to form solid nanoparticles at body temperature and room temperature. Compared with traditional liposomes, SLNs have lower cytotoxicity, higher stability, and bioavailability and can more effectively maintain drug stability and control drug release, as well as mature the industrial production process [96,101–104]. Although SLNs have many advantages as nanoscale drug carriers, its disadvantages should not be ignored. For example, SLNs still have the problem of gelation in the dispersed phase and expulsion of encapsulated antidepressants resulting from β-modification during storage, and low encapsulation rates and drug loads due to the fact that cavities are not allowed to occur within the lipoid nucleus during crystallization [104]. NLCs are the second generation of lipid nanomaterials. Different from SLNs, NLCs are composed of liquid lipids and solid lipids combined with surfactants as stabilizer to form nanoparticles that are also solid at room temperature and body temperature. NLCs significantly enhance drug loading due to the existence of liquid lipids, imperfect crystal order and loose structure of nanoparticles, and reduced drug leakage caused by β-modification during storage. The combined use of different carriers, or the functional modification of classical carriers, has aroused extensive interest in facilitating intranasal drug delivery and has promising prospects.

5. Intranasal Administration of Antidepressant Active Ingredients

5.1. Antidepressants

Venlafaxine (VLF) inhibits central 5-HT and NE neuronal reuptake, thereby increasing levels in the synaptic cleft between neurons in the brain. Oral VLF has low bioavailability, a short half-life, delayed onset of action, and significant systemic adverse reactions [105,106]. Cayero-Otero et al. prepared VLF-loaded PLGA-NPs and modified the nanoparticles with transferrin and specific peptide against transferrin receptor. Compared with plain NPs (around −25 mV), the amide reaction of the carboxylic acid terminus on the PLGA surface with the amino group of the peptide resulted in a less negative zeta potential value. The release rate of ordinary nanoparticles is higher than that of modified nanoparticles. Cell viability of h-CMEC/D3 cells is more than 85% in the MTT assay. In vivo biodistribution studies showed higher concentrations of plain fluorescent NPs than functionalized NPs in the brain after 30 min of administration. The authors hypothesize that the reason for this result is that plain NPs enter the brain through direct neural and olfactory epithelial pathways between the nose and brain. However, the speed and degree of functionalized NPs transport, which were mediated by the transferrin receptor expressed by the BBB, are not superior to the direct pathway between nose and brain [107]. Baboota S. et al. published a report about the new preparation of VLF-loaded CN-NPs by using CN and sodium tripolyphosphate ionic gel (Figure 4). CN is positively charged and has mucosal adhesion. It can inversely open TJs between nasal mucosal epithelial cells. The cumulative drug permeability after 24 h in VLF CN-NPs was nearly 3 times compared with VLF solution. VLF CN-NPs showed a more significant antidepressant effect than VLF solution on chronic depression rats by forced swimming method [108]. The properties of CN could be modified by functional groups, such as cross-linking, etherification, carboxymethylation, and graft copolymerization [109].

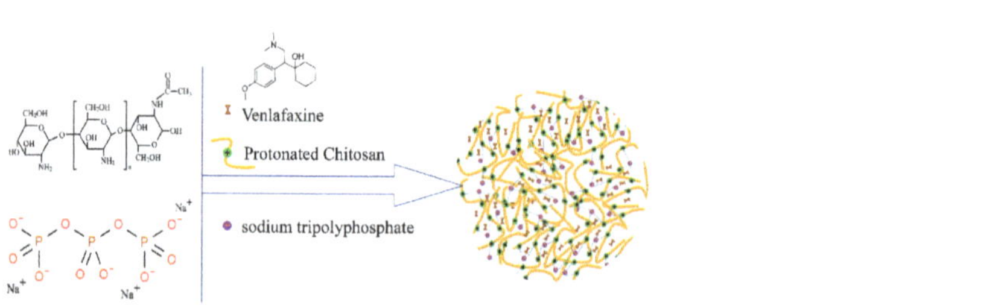

Figure 4. Chitosan is protonated under acidic conditions, and then ionically cross-linked with negatively charged sodium tripolyphosphate to prepare chitosan nanoparticles.

Desvenlafaxine (DVF) is a 5-HT and NE reuptake inhibitor. Systemic side effects were obvious after oral administration of DVF [110]. Tong et al. prepared desgavenlafaxine-loaded PLGA-CN-NPs. Intranasal PLGA-CN-NPs can increase the retention time and permeability of DVF in the nasal mucosa, avoiding degradation in the gastrointestinal tract, and liver intranasal PLGA-CN-NPs can increase the residence time and permeability of DVF in the nasal mucosa, avoiding degradation in the gastrointestinal tract and liver and bypassing BBB. In a rodent model of depression, compared with intranasal DVF solution and oral administration, increased levels of 5-HT and NE in the brain showed a more pronounced antidepressant effect. Pharmacokinetic parameters such as concentration, half-life, and AUC in the brain after intranasal administration were higher than those after intravenous [111].

As an atypical antidepressant, agomelatine has antidepressant effects by agonizing MT1, MT2 and antagonizing 5-HT2C. Agomelatine is well and rapidly absorbed (≥80%) after oral administration. Peak plasma concentrations are reached within 1–2 h after oral administration. However, the first-pass metabolism of agomelatine in the liver is obvious and the protein binding rate in the blood is high. Yasmin et al. prepared an intranasal formulation of agomelatine-loaded NE-thermosensitive in situ gel with CN. Pharmacokinetic study in Wistar rats showed that plasma concentration in the brain was 2.82 times higher than that of the intravenous suspension via the intranasal route [112]. The nasal solid lipid nanoparticles prepared by Ahmed et al. were superior to the oral suspension in brain concentration, $AUC_{0–360min}$, and absolute bioavailability (44.44%) [113]. Their other study combined drug-loaded SLNs-NPs with in situ gels, which also showed that intranasal administration has advantages over the oral route [114]. In the forced swim test, polymer nanoparticles receiving intranasal agomelatine significantly reduced immobility time in mice compared with suspension [115].

Duloxetine is a 5-HT and NE reuptake inhibitor. Duloxetine is acid intolerant, and its oral first-pass metabolism significantly affects bioavailability. Its concentration in cerebrospinal fluid is low. Baboota et al. loaded duloxetine into solid/liquid lipid-based nanostructured lipid carriers (DLX-NLCs). Intranasal DLX-NLCs showed higher concentrations in blood and brain compared with DLX solution and oral route, which showed the same results in behavioral tests in mice. Intranasal NLCs were 8-fold higher in brain concentrations than intravenous DLX. The controlled release provides the possibility for the sustained action of DLX in the brain [116,117]. Fares et al. designed a thiomer gel loaded with duloxetine proniosomes to increase the retention time, sustained release, and penetration of DLX in the nasal mucosa (1.86 times that of duloxetine proniosomes) [118]. Shah et al. designed intranasal DLX in situ cubo-gel by a central composite approach. Compared with the intranasal DLX solution, brain bioavailability was increased by 1.96 times [119].

Paroxetine, a phenylpiperidine derivative, selectively suppresses the 5-HT transporter, blocks the reuptake of 5-HT by the presynaptic membrane, and prolongs and increases the effect of 5-HT, thereby producing antidepressant effects. However, paroxetine has extensive first-pass metabolism and the BBB obstruction that limits its access to the brain [120]. Baboota S. et al. successfully developed an O/W type NE loaded with paroxetine for the study of intranasal treatment of depression. The permeability of paroxetine NEs was 2.57 times higher than that of its suspension via permeation studies. Results of behavioral studies in rats showed that intranasal administration of paroxetine NEs significantly improved behavioral activity in depressed rats compared with the oral suspension of paroxetine [121]. In addition, in nose-to-brain delivery for brain diseases, coating with adhesives such as CN is often added to NEs to reduce clarity and prolong the retention time of antidepressants in the nasal cavity [91,94]. Fortuna et al. prepared three nanostructured lipid carriers loaded with escitalopram and paroxetine. In vivo studies showed that intranasal delivery of the drug had similar pharmacokinetic parameters to intravenous administration, whereas escitalopram did not exhibit significant direct nasobrain delivery but reduced exposure elsewhere. In contrast, intranasal delivery of paroxetine-loaded borneol-NLCs significantly increased the concentration in the brain (5-fold), showing good brain targeting [122].

Trazodone is a tetracyclic atypical antidepressant that inhibits 5-HT reuptake and antagonizes 5-HT2 receptors in the presynaptic membrane to promote 5-HT release. Plasma concentrations and onset of oral trazodone are significantly affected by food, and there are significant systemic side effects. Sayyed et al. radiolabeled trazodone and compared the pharmacokinetic parameters of intranasal delivery of ^{131}I-TZ solution, ^{131}I-TZ microemulsion, and intravenous injection of ^{131}I-TZ solution. Intranasal ^{131}I-TZ microemulsion had sustained and higher brain uptake at any time tested than the other two formulations and routes. In addition, the blood exposure of intranasal ^{131}I-TZ microemulsion was lower than that of intravenous injection, reducing systemic toxicity [123].

Quetiapine fumarate can clinically treat manic episodes of bipolar disorder by antagonizing 5-HT2A receptors and D2 receptors. Oral quetiapine fumarate is also widely metabolized and has poor bioavailability. In addition, quetiapine fumarate is also a substrate of P-gp, and its absorption and brain distribution are limited. The intranasal NE prepared by Boche et al. enabled quetiapine fumarate to have higher brain-targeting efficiency and a shorter time to peak plasma concentration than intravenous injection [124]. In another comparative study, the brain bioavailability of quetiapine fumarate of CN-coated microemulsion was 3.8-fold and 2.7-fold higher than that of drug solution and CN-free microemulsion, respectively [125]. CN has an obvious promoting effect on the nasal mucosa.

Doxepin hydrochloride is a tricyclic antidepressant and anxiolytic. Doxepin hydrochloride has extensive first-pass metabolism and poor oral bioavailability (13–45%). Hema et al. prepared thermoreversible biogels based on CN and glycerophosphate loaded with doxepin hydrochloride. The formulation gelled rapidly at 37 °C, showing good drug permeability and long residence time. Compared with doxepin hydrochloride solution, the thermoreversible biogel showed more advantages in immobility time and swimming activity count in mice after 13 days of drug administration [126].

Intranasal administration is not appropriate for the type of antidepressant used. Some first-line antidepressants, such as escitalopram, have a good oral absorption effect, are not significantly metabolized, have high absolute bioavailability, and have a high ability to cross the BBB. Such antidepressants may be more suitable for oral formulations. Compared with intranasal administration, oral administration is undoubtedly more convenient and more easily accepted by the public who are now taking medication alone.

5.2. *Off-Label Drugs*

The rapid and potent antidepressant effect of ketamine for anesthesia is the most striking discovery in recent decades. The U.S. FDA approved the intranasal S-enantiomer of ketamine (Esketamine) in 2019 for the treatment of MDD in adults. The S-enantiomer of ketamine is more potent on NMDAR than the R-enantiomer and racemic ketamine [127,128]. R-enantiomer of ketamine exhibits more potent and longer-lasting antidepressant effects [129]. Studies have shown that in addition to antagonizing N-methyl-D-aspartate (NMDA) receptors, ketamine also inhibits possible mechanisms such as opioid receptors and monoamine transporters to exert antidepressant effects. It is now clear that ketamine, whether administered intravenously or intranasally, has a higher bioavailability than the oral route, and has a more rapid and significant effect than traditional antidepressants with delayed onset of action. Due to the plasma elimination half-life of ketamine of 2–4 h and the discomfort associated with invasive administration, delivery of ketamine directly to the brain via the nasal cavity is a more advantageous strategy (Figure 5).

Studies have shown that dopamine plays an indispensable role in the regulation of neural activity in states such as fear and anxiety. The ability of dopamine to cross the BBB is extremely poor, but intranasal dopamine exhibits antidepressant activity in the forced swim test [130]. Only dihydroxyphenylacetic acid produced by monoamine oxidase metabolism was detected in the CNS following intranasal dopamine delivery. However, the submucosal MAO saturates rapidly after administration, with little effect on the delivered dopamine content [131]. Liu et al. designed an interfering peptide capable of disrupting the interaction between dopamine D1 and D2 receptors. Due to the invasiveness of injection into the brain and cerebrospinal fluid and the restriction of the BBB to the peptide, they chose the route of intranasal administration. Intranasal administration showed similar antidepressant effects to imipramine, and disruptive effects of interfering peptides could be detected in the prefrontal cortex [132].

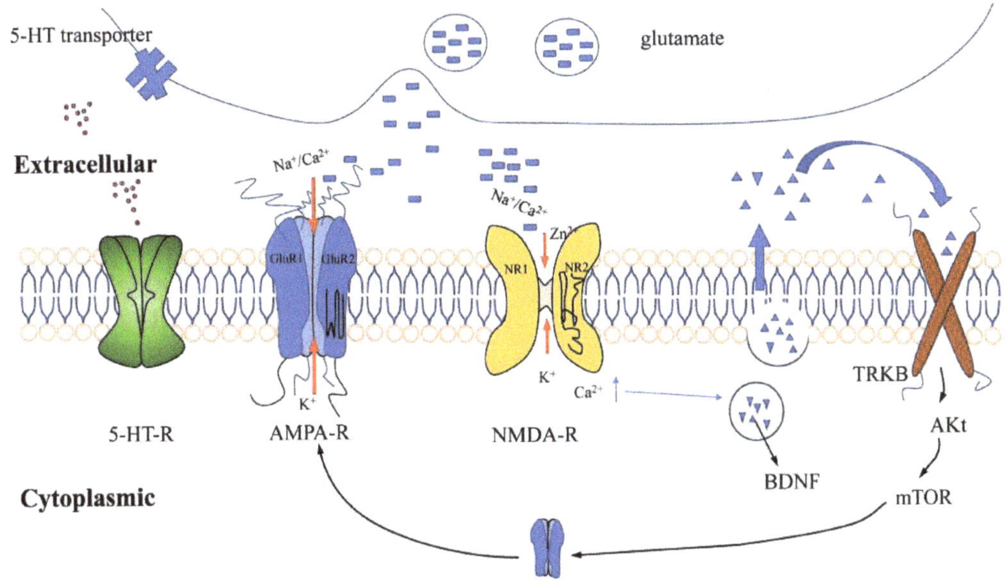

Figure 5. Receptors for antidepressant action.

Amisulpride is a second-generation atypical antipsychotic drug approved by the FDA for schizophrenia. Amisulpride selectively binds to D2/D3 dopaminergic receptors in the limbic system [133]. Multiple reports indicate that amisulpride blocks presynaptic dopamine D2/D3 receptors at low doses to promote dopamine transmission and antagonize 5-HT7 receptors to play an antidepressant role [134–136]. The absolute bioavailability of amisulpride after oral absorption is low (48%), and 25% to 50% of intravenous or oral administration will be eliminated with the original drug in the urine. Long-term use can also produce leukopenia or agranulocytosis [137,138]. Kokare et al. prepared amisulpride-loaded lipid-based poloxamer-gellan gum nanoemulgel for nose-to-brain delivery. Pharmacokinetic studies in Wistar rats showed that the intranasal Cmax of the brain was 3.39 times higher than that of the intravenous administration. Further, intranasal administration within one month did not affect blood leukocyte and granulocyte counts [139].

Aripiprazole—also a second-generation atypical antipsychotic; a partial agonist of dopamine D2, D3, and 5-HT1A receptors; and an antagonist of 5-HT2A receptors—has efficacy as a first-line antidepressant enhancer [140,141]. Although aripiprazole is well absorbed after oral administration, it undergoes extensive metabolism. Moreover, aripiprazole and its active metabolite dehydroaripiprazole are substrates for P-gp, limited by BBB permeability, resulting in low CNS bioavailability [142,143]. Hira et al. prepared aripiprazole-loaded mucoadhesive NE (ARP-MNE). Pharmacokinetic studies with single-dose administration showed that the plasma concentration in the brain of intranasal ARP-MNE was 1.44 and 6.03 times higher than that of intranasal and intravenous ARP-NE, respectively, and the Tmax was smaller than that of intravenously administered ARP-NE. Intranasal administration of ARP-MNE had higher values of drug targeting efficiency (96.90%) and drug targeting potential (89.73%) [144]. Aripiprazole-loaded poly(caprolactone) nanoparticles (APNPs) were prepared by Krutika et al. The $AUC_{0\text{-}8h}$ of Aripiprazole in the rat brain administered by the intranasal route of APNPs was approximately twice that of the intravenous route [145]. Somayeh et al. developed an ion-sensitive in situ gel containing Aripiprazole-loaded nanotransfersomes (APZ-TFS-Gel) for intranasal administration. Animal behavioral evaluations (swimming and climbing time, locomotor activity, and immobility time) showed that intranasal delivery of APZ-TFS-Gel was more effective for improving depression than oral or other formulation groups [146].

Selegiline is an irreversible type B monoamine oxidase (MAO-B) inhibitor clinically used to treat Parkinson's disease. High-dose selegiline is an irreversible inhibitor of MAO-A and MAO-B for the treatment of MDD. Selegiline transdermal patch has been clinically used to treat MDD. Selegiline can increase the content of dopamine in the hippocampus, activate D1 receptors, and regulate neuronal plasticity to play an antidepressant effect. However, oral selegiline has high first-pass metabolism and low bioavailability [147]. Wairkar et al. prepared selegiline-loaded CN-NPs and compared the pharmacokinetic parameters of selegiline delivered in the brain and plasma by different formulations and routes of administration. The C_{max} of plain solution of selegiline in the brain and plasma by intranasal administration (T_{max} = 5 min) was 20 and 12 times higher, respectively, compared with oral administration (T_{max} = 15 min). Furthermore, intranasal administration of selegiline-loaded CN-NPs and mucoadhesive thermosensitive gel showed superior formulation advantages compared with the $AUC_{0–24h}$ of plain solution [148,149]. There are also numerous studies on the enhancement of its bioavailability in the brain by intranasal delivery of selegiline-loaded vehicles, such as CN-NPs [150], thiolated CN-NPs [151], NE [147,152], nanostructured lipid carriers [153], etc. These studies have verified that the appropriate formulation delivered by the intranasal route is beneficial to reduce the dose of Selegiline and enhance the brain bioavailability for the treatment of MDD.

Racemic tramadol hydrochloride is used clinically as a central analgesic. The (-) enantiomer of tramadol hydrochloride can inhibit the synaptic reuptake of monoamine neurotransmitters such as NE and 5-HT, increasing the concentration outside neurons. The thermoreversible gel loaded with tramadol hydrochloride CN-NPs prepared by Goyal et al. showed antidepressant potential by intranasal administration [154].

5.3. Peptides

Growing evidence suggests that impaired insulin signaling may contribute to both depression and type 2 diabetes. Intranasal delivery of antihyperglycemic agents can avoid systemic exposure-induced abnormal blood glucose levels, the BBB, and metabolically induced reductions in brain-targeted content [155]. The brain is an insulin-sensitive site. There are high densities of insulin receptors in the olfactory bulb, hypothalamus, hippocampus, and limbic system. Numerous studies have shown that insulin signaling may modulate hypothalamic–pituitary–adrenal axis homeostasis; affect levels of central neurotrophic factors and monoamine neurotransmitters; interact with gastrointestinal microbes; and improve mood, learning, and memory functions [156,157]. Oral insulin bioavailability is low. Insulin injected under the skin can lower healthy blood sugar levels and can even be life-threatening. The study found that intranasal delivery of insulin showed a 2000-fold increased AUC in the brain: plasma ratio compared with subcutaneous administration, with no apparent effect on blood glucose levels [158]. Lixisenatide acts as a GLP-1 receptor agonist to control blood sugar levels. It has neuroprotective effects in degenerative diseases such as Parkinson's and Alzheimer's. Wu et al. explored the possible mechanism of action of lixisenatide in the improvement of olfactory function and mood [159]. Intranasal lixisenatide not only improved depressive and anxious behaviors in a chronic unpredictable mild stress model, but also improved olfactory system function. In addition, intranasal lixisenatide was demonstrated to play an antidepressant role by regulating cyclic-AMP response binding protein (CREB)-mediated neurogenesis. In addition, intraventricular injection of glucagon 2 (GLP-2) showed a good antidepressant effect in depression model mice. Yamashita et al. combined a cell-penetrating peptide and a penetration-accelerating sequence with GLP-2 to prepare a GLP-2 derivative (PAS-CPP-GLP-2). PAS-CPP-GLP-2 can be smoothly taken up and released by nasal mucosa epithelial cells to enter the CNS. Studies have found that intranasal PAS-CPP-GLP-2 exhibited antidepressant effects similar to intracerebroventricular injection in mouse models but not intravenous injection [56,160]. In another studies, they developed a GLP-2-loaded nasal formulation prepared from polyoxyethylene (25) lauryl ether and β-cyclodextrin, which similarly increased the brain targeting of GLP-2 effects and antidepressant effects in a rat model [161].

Oxytocin is a peptide neurohormone composed of 9 amino acids synthesized by the hypothalamus and secreted by the posterior pituitary. Intranasal delivery of oxytocin has potential value in regulating anxiety and depression. Oxytocin is secreted by the pituitary into the peripheral circulation. It is involved in regulating the balance of the hypothalamic–pituitary–adrenal axis (HPA) and reducing corticosterone levels [162,163]. In addition, oxytocin is distributed in the hippocampus, amygdala, prefrontal cortex, cingulate cortex, olfactory bulb, and other brain regions involved in emotion regulation. Oxytocin can directly stimulate oxytocin receptors in the center nucleus and striatum to promote the release of 5-HT and dopamine [164–166]. Oxytocin also inhibits microglial activation, reduces the release of pro-inflammatory cytokines, and promotes neuronal plasticity. However, synthetic exogenous oxytocin has a short half-life and poor BBB permeability.

Brain-derived neurotrophic factor (BDNF) in the hippocampus and prefrontal cortex, which is closely related to depressive symptoms, was significantly reduced in patients with depression. BDNF plays an important role in the adaptive regulation of neuronal function (neuronal plasticity) [167,168]. Increased expression of BDNF in the brain may be an effective treatment for depression. Intracranial injection, microinjection, and peripheral delivery cause tissue damage and low bioavailability. Liu et al. fused TAT, which can improve permeability, and HA2, a hydrophobic peptide sequence that promotes lipid membrane stability, with BDNF, and loaded them onto mitochondrial virus (AAV) to successfully construct BDNF-HA2TAT/AAV through intranasal route. Western-blotting analysis showed that the content of BDNF in the hippocampus increased [169]. Compared with the control group and the AVV group, the BDNF-HA2TAT/AAV group significantly reversed the depressive behavior of the rats [170].

NAP is a peptide derived from activity-dependent neuroprotective protein, consisting of eight amino acids (Asn-Ala-Pro-V al-Ser-Ile-Pro-Gln, NAPVSIPQ). NAP functions by protecting the structural stabilization of neuronal microtubules, neuroprotection, and the role of axonal transport. Dang et al. constructed NT4-NAP/AAV to promote the expression of NAP in the CNS via the intranasal route. Intranasal administration largely solves the limitations of synthetic NAPs with short plasma half-life and weak ability to cross the BBB. Experiments have shown that the depressive symptoms of female mice are improved after ten days of administration [171,172].

Using the $\alpha 2\delta$ auxiliary subunit of V-gated Ca^{2+} channels and GABAA receptors as targets, Sukhanova et al. screened LCGA-17 in silico. LCGA-17 restores hippocampal NE levels and exhibits favorable anxiolytic and antidepressant properties in behavioral evaluations in animal models of rats. The cyclic neuropeptide Cortistatin-14 (CST-14) mRNA was significantly decreased after exposure to stressed mice. CST-14 exerts a rapid antidepressant effect through the regulation of ghrelin and GABA(A) systems [173]. Intranasal delivery of neuropeptide Y (NPY) demonstrated lower behavioral impairment compared with controls in a single chronic stress animal model [174]. Melanin-concentrating hormone (MCH) exerts anxiolytic and antidepressant effects by regulating the target of rapamycin (mTOR) signaling pathway. MCH can also restore proteins downregulated by stress on the synaptic surface [175].

Depression lowers blood and brain levels of neurotrophic factors (NTFs), which are restored by antidepressant treatment. Nerve growth factor (NGF) was first discovered and has regulatory effects on the development, differentiation, growth, regeneration, and functional properties of neurons. Intranasal administration of NGF reduces immobility time in the forced swim test and tail suspension test in mice. In addition, intranasal NGF increases glucose intake and activity time, and NE and dopamine release in the frontal cortex and hippocampus in rats after UCMS [176]. It also increases neurogenesis to compensate for cellular damage from depression.

The cytokine hypothesis reveals an inextricable link between depression and inflammatory responses. Studies have shown that the levels of pro-inflammatory cytokines such as interleukin-6 (IL-6) and tumor necrosis factor (TNF) in patients with depression are significantly increased [177]. Pro-inflammatory cytokines promote increased corticosterone synthesis, decreased monoamine neurotransmitter synthesis, and increased reuptake [178–181]. Shevela et al. reduced anxiety responses in mice in the field by intranasal delivery of soluble factors in M2 macrophage culture medium [182]. Concurrent studies have shown that M2 macrophages exert antidepressant potential by downregulating pro-inflammatory cytokines in the hippocampus, prefrontal cortex, and striatum.

5.4. Natural Active Ingredients

Albiflorin is a safe natural active component of monoterpene glycoside extracted from the root of Radix Paeoniae Alba. Albiflorin can act on multiple targets such as hippocampal phospholipids, tryptophan, and dopamine to play anti-inflammatory and antioxidative stress effects [183]. However, albiflorin is easily metabolized to benzoic acid by microorganisms in the gastrointestinal tract. After oral administration, the absolute intracerebral drug concentration is low and the absolute bioavailability is low [184,185]. Wang et al. prepared albiflorin-loaded alginate nanogels for nasal administration to avoid gastrointestinal degradation. Fluorescent labeling showed that albiflorin could quickly reach the brain for distribution after intranasal administration ($\leq$30 min). The authors observed through tail suspension experiments in mice that low-dose intranasal administration significantly shortened the chronic unpredictable mild stress (CUMS) model of mice compared with intragastric gavage and intravenous injection of albiflorin solution. The reduction of pro-inflammatory cytokine levels and the repair of neuronal damage in CUMS rats further suggest that albiflorin has excellent potential for rapid antidepressant effects [186].

Berberine is a quaternary ammonium alkaloid biologically active ingredient extracted from Coptis chinensis with low toxicity and side effects. Berberine can reverse and improve the physiological changes caused by depression, such as monoamine neurotransmitter and dopamine levels, neuronal plasticity, and inflammatory responses [187,188]. Berberine has poor oral absorption, obvious intestinal first-pass metabolism, and limited absolute bioavailability [189]. Cui et al. improved the solubility of berberine by inclusion of hydroxypropyl-β-cyclodextrin (HP-β-CD). The encapsulated drug was then loaded into a thermosensitive hydrogel for intranasal administration. The relative intracerebral bioavailability of berberine showed that the intranasal formulation of berberine was 110 times higher than the oral inclusion complex of berberine–cyclodextrin. Pharmacological studies have found that the intranasal route, in addition to increasing the levels of monoamine neurotransmitters in the hippocampus compared with oral administration, exhibits a potential antidepressant mechanism by restoring sphingolipid and phospholipid abnormalities and mitochondrial dysfunction [190]. In another study, they combined two natural bioactive ingredients with antidepressant effects, berberine and evodiamine, into a nasal formulation. The bioavailability of intranasal hydrogels was more than 100 times higher than that of gavage drug solutions. Further, the intranasal formulation significantly improved behavioral despair by modulating monoamine levels and related metabolic pathways in mice [191].

Cang-ai volatile oil is a Chinese herbal volatile oil that has been used clinically as an intranasal inhaler. Studies have shown that Cang-ai volatile oil can inhibit microglia activation and kynurenine pathway to regulate 5-HT and play an antidepressant effect [192]. The forced swim test, open field test, sucrose preference test, etc. confirmed that intranasal delivery of Cang-ai volatile oil can effectively regulate the metabolism of dopamine and 5-HT in the brain of CUMS rats and improve depressive behavior [193].

Icariin (ICA), the main component of Epimedium grandiflorum, has shown therapeutic effects in osteoporosis, tumor, inflammation, cardiovascular disease, and even depression [194]. However, pharmacokinetic studies of ICA have shown poor oral absorption and brain bioavailability [195]. Wang Q. S. and Cui Y. L. et al. prepared an ICA nanogel loaded self-assembled thermosensitive hydrogel system (ICA-NGSTH) to treat depression by intranasal administration. ICA-NGSTH changes from a solid phase to a liquid phase in the nasal cavity and adheres to the nasal mucosa, thereby prolonging the residence time and releasing the drug slowly and continuously. They used fluorescence imaging of rhodamine B-labeled nanogels to study the in vivo distribution. ICA-NGSTH could be distributed in the brain in about half an hour and showed zero order kinetic release within 10 h. By comparing the oral route of ICA, intranasal ICA-NGSTH showed better behavior in an animal model of depression [196].

Olfactory dysfunction is a complication of a variety of CNS disorders, including depression. Gao et al. investigated the effects of white tea extracts on depressive behavior and olfactory disturbances in CUMS mice by intranasal administration. High and low levels of white tea extracts could effectively reverse depressive behavior in mice. Olfactory avoidance tests and olfactory sensitivity tests showed their relief of olfactory dysfunction. Pharmacological studies found that white tea reduced mitochondrial and synaptic damage in the olfactory bulb and enhanced the content of BDNF (Table 1) [197].

Table 1. Summary of studies on antidepressant active ingredients and new dosage forms for intranasal administration.

Types	Ingredients	Dosage Form	Characterization	Ex Vivo/In Vivo Studies	Relevant Outcomes	Ref.
Antidepressants	Venlafaxine	Poly lactic-co-glycolic acid nanoparticles (PLGA-NPs); Peptide-modified nanoparticles	PS = 206.3 ± 3.7 nm; PI = 0.041 ± 0.017; ZP = −26.5 ± 0.5 mV; around 200 nm after lyophilization process; DL = 10–12%; EE = 48–50%.	In vitro cell viability and cellular uptake (hCMEC/D3 cells); Permeability assay and transport studies; Biodistribution studies (C57/bl6 mice).	Cell viability of h-CMEC/D3 cells is more than 85% in the MTT assay. In vivo biodistribution studies showed higher concentrations of plain fluorescent NPs than functionalized NPs in the brain after 30 min of administration.	[107]
		Chitosan nanoparticles (CN-NPs)	PS = 167 ± 6.5 nm; PI = 0.367 ± 0.045; ZP = +23.83 ± 1.76 mV; DL = 32.25 ± 1.63%; EE = 79.3 ± 2.6%; Yield = 71.42 ± 3.24%.	Ex vivo permeation studies using porcine nasal mucosal membrane (Franz cells); Pharmacodynamic studies in Wistar rats (modified forced swim test, locomotor activity test); Qualitative localization and biodistribution studies by confocal laser scanning microscopy; Pharmacokinetic analysis.	The cumulative drug permeability after 24 h in VLF CN-NPs was nearly 3 times compared with VLF solution. VLF CN-NPs showed a more significant antidepressant effect than VLF solution on chronic depression rats by forced swimming method. DTE (%)/DTP (%): NPs = 508.59/80.34	[108]
	Desvenlafaxine	Chitosan-coated poly lactic-co-glycolic acid nanoparticles (PLGA-CN-NPs)	PS = 172.5 ± 10.2 nm; PI = 0.254 ± 0.02; ZP = +35.63 ± 8.25 mV; DL = 30.8 ± 3.1%; EE = 76.4 ± 4.2%; Release (24 h) = 77.21 ± 3.87% (pH 7.4) and 76.32 ± 3.54% (pH 6.0).	Ex vivo permeation studies on porcine nasal mucosa; Pharmacodynamic studies (Wistar rats); Stress-induced model (forced swimming test); Drug-induced model (reserpine reversal test); Biochemical estimation of serotonin, noradrenaline, and dopamine; Blood and brain pharmacokinetic studies.	In a rodent model of depression, compared with intranasal DVF solution and oral administration, increased levels of 5-HT and NE in the brain showed a more pronounced antidepressant effect. Pharmacokinetic parameters such as concentration, half-life, and AUC in the brain after intranasal administration were higher than those of through intravenous. DTE (%)/DTP (%) = 544.23/81.62 (DVF-NP s) DTE (%)/DTP (%) = 202.41/50.59 (DVF)	[111]

Table 1. *Cont.*

Types	Ingredients	Dosage Form	Characterization	Ex Vivo/In Vivo Studies	Relevant Outcomes	Ref.
	Agomelatine	Nanoemulsion thermosensitive in situ gel + 0.5% Chitosan	Gelling point = 28 ± 1 °C; Mucoadhesive strength = 6246.27 dynes/cm^2; NEs: PS = 206.3 ± 3.7 nm; Micelles of P-407: PS = 142.58 ± 4.21 nm; Ago-NE-gel + 0.5%chitosan: Viscosity = 2439 ± 23 cP (35 ± 1 °C); pH = 5.8 ± 0.2.	In vitro gel erosion study; Ex vivo drug permeation through the bovine nasal mucosa; Nasal toxicity study; Pharmacokinetic analysis: DTE (%) and DTP (%); Pharmacodynamic studies (Behavioral test; modified forced swim test and tail suspension test).	Pharmacokinetic study in Wistar rats showed plasma concentration in the brain was 2.82 times higher than that of the intravenous suspension via the intranasal route. DTE (%)/DTP (%) = 344.9/71.0	[112]
		Solid lipid nanoparticles (SLNs)	PS = 167.70 ± 0.42 nm; PI = 0.12 ± 0.10; ZP = −17.90 ± 2.70 mV; EE = 91.25 ± 1.70%; Release (1 h)/(8 h) = 35.40 ± 1.13%/80.87 ± 5.16%.	Pharmacokinetic study (rats): assay of agM in plasma and brain; Pharmacokinetic analysis: DTE (%) and DTP (%).	The nasal solid lipid nanoparticles prepared by Ahmed et al. were superior to the oral suspension in brain concentration, $AUC_{0\text{–}360min}$, and absolute bioavailability (44.44%) DTE (%)/DTP (%) = 190.02/47.37	[113]
	Duloxetine	Nanostructured lipid carriers (NLCs)	PS = 137.2 ± 2.88 nm; ZP = -31.53 ± 11.21 mV; DL = 9.73 ± 3.22%; EE = 79.15 ± 4.17%.	Biodistribution studies (Wistar rats); Pharmacokinetic study; Gamma-imaging study.	Intranasal DLX-NLCs showed higher concentrations in blood and brain compared with DLX solution and oral route, which showed the same results in behavioral tests in mice. Intranasal NLCs were 8-fold higher in brain concentrations than intravenous DLX. DTE (%)/DTP (%) = 757.74/86.80 (DLX-NLC) DTE (%)/DTP (%) = 287.34/65.12 (DLX)	[116, 117]
		Thiomer gel loaded with proniosomes	20% *w/v* PF127, 5% *w/v* PF68; 3.76 lipid ratio; PS = 265.13 ± 9.85 nm; GT = 32 ± 0.05 °C; EE = 98.13 ± 0.50%; Release (3 h) = 33%.	Pharmacokinetic analysis: DTE (%) and DTP (%); Stability study.	Thiomer gel loaded with duloxetine proniosomes increased the retention time and sustained release and penetration of DLX in the nasal mucosa (1.96 times that of duloxetine proniosomes). DTE (%)/DTP (%) = 137.77/10.5	[118]

Table 1. *Cont.*

Types	Ingredients	Dosage Form	Characterization	Ex Vivo/In Vivo Studies	Relevant Outcomes	Ref.
	Paroxetine	Nanoemulsion (NEs)	PS = 58.47 ± 3.02 nm; PDI = 0.339 ± 0.007; ZP = −33 mV; Transmittance = 100.60 ± 0.577%; Refractive index = 1.412 ± 0.003.	Ex vivo permeation studies using porcine nasal mucosal membrane (Franz cells); Pharmacodynamic studies (Wistar rats; forced swimming test, locomotor activity test); Biochemical estimation: GSH and TBARS.	The permeability of paroxetine NEs was 2.57 times higher than that of its suspension via permeation studies. Results of behavioral studies in rats showed that intranasal administration of paroxetine NEs significantly improved behavioral activity in depressed rats compared with the oral suspension of paroxetine.	[121]
	Trazodone	Microemulsion	labelling yield = 91.23 ± 2.12%; In vitro stability of ^{131}I-TZ = 6 h; Droplet size = 16.4 ± 2.5 nm; PDI = 0.11 ± 0.02; ZP = 3.83 ± 0.36; Viscosity (25 °C) = 261.7 ± 3.0; Viscosity (37 °C) = 157.3 ± 7.5.	Biodistribution of ^{131}I-TZ; The ^{131}I-TZ uptake in organs and body fluids.	Sayyed et al. radiolabeled trazodone and compared the pharmacokinetic parameters of intranasal delivery of ^{131}I-TZ solution, ^{131}I-TZ microemulsion, and intravenous injection of ^{131}I-TZ solution. Intranasal ^{131}I-TZ microemulsion had sustained and higher brain uptake at any time tested than the other two formulations and routes. In addition, the blood exposure of intranasal ^{131}I-TZ microemulsion was lower than that of intravenous injection, reducing systemic toxicity.	[123]
	Quetiapine fumarate	Microemulsion Chitosan microemulsion (CH-ME) methyl-β-cyclodextrin microemulsion (MeβCD-ME)	PS: QF-ME = 29.75 ± 0.99 nm; CH-ME = 35.31 ± 1.71 nm; MeβCD-ME = 46.55 ± 1.9 nm with; PDI: QF-ME = 0.221 ± 0.01; CH-ME = 0.249 ± 0.03; MeβCD-ME = 0.233 ± 0.02; ZP: QF-ME = 2.77 ± 0.51; CH-ME = 20.29 ± 1.23 MeβCD-ME = 8.43 ± 0.7; Viscosity: QF-ME = 17.5 ± 0.69 cP; CH-ME = 38.5 ± 1.26 cP; MeβCD-ME = 33.3 ± 0.93 cP.	Ex vivo mucoadhesive strength; Ex vivo nasal and intestinal diffusion study (goat nasal mucosa and small intestine); Nasal mucosal toxicity test; Pharmacokinetic analysis: DTE (%) and DTP (%).	The brain bioavailability of quetiapine fumarate of chitosan-coated microemulsion was 3.8-fold and 2.7-fold higher than that of drug solution and chitosan-free microemulsion, respectively. DTE (%)/DTP (%) = 371.20 ± 12.02/ 68.66 ± 6.84 (QF-ME) DTE (%)/DTP (%) = 453.69 ± 10.17/80.51 ± 6.46 (CH-ME)	[125]

Table 1. *Cont.*

Types	Ingredients	Dosage Form	Characterization	Ex Vivo/In Vivo Studies	Relevant Outcomes	Ref.
	Doxepin hydrochloride	Thermoreversible biogels	Gelation temperature = 37.4 °C; Gelation time = 7.32 min pH = 6.93.	In vitro penetration test on sheep nasal mucosa; Stress-induced model (forced swimming test).	Compared with doxepin hydrochloride solution, the thermoreversible biogel showed more advantages in immobility time and swimming activity count in mice after 13 days of drug administration.	[126]
Off-label drugs	Ketamine/Esketamine	Nasal spray	N/A	N/A	Ketamine, whether administered intravenously or intranasally, has a higher bioavailability than the oral route, and has a more rapid and significant effect than traditional antidepressants with delayed onset of action. Due to the plasma elimination half-life of ketamine of 2–4 h and the discomfort associated with invasive administration, delivery of ketamine directly to the brain via the nasal cavity is a more advantageous strategy.	[127, 128]
	Amisulpride	Lipid-based poloxamer-gellan gum nanoemulgel AMS nanoemulsion (AMS-NE) AMS in situ nanoemulgel (AMS-NG)	AMS-NE: PS = 92.15 ± 0.42 nm; PI = 0.46 ± 0.03; ZP = −18.22 mV; Transmittance = 99.57%; Mucoadhesive strength = 1.24 g; Release (4 h) = 99.99%; AMS-NG: PS = 106.11 ± 0.26 nm; PI = 0.51 ± 0.01; ZP = −16.01 mV; Transmittance = 98.47%; Mucoadhesive strength = 8.90 g; Release (4 h) = 98.96%.	Ex vivo drug permeation study on freshly isolated sheep nasal mucosa; In vivo animal experiments (pharmacokinetic study, AMS in brain and blood plasma samples); Animal behavioral studies (induced locomotor activity test, paw test); In vivo safety assessment.	Pharmacokinetic studies in Wister rats showed that the intranasal C(max) of the brain was 3.39 times higher than that of the intravenous administration and intranasal administration within one month did not affect blood leukocyte and granulocyte counts. DTE (%)/DTP (%) = 314.08/76.13 (AMS-NE) DTE (%)/DTP (%) = 1821.72/275.09 (AMS-NG)	[139]
	Aripiprazole	Mucoadhesive nanoemulsion	PS = 121.8 ± 1.5 nm; PI = 0.248 ± 0.05; ZP = −18.89 ± 3.47 mV; Viscosity = 187.79 ± 5.35 cP (25% Carbopol); Viscosity = 626.32 ± 8.63 cP (1% Carbopol); Release (8 h) = 84.92%.	Ex vivo permeation test and nasal ciliotoxicity on sheep nasal mucosa; In vitro cytotoxicity study (Vero cells, PC12 cells); In vivo pharmacokinetic study (DTE (%) and DTP (%)); Locomotor activity study.	Pharmacokinetic studies with single-dose administration showed that the plasma concentration in the brain of intranasal ARP-MNE was 1.44 and 6.03 times higher than that of intranasal and intravenous ARP-NE, respectively, and the Tmax was smaller than that of intravenously administered ARP-NE. DTE (%)/ DTP (%) = 96.90/89.73	[144]

Table 1. *Cont.*

Types	Ingredients	Dosage Form	Characterization	Ex Vivo/In Vivo Studies	Relevant Outcomes	Ref.
		Poly(caprolactone) nanoparticles	PS = 199.2 ± 5.65 nm; ZP = -21.4 ± 4.6 mV; EE = 69.2 ± 2.34%; Release (8 h) = 90 ± 2.69%.	Ex vivo diffusion studies on goat nasal mucosa; Nasal toxicity study (goat nasal mucosa); In vivo pharmacokinetics study (DTE (%) and DTP (%)).	The $AUC_{0\text{-}8h}$ of Aripiprazole in the rat brain administered by the intranasal route of APNPs was approximately twice that of the intravenous route. DTE (%)/DTP (%) = 64.11/74.34	[145]
	Selegiline	Chitosan nanoparticle	PS = 341.6 ± 56.91 nm; PI = 0.317 ± 0.29; ZP = −13.4 ± 0.04 mV; EE = 92.20 ± 7.15%; Release (8 h) = 90 ± 2.69%.	Ex vivo drug diffusion on sheep nasal mucosa; Pharmacokinetics and pharmacodynamics studies; Behavioral testing; Biochemical analyses: dopamine level, catalase activity, reduced glutathione (GSH) content.	The Cmax of plain solution of selegiline in the brain and plasma by intranasal administration (Tmax = 5 min) was 20 and 12 times higher, respectively, compared with oral administration (Tmax = 15 min). Furthermore, intranasal administration of selegiline-loaded CN-NPs and mucoadhesive thermosensitive gel showed superior formulation advantages compared with the $AUC_{0\text{-}24h}$ of plain solution.	[148, 149]
Peptides	Insulin	N/A	N/A	Pharmacokinetics study (insulin concentrations in brain and plasma via different delivery routes); $AUC_{brain:\ plasma}$ ratio; Repeated in insulin administration.	The study found intranasal delivery of insulin showed a 2000-fold increased $AUC_{brain:\ plasma}$ ratio compared with subcutaneous administration, with no apparent effect on blood glucose levels.	[158]
	Lixisenatide	N/A	N/A	Chronic unpredictable mild stress depression model (rats); Behavioral studies (forced swim test, tail suspension test, open field test); Cells were labeled with BrdU and neurogenesis in the olfactory bulb and hippocampus was observed.	Intranasal lixisenatide not only improved depressive and anxious behaviors in a chronic unpredictable mild stress model, but also improved olfactory system function. In addition, intranasal lixisenatide was demonstrated to play an antidepressant role by regulating cyclic-AMP response binding protein (CREB)-mediated neurogenesis.	[159]
	GLP-2	PAS-CPPs-GLP-2	N/A	Behavioral studies (forced swim test, tail suspension test, open field test); Distribution test (rats' brain).	Studies have found that intranasal PAS-CPP-GLP-2 exhibited antidepressant effects similar to intracerebroventricular injection in mouse models, but not intravenous injection.	[56, 160]

Table 1. *Cont.*

Types	Ingredients	Dosage Form	Characterization	Ex Vivo/In Vivo Studies	Relevant Outcomes	Ref.
	BDNF	BDNF-HA2TAT/AAV	Each step was qualified by specific restriction enzyme reactions and AGE; High expression of BDNF in infected Hela cells.	Chronic unpredictable mild stress depression model (rats); Behavioral assessment (forced swim test, sucrose preference test, open field test); Body weight; Western-blotting analysis; Expression of BDNF mRNA.	Western-blotting analysis showed that the content of BDNF in the hippocampus increased via intranasal administration. Compared with the control group and the AVV group, the BDNF-HA2TAT/AAV group significantly reversed the depressive behavior of the rats.	[169, 170]
	NAP	NT4-NAP/AAV	Each step was qualified by specific restriction enzyme reactions and AGE; Expression of BDNF in infected PC12 cells.	Behavioral assessment (forced swim test, sucrose preference test, open field test); Effect on plasma CORT; Expression of 5-HT and BNDF in hippocampus.	Experiments have shown that the depressive symptoms of female mice are improved after ten days of administration. Although the effect is not significant, it also proves that intranasal administration from different targets, such as microtubules, provide new ideas for the treatment of depression.	[171, 172]
	NPY/LCG-17/MCH/CST-14/NGF	N/A	N/A	Behavioral assessment (forced swim test, sucrose preference test, open field test); Biochemical studies.	These peptides bypass the blood–brain barrier via a non-invasive intranasal route of administration, improving bioavailability and brain targeting. The peptides both improve anxiety and depression behavior in animal models. The peptides also promote neuroplasticity in the central nervous system, especially the hippocampus and prefrontal cortex.	[173–176]
Natural active ingredients	Albiflorin	Alginate nanogels	PS = 45.6 ± 5.2 nm; PI < 0.20; ZP = −19.8 ± 0.9 mV; EE = ±7.15%; Release (12 h) = 99%; Gelling temperature = 28 °C.	In vivo fluorescence distribution analysis of alginate nanogels (rats); Pharmacodynamic study; Antidepressant behavioral studies: tail suspension test; Transcriptome studies: cAMP, calcium ion, and cGMP PKG signal pathway.	Fluorescent labeling showed that albiflorin could quickly reach the brain for distribution after intranasal administration (≤30 min). The authors observed through tail suspension experiments in mice that low-dose intranasal administration significantly shortened the chronic unpredictable mild stress model of mice compared with intragastric gavage and intravenous injection of albiflorin solution. Do not move time. The reduction of pro-inflammatory cytokine levels and the repair of neuronal damage in CUMS rats further suggest that albiflorin has an excellent potential for rapid antidepressant effects.	[186]

Table 1. *Cont.*

Types	Ingredients	Dosage Form	Characterization	Ex Vivo/In Vivo Studies	Relevant Outcomes	Ref.
	Berberine	Cyclodextrin + thermosensitive hydrogel	The berberine /HP-β-CD inclusion complex (^{1}H-NMR-NMR showed good degree of inclusion); Gelling temperature = 30 °C; Release (6 h) = 83.29 ± 3.98%; Loading efficiency = 22.86%.	Brain targeting of berberine study (Radioactive tracer of ^{125}I); Pharmacokinetic analysis: berberine in hippocampus; Monoamine neurotransmitters in rats (reserpine-induced model).	The relative intracerebral bioavailability of berberine showed that the intranasal formulation of berberine was 110 times higher than the oral inclusion complex of berberine–cyclodextrin. Pharmacological studies have found that the intranasal route, in addition to increasing the levels of monoamine neurotransmitters in the hippocampus compared with oral administration, exhibits a potential antidepressant mechanism by restoring sphingolipid and phospholipid abnormalities and mitochondrial dysfunction.	[190]
	Berberine and Evodiamine	Thermosensitive in situ hydrogels	P407/P188/HP-β-CD/PEG 8000 = 20/0/8/1; Release = 93% (berberine); Release = 43% (evodiamine); Gelling temperature = 28 °C.	Pharmacokinetic study (plasma and hippocampus); Antidepressant behavioral studies (open field test, tail suspension test); Monoamine neurotransmitters studies in rats.	The bioavailability of intranasal hydrogels was more than 135- and 112-fold higher than that of gavage berberine and evodiamine solutions. The intranasal formulation significantly improved behavioral despair by modulating monoamine levels and related metabolic pathways in mice.	[191]
	Cang-ai volatile oil	Intranasal inhaler	N/A	Chronic unpredictable mild stress depression model (rats); Behavioral studies (open field test, forced swim test, and sucrose preference test); Expression of pro-inflammatory cytokines and monoamine neurotransmitters studies in prefrontal cortex.	Studies have shown that Cang-ai volatile oil can inhibit microglia activation and kynurenine pathway to regulate 5-HT and play an antidepressant effect. The forced swim test, open field test, sucrose preference test, etc. confirmed that intranasal delivery of Cang-ai volatile oil can effectively regulate the metabolism of dopamine and 5-HT in the brain of CUMS rats and improve depressive behavior.	[192, 193]
	Icariin	Nanogel loaded thermosensitive hydrogel (NGSTH)	PS = 73.80 ± 2.34 nm; PI < 0.15; ZP = −19.2 ± 1.14 mV; Loading efficiency = 2.03%; Release (36 h) = 70% (nanogel); Gelling temperature = 30 °C; Release (36 h) = 100% (NGSTH).	In vivo distribution fluorescently labeled nanogels Behavioral testing (tail suspension test, forced swim test); Expression of pro-inflammatory cytokines and morphological changes in the hippocampus.	ICA-NGSTH could be distributed in the brain in about half an hour and showed zero order kinetic release within 10 h. By comparing the oral route of ICA, intranasal ICA-NGSTH showed better behavior improvement ability in an animal model of depression.	[196]

Table 1. *Cont.*

Types	Ingredients	Dosage Form	Characterization	Ex Vivo/In Vivo Studies	Relevant Outcomes	Ref.
	White tea	N/A	N/A	Chronic unpredictable mild stress depression model (rats); Behavioral testing (open-field test, sucrose preference test, buried food pellet test); Olfactory sensitivity test.	High and low levels of white tea extracts could effectively reverse depressive behavior in mice. Olfactory avoidance tests and olfactory sensitivity tests showed its relief of olfactory dysfunction. Pharmacological studies found that white tea reduced mitochondrial and synaptic damage in the olfactory bulb and enhanced the content of BDNF.	[197]

Abbreviations: PS, globule size; ZP, zeta potential; PI, polydispersity index; DTE (%), drug targeting efficiency; DTP (%), nose-to-brain direct transport percentage; BDNF, brain-derived neurotrophic factor; CORT, corticosterone; P407, poloxamer 407; P188, poloxamer 188.

6. Summary and Outlook

Under the action of different factors, the incidence of depression in today's society is becoming more and more serious. Drug treatment for depression is mainly oral administration, which is challenged by extensive first-pass metabolism, the BBB, and systemic side effects. Some antidepressant active ingredients, such as peptides, natural active ingredients, etc., cannot achieve their high brain bioavailability due to the limitations of their own physical and chemical properties. The direct pathway between the nasal cavity and the brain provides a reliable guarantee for improving the bioavailability of antidepressant active ingredients and reducing side effects. As the olfactory system highly overlaps with areas that process emotion and memory functions, intranasal administration (especially to the olfactory area) may be a potential route for treating depression. Antidepressant active ingredients can enter the brain directly through the olfactory sensory nerve, trigeminal nerve, and olfactory mucosal epithelial pathways, or indirectly through the rich capillaries in the respiratory area and lymphatic tissue.

With the continuous progress of medical technology and the pharmaceutical engineering industry, remarkable research achievements have been made in how to make more effective use of nose-to-brain drug delivery, but there are still some problems to be solved. The small size of the nasal space and olfactory area limits the amount of drug that needs to be delivered directly to the brain. Moreover, factors such as the close connection of olfactory epithelial cells, the clearance of nasal mucosa cilia and the degradation of antidepressant active ingredients by various enzymes all make fewer parts to be absorbed; so, the method of enhancing absorption is very important. At present, some remarkable achievements have been made in the research of promoting nasal mucous absorption. For example, some materials with good biocompatibility, biodegradation, and low toxicity, such as permeability enhancers, adhesives, enzyme inhibitors, and nanoparticles, have been well applied in promoting nose-to-brain delivery. Solvent enhancers, antioxidants, preservatives, moisturizers, buffers, and taste maskers are added to the formulation to ensure drug stability and patient compliance. However, the treatment of depression is often long-term, and these absorption-based materials can more or less cause irritation and side effects to the nasal cavity and other tissues and organs. On the other hand, the invention of some nasal delivery devices, such as Vianase™, has significantly improved the deposition of antidepressant active ingredients in the olfactory region of the upper nasal cavity compared with traditional delivery devices. However, the deposition rate in the olfactory area is still relatively low; the highest is only about 50%. Therefore, there is an urgent need to find better excipients and new devices for enhancing drug delivery in the nose and brain.

Current studies on intranasal administration of antidepressant active ingredients are mostly limited to the cellular level and model animal level. In addition, different stimulation methods may also lead to individual differences in the model. There are differences between the structure and physiological conditions of the human nasal cavity and experimental animals; thus, it is not effective in an animal model but in the human body. To further verify the drug, a clinical observation test is needed to evaluate its safety and effectiveness. In addition to intranasal administration, the development of new routes of administration, such as transdermal targeted administration, to overcome the disadvantages of oral and injectable administration is also one of the main exploration directions to improve the efficacy of antidepressant treatment in the future.

Author Contributions: Conceptualization, S.S.; methodology, S.S.; supervision, S.S.; funding acquisition, S.S. and B.H.; writing—reviewing and editing, Z.L., Y.J. and B.H.; data curation, Z.J.; writing—original draft preparation, Z.J.; visualization, Y.H. and D.Z.; investigation, Y.H. and D.Z. All authors have read and agreed to the published version of the manuscript.

Funding: This work was financially supported by the National Natural Science Foundation of China (No. U1803283 and No. 21878249), the Key R&D Program of Hebei Province (21377786D), and the Natural Science Foundation of Hebei Province (No. H2020208001).

Acknowledgments: This work was financially supported by the National Natural Science Foundation of China (No. U1803283 and No. 21878249), the Key R&D Program of Hebei Province (21377786D), and the Natural Science Foundation of Hebei Province (No. H2020208001). This work is dedicated to Xiaojun Peng on the occasion of his 60th birthday.

Conflicts of Interest: The authors declare that they have no known competing financial interest or personal relationship that could have appeared to influence the work reported in this paper.

References

1. Evans-Lacko, S.; Aguilar-Gaxiola, S.; Al-Hamzawi, A.; Alonso, J.; Benjet, C.; Bruffaerts, R.; Chiu, W.T.; Florescu, S.; de Girolamo, G.; Gureje, O.; et al. Socio-economic variations in the mental health treatment gap for people with anxiety, mood, and substance use disorders: Results from the WHO World Mental Health (WMH) surveys. *Psychol. Med.* **2017**, *48*, 1560–1571. [CrossRef] [PubMed]
2. Panek, M.; Kawalec, P.; Pilc, A.; Lasoń, W. Developments in the discovery and design of intranasal antidepressants. *Expert Opin. Drug Discov.* **2020**, *15*, 1145–1164. [CrossRef] [PubMed]
3. Depression and Other Common Mental Disorders: Global Health Estimates. Available online: https://www.who.int/publications/i/item/depression-global-health-estimates (accessed on 17 September 2022).
4. Illum, L. Transport of drugs from the nasal cavity to the central nervous system. *Eur. J. Pharm. Sci.* **2000**, *11*, 1–18. [CrossRef]
5. Kashyap, K.; Shukla, R. Drug Delivery and Targeting to the Brain Through Nasal Route: Mechanisms, Applications and Challenges. *Curr. Drug Deliv.* **2019**, *16*, 887–901. [CrossRef]
6. Mato, Y.L. Nasal route for vaccine and drug delivery: Features and current opportunities. *Int. J. Pharm.* **2019**, *572*, 118813. [CrossRef]
7. Giunchedi, P.; Gavini, E.; Bonferoni, M.C. Nose-to-brain delivery. *Pharmaceutics* **2020**, *12*, 138. [CrossRef]
8. O'Leary, O.F.; Dinan, T.G.; Cryan, J.F. Faster, better, stronger: Towards new antidepressant therapeutic strategies. *Eur. J. Pharmacol.* **2015**, *753*, 32–50. [CrossRef]
9. Pardridge, W.M. CSF, blood-brain barrier, and brain drug delivery. *Expert Opin. Drug Deliv.* **2016**, *13*, 963–975. [CrossRef]
10. Daneman, R.; Prat, A. The blood-brain barrier. *Cold Spring Harb. Perspect. Biol.* **2015**, *7*, a020412. [CrossRef]
11. Abbott, N.J.; Patabendige, A.A.K.; Dolman, D.E.M.; Yusof, S.R.; Begley, D.J. Structure and function of the blood-brain barrier. *Neurobiol. Dis.* **2010**, *37*, 13–25. [CrossRef]
12. Zheng, Y.; Chen, X.; Benet, L.Z. Reliability of In Vitro and In Vivo Methods for Predicting the Effect of P-Glycoprotein on the Delivery of Antidepressants to the Brain. *Clin. Pharmacokinet.* **2015**, *55*, 143–167. [CrossRef] [PubMed]
13. O'Brien, F.E.; Dinan, T.G.; Griffin, B.T.; Cryan, J.F. Interactions between antidepressants and P-glycoprotein at the blood-brain barrier: Clinical significance of in vitro and in vivo findings. *Br. J. Pharmacol.* **2011**, *165*, 289–312. [CrossRef] [PubMed]
14. Brückl, T.M.; Uhr, M. *ABCB1* genotyping in the treatment of depression. *Pharmacogenomics* **2016**, *17*, 2039–2069. [CrossRef] [PubMed]
15. Bicker, J.; Fortuna, A.; Alves, G.; Falcão, A. Nose-to-brain delivery of natural compounds for the treatment of central nervous system disorders. *Curr. Pharm. Des.* **2020**, *26*, 594–619.
16. Long, Y.; Yang, Q.; Xiang, Y.; Zhang, Y.; Wan, J.; Liu, S.; Li, N.; Peng, W. Nose to brain drug delivery—A promising strategy for active components from herbal medicine for treating cerebral ischemia reperfusion. *Pharmacol. Res.* **2020**, *159*, 104795. [CrossRef]
17. Shringarpure, M.; Gharat, S.; Momin, M.; Omri, A. Management of epileptic disorders using nanotechnology-based strategies for nose-to-brain drug delivery. *Expert Opin. Drug Deliv.* **2020**, *18*, 169–185. [CrossRef]
18. Wang, Z.; Xiong, G.; Tsang, W.C.; Schätzlein, A.G.; Uchegbu, I.F. Nose-to-brain delivery. *J. Pharmacol. Exp. Ther.* **2019**, *370*, 593–601. [CrossRef]
19. Kim, B.-Y.; Bae, J.H. Olfactory Function and Depression: A Meta-Analysis. *Ear Nose Throat J.* **2022**. [CrossRef]
20. Staszelis, A.; Mofleh, R.; Kocsis, B. The effect of ketamine on delta-range coupling between prefrontal cortex and hippocampus supported by respiratory rhythmic input from the olfactory bulb. *Brain Res.* **2022**, *1791*, 147996. [CrossRef]
21. Schwartz, J.S.; Tajudeen, B.A.; Kennedy, D.W. Diseases of the nasal cavity. *Handb. Clin. Neurol.* **2019**, *164*, 285–302. [CrossRef]
22. Gonçalves, J.; Alves, G.; Fonseca, C.; Carona, A.; Bicker, J.; Falcão, A.; Fortuna, A. Is intranasal administration an opportunity for direct brain delivery of lacosamide? *Eur. J. Pharm. Sci.* **2021**, *157*, 105632. [CrossRef] [PubMed]
23. Dhuria, S.V.; Hanson, L.R.; Frey, W.H., 2nd. Intranasal delivery to the central nervous system: Mechanisms and experimental considerations. *J. Pharm. Sci.* **2010**, *99*, 1654–1673. [CrossRef]
24. Samaridou, E.; Alonso, M.J. Nose-to-brain peptide delivery—The potential of nanotechnology. *Bioorg. Med. Chem.* **2018**, *26*, 2888–2905. [CrossRef] [PubMed]
25. Rottstädt, F.; Han, P.; Weidner, K.; Schellong, J.; Wolff-Stephan, S.; Strauß, T.; Kitzler, H.; Hummel, T.; Croy, I. Reduced olfactory bulb volume in depression-A structural moderator analysis. *Hum. Brain Mapp.* **2018**, *39*, 2573–2582. [CrossRef] [PubMed]
26. Rottstaedt, F.; Weidner, K.; Strauß, T.; Schellong, J.; Kitzler, H.; Wolff-Stephan, S.; Hummel, T.; Croy, I. Size matters—The olfactory bulb as a marker for depression. *J. Affect. Disord.* **2017**, *229*, 193–198. [CrossRef] [PubMed]

27. Cecon, E.; Ivanova, A.; Luka, M.; Gbahou, F.; Friederich, A.; Guillaume, J.; Keller, P.; Knoch, K.; Ahmad, R.; Delagrange, P.; et al. Detection of recombinant and endogenous mouse melatonin receptors by monoclonal antibodies targeting the C-terminal domain. *J. Pineal Res.* **2018**, *66*, e12540. [CrossRef]
28. Noseda, A.C.D.; Rodrigues, L.S.; Targa, A.D.S.; Ilkiw, J.L.; Fagotti, J.; Dos Santos, P.D.; Cecon, E.; Markus, R.P.; Solimena, M.; Jockers, R.; et al. MT(2) melatonin receptors expressed in the olfactory bulb modulate depressive-like behavior and olfaction in the 6-OHDA model of Parkinson's disease. *Eur. J. Pharmacol.* **2021**, *891*, 173722. [CrossRef]
29. Crowe, T.P.; Greenlee, M.H.W.; Kanthasamy, A.G.; Hsu, W.H. Mechanism of intranasal drug delivery directly to the brain. *Life Sci.* **2018**, *195*, 44–52. [CrossRef]
30. Renner, D.B.; Svitak, A.L.; Gallus, N.J.; Ericson, M.E.; Frey, W.H., 2nd; Hanson, L.R. Intranasal delivery of insulin via the olfactory nerve pathway. *J. Pharm. Pharmacol.* **2012**, *64*, 1709–1714. [CrossRef]
31. Tan, M.S.A.; Parekh, H.S.; Pandey, P.; Siskind, D.J.; Falconer, J.R. Nose-to-brain delivery of antipsychotics using nanotechnology: A review. *Expert Opin. Drug Deliv.* **2020**, *17*, 839–853. [CrossRef]
32. Altner, H.; Altner-Kolnberger, I. Freeze-fracture and tracer experiments on the permeability of the zonulae occludentes in the olfactory mucosa of vertebrates. *Cell Tissue Res.* **1974**, *154*, 51–59. [CrossRef] [PubMed]
33. Durante, M.A.; Kurtenbach, S.; Sargi, Z.B.; Harbour, J.W.; Choi, R.; Kurtenbach, S.; Goss, G.M.; Matsunami, H.; Goldstein, B.J. Single-cell analysis of olfactory neurogenesis and differentiation in adult humans. *Nat. Neurosci.* **2020**, *23*, 323–326. [CrossRef] [PubMed]
34. Trevino, J.; Quispe, R.; Khan, F.; Novak, V. Non-Invasive Strategies for Nose-to-Brain Drug Delivery. *J. Clin. Trials* **2020**, *10*, 439. [PubMed]
35. Lochhead, J.J.; Thorne, R.G. Intranasal delivery of biologics to the central nervous system. *Adv. Drug Deliv. Rev.* **2012**, *64*, 614–628. [CrossRef]
36. Croy, I.; Hummel, T. Involvement of nasal trigeminal function in human stereo smelling. *Proc. Natl. Acad. Sci. USA* **2020**, *117*, 25979. [CrossRef]
37. Lochhead, J.J.; Kellohen, K.L.; Ronaldson, P.T.; Davis, T.P. Distribution of insulin in trigeminal nerve and brain after intranasal administration. *Sci. Rep.* **2019**, *9*, 2621. [CrossRef]
38. Kumar, N.N.; Lochhead, J.; Pizzo, M.; Nehra, G.; Boroumand, S.; Greene, G.; Thorne, R.G. Delivery of immunoglobulin G antibodies to the rat nervous system following intranasal administration: Distribution, dose-response, and mechanisms of delivery. *J. Control. Release* **2018**, *286*, 467–484. [CrossRef]
39. Pang, Y.; Fan, L.-W.; Carter, K.; Bhatt, A. Rapid transport of insulin to the brain following intranasal administration in rats. *Neural Regen. Res.* **2019**, *14*, 1046–1051. [CrossRef]
40. Lochhead, J.; Wolak, D.J.; Pizzo, M.; Thorne, R.G. Rapid Transport within Cerebral Perivascular Spaces Underlies Widespread Tracer Distribution in the Brain after Intranasal Administration. *J. Cereb. Blood Flow Metab.* **2015**, *35*, 371–381. [CrossRef]
41. Iliff, J.J.; Wang, M.; Liao, Y.; Plogg, B.A.; Peng, W.; Gundersen, G.A.; Benveniste, H.; Vates, G.E.; Deane, R.; Goldman, S.A.; et al. A Paravascular Pathway Facilitates CSF Flow Through the Brain Parenchyma and the Clearance of Interstitial Solutes, Including Amyloid β. *Sci. Transl. Med.* **2012**, *4*, 147ra111. [CrossRef]
42. Pardeshi, C.V.; Rajput, P.V.; Belgamwar, V.S.; Tekade, A.R. Formulation, optimization and evaluation of spray-dried mucoadhesive microspheres as intranasal carriers for Valsartan. *J. Microencapsul.* **2011**, *29*, 103–114. [CrossRef]
43. Gänger, S.; Schindowski, K. Tailoring Formulations for Intranasal nose-to-brain delivery: A review on architecture, physico-chemical characteristics and mucociliary cearance of the nasal olfactory mucosa. *Pharmaceutics* **2018**, *10*, 116. [CrossRef] [PubMed]
44. Schwarz, B.; Merkel, O.M. Nose-to-brain delivery of biologics. *Ther. Deliv.* **2019**, *10*, 207–210. [CrossRef] [PubMed]
45. Smith, T.D.; Bhatnagar, K.P. Anatomy of the olfactory system. *Handb. Clin. Neurol.* **2019**, *164*, 17–28. [CrossRef] [PubMed]
46. Olivares, J.; Schmachtenberg, O. An update on anatomy and function of the teleost olfactory system. *PeerJ* **2019**, *7*, e7808. [CrossRef]
47. Palleria, C.; Roberti, R.; Iannone, L.F.; Tallarico, M.; Barbieri, M.A.; Vero, A.; Manti, A.; De Sarro, G.; Spina, E.; Russo, E. Clinically relevant drug interactions between statins and antidepressants. *J. Clin. Pharm. Ther.* **2019**, *45*, 227–239. [CrossRef]
48. Wyska, E. Pharmacokinetic considerations for current state-of-the-art antidepressants. *Expert Opin. Drug Metab. Toxicol.* **2019**, *15*, 831–847. [CrossRef]
49. Erdő, F.; Bors, L.A.; Farkas, D.; Bajza, Á.; Gizurarson, S. Evaluation of intranasal delivery route of drug administration for brain targeting. *Brain Res. Bull.* **2018**, *143*, 155–170. [CrossRef]
50. Ruigrok, M.J.; de Lange, E.C. Emerging insights for translational pharmacokinetic and pharmacokinetic-pharmacodynamic studies: Towards prediction of nose-to-brain transport in humans. *AAPS J.* **2015**, *17*, 493–505. [CrossRef]
51. Martins, P.P.; Smyth, H.D.; Cui, Z. Strategies to facilitate or block nose-to-brain drug delivery. *Int. J. Pharm.* **2019**, *570*, 118635. [CrossRef]
52. Iwasaki, S.; Yamamoto, S.; Sano, N.; Tohyama, K.; Kosugi, Y.; Furuta, A.; Hamada, T.; Igari, T.; Fujioka, Y.; Hirabayashi, H.; et al. Direct Drug Delivery of Low-Permeable Compounds to the Central Nervous System Via Intranasal Administration in Rats and Monkeys. *Pharm. Res.* **2019**, *36*, 76. [CrossRef] [PubMed]
53. Marttin, E.; Verhoef, J.C.; Merkus, F.W.H.M. Efficacy, Safety and Mechanism of Cyclodextrins as Absorption Enhancers in Nasal Delivery of Peptide and Protein Drugs. *J. Drug Target.* **1998**, *6*, 17–36. [CrossRef] [PubMed]

54. Li, Y.; Li, J.; Zhang, X.; Ding, J.; Mao, S. Non-ionic surfactants as novel intranasal absorption enhancers: In vitro and in vivo characterization. *Drug Deliv.* **2014**, *23*, 2272–2279. [CrossRef] [PubMed]
55. Rassu, G.; Soddu, E.; Cossu, M.; Brundu, A.; Cerri, G.; Marchetti, N.; Ferraro, L.; Regan, R.F.; Giunchedi, P.; Gavini, E.; et al. Solid microparticles based on chitosan or methyl-beta-cyclodextrin: A first formulative approach to increase the nose-to-brain transport of deferoxamine mesylate. *J. Control. Release* **2015**, *201*, 68–77. [CrossRef]
56. Akita, T.; Kimura, R.; Akaguma, S.; Nagai, M.; Nakao, Y.; Tsugane, M.; Suzuki, H.; Oka, J.-I.; Yamashita, C. Usefulness of cell-penetrating peptides and penetration accelerating sequence for nose-to-brain delivery of glucagon-like peptide-2. *J. Control. Release* **2021**, *335*, 575–583. [CrossRef]
57. Ozsoy, Y.; Güngör, S. Nasal route: An alternative approach for antiemetic drug delivery. *Expert Opin. Drug Deliv.* **2011**, *8*, 1439–1453. [CrossRef]
58. Espinoza, L.C.; Silva-Abreu, M.; Clares, B.; Rodriguez-Lagunas, M.J.; Halbaut, L.; Canas, M.A.; Calpena, A.C. Formulation dtrategies to improve nose-to-brain delivery of donepezil. *Pharmaceutics* **2019**, *11*, 64. [CrossRef]
59. Liu, L.; Tian, C.; Dong, B.; Xia, M.; Cai, Y.; Hu, R.; Chu, X. Models to evaluate the barrier properties of mucus during drug diffusion. *Int. J. Pharm.* **2021**, *599*, 120415. [CrossRef]
60. Wu, H.; Hu, K.; Jiang, X. From nose to brain: Understanding transport capacity and transport rate of drugs. *Expert Opin. Drug Deliv.* **2008**, *5*, 1159–1168. [CrossRef]
61. Graff, C.L.; Pollack, G.M. Nasal Drug Administration: Potential for Targeted Central Nervous System Delivery. *J. Curr. Chem. Pharm. Sci.* **2005**, *94*, 1187–1195. [CrossRef]
62. Shingaki, T.; Hidalgo, I.J.; Furubayashi, T.; Sakane, T.; Katsumi, H.; Yamamoto, A.; Yamashita, S. Nasal Delivery of P-gp Substrates to the Brain through the Nose–Brain Pathway. *Drug Metab. Pharmacokinet.* **2011**, *26*, 248–255. [CrossRef] [PubMed]
63. Bourganis, V.; Kammona, O.; Alexopoulos, A.; Kiparissides, C. Recent advances in carrier mediated nose-to-brain delivery of pharmaceutics. *Eur. J. Pharm. Biopharm.* **2018**, *128*, 337–362. [CrossRef] [PubMed]
64. Dhuria, S.V.; Hanson, L.R.; Frey, W.H., 2nd; Ii, W.H.F. Novel Vasoconstrictor Formulation to Enhance Intranasal Targeting of Neuropeptide Therapeutics to the Central Nervous System. *J. Pharmacol. Exp. Ther.* **2008**, *328*, 312–320. [CrossRef]
65. Pires, A.; Fortuna, A.; Alves, G.; Falcão, A. Intranasal drug delivery: How, why and what for? *J. Pharm. Pharm. Sci.* **2009**, *12*, 288–311. [CrossRef]
66. Perez-Caballero, L.; Torres-Sanchez, S.; Bravo, L.; Mico, J.A.; Berrocoso, E. Fluoxetine: A case history of its discovery and preclinical development. *Expert Opin. Drug Discov.* **2014**, *9*, 567–578. [CrossRef] [PubMed]
67. Suwała, J.; Machowska, M.; Wiela-Hojeńska, A. Venlafaxine pharmacogenetics: A comprehensive review. *Pharmacogenomics* **2019**, *20*, 829–845. [CrossRef] [PubMed]
68. Patel, R.G. Nasal Anatomy and Function. *Facial Plast. Surg.* **2017**, *33*, 3–8. [CrossRef] [PubMed]
69. Kumar, A.; Pandey, A.N.; Jain, S.K. Nasal-nanotechnology: Revolution for efficient therapeutics delivery. *Drug Deliv.* **2014**, *23*, 671–683. [CrossRef]
70. Costa, C.P.; Moreira, J.; Amaral, M.H.; Lobo, J.M.S.; Silva, A. Nose-to-brain delivery of lipid-based nanosystems for epileptic seizures and anxiety crisis. *J. Control. Release* **2019**, *295*, 187–200. [CrossRef]
71. Quintana, D.S.; Westlye, L.T.; Rustan, G.Ø.; Tesli, N.; Poppy, C.L.; Smevik, H.; Tesli, M.; Røine, M.; Mahmoud, R.A.; Smerud, K.T.; et al. Low-dose oxytocin delivered intranasally with Breath Powered device affects social-cognitive behavior: A randomized four-way crossover trial with nasal cavity dimension assessment. *Transl. Psychiatry* **2015**, *5*, e602. [CrossRef]
72. Mittal, D.; Ali, A.; Md, S.; Baboota, S.; Sahni, J.K.; Ali, J. Insights into direct nose to brain delivery: Current status and future perspective. *Drug Deliv.* **2013**, *21*, 75–86. [CrossRef] [PubMed]
73. Djupesland, P.G. Nasal drug delivery devices: Characteristics and performance in a clinical perspective—A review. *Drug Deliv. Transl. Res.* **2012**, *3*, 42–62. [CrossRef] [PubMed]
74. Tong, X.; Dong, J.; Shang, Y.; Inthavong, K.; Tu, J. Effects of nasal drug delivery device and its orientation on sprayed particle deposition in a realistic human nasal cavity. *Comput. Biol. Med.* **2016**, *77*, 40–48. [CrossRef] [PubMed]
75. Warnken, Z.N.; Smyth, H.D.; Watts, A.B.; Weitman, S.; Kuhn, J.G.; Williams, R.O. Formulation and device design to increase nose to brain drug delivery. *J. Drug Deliv. Sci. Technol.* **2016**, *35*, 213–222. [CrossRef]
76. Chen, Y.; Liu, Y.; Xie, J.; Zheng, Q.; Yue, P.; Chen, L.; Hu, P.; Yang, M. Nose-to-Brain Delivery by Nanosuspensions-Based in situ Gel for Breviscapine. *Int. J. Nanomed.* **2020**, *15*, 10435–10451. [CrossRef]
77. Cunha, S.; Forbes, B.; Lobo, J.M.S.; Silva, A.C. Improving drug delivery for alzheimer's disease through nose-to-brain delivery using nanoemulsions, nanostructured lipid carriers (NLC) and in situ hydrogels. *Int. J. Nanomed.* **2021**, *16*, 4373–4390. [CrossRef]
78. Agrawal, M.; Saraf, S.; Saraf, S.; Dubey, S.K.; Puri, A.; Gupta, U.; Kesharwani, P.; Ravichandiran, V.; Kumar, P.; Naidu, V.; et al. Stimuli-responsive In situ gelling system for nose-to-brain drug delivery. *J. Control. Release* **2020**, *327*, 235–265. [CrossRef]
79. Zahir-Jouzdani, F.; Wolf, J.D.; Atyabi, F.; Bernkop-Schnürch, A. In situ gelling and mucoadhesive polymers: Why do they need each other? *Expert Opin. Drug Deliv.* **2018**, *15*, 1007–1019. [CrossRef]
80. Kanwar, N.; Sinha, V.R. In Situ Forming Depot as Sustained-Release Drug Delivery Systems. *Crit. Rev. Ther. Drug Carr. Syst.* **2019**, *36*, 93–136. [CrossRef]
81. Mir, M.; Ahmed, N.; Rehman, A.U. Recent applications of PLGA based nanostructures in drug delivery. *Colloids Surf. B Biointerfaces* **2017**, *159*, 217–231. [CrossRef]

82. Kapoor, D.N.; Bhatia, A.; Kaur, R.; Sharma, R.; Kaur, G.; Dhawan, S. PLGA: A unique polymer for drug delivery. *Ther. Deliv.* **2015**, *6*, 41–58. [CrossRef] [PubMed]
83. Desai, K.G. Chitosan Nanoparticles Prepared by Ionotropic Gelation: An Overview of Recent Advances. *Crit. Rev. Ther. Drug Carr. Syst.* **2016**, *33*, 107–158. [CrossRef]
84. Prabaharan, M. Chitosan-based nanoparticles for tumor-targeted drug delivery. *Int. J. Biol. Macromol.* **2015**, *72*, 1313–1322. [CrossRef] [PubMed]
85. Mohebbi, S.; Nezhad, M.N.; Zarrintaj, P.; Jafari, S.H.; Gholizadeh, S.S.; Saeb, M.R.; Mozafari, M. Chitosan in Biomedical Engineering: A Critical Review. *Curr. Stem Cell Res. Ther.* **2019**, *14*, 93–116. [CrossRef] [PubMed]
86. Jana, S.; Sen, K.K.; Gandhi, A. Alginate Based Nanocarriers for Drug Delivery Applications. *Curr. Pharm. Des.* **2016**, *22*, 3399–3410. [CrossRef]
87. Tønnesen, H.H.; Karlsen, J. Alginate in Drug Delivery Systems. *Drug Dev. Ind. Pharm.* **2002**, *28*, 621–630. [CrossRef]
88. Severino, P.; Da Silva, C.F.; Andrade, L.N.; de Lima Oliveira, D.; Campos, J.; Souto, E.B. Alginate Nanoparticles for Drug Delivery and Targeting. *Curr. Pharm. Des.* **2019**, *25*, 1312–1334. [CrossRef]
89. Thai, H.; Nguyen, C.T.; Thach, L.T.; Tran, M.T.; Mai, H.D.; Nguyen, T.T.T.; Le, G.D.; Van Can, M.; Tran, L.D.; Bach, G.L.; et al. Characterization of chitosan/alginate/lovastatin nanoparticles and investigation of their toxic effects in vitro and in vivo. *Sci. Rep.* **2020**, *10*, 909–915. [CrossRef]
90. Reig-Vano, B.; Tylkowski, B.; Montané, X.; Giamberini, M. Alginate-based hydrogels for cancer therapy and research. *Int. J. Biol. Macromol.* **2020**, *170*, 424–436. [CrossRef]
91. Ahmad, E.; Feng, Y.; Qi, J.; Fan, W.; Ma, Y.; He, H.; Xia, F.; Dong, X.; Zhao, W.; Lu, Y.; et al. Evidence of nose-to-brain delivery of nanoemulsions: Cargoes but not vehicles. *Nanoscale* **2016**, *9*, 1174–1183. [CrossRef]
92. Bahadur, S.; Pardhi, D.M.; Rautio, J.; Rosenholm, J.M.; Pathak, K. Intranasal Nanoemulsions for Direct Nose-to-Brain Delivery of Actives for CNS Disorders. *Pharmaceutics* **2020**, *12*, 1230. [CrossRef] [PubMed]
93. Bonferoni, M.C.; Rossi, S.; Sandri, G.; Ferrari, F.; Gavini, E.; Rassu, G.; Giunchedi, P. Nanoemulsions for "nose-to-brain" drug delivery. *Pharmaceutics* **2019**, *11*, 84. [CrossRef]
94. Rinaldi, F.; Oliva, A.; Sabatino, M.; Imbriano, A.; Hanieh, P.N.; Garzoli, S.; Mastroianni, C.M.; De Angelis, M.; Miele, M.C.; Arnaut, M.; et al. Antimicrobial Essential Oil Formulation: Chitosan Coated Nanoemulsions for Nose to Brain Delivery. *Pharmaceutics* **2020**, *12*, 678. [CrossRef]
95. Pandey, V.; Kohli, S. Lipids and Surfactants: The Inside Story of Lipid-Based Drug Delivery Systems. *Crit. Rev. Ther. Drug Carr. Syst.* **2018**, *35*, 99–155. [CrossRef] [PubMed]
96. Urquhart, A.J.; Eriksen, A.Z. Recent developments in liposomal drug delivery systems for the treatment of retinal diseases. *Drug Discov. Today* **2019**, *24*, 1660–1668. [CrossRef] [PubMed]
97. Pattni, B.S.; Chupin, V.V.; Torchilin, V.P. New Developments in Liposomal Drug Delivery. *Chem. Rev.* **2015**, *115*, 10938–10966. [CrossRef]
98. Natsheh, H.; Touitou, E. Phospholipid Magnesome—A nasal vesicular carrier for delivery of drugs to brain. *Drug Deliv. Transl. Res.* **2018**, *8*, 806–819. [CrossRef]
99. Natsheh, H.; Touitou, E. Phospholipid Vesicles for Dermal/Transdermal and Nasal Administration of Active Molecules: The Effect of Surfactants and Alcohols on the Fluidity of Their Lipid Bilayers and Penetration Enhancement Properties. *Molecules* **2020**, *25*, 2959. [CrossRef]
100. Touitou, E.; Duchi, S.; Natsheh, H. A new nanovesicular system for nasal drug administration. *Int. J. Pharm.* **2020**, *580*, 119243. [CrossRef]
101. Tapeinos, C.; Battaglini, M.; Ciofani, G. Advances in the design of solid lipid nanoparticles and nanostructured lipid carriers for targeting brain diseases. *J. Control. Release* **2017**, *264*, 306–332. [CrossRef]
102. Garcês, A.; Amaral, M.H.; Lobo, J.M.S.; Silva, A.C. Formulations based on solid lipid nanoparticles (SLN) and nanostructured lipid carriers (NLC) for cutaneous use: A review. *Eur. J. Pharm. Sci.* **2018**, *112*, 159–167. [CrossRef] [PubMed]
103. Czajkowska-Kośnik, A.; Szekalska, M.; Winnicka, K. Nanostructured lipid carriers: A potential use for skin drug delivery systems. *Pharmacol. Rep.* **2018**, *71*, 156–166. [CrossRef] [PubMed]
104. Khosa, A.; Reddi, S.; Saha, R.N. Nanostructured lipid carriers for site-specific drug delivery. *Biomed. Pharmacother.* **2018**, *103*, 598–613. [CrossRef] [PubMed]
105. Chen, F.; Jiang, H.; Xu, J.; Wang, S.; Meng, D.; Geng, P.; Dai, D.; Zhou, Q.; Zhou, Y. In Vitro and In Vivo Rat Model Assessments of the Effects of Vonoprazan on the Pharmacokinetics of Venlafaxine. *Drug Des. Dev. Ther.* **2020**, *14*, 4815–4824. [CrossRef] [PubMed]
106. Schoretsanitis, G.; Haen, E.; Hiemke, C.; Endres, K.; Ridders, F.; Veselinovic, T.; Gründer, G.; Paulzen, M. Pharmacokinetic correlates of venlafaxine: Associated adverse reactions. *Eur. Arch. Psychiatry Clin. Neurosci.* **2019**, *269*, 851–857. [CrossRef]
107. Cayero-Otero, M.D.; Gomes, M.J.; Martins, C.; Álvarez-Fuentes, J.; Fernández-Arévalo, M.; Sarmento, B.; Martín-Banderas, L. In vivo biodistribution of venlafaxine-PLGA nanoparticles for brain delivery: Plain vs. functionalized nanoparticles. *Expert Opin. Drug Deliv.* **2019**, *16*, 1413–1427. [CrossRef]
108. Haque, S.; Md, S.; Fazil, M.; Kumar, M.; Sahni, J.K.; Ali, J.; Baboota, S. Venlafaxine loaded chitosan NPs for brain targeting: Pharmacokinetic and pharmacodynamic evaluation. *Carbohydr. Polym.* **2012**, *89*, 72–79. [CrossRef]
109. Rizeq, B.R.; Younes, N.N.; Rasool, K.; Nasrallah, G.K. Synthesis, Bioapplications, and Toxicity Evaluation of Chitosan-Based Nanoparticles. *Int. J. Mol. Sci.* **2019**, *20*, 5776. [CrossRef]

110. Norman, T.R.; Olver, J.S. Desvenlafaxine in the treatment of major depression: An updated overview. *Expert Opin. Pharmacother.* **2021**, *22*, 1087–1097. [CrossRef]
111. Tong, G.-F.; Qin, N.; Sun, L.-W. Development and evaluation of Desvenlafaxine loaded PLGA-chitosan nanoparticles for brain delivery. *Saudi Pharm. J.* **2016**, *25*, 844–851. [CrossRef]
112. Ahmed, S.; Gull, A.; Aqil, M.; Ansari, M.D.; Sultana, Y. Poloxamer-407 thickened lipid colloidal system of agomelatine for brain targeting: Characterization, brain pharmacokinetic study and behavioral study on Wistar rats. *Colloids Surfaces B Biointerfaces* **2019**, *181*, 426–436. [CrossRef] [PubMed]
113. Fatouh, A.M.; Elshafeey, A.H.; Abdelbary, A. Intranasal agomelatine solid lipid nanoparticles to enhance brain delivery: Formulation, optimization and in vivo pharmacokinetics. *Drug Des. Dev. Ther.* **2017**, *11*, 1815–1825. [CrossRef] [PubMed]
114. Fatouh, A.M.; Elshafeey, A.H.; Abdelbary, A. Agomelatine-based in situ gels for brain targeting via the nasal route: Statistical optimization, in vitro, and in vivo evaluation. *Drug Deliv.* **2017**, *24*, 1077–1085. [CrossRef] [PubMed]
115. Jani, P.; Vanza, J.; Pandya, N.; Tandel, H. Formulation of polymeric nanoparticles of antidepressant drug for intranasal delivery. *Ther. Deliv.* **2019**, *10*, 683–696. [CrossRef] [PubMed]
116. Alam, M.I.; Baboota, S.; Ahuja, A.; Ali, M.; Ali, J.; Sahni, J.K. Intranasal infusion of nanostructured lipid carriers (NLC) containing CNS acting drug and estimation in brain and blood. *Drug Deliv.* **2013**, *20*, 247–251. [CrossRef] [PubMed]
117. Alam, M.I.; Baboota, S.; Ahuja, A.; Ali, M.; Ali, J.; Sahni, J.K.; Bhatnagar, A. Pharmacoscintigraphic evaluation of potential of lipid nanocarriers for nose-to-brain delivery of antidepressant drug. *Int. J. Pharm.* **2014**, *470*, 99–106. [CrossRef] [PubMed]
118. Elsenosy, F.M.; Abdelbary, G.A.; Elshafeey, A.H.; Elsayed, I.; Fares, A.R. Brain Targeting of Duloxetine HCL via Intranasal Delivery of Loaded Cubosomal Gel: In vitro Characterization, ex vivo Permeation, and in vivo Biodistribution Studies. *Int. J. Nanomed.* **2020**, *15*, 9517–9537. [CrossRef]
119. Khatoon, M.; Sohail, M.F.; Shahnaz, G.; Rehman, F.U.; Din, F.U.; Rehman, A.U.; Ullah, N.; Amin, U.; Khan, G.M.; Shah, K.U. Development and Evaluation of Optimized Thiolated Chitosan Proniosomal Gel Containing Duloxetine for Intranasal Delivery. *AAPS PharmSciTech* **2019**, *20*, 288. [CrossRef]
120. Purgato, M.; Papola, D.; Gastaldon, C.; Trespidi, C.; Magni, L.R.; Rizzo, C.; Furukawa, T.A.; Watanabe, N.; Cipriani, A.; Barbui, C. Paroxetine versus other anti-depressive agents for depression. *Cochrane Database Syst. Rev.* **2014**. [CrossRef]
121. Pandey, Y.R.; Kumar, S.; Gupta, B.K.; Ali, J.; Baboota, S. Intranasal delivery of paroxetine nanoemulsion via the olfactory region for the management of depression: Formulation, behavioural and biochemical estimation. *Nanotechnology* **2015**, *27*, 25102. [CrossRef]
122. Silva, S.; Bicker, J.; Fonseca, C.; Ferreira, N.R.; Vitorino, C.; Alves, G.; Falcão, A.; Fortuna, A. Encapsulated Escitalopram and Paroxetine Intranasal Co-Administration: In Vitro/In Vivo Evaluation. *Front. Pharmacol.* **2021**, *12*, 751321. [CrossRef] [PubMed]
123. Motaleb, M.A.; Ibrahim, I.T.; Sayyed, M.E.; Awad, G.A.S. (131)I-trazodone: Preparation, quality control and in vivo biodistribution study by intranasal and intravenous routes as a hopeful brain imaging radiopharmaceutical. *Rev. Esp. Med. Nucl. Imagen Mol.* **2017**, *36*, 371–376. [PubMed]
124. Boche, M.; Pokharkar, V. Quetiapine Nanoemulsion for Intranasal Drug Delivery: Evaluation of Brain-Targeting Efficiency. *AAPS PharmSciTech* **2016**, *18*, 686–696. [CrossRef] [PubMed]
125. Shah, B.; Khunt, D.; Misra, M.; Padh, H. Non-invasive intranasal delivery of quetiapine fumarate loaded microemulsion for brain targeting: Formulation, physicochemical and pharmacokinetic consideration. *Eur. J. Pharm. Sci.* **2016**, *91*, 196–207. [CrossRef]
126. Naik, A.; Nair, H. Formulation and Evaluation of Thermosensitive Biogels for Nose to Brain Delivery of Doxepin. *BioMed Res. Int.* **2014**, *2014*, 847547. [CrossRef] [PubMed]
127. Eduardo, T.Q.; Angela, A.; Mateo, L.; Melanie, L.Z.; Valentina, P.F.; David, C.; Estefania, C.; Natalia, R.S.; Andrés, V.C.; Angel, R.O.; et al. Ketamine for resistant depression: A scoping review. *Actas Esp. Psiquiatr.* **2022**, *50*, 144–159. [PubMed]
128. Yavi, M.; Lee, H.; Henter, I.D.; Park, L.T.; Zarate, C.A., Jr. Ketamine treatment for depression: A review. *Discov. Ment. Health* **2022**, *2*, 9. [CrossRef]
129. Yao, W.; Cao, Q.; Luo, S.; He, L.; Yang, C.; Chen, J.; Qi, Q.; Hashimoto, K.; Zhang, J.C. Microglial ERK-NRBP1-CREB-BDNF signaling in sustained antidepressant actions of (R)-ketamine. *Mol. Psychiatry* **2022**, *27*, 1618–1629. [CrossRef]
130. Buddenberg, T.E.; Topic, B.; Mahlberg, E.D.; Silva, M.A.D.S.; Huston, J.P.; Mattern, C. Behavioral Actions of Intranasal Application of Dopamine: Effects on Forced Swimming, Elevated Plus-Maze and Open Field Parameters. *Neuropsychobiology* **2008**, *57*, 70–79. [CrossRef]
131. Chemuturi, N.V.; Donovan, M.D. Metabolism of Dopamine by the Nasal Mucosa. *J. Pharm. Sci.* **2006**, *95*, 2507–2515. [CrossRef]
132. Brown, V.; Liu, F. Intranasal Delivery of a Peptide with Antidepressant-Like Effect. *Neuropsychopharmacology* **2014**, *39*, 2131–2141. [CrossRef] [PubMed]
133. Zangani, C.; Giordano, B.; Stein, H.; Bonora, S.; D'Agostino, A.; Ostinelli, E.G. Efficacy of amisulpride for depressive symptoms in individuals with mental disorders: A systematic review and meta-analysis. *Hum. Psychopharmacol. Clin. Exp.* **2021**, *36*, e2801. [CrossRef] [PubMed]
134. Hopkins, S.C.; Wilkinson, S.; Corriveau, T.J.; Nishikawa, H.; Nakamichi, K.; Loebel, A.; Koblan, K.S. Discovery of Nonracemic Amisulpride to Maximize Benefit/Risk of 5-HT7 and D2 Receptor Antagonism for the Treatment of Mood Disorders. *Clin. Pharmacol. Ther.* **2021**, *110*, 808–815. [CrossRef] [PubMed]
135. Kishimoto, T.; Hagi, K.; Kurokawa, S.; Kane, J.M.; Correll, C.U. Efficacy and safety/tolerability of antipsychotics in the treatment of adult patients with major depressive disorder: A systematic review and meta-analysis. *Psychol. Med.* **2022**, 1–19. [CrossRef]

136. Yuan, B.; Yuan, M. Changes of Mental State and Serum Prolactin Levels in Patients with Schizophrenia and Depression after Receiving the Combination Therapy of Amisulpride and Chloroprothixol Tablets. *Comput. Math. Methods Med.* **2022**, *2022*, 6580030. [CrossRef]
137. Gadhave, D.G.; Tagalpallewar, A.A.; Kokare, C.R. Agranulocytosis-Protective Olanzapine-Loaded Nanostructured Lipid Carriers Engineered for CNS Delivery: Optimization and Hematological Toxicity Studies. *AAPS PharmSciTech* **2019**, *20*, 22. [CrossRef]
138. El Assasy, A.E.I.; Younes, N.F.; Makhlouf, A.I.A. Enhanced oral absorption of amisulpride via a nanostructured lipid carrier-based capsules: Development, optimization applying the desirability function approach and in vivo pharmacokinetic study. *AAPS PharmSciTech* **2019**, *20*, 82. [CrossRef]
139. Gadhave, D.; Tupe, S.; Tagalpallewar, A.; Gorain, B.; Choudhury, H.; Kokare, C. Nose-to-brain delivery of amisulpride-loaded lipid-based poloxamer-gellan gum nanoemulgel: In vitro and in vivo pharmacological studies. *Int. J. Pharm.* **2021**, *607*, 121050. [CrossRef]
140. Lenze, E.J.; Mulsant, B.H.; Blumberger, D.M.; Karp, J.F.; Newcomer, J.W.; Anderson, S.; Dew, M.A.; Butters, M.A.; Stack, J.A.; Begley, A.E.; et al. Efficacy, safety, and tolerability of augmentation pharmacotherapy with aripiprazole for treatment-resistant depression in late life: A randomised, double-blind, placebo-controlled trial. *Lancet* **2015**, *386*, 2404–2412. [CrossRef]
141. Nelson, J.C.; Papakostas, G.I. Atypical Antipsychotic Augmentation in Major Depressive Disorder: A Meta-Analysis of Placebo-Controlled Randomized Trials. *Am. J. Psychiatry* **2009**, *166*, 980–991. [CrossRef]
142. Kirschbaum, K.M.; Uhr, M.; Holthoewer, D.; Namendorf, C.; Pietrzik, C.; Hiemke, C.; Schmitt, U. Pharmacokinetics of acute and sub-chronic aripiprazole in P-glycoprotein deficient mice. *Neuropharmacology* **2010**, *59*, 474–479. [CrossRef] [PubMed]
143. Piazzini, V.; Landucci, E.; Urru, M.; Chiarugi, A.; Pellegrini-Giampietro, D.E.; Bilia, A.R.; Bergonzi, M.C. Enhanced dissolution, permeation and oral bioavailability of aripiprazole mixed micelles: In vitro and in vivo evaluation. *Int. J. Pharm.* **2020**, *583*, 119361. [CrossRef] [PubMed]
144. Kumbhar, S.A.; Kokare, C.R.; Shrivastava, B.; Gorain, B.; Choudhury, H. Antipsychotic Potential and Safety Profile of TPGS-Based Mucoadhesive Aripiprazole Nanoemulsion: Development and Optimization for Nose-To-Brain Delivery. *J. Pharm. Sci.* **2021**, *110*, 1761–1778. [CrossRef] [PubMed]
145. Sawant, K.; Pandey, A.; Patel, S. Aripiprazole loaded poly(caprolactone) nanoparticles: Optimization and in vivo pharmacokinetics. *Mater. Sci. Eng. C Mater. Biol. Appl.* **2016**, *66*, 230–243. [CrossRef]
146. Taymouri, S.; Shahnamnia, S.; Mesripour, A.; Varshosaz, J. In vitro and in vivo evaluation of an ionic sensitive in situ gel containing nanotransfersomes for aripiprazole nasal delivery. *Pharm. Dev. Technol.* **2021**, *26*, 867–879. [CrossRef] [PubMed]
147. Singh, D.P.; Rashid, M.; Hallan, S.; Mehra, N.K.; Prakash, A.; Mishra, N. Pharmacological evaluation of nasal delivery of selegiline hydrochloride-loaded thiolated chitosan nanoparticles for the treatment of depression. *Artif. Cells Nanomed. Biotechnol.* **2015**, *44*, 865–877. [CrossRef]
148. Sridhar, V.; Gaud, R.; Bajaj, A.; Wairkar, S. Pharmacokinetics and pharmacodynamics of intranasally administered selegiline nanoparticles with improved brain delivery in Parkinson's disease. *Nanomedicine* **2018**, *14*, 2609–2618. [CrossRef] [PubMed]
149. Sridhar, V.; Wairkar, S.; Gaud, R.; Bajaj, A.; Meshram, P. Brain targeted delivery of mucoadhesive thermosensitive nasal gel of selegiline hydrochloride for treatment of Parkinson's disease. *J. Drug Target* **2018**, *26*, 150–161. [CrossRef]
150. Rukmangathen, R.; Yallamalli, I.M.; Yalavarthi, P.R. Biopharmaceutical potential of selegiline loaded chitosan nanoparticles in the management of parkinson's disease. *Curr. Drug Discov. Technol.* **2019**, *16*, 417–425. [CrossRef]
151. Kumar, S.; Ali, J.; Baboota, S. Design Expert® supported optimization and predictive analysis of selegiline nanoemulsion via the olfactory region with enhanced behavioural performance in Parkinson's disease. *Nanotechnology* **2016**, *27*, 435101. [CrossRef]
152. Kumar, S.; Dang, S.; Nigam, K.; Ali, J.; Baboota, S. Selegiline nanoformulation in attenuation of oxidative stress and upregulation of dopamine in the brain for the treatment of parkinson's disease. *Rejuvenation Res.* **2018**, *21*, 464–476. [CrossRef]
153. Mishra, N.; Sharma, S.; Deshmukh, R.; Kumar, A.; Sharma, R. Development and Characterization of Nasal Delivery of Selegiline Hydrochloride Loaded Nanolipid Carriers for the Management of Parkinson's Disease. *Central Nerv. Syst. Agents Med. Chem.* **2019**, *19*, 46–56. [CrossRef] [PubMed]
154. Kaur, P.; Garg, T.; Vaidya, B.; Prakash, A.; Rath, G.; Goyal, A.K. Brain delivery of intranasal in situ gel of nanoparticulated polymeric carriers containing antidepressant drug: Behavioral and biochemical assessment. *J. Drug Target* **2015**, *23*, 275–286. [CrossRef] [PubMed]
155. Woo, Y.S.; Lim, H.K.; Wang, S.-M.; Bahk, W.-M. Title Clinical Evidence of Antidepressant Effects of Insulin and Anti-Hyperglycemic Agents and Implications for the Pathophysiology of Depression—A Literature Review. *Int. J. Mol. Sci.* **2020**, *21*, 6969. [CrossRef] [PubMed]
156. Hamer, J.A.; Testani, D.; Mansur, R.B.; Lee, Y.; Subramaniapillai, M.; McIntyre, R.S. Brain insulin resistance: A treatment target for cognitive impairment and anhedonia in depression. *Exp. Neurol.* **2019**, *315*, 1–8. [CrossRef] [PubMed]
157. Zou, X.H.; Sun, L.H.; Yang, W.; Li, B.J.; Cui, R.J. Potential role of insulin on the pathogenesis of depression. *Cell Prolif.* **2020**, *53*, e12806. [CrossRef]
158. Nedelcovych, M.T.; Gadiano, A.J.; Wu, Y.; Manning, A.A.; Thomas, A.G.; Khuder, S.S.; Yoo, S.-W.; Xu, J.; McArthur, J.C.; Haughey, N.J.; et al. Pharmacokinetics of Intranasal versus Subcutaneous Insulin in the Mouse. *ACS Chem. Neurosci.* **2017**, *9*, 809–816. [CrossRef]
159. Ren, G.; Xue, P.; Wu, B.; Yang, F.; Wu, X. Intranasal treatment of lixisenatide attenuated emotional and olfactory symptoms via CREB-mediated adult neurogenesis in mouse depression model. *Aging* **2021**, *13*, 3898–3908. [CrossRef]

160. Sasaki-Hamada, S.; Nakamura, R.; Nakao, Y.; Akimoto, T.; Sanai, E.; Nagai, M.; Horiguchi, M.; Yamashita, C.; Oka, J.-I. Antidepressant-like effects exerted by the intranasal administration of a glucagon-like peptide-2 derivative containing cell-penetrating peptides and a penetration-accelerating sequence in mice. *Peptides* **2017**, *87*, 64–70. [CrossRef]
161. Nakao, Y.; Horiguchi, M.; Nakamura, R.; Sasaki-Hamada, S.; Ozawa, C.; Funane, T.; Ozawa, R.; Oka, J.-I.; Yamashita, C. LARETH-25 and β-CD improve central transitivity and central pharmacological effect of the GLP-2 peptide. *Int. J. Pharm.* **2016**, *515*, 37–45. [CrossRef]
162. Dickens, M.J.; Pawluski, J.L. The HPA Axis During the Perinatal Period: Implications for Perinatal Depression. *Endocrinology* **2018**, *159*, 3737–3746. [CrossRef] [PubMed]
163. Riem, M.; Kunst, L.; Bekker, M.; Fallon, M.; Kupper, N. Intranasal oxytocin enhances stress-protective effects of social support in women with negative childhood experiences during a virtual Trier Social Stress Test. *Psychoneuroendocrinology* **2019**, *111*, 104482. [CrossRef] [PubMed]
164. Yoon, S.; Kim, Y.-K. Possible oxytocin-related biomarkers in anxiety and mood disorders. *Prog. Neuro-Psychopharmacol. Biol. Psychiatry* **2022**, *116*, 110531. [CrossRef] [PubMed]
165. Chen, Q.; Zhuang, J.; Zuo, R.; Zheng, H.; Dang, J.; Wang, Z. Exploring associations between postpartum depression and oxytocin levels in cerebrospinal fluid, plasma and saliva. *J. Affect. Disord.* **2022**, *315*, 198–205. [CrossRef] [PubMed]
166. Ross, H.; Cole, C.; Smith, Y.; Neumann, I.; Landgraf, R.; Murphy, A.; Young, L. Characterization of the oxytocin system regulating affiliative behavior in female prairie voles. *Neuroscience* **2009**, *162*, 892–903. [CrossRef]
167. Zhao, J.L.; Jiang, W.T.; Wang, X.; Cai, Z.D.; Liu, Z.H.; Liu, G.R. Exercise, brain plasticity, and depression. *CNS Neurosci. Ther.* **2020**, *26*, 885–895. [CrossRef]
168. Castrén, E.; Monteggia, L.M. Brain-Derived Neurotrophic Factor Signaling in Depression and Antidepressant Action. *Biol. Psychiatry* **2021**, *90*, 128–136. [CrossRef]
169. Ma, X.-C.; Liu, P.; Zhang, X.-L.; Jiang, W.-H.; Jia, M.; Wang, C.-X.; Dong, Y.-Y.; Dang, Y.-H.; Gao, C.-G. Intranasal Delivery of Recombinant AAV Containing BDNF Fused with HA2TAT: A Potential Promising Therapy Strategy for Major Depressive Disorder. *Sci. Rep.* **2016**, *6*, 22404. [CrossRef]
170. Chen, C.; Dong, Y.; Liu, F.; Gao, C.; Ji, C.; Dang, Y.; Ma, X.; Liu, Y. A Study of Antidepressant Effect and Mechanism on Intranasal Delivery of BDNF-HA2TAT/AAV to Rats with Post-Stroke Depression. *Neuropsychiatr. Dis. Treat.* **2020**, *16*, 637–649. [CrossRef]
171. Liu, F.; Liu, Y.-P.; Lei, G.; Liu, P.; Chu, Z.; Gao, C.-G.; Dang, Y.-H. Antidepressant effect of recombinant NT4-NAP/AAV on social isolated mice through intranasal route. *Oncotarget* **2016**, *8*, 10103–10113. [CrossRef]
172. Ma, X.-C.; Chu, Z.; Zhang, X.-L.; Jiang, W.-H.; Jia, M.; Dang, Y.-H.; Gao, C.-G. Intranasal Delivery of Recombinant NT4-NAP/AAV Exerts Potential Antidepressant Effect. *Neurochem. Res.* **2016**, *41*, 1375–1380. [CrossRef] [PubMed]
173. Jiang, J.; Peng, Y.; Liang, X.; Li, S.; Chang, X.; Li, L.; Chang, M. Centrally administered cortistation-14 induces antidepressant-like effects in mice via mediating ghrelin and GABA(A) receptor signaling pathway. *Front. Pharmacol.* **2018**, *9*, 767. [CrossRef] [PubMed]
174. Serova, L.; Laukova, M.; Alaluf, L.; Pucillo, L.; Sabban, E. Intranasal neuropeptide Y reverses anxiety and depressive-like behavior impaired by single prolonged stress PTSD model. *Eur. Neuropsychopharmacol.* **2014**, *24*, 142–147. [CrossRef]
175. Oh, J.-Y.; Liu, Q.F.; Hua, C.; Jeong, H.J.; Jang, J.-H.; Jeon, S.; Park, S.J.A.H.-J. Intranasal Administration of Melanin-Concentrating Hormone Reduces Stress-Induced Anxiety- and Depressive-Like Behaviors in Rodents. *Exp. Neurobiol.* **2020**, *29*, 453–469. [CrossRef]
176. Shi, C.-G.; Wang, L.-M.; Wu, Y.; Wang, P.; Gan, Z.-J.; Lin, K.; Jiang, L.-X.; Xu, Z.-Q.; Fan, M. Intranasal Administration of Nerve Growth Factor Produces Antidepressant-Like Effects in Animals. *Neurochem. Res.* **2010**, *35*, 1302–1314. [CrossRef] [PubMed]
177. Beurel, E.; Toups, M.; Nemeroff, C.B. The Bidirectional Relationship of Depression and Inflammation: Double Trouble. *Neuron* **2020**, *107*, 234–256. [CrossRef]
178. Obermanns, J.; Krawczyk, E.; Juckel, G.; Emons, B. Analysis of cytokine levels, T regulatory cells and serotonin content in patients with depression. *Eur. J. Neurosci.* **2021**, *53*, 3476–3489. [CrossRef]
179. Troubat, R.; Barone, P.; Leman, S.; Desmidt, T.; Cressant, A.; Atanasova, B.; Brizard, B.; El Hage, W.; Surget, A.; Belzung, C.; et al. Neuroinflammation and depression: A review. *Eur. J. Neurosci.* **2021**, *53*, 151–171. [CrossRef]
180. Comai, S.; Melloni, E.; Lorenzi, C.; Bollettini, I.; Vai, B.; Zanardi, R.; Colombo, C.; Valtorta, F.; Benedetti, F.; Poletti, S. Selective association of cytokine levels and kynurenine/tryptophan ratio with alterations in white matter microstructure in bipolar but not in unipolar depression. *Eur. Neuropsychopharmacol.* **2021**, *55*, 96–109. [CrossRef]
181. Rengasamy, M.; Brundin, L.; Griffo, A.; Panny, B.; Capan, C.; Forton, C.; Price, R.B. Cytokine and Reward Circuitry Relationships in Treatment-Resistant Depression. *Biol. Psychiatry Glob. Open Sci.* **2021**, *2*, 45–53. [CrossRef]
182. Markova, E.V.; Shevela, E.Y.; Knyazeva, M.A.; Savkin, I.V.; Serenko, E.V.; Rashchupkin, I.M.; Amstislavskaya, T.G.; Ostanin, A.A.; Chernykh, E.R. Effect of M2 Macrophage-Derived Soluble Factors on Behavioral Patterns and Cytokine Production in Various Brain Structures in Depression-Like Mice. *Bull. Exp. Biol. Med.* **2022**, *172*, 341–344. [CrossRef] [PubMed]
183. Wang, Y.-L.; Wang, J.-X.; Hu, X.-X.; Chen, L.; Qiu, Z.-K.; Zhao, N.; Yu, Z.-D.; Sun, S.-Z.; Xu, Y.-Y.; Guo, Y.; et al. Antidepressant-like effects of albiflorin extracted from Radix paeoniae Alba. *J. Ethnopharmacol.* **2015**, *179*, 9–15. [CrossRef] [PubMed]
184. Zhao, Z.-X.; Fu, J.; Ma, S.-R.; Peng, R.; Yu, J.-B.; Cong, L.; Pan, L.-B.; Zhang, Z.-G.; Tian, H.; Che, C.-T.; et al. Gut-brain axis metabolic pathway regulates antidepressant efficacy of albiflorin. *Theranostics* **2018**, *8*, 5945–5959. [CrossRef] [PubMed]

185. Lindqvist, D.; Dhabhar, F.S.; James, S.J.; Hough, C.M.; Jain, F.A.; Bersani, F.S.; Reus, V.I.; Verhoeven, J.E.; Epel, E.S.; Mahan, L.; et al. Oxidative stress, inflammation and treatment response in major depression. *Psychoneuroendocrinology* **2016**, *76*, 197–205. [CrossRef] [PubMed]
186. Xu, D.; Qiao, T.; Wang, Y.; Wang, Q.-S.; Cui, Y.-L. Alginate nanogels-based thermosensitive hydrogel to improve antidepressant-like effects of albiflorin via intranasal delivery. *Drug Deliv.* **2021**, *28*, 2137–2149. [CrossRef]
187. Zhu, W.-Q.; Wu, H.-Y.; Sun, Z.-H.; Guo, Y.; Ge, T.-T.; Li, B.-J.; Li, X.; Cui, R.-J. Current Evidence and Future Directions of Berberine Intervention in Depression. *Front. Pharmacol.* **2022**, *13*, 824420. [CrossRef]
188. Lee, B.; Shim, I.; Lee, H.; Hahm, D.-H. Berberine alleviates symptoms of anxiety by enhancing dopamine expression in rats with post-traumatic stress disorder. *Korean J. Physiol. Pharmacol.* **2018**, *22*, 183–192. [CrossRef]
189. Fan, J.; Zhang, K.; Jin, Y.; Li, B.; Gao, S.; Zhu, J.; Cui, R. Pharmacological effects of berberine on mood disorders. *J. Cell. Mol. Med.* **2018**, *23*, 21–28. [CrossRef]
190. Wang, Q.-S.; Li, K.; Gao, L.-N.; Zhang, Y.; Lin, K.-M.; Cui, Y.-L. Intranasal delivery of berberine via in situ thermoresponsive hydrogels with non-invasive therapy exhibits better antidepressant-like effects. *Biomater. Sci.* **2020**, *8*, 2853–2865. [CrossRef]
191. Xu, D.; Qiu, C.; Wang, Y.; Qiao, T.; Cui, Y.-L. Intranasal co-delivery of berberine and evodiamine by self-assembled thermosensitive in-situ hydrogels for improving depressive disorder. *Int. J. Pharm.* **2021**, *603*, 120667. [CrossRef]
192. Zhang, K.; Lei, N.; Li, M.; Li, J.; Li, C.; Shen, Y.; Guo, P.; Xiong, L.; Xie, Y. Cang-Ai Volatile Oil Ameliorates Depressive Behavior Induced by Chronic Stress Through IDO-Mediated Tryptophan Degradation Pathway. *Front. Psychiatry* **2021**, *12*, 791991. [CrossRef] [PubMed]
193. Chen, B.; Li, J.; Xie, Y.; Ming, X.; Li, G.; Wang, J.; Li, M.; Li, X.; Xiong, L. Cang-ai volatile oil improves depressive-like behaviors and regulates DA and 5-HT metabolism in the brains of CUMS-induced rats. *J. Ethnopharmacol.* **2019**, *244*, 112088. [CrossRef] [PubMed]
194. He, C.; Wang, Z.; Shi, J. Pharmacological effects of icariin. *Adv. Pharmacol.* **2020**, *87*, 179–203. [PubMed]
195. Xu, S.; Yu, J.; Zhan, J.; Yang, L.; Guo, L.; Xu, Y. Pharmacokinetics, Tissue Distribution, and Metabolism Study of Icariin in Rat. *BioMed Res. Int.* **2017**, *2017*, 4684962. [CrossRef] [PubMed]
196. Xu, D.; Lu, Y.-R.; Kou, N.; Hu, M.-J.; Wang, Q.-S.; Cui, Y.-L. Intranasal delivery of icariin via a nanogel-thermoresponsive hydrogel compound system to improve its antidepressant-like activity. *Int. J. Pharm.* **2020**, *586*, 119550. [CrossRef]
197. Hu, W.; Xie, G.; Zhou, T.; Tu, J.; Zhang, J.; Lin, Z.; Zhang, H.; Gao, L. Intranasal administration of white tea alleviates the olfactory function deficit induced by chronic unpredictable mild stress. *Pharm. Biol.* **2020**, *58*, 1230–1237. [CrossRef]

pharmaceutics

Article

Sporopollenin Microcapsule: Sunscreen Delivery System with Photoprotective Properties

Silvia Tampucci [1,*], Giorgio Tofani [1,2,*], Patrizia Chetoni [1], Mariacristina Di Gangi [1], Andrea Mezzetta [1], Valentina Paganini [1], Susi Burgalassi [1], Christian Silvio Pomelli [1] and Daniela Monti [1]

1 Department of Pharmacy, University of Pisa, Via Bonanno 6, 56126 Pisa, Italy
2 Department of Physics, University of Pisa, Largo B. Pontecorvo 3, 56127 Pisa, Italy
* Correspondence: silvia.tampucci@unipi.it (S.T.); giorgio.tofani@df.unipi.it (G.T.)

Abstract: In recent years, the demand for high-quality solar products that combine high efficacy with environmentally friendly characteristics has increased. Among the coral-safe sunscreens, ethylhexyl triazone (Uvinul® T150) is an effective organic UVB filter, photostable and practically insoluble in water, therefore difficult to be formulated in water-based products. Oil-free sunscreens are considered ideal for most skin types, as they are not comedogenic and do not leave the skin feeling greasy. Recent studies reported that pollen grains might represent innovative drug delivery systems for their ability to encapsulate and release active ingredients in a controlled manner. Before being used, the pollen grains must be treated to remove cellular material and biomolecules, which could cause allergic reactions in predisposed subjects; the obtained hollow structures possess uniform diameter and a rigid wall with openings that allow them to be filled with bioactive substances. In the present work, pollen from *Lycopodium clavatum* has been investigated both as a delivery system for ethylhexyl triazone and as an active ingredient by evaluating its photoprotective capacity. The goal is to obtain environmentally friendly solar aqueous formulations that take advantage of both sunscreen and sporopollenin microcapsules' UV protection with a relatively low cost, as these pollen grains are widely available.

Keywords: photoprotection; sporopollenin; sunscreen; ethylhexyl triazone; skin delivery

Citation: Tampucci, S.; Tofani, G.; Chetoni, P.; Di Gangi, M.; Mezzetta, A.; Paganini, V.; Burgalassi, S.; Pomelli, C.S.; Monti, D. Sporopollenin Microcapsule: Sunscreen Delivery System with Photoprotective Properties. *Pharmaceutics* **2022**, *14*, 2041. https://doi.org/10.3390/pharmaceutics14102041

Academic Editor: Bozena B. Michniak-Kohn

Received: 17 August 2022
Accepted: 20 September 2022
Published: 24 September 2022

1. Introduction

Sunscreens play a strategic role in the prevention of photo-induced damage. In recent years, the demand for high-quality solar products that combine high efficacy with excellent sensory properties, but are also environmentally friendly, has increased [1]. In particular, the impact of solar products on marine ecosystems with the possibility of bioaccumulation of organic filters and consequent toxicity towards the coral reef has received attention [2–4].

Several strategies can be adopted to limit the environmental damage caused by UV filters. Among these, the use of natural photoprotectors in association with UV filters in solar products has been proposed for an increase in solar protection factor (SPF) combined with an antioxidant activity, which therefore allows reduction in the amount of chemical filters in solar formulations [5]. Moreover, the formulation also has a role, as the different excipients employed contribute to some ecotoxicological hazards [1].

In the last decade, the research in drug delivery systems deriving from substances of natural origin has been intensified and pollen grains have attracted a lot of interest as microencapsulation materials due to their ecological nature, uniform size distribution in the micrometric scale and chemical–physical stability [6–13]. Indeed, the outermost layer (exine) of the pollen grain wall is mainly composed of sporopollenin, an extremely robust mixture of biopolymers resistant to physical, biological and chemical non-oxidative degradation. Conversely, the innermost layer (intina) of the pollen grain walls consists of cellulose, pectin, proteins and polysaccharides susceptible to acidic hydrolysis and alkaline treatments [7]. Due to their robustness, sporopollenin microcapsules (SPMs) have been proposed for various applications, including drug delivery [8,14,15]. SPMs with different sizes and morphology can be obtained from a variety of plants, but even though many recent studies have demonstrated the capability of SPMs to control and enhance the delivery of drugs [16–18], only a few of them concern cutaneous and cosmetical applications [19,20].

Among the different plant sources that have been investigated throughout the years, SPMs from *Lycopodium clavatum* have been shown to possess a uniform diameter of about 25 μm [21] and to be endowed with antioxidant properties [22]. Furthermore, *L. clavatum* pollen grains are easily available and not expensive, therefore representing an ideal material for new excipient research.

In the present work, pollen from *L. clavatum* has been investigated as a delivery system for the organic solar filter ethylhexyl triazone (ETZ, Uvinul® T150), which is authorized for use in Europe up to a 5% concentration (Annex VI, Regulation (EC) No. 1223/2009) and is considered environmentally friendly [1] and reef-safe [23]. A particularly interesting property is its high molecular weight (823.07 g mol^{-1}), which impairs skin penetration. It shows a broad absorption profile in the UVB region and it is considered not only highly photostable, but it acts as a photostabilizer for other organic filters such as avobenzone [24,25]. The water insolubility makes this filter easily embeddable in water-resistant formulations, while the formulation in aqueous media presents some difficulties. On the other hand, oil-free sunscreens are considered ideal for most skin types, as they are not comedogenic and do not leave the skin feeling greasy; in particular, for people prone to acne it is recommended to avoid oil-based formulations [26].

The aim of the present work is to obtain an aqueous formulation containing ETZ encapsulated in the hydrophobic cavity of *L. clavatum*, which should be able to deliver the solar filter to the first layers of the skin, in order to exploit the solar filter substantivity (i.e., the capability to adhere well to the skin, without being easily removed by simple washing; [27]). The SPMs were extracted from *L. clavatum* pollen grains through a standardized protocol that includes alkaline and acidolysis steps and the obtained microcapsules have been characterized through different methods. Moreover, the capability of SPMs to act as a photoprotective agent has been investigated, to take advantage both of the sunscreen and sporopollenin microcapsules' UV protection. Finally, the in vitro permeation and penetration through and into the skin layers after cutaneous application of an aqueous dispersion of ETZ-SPMs were studied using porcine ear skin as a model.

2. Materials and Methods

2.1. Materials

The following materials were used: pollen from *Lycopodium clavatum* L. (Merk Life Science S.r.l., Milano, Italy); ethylhexyl triazone (ETZ, Uvinul® T150; BASF, Ludwigshafen, Germany); polyoxyethylene-20-oleylether (Brij98, Sigma Aldrich (Darmstadt, Germany); xanthan gum (Xantural 75; CP Kelko, Atlanta, GA, USA).

All other chemicals and solvents were of analytical grade. Ultrapure water was obtained using MilliQ® plus apparatus (Millipore, Milan, Italy).

2.2. Extraction of Sporopollenin of Lycopodium Pollen

Lycopodium clavatum pollen grains (100 g) were refluxed in 500 mL of acetone for 6 h, and then the solution was filtered under vacuum with glass fiber filters and air-dried overnight. Pollen grains were refluxed in 500 mL of KOH (6% *w/v*) for 6 h, filtered under vacuum, rinsed with distilled water (5 × 500 mL) and with hot ethanol (500 mL) and air-dried overnight.

Subsequently, the pollen grains were treated with 500 mL phosphoric acid (85% *v/v*) at 70 °C for 6 h, filtered under vacuum, rinsed with hot water (5 × 800 mL), acetone (600 mL), hydrochloric acid 2 M (600 mL), sodium hydroxide 2 M (600 mL) and washed again with hot water (5 × 500 mL), acetone (600 mL) and ethanol (600 mL).

Finally, the SPMs obtained were air-dried until reaching 30.0 g weight.

2.3. Preparation of Encapsulated ETZ-Loaded Sporopollenin Microcapsules (ETZ-SPMs)

Preliminary studies for the selection of the best method for encapsulation of ETZ inside sporopollenin microcapsules have been performed, varying both the solvent in which to solubilize the sunscreen and the treatment to eliminate the non-incorporated ETZ fraction. In all cases, the passive loading technique was employed and finally acetone was selected as the best solvent.

Briefly, 150 mg of ETZ was solubilized in 4 mL of acetone and subsequently an exactly weighed amount (300 mg) of sporopollenin microcapsules was added, vortexed for 7 min and stirred at 4 °C for 2 h. To remove the non-incorporated ETZ fraction, the suspension was subjected to centrifugation at 4000 rpm for 7 min (an additional 2 mL was used to wash the vial to recover all the material).

For further removal of the non-encapsulated ETZ from the recovered microcapsules, the samples were subjected to 3 more consecutive cycles of solvent addition and centrifugation (6 mL acetone and 4000 rpm for 7 min). For each cycle, the supernatant was collected and analyzed by spectrophotometry.

Finally, the residue, representing the ETZ-loaded sporopollenin microcapsules (ETZ-SPMs), was collected and dried at 37 °C until a constant weight was reached. The amount of sunscreen encapsulated was spectrophotometrically determined. In order to control the reproducibility of the production method, six different batches were prepared.

2.4. Encapsulation Efficiency

The quantitative determination of ETZ encapsulated into the sporopollenin microcapsules was performed with a UV–Vis spectrophotometer (UV-2600 Shimadzu Europe GmbH, Milan, Italy) for comparison with an external calibration curve ($R^2 = 0.998$) at the maximum absorption wavelength of 315 nm.

Basically, to ensure complete release of the incorporated solar filter, 5 mg ETZ-SPMs were suspended in 10 mL ethanol. The suspension was stirred for 30 min at room temperature and then centrifuged at 4000 rpm for 7 min. The supernatant was collected and analyzed.

The entrapment efficiency of ETZ was calculated using the following equation:

$$Entrapment\ efficiency, EE\% = \frac{mass\ of\ solar\ filter\ in\ the\ microcapsule}{mass\ of\ solar\ filter\ added\ to\ the\ microcapsule} \times 100.$$

2.5. FTIR Spectral Analysis and Imaging Techniques

ATR-FTIR spectra of *Lycopodium clavatum* L. untreated pollen grains, SPMs, ETZ and ETZ-SPMs were recorded with an IR Cary 660 FTIR spectrometer (Agilent Technologies Santa Clara, CA, USA) using a macro-ATR accessory with a Zn/Se crystal. The spectra were measured as previously described by [9] in a range from 4000 to 500 cm^{-1}, with 32 scans both for background and samples. Chemical imaging data were collected by Agilent's ATR-imaging technique using an FTIR Cary 620 imaging system equipped with a 64 × 64 focal plane array (FPA) detector cooled by liquid nitrogen and a Ge crystal. Spectra were measured, from 3300 to 900 cm^{-1}, with 256 scans for both background and pollen samples. For all the FTIR analyses, a few milligrams of sample were used.

The same samples were also observed with the MicroStar120 optical microscope, (Reichert-Jung, Buffalo, NY, USA) at 40× magnification.

2.6. Differential Scanning Calorimetry

The thermal behavior of ETZ, SPMs, ETZ-SPMs was analyzed by a differential scanning calorimeter (DSC 4000, Perkin Elmer, Milan, Italy) following the procedure reported in [28]. The method involves two steps: the sample was maintained at 10 °C for 1 min and then exposed to heating from 10 °C to 300 °C with a constant heating rate of 10 °C/min under nitrogen atmosphere (20 mL/min).

2.7. Thermogravimetric Analysis

The thermal stability was investigated by thermal gravimetric analysis (TG) (TA Instruments Q500 TGA, New Castle, USA, Delaware; weighing precision ± 0.01%, sensitivity 0.1 μg, baseline dynamic drift < 50 μg). The temperature calibration was performed using the Curie point of nickel and Alumel standards and for mass calibration weight standards of 1 g, 500 mg and 100 mg were used. All the standards were supplied by TA Instruments Inc. In a platinum crucible as a sample holder, 12–15 mg of each sample was heated. First, the heating mode was set to isothermal at 60 °C in N_2 (80 cm^3/min) for 30 min. Then, the sample was heated from 40 °C to 500 °C at 10 °C/min under nitrogen (80 cm^3/min) and maintained at 800 °C for 3 min. Mass change was recorded as a function of temperature and time. TGA experiments were carried out in duplicate.

2.8. In Vitro Release Study

In vitro release studies through a dialysis membrane (regenerated cellulose, Spectra/Por molecular porous membrane tubing, 12–14 KDa; Spectrum Laboratories, Inc., Piscataway, NJ, USA) were performed with Gummer-type diffusion cells.

Briefly, 200 mg of ETZ-SPMs was added in the donor compartment in direct contact with the membrane. The receptor compartment consisted of 5.0 mL of pH 7.4 (phosphate buffered saline, PBS) added to 0.01% Brij 98 maintained at 37 °C and stirred at 600 rpm. Brij 98 was added to improve the ETZ solubility in the receiving fluid [29].

At predetermined time intervals, 5 mL of the samples was withdrawn and replaced with the same amount of fresh receiving fluid. All the experiments lasted 24 h and were performed in triplicate. The amount of ETZ in the samples was determined by HPLC after filtration through cellulose acetate filters (0.22 um pore size, Minisart® NML Syringe filters, Sartorius, Florence, Italy).

2.9. Sun Protection Factor (SPF) Determination

The SPF quantifies the filtering capacity of a sun product, providing a correct indication of the level of protection and therefore of the duration of exposure to UV rays without having erythematogenic effects.

The sun protection factor was determined using the Labsphere 2000S spectrophotometer (Labsphere UV-2000S UV Transmittance Analyzer, Labsphere, Inc., North Sutton, NH 03260, USA) and the substrate used for the determination of the SPF was surgical tape (3M™ Health Care, St, Paul, MN, USA).

Briefly, an exactly weighed amount of the sample was uniformly applied with a latex glove-coated finger on the Transpore™ membrane to obtain a 2 mg/cm^2 product/surface ratio. Then, it was allowed to rest for 15 min at room temperature, protected from light.

For the SPF determination, an aqueous dispersion of xanthan gum 0.8% *w/w* was prepared and added to an appropriate amount of ETZ-SPMs corresponding to ETZ 1.0% *w/w*. As a reference, both SPMs and natural pollen grains in the same semisolid formulation were analyzed.

Before reading the substrate on which the solar product was spread, the substrate was read as it was (blank scan), 12 scans were carried out.

The estimated value of the SPF is:

$$\textbf{Rated SPF} = \textbf{SPF} - \textbf{E},$$

where E is the error that depends on the number of measurements (scans) collected.

The absorbance for the two radiations was calculated as follows:

$$\text{a UVA} = \frac{1\text{nm}}{(400\text{ nm} - 320\text{ nm})}\left(\frac{A_{320}}{2} + A_{321} + A_{322} + \ldots + A_{399} + \frac{A_{400}}{2}\right),$$

$$\text{a UVB} = \frac{1\text{ nm}}{(320\text{ nm} - 290\text{ nm})}\left(\frac{A_{290}}{2} + A_{291} + A_{292} + \ldots + A_{319} + \frac{A_{320}}{2}\right).$$

The *Boots Star Rating* and the *Star Category* are based on the *UVA ratio* parameter and classified as reported in Table 1:

$$\textbf{Ratio UVA} = \frac{\textbf{a UVA}}{\textbf{a UVB}}$$

Table 1. *Star Category* and *Star Rating values* according to the UVA ratio (2004 *revision*).

UVA Ratio	Star Category	Star Rating
0.0 to 0.19	-	Too low for UVA claim
0.2 to 0.39	*	Minimum
0.4 to 0.59	**	Moderate
0.6 to 0.79	***	Good
0.8 to 0.89	****	Superior
≥0.90	*****	Ultra

2.10. *In Vitro Cutaneous Permeation and Distribution Studies*

Porcine ear skin, used as a model, was obtained from freshly sacrificed animals in a local slaughterhouse and treated as described in [30]. The pig hairs were abscised and the skin (thickness: 1.46 ± 0.06 mm and area available for permeation: 1.23 ± 0.99 cm^2) was placed in the Gummer-type diffusion cells.

The donor phase consisted of 50 mg (or 200 mg) of ETZ-SPMs added to 100 µL (or 400 µL) of PBS + 0.01% Brij 98 in order to improve the contact of the formulation with the skin.

The receiving phase consisted of pH 7.4 (phosphate buffered saline, PBS) added to 0.01% Brij 98 maintained at 37 °C and stirred at 600 rpm.

At predetermined time intervals, the receiving phase (5 mL) was withdrawn for HPLC analysis and replaced with the same volume of fresh fluid, and sink conditions were maintained throughout the entire study. All experiments lasted 24 h and were replicated four times.

At the end of the permeation experiments, the skin was collected and treated following the procedure reported and validated in [31].

Briefly, at the end of the in vitro permeation experiments the skin was removed from the diffusion cells, rinsed with distilled water to eliminate excess formulation from the skin surface and gently wiped with cotton-wool wipes.

Afterwards, the stratum corneum was removed using the tape-stripping method as described in [31]. The skin was stripped using an adhesive tape (Tesa film N. 5529; Kristall-Klar, Beiersdorf, Hamburg, Germany) and the tape strips were pressed on the skin by applying uniform pressure in order to obtain intimate contact between the film and the skin. The first tape strip was discarded, as this represents unabsorbed material only. Then, the subsequent strips were carefully removed and the entire procedure was repeated 25 times (tape strips no. 1–25). The strips were collected following a predetermined scheme in seven glass vials, each containing 5 mL of ethanol, sonicated for 10 min and subjected to centrifugation (15 min, 4000 rpm). The supernatant was collected for HPLC analysis to determine the amount of ETZ in the SC. To evaluate the presence of ETZ in the viable epidermis, the samples were treated with 2 mL of 2% sodium dodecyl sulfate (SDS) for 23 h under magnetic stirring and then 4 mL of methanol was added and stirred for 1 h.

The mixture was centrifuged at 4000 rpm for 15 min and then supernatant was collected for HPLC analysis.

2.11. HPLC Analysis

The quantitative determination of ETZ in the receiving fluid and in skin samples was determined by a reverse-phase HPLC method with apparatus consisting of an LC-10AD pump and 20 μL Rheodyne injector, SPD-10AV detector and C-R4A computer integrating system (Shimadzu Corp., Kyoto, Japan). A reverse-phase C18 column (Synergi 4u Fusion-RP 80A, 150 × 4.6 mm, Phenomenex, Torrance, CA, USA) was used. Isocratic elution was performed using a mobile phase consisting of a mixture of CH_3OH:H_2O acidified with 1% glacial acetic acid (98:2 *v/v*), filtered through a 0.45 μm pore size membrane filter. The detection wavelength was 310 nm, the flux was 1.0 mL/min and the retention time was 11.3 min.

The amount of ETZ in the samples was determined by comparison with appropriate external standard curves obtained applying the least square linear regression analysis. For in vitro studies, the calibration curves were obtained by dissolving the ETZ in acetonitrile and then diluting with PBS pH 7.4 added to 0.01% Brij 98. In the case of biological materials, a standard curve was obtained by adding increasing amounts of ETZ to a blank biological matrix.

2.12. Statistical Analysis

Statistical differences in the exploited SPF between empty and ETZ-loaded SPMs were evaluated applying Student's two-tailed unpaired *t*-test (GraphPad Prism Software, version 8.4.3, San Diego, CA, USA).

In the case of cutaneous permeation and distribution studies, statistical differences between the different doses of product applied on the skin have been determined by Student's two-tailed unpaired *t*-test using GraphPad Prism software, version 8.0.

The evaluation included calculation of the mean and standard error (S.E). In all cases, differences were considered statistically significant at $p < 0.05$.

3. Results and Discussion

3.1. Preparation of Encapsulated ETZ-Loaded Sporopollenin Microcapsules (ETZ-SPMs)

To verify that the sporopollenin microcapsules obtained after the extraction procedure can be used as a vehicle for the skin application of a sunscreen, the passive loading method was employed to encapsulate the solar filter ETZ, after its solubilization in acetone.

The results obtained are reported in Table 2 in terms of the amount of ETZ encapsulated by grams of SPMs (ETZ, mg/g_{SPMs}) and EE%, for the different batches prepared. The prepared ETZ-SPMs exhibited an inter-batch variability with a CV% < 20, that can be considered acceptable for an attempt on a laboratory scale, which will in any case be subjected to further refinements. UV–Vis analysis of ETZ-SPMs showed ETZ loading content in the range 36.99–62.12 mg for grams of sporopollenin microcapsules and encapsulation efficiency in the range 6.49–11.91, suggesting that *L. clavatum* sporopollenin microcapsules may represent a potential system for the encapsulation of active substances.

Table 2. Physical–chemical characterization of ETZ-SPMs produced with acetone and four consecutive centrifugations with solvent replacement after supernatant removal (n = 2).

Batch	ETZ, mg/g_{SPMs}	EE%
1	61.28 ± 0.39	11.91 ± 0.33
2	59.44 ± 0.37	10.39 ± 0.38
3	43.34 ± 4.32	7.76 ± 0.77
4	62.12 ± 8.56	11.34 ± 1.58
5	55.82 ± 7.46	9.93 ± 1.33
6	36.99 ± 1.60	6.49 ± 0.28

3.2. FTIR Spectra Analysis and Optical Microscopy

Lycopodium clavatum L. untreated pollen grains, SPMs, ETZ and ETZ-SPMs were analyzed by infrared spectroscopy and FTIR infrared microscopy.

In Figure 1, the visible image and the FPA image, together with the spectra obtained with the microscope and the spectrometer for the untreated pollen grains, are shown.

In accordance with the literature [32], the infrared spectrum allows the identification of the three main components of pollen based on the different vibrational bands:

1. Lipids: 1700–1760 cm^{-1} (C = O stretching), 1440–1450 cm^{-1} (CH_2 deformation) and 1280 cm^{-1} (C – O stretching, not visible due to the lower resolution);
2. Proteins: 1600–1650 cm^{-1} (amide I) and 1515–1525 cm^{-1} (amide II);
3. Carbohydrates: 1000–1200 cm^{-1} (C – O – C and C – OH stretching).

In addition, there are two other bands attributable to the three components, ≈3300 cm^{-1} stretching of the – O – H and 2950–2850 cm^{-1} stretching of the aliphatic C – H.

The different resolutions between the two spectra reported in Figure 1b,c are correlated by the use of different ATR crystals (Ge vs. Zn/Se) and the major sensitivity of the ATR microscope to the humidity.

If the data of the untreated pollen grains are compared with the ones obtained for SPMs, shown in Figure 2, the major highlighted differences are represented by a reduction in intensity of the bands corresponding to lipids and proteins. The carbohydrate bands appear to show a less marked reduction in their intensity. Unlike the literature [33], it is difficult to define a greater isolation of sporopollenin as its bands are almost superimposable to those of proteins.

On the other hand, the infrared spectrum of the solar filter (Figure 3) shows four characteristic bands that differentiate it from pollen; ≈3300 cm^{-1} (N-H stretching of secondary amines), ≈1700 cm^{-1} (C = O stretching of esters), ≈1410 cm^{-1} (bending of C – H bonds) and ≈1270 cm^{-1} (C – N stretching of aromatic amines).

In Figure 4, the visible image and the FPA image of the ETZ-SPMs before the washing cycles necessary to remove the non-encapsulated solar filter, along with the spectra obtained with the microscope and the spectrometer, are shown.

The three characteristic bands of the solar filter are very evident in the spectrum obtained with the spectrometer, in particular the bands at ≈3300 cm^{-1} and ≈1700 cm^{-1}, while with the microscope the band at ≈3300 cm^{-1} is more difficult to identify due to the sensitivity of the microscope to humidity, but the intense signal around 1700 cm^{-1} confirms the presence of the solar filter, suggesting that a non-negligible amount of ETZ is deposited outside the pollen grains.

Conversely, when ETZ-SPMs were subjected to the procedure for the removal of the non-encapsulated filter, the band at ≈3300 cm^{-1} is no longer observable and the ester band at ≈1700 cm^{-1} together with other characteristic bands of the solar filter, such as the one at 1410 cm^{-1}, has been significantly reduced, indicating that a substantial amount of sunscreen has been removed from outside the pollen (Figure 5).

Furthermore, the samples were also observed by using the optical microscope.

In Figure 6, the photomicrographs obtained for the natural pollen, the SPMs and the ETZ-SPMs are shown. In all cases, the size of the granules, of about 30 μm, confirms the literature data [34]. It is also interesting to note that both the emptying treatment and the filling procedure with the solar filter do not cause breakage of the granules or affect the homogeneity of the sample and its size.

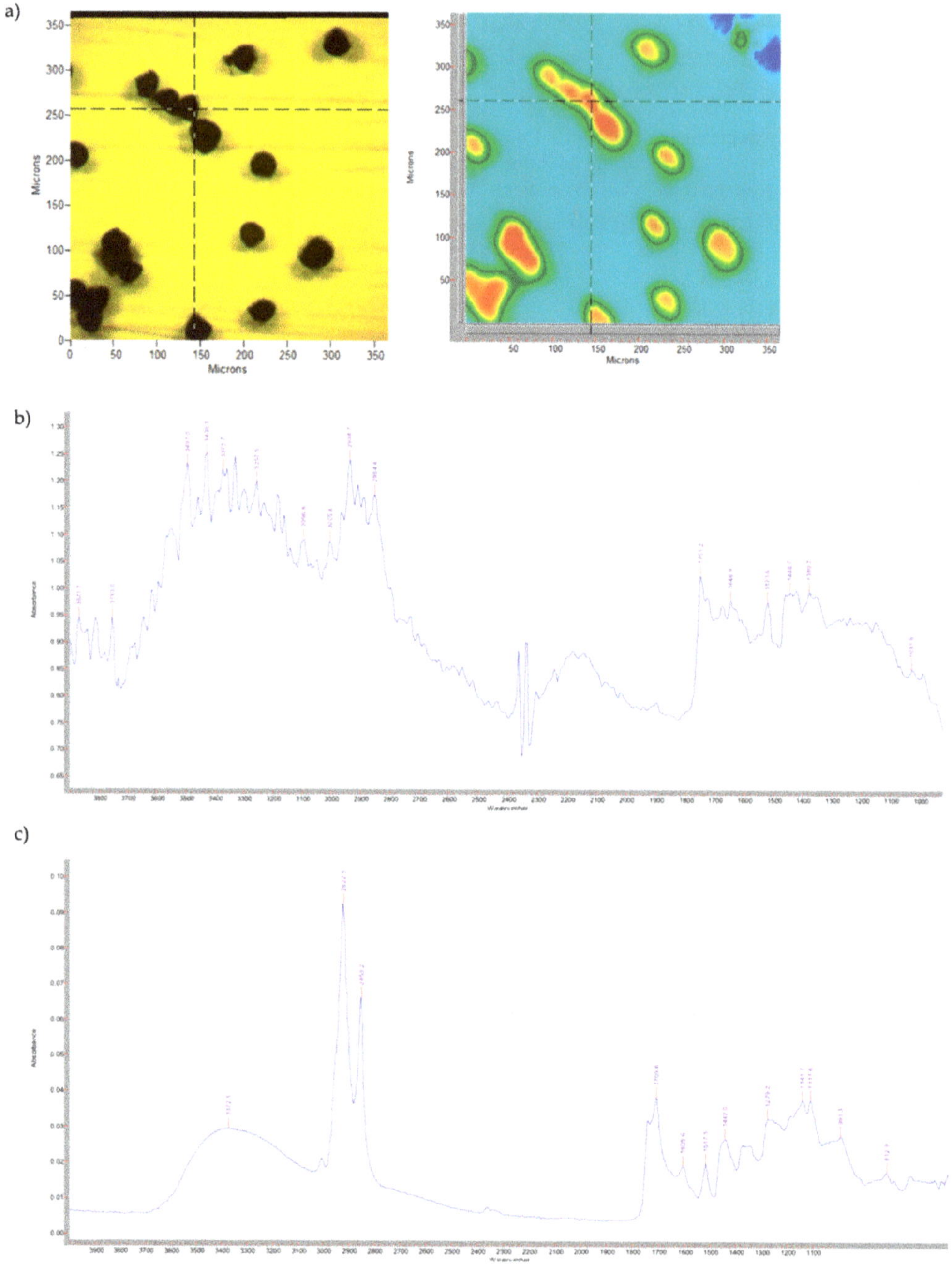

Figure 1. Untreated pollen grains. (**a**) Visible and FPA images; (**b**) spectrum obtained with the microscope; (**c**) spectrum obtained with the spectrometer.

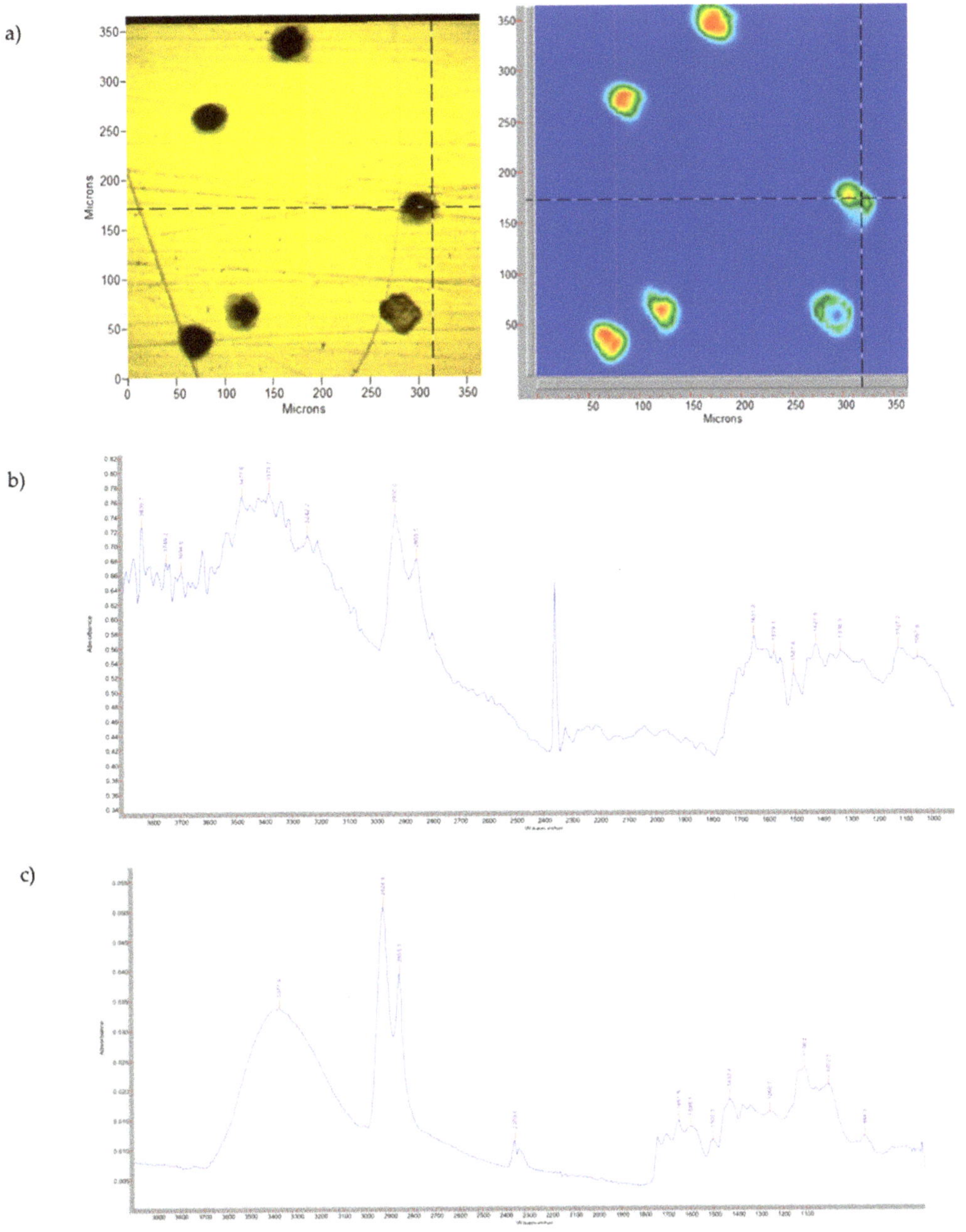

Figure 2. SPMs: (**a**) visible and FPA images; (**b**) spectrum obtained with the microscope; (**c**) spectrum obtained with the spectrometer.

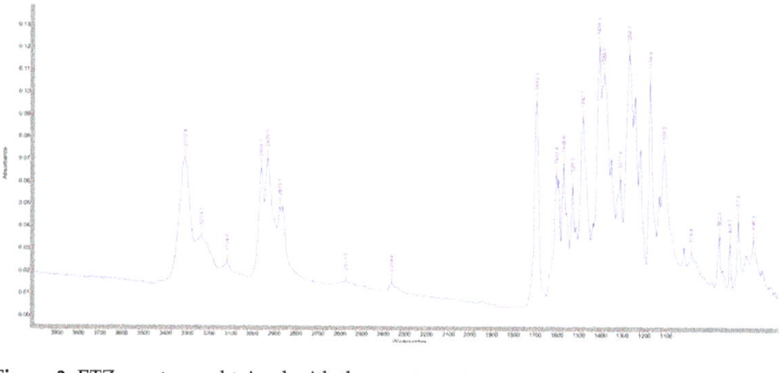

Figure 3. ETZ spectrum obtained with the spectrometer.

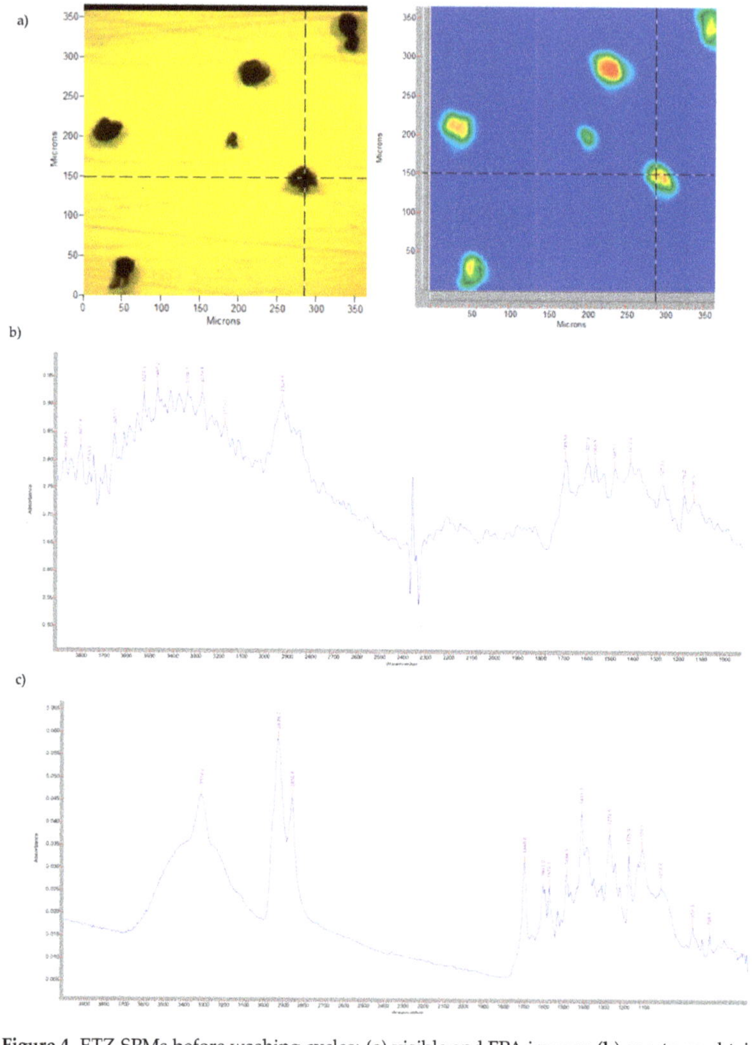

Figure 4. ETZ-SPMs before washing cycles: (**a**) visible and FPA images; (**b**) spectrum obtained with the microscope; (**c**) spectrum obtained with the spectrometer.

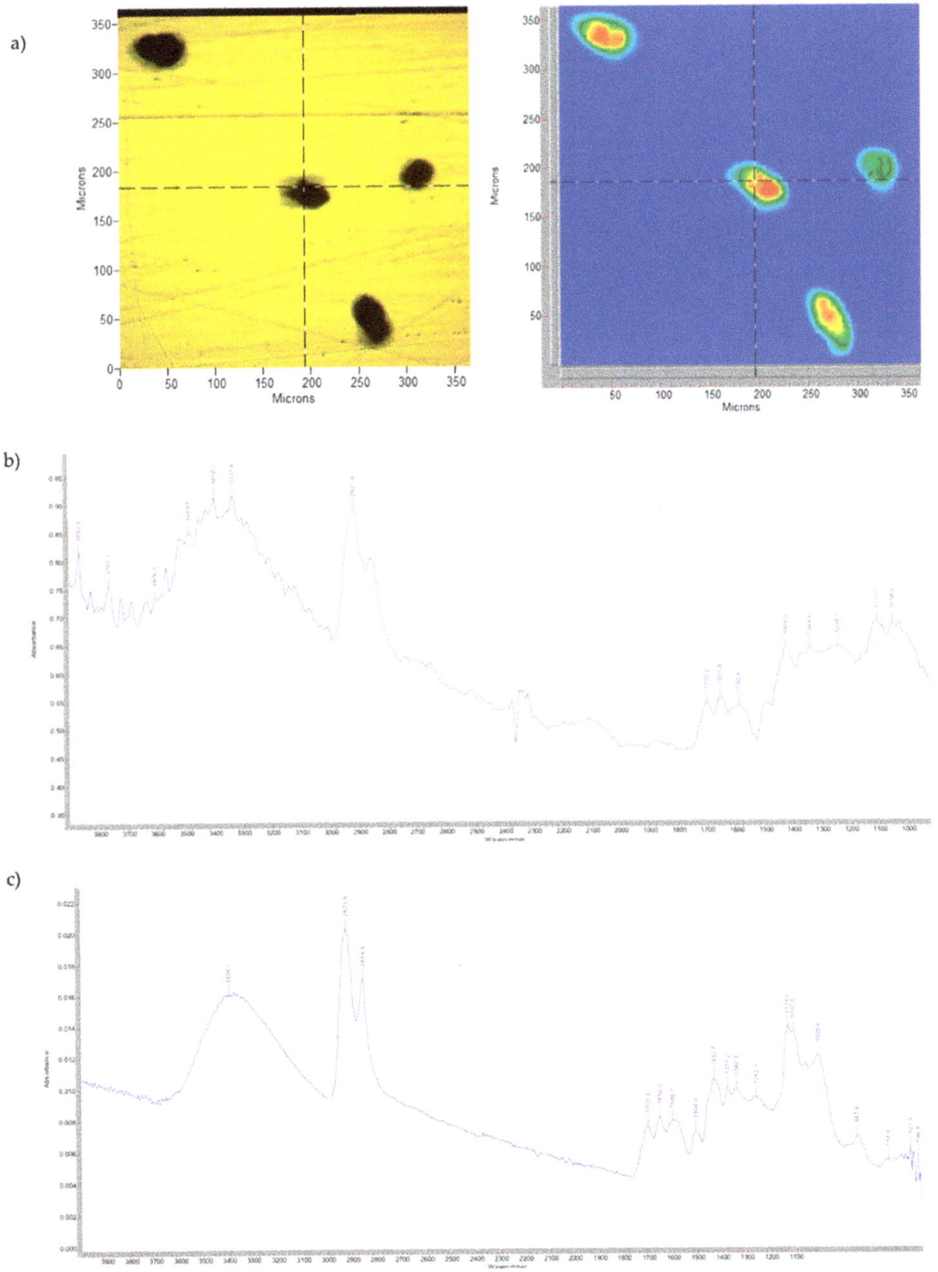

Figure 5. ETZ-SPMs after washing cycles: (**a**) visible and FPA images; (**b**) spectrum obtained with the microscope; (**c**) spectrum obtained with the spectrometer.

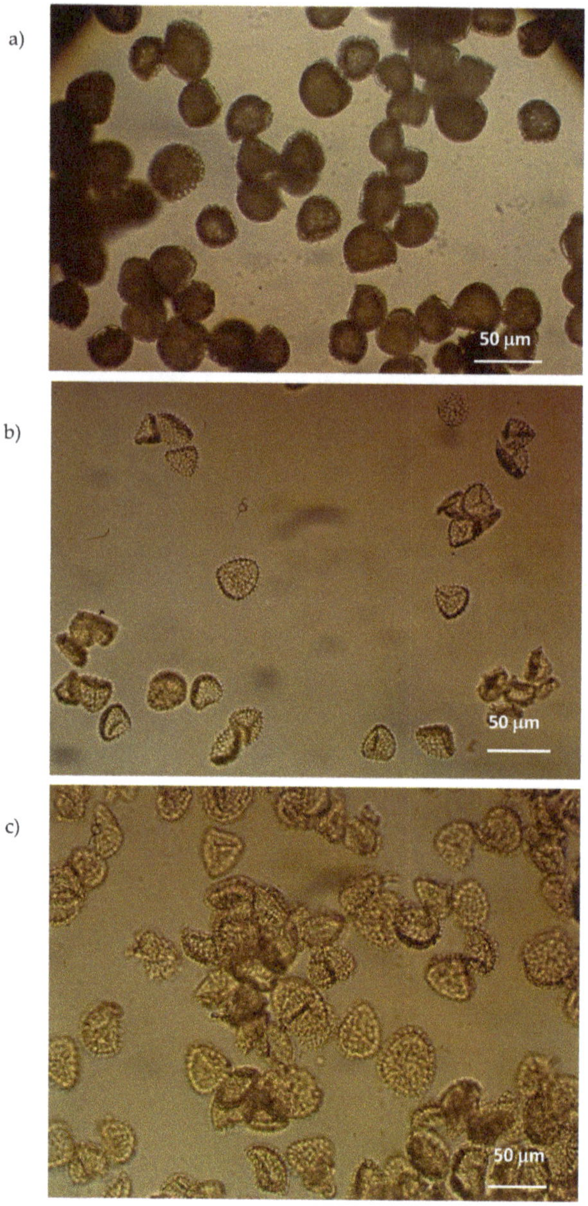

Figure 6. Photomicrographs of (**a**) untreated pollen grains, (**b**) SPMs and (**c**) ETZ-SPMs obtained by optical microscopy.

3.3. *Differential Scanning Calorimetry and Thermogravimetric Analysis*

A further characterization of ETZ-SPMs was conducted by analyzing their thermal behavior using a differential scanning calorimeter (DSC), in order to verify the effective encapsulation of the solar filter within the microcapsules. Therefore, both SPMs and ETZ-SPMs were analyzed. For comparison, the ETZ alone and the physical mixture given by

ETZ and empty SPMs in the same weight ratio present in the microcapsules loaded with solar filter were also analyzed.

The thermograms obtained by DSC are shown in Figure 7. The melting peak at about 120 °C relative to ETZ confirms the literature data (123.27 °C) [35] and reflects its crystalline structure. It is possible to notice a shift at higher temperatures of the peak relative to the ETZ-SPMs compared to the empty SPMs. Furthermore, the melting peak of the solar filter disappears in the thermogram of ETZ-SPMs, indicative of a change from a crystalline state to a dispersed amorphous state. The physical mixture shows the peak at about 140 °C characteristic of sporopollenin microcapsules, but the spike disappears at about 120 °C and the ETZ melting peak is not detected, a sign of a surface interaction between the two products.

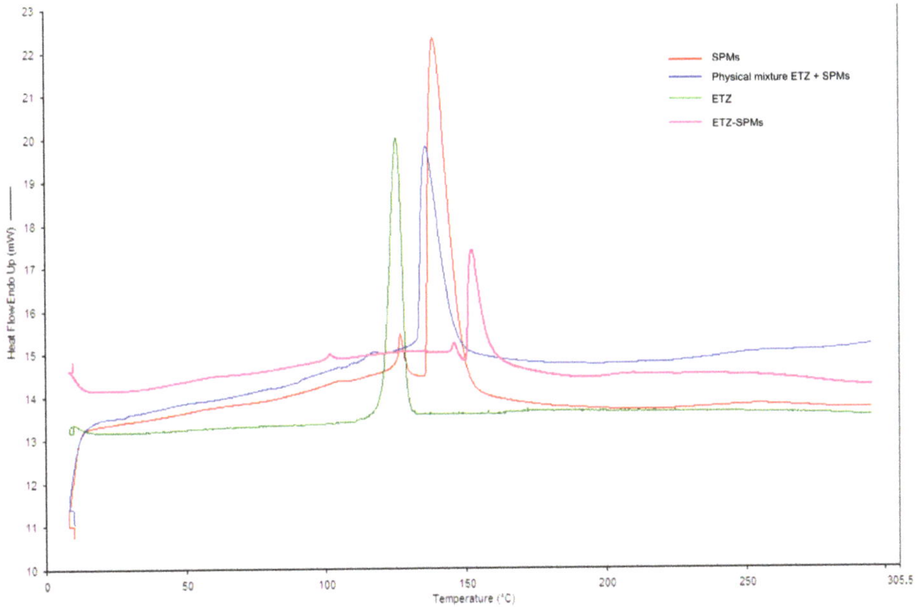

Figure 7. DSC of empty SPMs, ETZ, ETZ-SPMs and physical mixture ETZ + SPMs.

Thermograms of *Lycopodium* pollen grains, empty SPMs, ETZ and ETZ- SPMs (batch no. 6, Table 1) from thermogravimetric analysis are shown in Figure 8. As expected, empty SPMs show a higher thermal stability than the *Lycopodium* pollen grains with a starting degradation temperature (T_{start}) of 265 °C and 172 °C, respectively. This is mainly due to the removal of the biological material within the pollen, which preserves the thermally stable exine and intine components. The ETZ compound shows higher thermal stability than empty SPMs, with a T_{start} value of 400 °C. On the other hand, the ETZ compound itself shows higher thermal stability than empty SPMs, with a T_{start} value of 400 °C and, for this reason, there is no significant effect of ETZ on the stability of the filled versus unfilled sporopollenin. In the temperature range around 400 °C, the difference in mass loss between the ETZ-loaded and empty sporopollenin is consistent with the previously determined EE% for the same batch (6.49 ± 0.28).

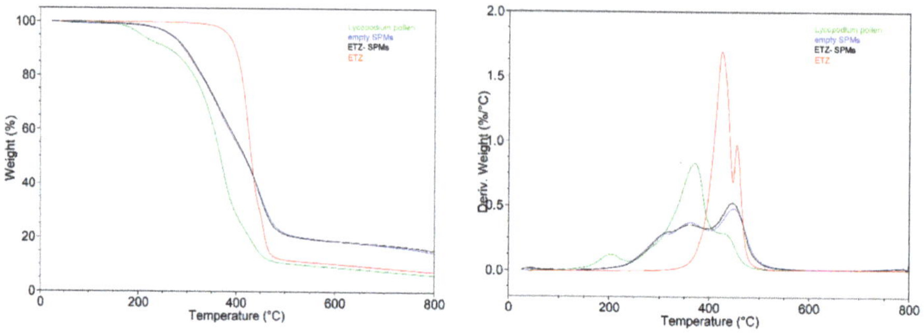

Figure 8. TG (**left**) and DTG (**right**) curves of *Lycopodium* pollen (green), empty SPMs (blue), ETZ-SPMs (black) and ETZ (red).

3.4. In Vitro Release Studies

In vitro release studies through the dialysis membrane were carried out. In any case, the spectrophotometric analysis of the receiving phase samples did not show a detectable amount of ETZ for each performed withdrawal (LOQ = 0.0433 μg/mL). Previous studies reporting on the topical application of sporopollenin microcapsules suggested that the release of actives could only take place after applying a moderate pressure, such as that applied by spreading a formulation on the skin [19]. Therefore, it can be apparent that the ETZ-SPMs are stable in water solution at least for the 24 h of the in vitro release study, without allowing the solar filter to exit from the microcapsules. For further investigation, ETZ-SPMs were subjected to a stability study after having dispersed them in purified water or in lactate buffer at pH 5.5 for a week, in order to verify that the release of solar filter from the microcapsule in an aqueous environment did not occur, an important factor for the formulation development of the product in anticipation of a skin application. In no case was the characteristic peak of ETZ at the wavelength of maximum absorption of the solar filter (303 nm) revealed and therefore the product can be considered stable for the observed period.

3.5. SPF Determination

In order to evaluate a possible cosmetic application of ETZ-SPMs, the sun protection factor (SPF), which gives an indication of the filtering capacity of a solar product and provides a correct indication of the level protection and therefore of the duration of exposure to UVB radiation without having erythematogenic effects, was determined. The results obtained did not show a high protective capacity against UVB radiation by the SPMs, although ETZ-SPMs possessed a higher SPF with respect to empty SPMs with a statistically significant difference. Indeed, the mean SPF values measured were 1.20 ± 0.06 and 1.50 ± 0.09 for the empty SPMs and ETZ-SPMs, respectively; the values found are lower than the value of 3.4 calculated with the program Sunscreen-simulator (BASF) for the ETZ solar filter at the same concentration (1% *w*/*w*). This result seems to confirm that the solar filter is encapsulated inside the microcapsules and is not released in the aqueous medium of the formulation until use.

The Cosmetic Regulation (EC) No. 1223/2009 also establishes that sunscreen products must also contain filters that are able to protect the skin from UVA radiation and therefore this feature has also been evaluated. In this case, the results reported in Tables 3 and 4 show a UVA ratio in both cases ≥ 0.90 with the highest level of Star Rating (5 stars) definable as "ultra" protection, the highest level of protection according to the "Boots Star Rating System". It can therefore be considered that the SPMs have a high protective power against UVA radiation.

Table 3. Calculation of SPF and UVA ratio of empty SPMs (transpore 2 mg/cm^2).

Mean	STD	COV (%)	UVA/UVB Ratio	Star Category	Λ_{cr} (nm)
1.23	0.01	0.42	1.33	Ultra	390
1.22	0.01	0.48	1.33	Ultra	390
1.23	0.01	0.55	1.37	Ultra	390
1.10	0.01	1.31	1.94	Ultra	392
1.22	0.01	1.02	1.39	Ultra	391
SPF (mean ± SD): 1.20 ± 0.06					

Table 4. Calculation of SPF and UVA ratio of ETZ-SPMs (transpore 2 mg/cm^2).

Mean	STD	COV (%)	UVA/UVB Ratio	Star Category	Λ_{cr} (nm)
1.46	0.02	1.40	1.06	Ultra	389
1.66	0.02	1.46	1.02	Ultra	389
1.46	0.01	0.80	1.07	Ultra	389
1.49	0.00	0.21	1.07	Ultra	389
1.46	0.01	0.30	1.11	Ultra	390
SPF (mean ± SD): 1.50 ± 0.09					

3.6. Cutaneous Permeation and Distribution Studies

A further step was to verify the ability of the solar filter encapsulated in the SPMs to interact with the skin and permeate through excised porcine ear skin. The in vitro permeation experiments were performed using as donor phase two different amounts of ETZ-SPMs, namely 50 mg and 200 mg, containing about 2.7 and 10.6 mg of ETZ, respectively, to investigate the influence of the applied amount on both the permeation and the accumulation of the sun filter in the skin. In any case, the transcutaneous permeation of ETZ was observed throughout the 24 h experiment, during which the permeation was monitored. These data were quite expected, since it is widely known that ETZ is not prone to skin permeation due to its high molecular weight (823 Da) and log $p > 8.10$ [36,37]. Anyway, it is crucial to assess the role of the formulation in not enhancing the skin flux of a permeant.

The in vitro penetration data after skin application of ETZ-SPMs in different amounts are reported in Table 5 and illustrated in Figure 9a,b for the recovery of ETZ in the tape strips (stratum corneum) and in the skin depth (epidermis and dermis), respectively. The obtained results showed that the application of either 50 or 200 mg of ETZ-SPMs produced a similar recovery of solar filter in all the skin layers, with a total amount of sun filter retained accounting for 6.195 µg and 8.782 µg, respectively.

It is evident that the amount of ETZ accumulated in the skin does not depend on the amount of filter applied, as there are no statistically significant differences between the results of the two tests. The observed behavior could be related to the finding reported above (Section 3.4) that the SPMs do not release the active ingredient in the aqueous medium, therefore only the preparation in direct contact with the skin allows the release of the filter for interaction of the sporopollenin with the skin itself, with an unknown mechanism that should be further elucidated. Degradation mechanisms of pollen grains are not yet fully understood, but it has been reported that *L. clavatum* microcapsules undergo degradation in human blood plasma and the release of the contents occurs through an enzymatic pathway [38–40]. Even though the literature lacks a characterization of the enzymatic species capable of degrading exine microcapsules [6], some speculation could be performed. Since it has been reported that the skin itself possesses a high enzymatic biotransformation activity [41,42], it can be hypothesized that some of the enzymes that are present in the skin (esterase, dehydrogenase, monooxygenase, etc.), and represent a further barrier together with the stratum corneum, could degrade the sporopollenin and allow ETZ release.

Table 5. Recovery of ETZ in both the stratum and the skin depth, when different amounts of ETZ-SPMs are applied on the porcine ear skin (mean ± standard error).

Skin Depth	**ETZ (µg) 24 h**	
	Donor Phase 50 mg	**Donor Phase 200 mg**
Stratum corneum		
Strip 1	0.624 ± 0.312	1.554 ± 0.713
Strip 2–4	2.212 ± 1.300	0.872 ± 0.357
Strip 5–8	0.370 ± 0.148	1.937 ± 0.966
Strip 9–12	0.617 ± 0.355	0.722 ± 0.465
Strip 13–15	0.295 ± 0.088	0.494 ± 0.212
Strip 16–20	1.063 ± 0.931	1.745 ± 0.893
Strip 21–25	0.463 ± 0.351	1.131 ± 0.918
Viable epidermis/dermis	0.551 ± 0.310	0.327 ± 0.210

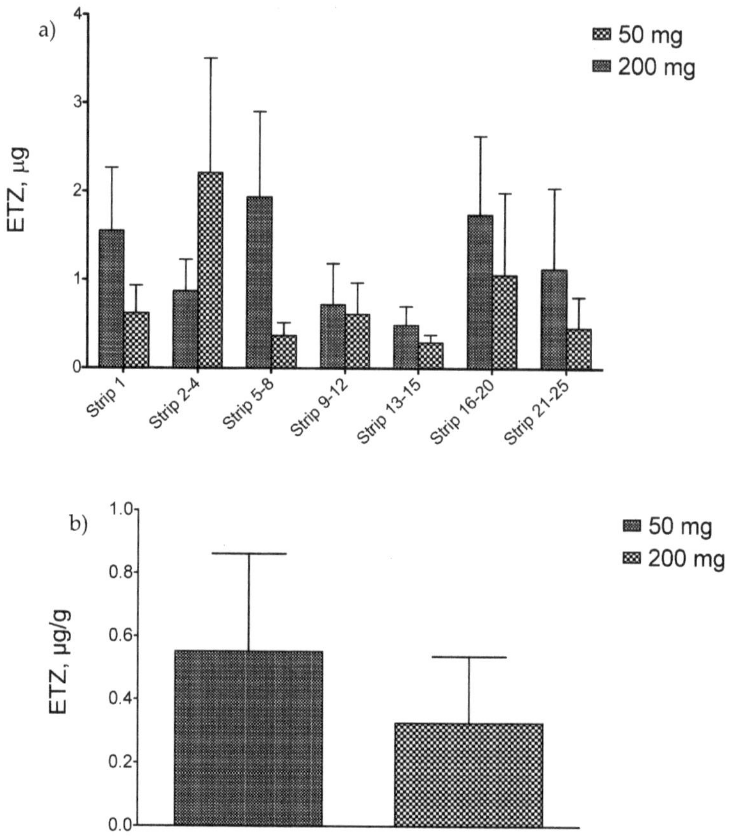

Figure 9. Recovery of ETZ (**a**) in the different tape strips (stratum corneum) and (**b**) in the skin depth (epidermis and dermis) when different amounts of ETZ-SPMs are applied on the porcine ear skin.

Anyway, if we try to analyze the obtained results in terms of sun protection, we have to take into account that when we performed the SPF and UVA protection experiments, we applied 2 mg/cm^2 of a 1% ETZ semisolid formulation, corresponding to ≅0.38 mg/cm^2 ETZ-SPMs. In this context, it seems that the application of a dose of 50 mg ETZ-SPMs could be enough to exploit the property of the pollen grains in protecting from UVA radiation and to obtain a complete photoprotection.

Moreover, the results obtained are consistent with the very few studies reported in the literature for ETZ investigating the skin penetration of the sun filter from different formulations at different concentrations of the active ingredient [36,43]. In particular, Ref. [43] studied the penetration profile of ETZ into rat skin or reconstituted skin from O/W and W/O emulsions containing 6% sun filter. In every case, the amount of ETZ penetrating into the first skin layers was of the same order of magnitude (a few micrograms) as in the present study, confirming our hypothesis that SPMs could represent an alternative to conventional formulations for sunscreen development.

4. Conclusions

In recent years, one of the most important challenges of cosmetic research has been to find innovative formulations that are functional and respectful of the environment at the same time. In this perspective, SPMs deriving from natural pollen, capable of encapsulating active ingredients by passive loading and therefore a method that is ecological, cost-effective, simple and relatively quick, can represent a valid alternative to conventional delivery systems. Furthermore, the possibility of functioning both as a release system and as an active ingredient provides these products with added value, together with easy availability and good reproducibility in the production phase.

In the present work, *L. clavatum* SPMs have been proposed as an alternative delivery system for a solar filter, ETZ, considered safe for both humans and the ecosystem; the data obtained from the imaging and FTIR analysis demonstrated the intact and monodisperse nature of the *L. clavatum* SPMs and the removal of foreign materials from the microcapsule. Moreover, the investigations on the thermal behavior of ETZ-SPMs confirmed that the solar filter was successfully loaded into the microcapsules and the aqueous dispersion remain stable over time, supporting the idea that SPMs may represent a promising material for the encapsulation of active ingredients that need to be protected from the external environment and subsequently released in the tissues of interest. Additionally, ETZ-SPMs are able to deliver the solar filter to the first layers of the skin in order to let ETZ protect from UVB light, without permeating through the skin as requested by the Cosmetic Regulation. Finally, for the first time in the literature, *L. clavatum* SPMs have been demonstrated to possess a very high UVA protection at the concentration of use in the cosmetic field, accordingly with the *Boots Star Rating* and the *Star Category* based on the *UVA ratio* parameter.

Therefore, the results of this work seem very interesting for the cosmetic field and can be prodromal for future studies on the use of sporopollenin microcapsules from different plant species, with the dual function of vehicle and active ingredient with photoprotective properties.

Author Contributions: Conceptualization: S.T. and D.M.; methodology: S.T., D.M. and C.S.P.; validation: S.T., D.M., P.C. and S.B.; formal analysis: V.P., M.D.G., A.M. and G.T.; investigation: S.T., D.M., V.P. and M.D.G. performed the solar filter loading in the sporopollenin microcapsules, the DSC analysis, the in vitro release and skin permeation/penetration studies and the preparation of biological materials for the instrumental analysis; A.M. performed the TGA analysis; C.S.P. and G.T. performed the FTIR analysis and evaluation; funding acquisition: P.C. and S.B.; writing—original draft preparation: S.T.; writing—review and editing: D.M. and P.C.; supervision: S.T. and D.M. All authors have read and agreed to the published version of the manuscript.

Funding: This research received no external funding.

Informed Consent Statement: Not applicable.

Acknowledgments: The authors gratefully acknowledge Marisanna Centini, Department of Biotechnology, Chemistry and Pharmacy, University of Siena for supporting in the UV protection determination.

Conflicts of Interest: The authors declare no conflict of interest.

References

1. Pawlowski, S.; Herzog, B.; Sohn, M.; Petersen-Thiery, M.; Acker, S. EcoSun Pass: A tool to evaluate the ecofriendliness of UV filters used in sunscreen products. *Int. J. Cosmet. Sci.* **2021**, *43*, 201–210. [CrossRef] [PubMed]
2. Danovaro, R.; Bongiorni, L.; Corinaldesi, C.; Giovannelli, D.; Damiani, E.; Astolfi, P.; Greci, L.; Pusceddu, A. Sunscreens cause coral bleaching by promoting viral infections. *Environ. Health Perspect.* **2008**, *116*, 441–447. [CrossRef] [PubMed]
3. He, T.; Tsui, M.M.P.; Tan, C.J.; Ma, C.Y.; Yiu, S.K.F.; Wang, L.H.; Chen, T.H.; Fan, T.Y.; Lam, P.K.S.; Murphy, M.B. Toxicological effects of two organic ultraviolet filters and a related commercial sunscreen product in adult corals. *Environ. Pollut.* **2019**, *245*, 462–471. [CrossRef] [PubMed]
4. Da Silva, A.C.P.; Santos, B.A.M.C.; Castro, H.C.; Rodrigues, C.R. Ethylhexyl methoxycinnamate and butyl methoxydibenzoylmethane: Toxicological effects on marine biota and human concerns. *J. Appl. Toxicol.* **2022**, *42*, 73–86. [CrossRef]
5. Da Silva, A.C.P.; Paiva, J.P.; Diniz, R.R.; Dos Anjos, V.M.; Silva, A.B.S.M.; Pinto, A.V.; Dos Santos, E.P.; Leitão, A.C.; Cabral, L.M.; Rodrigues, C.R.; et al. Photoprotection assessment of olive (*Olea europaea* L.) leaves extract standardized to oleuropein: In vitro and in silico approach for improved sunscreens. *J. Photochem. Photobiol. B* **2019**, *193*, 162–171. [CrossRef]
6. Mackenzie, G.; Boa, A.N.; Diego-Taboada, A.; Atkin, S.L.; Sathyapalan, T. Sporopollenin, the least known yet toughest natural biopolymer. *Front. Mater.* **2015**, *2*, 66. [CrossRef]
7. Sargin, I.; Akyuz, L.; Kaya, M.; Tan, G.; Ceter, T.; Yildirim, K.; Ertosun, S.; Aydin, G.H.; Topal, M. Controlled release and anti-proliferative effect of imatinib mesylate loaded sporopollenin microcapsules extracted from pollens of Betula pendula. *Int. J. Biol. Macromol.* **2017**, *105 Pt 1*, 749–756. [CrossRef]
8. Fan, T.; Park, J.H.; Pham, Q.A.; Tan, E.L.; Mundargi, R.C.; Potroz, M.G.; Jung, H.; Cho, N.J. Extraction of cage-like sporopollenin exine capsules from dandelion pollen grains. *Sci. Rep.* **2018**, *8*, 6565. [CrossRef]
9. Palazzo, I.; Mezzetta, A.; Guazzelli, L.; Sartini, S.; Pomelli, C.S.; Parker, W.O., Jr.; Chiappe, C. Chiral ionic liquids supported on natural sporopollenin microcapsules. *RSC Adv.* **2018**, *8*, 21174–21183. [CrossRef]
10. Pomelli, C.S.; D'Andrea, F.; Mezzetta, A.; Guazzelli, L. Exploiting pollen and sporopollenin for the sustainable production of microstructures. *New J. Chem.* **2020**, *44*, 647–652. [CrossRef]
11. Gonzalez-Cruz, P.; Uddin, M.J.; Atwe, S.U.; Abidi, N.; Gill, H.S. A chemical treatment method for obtaining clean and intact pollen shells of different species. *ACS Biomater. Sci. Eng.* **2018**, *4*, 2319–2329. [CrossRef] [PubMed]
12. Uddin, M.J.; Liyanage, S.; Abidi, N.; Gill, H.S. Physical and Biochemical Characterization of Chemically Treated Pollen Shells for Potential Use in Oral Delivery of Therapeutics. *J. Pharm. Sci.* **2018**, *107*, 3047–3059. [CrossRef] [PubMed]
13. Ageitos, J.M.; Robla, S.; Valverde-Fraga, L.; Garcia-Fuentes, M.; Csaba, N. Purification of Hollow Sporopollenin Microcapsules from Sunflower and Chamomile Pollen Grains. *Polymers* **2021**, *13*, 2094. [CrossRef] [PubMed]
14. Iravani, S.; Varma, R.S. Plant Pollen Grains: A Move Towards Green Drug and Vaccine Delivery Systems. *Nanomicro. Lett.* **2021**, *13*, 128. [CrossRef]
15. Atalay, F.E.; Culum, A.A.; Kaya, H.; Gokturk, G.; Yigit, E. Different Plant Sporopollenin Exine Capsules and Their Multifunctional Usage. *ACS Appl. Bio. Mater.* **2022**, *5*, 1348–1360. [CrossRef]
16. Diego-Taboada, A.; Beckett, S.T.; Atkin, S.L.; Mackenzie, G. Hollow pollen shells to enhance drug delivery. *Pharmaceutics* **2014**, *6*, 80–96. [CrossRef]
17. Alshehri, S.M.; Al-Lohedan, H.A.; Chaudhary, A.A.; Al-Farraj, E.; Alhokbany, N.; Issa, Z.; Alhousine, S.; Ahamad, T. Delivery of ibuprofen by natural macroporous sporopollenin exine capsules extracted from *Phoenix dactylifera* L. *Eur. J. Pharm. Sci.* **2016**, *88*, 158–165. [CrossRef]
18. Meligi, N.M.; Dyab, A.; Paunov, V.N. Sustained In Vitro and In Vivo Delivery of Metformin from Plant Pollen-Derived Composite Microcapsules. *Pharmaceutics* **2021**, *13*, 1048. [CrossRef]
19. Atkin, S.A.; Beckett, S.T.; Mackenzie, G. Topical Formulations Containing Sporopollenin. WO2007012857A1, 1 February 2007.
20. Sporomex Ltd. Available online: https://www.sporomex.co.uk/applications/cosmetics (accessed on 26 July 2022).
21. Harris, T.L.; Wenthur, C.J.; Diego-Taboada, A.; Mackenzie, G.; Corbittc, T.S.; Janda, K.D. *Lycopodium clavatum* exine microcapsules enable safe oral delivery of 3,4-diaminopyridine for treatment of botulinum neurotoxin A intoxication. *Chem. Commun.* **2016**, *52*, 4187–4190. [CrossRef]
22. Thomasson, M.J.; Diego-Taboada, A.; Barrier, S.; Martin-Guyout, J.; Amedjou, E.; Atkin, S.L.; Queneau, Y.; Boa, A.N.; Mackenzie, G. Sporopollenin exine capsules (SpECs) derived from *Lycopodium clavatum* provide practical antioxidant properties by retarding rancidification of an ω-3 oil. *Ind. Crops Prod.* **2020**, *154*, 112714. [CrossRef]
23. Miller, I.B.; Pawlowski, S.; Kellermann, M.Y.; Petersen-Thiery, M.; Moeller, M.; Nietzer, S.; Schupp, P.J. Toxic effects of UV filters from sunscreens on coral reefs revisited: Regulatory aspects for "reef safe" products. *Environ. Sci. Eur.* **2021**, *33*, 1–13. [CrossRef]
24. Baker, L.A.; Clark, S.L.; Habershon, S.; Stavros, V.G. Ultrafast Transient Absorption Spectroscopy of the Sunscreen Constituent Ethylhexyl Triazone. *J. Phys. Chem. Lett.* **2017**, *8*, 2113–2118. [CrossRef] [PubMed]
25. Herzog, B.; Wehrle, M.; Quass, K. Photostability of UV absorber systems in sunscreens. *Photochem. Photobiol.* **2009**, *85*, 869–878. [CrossRef] [PubMed]

26. National Guideline Alliance (UK). *Skin Care Advice for People with Acne Vulgaris: Acne Vulgaris: Management: Evidence Review B*; NICE Guideline; No. 198; National Institute for Health and Care Excellence (NICE): London, UK, 2021. Available online: https://www.ncbi.nlm.nih.gov/books/NBK573057/ (accessed on 13 August 2022).
27. Dario, M.F.; Baby, A.R.; Velasco, M.V. Effects of solar radiation on hair and photoprotection. *J. Photochem. Photobiol. B* **2015**, *153*, 240–246. [CrossRef] [PubMed]
28. Terreni, E.; Chetoni, P.; Tampucci, S.; Burgalassi, S.; Al-Kinani, A.A.; Alany, R.G.; Monti, D. Assembling Surfactants-Mucoadhesive Polymer Nanomicelles (ASMP-Nano) for Ocular Delivery of Cyclosporine-A. *Pharmaceutics* **2020**, *12*, 253. [CrossRef]
29. Berben, P.; Bauer-Brandl, A.; Brandl, M.; Faller, B.; Flaten, G.E.; Jacobsen, A.C.; Brouwers, J.; Augustijns, P. Drug permeability profiling using cell-free permeation tools: Overview and applications. *Eur. J. Pharm. Sci.* **2018**, *119*, 219–233. [CrossRef]
30. Caon, T.; Costa, A.C.; de Oliveira, M.A.; Micke, G.A.; Simões, C.M. Evaluation of the transdermal permeation of different paraben combinations through a pig ear skin model. *Int. J. Pharm.* **2010**, *391*, 1–6. [CrossRef]
31. Tampucci, S.; Burgalassi, S.; Chetoni, P.; Lenzi, C.; Pirone, A.; Mailland, F.; Caserini, M.; Monti, D. Topical formulations containing finasteride. Part II: Determination of finasteride penetration into hair follicles using the differential stripping technique. *J. Pharm. Sci.* **2014**, *103*, 2323–2329. [CrossRef]
32. Jessop, P.; Leitner, W. Green Chemistry in 2017. *Green Chem.* **2017**, *19*, 15–17. [CrossRef]
33. Becherini, S.; Mitmoen, M.; Tran, C.D. Natural Sporopollenin Microcapsules Facilitated Encapsulation of Phase Change Material into Cellulose Composites for Smart and Biocompatible Materials. *ACS Appl. Mater. Interfaces* **2019**, *11*, 44708–44721. [CrossRef]
34. Atkin, S.L.; Barrier, S.; Cui, Z.; Fletcher, P.D.; Mackenzie, G.; Panel, V.; Sol, V.; Zhang, X. UV and visible light screening by individual sporopollenin exines derived from *Lycopodium clavatum* (club moss) and Ambrosia trifida (giant ragweed). *J. Photochem. Photobiol. B* **2011**, *102*, 209–217. [CrossRef] [PubMed]
35. Chu, C.C.; Tan, C.P.; Nyam, K.L. Development of Nanostructured Lipid Carriers (NLCs) Using Pumpkin and Kenaf Seed Oils with Potential Photoprotective and Antioxidative Properties. *Eur. J. Lipid Sci. Technol.* **2019**, *121*, 1900082. [CrossRef]
36. Puglia, C.; Damiani, E.; Offerta, A.; Rizza, L.; Tirendi, G.G.; Tarico, M.S.; Curreri, S.; Bonina, F.; Perrotta, R.E. Evaluation of nanostructured lipid carriers (NLC) and nanoemulsions as carriers for UV-filters: Characterization, in vitro penetration and photostability studies. *Eur. J. Pharm. Sci.* **2014**, *51*, 211–217. [CrossRef]
37. Hojerová, J.; Peráčková, Z.; Beránková, M. Margin of safety for two UV filters estimated by in vitro permeation studies mimicking consumer habits: Effects of skin shaving and sunscreen reapplication. *Food Chem. Toxicol.* **2017**, *103*, 66–78. [CrossRef]
38. Fan, T.F.; Potroz, M.G.; Tan, E.L.; Ibrahim, M.S.; Miyako, E.; Cho, N.J. Species-Specific Biodegradation of Sporopollenin-Based Microcapsules. *Sci. Rep.* **2019**, *9*, 9626. [CrossRef] [PubMed]
39. Fan, T.F.; Potroz, M.G.; Tan, E.L.; Park, J.H.; Miyako, E.; Cho, N.J. Human blood plasma catalyses the degradation of Lycopodium plant sporoderm microcapsules. *Sci. Rep.* **2019**, *9*, 2944. [CrossRef] [PubMed]
40. Lorch, M.; Thomasson, M.J.; Diego-Taboada, A.; Barrier, S.; Atkin, S.L.; Mackenzie, G.; Archibald, S.J. MRI contrast agent delivery using spore capsules: Controlled release in blood plasma. *Chem. Commun. Camb* **2009**, *42*, 6442–6444. [CrossRef]
41. Pyo, S.M.; Maibach, H.I. Skin Metabolism: Relevance of Skin Enzymes for Rational Drug Design. *Ski. Pharmacol. Physiol.* **2019**, *32*, 283–294. [CrossRef]
42. Svensson, C.K. Biotransformation of drugs in human skin. *Drug Metab. Dispos.* **2009**, *37*, 247–253. [CrossRef]
43. Monti, D.; Brini, I.; Tampucci, S.; Chetoni, P.; Burgalassi, S.; Paganuzzi, D.; Ghirardini, A. Skin permeation and distribution of two sunscreens: A comparison between reconstituted human skin and hairless rat skin. *Ski. Pharmacol. Physiol.* **2008**, *21*, 318–325. [CrossRef]

MDPI
St. Alban-Anlage 66
4052 Basel
Switzerland
www.mdpi.com

Pharmaceutics Editorial Office
E-mail: pharmaceutics@mdpi.com
www.mdpi.com/journal/pharmaceutics

www.ingramcontent.com/pod-product-compliance
Lightning Source LLC
LaVergne TN
LVHW070241100526
838202LV00015B/2162